Carbohydrate Chemistry

Volume 34

A Specialist Periodical Report

Carbohydrate Chemistry
Monosaccharides, Disaccharides and Specific Oligosaccharides
Volume 34

A Review of the Literature Published during 2000

Senior Reporter
R.J. Ferrier, *Industrial Research Limited, Lower Hutt, New Zealand*

Reporters
P. Benjes, *Industrial Research Limited, Lower Hutt, New Zealand*
R. Blattner, *Industrial Research Limited, Lower Hutt, New Zealand*
R.A. Field, *University of St. Andrews, St. Andrews, UK*
R.H. Furneaux, *Industrial Research Limited, Lower Hutt, New Zealand*
C. Hamilton, *University of East Anglia, Norwich, UK*
J.O. Hoberg, *Victoria University of Wellington, Wellington, New Zealand*
K.P.R. Kartha, *University of St. Andrews, St. Andrews, UK*
P.C. Tyler, *Industrial Research Limited, Lower Hutt, New Zealand*
R.H. Wightman, *Heriot-Watt University, Edinburgh, UK*

advancing the chemical sciences

NEW FROM 2003

If you buy this title on standing order, you will be given FREE access to the chapters online. Please contact sales@rsc.org with proof of purchase to arrange access to be set up.

Thank you.

ISBN 0-85404-238-5

ISSN 0951-8428

A catalogue record for this book is available from the British Library

Published by The Royal Society of Chemistry,
Thomas Graham House, Science Park, Milton Road, Cambridge CB4 0WF, UK

Registered Charity Number 207890

For further information see our web site at www.rsc.org

Typeset by Vision Typesetting, Manchester, UK
Printed and bound by Athenaeum Press Ltd, Gateshead, Tyne & Wear, UK

Preface

In order to address the ever-increasing significance of carbohydrates in biology the presentation of material in this volume has been somewhat modified. A new introductory chapter (1) highlights some of the reviews published during the year that give an overview of the glycobiological aspects of carbohydrate chemistry, and another new chapter (20) brings together reports of the use of enzymic methods in the field – particularly those employed in the synthesis of glycosides and especially di- and oligosaccharides. There is consequently some inevitable overlap between Chapter 20 and Chapters 3 and 4. Antibiotics no longer have their separate chapter, rather material on this topic is distributed between the chapters on *e.g.* glycosides and nucleosides. The various analytical and separatory methods, which in previous volumes had their own chapters, are now covered together in Chapter 21.

As was foreshadowed in the Preface to Volume 33 the present team completes its writing responsibilities with Volume 34, the decision having been forced by the ever-increasing workload involved in covering not just an expanding field but one of appreciably greater complexity. While it used to be a relatively straightforward task to abstract a paper and assign the précis to one of the chapters, it is now commonplace to find important features of a report which fall under several headings. All too often, it has appeared to the writers, important material was being buried in the mass, and highly significant and long papers have had to be condensed beyond an appropriate degree. Our procedures do not allow automatic searching, and in the face of powerful alternatives, it has to be conceded that this is a fundamental weakness. It also has to be accepted that, despite much effort to reduce the delay in publication, we have not achieved our objectives. Nevertheless, the approach adopted over 34 years has provided a search method with unique advantages, bringing together related chemistry in a way that facilitates searching within categories of particular compounds and which affords a 'snapshot' of the state of each category at a particular point in time.

Highlights this year are the major syntheses of everninomycin and of glycoconjugate-based anti-cancer antigens mentioned at the beginning of Chapter 4, and also the less spectacular but highly significant developments in the vital area of synthesis of oligosaccharides. The further recognition of the desirability of using complementary and tuned donors and acceptors, the advances in 'one-pot', biochemical and automated procedures indicate very real progress, but nevertheless, each oligosaccharide still appears to presents its own challenges.

This year Paul Benjes and Chris Hamilton contributed to the abstracting and writing tasks. Their help has been much appreciated, and their longer serving colleagues are thanked warmly for all their efforts over, in some cases, many years. Richard Wightman has been a particular tower of strength during the production of 17 volumes, always providing many of the most complex and detailed abstracts from the most difficult sources and writing key and highly involved chapters with the greatest commitment and professionalism. He seemed to check everything – even after publication – and was never slow to point out deficiencies with a directness that could only be forgiven by the succeeding friendly references to British/New Zealand sporting interactions.

Over many years Janet Freshwater and Alan Cubitt, of the Royal Society of Chemistry, provided us with invaluable help by administering the projects and smoothing the paths of our manuscripts from delivery to publication. Sincere thanks are extended for all their work on behalf of all the contributors.

R. J. Ferrier
December 2002

Contents

Abbreviations

The following abbreviations have been used:

Ac	acetyl
Ade	adenin-9-yl
AIBN	2,2-azobisisobutyronitrile
All	allyl
Ar	aryl
Ara	arabinose
Asp	aspartic acid
BBN	9-borabicyclo[3.3.3]nonane
Bn	benzyl
Boc	*t*-butoxycarbonyl
Bu	butyl
Bz	benzoyl
CAN	ceric ammonium nitrate
Cbz	benzyloxycarbonyl
CD	circular dichroism
Cer	ceramide
CI	chemical ionization
Cp	cyclopentadienyl
Cyt	cytosin-1-yl
Dahp	3-deoxy-D-*arabino*-2-heptulosonic acid 7-phosphate
DAST	diethylaminosulfur trifluoride
DBU	1,8-diazabicyclo[5.5.0]undec-5-ene
DCC	dicyclohexylcarbodi-imide
DDQ	2,3-dichloro-5,6-dicyano-1,4-benzoquinone
DEAD	diethyl azodicarboxylate
DIBALH	di-isobutylaluminium hydride
DMAD	dimethyl acetylenedicarboxylate
DMAP	4-(dimethylamino)pyridine
DMF	*N,N*-dimethylformamide
DMSO	dimethyl sulfoxide
Dmtr	dimethoxytrityl
e.e.	enantiomeric excess
Ee	1-ethoxyethyl
ESR	electron spin resonance
Et	ethyl

FAB	fast-atom bombardment
Fmoc	9-fluorenylmethylcarbonyl
Fru	fructose
FTIR	Fourier transform infrared
Fuc	fucose
Gal	galactose
GalNAc	2-acetamido-2-deoxy-D-galactose
GLC	gas–liquid chromatography
Glc	glucose
GlcNAc	2-acetamido-2-deoxy-D-glucose
Gly	glycine
Gua	guanin-9-yl
Hep	L-*glycero*-D-*manno*-heptose
HMPA	hexamethylophosphoric triamide
HMPT	hexamethylphosphorous triamide
HPLC	high performance liquid chromatography
IDCP	iodonium dicollidine perchlorate
Ido	idose
Im	imidazolyl
IR	infrared
Kdo	3-deoxy-D-*manno*-2-octulosonic acid
LAH	lithium aluminium hydride
LDA	lithium di-isopropylamide
Leu	leucine
LTBH	lithium triethylborohydride
Lyx	lyxose
Man	mannose
*m*CPBA	*m*-chloroperbenzoic acid
Me	methyl
Mem	(2-methoxyethoxy)methyl
Mmtr	monomethoxytrityl
Mom	methoxymethyl
Ms	methanesulfonyl (mesyl)
MS	mass spectrometry
NAD	nicotinamide adenine dinucleotide
NBS	*N*-bromosuccinimide
NeuNAc	*N*-acetylneuraminic acid
NIS	*N*-iodosuccinimide
NMNO	*N*-methylmorpholine *N*-oxide
NMR	nuclear magnetic resonance
NOE	nuclear Overhauser effect
ORD	optical rotatory dispersion
PCC	pyridinium chlorochromate
PDC	pyridinium dichromate
Ph	phenyl
Phe	phenylalanine

Piv	pivaloyl
Pmb	*p*-methoxybenzyl
Pr	propyl
Pro	proline
p.t.c.	phase transfer catalysis
Py	pyridine
Rha	rhamnose
Rib	ribose
Ser	serine
SIMS	secondary-ion mass spectrometry
TASF	tris(dimethylamino)sulfonium(trimethylsilyl)difluoride
Tbdms	*t*-butyldimethylsilyl
Tbdps	*t*-butyldiphenylsilyl
Tipds	tetraisopropyldisilox-1,3-diyl
Tips	triisopropylsilyl
Tf	trifluoromethanesulfonyl (triflyl)
Tfa	trifluoroacetyl
TFA	trifluoroacetic acid
THF	tetrahydrofuran
Thp	tetrahydropyranyl
Thr	threonine
Thy	thymin-1-yl
Tips	1,1,3,3-tetraisopropyldisilox-1,3-diyl
TLC	thin layer chromatography
Tms	trimethylsilyl
TPP	triphenylphosphine
Tps	tri-isopropylbenzenesulfonyl
Tr	triphenylmethyl (trityl)
Ts	toluene-*p*-sulfonyl (tosyl)
Ura	uracil-1-yl
UDP	uridine diphosphate
UDPG	uridine diphosphate glucose
UV	ultraviolet
Xyl	xylose

1
Introduction and General Aspects

This year saw the publication of a triple issue (numbers 7–9) of *Glycoconjugate Journal* highlighting 'Glycobiology at the Millennium, a look back and a glance ahead'.[1] Topics covered range from an appreciation of A. Kabat (Feizi and Lloyd) and affinity enhancement of lectin–carbohydrate interactions (Lee and Lee) to carbohydrates as future anti-adhesion drugs for bacterial disease (Sharon and Ofek). A symposium issue of *Journal of Carbohydrate Chemistry* features a collection of articles from the First Euroconference on Carbohydrates in Drug Research.[2] A special issue of *Chemical Reviews* has been published entitled 'Frontiers in Carbohydrate Research', and sub-titled by the guest editor (J.K. Bashkin) 'Carbohydrates – A Hostile Scientific Frontier Becomes Friendlier'.[3] Amongst other topics, many of which are referred to at the beginning of relevant chapters, this issue covers solid-phase oligosaccharide synthesis and combinatorial carbohydrate libraries (Seeberger and Haase),[4] intramolecular *O*-glycoside formation (Schmidt and co-workers),[5] enzyme-based and programmable one-pot strategies for synthesis of complex carbohydrates and glycoconjugates (Koeller and Wong,[6] who have written a further review on this topic[7]) and theoretical approaches and experimental validation of studies on the structure, conformation and dynamics of bioactive oligosaccharides (Imberty and Perez).[8]

A review from Zechel and Withers, entitled 'Glycosidase Mechanisms: Anatomy of a Finely-tuned Catalyst', addresses issues of transition state structure, substrate distortion, acid–base catalysis and trapping of covalent intermediates.[9] Winchester and Fleet have reviewed modification of glycosylation of glycoconjugates as a therapeutic strategy;[10] the use of glycosphingolipid synthesis inhibitors as therapy for glycolipid storage disorders has been also reviewed.[11] The glycan repertoire of genetically modified mice has been analysed by nano-NMR spectroscopy – a key step on the way to understanding the role of glycosylation *in vivo*.[12] At the whole-cell level, synthetic *N*-glycolylmannosamine pentaacetate has been used to prime *N*-glycolylneuraminic acid formation in neural cell cultures and hence alter cell phenotype.[13] Synthetic heparin–diazeniumdiolate conjugates have been shown to act as inhibitors of thrombin-induced blood coagulation by virtue of their ability to generate nitric oxide.[14] A comprehensive, and comprehensible, survey of topical issues in glycobiology appears in *Essentials of Glycobiology*.[15]

The synthesis and biological activity of glycolipids, with a focus on ganglio-

Carbohydrate Chemistry, Volume 34
© The Royal Society of Chemistry, 2003

sides and sulfatide, has been reviewed.[16] Extensive studies at the forefront of chemical synthesis, detailing the successful total synthesis of the oligosaccharide antibiotic everninomycin **1**, have been reported by Nicolaou and co-workers.[17,18] The use of olefin metathesis in carbohydrate chemistry has been reviewed.[19,20]

1

Combinatorial methods have been used to generate penta- and hexa-peptides with monosaccharide-recognition ability.[21] A number of review articles have appeared concerning solution and solid-phase approaches to the generation of carbohydrate-based combinatorial libraries.[22-25] The use of anomeric radicals in the synthesis of *O*- and *C*-glycosides has been reviewed,[26] as have the conformations of the radicals and their reactions under reductive conditions.[27] An extensive review of iodine and iodine-based reagents in carbohydrate chemisry has appeared.[28]

Last year saw the death of Professor Guy Dutton, well known for his leading work on polysaccharide structural analysis; his obituary has appeared.[29]

References

1.   Various authors, *Glycoconjugate J.*, 2000, **17**, 437–668.
2.   Various authors, *J. Carbohydr. Chem.*, 2000, **19**, 419–657.
3.   Various authors, *Chem. Rev.*, 2000, **100**, 4265–4711.
4.   P.H. Seeberger and W.-C. Haase, *Chem. Rev.*, 2000, **100**, 4349.
5.   K.-H. Jung, M. Muller and R.R. Schmidt, *Chem. Rev.*, 2000, **100**, 4423.
6.   K.M. Koeller and C.-H. Wong, *Chem. Rev.*, 2000, **100**, 4465.
7.   K.M. Koeller and C.-H. Wong, *Glycobiology*, 2000, **10**, 1157.
8.   A. Imberty and S. Perez, *Chem. Rev.*, 2000, **100**, 4567.
9.   D.L. Zechel and S.G. Withers, *Acc. Chem. Res.*, 200, **33**, 11.
10.   B. Winchester and G.W.J. Fleet, *J. Carbohydr. Chem.*, 2000, **19**, 471.
11.   C.J. Tifft and R.L. Proia, *Glycobiology*, 2000, **10**, 1249.
12.   A.E. Manzi, K. Norgard-Sumnicht, S. Argade, J.D. Marth, H. van Halbeek and A. Varki, *Glycobiology*, 2000, **10**, 669.
13.   B.E. Collins, T.J. Fralich, S. Itonori, Y. Ichikawa and R.L. Schnaar, *Glycobiology*, 2000, **10**, 11.
14.   J.E. Saavedra, D.L. Mooradian, K.A. Mowery, M.H. Schoefisch, M.L. Citro, K.M. Davies, M.E. Meyerhoff and L.K. Keefer, *Bioorg. Med. Chem. Lett.*, 2000, **10**, 751.
15.   *Essentials of Glycobiology*, 1999, ed. A. Varki, Cold Spring Harbor Laboratory Press, Cold Spring Harbor, New York.
16.   T. Ikami, H. Ishida and M. Kiso, *Methods Enzymol.*, 2000, **311**, 547 (*Chem. Abstr.*, 2000, **132**, 265 348).

17. K.C. Nicolaou, H.J. Mitchell, H. Suzuki, R.M. Rodriguez, O. Baudoin and K.C. Fylaktakidou, *Angew. Chem. Int. Ed. Engl.*, 2000, **39**, 3334, 3340, 3345.
18. K.C. Nicolaou, H.J. Mitchell, R.M. Rodriguez, K.C. Fylaktakidou, H. Suzuki and S.R. Conley, *Chem. Eur. J.*, 2000, **6**, 3149.
19. N. Sugimoto, D. Miyoshi and J. Zou, *Chem. Commun.*, 2000, 2295.
20. K. Fukase, *Kagaku Kogyo*, 2000, **51**, 232 (*Chem. Abstr.*, 2000, **132**, 265–351).
21. T. Kanemitsu and O. Kanie, *Trends Glycosci. Glycotechnol.*, 1999, **11**, 267 (*Chem. Abstr.*, 2000, **132**, 237 246).
22. M.J. Sofia, *Annu. Rep. Comb. Chem. Mol. Diversity*, 1999, **2**, 41 (*Chem. Abstr.*, 2000, **132**, 208 007).
23. H. An and P.D. Cook, *Chem. Rev.*, 2000, **100**, 3311.
24. R. Roy and S.K. Das, *Chem. Commun.*, 2000, 519.
25. A. Furstner, *Angew. Chem. Int. Ed. Engl.*, 2000, **39**, 3013.
26. J. Jimenez-Barbero, J. F. Espinosa, J.L. Asensio, F.J. Canada and A. Poveda, *Adv. Carbohydr. Chem. Biochem.*, 2000, **56**, 235.
27. J.P. Praly, *Adv. Carbohydr. Chem. Biochem.*, 2000, **56**, 65.
28. A.R. Vaino and W.A. Szarek, *Adv. Carbohydr. Chem. Biochem.*, 2000, **56**, 9.
29. H. Parolis, *Adv. Carbohydr. Chem. Biochem.*, 2000, **56**, ix.

2
Free Sugars

1 Synthesis

1.1 Tetroses and Pentoses. – A new synthesis of 2-*O*-benzyl-3,4-*O*-isopropylidene-D-erythrose (**2**) from 2,3-*O*-isopropylidene-D-glyceraldehyde involved chain-extension by use of methyl tolyl sulfoxide, followed by benzylation of the new hydroxyl group to give **1**. Quantitative transformation of the sulfoxide to a formyl group (**1→2**) was achieved by exposure to lutidine-trifluoroacetic anhydride, then *aq.* sodium hydrogen carbonate.[1]

Compound **3** was prepared from commercially available (*R*)-(+)-5-hydroxymethyl-5*H*-furan-2-one by *O*-benzylation and subsequent conjugate addition of (PhMe$_2$Si)$_2$ Cu(CN)Li$_2$, and converted to 2-deoxy-L-ribose (**5**) *via* the 2-deoxy-L-ribonolactone derivative **4**.[2] 2′-Deoxy-D-ribose 5-phosphates ^{13}C-labelled at C-3 and C-4, and/or at C-5, were prepared in a chemoenzymatic approach by cyclizing appropriately labelled dihydroxyacetone monophosphates with unlabelled acetaldehyde. By use of [[13]C$_2$]-, [1-^{13}C]- or [2-^{13}C]-acetaldehyde, labels were also introduced at C-1 and/or C-2.[3]

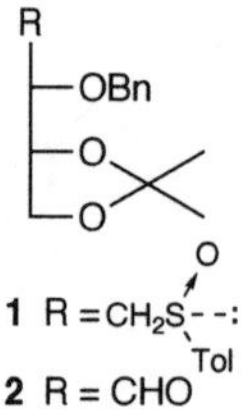

1 R = CH$_2$S--: Tol
2 R = CHO

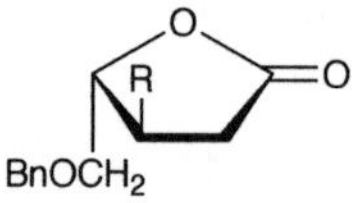

3 R = SiMe$_2$Ph
4 R = OH

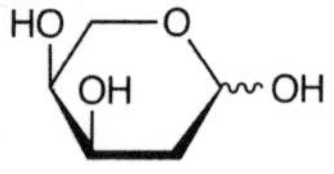

5

1.2 Hexoses. – The *de novo* syntheses of enantiopure D- as well as L-hexoses from vinylfuran (see Vol. 33, Chapter 2, Ref. 10) were greatly improved by use of optimal conditions for the required Sharpless catalysed asymmetric dihydroxylation.[4,5]

Hetero Diels–Alder cycloaddition of α,β-unsaturated carbonyl compounds and dioxygenated alkenes in the presence of a chiral bisoxazoline-Cu(OTf)$_2$ complex as Lewis acid catalyst furnished hexopyranose precursors in good yields and high enantiomeric excess. In the synthesis of the precursor **6** of ethyl tetra-*O*-acetyl-β-D-mannopyranose outlined in Scheme 1, for example, a 69% overall yield and 99% ee were achieved.[6] A new route to hex-2-uloses involving boron- or, preferably, lithium-enolates is exemplified in Scheme 2. Only 3,4-*trans*-

Carbohydrate Chemistry, Volume 34
© The Royal Society of Chemistry, 2003

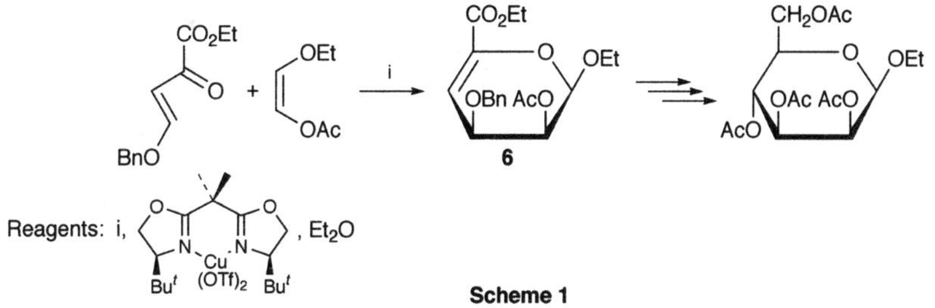

Scheme 1

products were formed when the chiral lithium amide **7** was used to generate the enolate, with the D-tagatose derivative **8** as the major (69%) and its D-psicose isomer **9** as the minor (8%) products.[7]

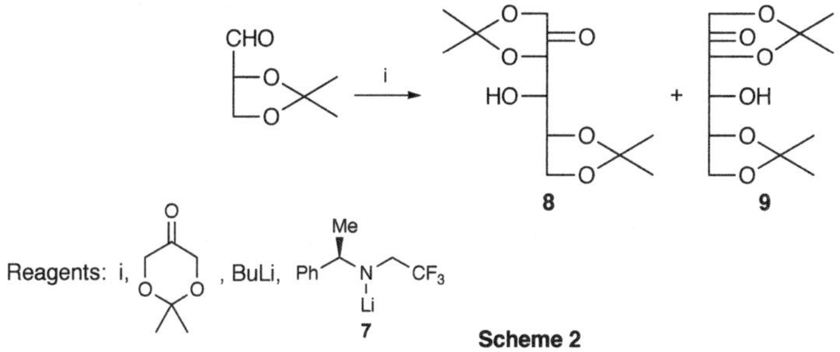

Scheme 2

A new approach to multiply-protected aminodeoxyhexoses using an $Sn(OTf)_2$-catalysed cross-aldol condensation between lactaldehyde and a tricar-bonyliron/α-aminoheptadiene complex is referred to in Chapter 9.

The synthesis of L-sugars has received considerable attention: the furfural-derived, optically active, bicyclic enone **10** served as common precursor of L-galactose, L-gulose and L-idose.[8] Compound **10** was converted in a four step 1,3-enone transposition to its isomer **11**, from which the remaining five L-aldohexoses (L-allose, L-altrose, L-glucose, L-mannose and L-talose) were obtained.[9] Several L-hexoses have been synthesized from the appropriate perbenzylated D-hexono-1,5-lactones by exposure to $BnONH_2$-Me_3Al to give acyclic intermediates (*e.g.* **12**) which ring-closed with inversion at C-5 on treatment with DEAD-TPP. The resulting oximes (*e.g.* **13**) were readily hydrolysed and reduced to the perbenzylated free L-sugars.[10]

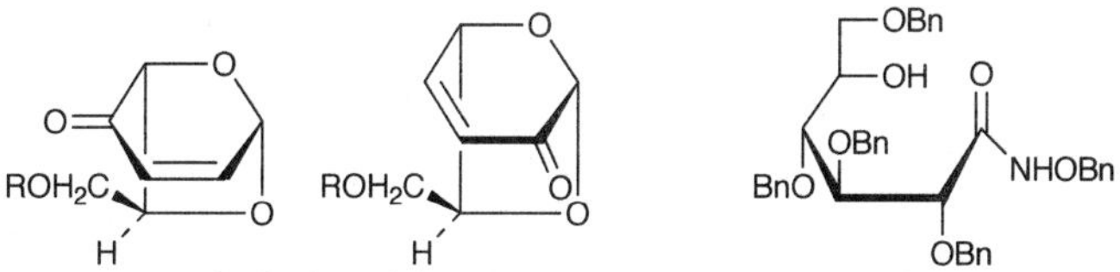

D-Glucono-1,4-lactone derivative **14** was converted to L-gulose by reductive opening, followed by persilylation and acetal hydrolysis to give diol **15**, then primary oxidation and deprotection. Acetal hydrolysis in the presence of silyl

groups was achieved by use of BCl$_3$.[11] In another, as yet incompleted synthesis of
L-gulose starting from 2,3,4,6-tetra-*O*-Tbdms-D-gulono-1,5-lactone, use was
made of BCl$_3$ in THF for the selective cleavage of the primary Tbdms ether
(**16→17**), to allow the necessary C-6 oxidation (**17→18**).[12] 1,2:5,6-Di-*O*-isopro-
pylidene-β-L-idofuranose, formed on acid treatment of the 1,2:3,5-di-*O*-acetal
(see Chapters 6 and 13 for synthesis), gave the elimination products **19** and **20** on
treatment with DAST-pyridine and PDC-Ac$_2$O-pyridine, respectively. The former underwent stereo- and regioselective hydroboration to furnish L-altrose,
after acid hydrolysis; the latter was converted to L- mannose by catalytic hydrogenation, then deprotection.[13,14] An immobilized L-rhamnose isomerase of
Pseudomonas sp. was used to produce L-talose from L-tagatose and D-gulose
from D-sorbose in 12 and 10% crystalline yield, respectively.[15] A new formal
synthesis of 4-deoxy-L-hexoses from (*R*)-benzylglycidyl ether is covered in Chapter 12.

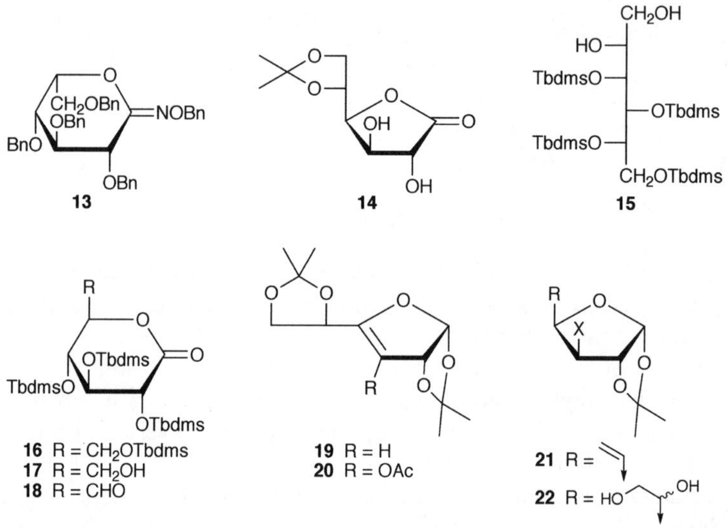

13   14   15

16 R = CH$_2$OTbdms   19 R = H   21 R =
17 R = CH$_2$OH   20 R = OAc
18 R = CHO   22 R = HO

Asymmetric dihydroxylation experiments with α-D-*xylo*-hex-5-enofuranose
derivatives (**21→22**) under a variety of conditions showed that the C-3 substituent (X) plays an important role. 3-Esters **21** (X = OAc or OBz) and the 3-deoxy
compound **21** (X = H) gave D-gluco- and 3-deoxy-L-ido-products, respectively,
with good selectivity when 'Admix α' was used as the reagent.[16]

1.3   Chain-extended Sugars. – *1.3.1   Chain-extension at the 'Non-reducing
End'*. Methyl 2,3,4-tri-*O*-benzyl-D-*erythro*-α-D-*gluco*-oct-1,5-pyranoside and its
L-*erythro*-β-D-*gluco*-isomer have been synthesized for the first time by *cis*-hy-
droxylation of the appropriate 6-*C*-vinyl-D- and -L-*glycero*-D-*gluco*-pyranose
derivatives, respectively.[17]

    Homologation of protected dialdohexoses **23** with α-D-gluco-, α-D-galacto-
and α-D-allo-configurations by use of various (substituted-methyl)magnesium
chlorides, has been undertaken. Good L-selectivity was achieved with

PhMe$_2$SiCH$_2$MgCl (*e.g.* **23→24**).[18] The *glycero*-D-*manno*-heptopyranoside-7-phosphate analogues **25** were obtained from a dialdehydo-mannopyranoside precursor by reaction with MePO(OEt)$_2$-*n*-BuLi, followed by deprotection.[19]

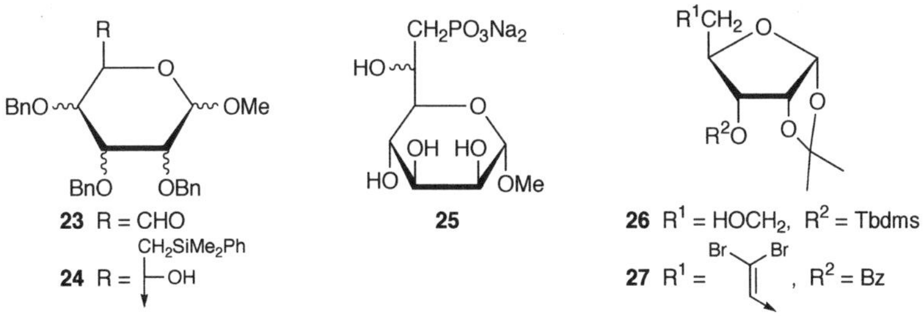

23 R = CHO
24 R = $\overset{CH_2SiMe_2Ph}{\underset{}{\mid}}$—OH

25

26 R^1 = HOCH$_2$, R^2 = Tbdms
27 R^1 = (Br Br), R^2 = Bz

Dibromide **27**, required for the preparation of doubly homologated analogues of adenosine (see Chapter 19), was obtained by oxidation of 5-deoxy-α-D-allofuranose derivative **26** to the corresponding 6-aldehyde, treatment with Br$_2$C=PPh$_3$ and replacement of the silyl by a benzoyl group.[20] The synthesis of the bridged α,β-unsaturated lactone **30** involved Wittig-extension of the dispiroketal-protected 6-aldehydo-D-mannopyranoside **28** to furnish **29**, followed by debenzylation with concomitant cyclization by use of FeCl$_3$ in dry dichloromethane.[21]

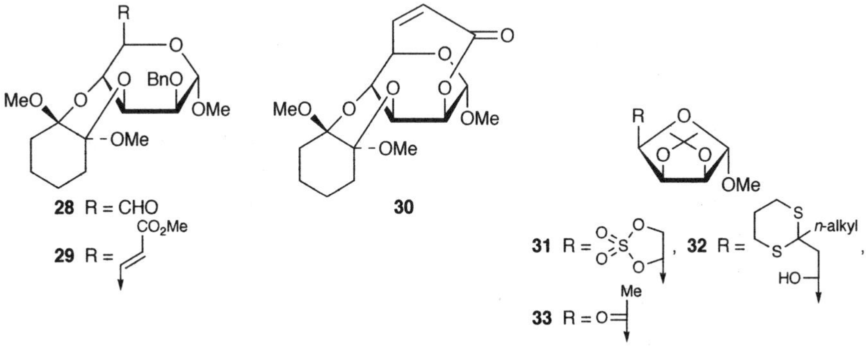

28 R = CHO
29 R = $\overset{CO_2Me}{\underset{}{}}$

30

31 R = (cyclic sulfate), **32** R = (dithiane, *n*-alkyl, HO)

33 R = O=$\langle$ Me

C-Alkylation of mannofuranoside 5,6-cyclic sulfate **31** with *n*-alkylated lithium dithianes gave chain-extended 7-osulose precursors **32**, accompanied by by-products **33** in increasing amounts (0–42%) with increasing alkyl chain length (*n* = 0–12).[22] Cyclic sulfate **34** was opened with lithium trimethylsilyldithiane to furnish ketone **35** after dethioacetalation. Further processing gave the 3-trimethylsilyl-5-(threos-4-yl)-pyrazole derivatives **36**, as shown in Scheme 3.[23]

3-*O*-Benzyl-6-deoxy-1,2-*O*-isopropylidene-α-D-*xylo*-hexofuranos-5-ulose (**37**) on treatment with carbon disulfide and methyl iodide under basic conditions afforded α-oxoketene dithioacetal **38**, which was transformed to pyrazole derivative **39** by exposure to hydrazine hydrate. A 3-deoxy-3,4-unsaturated analogue was similarly prepared.[24] Conversion of ketone **37** to the Knoevenagel product **40** prior to treatment with CS$_2$-MeI-NaH led to the formation of the sugar 'push-pull-butadiene' **41**.[25]

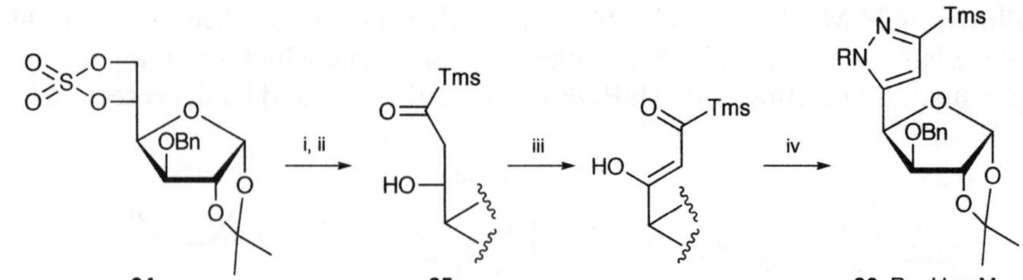

34   **35**   **36** R = H or Me

Reagents: i, $\overset{Tms}{\underset{Li}{\big\rangle}}\hspace{-0.3em}\big\langle^{S}_{S}$, H$^+$; ii, I$_2$, CaCO$_3$; iii, DMSO, TFAA; iv, RNHNH$_2$

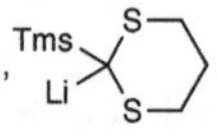

Scheme 3

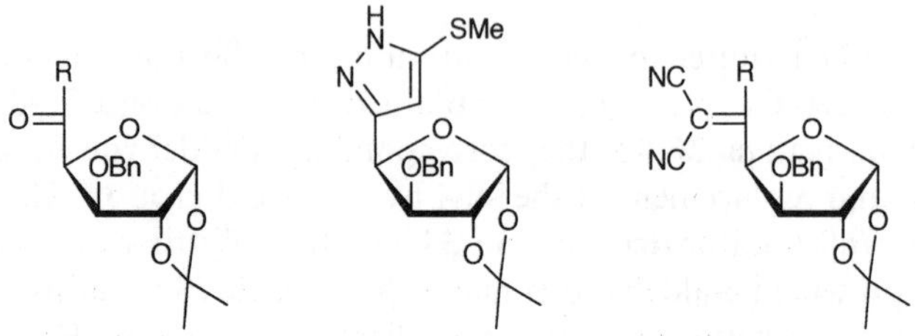

37 R = Me   **39**   **40** R = Me

38 R = C=C$\overset{SMe}{\underset{SMe}{}}$ **41** R = C=C$\overset{SMe}{\underset{SMe}{}}$

In the presence of chloramine T, the non-reducing end oxime **42** underwent 1,3-polar cycloadditions with terminal alkynes to give 3-glycosyl-5-substituted isoxazoles **43** in moderate yields.[26] Reductive alkylation of pyrroline derivative **44** with methyl 2,3,4-tri-*O*-benzyl-6-deoxy-6-iodo-α-D-glucopyranoside gave, after ester reduction, osmylation and acetylation-deacetylation, the novel poly-hydroxypyrrolidine **45**, containing a methyl glucoside moiety.[27]

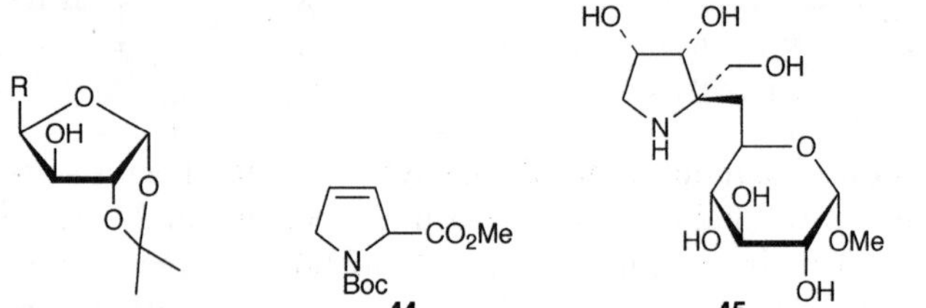

42 R = HC$\overset{NOH}{}$

43 R =

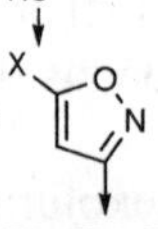

X = Ph, CH$_2$OH, *etc.*

44

45

Chain-extension has been brought about intramolecularly by ring-closing metathesis of glycoside dienes. The 3-*O*-allyl-hex-5-enofuranose derivative **46**, for example, afforded the 3,7-anhydroheptofuranose **47**.[28]

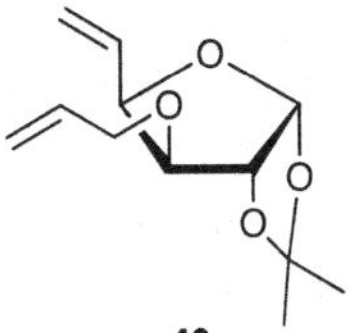

46 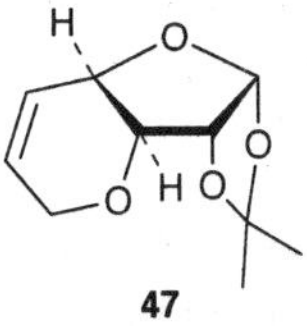   **47**

Crossed-aldol condensations between 1-deoxy-3,4:5,6-di-*O*-isopropylidene-L-fructose and various protected aldehydo-pentoses afforded C-11 sugars, mostly as diastreomeric mixtures, in modest yields.[29] Higher analogues of sucrose were obtained by oxidation and one-carbon Wittig extension at C-6 of an appropriately protected starting compound and subsequent *cis*-dihydroxylation of the 6,7-double bond.[30] Radical addition reactions furnishing D-galactofuranosyl-containing *C*-disaccharides are covered in Chapter 3.

1.3.2 Chain-extension at the 'Reducing End'. Mercuricyclization of hept-dienitol **48** and subsequent reductive demercuration in the presence of oxygen afforded a 3:1 mixture of 1,2-unsaturated hept-3-ulofuranosides **49**.[31] Reaction of 5,6-*O*-isopropylidene-2,3-di-*O*-Tbdms-D-allono-1,4-lactone with allyl magnesium chloride, followed by acetylation, then ozonolysis or epoxidation gave octos-3-ulose derivative **50** and non-4-ulose derivatives **51**, respectively.[11]

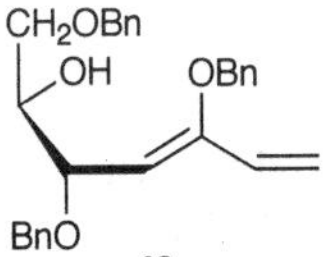

48 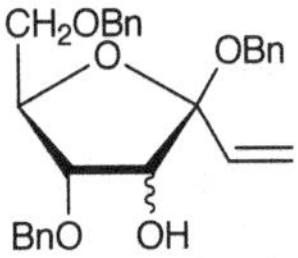   **49**

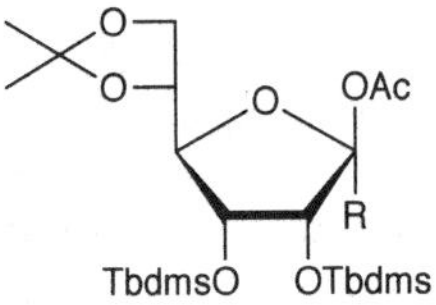

50 R = CHO

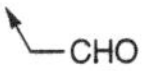

51 R =

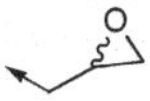

The differentially protected F-ring moiety **53** of the altohyrtin group of anticancer macrolides has been prepared using intramolecular opening of epoxide **52** to form the pyranose ring. Compound **52** was obtained by epoxidation of the corresponding unsaturated octitol, which in turn was constructed from small molecules in a multistep asymmetric synthesis.[32]

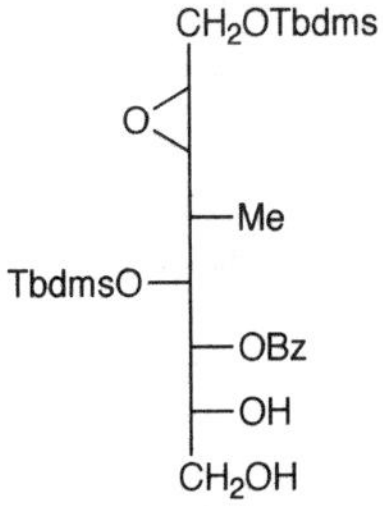

52 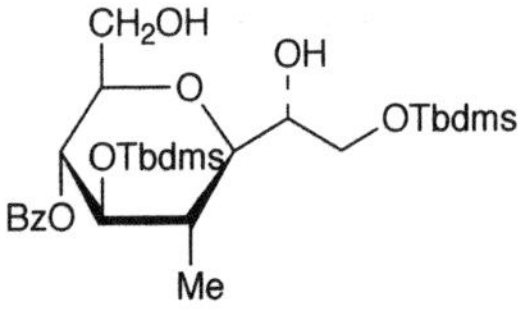   **53**

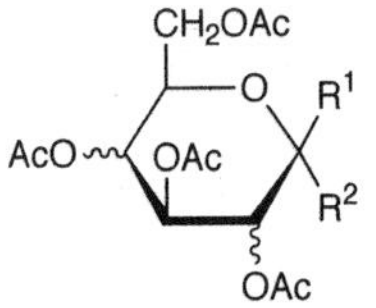

54 R^1 = Cl, R^2 = Br
55 R^1 = , R^2 = Cl
56 R^1 = , R^2 = H
57 R^1,R^2 =

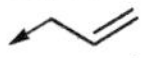

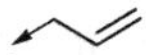

Radical allylation of peracetylated 1-bromo-β-D-glycopyranosyl chlorides **54** (β-D-gluco-, β-D-manno- or β-D-galacto-configuration) with allyltributyl tin under photolytic conditions gave 4-ulopyranosyl chlorides **55** in moderate to excellent yields. Radical dechlorination then furnished the expected 3-(β-D-glycopyranosyl)-1-propenes **56**. Alternatively, base-induced dehydrochlorination led to the new glycopyranosylidene dienes **57**.[33] Insertion of glycosylidene carbene derived from **58** into boron-alkyl bonds of trialkylborons or *B*-alkyl-9-oxo-10-borabicyclo[3.3.2]decanes gave the base-stable glycosyl boranes **59** and glycosyl borinates **60**, respectively, with low stereoselectivity. Oxidation of **59** and **60** (H_2O_2-NaOH) afforded exclusively hemiketals **61** and **61a**, respectively, with axial OH.[34]

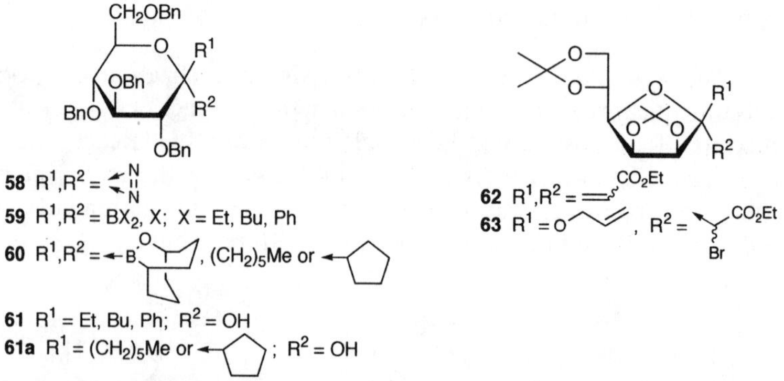

α-Bromoacetals **63**, available by 1,4-addition of allyl alcohol to the D-mannono-1,4-lactone-derived enol ester **62** in the presence of NBS, underwent radical cyclization on treatment with Bu₃SnH to afford spiroacetal **64**.[35] The spiroketal moieties of spirophostins **66**, conformationally restricted analogues of the adenophostin-type nucleoside antibiotics, were formed by acid-promoted cyclizations of precursor non-4-uloses **65**.[36] On exposure to DIBAL, the known spiroketal mesylate **67** rearranged to give the hydroxylated, *cis*-fused 1,6-dioxadecalin **68**. Formation of a cyclic intermediate **67a** by participation of the C-3 benzyloxy group in the displacement is proposed, the sulfonate being displaced intramolecularly by the tetrahydrofuranyl oxygen atom.[37]

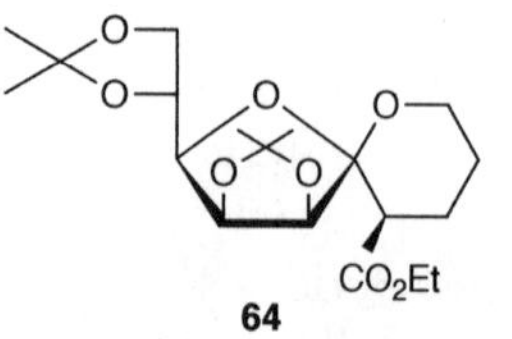

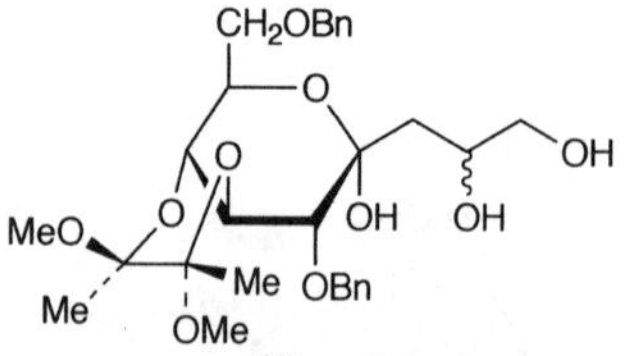

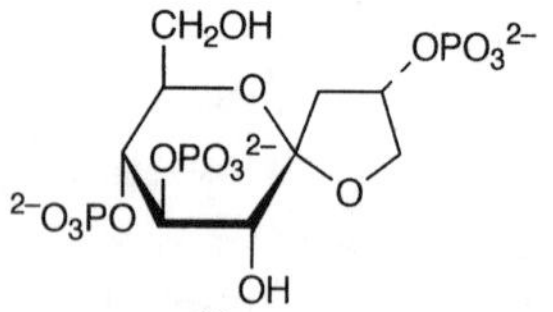

2 Reactions

A short summary of the Bilik reaction (molybdic acid-catalysed epimeric interconversion of aldoses with skeletal rearrangement) has been published.[38] The molybdic acid-catalysed isomerizations of D-*erythro*- and D-*threo*-pentulose to the corresponding 2-*C*-(hydroxymethyl)-D-tetroses is covered in Chapter 14.

The platinum-catalysed oxidation and carboxy-alkylation of free sugars has been reviewed (11 pp., 44 refs.).[39] A kinetic study on the oxidation of aldopentoses in *aq.* sulfuric acid by electrolytically generated manganese dioxide has been undertaken.[40] The iridium(III)-catalysed oxidation of maltose and lactose by NBS in perchloric acid has been investigated, with particular attention to the effects of various additives, such as acetamide, acetic acid, KCl and Hg(OAc)$_2$ on the reaction rates.[41] In order to locate the reaction sites in starch oxidation processes, the characteristics of the reactions of the model compounds methyl α-D-glucopyranoside and 1,2-*O*-isopropylidene-α-D-glucofuranose with oxygen in alkaline solution in the presence of copper phenanthroline have been studied.[42]

The highly efficient catalytic activity of lanthanide(III) ions in the thermal degradation of the most common monosaccharides as well as several fructose-containing di- and tri-saccharides in organic solvents, especially in DMSO, with formation of 5-hydroxymethylfurfural has been investigated.[43]

Addition of glycine during the conversion of D-glucose by aqueous alkali greatly increased the rate of formation of carbonyl and dicarbonyl degradation intermediates, whereas their subsequent transformation to carboxylic products was inhibited.[44] In the presence of thiols, browning was accelerated, but when both glycine and a thiol were added, it was slowed down.[45] The effects of constant reaction pH on sucrose degradation has been assessed by use of simulated industrial model systems. It was shown that minimum sucrose degradation occurred between pH 6.45 and 8.50.[46]

3   Other Aspects

The relative proportions of the acyclic aldehydic forms of eight hexoses, four pentoses and three deoxyaldoses has been estimated from the rate constants for their reactions with urazole. The absolute values were obtained by correlation with literature NMR measurements.[47]

Infinite-dilution apparent molar volumes for monosaccharides (D-xylose, D-arabinose, D-glucose and D-galactose) in aqueous NaCl and those for NaCl in aqueous saccharide solutions and related volumetric properties of sugar/NaCl/water systems have been determined and the results have been evaluated in terms of structural interaction and stereochemistry.[48]

When monosaccharides (Glc, Gal, Man, GlcNAc) were doped with the corresponding alkyl glycosides, glasses were obtained that were able to solubilize hydrophobic compounds such as phenols, terpenols, arylalkyl alcohols and lipids) to form isotropic liquids. The use of these materials in enzymic glycosylations is referred to in Chapter 20.[49,50]

References

1. Y. Arroyo-Gómez, J.F. Rodríguez-Amo, M. Santos-García and M.A. Sanz-Tejedor, *Tetrahedron: Asymm.*, 2000, **11**, 789.
2. F. Fazio and M.P. Schneider, *Tetrahedron: Asymm.*, 2000, **11**, 1869.
3. N. Ouwerkerk, J.H. van Boom, J. Lugtenburg and J. Raap, *Eur. J. Org. Chem.*, 2000, 861.
4. J.M. Harris, M.D. Keränen, H. Nguyen, V.G. Young and G.A. O'Doherty, *Carbohydr. Res.*, 2000, **328**, 17.
5. J.M. Harris and G.A. O'Doherty, *Tetrahedron Lett.*, 2000, **41**, 183.
6. H. Audrain, J. Thorhauge, R.G. Hazell and K.A. Jorgensen, *J. Org. Chem.*, 2000, **65**, 4487.
7. M. Majewski and P. Nowak, *J. Org. Chem.*, 2000, **65**, 5152.
8. M. Takeuchi, T. Taniguchi and K. Ogasawara, *Synthesis*, 1999, 341
9. M. Takeuchi, T. Taniguchi and K. Ogasawara, *Chirality*, 2000, **12**, 338 (*Chem. Abstr.*, 2000, **133**, 177 360).
10. H. Takahashi, Y. Hitomi, Y. Iwai and S. Ikegami, *J. Am. Chem. Soc.*, 2000, **122**, 2995.
11. W.-B. Yang, C.-H. Tsai and C.-H. Lin, *Tetrahedron Lett.*, 2000, **41**, 2569.
12. Y.-Y. Yang, W.-B. Yang, C.-F. Teo and C.-H. Lin, *Synlett*, 2000, 1634.
13. S.-C. Hung, R. Puranik and F.-C. Chi, *Tetrahedron Lett.*, 2000, **41**, 77.
14. S.-C. Hung, C.-C. Wang and S.R. Thopate, *Tetrahedron Lett.*, 2000, **41**, 3119.
15. S.H. Bhuiyan, Y. Itami, G. Takada and K. Izumori, *J. Biosci. Bioeng.*, 1999, **86**, 567 (*Chem. Abstr.*, 2000, **132**, 222 721).
16. H.B. Mereyala, P.M. Goud, R.R. Gadikota, R.K. Maddala and K.R. Reddy, *J. Carbohydr. Chem.*, 2000, **19**, 1201.
17. S. Jarosz, S. Skora and S. Kosciolowska, *Pol. J. Chem.*, 1999, **73**, 1797 (*Chem. Abstr.*, 2000, **132**, 78 764).
18. M. Kim, B. Grzeszczyk and A. Zamojski, *Tetrahedron*, 2000, **56**, 9319.
19. S. Vidal, C. Vidil, A. Morère, M. Garcia and J.-L. Montero, *Eur. J. Org. Chem.*, 2000, 3433.
20. S.F. Wauk, C.A. Valdez, J. Khan, P. Moutinho, M.J. Robins, X. Yang, R.T. Borchardt, J. Balzarini and E. De Clercq, *J. Med. Chem.*, 2000, **43**, 1180.
21. H.H.A.M. Hassan, *Heterocycles*, 2000, **53**, 397(*Chem. Abstr.*, 2000, **132**, 251 350).
22. T. Gourlain, A. Wadouachi and D. Beaupère, *Carbohydr. Res.*, 2000, **324**, 66.
23. S. Gérard, R. Plantier-Royon, J.-M. Nuzillard and C. Portella, *Tetrahedron Lett.*, 2000, **41**, 9791.
24. D. Michalik and K. Peseke, *J. Carbohydr. Chem.*, 2000, **19**, 1049.
25. D. Michalik and K. Peseke, *J. Carbohydr. Chem.*, 2000, **19**, 141.
26. M.L. Fascio, V.J. Montesano and N.B. D'Accorso, *J. Carbohydr. Chem.*, 2000, **19**, 393.
27. F. Schieweck and H.-J. Altenbach, *J. Carbohydr. Chem.*, 2000, **19**, 647.
28. M.K. Gurjar, A.M.S. Murugaiah, J. Cherian and M.S. Chorghade, *Carbohydr. Lett.*, 1999, **3**, 343 (*Chem. Abstr.*, 2000, **132**, 166 407).
29. A.H. Haines and A.J. Lamb, *Carbohydr. Res.*, 2000, **325**, 323.

30. S. Jarosz, M. Mach and J. Frelek, *J. Carbohydr. Chem.*, 2000, **19**, 693.
31. C.K. Lee and H. Jiang, *Carbohydr. Lett.*, 1999, **3**, 309 (*Chem. Abstr.*, 2000, **132**, 166 404).
32. J.C. Anderson, B.P. McDermott and E.J. Griffin, *Tetrahedron*, 2000, **56**, 8747.
33. J.-P. Praly, G.-R. Chen, J. Gola and G. Hetzer, *Eur. J. Org. Chem.*, 2000, 2831.
34. W. Wenger and A. Vasella, *Helv. Chim. Acta*, 2000, **83**, 1542.
35. G.V.M. Sharma, A. Chander, V.G. Reddy, K. Krishnudu, M.H.V.R. Rao and A.C. Kunwar, *Tetrahedron Lett.*, 2000, **41**, 1997.
36. M. de Kort, A.D. Regenbogen. A.R.P.M. Valentijn, R.A.J. Challiss, Y. Iwata, S. Miyamoto, G.A. van der Marel and J.H. van Boom, *Chem. Eur. J.*, 2000, **6**, 2696.
37. Betancor, R.L. Dorta, R. Freire, T. Prangé and E. Suárez, *J. Org. Chem.*, 2000, **65**, 8822.
38. L. Petrus, *Enantiomer*, 1999, **4**, 369 (*Chem. Abstr.*, 2000, **132**, 64 454).
39. H. Van Bekkum, *Stud. Surf. Sci. Catal.*, 1999, **121**, 117 (*Chem. Abstr.*, 2000, **132**, 108 157).
40. M.A. Nikath, N. Anitha, K.M.L. Rai and K.S. Rangappa, *Trends Carbohydr. Chem.*, 1999, **4**, 109 (*Chem. Abstr.*, 2000, **132**, 108 200).
41. P.K. Tandon, J.P. Singh and M.P. Singh, *Oxid. Commun.*, 1999, **22**, 424 (*Chem. Abstr.*, 2000, **132**, 50 168).
42. N.A. Borovko, S.S. Chernaya and S.R. Trusov, *Zh. Prikl. Khim.*, 1999, **72**, 1842 (*Chem. Abstr.*, 2000, **132**, 251 308).
43. K.-i Seri, Y. Inoue and H. Ishida, *Chem. Lett.*, 2000, 22.
44. V.N. Rudenko, *Izv. Vyssh. Uchebn. Zaved., Pishch. Tekhnol.*, 1999, 30 (*Chem. Abstr.*, 2000, **132**, 194 554).
45. V.N. Rudenko, *Izv. Vyssh. Uchebn. Zaved., Pishch. Tekhnol.*, 1999, 33 (*Chem. Abstr.*, 2000, **132**, 194 555).
46. G. Eggleston and J.R. Vercellotti, *J. Carbohydr. Chem.*, 2000, **19**, 1305.
47. J.P. Dworkin and S.L. Miller, *Carbohydr. Res.*, 2000, **329**, 359.
48. K. Zhuo, J. Wang, Y. Yue and H. Wang, *Carbohydr. Res.*, 2000, **328**, 383.
49. I. Gill and Valivety, *Angew. Chem., Int. Ed. Engl.*, 2000, **39**, 3802.
50. I. Gill and Valivety, *Angew. Chem., Int. Ed. Engl.*, 2000, **39**, 3804.

3
Glycosides and Disaccharides

1 *O*-Glycosides

1.1 Synthesis of Monosaccharide Glycosides. – Relevant reviews to have appeared have dealt with glycosides bearing highly fluorinated aglycons, their biological activities and surface activity properties also being discussed,[1] glycosyl fluorides as glycosylating agents,[2] and the chemistry and biology of glycoside-controlled plant leaf movements.[3] Of particular important is a *Chemical Review* on intramolecular *O*-glycosylation by R.R. Schmidt and co-workers who dealt with the various approaches to this important technology and several applications.[4] Two new papers on this subject have dealt with compounds having the donor and acceptor parts linked by malonate or succinate[5] or isophthaloyl diester[6] tethers. In all cases the donor units were in the form of 1-thioglycosides. A relatively simple application of the intramolecular principle affords efficient access to β-mannosides and α-glucosides from[1,2] 1,2-*trans*-related 2-*O*-allyl 1-thioglycosides (Scheme 1). The reactions are stereospecific, and the critical last step occurs with 60–80% efficiency for the formation of simple glycosides and disaccharides.[7]

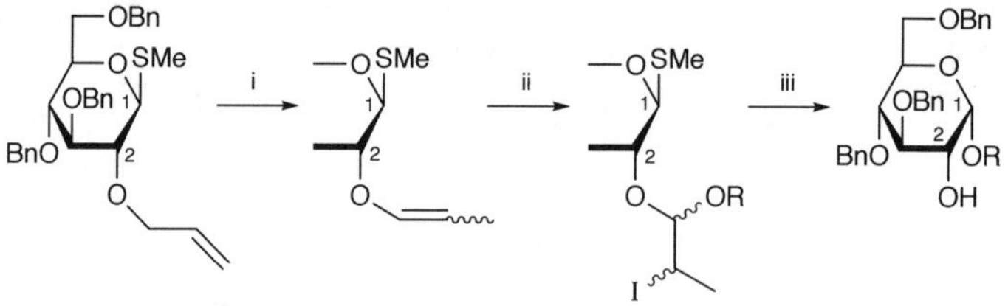

Reagents: i, (Ph₃P)₃RhCl, Bu^{n}Li; ii, ROH, NIS, AgOTf, base; iii, MeOTf, base

Scheme 1

A review on the use of triflic anhydride in organic chemistry has surveyed the activation of glycosyl donors with this reagent.[8]

1.1.1 Methods of Synthesis of Glycosides. Much material in this section overlaps reports in the synthesis of disaccharides, Section 1.4.

Hanessian and Lou have reviewed the 'remote activation' concept applied with *O*-unprotected glycosyl donors *e.g.* **1** which, with methyl triflate, give access to α-glucosides. This approach can also be used to make glycosyl carboxylates,

Carbohydrate Chemistry, Volume 34
© The Royal Society of Chemistry, 2003

phosphates, azides and nucleotides such as UDPG.[9] D-Glucose, D-mannose, D-ribose and L-arabinose, *O*-benzylated but with the anomeric hydroxyl groups free, on treatment with diphenylsulfoxide and triflic anhydride give sulfonium species, *e.g.* **2**, which can be used as glycosylating agents to give good yields of glycosyl derivatives of aliphatic and aromatic hydroxy compounds, thiols, azides and amides. The reactions are not highly stereoselective – especially when non-participating substituents at O-2 are present.[10] Contrastingly, the anticancer etoposide **3** was made in 79% overall yield by use of the relevant benzylic alcohol together with 2,3-di-*O*-benzyl-4,6-*O*-ethylidene-D-glucose. The coupling was promoted by use of BF₃.Et₂O in acetonitrile at − 10 °C, and the unusual but usefully high selectivity was driven by product crystallization. Debenzylation was effected by standard hydrogenolysis.[11]

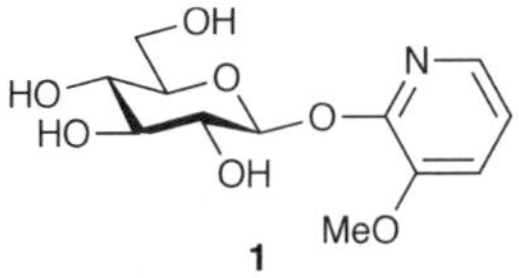

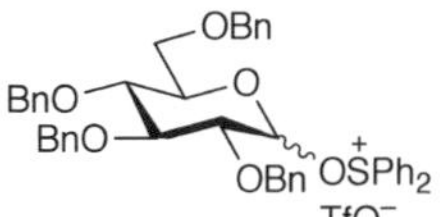

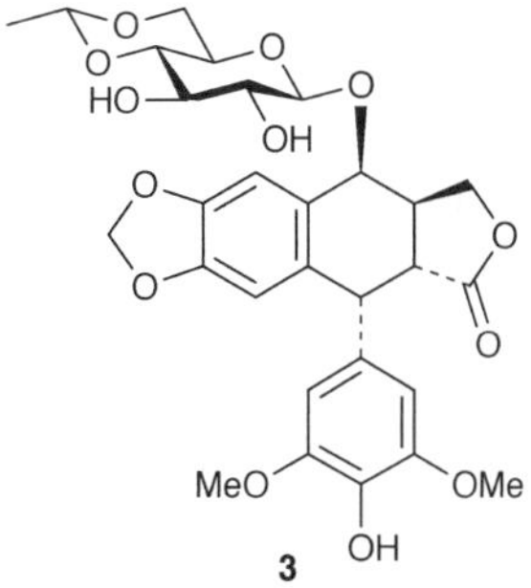

Boons has continued work with the vinylic 2-buten-2-yl glycosides (made from allylic 3-buten-2-yl isomers) as glycosyl donors, making glucosaminides and corresponding disaccharides from donors such as **4** in yields of about 80%.[12]

Standard conversions of glycal esters to 2,3-unsaturated glycosides can be promoted by use of indium triflate[13] (see Chapter 13 for other examples). For the synthesis of 2,6-dideoxyhexosyl steroid glycosides additions to glycals were effected on both soluble and insoluble polymers by use of triphenylphosphonium bromide without formation of allylically rearranged products.[14] Diene **5**, made from the 2-*C*-formylglycal, undergoes 1,4-addition of alcohols to give β-glyco-sides **6** in the presence of various acid catalysts.[15]

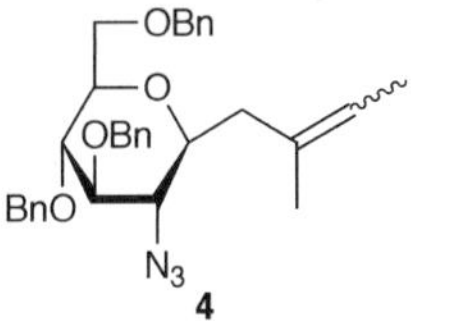

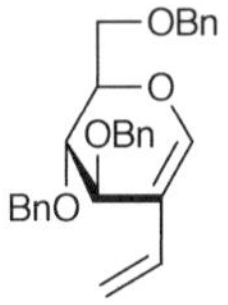

An investigation with altrosyl donors **7** [R = Ac, Me; X = Br, OC(N = H)CCl₃] indicated that participation by O-3 substituents was less important in determining the anomeric configurations of the products than the reactivities and bulkiness of the nucleophiles.[16]

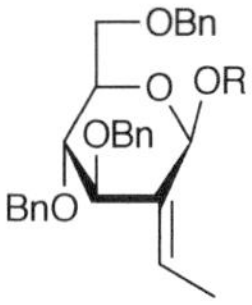

During TMSOTf-catalysed reactions of mixtures of D-glucose-and D-galac-tose-based 1,2-orthoacetates S_1R_1 and S_2R_2 (implying derivatives of sugars S_1 and

S_2 and aglycon groups R_1 and R_2), all four 2-*O*-acyl-β-glycosides S_1R_1, S_1R_2, S_2R_1 and S_2R_2 were formed showing that cross-over as well as intramolecular processes were involved.[17] 1,2,5-Orthoesters of D-arabinose **8–10** can be converted efficiently and selectively to 2-*O*-acyl-*O*- or *S*-linked α-D-arabinofuranosides (*e.g.* **11** from **9**) by treatment with alcohols on thiols in the presence of Lewis acid catalysts. Application of this method to the synthesis of an arabinofuranopentaose is noted in Chapter 4.[18]

7

8 R = Ph, R^1 = H
9 R = Ph, R^1 = Bn
10 R = CHCl$_2$, R^1 = Bn

11

Tetra-*O*-benzyl-D-mannopyranosyl propane-2,3-diyl phosphate together with trimethylsilyl triflate affords glycosides and disaccharides in yields of 46–90% with α-products either dominating or being formed exclusively.[19] Phosphinate donor **12** affords the corresponding α-glycosides with good selectivity on treatment with alcohols in the presence of trimethylsilyl triflate. Yields are about 50%.[20] Further methods of preparing α-D-mannopyranosides with very high selectivity have involved the use of *S*-phenyl 2,6-di-*O*-benzyl-3,4-*O*-(2',3'-dimethoxybutane-2',3'-yl)-1-thio-α-D-mannopyranoside or its sulfoxide together with triflic anhydride, the reactions proceeding *via* the α-glycosyl triflate. Modelling studies have provided a stereochemical rationalization for this discovery.[20a]

The use of thioglycosides as glycosylating agents has frequently been reported. In an indirect application, they are converted to the corresponding glycosyl chlorides by use of chlorosulfonium chlorides (*e.g.* Me$_2$S$^+$Cl$_2^-$).[21] With iodine as catalyst the reactivities of various *O*-benzylated 1-thio-β-D-galactopyranosides correlate with the electronic characteristics of the *S*-substituents. Methanolysis reaction rates are in the order Me>Bn>Ph>PhNO$_2$. For the *p*-nitrophenyl thioglycoside the reaction rate depends to a large extent on the promoter used. Iodine in dichloromethane, for example, was very much less effective than IBr in acetonitrile which was identified as the most practical reagent.[22]

12

13

14

Thioglycoside **13** (R = Bn, R^1 = Et) was designed for delivering the ethoxy group to give the ethyl glycoside. When activation was effected, however, in the presence of the analogue **13** (R = OPmb, R^1 = *o*-bromobenzyl) appreciable cross-over took place and led, for example, to *o*-bromobenzyl tetra-*O*-benzyl-D-glucoside, showing that the reaction was not exclusively intramolecular.[23] L-Rhamnose-derived **14**, on treatment with trimethylsilyl triflate in the presence of alcohols (including monosaccharide secondary alcohols), was activated at the orthoester centre and the ethylthio group migrated to C-2 *via* an episulfonium ion that was opened by attack of the alcohols at the anomeric centre. The yields of the products **15** were about 50%, and a novel route to 2-thio- and hence 2-deoxy-and 2,6-dideoxy-β-L-glucosides was opened.[24] Yet another example of unexpected reaction occurred on attempted glycosylation of alcohol **17** with tetra-*O*-acetyl-α-D-galactopyranosyl trichloroacetimidate. Cross-over occurred to give the acetylated galactosyl thioglycoside and the partially substituted amino-sugar trichloroacetimidate **16**.[25] This can be inhibited by careful choice of protecting groups. Compound **18** and the 2,3,4-tribenzoyl analogue have been used in solid phase oligosaccharide synthesis, the substituted O-6 benzyl group being stable to the glycosylation conditions with Lewis acids, and readily removed by treatment with PPh₃ followed by DDQ.[26]

15   **16**   **17**

Enzymic glycosylations are becoming increasingly important. Traditional methods as well as some novel approaches designed to improve methods applicable with hydrophobic alcohols have been reviewed.[27] Glycoside syntheses carried out in plasticized glass phases, which support high concentrations of donors and acceptors, allow increased substitution rates and yields in the 40–60% range. Various β-glucosidases and α- and β-galactosidases were examined.[28] A computational assessment of the ease of glycosylation of various alcohols with β-glucosidases has been reported. The nucleophilicities of the alcohols seem to determine the ease of substitution, and phenols are not sufficiently nucleophilic.[29]

1.1.2 Classes of Glycosides – Furanosides. Galactofuranosyl sulfoxide **19** (both stereoisomers) have been made from the anomeric octyl galactofuranosides and tested as glycosylating agents. With cyclohexanol and trifluoroacetic anhydride this gave a mixture of the cyclohexyl β-furanoside, the corresponding 1,2-orthoester and the 1,1′-linked non-reducing disaccharide, this last product conceivably pointing to the possibility that water was present in the system. In the

18   **19**   **20**

course of the work the *R*- and *S*-isomers of the benzoylated analogue of **19** were subjected to X-ray analysis.[30] In an extension of this work the same group has reported the formation of α-galacto- and manno-furanosyl disaccharides from the ethylthio- and phenylthio-glycofuranoside analogues of **19** with various primary and secondary sugar alcohols as acceptors. Yields were about 80% with good to moderate 1,2-*cis* selectivity in the products.[31] In Scheme 2 the formation of a 6-*O*-acetyl-β-D-galactofuranosyl acetate and hence a 6-*O*-acetyl-thioglycoside formed *via* the 1,6-anhydride **21** is illustrated. The acetate and thioglycoside were then used as galactofuranosyl donors.[32] Alternatively, the 1,5-anhydro-galactofuranose derivative **20** has been ring opened to give access mainly to 2,3,6-tri-*O*-substituted β-galactofuranosides.[33]

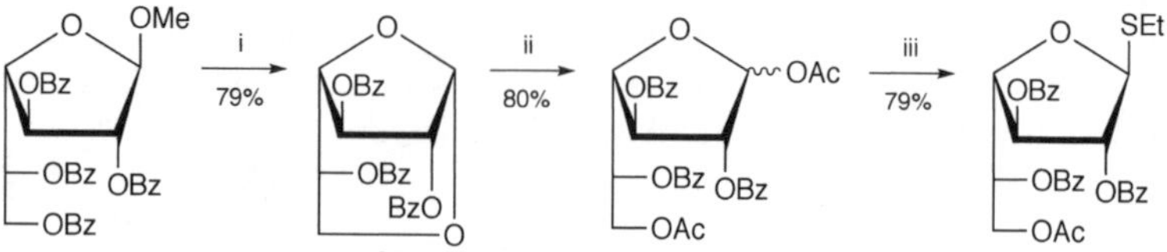

Reagents: i, SnCl₄; ii, Ac₂O, H₂SO₄; iii, EtSH, BF₃·Et₂O

Scheme 2

Otherwise β-galactofuranosides can be made from the galactono-γ-lactone. By standard chemistry the D-fucoside **22** was made, and the *O*-benzoylated glycosyl acetate was used to prepare a 6-deoxy-β-D-galactofuranosyl-(1→3)-D-mannose.[34] By standard synthetic steps 2-deoxy-pentofuranoside **23** was converted to the furanoid **24** and hence to the pyranoid thioglycoside **25** for introducing L-axenose into various antibiotics.[35]

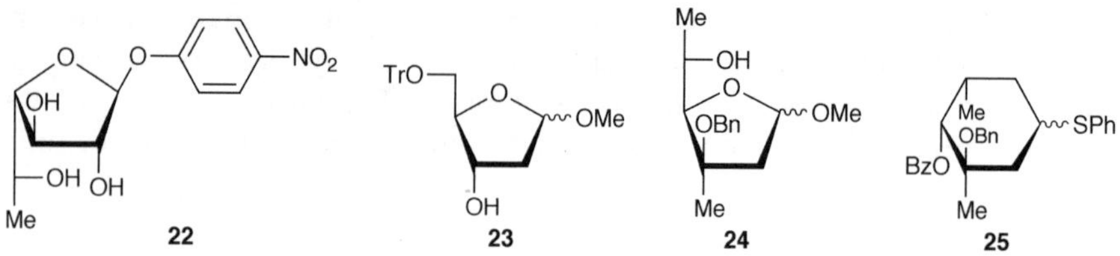

β-Mannosides. One interesting, novel method for making glycosides of this category has already been referred to,[7] and a review on the synthesis of such glycosides and β-D-mannosaminides has been published.[36] Weingart and Schmidt have enriched the topic with their report on the use of 4,6-*O*-benzylidene mannopyranosyl sulfoxides and trichloroacetimidates. Compound **26** gives β-mannosides with high selectivity when catalytic quantities of TMSOTf are used. Examples of disaccharide- and β-D-mannosyl-serine and threonine syntheses were reported.[37] The 2,3-di-*O*-benzyl-1-phenylthio analogue of compound **26** can be converted to the glycosyl triflate by use of compound **27** and triflic anhydride as thiophile. The triflate with 2,6-di-*tert*-butyl-4-methylpyridine then reacts with alcohols to give β-mannosides with good yields and selectivities.[38] A further but well known 'stereospecific' synthesis involves a four-step process from α-D-*gluco*-1,2-orthoesters which are used to make β-glucosides before inverson at C-2.[39]

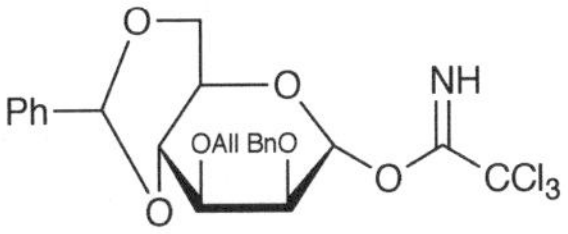

26 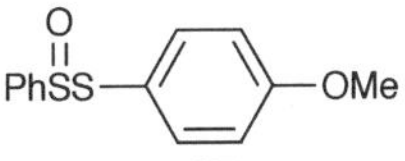 **27** 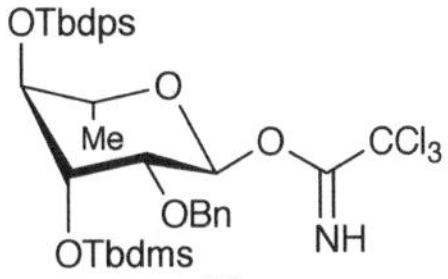 **28**

2,3-Diazido-2,3-dideoxy-β-D-mannopyranosides are referred to in Chapter 10.

Deoxy sugar glycosides and pentopyranosides. The L-rhamnosyl α-trichloroacetimidate **28**, which adopts the illustrated conformation that exemplifies the propensity for large silyloxy groups to favour the axial orientation, reacts with cyclohexylmethanol to give mainly the β-glycosidic product.[40]

By standard chemistry paratose and abequose (3,6-dideoxy-D-*ribo*- and -*xylo*-hexopyranose, respectively) have been made from 4,6-*O*-benzylidene-1,2-*O*-propylidene-α-D-glucose and -galactose, and their phenyl 1-thioglycosides are suitable as glycosylating agents.[41] The synthetic anthraquinone α- and β-glycosides **29** bind to DNA and cleave it on UV irradiation,[42] and the δ-lactam xylopyranoside **30** is a high affinity glycosidase inhibitor.[43]

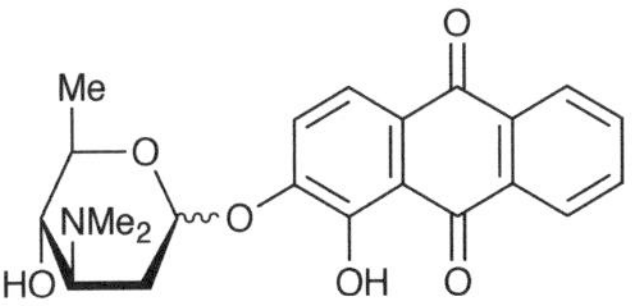

29 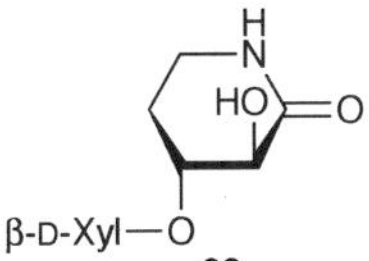 **30** 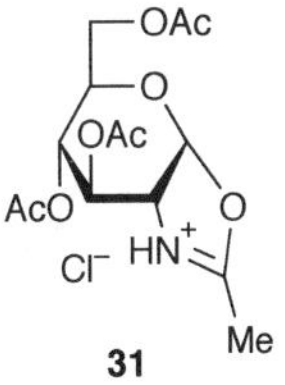 **31**

Amino-sugar glycosides. The torsionally constrained 2-azidoglycosyl donor **32** was used in the preparation of the α-linked glucosaminyl inositol (Scheme 3), and in the course of the same work alternative access to α-linked compounds of this type was obtained *via* a β-orthoester as indicated in Scheme 4.[44] New work has been reported on the preparation of β-D-glucosaminides from the 1,2-oxazoline hydrochloride **31** made from the β-glycosyl chloride. *N,N'*-Dicyclohexylurea inhibits the anomerization of the 1,2-*cis*-glycosides.[45] More specifically, compounds **33** were made from the corresponding allyl glycosides by

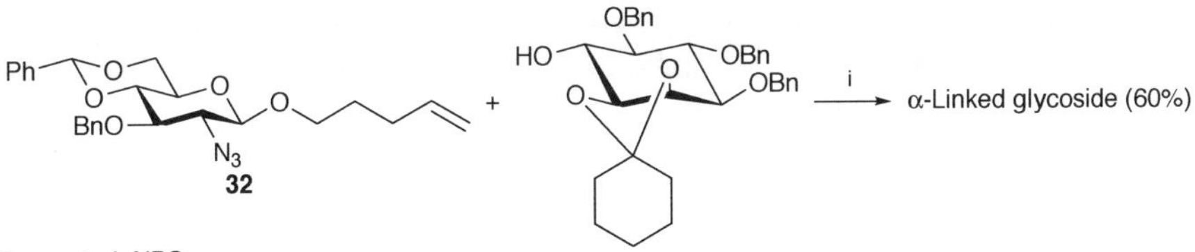

Reagent: i, NBS

Scheme 3

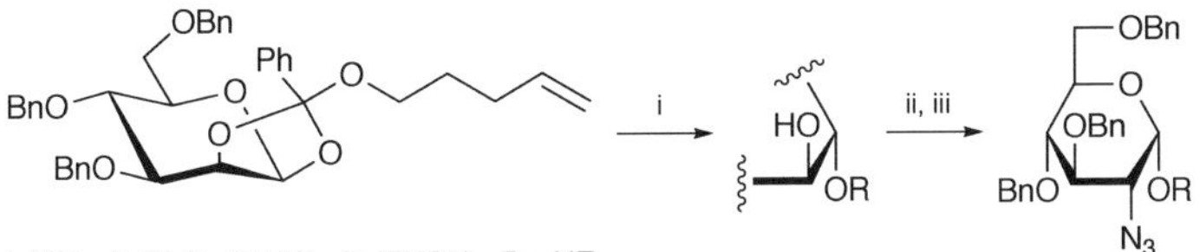

Reagents: i, NIS; ii, Tf₂O, DMAP; iii, TMSN₃, Bu₄NF

Scheme 4

initial epoxidations and nucleophilic epoxide ring openings,[46] and diosgenyl 2-amino-2-deoxy-β-D-glucopyranosides with trichloroacetyl and tetrachlorophthaloyl *N*-protection were made from the corresponding 2-acetylated glycosyl bromides.[47] Muramic acid having a pentapeptide amide linked to the carboxylic acid group has been made as its methyl and phenylthio α-glycosides in studies of the role of the corresponding glycosyl diphosphate in the recognition of Lipid 1 by the enzyme MurG.[48]

Compound **34** was made from the corresponding allyl 4,6-*O*-benzylidene glycoside which was acetal-ring opened with $BH_3.NMe_3/Me_2BBr$ to give the 4-*O*-benzyl-6-hydroxy compound. The polyfunctional azido product was used as a 'scaffold' for generation of broad spectrum libraries. It was amenable to use on a multi-gram scale and gave access to 12 000 compounds.[49] Also in the area of 2,6-di-*N*-substituted glycosides compound **35** was made for study of the binding of aminoglycosides to nucleic acids.[50]

33 R = Me, C_5H_{11}, C_7H_{15}; X = N_3, CN, SPh *etc.*   **34**   **35**

Glycosides of sugar acids. Compounds **36–38**, for use in glycopeptide synthesis, have been prepared from corresponding thioglycosides.[51] Otherwise compound **39** with monohydroxy sugar alcohols has been used to make α-linked sialyl disaccharides (Section 1.4).[51a] Uronic acids glycosidically *O*-linked to the nitrogen atoms of amides[51b] are referred to in Chapter 10.

Glycosides having aromatic rings in the aglycons. A series of *p*-iodophenyl β-D-glucopyranosides with various acyl groups at O-2, alkyl groups at O-3 and arylsulfate ester groups at O-6 have been made by standard procedures and found to inhibit two carcinoma cell lines.[52] Photo-labile *p*-substituted phenyl α-D-galactosides having azido, diazonium (trifluoroacetate) or (1-diaziridino-2,2,2-trifluoro)ethyl groups on the aromatic ring have been used to probe the sugar binding site of melibiose permease.[53] The β-glucuronoside **40** of nitrogen mustard has been made,[54] and *p*-nitrophenyl α-glycosides of NeuAc with various groups at C-3 have been prepared from a 2,3-epoxide. Some potently and selectively inhibited a sialidase, and liposome-like conjugates containing the glycoside with OH or F at C-3 selectively inhibited haemagglutination.[55] β-Glucosylation of the dopamine derivative **41** at either of the phenolic positions gave products made as potential anti-Parkinson's agents,[56] and α-deamino-kynurenine glucoside **42**, present in human lens, has been made by standard methods.[57] Amongst aryl glycosides made in the course of work on the bioavailability of anti-inflammatory compounds were the β-glucosides of gallic and gentisic acid, α-tocopherol and ibuprofen.[58]

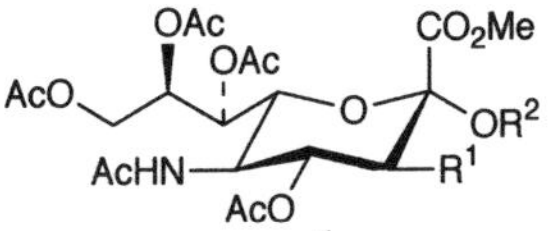

36 R^1 = H, R^2 =

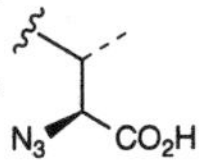

37 R^1 = SPh, R^2 =

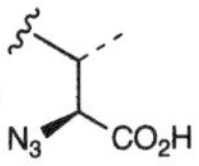

38 R^1 = H, R^2 =

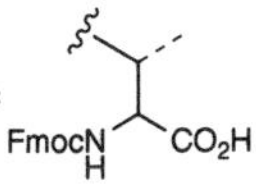

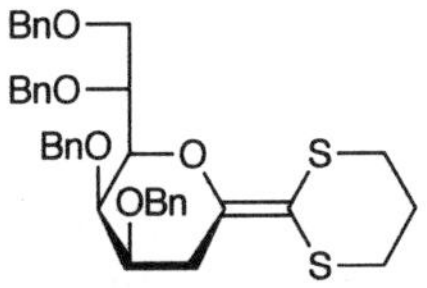

39

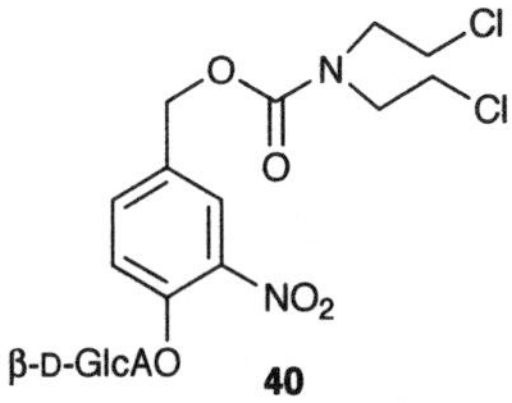

β-D-GlcAO **40**

In the area of compounds having *O*-glycosyl groups at benzylic (or equivalent) positions the polymer system **43** was made to illustrate the orthogonal nature of the *O*-protecting groups involved.[59] Enzymic transglycosylation effected the synthesis of *o*-, *m*- and *p*-hydroxybenzyl α-glucosides from soluble starch as sugar source,[60] and two publications in Japanese report the enzymic synthesis of pyridoxine β-glycosides *e.g.* **44** with maltodextrin as source.[61] Compound **45** was made as a fluorescent probe for studies of chemically induced leaf closing and opening,[62,63] and the somewhat related **46** by standard glycosylation of the enol form of the conjugated enone.[64]

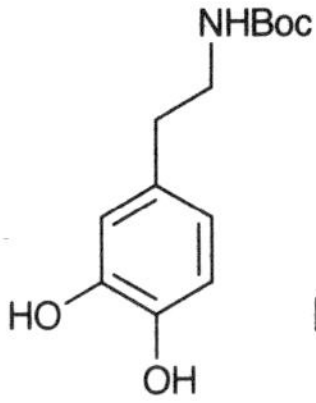

41

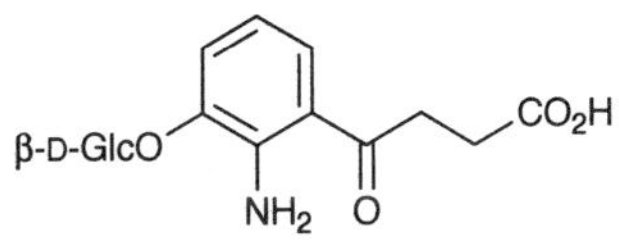

β-D-GlcO **42**

O(CH₂)₅CO₂PEG **43**

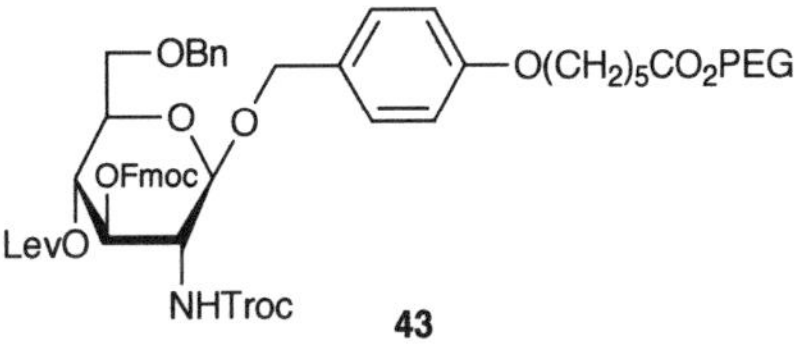

β-D-GlcO **44**

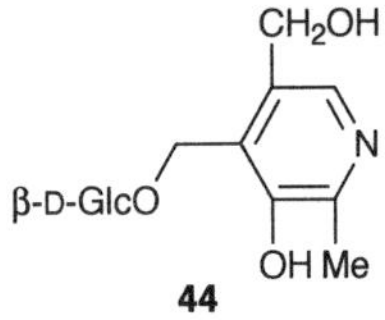

NHfluorescein **45**

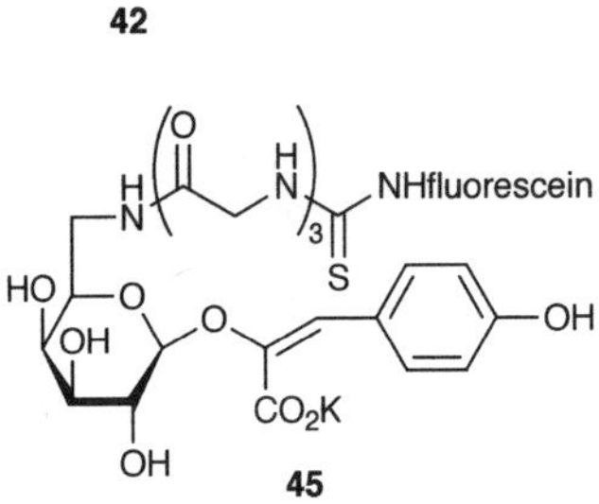

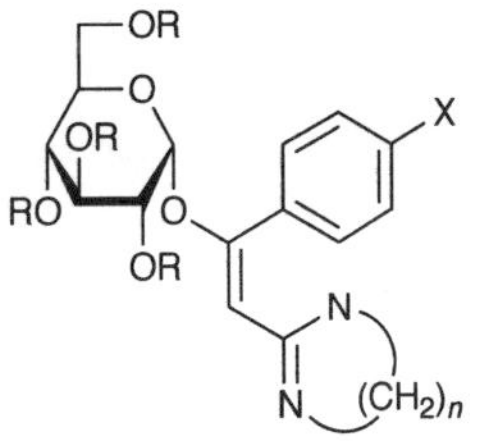

46 R = Ac, Bn; X = H, Me, OMe Cl; n = 2, 3

Other glycosides with aryl-containing aglycons are the β-glucosides and -galactosides of the *cis*- and *trans*-isomers of compound **47**, made enzymically as models from juvenogens,[65] and the polymer-bound **48** in which the polymer is TentaGel or PEGA resin, and from which the benzyl groups were removed under mild conditions by hydrogenolysis over palladium nanoparticles. The PEGA-based product was much more reactive, and hydrogenolysis proceeded smoothly provided the pore size of the support was carefully controlled.[66] Amongst compounds containing the common fluorescence label 2-aminobenz-amide to have been made is the glucoside **49**.[67] A set of compounds found in traditional Japanese health tonics and screened for prolyl endopeptidase inhibi-ting activity included (−)-epicatechin 3-*O*-gallate which is amongst the most potent.[68] Compound **50**, synthesized by use of a galactosyl phosphate, is an excellent fluorophore.[69]

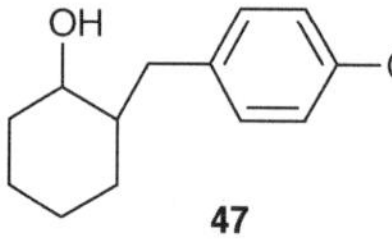

47 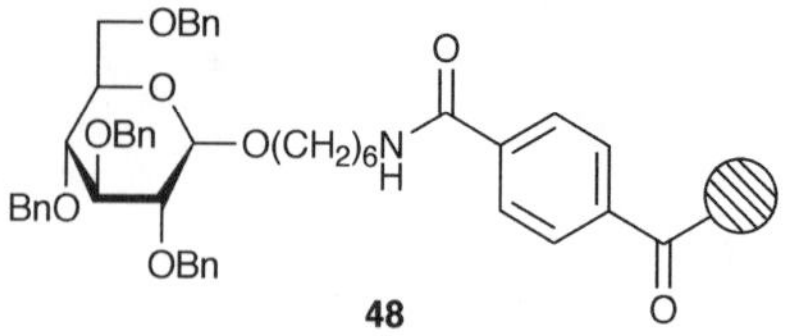   48

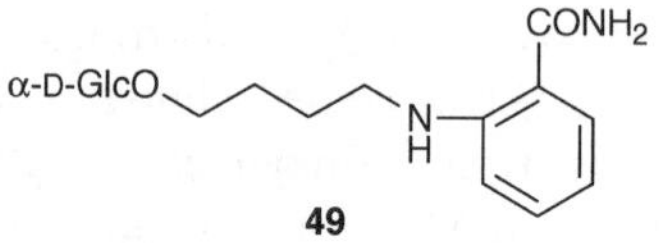

49

Syntheses of the following glycosylated polycyclic phenols have been reported: [2-^{13}C]-quercitin-4′-β-D-glucopyranoside (made by β-glucosylation of 2-ben-zyloxy-4-iodophenol),[70] novobiocin-related compounds **51**,[71] and the alizarin blue glycoside **52** which inhibits topoisomerease II and induces apoptosis.[72] In the first two of these syntheses trichloroacetimidate technology was used, while in the third case glycosylation was by way of a glycosyl bromide.

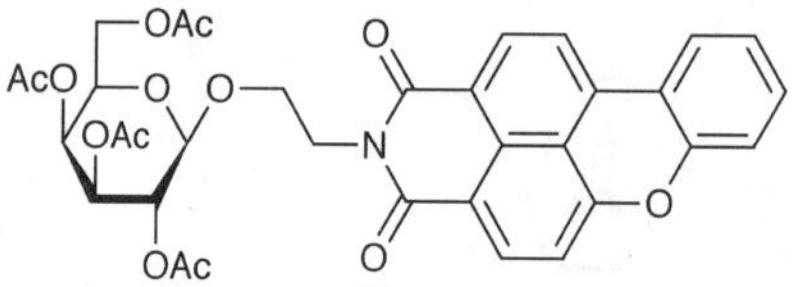

50 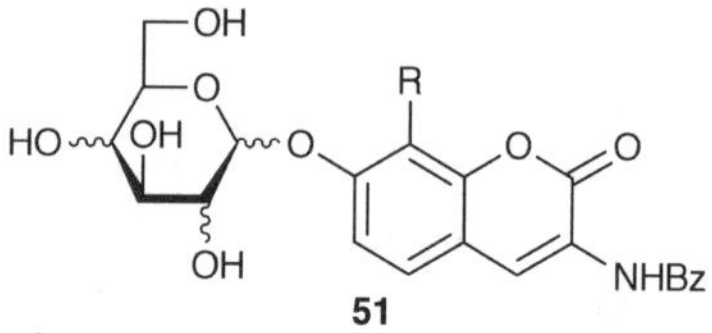   51

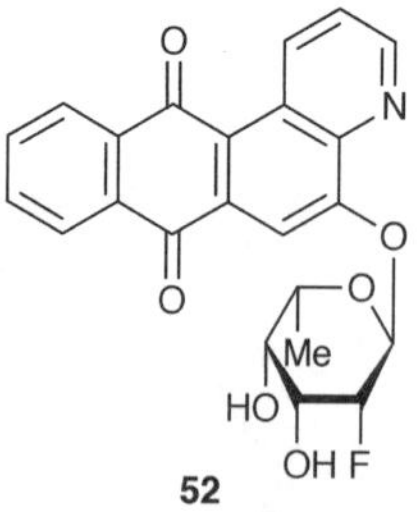

52

Glycosylated compounds containing more than one sugar unit. Interest is devel-oping rapidly in this area, and several new cluster glycosides have been reported. 6-Aminohexyl β-D-glucopyranoside, serving as the 'core', has been tetra-*O*-substituted with (α-D-mannopyranosyl)oxypropyl groups, and the analogues product with trehalose as the core was also made. In the first case the product has a functionalized tether to allow incorporation into more complex systems.[73] Lindhorst's group have also used pentaerythritol as core and made the tetra-3-

hydroxypropyl ether and hence the tetra-α-D-mannopyranosyl glycoclusters aimed at the inhibition of adhesion of *E. coli* to host cells.[74] Related work has led to pentaerythritol with four different *O*-substituents, one being β-D-galactosyl and another β-D-glucopyranosyl,[75] and has resulted in the linking of one and two copies of the triglycosidic amine **53** by amide tethers to C_{60}-based carboxylic acids.[76]

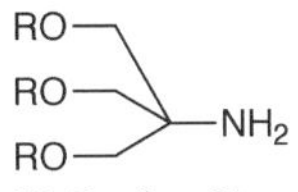

53 R = β-D-Glc*p*

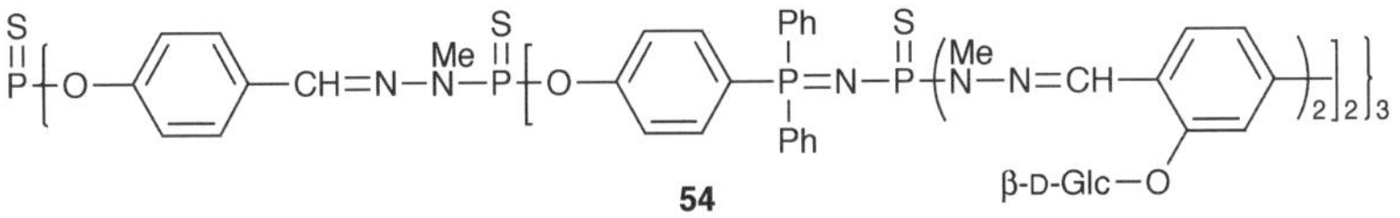

54

O-Glycosylated phenolic glycosides have been substituted into porphyrin dimers to provide neutral and cationic products for studies of photocytotoxicity against cancer cells.[77] The polyfunctional **54** represents a 'carbohydrate-capped' water soluble dendrimer. It and more complex related compounds have been described.[78] Several other dendrimers based on α-D-Man linked through $O(CH_2)_2NHCOCH_2NHCOCH_2NH$ to benzene-1,3,5-tricarboxylic acid and second generation acids of this type, have been reported.[78a] Three copies of the bipyridyl diglycoside **55** self assemble around a Fe(II) ion to give a hexa-glycosidic complex. The α-glucosylated analogue exhibits enhanced affinity for concanavalin A.[79] Base-catalysed condensation of 2,3-di-*O*-benzyl-4-*O*-*p*-methoxybenzyl-β-D-glucopyranosyl azide and two mols of *N*-(tetra-*O*-acetyl-β-D-glucopyranosyloxyethyl)-*p*-nitrophenyl carbamate gave the allophanate **56** with the inter-sugar linkages rigidified by the H-bonded heterocyclic ring.[80]

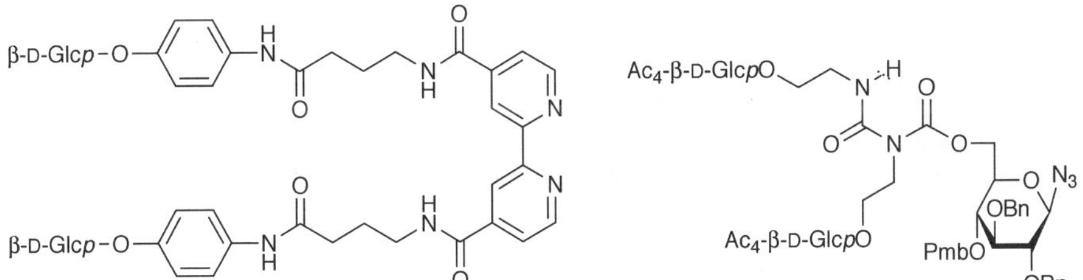

55

56

Orthogonally protected **57** has been selectivity amide condensed to give a trimeric branched product containing three sugar units.[81] Iteration then gave a compound with seven sugars, and the third generation has 15 such units.[82] Complex polyamide **58** has been made by solid phase methods as a high affinity ligand for asialoglycoprotein receptor.[83]

One pot coupling of chlorosilane **59** (X = Cl) with 3-thiopropyl glycosides has given access to the hexaglycosides **59** [X = sugar-$O(CH_2)_3S$, with GlcA, GlcNAc and NeuNAc as the sugars (S)].[84]

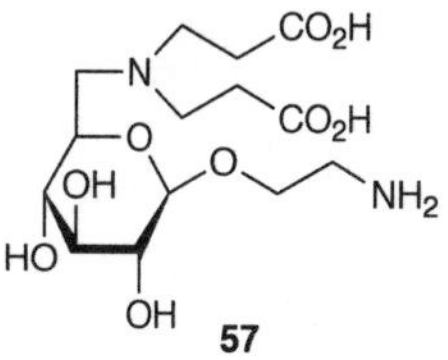

57   RO   58 R = β-D-GalNAc

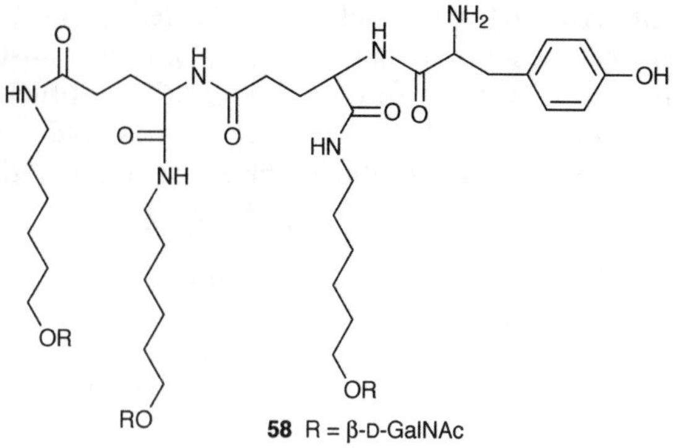

Other glycosides. Pent-4-enyl β-D-glucopyranosides have been converted in high yield to β-glucosides with functionalized alkyl aglycons that can be used as spacer linkages in the preparation of neoglycoconjugates *etc.* The aglycons were, for example, 5-alkylthio- or 5-acylthio-pentyl or -4-butanoyl groups.[85]

α-Mannoside **60** was made as a mimetic of the β-turn of the tetrasaccharide SiaLe^x.[86]

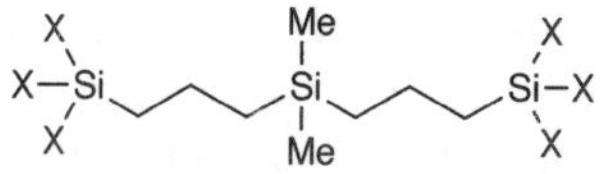

59 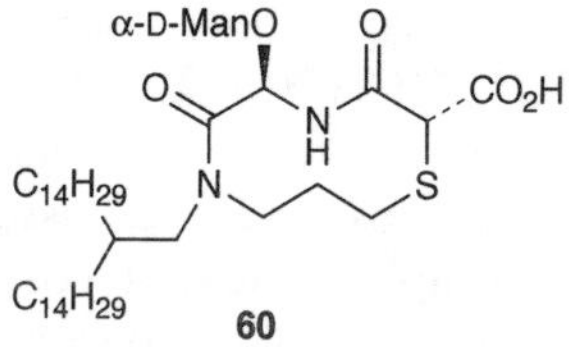   60

1.2 Synthesis of Glycosylated Natural Products and Their Analogues. – In the field of *O*-glycosylated sphingosines 1-*O*-β-D-glucopyranosyl-*N*-stearoyl-D-*erythro*-sphingosine ^{14}C labelled throughout the sugar, and 1-*O*-β-D-glucopyranosyl-sphingosine ^{13}C labelled at C-6 of the sugar have been made for pharmacological and mass spectrometric purposes, respectively.[87] Asymmetic allylboration and cyclizing metathesis applied to the aglycon are key steps in a new synthesis of glucosphingolipids.[88] 4,6-*O*-Benzylidene-D-galactose was the source of both the glycosyl trichloroacetimidate and triol **61** used in the synthesis of α-galactosyl cerebroside **62**.[89] Total syntheses have been reported of phlakosides A and B which have 2-*O*-(4-methylbut-2-enyl)- β-D-galactosyl groups bonded to aglycons **63** in which the stereochemistry at the cyclopropa centres of the main chains differ.[90]

Considerable activity continues with *O*-glycosyl amino-acids. Two reports have appeared on β-D-galactopyranosyl L-hydroxylysine derivatives,[91,92] the first in connection with the development of assays for osteoporosis. An extensive range of related compounds **64** (with R = alkyl and fluoroalkyl), as well as related thioglycoside and 2-*N*-acetyl-2-deoxy compounds have been made as GalCer analogues for anti-HIV testing.[93]

Compounds **65**, with R = H or tetra-*O*-acetyl-α-D-glucopyranosyl, have been made as *O*-β-D-glucosylserine analogues for studies of glycosylated α-aminooxy pseudo peptides. *N*-β-D-Glycosylated oxyamino modifications of amino acids were also made in the course of this work and are referred to in Chapter 10.[94]

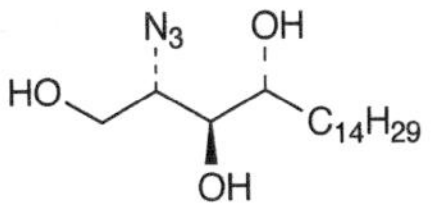

Considerable attention has also been given to the development of *O*-glycosylated peptides and related compounds. For example glucose (and other hexoses) were incorporated into triamide **66** by a simple solid phase method suitable for preparing a range of such glycolipid compounds.[95] By a 1-pot, convergent method α-mannose-bearing cluster compounds were produced by coupling the thiol group of 2-thioethyl α-D-mannosides with *N*-(chloro-acetyl)lysine-based peptides.[96]

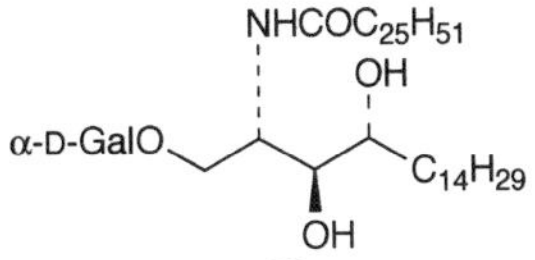

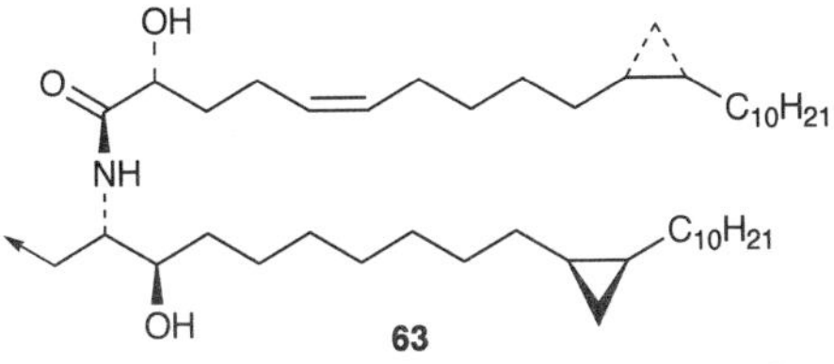

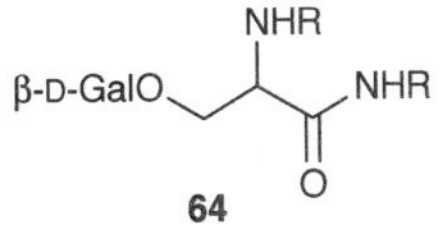

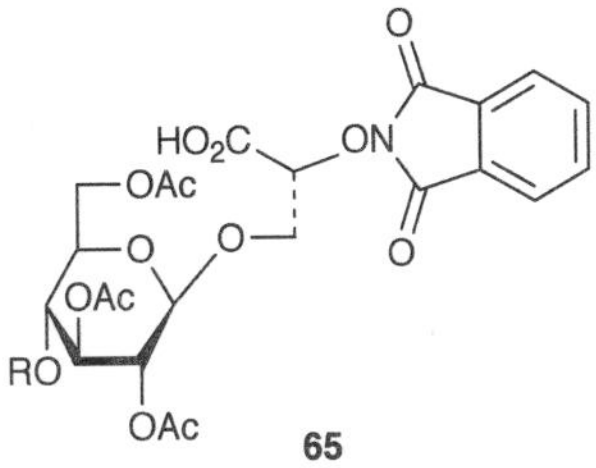

O-β-D-Galactosylation of hydroxylysine gave a product that was incorporated by solid phase methods into a type IV collagen peptide,[97] and 3-aminopropyl β-D-galactoside was coupled with a tetrapeptide coupled in turn to a complex hydrophobic acid to give a glycopeptidolipid as a potential antigen delivery type of product.[98]

Glycosidic group **67** has been introduced into cyclic peptides based on di-amino acids by use of the corresponding *p*-nitrophenyl carbonate in solid-phase synthesis designed to give libraries of cyclic neoglycopeptides containing differ-ent densities of sugars spaced at different distances to each other.[99] Otherwise GlcNAc has been converted to the serinyl glycoside **68** for incorporation into a heptapeptide phosphate from human serum response factor. The elegant chemis-try used in this work involved sequential extensions from the *C*- and *N*-termini of **68** and selective enzymic deesterification at several points.[100] Otherwise sugars, including GlcNAc, can be incorporated into proteins (*e.g.* bovine serum albumin, horseradish peroxidase) directly by application of the Ugi reaction. GlcNAcO(CH₂)₅NH₂, together with acetone and cyclohexylisocyanide, give the bioconjugates **69**.[101]

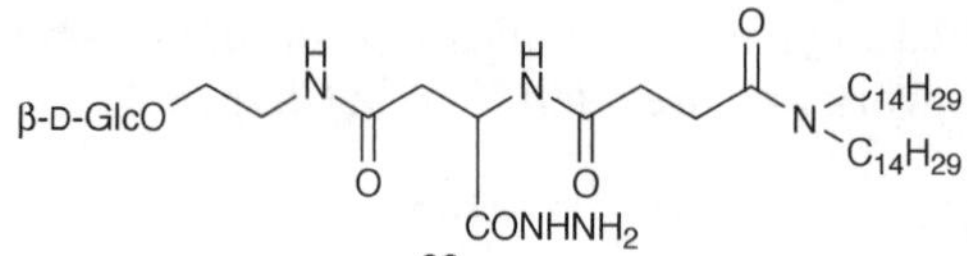

66

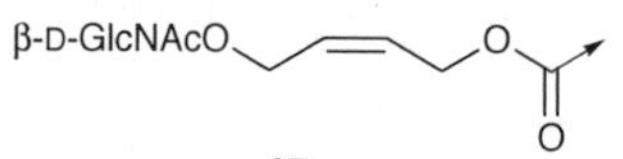

67

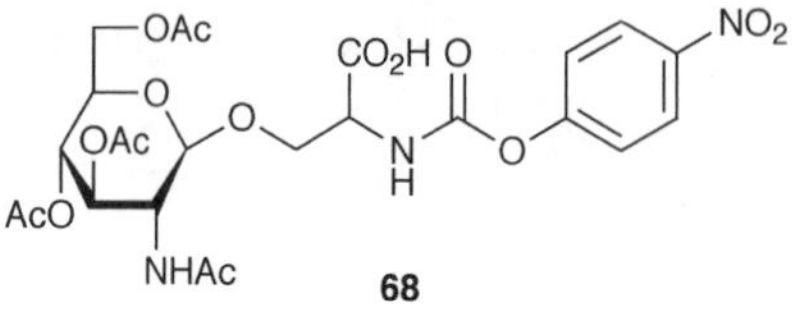

68

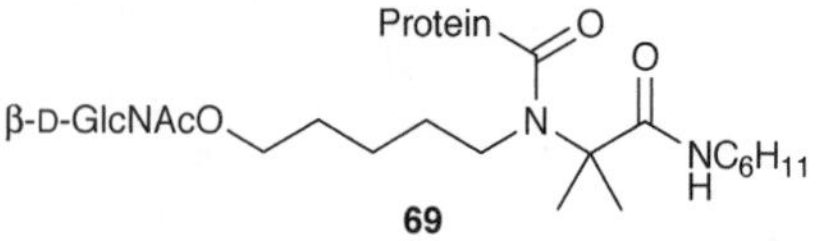

69

A new block synthesis of peptides can be applied with *O*-glycosylated derivatives, and in this way the 82-mer diptericin **70** has been made by methods
involving the coupling of cysteine to a benzylthio ester.[102] Solid phase methods
were used to make serglycin glycopeptides *e.g.* **71**.[103]

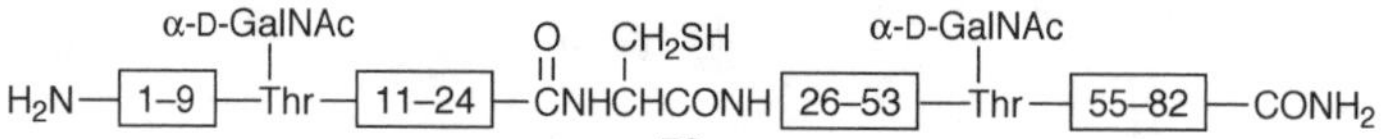

70

Interest in glycosyl inositols is now extremely high. β-L-Arabinopyranose,
α-D-galactopyranose and α-D-Gal-(1→6)-α-D-Gal have been bonded to O-2 of
myo-inositol,[104] and the 2-*O*-α-D-mannopyranosyl compound, the LAM antigenic component of *Mycobacterium tuberculosis*, was made by use of the pentenyl glycosylation method.[105] A new synthesis of 1-D-6-*O*-(2-amino-2-deoxy-D-
glucopyranosyl)-*chiro*-inositol 1-phosphate and 1,2-cyclicphosphate depended
on *trans*-diaxial opening of an epoxide followed by 2-azido-2-deoxy glucosyl
trichloracetimidate glycosylation.[106]

In the area of glycosylated deoxyinosamines a series of neamine analogues
acylated at N-6′ with aroyl groups, *e.g.* **72**, were tested for their binding affinity to
RNA.[107] Sinay's method for converting 6-deoxyhex-5-enopyranosides into
deoxyinositols with retained aglycons has been extended to the conversion of
diunsaturated disaccharides to di-(deoxyinositols) (Chapter 18). Diels–Alder
chemistry has been used to develop the carbocycle of a derivative of 4-*O*-β-D-
glucopyranosyl-5a-carba-α-L-idopyranose (Chapter 18).

Glycosides containing aromatic rings in the aglycons to have been made are:
α-L-rhamnosides of *N*-substituted *p*-(thiocarbamoylmethyl)phenol which are
hypertensive compounds found in *Moningaoleifera*;[108] astilbin (**73**) an aldose
reductase inhibitor used in Chinese folk medicine;[109] noviose-based inhibitors of
gyrase B *e.g.* **74**[110] and the prodrug **75** which releases daunomycin on treatment
with a β-galactosidase.[111]

Several reports appeared on the preparation of nonaromatic, cyclic natural
products. The trichloroacetimidate procedure has been applied to make the
β-glucuronoside of steroid derivative **76**,[112] and the morphinyl 3,6-diglycoside of
the same sugar.[113] Morphine 6-α-glucuronoside was made using the *O*-acetyl
glycosyl bromide methyl ester as a reference marker for testing the purity and
stability of the pharmaceutically important β-anomer.[114] Analogues of the insec-

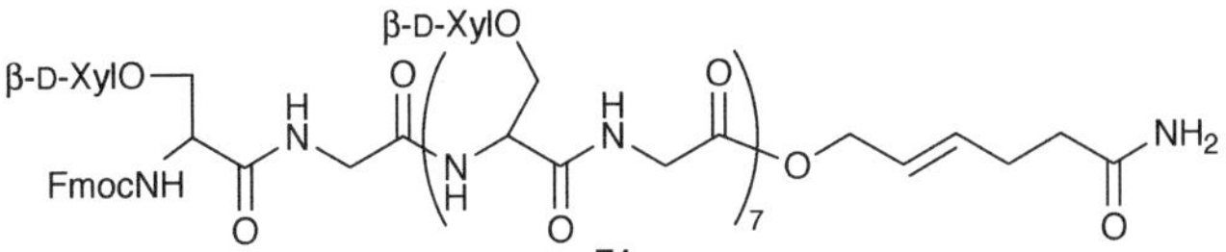

ticide spinosyn (**77**) without methyl and also without methoxy groups at positions 2′, 3′ and 4′ have been biologically tested.[115]

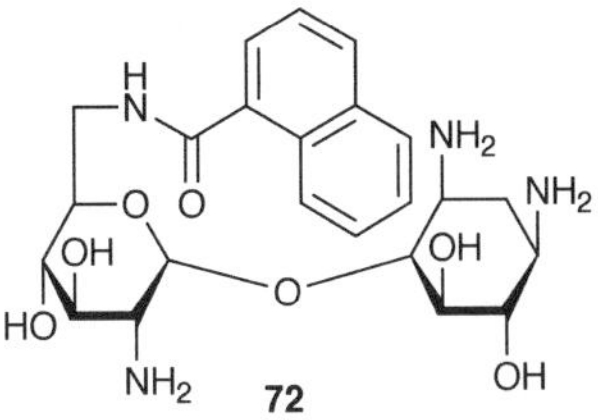

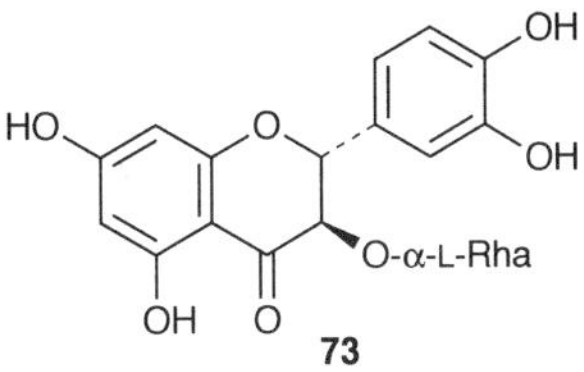

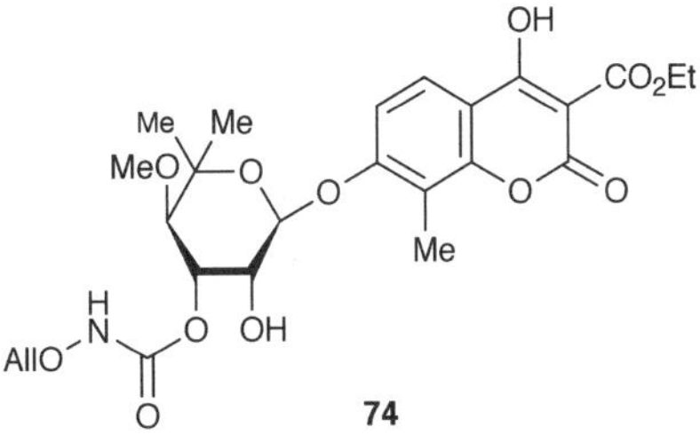

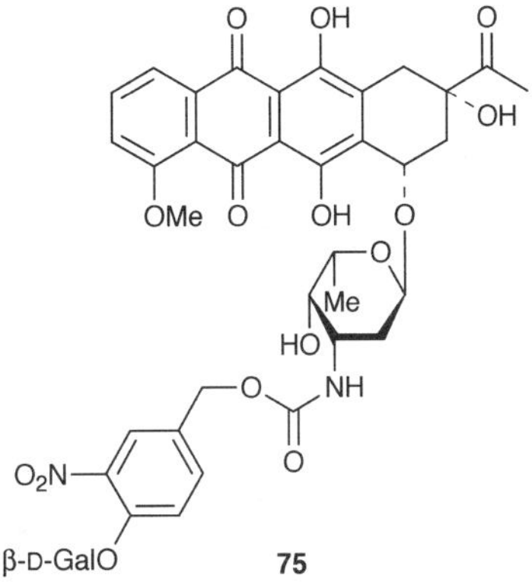

The 3β-hydroxy triterpenes of the oleanane, ursane and lupane series have been converted to the 2-deoxy- and 2,6-dideoxy-α-L-*arabino*-hexopyranosyl derivatives by addition processes applied to L-glucal and L-rhamnal.[116] β-Glucosides of diol **78** (R = R′ = H) have been made with the sugar bonded separately to each of the hydroxyl groups.[117]

1.3 Glycosides Isolated from Natural Products. – As usual in this series, only a selection of papers published on this aspect of glycoside chemistry are dealt with: those describing compounds with notable biological properties or structural features.

The leaf-closing substance **79** has been isolated from a nyctinastic plant,[118] and compound **80** is a new lignan glycoside isolated in tiny proportions from the roots of *Saussurea lappa*.[119] Xyloside **81** is a new glycoside isolated from the mycelia of *Hericium erinaceum* which contains compounds that stimulate nerve growth factor synthesis. Two derivatives were made by chemical transformations.[120]

3-*O*-α-D-Galactopyranosyl-D-*chiro*-inositol is a novel product found in buckwheat seeds,[121] and the further glycosylated inositol derivative **82** has been isolated from the roots of *P. acidus* as a derivative bonded to a terpene at the indicated hydroxyl group. Extracts of the plant have been used in alcohol rehabilitation in Thailand.[122]

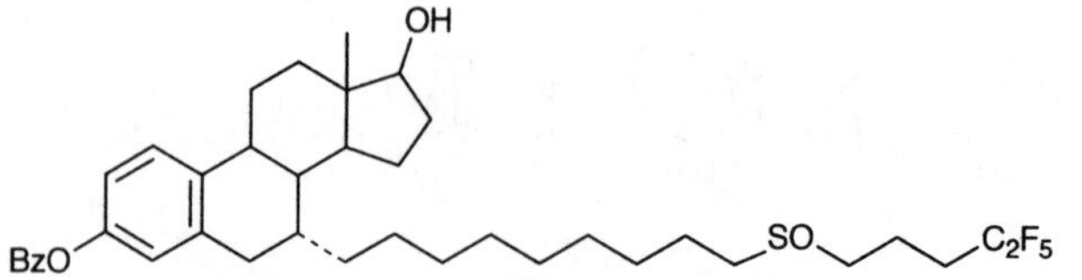

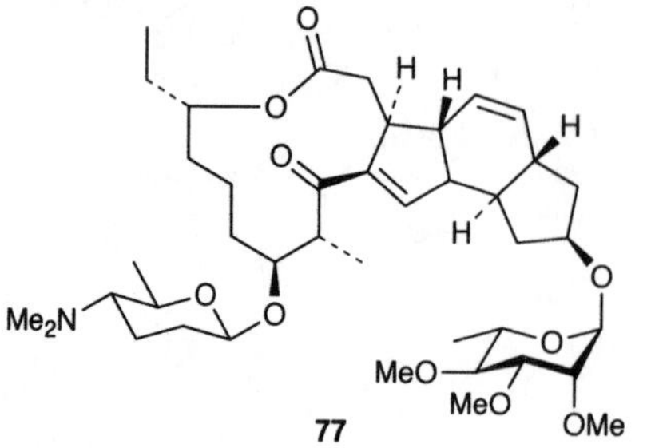

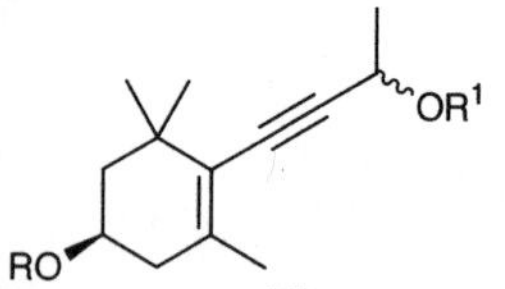

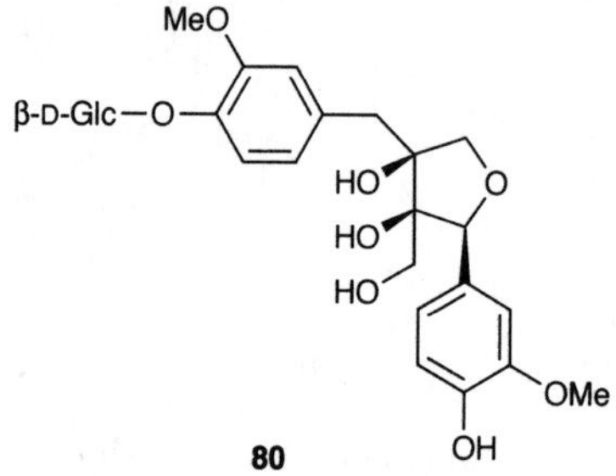

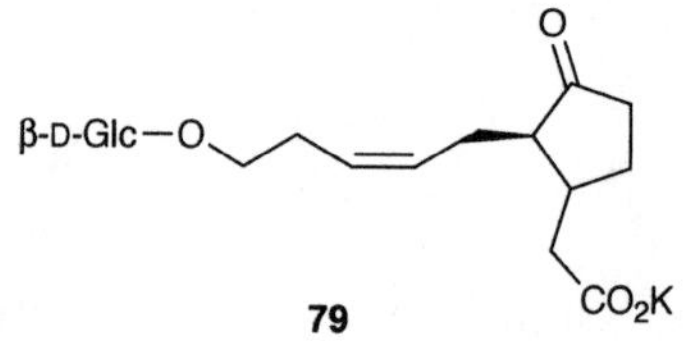

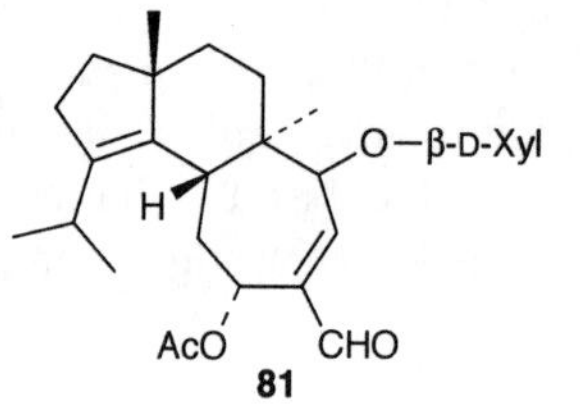

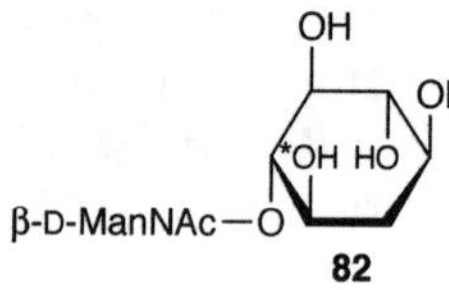

The absolute stereochemistry of oceanapiside, the antifungal compound **83** isolated from a marine sponge, has been determined.[123] Four new cerebrosides which induce neuronal differentiation in rat PC12 cells have been isolated from an edible Chinese mushroom and found to have structures **84–87**.[124]

Further examples of the family of fattiviracin antiviral antibiotics with different lengths of fatty acid chains (see Vol. 32, p. 244) include compound **88**.[125] Compound **89**, with R = 2,3-di-*O*-methyl-α-L-fucosyl-(1→3)-2,4-dimethyl-β-D-xylose, is a powerful toxic metabolite of the red alga *Polycavernosa tsudai* and its total synthesis has been reported.[126] The biosynthesis pathways to vicenistatin 1, **90** (R = 2,4,6-trideoxy-4-methylamino-β-D-*ribo*-hexose) and vicenistatin 2, **90**

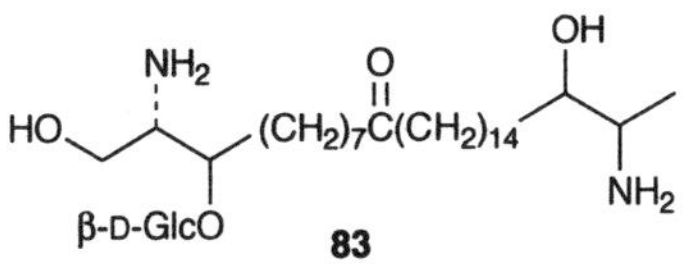

The following structures are shown:

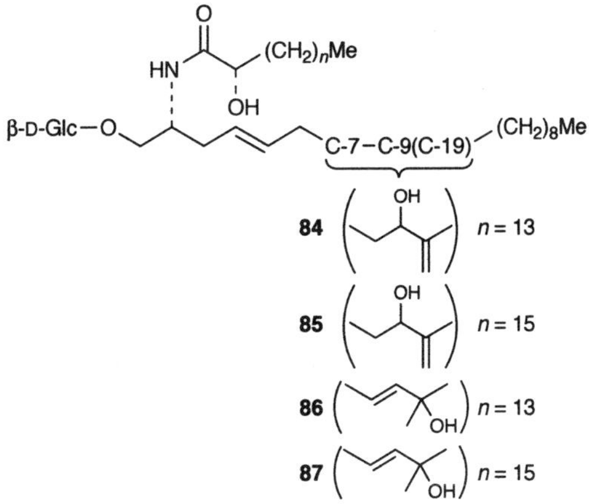

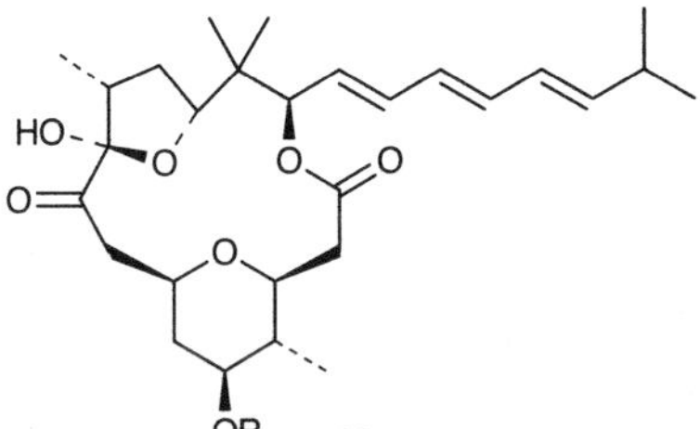

Structure **83** (β-D-GlcO), with groups NH$_2$, (CH$_2$)$_7$C(CH$_2$)$_{14}$, OH, NH$_2$.

Structures **84**, **85**, **86**, **87** (β-D-Glc—O) with:
- **84** n = 13
- **85** n = 15
- **86** n = 13
- **87** n = 15

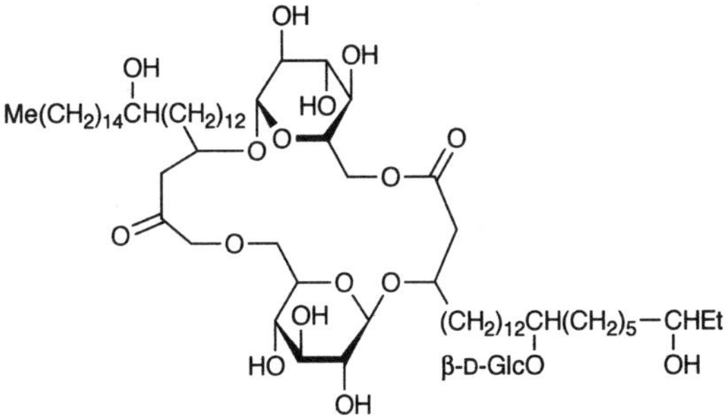

Structure **88**: Me(CH$_2$)$_{14}$CH(CH$_2$)$_{12}$, β-D-GlcO, (CH$_2$)$_{12}$CH(CH$_2$)$_5$—CHEt, OH.

Structure **89** (OR, R = β-mycarose).

(R = β-mycarose) have been established, the sugar units being derived from a common precursor.[127]

Gentiobiose glycosides of (*R*)- and (*S*)-terosachrysone and anthraquinones have been isolated from toadstools,[128,129] and a sophorose-containing flavanone has been found in the stem bark of *Ochna calodendron*.[130] An unusual disacchar- ide derivative, a tricyclic bisglycoside of β-D-glucopyranose and 2-*C*-hydro- methyl-D-threose has been isolated from a plant source. (See this chapter, Section 1.4.9).

1.4 Synthesis of Disaccharides and Their Derivatives. – *1.4.1   General*. Much material in Section 1.1.1 (Methods of synthesis of glycosides) is also relevant to the methods developed for the synthesis of disaccharides.

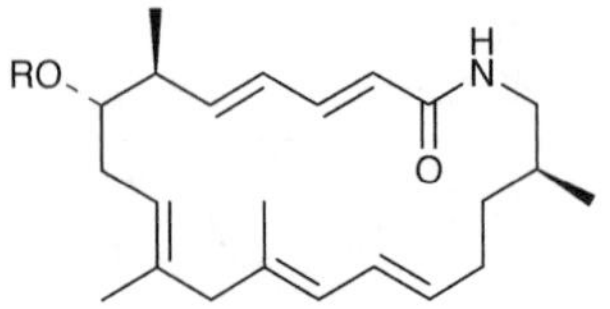

90 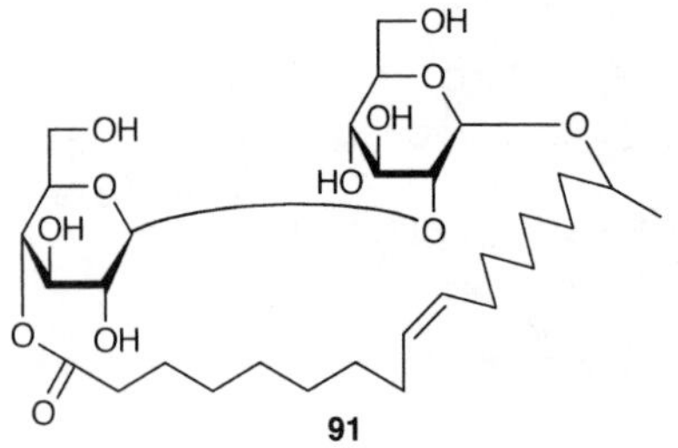   91

1.4.2 Non-reducing Disaccharides. An elegant stereospecific synthesis of sucrose, dependent on the novel β-D-fructofuranosyl donor **92**, is illustrated in Scheme 5. The donors were produced in about 30% yield, the coupling steps were about 70% efficient and deprotection was quantitative.[131] A detailed conformational analysis has been conducted on α-D-mannopyranosyl β-D-galactopyranoside and on the analogue with a methylene group replacing the inter-unit oxygen atom. The former adopted the *exo-syn/exo-syn* conformation as required by the exo-anomeric effect, whereas the latter was much more flexible.[132] Condensation of 3,4,6-tri-*O*-benzyl-1,2-dibutylstannylene-β-D-mannose with 2-*O*-acetyl-3,4,6-tri-*O*-benzyl-D-glucosyl trichloroacetimidate gave access to β-D-glucopyranosyl β-D-mannopyranoside in 68% yield in a novel approach to the synthesis of non-reducing disaccharides.[132a]

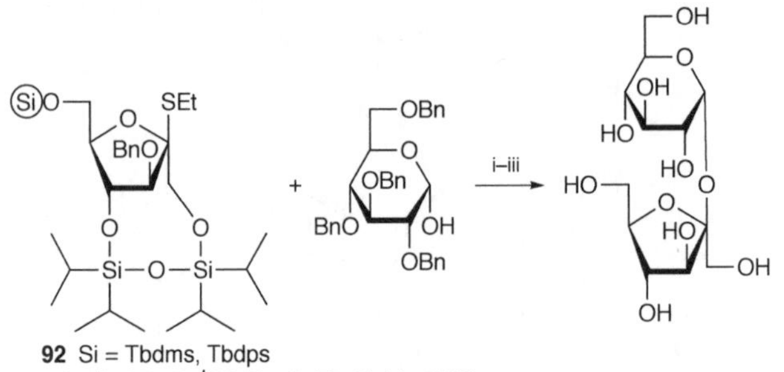

92 Si = Tbdms, Tbdps
Reagents: i, Me$_2$(MeS)S$^+$ $^-$OTf, Bu$^t_2$MePy; ii, F$^-$; iii, H$_2$, Pd/C
Scheme 5

1.4.3 Glucosyl Disaccharides. The sophorolipid lactone **91**, which is a microbial surfactant, has been made from a sophoroside having an unsaturated C$_{11}$ aglycon and a related O-4′ acyl substituent by metathetical cyclization.[133] A concise synthesis of 6-*O*-acetyl-α-D-kojibiose [α-Glc-(1→2)-Glc] bonded to O-3 of 1,2-di-*O*-acyl-sn-glycerols (acyl = C$_{14}$, C$_{16}$ groups) has been described.[134] The syntheses of 1,2-linked diglucosides have been effected by use of tetra-*O*-benzyl-α-D-glucopyranosyl trichloroacetimidate and methyl 3-*O*-acetyl-4,6-*O*-benzylidene-α-D-glucopyranoside. With Sm(OTf)$_3$ as catalyst and acetonitrile as solvent 4:1 β-selectivity was obtained; in ether/dioxane the selectivity favoured the α-product by 3.5:1. Similar methods were applied to the preparation of the 1,3-linked glucobioses.[135]

Reduction of glycosylidene acetals, made by condensing aldono-δ-

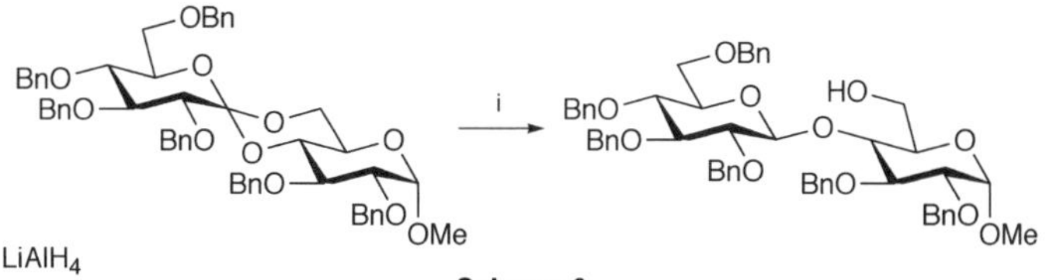

Reagent: i, LiAlH₄

Scheme 6

lactones with 4,6-diols of hexopyranosides, has been used efficiently in the synthesis of methyl α -D-cellobioside (Scheme 6). Otherwise orthoesters derived from sugars with axial C-4 hydroxyl groups, on reduction with NaBH₃CN/AlCl₃, give β-1,6-linked products.[136] Condensation of 3,4,6-tri-*O*-benzyl-2-*O*-benzoyl-β-D-glucopyranosyl fluoride with methyl 2,3,6-tri-*O*-benzyl-α-D-glucoside and with triflic acid as catalyst gave the cellobioside product in good yield.[137] Better yields with good β-selectivity were observed when tetra-*O*-benzyl-β-D-glucosyl fluoride was the donor, and similar results were obtained with analogous galactopyranosyl β-fluorides.[138] In closely related work methyl 2,3,6-tri-*O*-benzyl-α-D-glucopyranoside treated with lauryl or stearyl tetra-*O*-benzoyl-1-thio-β-D-glucopyranoside in the presence of NIS/TfOH gave the β-disaccharide in 94% yield. The advantage of these donors is that they are made from almost odourless thiols.[139] Reference to a further Mukaiyama β-glucosylating agent is made in reference 147.

Amongst a set of glycosyl methanethiosulfonates, made to permit attachment of sugars to proteins *via* disulfide linkages, was a cellobiosyl compound.[140] Similarly, selectively *O*-substituted maltose thioglycosides have been produced as building blocks for the stepwise synthesis of starch-derived oligosaccharides,[141] and aryl maltosides have been made for assaying α-glucosidases.[142]

By solid phase methods, with the aglycon selenium-bonded to a resin, a gentiobiosyl benzopyran (chromene) glycoside was made.[143] Methyl β-D-gentiobioside, amongst several other glycosides, has been produced in about 20% yield by enzymic transglycosylation from *p*-nitrophenyl β-D-glucopyranoside.[144]

In the area of glucosylated heterodisaccharides α-D-Glc-(1→3)-α-D-Man has been made as its 4-methylumbelliferone glucoside as a substrate for an endo-1,2-mannosidase.[145]

Methyl β-D-galactopyranoside, activated with dibutyltin oxide and treated with tetra-*O*-pivaloyl-α-D-glucopyranosyl bromide, gave the 6-β-linked product in 72% yield. In this work the term 'open glycosylation' is used to imply an unprotected acceptor strategy.[146] Related work with the glucuronic acid analogue as donor is reported in Section 1.4.8.

3,4,6-Tri-*O*-benzyl-2-*O*-toluoyl-β-D-glucopyranosyl phenyl carbonate activated by TrB(C₆H₅)₄ is a new β-glycosyl donor and has provided means of obtaining 2-amino-2-deoxy-D-glucose derivatives having β-D-Glc linked separately to O-6 and O-4.[147] 6-*O*-α-D-Glucosyl-D-glucosaminol and the mannosaminol analogue were made with *N*-methacroyl substituents and copolymerized. The physical properties of the products were examined.[148]

α-D-Glc-(1→4)-α-L-Rha-OMe and higher fragments of an O-specific *Shigella* polysaccharide have been synthesized.[149]

Considerable effort has gone into the synthesis and study of adenophostin A (**93**, see Vol. 31, p. 31–32, 258 for preliminary accounts) and analogues which are agonists of IP_3 receptor. Several analogues carrying different aromatic aglycons were made from an *O*-substituted glucosyl-ribofuranosyl acetate,[150] and compounds with alkoxy groups in place of the heterocyclic base, including MeO (**94**), PrO, and $PhCH_2CH_2CH_2O$ were described. In this work cluster compounds involving up to four of the groups **95** bonded to a central benzene ring were produced. Cooperative binding to the tetravalent IP_3 receptor was not observed.[151] Cyclophostin, which has the 5 and 6' hydroxyl groups of adenophostin A tethered together with a but-1,4-diyl group, has been synthesized, conformationally analysed by NMR and modelling methods and biologically evaluated.[152]

1.4.4 Mannosyl Disaccharides. α-D-Man-(1→2)- and α-D-Man-(1→4)-D-Man have been made by standard glycosylations and suitable for chain extensions as the antigenic factors of *Candida guillier-mondii*.[153] The former disaccharide has also been made as its propyl glycoside 6'-phosphate which represents a partial structure of the glycosyl phosphatidylinositol anchor.[154] The α- and β-(1→3)-linked mannobioses were obtained selectively on mannosylation of methyl 6-*O*-acetyl-α-D-mannopyranoside using an ethylthio glycosyl donor. The same couplings were then repeated with a peptide linker joining the two mannose units.[155]

By use of trichloroacetimidate coupling and preparation of monohydroxy acceptors made with the aid of specifically removable 4-bromobenzyl ether groups α-D-Man-(1→2)-D-Glc was made in multi protected form for use in automated synthesis of specific oligosaccharides.[156] Seeberger's group have reported the synthesis of β-D-Man-(1→2)-D-Glc by direct use of tetra-*O*-benzyl-α-D-mannopyranosyl diphenyl phosphate as donor. With TMSOTf as activator and in CH_2Cl_2 as solvent 88% yield of the disaccharide (α:β 3:2) was achieved.[157]

An interesting survey has appeared of the synthesis of D-Man-(1→4)-D-Glc using intramolecular strategies with succinate or malonate tethers between O-2,3,4 or 6 of the thiomannoside donor and O-2,3 or 6 of the acceptor. Various conditions were found that enabled the synthesis of α- or β-(1→4) linked products exclusively.[158] The same authors published independently on a related study of the synthesis of D-Man-(1→4)-D-GlcNH$_2$ compounds. Malonate linking between O-3 of the aminosugar and O-6 of the mannose led to a method for an exclusive β-linking.[159] An intermolecular method for mannosylating O-4 of a 2-azido-2-deoxyglucose derivative using a sulfoxide donor also described details that allowed the preparation of the β-linked disaccharide exclusively, and others that favoured the production of the α with modest selectivity. Steric buttressing was used to assist the direction of coupling.[160]

α-Linked products were strongly favoured when tetra-*O*-benzyl-α-D-mannopyranosyl phenyl sulfinate was used with $H_3PW_{12}O_{40}$ as a cheap and otherwise desirable activator. α-D-Man-(1→6)-D-Glc was made with >99% α-selectivity by this means.[161]

1.4.5 Galactosyl Disaccharides. α-D-Gal-(1→3)-D-Gal was made following galactosylation of 1,2:5,6-di-*O*-cyclohexylidene-D-galactofuranose with phenyl 2,3,4,6-tetra-*O*-benzyl-1-thio-β-D-galactopyranoside activated with NIS/TfOH,[162] and the same disaccharide as its *O*-allyl glycoside has been converted to the 3-aminopropyl glycoside which was bound to *N*-hydroxysuccinimide-activated agarose gel columns, the product being used for immunoabsorption of anti-α-Gal antibodies from human plasma.[162a] Phenyl 1-thio-galactoside technology has also been used to make α-D-Gal-(1→4)-D-Gal as its *O*-protected tolyl 1-thio-β-glycoside, α-trichloroacetimidate and bromide as donors for the synthesis of galactobiosides.[163]

Enzymic methods are significant in this area and bovine β-1,4-galactosyltransferase transfers to C-4-OH of acceptor D-sugars. In some circumstances, however, L-sugars accept β-galactosyl residues at O-3.[164] Enzymic procedures have also led to α-D-Gal-(1→3)-α-D-GalOC$_6$H$_4$NO$_2$(*p*),[165] β-D-Gal-(1→6)-α-D-GalOC$_6$H$_4$NO$_2$(*p*), its β,β-linked isomer, β-D-Gal-(1→6)-α-D-GalNAc-OC$_6$H$_4$NO$_2$(*p*), its β,β-isomer, β-D-Gal-(1→6)-α-GlcNAcOC$_6$H$_4$NO$_2$(*p*) and to its β,β-isomer.[166] Related work has yielded (57%) β-D-Gal-(1→3)-α-D-GlcNAc glycosides.[167]

Lactosylsphingolipids have been prepared[168] as have several β-lactosides with carboranyl residues within the aglycons.[169] Octaacetyllactose is a donor for the synthesis of several α, ω-diol lactosides and bislactosides.[170]

Several new references have been made to lactosamine chemistry. An enzymic procedure using the β-galactosidase from *Bifidbacterium bifidum* immobilized on a nylon powder column permitted efficient recycling of the enzyme and higher yields of LacNAc than were obtained by earlier methods. Yields up to 41% were obtained with good selectivity by use of *p*-nitrophenyl β-D-galactopyranoside as donor.[171] β-D-Gal-(1→3)-α-D-GalNAc glycosides of serine and theonine have been made by use of a phenyl 2-azido-2-deoxy-1-selenyl-α-D-galactoside,[172] as have a range of selectively acylated β-D-Gal-(1→3)- and-(1→4)-β-D-GlcNH$_2$ derivatives by selective enzymic acylations.[173] 4-*O*-β-D-Galactosyl-D-gluconamide and 4-*O*-α-D-glucosyl-D-gluconamide have been linked *via* the amino nitrogen atoms to the aromatic rings of tetraaryl-substituted porphyrins for studies of saccharide-directed cell recognition and molecular delivery.[174] Chapter 4 contains many references to oligosaccharides containing lactose and lactosamine components.

93 R = Ade
94 R = OMe
95 R = OCH$_2$CH$_2$CH$_2$→

96

97

1.4.6 Amino-sugar Disaccharides. Oximinoglycosyl donor **96** has been coupled with four different sugar acceptor with NIS/TfOH as activator to give mainly α-linked products when the reactions were conducted in dichloromethane. In acetonitrile β-glycosidic bond formation is favoured as it is if the thioglycoside is converted *in situ* to a glycosyl triflate as donor. Reduction of the α- and β-linked oximino disaccharides give access to α-linked GlcNAc and β-linked ManNAc products, respectively.[175] On the other hand thioglycoside **97** activated with TrBPh$_4$, I$_2$ and DDQ leads with high selectivity to β-linked GlcNH$_2$ glycosides, β-D-GlcNH$_2$-(1→6)-D-Glc being linked with 94% efficiency by this method.[176]

O- And *C*-glycosides of β-D-GlcNAc-(1→6)-α-D-Gal have been tested as acceptors by two human GlcNAc-6-sulfotransferases,[177] and a report has appeared on the synthesis of a β-D-GlcNH$_2$-(1→3)-D-Gal derivative by use of a thioglycoside coupling reaction.[178] To prepare β-D-GlcNAc-(1→3)-α-Fuc 3'-sulfate glycosyl trichloroacetimidate coupling was used, the product as a 3-(2-aminoethylthio)-propyl glycoside being coupled to bovine serum albumin.[179]

Lipid A analogues lacking the phosphate group at O-4' ,[180] and with a *C*-linked carboxylic acid group replacing the phosphate at C-1[181,182] have been made. In the latter case the alteration did not affect the LPS-antagonistic activity, but the substituent on nitrogen of the non-reducing end moiety is important to the biological properties.[181,182]

Compounds having GlcNAc α-(1→4) linked to 2,3-diacetamido-2,3-dideoxy-α-D-mannose, and to the corresponding mannuronic acid, have been made as 2-phenylethyl glycosides to represent spacer group-containing disaccharide fragments of the lipopolysaccharide of *Bordetella pertussis*.[183]

β-D-GalNAc-(1→4)-β-D-GalO(CH$_2$)$_8$CONH(CH$_2$)$_2$NH$_2$ has been amide-linked to a symmetrical cyclohexane tricarboxylic acid to give one of five cluster compounds to bind to adhesion receptors,[184] and β-D-GalNAc-(1→4)-L-IdoA 2,4'-disulfate, a disaccharide fragment of dermatan sulfate, has been made together with its methyl α-glycoside.[185]

In the area of diamino-sugar disaccharides compound **98**, corresponding to the repeating unit of the O-specific polysaccharide of *Shigella sonnei* has been made,[186] as well as the analogue containing the D-enantiomer of the illustrated acid and the latter compound with the two sugar units interchanged.[187] In Chapter 4, Section 2.2.6 reference is made to a neamine derivative with a spacer-linked 5-amino-5-deoxy-β-D-ribofuranose substituent.

98

1.4.7 Deoxy-sugar Disaccharides. Illustrated in Scheme 7 is the conversion of D-galactal to a β-ketolactone which, in a 1-pot process, was transformed to the reactive dioxothine and trapped by tri-*O*-benzyl-D-glucal in a 4 + 2 cycloaddition step to give a tricyclic product which on desulfurization afforded 2,2'-dideoxy-2,3-unsaturated disaccharide lactone **99**.[188] Disaccharides consisting of β-D-glucose and α- and β-D-galactose linked 1→4 or 1→6 to 2-deoxy-D-glucose and

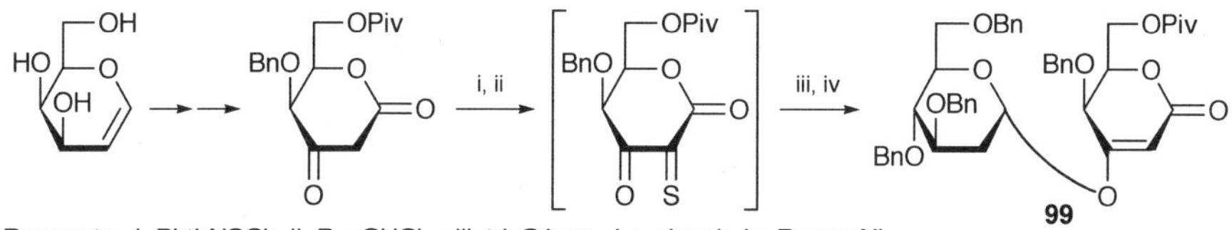

Reagents: i, PhthNSCl; ii, Py, CHCl₃; iii, tri-*O*-benzyl-D-glucal; iv, Raney Ni

Scheme 7

2-deoxy-D-galactose have given products which, by Wittig coupling and further elaboration, have been converted to glycosylphytosphingosines.[189] Compound **100**, a potential disaccharide donor, has been made by direct glycosylation with the deoxy-sugar glycosyl dithiocarbamate.[190] The anti-tumour antibiotic AT 2433-A1 (**101**) has been synthesized by acid-catalysed coupling of the *N*-protected disaccharide to the base followed by treatment with iodine and DBU.[191]

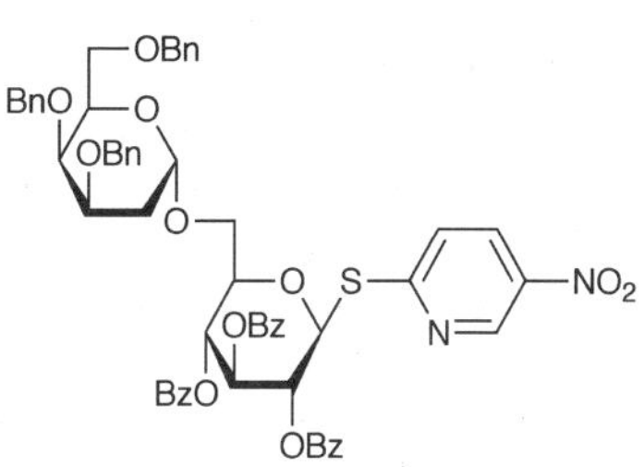

100

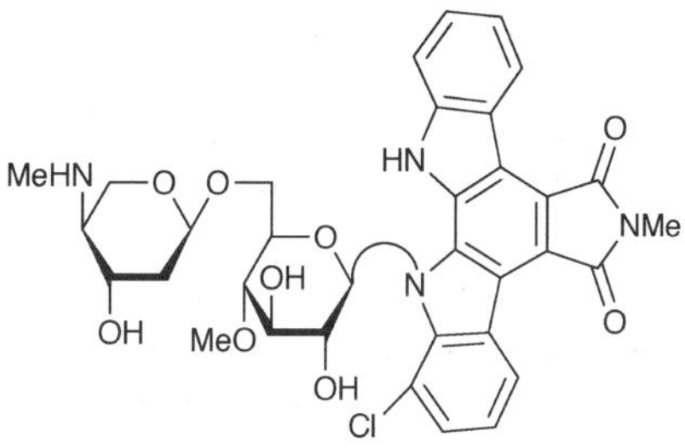

101

Several sulfate and succinate esters of rutin, a flavanol glycoside of α-L-Rha-(1→6)-β-D-Glc, have been made. These have antioxidant activities and also complex with a pigment of the anthocyanin family to modify the natural colours.[192]

α-L-Fuc-(1→2)- and -(1→3)-β-D-GalO-4-methylumbelliferone have been synthesized,[193] and enzymic means have been employed to obtain α-L-Fuc-(1→3)-R where R is D-Glc, D-GlcNAc and various glycosides of the latter.[194] The four α-L-Fuc-(1→X)-β-<F4·d-GlcNAcOMe, X = 2, 3, 4 or 6, were made for NMR conformational studies using 2,3,4-trideuterio-L-fucose.[195]

1.4.8 Sugar Acid Disaccharides. β-D-GlcA-(1→3)-GalNAc 6,2'-disulfate, which represents the repeating unit of chondroitin sulfate D, has been made.[196] The *O*-pivaloated glycosyl bromide of methyl glucuronate can be used to effect 6-β-glycosylation of methyl β-D-galactopyranoside with 78% efficiency provided the acceptor is used after treatment with dibutyltin oxide.[146] A valuable observation was made when thioglycoside **102** was condensed with a 1:1 mixture of α- and β-acceptors **103.** The β-anomer was the more reactive, the products **104** having the α,β ratio 1:4 (Scheme 8), which was interpreted as implying that the hydroxyl group of the latter anomer was more strongly hydrogen bonded intramolecularly to the ring oxygen atom. This in turn is consistent with the weakening of this H-bonding by the endo-anomeric effect in the case of the α-anomer.[197]

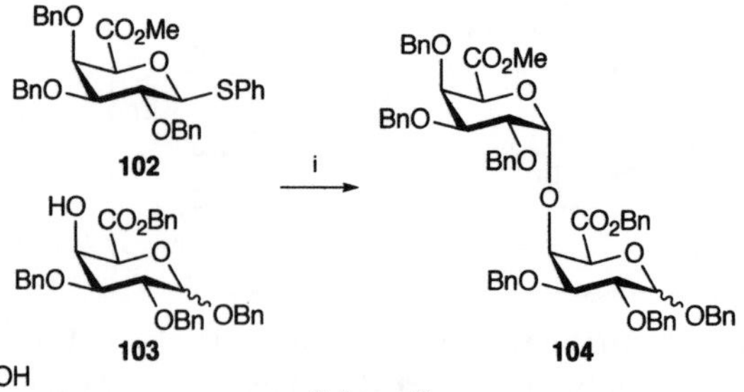

Reagents: i, NIS, TfOH

Scheme 8

A review, *Recent Advances in O-Sialylation*, covers the synthesis of disaccharides including sialic acid dimers and deals with NMR criteria for determining anomeric configuration.[198]

Efficient chemoenzymic syntheses of ganglioside GM4 analogues with potent immunosuppressive activity have yielded α-Neu5Ac-(2→3)- α-D-GalOR with R = $(CH_2)_nCH_3$, $n = 7, 11, 15$; CH_2CH_2TMS, SPh, $CH_2CH[(CH_2)_{13}CH_3]_2$.[199] α-Neu 5Ac-(2→6)-α-D-GalNAc has been made as a spacer group glycoside.[200] Neoglycoproteins containing D-*glycero*-D-*talo*-oct-2-ulosonic acid-based disaccharides, corresponding to the core units of *Burkholderia* and *Acinetobacter* lipopolysaccharides, have been made.[201] This ulosonic acid and its 3-deoxy analogue have been linked to O-6 of GlcNAc.[201a] Ketene dithioacetals derived from aldonolactones react with monohydroxy sugar derivatives in the presence of TMSOTf to give aldos-2-ulosides dithioacetals and these can be converted into glycosides of 2-ulosonic acids. Kdo-based disaccharides have been made in this way.[51]

1.4.9 Other Disaccharides. Several reports relating to pentose-containing disaccharides have appeared. β-Xylobiose has been linked to 4-methylumbellifone and 5-bromo-3-hydroxyindole to give compounds for the sensitive detection of xylanase activity.[202] Twelve steroidal glycosides of β-D-Xyl-(1→3)-α-L-Ara or the 1,4-linked isomer carrying aroyl substitutes have been made and tested for antitumour activity.[203] A 2-(dihydroxyphenyl)ethyl glycoside (conandroside) of β-D-Xyl-(1→3)-D-Glc, also carrying an aroyl substituent, has been synthesized.[204]

Model glycosylations aimed at the synthesis of pradimicin-benanomicin antibiotics have used fluoride **105** as a donor in a range of acceptors including anthracene derivative **106**.[205]

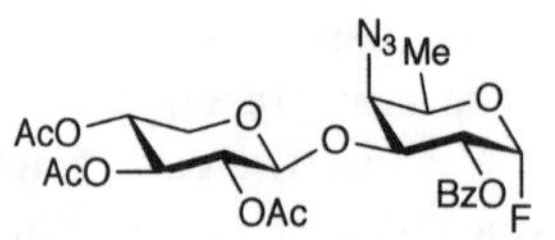

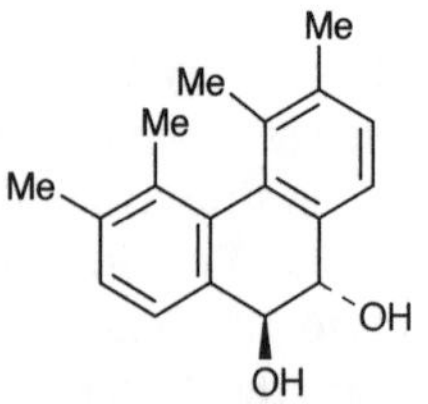

3-*O*-β-D-Apiofuranosyl-D-xylose has been made[206] and the further novel branched-chain dimer **107** has been isolated from rhizomes of *Hemsleya amabilis*.

It is a tricyclic bisglycoside derived from β-D-glucopyranose and the uncommon branched-chain 2-*C*-hydroxymethyl-D-threose.[207]

1.5 Disaccharides Containing Sugar Analogues or with Anomalous Linking.

– Several compounds belonging to these categories are noted elsewhere, *e.g.* in Chapters 3 (*S*- and *C*-glycosides), 11, 14, 18; others are as follows.

107   108

Compounds described as 'sugar rods' and with phytohemagglutinin cross-linking properties, *e.g.* compound **108**, have been made by a Pd-induced coupling reaction of propargyl and *p*-iodophenyl α-D-mannoside tetraacetate, and *p*-diiodobenzene led to the linear product with two propargyl α-D-mannoside substituents on the benzene rings.[208]

In the area of compounds containing *C*-bonded linkages 6(*S*)-methyl tri-*O*-benzyl-6-*C*-vinyl-α-D-glucopyranoside was dimerized by initial coupling two molecules by use of Ph$_2$SiCl$_2$ and then by metathesis to give tetradecaose derivative **109**. The (6*R*)-isomer was similarly converted.[209] A dimer formed by linking ethyl 2,3-dideoxy-β-D-*ribo*-hexopyranoside and 3-deoxy-1,2:5,6-di-*O*-isopropylidene-α-D-allose together at positions 3 and 3′ *via* a methylene group has been described.[210]

The 5a-carba-sugar analogues of α-D,L-GalNAc, α- and β-D,L-GlcNAc and α- and β-D,L-ManNAc were examined as acceptors for β-(1→4)-galactosyltransferase. Both anomers of 5-carba-D-GlcNAc were substrates and gave access to β-D-Gal-(1→4)-5a-carba-α- and β-D-GlcNAc.[211] The octyl glycosides of these were good acceptor substrates for α-(1→3)-fucosyltransferase.[212]

1.6 Reactions and Other Features of *O*-Glycosides.

– The study of the spontaneous hydrolysis of a series of deoxy and deoxyfluoro 2,4-dinitrophenyl β-D-glucopyranosides led to the conclusion that the rates correlate with the stabilities of the oxocarbenium ions involved. Of the deoxy derivatives the 2-deoxy compound reacted fastest, whereas the 2-deoxy-2-fluoro compound was most stable of the deoxyfluoro set.[213] Linkages of sugars (R) to Wang resins of the kind illustrated in **110** are stable to Lewis acid glycosylation conditions. However reduction of the nitro group by use of Sn(SPh)$_2$, PhSH, Et$_3$N is accompanied by cyclization to give the hydroxamic acid **111** which can be cleaved under acidic conditions.[214]

Various tripeptides (**112**) containing serine β-D-galactosides have been examined for base stability. Replacing H by Me at R^2 had no effect, but a similar change at R^1 caused appreciable increase in the rate of β-elimination, suggesting that the carbonyl group labelled A acts as a base to remove the illustrated proton and induce β-elimination of the sugar.[215]

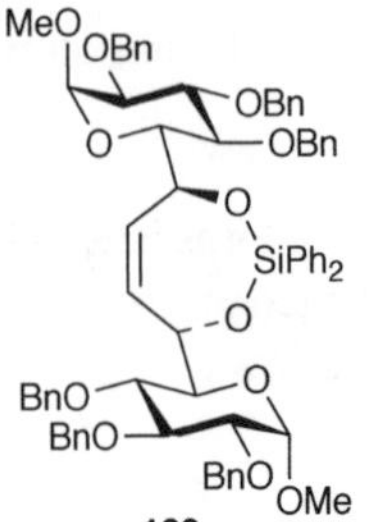

109 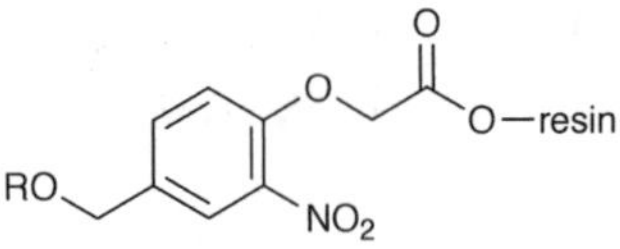   **110** 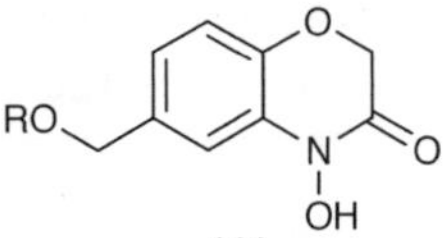   **111**

The dicarboxylate **113** and its epimers at C-2 and at the aglycon chiral centre have been isolated from *Sambucus nigra* suggesting that they are derived biogenically from the corresponding cyanogenic hexosides.[216]

Hydrogen peroxide oxidation of palatinose [α-D-Glc-(1→6)-D-Fru] or trehalulose [α-D-Glc-(1→1)-D-Fru] affords carboxymethyl α-D-glucopyranoside in about 40% yield. Various reactions of this product, *e.g.* its conversion to the bicyclic lactone involving O-2 on treatment under acetylating conditions, were described.[217]

Physical studies on glycosides have included the X-ray diffraction analysis of methyl α-D-galactopyranosides and methyl β-D-glucopyranosides with fatty acid esters at C-6 which showed bilayer structures with zig-zag alkyl chains.[218] The conformational behaviour of several non-ionisable lactose analogues have been examined by NMR and molecular mechanics methods,[219] and the energy profile of methyl β-D-arabinofuranoside as a function of ring conformations has been determined by gas phase computations.[220]

2   S- and Se- Glycosides

Direct syntheses of the *tert*-butyl 1-thiopyranosides can be carried out directly from D-Glc, D-Gal, D-Man, L-Rha, maltose and 2-deoxy-D-Glc by direct reaction at room temperature with butanethiol in the presence of trifluoroacetic acid as catalyst. Except with the last sugar the 1,2-*cis* anomers are formed slectively.[221] β-Fructofuranosidases solvolyse sucrose with 2-thioethanol to give glucose and *S*-hydroxyethyl 2-thio-β-D-fructofuranoside,[222] and the same workers have reported that β-galactosidases catalyse the transfer of D-galactose to 2-thioethanol to give the *S*- and *O*- bonded glycosides in the ratio 2:1.[223]

Many examples, as usual, of the use of thioglycosides in *O*-glycoside synthesis

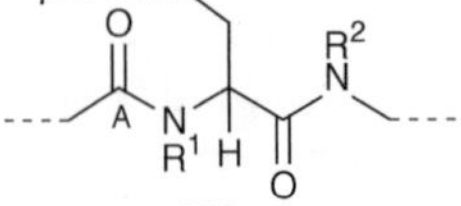

112 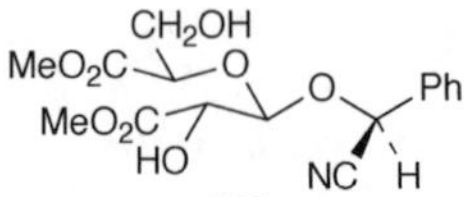   **113** 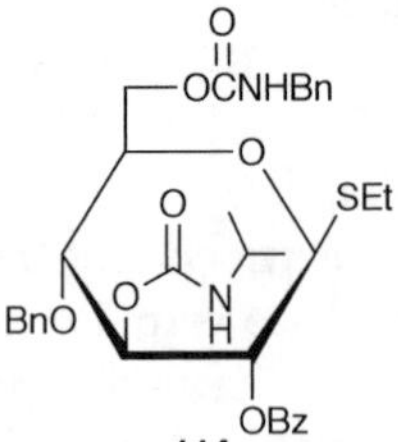   **114**

have been reported; several are noted earlier in this chapter. Compound **114** has been made to offer routes to products which require selective substitution at all five hydroxyl groups of β-D-glucose, and hence it opens a new approach to the preparation of carbohydrate-based libraries.[224]

Sulfur-linked molecular rod-like thioglycosides which are analogues of **108**, and symmetrical diacetylenic compounds of this type, have been reported.[225] *p*-Aminobenzyl thioglycosides of sugars and oligosaccharides have been *N*-linked to biotin and used in the study of the interactions of carbohydrates with proteins.[226] Thioglycoside **115**, the major glucosinolate of the seeds of *Moringa oleifera*, has been synthesized from *p*-(nitroethenyl)phenyl α-L-rhamnoside triacetate which was reduced to the hydroximoyl chloride and coupled to tetra-*O*-acetyl-1-thio-β-D-glucose.[227] Glycosylations of structure **116** with acetohalogeno sugars give thioglycosides rather than *N*-linked nucleosides,[228] and likewise compounds of type **117** react in the thioenolic form also to give thioglycosides.[229]

In the area of thioglycosyl derivatives of amino-acids tri-*O*-acetyl-2-deoxy-2-nitroso-α-D-galactopyranosyl chloride has been treated with the methyl ester of *N*-benzoylcysteine to give the 2-deoxy-2-oximo thioglycosides,[230] and GalNAc has been α-linked to cysteine and converted *via* the carboxylic acid group to the 3-carboxypropylamine amide which is the *S*-analogue of a tumour-associated antigen.[231] In related work two methods were used to prepare *S*-linked β-D-GlcNAc-cysteine, the preferred procedure involving Mitsunobu condensation of 1-thioGlcNAc with a protected derivative of serine.[232] Solid phase coupling of 1-thioGlcNAc (O-6 linked to the resin) to *t*-butyl 3-Fmoc-amino-4-iodo-butanoate gave access to thioglycoside **118**.[233] Resin-bound thioglycosides *e.g.* **119** can be selectively and quantitatively deallylated by use of toluenesulfinic acid with Pd(PPh₃)₄ as catalyst.[234]

S-Linked disaccharides to have been made include the analogues of the methyl glycosides of kojibiose and sophorose which were produced from a non-reducing disaccharide derivative by the rearrangement process illustrated in Scheme 9.[235] Unsaturated thio-linked disaccharide **120** was made by Pd-catalysed coupling of *O*-acetylated β-D-GlcSH with the 4-methylcarbonate of the 2,3-unsaturated ethyl glycoside.[236] The same thioglucose derivative and the corresponding thiols

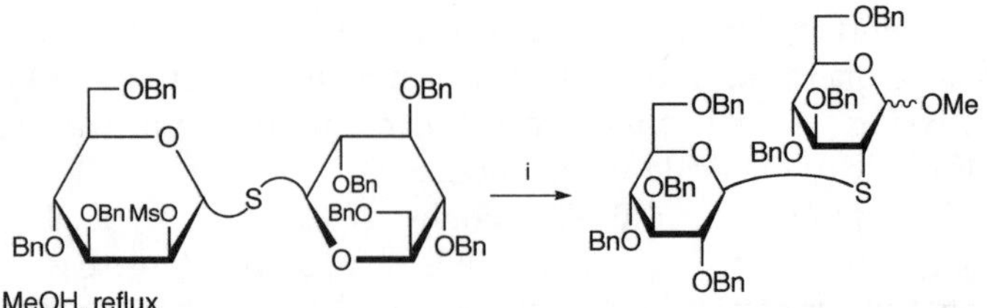

Reagent: i, MeOH, reflux

Scheme 9

of β-D-Gal and α-L-Fuc, added to isolevoglucosenone, gave access to β-D-Glc-(1→2)-3-deoxy-2-thio-D-*ribo*-hexose and the β-D-Gal and α-L-Fuc analogues.[237]

The thio-linked analogue of β-D-Man-(1→4)-D-GlcOMe has been made by coupling a mannopyranosyl sulfoxide with a 4-thioglucose derivative,[238] and a NeuNAc-2SAc compound used to displace the triflate of 1,2:5,6-di-*O*-isopropylidene-D-gulose 3-triflate gave NeuNAcS-(2→3)-D-Gal.[239] A set of unusual *S*-linked branched-chain sugar disaccharides is noted in Chapter 11.

Chemoenzymic methods have been used to make polyoxyalkyl dendimers carrying six copies of β-*S*-linked β-D-Lac and -β-D-GlcNAc which were used as precursors for the synthesis of lactose 3'-sulfate (chemical methods) and LacNAc (enzymic methods), respectively.[240]

N,*N*-Dialkyl-*S*-glycosylsulfenamides *e.g.* **121** were made from the corresponding glycosyl *S*-acetate by treatment with secondary amines in the presence of NBS.[241]

Studies of the reactions of iodine, TmsI, IBr and ICl with *O*-substituted 1-thiogalactosides show them to be dependent on the activating or deactivating nature of the sugar O-2 substituents and the activator used, and may give α- or β-glycosyl halides or cause epimerization. It was suggested that choice of the promoter may allow control of the stereochemical outcome of thioglycoside-based glycosidation reactions.[242]

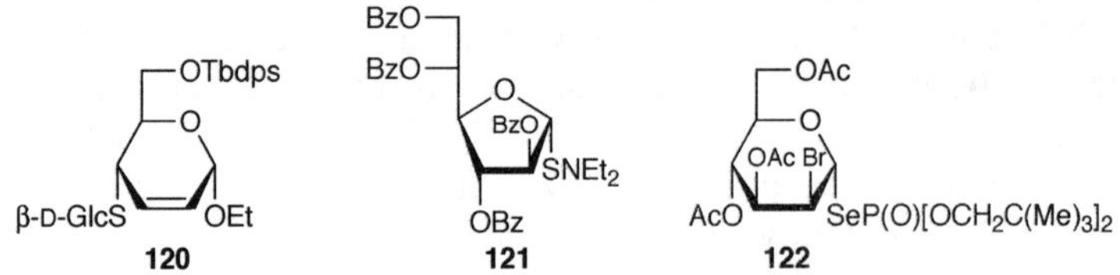

β-D-GlcS   **120**

121

122

Compounds such as the seleno-linked **122** with the α-*manno*- or α- or β-*gluco*-configurations, together with carboxylic acids and in the presence of silver salts, give the glycosyl carboxylates in quantitative yield.[243]

3   *C*-Glycosides

3.1   General. – A review has been written on the use of titanocene(III) and zirconocene(III) chlorides as mild reagents for generating glycosyl radicals from corresponding halides and converting them to their *C*-glycosides,[244] and another dealt with the use of palladium complexes in the synthesis of alkyl or aryl

C-glycosides.[245] Vasella has reviewed his work on glycosylidene carbenes which is of relevance to the subject,[246] and the use of glycosyl carbanions in *C*-glycoside synthesis has also been surveyed.[247]

A further review of compounds bearing CF_2X substituents includes references to relevant *C*-glycosides,[248] and one on acetylenic coupling covers Vasella's work on couplings involving acetylenic *C*-glycosides.[249]

3.2 Pyranoid Compounds. – *3.2.1 Compounds with Short-chain 'Aglycons'.* Reaction of 2-*C*-formylglycals with organocopper reagents can occur in different ways to give saturated or unsaturated alkyl *C*-glycosides as shown in Scheme 10.[250]

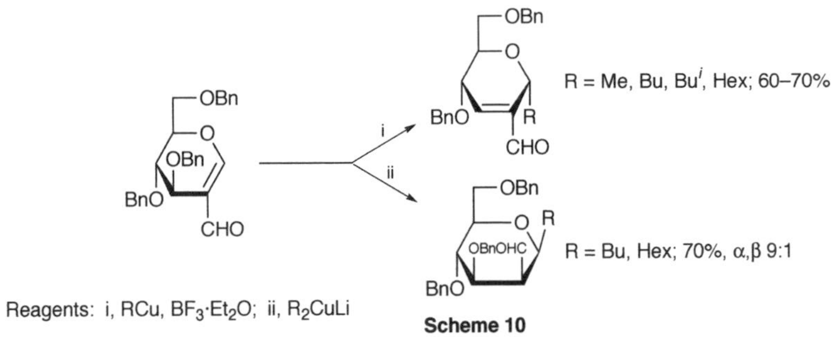

$R = Me, Bu, Bu^i, Hex; 60–70\%$

$R = Bu, Hex; 70\%, \alpha,\beta \ 9:1$

Reagents: i, RCu, $BF_3 \cdot Et_2O$; ii, R_2CuLi

Scheme 10

O-Acetylated *S*-glycosyl phosphorothioates treated with TmsCN and an acid catalyst give glycosyl cyanides,[251] and 3-ketoglycals similarly afford 2-deoxy-2-keto-glycosyl cyanides.[252] From tri-*O*-benzyl-D-galactal 2-tolylthiogalactosyl cyanides and various related *C*-glycosides can be made *via* the corresponding glycosyl chlorides with good selectivity.[253] A useful method for making 1,2-*cis*-related glycosyl cyanides involves reduction of readily available nitromethyl *C*-glycosides to the oximes followed by dehydration with acetic anhydride.[254] 2,6-Anhydro-3-deoxy-D-*erythro*-L-*gluco*-nonate, a *C*-glycosidic analogue of Kdn, has been made from D-isoascorbic acid.[254a]

C-Hydroxymethyl α-D-glucopyranoside, a 2,6-anhydroheptitol made from the aldonolactone, was converted to the 1,4,5-trisphosphate as a mimic of the analogues D-*myo*-inositol triester (*cf.* **125**),[255] and from α-D-galactopyranosyl formic acid a C_{18} amido substituent was introduced to give an analogue of galactosyl ceramide.[256] Compound **123** represents a modification of a cyclic peptide with Phe-Val replaced by the *C*-glycosidic amino uronic acid moiety and was made for studies of binding to integrins.[257]

In the area of *C*-glycosides with C_2 aglycons tetra-*O*-benzyl-β-D-mannopyranosylacetylene, treated with butyllithium, gives the glycal derivative **124** in better yields than do the α-*manno* or β-*gluco* isomers.[258] Further work on inositol trisphosphate mimics has yielded compound **125** made by the ingenious use of Tbdms protecting groups to hold the starting material in the unusual $^1C_4$ conformation (Scheme 11).[259] Similar work with an analogous allylsilyl intermediate gave access to 3-hydroxypropyl α-*C*-glycoside exclusively (85%), whereas the same chemistry carried out with *O*-benzyl protection of the phenylselenyl

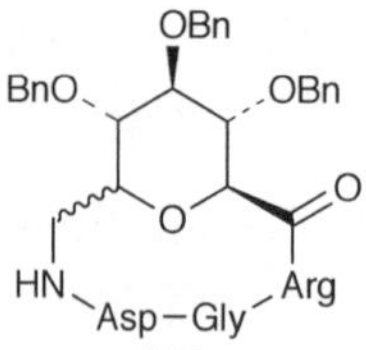

Scheme 11

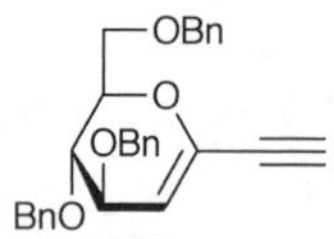

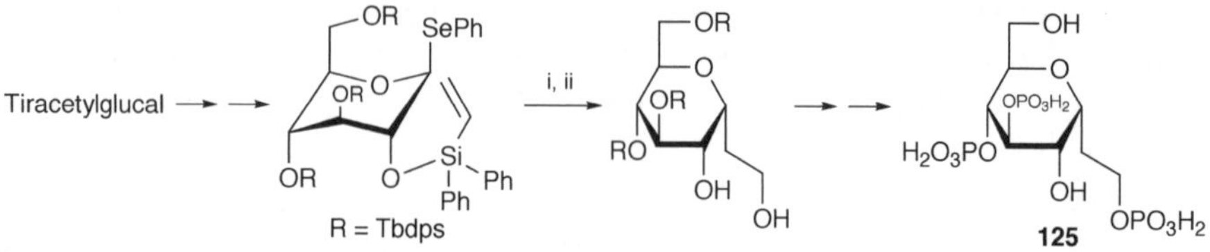

Reagents: i, Bu₃SnH, AlBn; ii, H₂O₂, NaHCO₃, MeOH, H₂O

donor afforded the anomeric β-compound in 80% yield in a 4:1 mixture with the α-isomer.[260] α-GlcNAc-*C*-allyl has been used to make the α-*C*-2-hydroxyethyl glycoside as the phosphate ester. Also, by way of the 2-bromoethyl glycoside, the corresponding phosphonate was produced.[261]

Various *C*-glycosides of NeuNAc have been made from the glycosyl phenyl sulfone which, with samarium iodide, gave the glycosyl anion which added to simple carbonyl electrophiles.[262] In somewhat related manner the aldehyde derived from 2,3:4,5-di-*O*-isopropylidene-D-fructose gives related adducts with *e.g.* Grignard reagents (although the products are technically not *C*-glycosides but rather extended chain ketoses).[263]

Cyanomethyl *C*-glycosides can be made from aldonolactones *via* the Wittig-derived cyanomethylene adducts,[264] and carboxymethyl β-*C*-glucopyranoside has been condensed with a series of primary amines to give *C*-glycosyl amide derivatives.[265] The 2-phenylethyl α-*C*-glycoside of GalNAc was made by a radical procedure from a phenylseleno glycosidic precursor.[266]

Homodimerization of allyl tetra-*O*-benzyl-α-*C*-mannopyranoside with Grubbs' catalyst gave access to the hexadecitol derivative.[267] Allyl *C*-glycosides have been made by Et₃SiH reduction of the hemiacetals made by allyl additions to aldonolactones. In this way allyl β-*C*-lactoside was produced.[268] Treatment of 1,2-anhydro-3,4,6-tri-*O*-benzyl-α-D-glucopyranose with *e.g.* triallylaluminium occurs by a *cis*-opening procedure to give the allyl α-*C*-glucoside.[269] α-D-Gal-NAc-*C*-All has been made from the *gluco*-isomer by a displacement with inversion at C-4, and oxidized to the *C*-acetonyl compound **126**.[270] The same type of acetonyl products, except the β-anomers, were made by Knoevenagel condensation of pentane-2,4-dione with unprotected sugars *e.g.* glucose, mannose or cellobiose in alkaline conditions.[271] *C*-Linked glucosinolate analogue **127**, which is a substrate for the enzyme that desulfates natural *O*-glucosinolates, was made by oximation from the carbonyl precursor, itself obtained from 4,6-*O*-benzylidene-D-glucose by treatment with dimethyl 2-oxo-3-phenylpropan-1-yl phosphonate.[272]

Free radicals derived from *O*-acetylated glycosyl bromides with nickel catalysis add to acrylates and acrylonitrile to give *C*-glycosides with axial C-1 substituents as expected.[273]

A sulfoxide derived from phenyl 3,4-*O*-isopropylidene-1-thio-β-L-fucoside, on treatment with MeLi followed by *tert*-BuLi, gave the glycosyl carbanion which added to *iso*-butanal to give β-*C*-glycosides as an epimeric pair, and the α-thioglycoside led to the α-*C*-glycosides.[274]

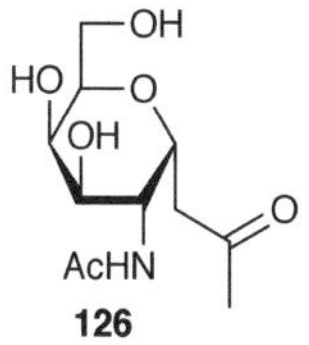

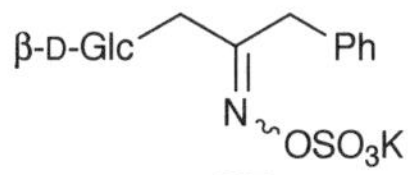

Allyl *C*-glycosides can give access to compounds with longer chain aglycons. For example compound **128** is obtainable by a Grubbs' reagent condensation applied to the α-*C*-allyl compound, and more complex compounds made by this approach include a peptidomimetic (see Section 3.2.6) and aryl glycoside cluster compounds.[275] An interesting way of conducting chain extension on *C*-allyl glycosides uses metathesis involving an unsaturated substituent on O-6. For example, compound **129** was made in this way and hydrolysed to give a *C*-glycoside with an extended aglycon.[276]

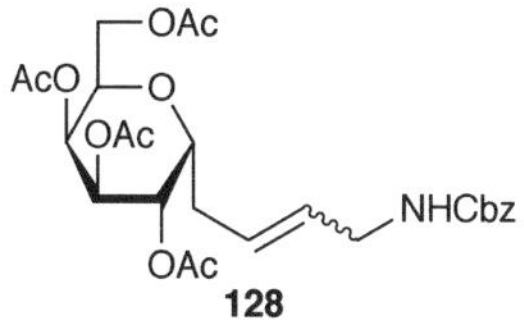

Compound **130**, made by oxidative cleavage of the GlcNAc allyl *C*-glycoside followed by Wittig extension, has been used to make the nucleoside derivative **131** for studies of potential inhibitors of chitin synthases.[277]

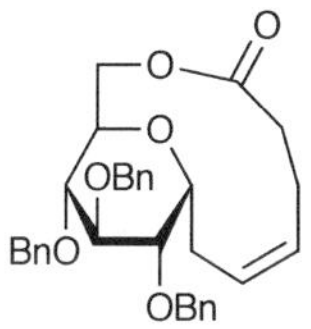

3.2.2 Compounds with Alkynyl 'Aglycons'. A set of oligosaccharide analogues containing two sugar units and four alkyne bonds have been made as illustrated in Scheme 12.[278]

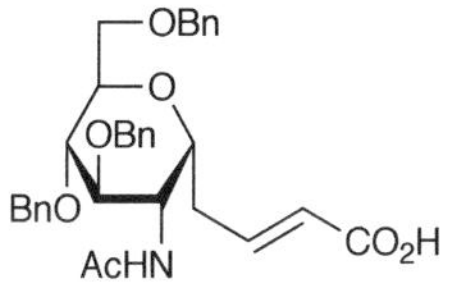

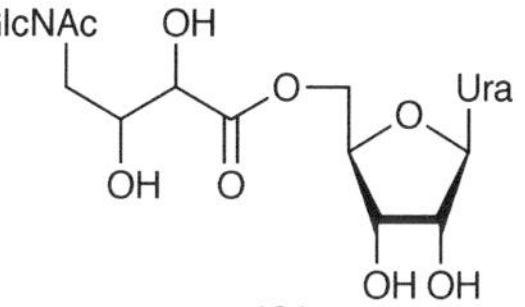

3.2.3 Compounds with Aromatic Aglycons. A variety of ways of making aryl *C*-glycosides have been illustrated. For example, *O*-benzylated D-xylopyranosyl and L-arabinopyranosyl fluorides have been condensed in the presence of BF₃.Et₂O with 2,4,6-tribenzyloxyacetophenone to give the β-glycosides as thermodynamic products. In the case of the latter sugar the kinetic α-anomer was also obtained.[279] The 1-lithio derivative of the glycal added to 2-benzyloxy-6-benzyloxymethylquinone, and the product reductively aromatized, gave the glycal *C*-glycoside **132** which was converted to the β-*C*-glucoside by hydroboration and to a spiroketal following epoxidation.[280]

2-Hydroxyglycal esters condensed with substituted benzenes in the presence of

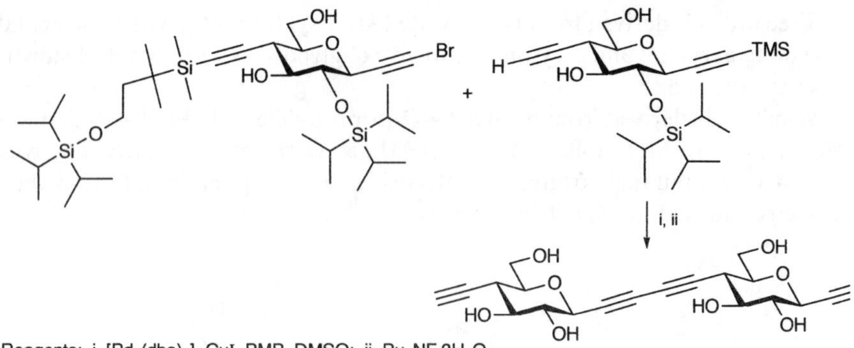

Reagents: i, [Pd$_2$(dba)$_3$], CuI, PMP, DMSO; ii, Bu$_4$NF·3H$_2$O

Scheme 12

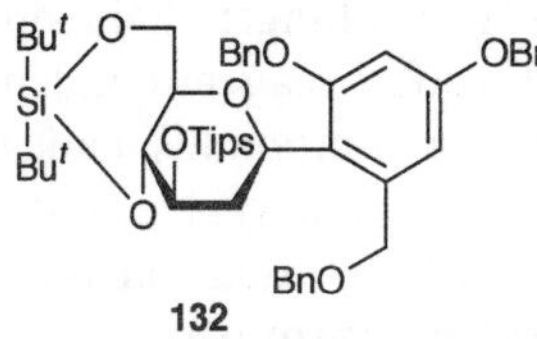

132 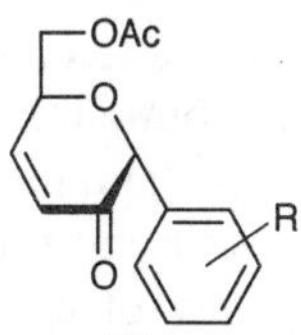   133 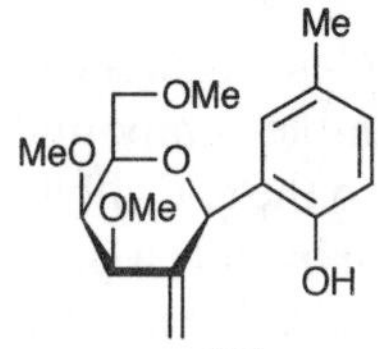   134

HF-pyridine gave enone *C*-glycosides *e.g.* **133** from tetra-*O*-acetyl-2-hydroxy-D-glucal.[281] Reaction of the branched chain 2-acetoxymethyl-tri-*O*-methyl-D-galactal with *p*-cresol in the presence of Montmorillonite gives *C*-glycoside **134** in 62% yield.[282] In a related example reaction of cyclopropa-glycal adduct **135** with phenols in the presence of a Pt catalyst gave 2-*C*-aryl-2-deoxy-*C*-methyl mannosides, both anomeric configurations being produced.[283]

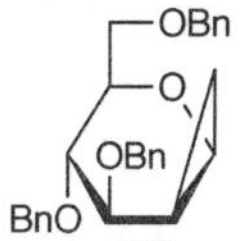

135 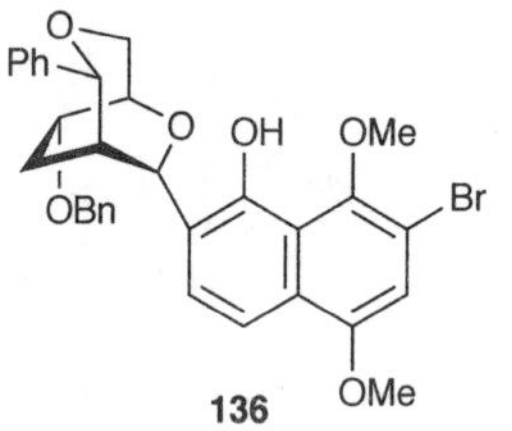   136

Reaction of 3,4,6-tri-*O*-benzyl-2-deoxy-D-glucosyl acetate with the appropriate naphthol in the presence of BF$_3$.Et$_2$O gave the β-*C*-glycoside, but only as the minor product, the major one being **136** which was formed by rearrangement of the sugar moiety prior to substitution. This process, which involves hydride abstraction from the O-6 benzylic methylene group by a C-3 carbocation, has been observed before in a reaction of tri-*O*-benzyl-D-glucal (*J. Org. Chem.* 1998, **63**, 193) which suggests, therefore, that it was this glycal, formed as an intermediate, which reacted with the naphthol.[284] A closely related reaction of tetra-*O*-acetyl-2-deoxyglucose with the same naphthol without the bromine substituent proceeded without complication to give the β-*C*-glycoside in high yield.[285]

Compound **137** was converted to **138** *via* an aryne intermediate,[286] and the latter (as the tri-*O*-benzyl analogue) was used in the synthesis of aquayamycin

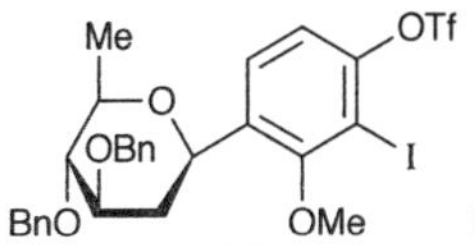

137 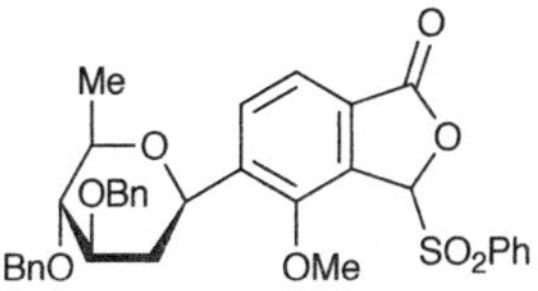   **138** 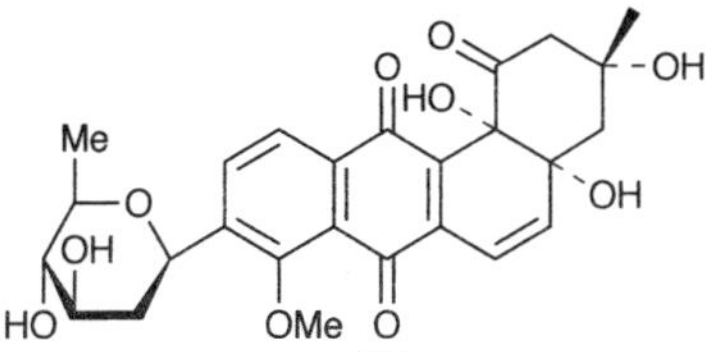   **139**

139.[287] Isoorientin , a 6-*C*-glucosylflavone, has been made by use of glycosyl fluoride technology,[288] and a range of related flavonoid compounds, apigenin 6-*C*-β-D-glucopyranoside amongst them, have been isolated from *Ziziphus jujuba*, a sedative source used in Chinese medicine.[289] Compounds **140–142** have been synthesized with the following *C*-linked sugar substituents, respectively, at the asterisked carbon atoms: 4-*O*-benzoyl-2,6-dideoxy-3-*O*-mesyl-β-D-*ribo*-hexopyranosyl,[290] 2-deoxy-β-D-*arabino*-hexopyranosyl,[291] and 3,6-dideoxy-3-dimethylamino-β-L-galactopyranosyl (the L-enantiomer of the naturally occurring radivomycin).[292] In addition, compounds **143** and **144** have been isolated from two plant sources as *C*-bonded β-D-glucopyranosyl derivatives,[293,294] substituted at the indicated positions. Two *C*-glycosides with 2,6-dideoxy-β-D-*lyxo*- and β-D-*arabino*-hexopyranosyl bonded as indicated in **145** have been isolated from reactions of the *O*-glycoside mithramycin under acid conditions. They inhibit MDRI gene expression in a tumour cell line and thus have potential to overcome multidrug resistance towards antitumour agents.[295]

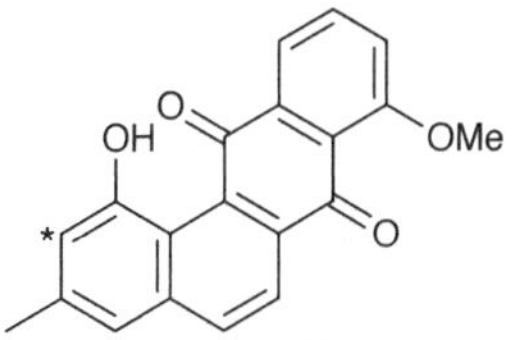

140 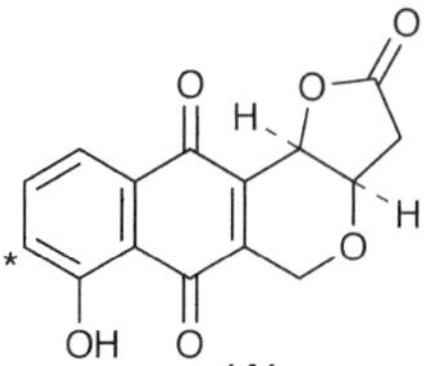   **141** 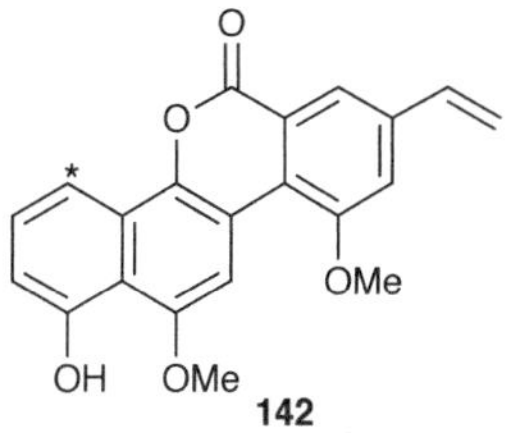   **142**

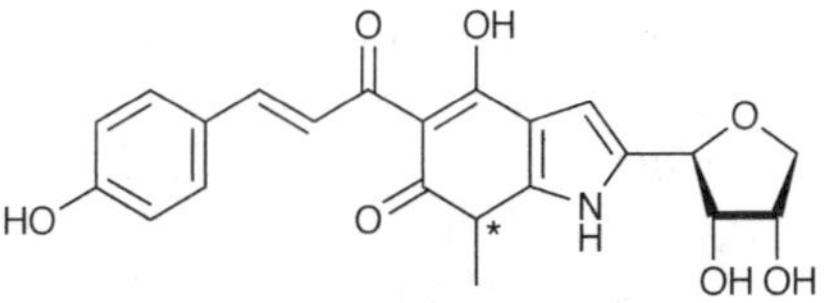

143 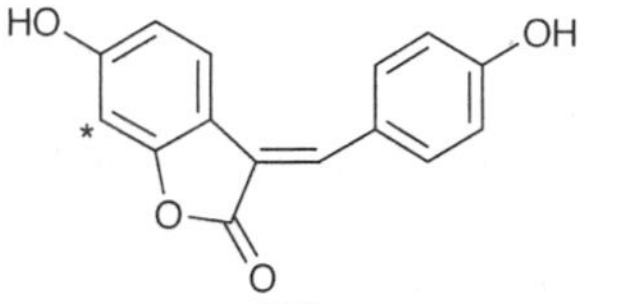   **144**

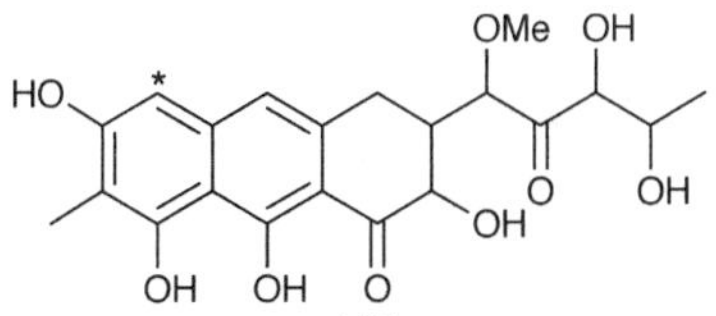

145

3.2.4   Spiro- and 1,2-fused Bicyclic Systems. Spiroketals involving *C*-glycosidic structures can more correctly be considered to be chain-extended sugars (Chapter 2, Section 2.3). Sharpless asymmetric dihydroxylation of 5-aryl-2-vinylfuran

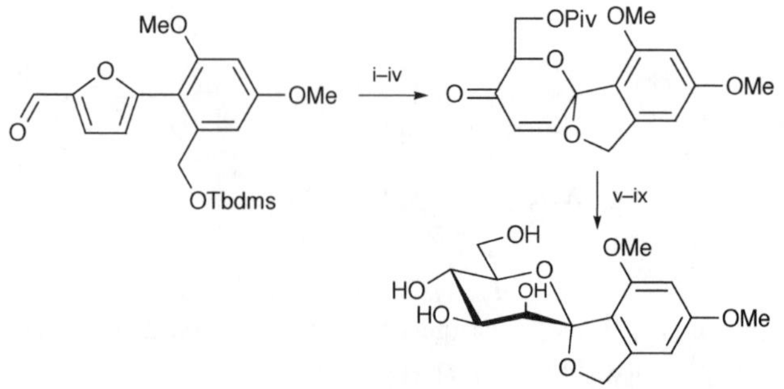

Reagents: i, Ph$_3$P=CH$_2$; ii, ADmixα; iii, PivCl; iv, NBS then HCl; v, NaBH$_4$; vi, TbdmsCl; vii, OsO$_4$; viii, DIBALH; ix, Bu$_4$NF

Scheme 13

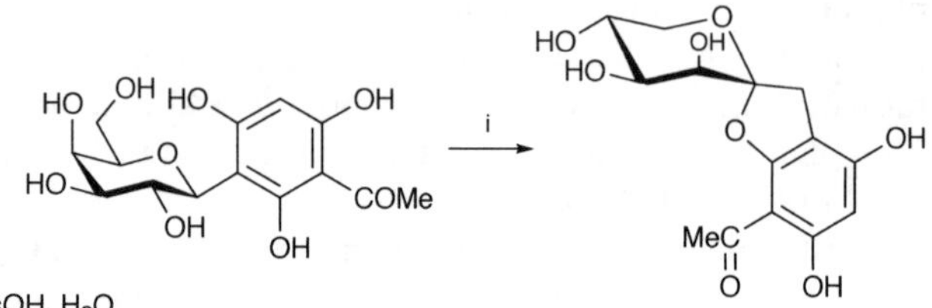

Reagents: i, TsOH, H$_2$O

Scheme 14

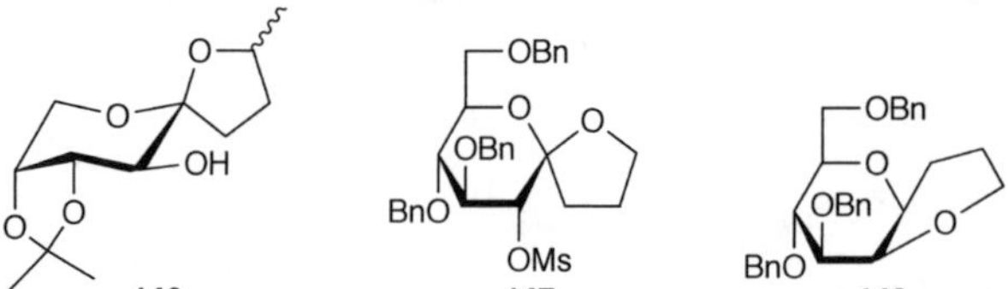

146   147   148

has been applied in the synthesis of papulacandin D, Scheme 13.[296] An interesting rearrangement to give pentopyranosyl analogues of the compound occurs when *C*-galactopyranosyl phloroacetophenone is treated with aqueous acid (Scheme 14).[297] Epimers **146** have been obtained separately from a mixture of 1,3,4-trideoxy-5,6:7,8-di-*O*-isopropylidene-β-D-*manno*- and -*gluco*-non-5-ulo-5,9-pyranose following selective enzymic acetylation of the former at the unsubstituted C-2 OH group.[298] Treatment of *spiro*-ketal **147** with DIBALH gave the 1,2-fused ring compound **148** in 54% yield together with the β-*C*-propyl glycoside.[299]

Reaction of the 2-trimethoxybenzyl glycosylating agent **149** with IDCP gave the fused-ring compound **150** which with BF$_3$.Et$_2$O isomerized to the *trans*-fused β-anomer which was taken through to lactone **151**, the triacetate of tri-*O*-methylnorbergenin (Scheme 15).[300]

Glycosylation of the protected octosyl nucleoside ezomycin at O-6′ led to 4-desamino-4-oxoezomycin A$_2$ (**152**).[301]

3.2.5 C-Linked Disaccharides. Compounds in this section will be described in the abbreviated form used for *O*-linked disaccharides with the group (usually

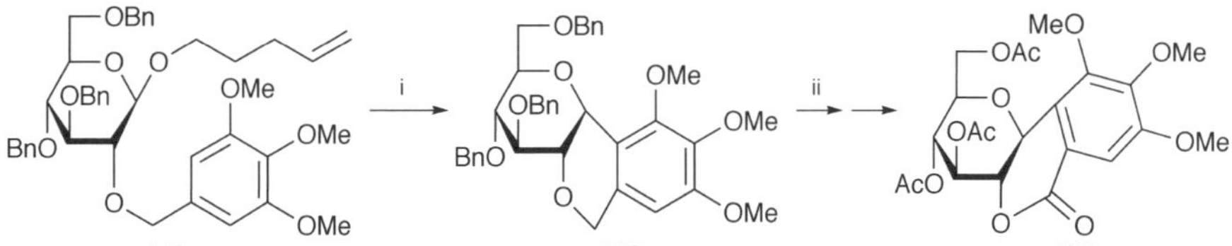

149   **150**   **151**

Reagents: i, IDCP; ii, BF₃·Et₂O

Scheme 15

methylene) that replaces the linking oxygen atom being specified.

O-Benzylated α-allyl *C*-glycosides derived from D-glucose, D-galactose and D-mannose have been linked in pairs by O-6 diester tethers. Ring closure methathesis then gave in high yields cyclic compounds also joined 1→1 by *C*-but-2-enyl linkages.[302] β-D-Glc-(1-CH₂-6)-D-Glc has also been made by a metathesis key step by way of a 1-*C*-substituted D-glucal derivative.[303] Cellobiose has been *C*-linked to C-6 of D-glucose by use of the β-*C*-nitromethyl glycoside of the former.[304]

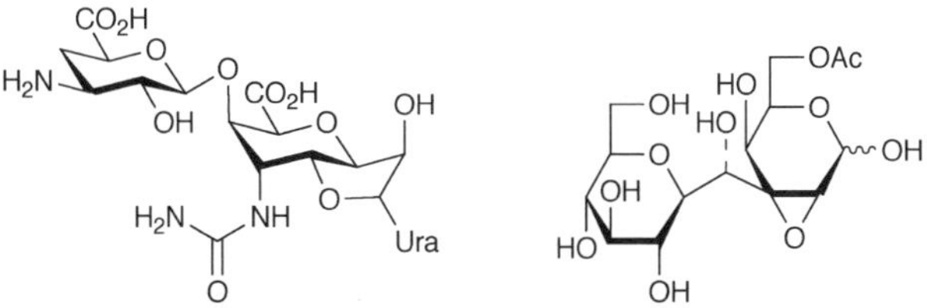

152   **153**

Isolevoglucosenone was the starting material from which the unusual β-D-Glc-(1-CHOH-3)-D-Gal epoxide derivative **153** was made.[305] β-D-Glc-(1-CHOH-6)-α-D-ManOMe was synthesized from the *O*-benzylated glucosyl iodide and the *O*-benzylated 6-deoxyheptose dialdoside.[306]

The *C*-linked analogue **154** of adenophostin A, a strong inhibitor of IP₃ receptor ligand from a *Penicillium*, was made by an intramolecular technique as indicated in Scheme 16.[307,308]

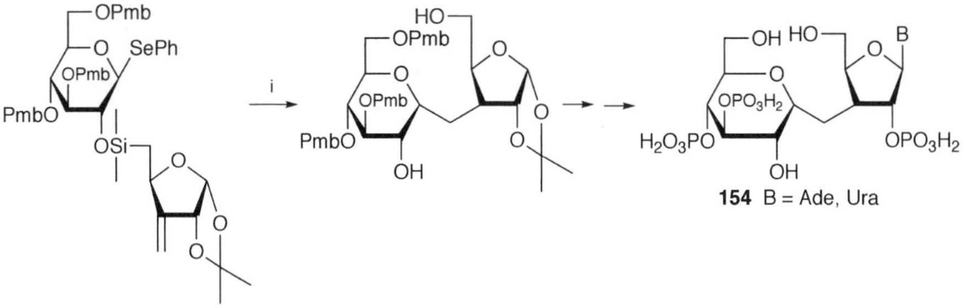

154 B = Ade, Ura

Reagents: i, Bu₃SnH, AIBN

Scheme 16

3-*O*-Carboxymethyl-β-D-Gal-(1-CH₂-1)-D-Man was produced in 8-steps as a hydrolytically stable analogue of SiaLeˣ, the galactose moiety being constructed from a threose derivative,[309] and β-D-Gal-(1-CHOH-1)-D-Gal has been reported.[310]

α-D-Man-(1-CH$_2$-3)-D-GalNAc and α-D-Man-(1-CH$_2$-3)-D-TalNAc have been tested as inhibitors of several glycosidases and human α-1,3-fucosyltransferase. While the former is active against these enzymes, the latter is not.[311]

Ester **155** was converted by the Tebbe method to the methylene analogue (92%) which underwent Claisen rearrangement to give the *C*-linked uronic acid derivative **156** (56%).[312]

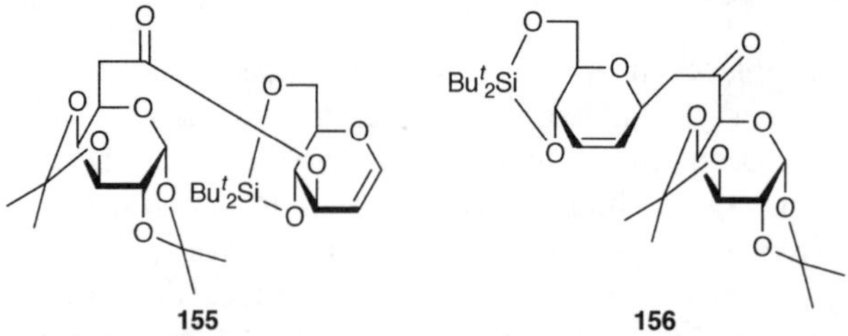

155   **156**

3.2.6 C-Glycosylated Amino-acids and Related Compounds. A review which covers compounds of the category has appeared.[313] The cyclic glycopeptide **157**, based on an α-L-galactosyl carboxylic acid, is 1000 times more active than SiaLe^x in binding P-selectin. Several other SiaLe^x mimetics were also made in the course of the work.[314] Wong's group has also prepared compound **158** and Gal, Man, Fuc analogues by application of asymmetric Strecker reactions.[315]

L-Serine *C*-linked to C-1 of nojirimycin has been made from the α-*C*-allyl glycoside,[316] and a related α-*C*-glucosyl derivative, made from the formylmethyl *C*-glycoside, has CH$_2$CH(OH)CH(NH$_2$)CO$_2$H as the basis of the aglycon. It is therefore a *C*-linked glycosyl threonine.[316a] A *C*-glycosidic isostere of β-D-glucosylasparagine having the (CH$_2$)$_3$CH(NH$_2$)CO$_2$H aglycon has been described,[317] as have homologues with butyl and but-2-enyl linkages.[275]

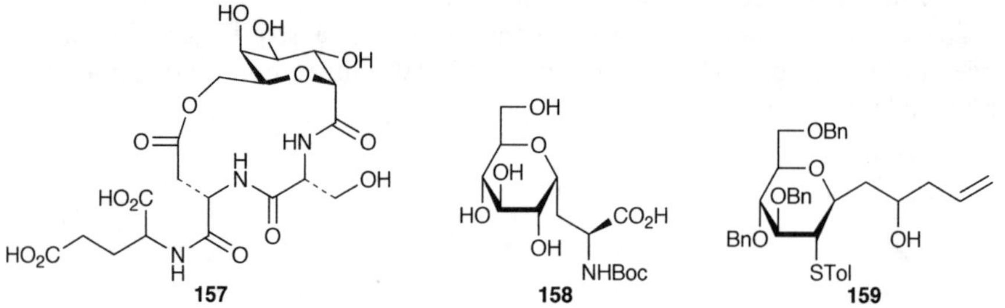

157   **158**   **159**

3.2.7 Other C-glycosides. Compound **159** was the product formed on reaction of the β-formylmethyl *C*-glycoside with pent-4-en-4-ol in the presence of tin tetrachloride rather than the expected tetrahydropyran derivative.[318]

The tertiary alcohol **160**, made as indicated in Scheme 17, was the exclusive product whereas, without the nickel salt, the reaction gave high proportions of the glycal.[319]

3.3   Furanoid (and Pyrrolidine) Compounds. – For making 2,3,5-tri-*O*-benzoyl-β-D-ribofuranosyl cyanide it is reported to be better to use KCN rather than

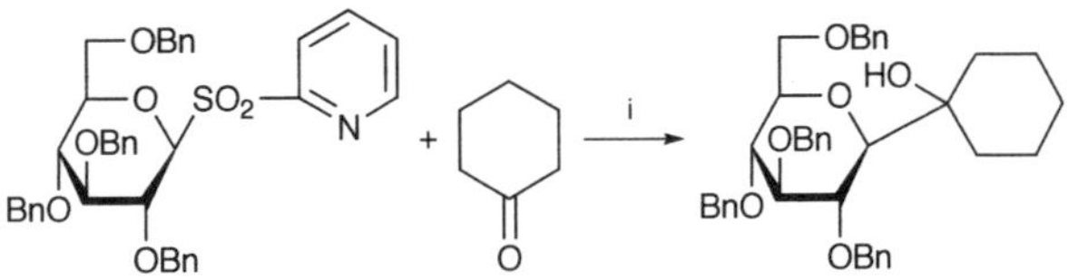

Reagents: i, SmI$_2$ + 1 mol% NiI$_2$

Scheme 17

TmsCN.[320] Treatment of 1,2-*O*-isopropylidene-3,5-di-*O*-mesyl-6-*O*-trityl-D-glucose with ethanediol and TsOH gives the *C*-formyl compound **161** in 53% yield from which (+)-muscarine iodide **162** and its epimer at the alcohol centre were made.[321] Compounds **163** and **164** have been made from 2-deoxy-D-ribose donors by reaction with a benzyl Grignard reagent and allyltrimethylsilane, respectively, in the key steps. The products were incorporated into oligonucleotides during a study of the interactions of DNA with peptides.[322]

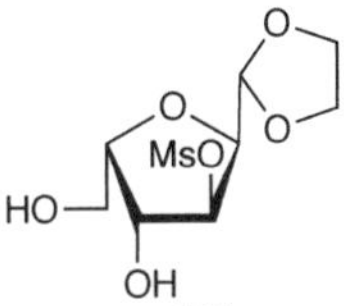

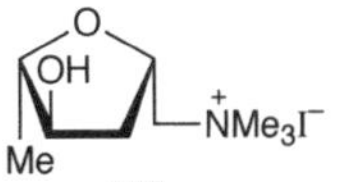

A long paper has appeared on synthetic and conformational studies of compound **165** (and its epimers at C-2 and C-5), made from 6-azido-2,3,4-tri-*O*-benzyl-6-deoxy-D-glucitol. These were amide linked to amino-acids at both C-1 and C-6.[323]

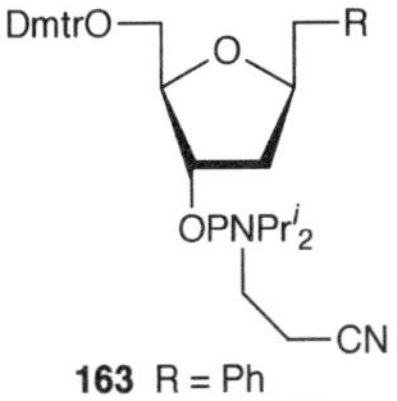

Condensation of the corresponding aldonolactone derivative with 1,1,1-trichlorotoluene and trimethyl phosphite gives the *exo*-alkenes **166** from which *C*-glycosides are available.[324] Reaction of *O*-substituted 1,2-*O*-isopropylidene-α-D-xylofuranose with allyltrimethylsilane and a Lewis acid gives the 2-*O*-unprotected β-*C*-allyl *C*-glycosides.[325]

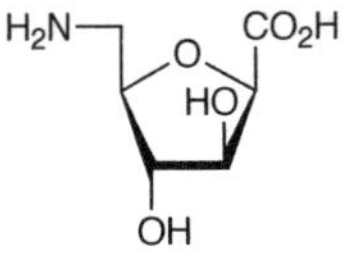

Conversion of the corresponding lactone to the ethynyl hemiacetal followed by appropriate *C*-substitution gave dialkyne **167** which, on [1,2] Wittig rearrangement, was converted to the doubly C-1 *C*-substituted **168** (Scheme 18).[326]

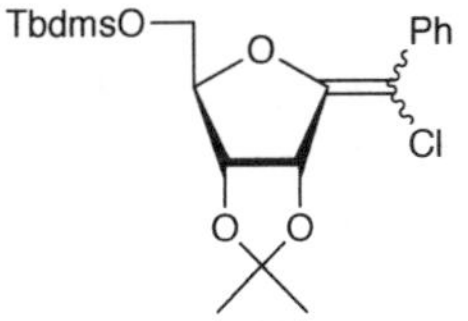

Considerable attention has been given to *C*-furanosyl derivatives having aryl aglycons, and several synthetic approaches have been adopted. The simplest involves the addition of aryllithiums to aldonolactones and reduction of the derived alcohol group using triethylsilane. In this way compound **169** and other

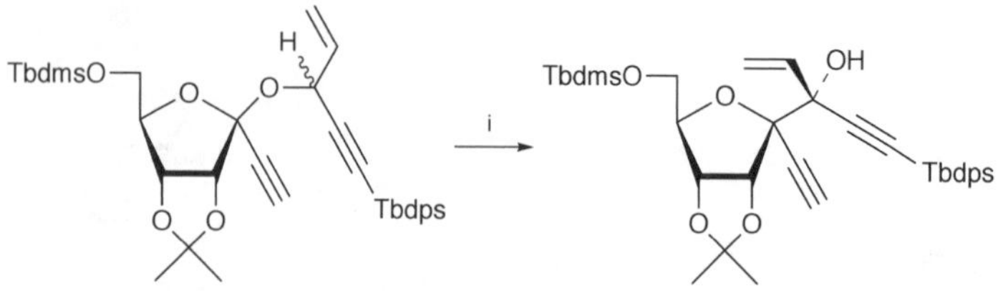

Reagent: i, Bu^{n}Li

167   **168**

Scheme 18

fluorophenyl compounds have been made and incorporated into RNA. This does not result in conformational change to the latter.[327] A new synthesis of tiazofurin (**170**), which is active against several human cancers, was based on development of the heterocycle from the cyano group,[328] and a related approach to heterocyclic *C*-glycosides, based on the formation of the precursor to the heterocycle by intramolecular transfer, is illustrated in Scheme 19.[329]

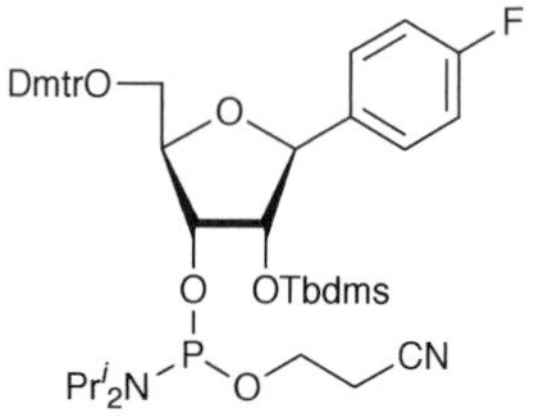

169 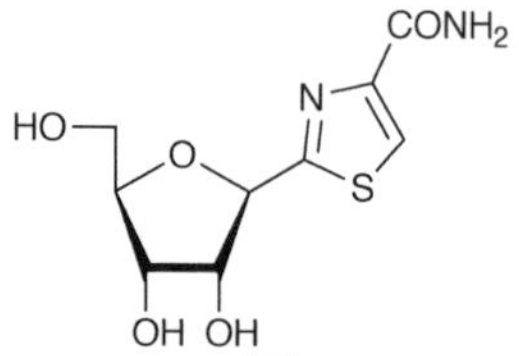   **170**

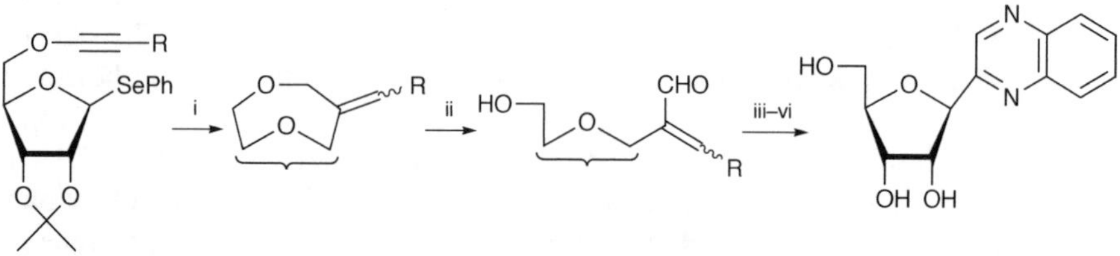

R = Me, Tms

Reagents: i, Bu$_3$SnH; ii, SeO$_2$, AcOH; iii, O$_3$; iv, Me$_2$S; v, [aminobenzene structure]; vi, H$^+$

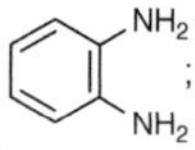

Scheme 19

Reactions involving use of aldehydo-sugars for introducing the aglycons have given the conformationally locked bicyclic system **171** required for the study of the triplex-forming ability of oligonucleotide analogues.[330,331] By a related approach 2′-α-*C*-mannofuranosyl-2-formylthiophene was made from the 2,3:5,6-di-*O*-isopropylidene-D-mannose and used as a source of various short chain alkyl and functionalized alkyl *C*-glycosides following desulfurization.[332]

O-Protected methyl 2-deoxy-D-ribofuranosides have been condensed with *o*-disubstituted benzenes to give compounds **172** (X,Y = OH, NH$_2$) for incorporation into DNA which permits 'base-pairing' by means other than H-bonding.[333] For example, borate anion coupling was examined.[334]

3,5-*O*-Silylated-D-arabinal coupled by the Pd-promoted Heck procedure with substituted benzenes, followed by fluoride-promoted deprotection, gave, for example, the 3-uloses from which compounds **173** were obtained by carbonyl

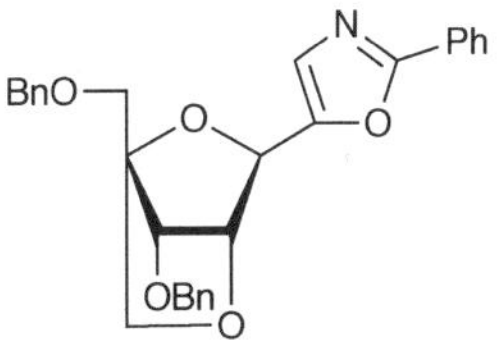

171 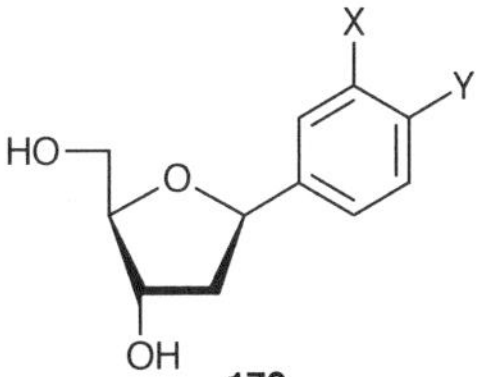 **172** 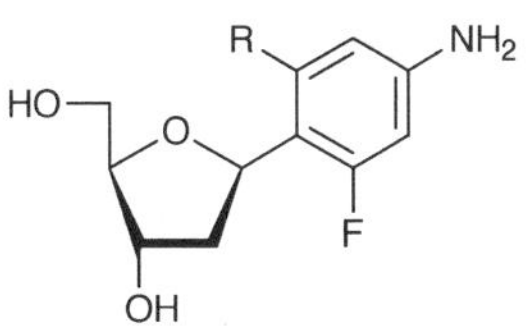 **173** R = H, Me, F

reduction.[335] Analogues of cytidine were then made by the same approach, but from L-arabinal, to give mimics of the unnatural L-nucleosides.[336]

C-Linked furanosyl disaccharides and analogues to have been made are α-D-Galf-(1-CF$_2$-3)-β-D-GlcOMe[337] and compounds **174**[338] and **175**.[339] The former was made together with several isomers by a Wittig–Michael approach, and the latter by a 1-C-formyl glycoside, 6'-Wittig reagent strategy applied to an enal at C-5–C-7 of a heptofuranose derivative.

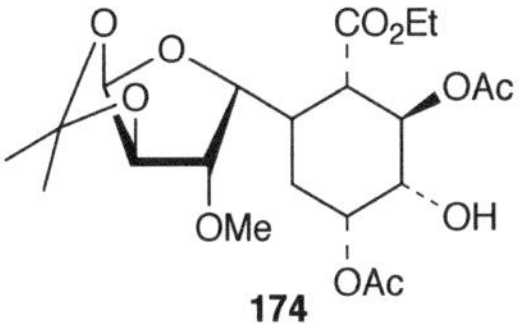

174 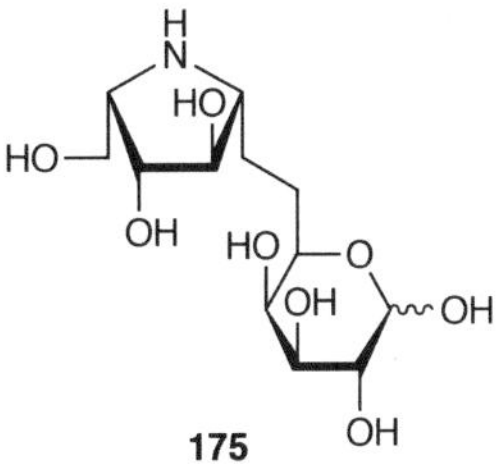 **175**

References

1. J.G. Riess and J. Greiner, *Carbohydr. Res.*, 2000, **327**, 147.
2. K. Toshima, *Carbohydr. Res.*, 2000, **327**, 15.
3. K. Ueda and S. Yamamura, *Angew. Chem., Int. Ed. Engl.*, 2000, **39**, 1400.
4. K.-H. Jung, M. Müller and R.R. Schmidt, *Chem. Rev.*, 2000, **100**, 4423.
5. G. Lemanski and T. Ziegler, *Helv. Chim. Acta*, 2000, **83**, 2676.
6. S. Valverde, M. García, A.M. Gómez and J.C. López, *Synlett*, 2000, 22.
7. C.M.P. Seward, I. Cumpstey, M.A. Loui, S.C. Ennis, A.J. Redgrave and A.J. Fairbanks, *Chem. Comm.* 2000, 1409.
8. I.L. Baraznenok, V.G. Nenajdenko and E.S. Balenkova, *Tetrahedron*, 2000, **56**, 3077.
9. S. Hanessian and B. Lou, *Chem. Rev.*, 2000, **100**, 4443.
10. B.A. Garcia and D.Y. Gin, *J. Am. Chem. Soc.*, 2000, **122**, 4269.
11. L.J. Silverberg, S. Kelly, P. Vemishetti, D.H. Vipond, F.S. Gibson, B. Harrison, R. Spector and J.L. Dillon, *Org. Lett.* 2000, **2**, 3281.
12. Y. Bai, G.-J. Boons, A. Burton, M. Johnson and M. Haller, *J. Carbohydr. Chem.*, 2000, **19**, 939.
13. K.K. Chauchan and C.G. Frost, *J. Chem. Soc. Perkin Trans. 1*, 2000, 3015.
14. M. Jesberger, J. Jaunzems, A. Jung, G. Jas, A. Schönberger and A. Kirschning, *Synlett*, 2000, 1289.
15. B.-A. Feit, I.K. Kelson, A. Gerull, S. Abramson and R.R. Schmidt, *J. Carbohydr. Chem.*, 2000, **19**, 661.
16. Z. Surma and K. Agoston, *Carbohydr. Lett.* 2000, **3**, 457.

17.   Z. Yang, W. Lin and B. Yu, *Carbohydr. Res.* 2000, **329**, 879.
18.   S. Sanchez, T. Bamhaoud and J. Prandi, *Tetrahedron Lett.*, 2000, **41**, 7447.
19.   G. Singh and H. Vankayalapati, *Tetrahedron: Asymmetry*, 2000, **11**, 125.
20.   J.-i. Kadokawa, T. Nagaoka, J. Ebana, H. Tagaya and K. Chiba, *Carbohydr. Res.*, 2000, **327**, 341.
20a. D. Crich, W. Cai and Z. Dai, *J. Org. Chem.*, 2000, **65**, 1291.
21.   S. Sugiyama and J.M. Diakur, *Org. Lett.*, 2000, **2**, 2713.
22.   P. Cura, M. Aloui, K.P.R. Kartha and R.A. Field, *Synlett*, 2000, 1279.
23.   G. Scheffler, M.E. Behrondt and R.R. Schmidt, *Eur. J. Org. Chem.*, 2000, 3527.
24.   B. Yu and Z. Yang, *Tetrahedron Lett.*, 2000, **41**, 2961.
25.   T. Zhu and G.-J. Boons, *Carbohydr. Res.*, 2000, **329**, 709.
26.   K. Egusa, K. Fukase, Y. Nakai and S. Kusumoto, *Synlett*, 2000, 27.
27.   J.A. Khan, A. Millquist-Fureby and E.N. Vulfson, *Methods Biotechnol.*, 1999, **10**, 313 (*Chem. Abstr.*, 2000, **132**, 108 149).
28.   I. Gill and R. Valivety, *Angew. Chem., Int. Ed. Engl.*, 2000, **39**, 3804.
29.   B.M. de Roode, H. Zuilhuf, M.C.R. Frannsen, A. van der Padt and A. de Groot, *J. Chem. Soc., Perkin Trans. 2*, 2000, 2217.
30.   V. Ferrières, J. Joutel, R. Boulch, M. Roussel, L. Toupet and D. Plusquellec, *Tetrahedron Lett.*, 2000, **41**, 5515.
31.   M. Gelin, V. Ferrières and D. Plusquellec, *Eur. J. Org. Chem.*, 2000, 1423.
32.   S.K. Sarkar, A.K. Choudhury, B. Mukhapadhyay and N. Roy, *J. Carbohydr. Chem.*, 1999, **18**, 1121.
33.   J. Kovensky and P. Sinaÿ, *Eur. J. Org. Chem.*, 2000, 3523.
34.   A. Chiocconi, C. Marino and R.M. de Lederkremer, *Carbohydr. Res.*, 2000, **323**.
35.   G.R. Smith, F.J. Villani Jr., L. Failli and R.M. Guiliano, *Tetrahedron Asym.*, 2000, **11**, 139.
36.   J.J. Gridley and H.M.I. Osborn, *J. Chem. Soc. Perkin Trans. 1*, 2000, 1471.
37.   R. Weingart and R.R. Schmidt, *Tetrahedron Lett.*, 2000, **41**, 8753.
38.   D. Crich and M. Smith, *Org. Lett.*, 2000, **2**, 4067.
39.   E. Kaji and Y. Hosakawa, *Heterocycles*, 2000, **52**, 579 (*Chem. Abstr.*, 2000, **132**, 279 407).
40.   T. Ikeda, H. Yamada, *Carbohydr. Res.*, 2000, **329**, 889.
41.   H.N. Yu, P. Zhang, C.-C. Ling and D.R. Bundle, *Tetrahedron: Asymmetry*, 2000, **11**, 465.
42.   K. Toshima, Y. Maeda, H. Ouchi, A. Asai and S. Matsumura, *Bioorg. Med. Chem. Lett.*, 2000, **10**, 2163.
43.   S.J. Williams, V. Notenboom, J. Wicki, D.R. Rose and S.G. Withers, *J. Am. Chem. Soc.*, 2000, **122**, 4229.
44.   G. Anilkamar, L.G. Nair, L. Olsson, J.K. Daniels and B. Fraser-Reid, *Tetrahedron Lett.*, 2000, **41**, 7605.
45.   S.P. Sergey, V.Y. Chirva, A.L. Kadun, E.S. Kakayan, *Carbohydr. Res.*, 2000, **329**, 895.
46.   J.M. Vega-Pérez. J.I. Candela, I. Romero, E. Blanco and F. Iglesias-Guerra, *Eur. J. Org. Chem.*, 2000, 3949.
47.   D. Bednarczik, W. Kaca, H. Myszka, L. Serwecinska, Z. Smiatacz and A. Zaborowski, *Carbohydr. Res.*, 2000, **328**, 249.
48.   D.J. Silva, C.L. Bowe, A.A. Branstrom, E.R. Baizman and M.J. Sofia, *Bioorg. Med. Chem. Lett.*, 2000, **10**, 2811
49.   M. Ghosh, R.G. Dulina, R. Kakarla and M.J. Sofia, *J. Org. Chem.*, 2000, **65**, 8387.
50.   R. Tona, R. Bertolini and J. Hunziker, *Org. Lett.*, 2000, **2**, 1693.

51. K.M. Halkes, P.M. St Hilaire, A.H. Jansson, C.H. Gotfredsen and M. Meldal, *J. Chem. Soc., Perkin Trans. 1*, 2000, 2127.

51a. J. Mynarski and A. Banaszek, *Tetrahedron: Asymmetry*, 2000, **11**, 3737.

51b. M.B. Mitchell and I.W.A. Whitcombe, *Tetrahedron Lett.*, 2000, **41**, 8829.

52. S. Hanessian, O.M. Saavedra, F. Xie, N. Amboldi and C. Battistini, *Bioorg. Med. Chem. Lett.*, 2000, **10**, 439.

53. Y. Ambroise, C. Mioskowski, G. Leblanc and B. Rousseau, *Bioorg. Med. Chem. Lett.*, 2000, **10**, 1125.

54. S. Papot, D. Combaud, K. Bosslet, M. Gerken, J. Czech and J.-P. Gesson, *Bioorg. Med. Chem. Lett.*, 2000, **10**, 1835.

55. X.-L. Sun, Y. Kanie, C.-T. Guo, O. Kanie, Y. Suzuki and C.-H. Wong, *Eur. J. Org. Chem.*, 2000, 2643.

56. C. Fernández, O. Nieto, E. Rivas, G. Montenegro, J.A. Fontenla and A. Fernández-Mayoralas, *Carbohydr. Res.*, 2000, **327**, 353.

57. H.K. Chenault, J. Yang and D.F. Talser, *Tetrahedron*, 2000, **56**, 3673.

58. R.K. Uhrig, M.A. Picard, K. Beyreuther and M. Wiessler, *Carbohydr. Res.*, 2000, **325**, 72.

59. T. Zhu and G.-J. Boons, *Tetrahedron: Asymmetry*, 2000, **11**, 199.

60. H.K. Shin, J.Y. Kong, J.D. Lee and T.H. Lee, *Biotechnol. Lett.*, 2000, **22**, 321.

61. T. Yamamoto, T. Hosokawa, S. Kishihara, H. Tamaki, N. Wakiuchi and Y. Suzuki, *Seito Gijutsu Kenkyu Kaishi*, 1998, **76**, 49, 57 (*Chem. Abstr.*, 2000, **132**, 108 169).

62. M. Ueda, Y. Sawai and S. Yamamura, *Tetrahedron Lett.*, 2000, **41**, 3433.

63. M. Ueda, Y. Sawai, Y. Wada and S. Yamamura, *Tetrahedron*, 2000, **56**, 5123.

64. X.-M. Chen, Z.-J. Li and Z.-T. Huang, *Carbohydr. Res.*, 2000, **328**, 253.

65. M. Zarevucka, M. Vacek, Z. Wimmer, C. Brunet and M.-D. Legoy, *Biotechnol. Lett.*, 1999, **21**, 785 (*Chem. Abstr.*, 2000, **132**, 12 457).

66. O. Kanie, G. Grotenbreg and C.-H. Wong, *Angew. Chem., Int. Ed. Engl.*, 2000, **39**, 4545.

67. R.R. France, I. Cumpstey, T.D. Butters, A.J. Fairbanks and M.R. Wormald, *Tetrahedron: Asymmetry*, 2000, **11**, 4985.

68. W. Fan, Y. Tezuka and S. Kadota, *Chem. Pharm. Bull.*, 2000, **48**, 1055.

69. M.P. Prickett, G. Singh and H. Vankayalapati, *Tetrahedron Lett.*, 2000, **41**, 2987.

70. S.T. Caldwell, A. Crozier and R.C. Hartley, *Tetrahedron*, 2000, **56**, 4101.

71. M. Ješelnik, J. Plavec, S. Polanc and M. Kočevar, *Carbohydr. Res.*, 2000, **328**, 591.

72. T. Tsuchiya, Y. Takagi and H. Yamada, *Bioorg. Med. Chem. Lett.*, 2000, **10**, 203.

73. M. Dubber and T.B. Lindhorst, *J. Org. Chem.*, 2000, **65**, 5275.

74. T.K. Lindhorst, M. Dubber, U. Krallmann-Wenzel and S. Ehlers, *Eur. J. Org. Chem.*, 2000, 2027.

75. M. Schmidt, B. Dobner and P. Nuhn, *Synlett*, 2000, 1157.

76. J.-F. Nierengarten, *Chem. Eur. J.*, 2000, **6**, 3667.

77. C. Kaldapa, J.C. Blais, V. Carré, R. Granet, V. Sol, M. Guilloton, M. Spiro and P. Krausz, *Tetrahedron Lett.*, 2000, **41**, 331.

78. R.-M. Sebastian, G. Magro, A.-M. Caminade and J.-P. Majoral, *Tetrahedron*, 2000, **56**, 6269.

78a. J.B. Corbell, J.J. Lundquist and E.J. Toone, *Tetrahedron: Asymmetry*, 2000, **11**, 95.

79. T. Hasegawa, K. Matsuura and K. Kobayashi, *Chem. Lett.*, 2000, 466.

80. P.Y. Chong and P.A. Petillo, *Org. Lett.*, 2000, **2**, 2113.

81. C. Kieburg, K. Sadalapure and T.K. Linkhorst, *Eur. J. Org. Chem.*, 2000, 2035.

82. K. Sadalapure and T.K. Lindhorst, *Angew. Chem., Int. Ed. Engl.*, 2000, **39**, 2010.

83. W.-T. Jiaang, P.-H. Tseng and S.-T. Chen, *Synlett*, 2000, 797.

84. K. Matsuoka, H. Kurosawa, Y. Esumi, D. Terunuma and H. Kuzuhara, *Carbohydr. Res.*, 2000, **329**, 765.

85. T. Buskas, E. Söderberg, P. Konradsson and B. Fraser-Reid, *J. Org. Chem.*, 2000, **65**, 658.

86. K. Kurokawa, H. Kumihara and H. Kondo, *Bioorg. Med. Chem. Lett.*, 2000, **10**, 1827.

87. G.R. Duffin, G.J. Ellames, S. Hartmann, J.M. Herbert and D.I. Smith, *J. Chem. Soc., Perkin Trans.1*, 2000, 2237.

88. A.G.M. Barrett, J.C. Beall, D.C. Braddock, K. Flack, V.C. Gibson and M.M. Salter, *J.Org. Chem.*, 2000, **65**, 6508.

89. S. Figueroa-Pérez and R.R. Schmidt, *Carbohydr. Res.*, 2000, **328**, 95.

90. K.C. Nicolaou, J. Li and G. Zenke, *Helv. Chim. Acta*, 2000, **83**, 1977.

91. M. Adamczyk, R. E. Reddy and S.D. Rege, *Synth. Commun.*, 2000, **30**, 3281.

92. B. Holm, J. Broddefalk, S. Flodell, E. Wellner and J. Kihlberg, *Tetrahedron*, 2000, **56**, 1579.

93. B. Faroux-Corlay, L. Clary, C. Gadras, D. Hammache, J. Greiner, C. Santaella, A.-M. Aubertin, P. Vierling and J. Fatini, *Carbohydr. Res.*, 2000, **327**, 223.

94. I. Shin and J. Lee, *Synlett*, 2000, 1297.

95. I. Hanachi, S. Kiyonaka and S. Shinkai, *Chem. Commun.*, 2000, 1281.

96. C. Grandjean, C. Rommens, H. Gras-masse and O. Melnyk, *Angew. Chem., Int. Ed. Engl.*, 2000, **39**, 1068.

97. N.B. Malkar, J.L. Lauer-Fields and G.B. Fields, *Tetrahedron Lett.*, 2000, **41**, 1137.

98. F. Reichel, A.M. Roelofsen, H.P.M. Geurts, S.J. van der Gaast, M.C. Feiters and G-J. Boons, *J. Org. Chem.*, 2000, **65**, 3357.

99. V. Wittmann and P.H. Seeberger, *Angew Chem., Int. Ed. Engl.*, 2000, **39**, 4348.

100. J. Sander and H. Waldmann, *Chem. Eng. J.*, 2000, **6**, 1564.

101. T. Ziegler, S. Gerling and M. Lang, *Angew. Chem., Int. Ed. Engl.*, 2000, **39**, 2109.

102. D. Macmillan and C.R. Bertozzi, *Tetrahedron*, 2000, **56**, 9515.

103. Y. Nakahara, S. Ando, M. Itakura, N. Kumabe, H. Hojo, Y. Ito and Y. Nakahara, *Tetrahedron Lett.*, 2000, **41**, 6489.

104. S. Koto, M. Hirooka, T. Yoshida, K. Takenaka, C. Asai, T. Nagamitsu, H. Sakuma, M. Sakurai, S. Masuzawa, M. Komiya, T. Sato, S. Zen, K. Yago and F. Tomonaga, *Bull. Chem. Soc. Jpn.*, 2000, **73**, 2521.

105. G. Anilkumar, M.R. Gilbert and B. Fraser-Reid, *Tetrahedron*, 2000, **56**, 1993.

106. M. Martin-Lomas, M. Flores-Mosquera and N. Khiar, *Eur. J. Org. Chem.*, 2000, 1539.

107. K. Hamasaki, M.-C. Woo and A. Veno, *Tetrahedron Lett.*, 2000, **41**, 8327.

108. R. Saleem and J. Meinwald, *J. Chem. Soc., Perkin Trans. 1*, 2000, 391.

109. K. Ohmori, H. Ohrui and K. Suzuki, *Tetrahedron Lett.*, 2000, **41**, 5537.

110. A.-M. Periers, P. Laurin, D. Ferroud, J.-L. Haesslein, M. Klich, C. Dupuis-Hamelin, P. Mavais, P. Lassaigne, A. Bonnefoy and B. Musicki, *Bioorg. Med. Chem. Lett.*, 2000, **10**, 161.

111. A.K. Ghosh, S. Khan, F. Marini, J.A. Nelson and D. Farquhar, *Tetrahedron Lett.*, 2000, **41**, 4871.

112. J.R. Ferguson, J.R. Harding, K.W. Lumbard, F. Scheinmann and A.V. Stachulski, *Tetrahedron Lett.*, 2000, **41**, 389.

113. R.T. Brown, N.E. Carter, S.P. Mayalarp and F. Scheinmann, *Tetrahedron*, 2000, **56**, 7591.

114. I. Rukhman and A.L. Gutman, *Tetrahedron Lett.*, 2000, **41**, 6889.

115. L.C. Creemer, H.A. Kirst, J.W. Paschal and T.V. Worden, *J. Antibiotics*, 2000, **53**, 171.

116. O.B. Flekhter, L.A. Baltina and G.A. Tolstikov, *J. Nat. Prod.*, 2000, **63**, 992.

117. Y. Yamano, Y. Watanabe, N. Watanabe and M. Ito, *Chem. Pharm. Bull.*, 2000, **48**, 2017.

118. M. Ueda, M. Okazaki, K. Ueda and S. Yamamura, *Tetrahedron*, 2000, **56**, 8101.

119. H. Matsuda, T. Kageura, Y. Inoue, T. Morikawa and M. Yoshikawa, *Tetrahedron*, 2000, **56**, 7763.

120. H. Kenmoku, T. Sassa and N. Kato, *Tetrahedron Lett.*, 2000, **41**, 4389.

121. R.L. Obendorf, K.J. Steadman, D.J. Fuller, M. Horbowicz and B.A. Lewis, *Carbohydr. Res.*, 2000, **328**, 623.

122. N. Vongvanich, P. Kittakoop, J. Kramyu, M. Tanticharoen and Y. Thebtaranonth, *J. Org. Chem.*, 2000, **65**, 5420.

123. G.M. Nicholas and T.F. Mounski, *J. Am. Chem. Soc.*, 2000, **122**, 4011.

124. T.Qi, M. Ojika and Y. Sakagami, *Tetrahedron*, 2000, **56**, 5835.

125. E.-S.E. Habib, K. Yokomizo, K. Murata and M. Uyeda, *J. Antibiotics*, 2000, **53**, 1420.

126. L.A. Paquette, L. Barriault, D. Pissarnitski and J.N. Johnston, *J. Am. Chem. Soc.*, 2000, **122**, 619.

127. M. Otsuka, M. Fujita, Y. Matsushima, T. Eguchi, K. Shindo and K. Kakinuma, *Tetrahedron*, 2000, **56**, 8281.

128. M. Gill, S. Saubern and J. Yu, *Aust. J. Chem.*, 2000, **53**, 213.

129. M. Gill and A.F. Smidel, *Aust. J. Chem.*, 2000, **53**, 47.

130. B.B. Messanga, B.L. Sondengam and B. Bodo, *Can. J. Chem.*, 2000, **78**, 487.

131. S. Oscarson and F.W. Sehgelmeble, *J. Am. Chem. Soc.*, 2000, **122**, 8869.

132. J.L. Asensio, F.J. Cañada, X. Cheng, N. Khan, D.R. Mootoo and J. Jiménez-Barbero, *Chem. Eur. J.*, 2000, **6**, 1035.

132a. K.C. Nicolaou, K.C. Fylaktakidon, H.J. Mitchell, F.L. van Delft, R.M. Rodriguez, S.R. Conley and Z. Jin, *Chem. Eur. J.*, 2000, **6**, 3166.

133. A. Fürstner, K. Radkowski, J. Grabowski, C. Wirtz and R. Mynott, *J. Org. Chem.*, 2000 **65**, 8758.

134. H. Imai, T. Oishi, T. Kikuchi and M. Hirama, *Tetrahedron*, 2000, **56**, 8451.

135. M. Adinolfi, G. Barone, L. Guariniello and A. Iadonisi, *Tetrahedron Lett.*, 2000, **41**, 9005.

136. H. Ohtake, N. Ichiba and S. Ikegami, *J. Org. Chem.*, 2000, **65**, 8171.

137. T. Makaiyama, H. Jona and K. Takeuchi, *Chem. Lett.*, 2000, 696.

138. T. Mukaiyama, K. Takeuchi, H. Jona, H. Maeshima and T. Saitoh, *Helv. Chim. Acta*, 2000, **83**, 1901.

139. H. Matsui, J. Furukawa, T. Awano, N. Nishi and N. Sakairi, *Chem. Lett.*, 2000, 326.

140. B.G. Davis, M.A.T. Maughan, M.P. Green, A. Ullman and J.B. Jones, *Tetrahedron: Asymmetry*, 2000, **11**, 245.

141. M.S. Motawia, K. Larsen, C.E. Olsen and B.L. Moller, *Synthesis*, 2000, 1547.

142. Ya.V. Voznyi, I.S. Lukomskaya, I.M. Lanskaya and E.I. Podkidysheva, *Vopr. Med. Khim.*, 1996, **42**, 348 (*Chem. Abstr.*, 2000, **132**, 237 265).

143. K.C. Nicolaou, J.A. Pfefferkorn and G.-Q. Cao, *Angew. Chem., Int. Ed. Engl.*, 2000, K**39**, 734.

144. J. Ortner, M. Albert, K. Terler, W. Steiner and K. Dax, *Carbohydr. Res.*, 2000, **327**, 483.

145. C. Vogel and G. Pohlentz, *J. Carbohydr. Chem.*, 2000, **19**, 1247.

146. E. Kaji and N. Harita, *Tetrahedron Lett.*, 2000, **41**, 53.

147. K. Takeuchi, T. Tamura and T. Makaiyama, *Chem. Lett.*, 2000, 122.
148. Y. Hirata and T. Nakagawa, *Bull. Chem. Soc. Jpn.*, 2000, **73**, 1905.
149. L.A. Mulard, C. Costachel and P.J. Sansonetti, *J. Carbohydr. Chem.*, 2000, **19**, 849.
150. R.D. Marwood, S. Shuto, D.J. Jenkins and B.V.L. Potter, *Chem. Commun.*, 2000, 219.
151. M. de Kort, V. Correa, A.R.P.M. Valentijn, G.A. van der Marel, B.V.L. Potter, C.W. Taylor and J.H. van Boom, *J.Med. Chem.*, 2000, **43**, 3295.
152. M. de Kortf, A.D. Regenbogen, H.S. Overkleeft, R.A.J. Challiss, Y. Iwata, S. Miyamoto, G.A. van der Marel and J.H. van Boom, *Tetrahedron*, 2000, **56**, 5915.
153. S.S. Srivastav, R. Saxena, D. Deepak, N.K. Khare and A. Khare, *Trends Carbohydr. Chem.*, 1999, **4**, 51 (*Chem. Abstr.*, 2000, **132**, 108 165).
154. T. Yananoi, J. Inagaki, M. Mizuno, K. Yamashita and T. Inazu, *Heterocycles*, 2000, **52**, 921 (*Chem. Abstr.*, 2000, **132**, 279 417).
155. R.J. Tennant-Eyles, B.G. Davis and A.J. Fairbanks, *Tetrahedron: Asymmetry*, 2000, **11**, 231.
156. O.J. Plante, S.L. Buchwald and P.H. Seeberger, *J. Am. Chem. Soc.*, 2000, **122**, 71.
157. O.J. Plante, E.R. Palmacci and P.H. Seeberger, *Org. Lett.*, 2000, **2**, 3841.
158. G. Lemanski and T. Ziegler, *Helv. Chim. Acta*, 2000, **83**, 2655.
159. G. Lemanski and T. Ziegler, *Tetrahedron*, 2000, **56**, 563.
160. D. Crich and V. Dudkin, *Tetrahedron Lett.*, 2000, **41**, 5643.
161. H. Nagai, S. Matsumura and K. Toshima, *Tetrahedron Lett.*, 2000, **41**, 10233.
162. I. Sakamoto and H. Ohrui, *Biosci. Biotechnol. Biochem.*, 2000, **64**, 1974.
162a. J. Liaigre, D. Dubreuil, J.-P. Pradère and J.-F. Bonhours, *Carbohydr. Res.*, 2000, **325**, 265.
163. J. Ohlsson, G. Magnusson, *Carbohydr. Res.*, 2000, **329**, 49.
164. Y. Nishida, H. Tamakoshi, Y. Kitagawa, K. Kobayashi and J. Thiem, *Angew. Chem., Int. Ed. Engl.*, 2000, **39**, 2000.
165. L. Haga, M. Satake, H. Ohrui, N. Ohkohchi and S. Satomi, *Transplant Proc.*, 1999, **31**, 2826 (*Chem. Abstr.*, 2000, **132**, 334 684).
166. X. Zeng, R. Yoshino, T. Murata, K. Ajisaka and T. Usui, *Carbohydr. Res.*, 2000, **325**, 120.
167. E. Farkas, J. Thiem, F. Krzewinski and S. Bouquelet, *Synlett*, 2000, 728.
168. J. Ohlsson and G. Magnusson, *Tetrahedron*, 2000, **56**, 9975.
169. L.F. Tietze, V. Bothe and I. Schuberth, *Chem. Eur. J.*, 2000, **6**, 836.
170. Q. Li, M.-S. Cai, Z.-J. Li and R.-L. Zhou, *Gaodong Xuexiao Huaxue Xuebao*, 2000, **21**, 70 (*Chem. Abstr.*, 2000, **132**, 194 575).
171. J.H. Yoon and J.S. Rhee, *Carbohydr. Res.*, 2000, **327**, 377.
172. W.-T. Jiaang, M.-Y. Chang, P.-H. Tseng and S.-T. Chen, *Tetrahedron Lett.*, 2000, **41**, 3127.
173. B. La Ferla, L. Lay, G. Russo and L. Panza, *Tetrahedron: Asymmetry*, 2000, **11**, 3647.
174. K. Fujimoto, T. Miyata and Y. Aoyama, *J. Am. Chem. Soc.*, 2000, **122**, 3558.
175. S.C. Ennis, J.J. Gridley, H.M.I. Osborn and D.G. Spackman, *Synlett*, 2000, 1593.
176. K. Takeuchi, T. Tamura, H. Jona and T. Mukaiyama, *Chem. Lett*, 2000, 692.
177. B.N. Cook, S. Bhakta, T. Biegel, K.G. Bowman, J.I. Armstrong, S. Hemmerich and C.R. Bertozzi, *J. Am. Chem. Soc.*, 2000, **122**, 8612.
178. U.S. Chowdhury, *Synth. Commun.*, 2000, **30**, 3785.
179. H.J. Vermeer, J.P. Kamerling and J.F.G. Vliegenthart, *Tetrahedron: Asymmetry*, 2000, **11**, 539.
180. Y. Sakai, M. Oikawa, H. Yoshizaki, T. Ogawa, Y. Suda, K. Fukase and S.

Kusumoto, *Tetrahedron Lett.*, 2000, **41**, 6843.

181. T. Mochizuki, Y. Iwano, M. Shiozaki, S.-I. Kurakata, S. Kanai and M. Nishijima, *Tetrahedron*, 2000, **56**, 7691.

182. T. Mochizuki, Y. Iwano, M. Shiozaki, S.-I. Kurakata, S. Kanai and M. Nishijima, *Carbohydr. Res.*, 2000, **324**, 225.

183. M. Nilsson and T. Norberg, *Carbohydr. Res.*, 2000, **327**, 261.

184. H. Jiao and O. Hindsgaul, *J. Carbohydr. Chem.*, 1999, **18**, 499.

185. N. Barroca, J.-C. Jacquinet, *Carbohydr. Res.*, 2000, **329**, 667.

186. A. Tóth, A. Medgyes, I. Bajza, A. Lipták, G. Batta, T. Kontrohr, K. Péterffy and V. Pozgay, *Biorg. Med. Chem. Lett.*, 2000, **10**, 19.

187. A. Medgyes, I. Bajza, E. Farkas, V. Pozsgay and A. Liptak, *J. Carbohydr. Chem.*, 2000, **19**, 285.

188. B. Alessandra, G. Capozzi, S. Menichetti and C. Nativi, *Org. Lett.* 2000, **2**, 251.

189. A. Graziani, P. Passacamtilli, G. Piancatelli and S. Tani, *Tetrahedron: Asymmetry*, 2000, **11**, 3921.

190. G. Pastuch, I. Wandzik and W. Szeja, *Tetrahedron Lett.*, 2000, **41**, 9923.

191. J.D. Chisholm and D.L. Van Vranken, *J. Org. Chem.*, 2000, **65**, 7541.

192. B. Alluis, N. Pérol, H. El Hajji and O. Dangles, *Helv. Chim. Acta*, 2000, **83**, 428.

193. H. Vankayalapati and G. Singh, *J .Chem. Soc., Perkin Trans. 1*, 2000, 2187.

194. E. Farkas, J. Thiem and K. Ajisaka, *Carbohydr. Res.*, 2000, **328**, 293.

195. W. Hellebrandt, T. Haselhorst, T. Köhli, E. Bäuml and T. Peters, *J. Carbohydr. Chem.*, 2000, **19**, 769.

196. N. Karst and J.-C. Jacquinet, *J. Chem. Soc., Perkin Trans. 1*, 2000, 2709.

197. Y.L. Dory, P. Deslongchamps, D. Magaud, R. Dolmazon, D. Anker and A. Doutheau, *Org. Lett.*, 2000, **2**, 2225.

198. G.-J. Boons and A.V. Demchenko, *Chem. Rev.*, 2000, **100**, 4539.

199. X. Zhang, T. Kamiya, N. Otsubo, H.K. Ishida and M. Kiso, *J. Carbohydr. Chem.*, 1999, **18**, 225.

200. L.A. Simeoni, N.E. Byramova and N.V. Bovin, *Russ. J. Bioorg. Chem.*, 2000, **26**, 183. (*Chem. Abstr.*, 2000, **133**, 252 642).

201. N. Wimmer, H. Brade, P. Kosma, *Carbohydr. Res.*, 2000, **329**, 549.

201a. M. Reiner and R.R. Schmidt, *Tetrahedron: Asymmetry*, 2000, **11**, 319.

202. S. Kaneko, M. Kitaoka, A. Kuno and K. Hayashi, *Biosci. Biotechnol. Biochem.*, 2000, **64**, 741.

203. X. Ma, B. Yu, Y. Hui, D. Xiao, J. Ding, *Carbohydr. Res.*, 2000, **329**, 495.

204. T. Kawada, R. Asano, K. Makino and T. Sakuno, *Eur. J. Org. Chem.*, 2000, 2723.

205. H. Kato, K. Ohmori and K. Suzuki, *Tetrahedron Lett.*, 2000, **41**, 6827.

206. X.-M. Zhu and B.Yu, *Chin. J. Chem.*, 2000, **18**, 72. (*Chem. Abstr.*, 2000, **132**, 208 028).

207. Y.-K. Yang, M.-H. Chiu, C.-W. Gao, R.-L. Nie, Y. Lu and Q.-T. Zheng, *Tetrahedron*, 2000, **56**, 7433.

208. R. Roy, S.K. Das, F. Santoyo-González, F. Hernández-Matteo, T.K. Dam and C.F. Brewers, *Chem. Eur. J.*, 2000, **6**, 1757.

209. M. Lobbel and P. Köll, *Tetrahedron: Asymmetry*, 2000, **11**, 393.

210. L. Kniezo, M. Budesinski, P. Voitisek and M. Martinkova, *Enantiomer*, 1999, **4**, 351 (*Chem. Abstr.*, 2000, **132**, 64 453).

211. Y. Kajihara, H. Hashimoto and S. Ogawa, *Carbohydr. Res.*, 2000, **323**, 44.

212. S. Ogawa, K. Gamou, Y. Kugimiya, Y. Senba, A. Lu and M.M. Palcic, *Carbohydr. Lett.*, 2000, **3**, 451.

213. M.N. Namchuk, J.D. McCarter, A. Becalski, T. Andrews and S.G. Withers, *J. Am. Chem. Soc.*, 2000, **122**, 1270.

214. S. Manabe, Y. Nakahara and Y. Ito, *Synlett*, 2000, 1241.
215. P Sjölin and J. Kihlberg, *Tetrahedron Lett.*, 2000, **41**, 4435.
216. M. DellaGreca, A. Fiorentino, P. Monaco, L. Previtera and A.M. Simonet, *Tetrahedron Lett.*, 2000, **41**, 6507.
217. S. Trombotto, A. Bouchu, G. Descotes and Y. Queneau, *Tetrahedron Lett.*, 2000, **41**, 8273.
218. Y. Abe, M. Fujiwara, K. Ohbu and K. Harata, *J. Chem. Soc., Perkin Trans. 2*, 2000, 341.
219. E. Montero, A. García-Herrero, J.L. Asonsia, K. Hirai, S. Ogawa, F. Santoyo-Gonzales, F.J. Cañada and J. Jiménez-Barbero, *Eur. J. Org. Chem.*, 2000, 1945.
220. M.T. Gordon, T.L. Lowary and C.M. Hadad, *J. Org. Chem.*, 2000, **65**, 4954.
221. M. Yanase and M. Funabashi, *J. Carbohydr. Chem.*, 2000, **19**, 53.
222. H. Nakano, H. Murakami, M. Shizuma, T. Kiso, T.L. de Aranjo and S. Kitahata, *Biosci. Biotechnol. Biochem.*, 2000, **64**, 1472.
223. H. Nakano, M. Schizuma, T. Kiso and S. Kitahata, *Biosci. Biotechnol. Biochem.*, 2000, **64**, 735.
224. R. Hirschmann, L. Ducry and A.B. Smith, *J. Org. Chem.*, 2000, **65**, 8307.
225. Z. Gan and R. Roy, *Tetrahedron Lett.*, 2000, **41**, 1155.
226. D.I., Angus, M.J. Kiefel and M. von Itzstein, *Bioorg. Med. Chem.*, 2000, **8**, 2709.
227. D. Gueyrard, J. Barillari, R. Iori, S. Palmieri and P. Rollin, *Tetrahedron Lett.*, 2000, **41**, 8307.
228. G.E.H. Elgemeie, A.M.E. Attia and S.S. Alkabai, *Nucleosides, Nucleotides, Nucleic Acids*, 2000, **19**, 723.
229. M.S. Chande and U.S. Bhat, *Indian J. Chem.*, 1999, **38B**, 932 (*Chem. Abstr.*, 2000, **132**, 194 566).
230. B. Liberek, A. Konitz, R. Frankowski and Z. Smiatacz, *Carbohydr. Res.*, 2000, **326**, 151.
231. D. Bousquet, A. Spadaro, M.S. Pappalardo, R. Bernardini, R. Romeo, L. Panza and G. Ronsisvalle, *J. Carbohydr. Chem.*, 2000, **19**, 527.
232. Y. Ohnishi, M. Ichikawa and Y. Ichikawa, *Bioorg. Med. Chem. Lett.*, 2000, **10**, 1289.
233. L. Jobron and G. Hummel, *Org. Lett.*, 2000, **2**, 2265.
234. T. Opatz and H. Kunz, *Tetrahedron Lett.*, 2000, **41**, 10185.
235. B.D. Johnston and B.M. Pinto, *J. Org. Chem.*, 2000, **65**, 4607.
236. A. Zawisza, B. Kryczka, P. Lhoste, S. Porwanski and D. Sinou, *J. Carbohydr. Chem.*, 2000, **19**, 795.
237. Z.J. Witczak, H. Chen and P. Kaplon, *Tetrahedron: Asymmetry*, 2000, **11**, 519.
238. D. Crich and H. Li, *J. Org. Chem.*, 2000, **65**, 801.
239. W.B. Turnbull and R.A. Field, *J. Chem. Soc., Perkin Trans. 1*, 2000, 1859.
240. A. Lubineau, A. Malleron and C. LeNarvor, *Tetrahedron Lett.*, 2000, **41**, 8887.
241. D.J. Owen and M. von Itzstein, *Carbohydr. Res.*, 2000, **328**, 287.
242. K.P.R. Kartha, P. Cura, M. Aloui, S.K. Readman, T.J. Rutherford and R.A. Field, *Tetrahedron: Asymmetry*, 2000, **11**,581.
243. J. Borowiecka, *Heteroat. Chem.*, 2000, **11**, 292; (*Chem Abstr.*, 2000, **133**, 177 367).
244. R.P. Spencer and J. Schwartz, *Tetrahedron*, 2000, **56**, 2103.
245. D. Sinou and C. Moineau, *Recent Res. Dev. Org. Chem.*, 1999, **3**, 1 (*Chem. Abstr.*, 2000, **132**, 237 247).
246. A. Vasella, *Bioorg. Chem.: Carbohydr.*, 1999, **56**, 556 (*ChemAbstr.*, 2000, **132**, 279 399).
247. B. Becker, *J. Carbohydr. Chem.*, 2000, **19**, 253.
248. R. Plantier-Royon and C. Portella, *Carbohydr. Res.*, 2000, **327**, 119.

249. P. Siemsen, R.C. Livingston and F. Diederich, *Angew. Chem., Int. Ed. Engl.*, 2000, **39**, 2633.
250. J. Cossy and H. Rakotoarisoa, *Synlett*, 2000, 734.
251. W. Kudelska, *Carbohydr. Res.*, 2000, **329**, 687.
252. M. Hayashi, H. Kawabata, S. Shimono and A. Kakehi, *Tetrahedron Lett.*, 2000, **41**, 2591.
253. M. Han, I.P. Smoliakova and L.N. Koikov, *Carbohydr. Res.*, 2000, **323**, 202.
254. D.-P. Pham-Huu, M. Petruskova, J.N. BeMiller and L. Petrus, *J. Carbohydr. Chem.*, 2000, **19**, 93.
254a. X. Shen, Y.-L. Wu and Y. Wu, *Helv. Chim. Acta.*, 2000, **83**, 943.
255. H.J. Rosenberg, A.M. Riley, V. Correa, C.W. Taylor and B.V.L. Potter, *Carbohydr. Res.*, 2000, **329**, 7.
256. K.T. Weber, D. Hammache, J. Fantini and B. Ganem, *Bioorg. Med. Chem. Lett.*, 2000, **10**, 1011.
257. E. Lohof, E. Planker, C. Mang, F. Burkhart, M.A. Dechantsreiter, R. Haubner, H.-J. Wester, M. Schwaiger, G.Hölzemann, S.L. Goodman and H. Kessler, *Angew. Chem., Int. Ed. Engl.*, 2000, **39**, 2761.
258. T.M. Davis and T.L. Lowary, *Carbohydr. Res.*, 2000, **324**, 210.
259. S. Shuto, Y. Yahiro, S. Ichikawa and A. Matsuda, *J. Org. Chem.*, 2000, **65**, 5.
260. S. Shuto, M. Terauchi, Y. Yahiro, H. Abe, S. Ichikawa and A. Matsuda, *Tetrahedron Lett.*, 2000, **41**, 4151.
261. O. Gaurat, J. Xie and J.-M. Valéry, *Tetrahedron Lett.*, 2000, **41**, 1187.
262. Q. Wang, M. Wolff, T. Polat, Y. Du and R.J. Linhardt, *Biorg. Med. Chem. Lett.*, 2000, **10**, 941.
263. M.J. Costanzo, L. Jaroskova, D.A. Gauthier and B.E. Maryanoff, *Tetrahedron: Asymmetry*, 1999, **10**, 689.
264. Y. Lakhrissi, C. Taillefumier, M. Lakhrissi and Y. Chapleur, *Tetrahedron: Asymmetry*, 2000, **11**, 417.
265. T. Ziegler and C. Jurisch, *Tetrahedron: Asymmetry*, 2000, **11**, 3403.
266. R. SanMartin, B. Tavassoli, K.E. Walsh, D.S. Walter and T. Gallagher, *Org. Lett.*, 2000, **2**, 4051.
267. S.K. Das, R. Dominique, C. Smith, J. Nahro and R. Roy, *Carbohydr. Lett.*, 1999, **3**, 361 (*Chem. Abstr.* 2000, **132**, 166 415).
268. S.D. Debenham and E.J. Toone, *Tetrahedron: Asymmetry*, 2000, **11**, 385.
269. J.D. Rainer and J.M. Cox, *Org. Lett.*, 2000, **2**, 2707.
270. L. Cipolla, B. La Ferla, L. Lay, F. Peri and F. Nicotra, *Tetrahedron: Asymmetry*, 2000, **11**, 295.
271. P. Rodrigues, Y. Canac and A. Lubineau, *Chem. Commun.*, 2000, 2049.
272. V. Aucagne, D. Gueyard, A. Tatibouët, A. Quinsac and P. Rollin, *Tetrahedron*, 2000, **56**, 2647.
273. S.K. Readman, S.P. Marsden and A. Hodgson, *Synlett*, 2000, 1628.
274. M. Carpintero, C. Jaramillo and A. Fernández Mayoralas, *Eur. J. Org. Chem.*, 2000, 1285.
275. R. Dominique, B. Liu, S.K. Das and R. Roy, *Synthesis*, 2000, 862.
276. A. Walter and B. Westermann, *Synlett*, 2000, 1682.
277. J. Grugier, J. Xie, I. Duarte and J.-M. Valéry, *J. Org. Chem.*, 2000, **65**, 979.
278. T.V. Bohner, R. Beaudegnies and A. Vasella, *Helv. Chim. Acta*, 1999, **82**, 143.
279. T. Kumazawa, T. Saito, S. Matsuba, S. Sato and J.-i. Onodera, *Carbohydr. Res.*, 2000, **329**, 855.
280. K.A. Parker and A.T. Georges, *Org. Lett.*, 2000, **2**, 497.

281. N. Hayashi, S.-Z. Nakayama and M. Kawabata, *Chem. Commun.*, 2000, 1329.
282. A. Gupta and Y.D. Vankar, *Tetrahedron*, 2000, **56**, 8525.
283. J. Beyer, P.R. Skaanderup and R. Madsen, *J. Am. Chem. Soc.*, 2000, **122**, 9575.
284. M.A. Brimble and T.J. Brenstrum, *Tetrahedron Lett.*, 2000, **41**, 1107.
285. F.L. Andrews, D.S. Larsen and L. Larsen, *Aust. J. Chem.*, 2000, **53**, 15.
286. T. Matsumoto, H. Yamaguchi, T. Hamura, M. Tanabe, Y. Kuriyama and K. Suzuki, *Tetrahedron Lett.*, 2000, **41**, 8383.
287. T. Matsumoto, H. Yamaguchi, M. Tanabe, Y. Yasui and K. Suzaki, *Tetrahedron Lett.*, 2000, **41**, 8393.
288. T. Kumazawa, T. Minatagawa, S. Matsuba, S. Sato and J.-i. Onodera, *Carbohydr. Res.*, 2000, **329**, 507.
289. G. Cheng, Y. Bai, Y. Zhao, J. Tao, Y. Liv, G. Tu, L. Ma, N. Liao and X. Xu, *Tetrahedron*, 2000, **56**, 8915.
290. K.A. Parker and G.-J. Ding, *Tetrahedron*, 2000, **56**, 10255.
291. M.A. Brimble and T.J. Brenstrum, *Tetrahedron Lett.*, 2000, **41**, 2991.
292. S. Futagami, Y. Ohashi, K. Imura, T. Hosoya, K. Ohmori, T. Matsumoto and K. Suzuki, *Tetrahedron Lett.*, 2000, **41**, 1063.
293. H.-B. Yin and Z.-S. He, *Tetrahedron Lett.*, 2000, **41**, 1955.
294. S.S. Handa, R. Singh, R. Maurya, N.K. Satti, K.A. Suri and O.P. Suri, *Tetrahedron Lett.*, 2000, **41**, 1579.
295. M. Tagashira, T. Kitagawa, N. Nozato, S. Isonishi, A. Okamoto, K. Ochiai and Y. U. Ohtake, *Chem. Pharm. Bull.*, 2000, **48**, 575.
296. D. Balachari and G.A. O'Doherty, *Org. Lett.*, 2000, **2**, 863.
297. T. Kumazawa, M. Chiba, S. Matsuba, S. Sato and J.-i. Onodera, *Carbohydr. Res.*, 2000, **328**, 599.
298. I. Izquierdo, M.T. Plaza, M. Rodríguez and J. Tamayo, *Tetrahedron: Asymmetry*, 2000, **11**, 1749.
299. C. Betancor, R.L. Dorta, R. Freire, T. Prangé and E. Suárez, *J. Org. Chem.*, 2000, **65**, 8822.
300. C. Rousseau and O.R. Martin, *Tetrahedron: Asymmetry*, 2000, **11**, 409.
301. S. Knapp and V.K. Gore, *Org. Lett.*, 2000, **2**, 1391.
302. G.-T. Fan, T.-S. Hus, C.-C. Lin and C.-C. Lin, *Tetrahedron Lett.*, 2000, **41**, 6593.
303. M.H.D. Postema, D. Calimente, L. Liu and T.L. Behrmann, *J. Org. Chem.*, 2000, **65**, 6061.
304. S.J. Spak and O.R. Martin, *Tetrahedron*, 2000, **56**, 217.
305. Y.-H. Zhu, R. Demange and P. Vogel, *Tetrahedron: Asymmetry*, 2000, **11**, 263.
306. N. Miquel, G. Doisneau and J.-M. Beau, *Chem. Commun.*, 2000, 2347.
307. H. Abe, S. Shuto and A. Matsuda, *Tetrahedron Lett.*, 2000, **41**, 2391.
308. H. Abe, S. Shuto and A. Matsuda, *J. Org. Chem.*, 2000, **65**, 4315.
309. X. Cheng, N. Khan and D.R. Mootoo, *J. Org. Chem.*, 2000, **65**, 2544.
310. B. Patro and R.R. Schmidt, *J. Carbohydr. Chem.*, 2000, **19**, 817.
311. C. Pasquarello, S. Picasso, R. Demange, M. Malissard, E.G. Berger and R. Vogel, *J. Org. Chem.*, 2000, **65**, 4251.
312. H. Godage and A.J. Fairbanks, *Tetrahedron Lett.*, 2000, **41**, 7589.
313. A. Dondoni and A. Marra, *Chem. Rev.*, 2000, **100**, 4395.
314. C.-Y. Tsai, X. Huang and C.-H. Wong, *Tetrahedron Lett.*, 2000, **41**, 9499.
315. S.P. Vincent, A. Schleyer and C.-H. Wong, *J. Org. Chem.*, 2000, **65**, 4440.
316. T. Fuchss and R.R. Schmidt, *Synthesis*, 2000, 259.
316a. M. Ricci, L. Madariaga and T. Skrydstrup, *Angew. Chem., Int. Ed. Engl.*, 2000, **39**, 242.

317. A. Dondoni, G. Mariotti and A. Marra, *Tetrahedron Lett.*, 2000, **41**, 3483.
318. V.V. Samoshin, I.P. Smoliakova, M. Han and P.H. Gross, *Mendeleev Commun.*, 1999, 219 (*Chem. Abstr.*, 2000, **132**, 194 563).
319. N. Miquel, G. Doisneau and J.-M. Beau, *Angew. Chem., Int. Ed. Engl.*, 2000, **39**, 4111.
320. K. Lee, B. Soh and Y. Kong, *Bull. Korean Chem. Soc.*, 2000, **21**, 449.
321. V. Popsavin, O. Beric, M. Popsavin, L. Radic, J. Csanadi and V. Cirin-Novta, *Tetrahedron*, 2000, **56**, 5929.
322. U. Diederichsen and C.M. Biro, *Bioorg. Med. Chem. Lett.*, 2000, **10**, 1417.
323. T.R. Chakraborty, S. Ghosh, S. Jayaprakash, J.A.R.P. Sarma, V. Ravikanth, P.V. Diwan, R. Nagaraj and A.C.Kunwar, *J. Org. Chem.*, 2000, **65**, 6441.
324. C. Lamberth, *Carbohydr. Lett.*, 1999, **3**, 375 (*Chem. Abstr.*, 2000, **132**, 166 339).
325. F. García-Tellado, P. de Armas and J.J. Marrero-Tellado, *Angew. Chem., Int. Ed. Engl.*, 2000, **39**, 2727.
326. K. Tomooka, H. Yamamoto and T. Nakai, *Angew. Chem., Int. Ed. Engl.*, 2000, **39**, 4500.
327. J. Parsch and J.W. Engels, *Helv. Chim. Acta*, 2000, **83**, 1791.
328. K.S. Ramasamy, R. Bandaru and D. Averett, *J. Org. Chem.*, 2000, **65**, 5849.
329. G. Kim and H.S. Kim, *Tetrahedron Lett.*, 2000, **41**, 225.
330. S. Obika, Y. Hari, K.-i. Morio and T. Imanishi, *Tetrahedron Lett.*, 2000, **41**, 215.
331. S. Obika, Y. Hari, K.-i. Morio and T. Imanishi, *Tetrahedron Lett.*, 2000, **41**, 221.
332. P.R. Krishna, B. Laranya, A. Ilangaran and G.V.M. Sharma, *Tetrahedron: Asymmetry*, 2000, **11**, 4463.
333. M. Shionoyia and K. Tanaka, *Bull. Chem. Soc. Jpn.*, 2000, **73**, 1945.
334. H. Cao, K. Tanaka and M. Shionoya, *Chem. Pharm. Bull.*, 2000, **48**, 1745.
335. Z.-X. Wang, L.I. Wiebe, E. De Clercq, J. Balzarini and E.E. Knaus, *Can. J. Chem.*, 2000, **78**, 1081.
336. Z.-X. Wang, L.I. Wiebe, J. Balzarini, E. De Clercq and E.E. Knaus, *J. Org. Chem.*, 2000, **65**, 9214.
337. J. Kovensky, D. Burrieza, V. Colliou, A. Fernández-Cirelli and P. Sinaÿ, *J. Carbohydr. Chem.*, 2000, **19**, 1.
338. G.V.M. Sharma, A. Subhash Chander, P. Radha Krishna, K. Krishnudu, M.H.V. Ramana Rao and A.C. Kunwar, *Tetrahedron: Asymmetry*, 2000, **11**, 2643.
339. A. Dondoni, P.P. Biovannini and A. Marra, *Tetrahedron Lett.*, 2000, **41**, 6195.

4
Oligosaccharides

1 General

Before mention is made of many reviews relevant to oligosaccharide synthesis attention is called to two contributions which illustrate the 'state of the art' in this fast developing but extremely demanding subject. A survey of the work of Danishefsky's laboratory gives an account of highly complex syntheses of oligosaccharide and glycoconjugate antigens and the development of anti-cancer vaccines from them.[1] Also monumentally impressive is Nicolaou's synthesis of everninomicin (1) which was described in three massive papers each dealing with the synthesis of one part of the molecule as indicated by the dotted lines.[2-4] Briefer accounts of the work on this orthosomicin class antibiotic, which describe syntheses of a 1-1' disaccharide linkage, a branched-chain nitrosugar, orthoester acetal formation and several specific deoxy and O-substituted sugars, 13 rings and 35 stereogenic centres have appeared separately.[5]

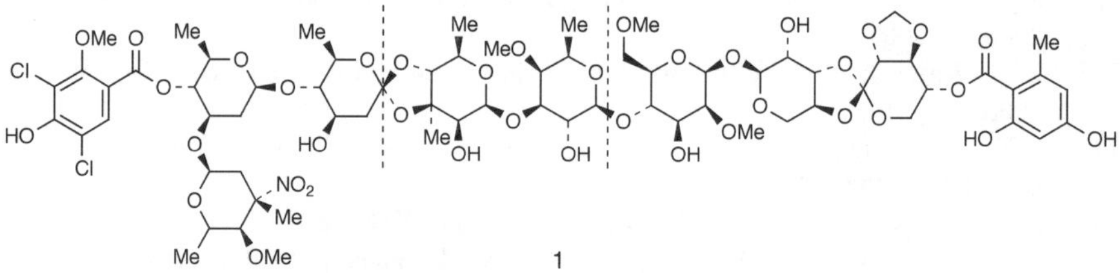

1

A synopsis covering the literature from October 1995 to December 1999 on developments in oligosaccharide synthesis[6] and a further important one dealing with work utilizing intramolecular methods[7] have appeared. A more selective survey on the use of glycosyl halides in solid phase work was written as part of a major survey of bioorganic aspects of the subject.[8]

Wong has described a one-pot oligosaccharide synthesis for application to the creation of libraries. Thioglycosides containing one free hydroxyl group the glycosylating powers of which are variable were used such that the hydroxyl group could be glycosylated and the product then used as a di-(or higher) saccharide glycosylating agent.[9] (See later in this chapter for further reports on this approach to tunable glycosylations; especially refs. 40, 41, 59). Methods involving the use of both solid and liquid phase techniques to produce oligosac-

Carbohydrate Chemistry, Volume 34
© The Royal Society of Chemistry, 2003

charide libraries have been reviewed,[10,11] the latter also dealing with O- and N-linked glycopeptides. A further important one limited to solid phase methodology covers Seeberger's new methods and deals with strategies, supports, linkers, protecting groups, glycosylations, 'on-bead' analysis, release from the resins and specific applications and generations of libraries. Automated procedures are also dealt with.[12]

An extensive review covers the synthesis of glycopeptides, mostly by use of O- or N-glycosylated amino-acid building blocks in solid phase techniques.[13] An N-terminal glycodecapeptide fragment of interleukin has been made by use of a new solid phase linker.[14] Homooligomers of glycamino-acids (cyclic carbohydrate derivatives bearing a carboxylic acid group and an amino group replacing one of the hydroxyl functions) have been made with inter-unit amide bonding between different positions. Hybrids of these and amino acids were also made and, after sulfation, bioactive products were obtained.[15]

A review 'Recent Advances in O-Sialylation' includes treatment of the synthesis of sialyl dimers with α-(2→8) and α-(2→9) linkages and the sialylation of oligosaccharides. NMR criteria for characterizing α- and β-sialyl linkages are dealt with.[16] A shorter review deals with the synthesis of Sia Le^x gangliosides and blood group oligosaccharides.[17] *In vivo* lactonization within polysialic acids, has been critically assessed.[18]

In the area of glycolipids reviews have been published on the synthesis of bacterial lipopolysaccharides,[19] glycolipid oligosaccharides and Sia Le^x and sulfated analogues as selectin receptors,[20] and glycosphingolipids containing poly-N-acetylactosamine and polyglycosylceramides.[21]

Interest in the field of enzymic procedures for making oligosaccharides is illustrated by the number of reviews that have dealt with the topic. Of particular significance is the review from Wong's laboratory 'Syntheses of Complex Carbohydrates and Glycoconjugates: Enzyme-based and Programmable One-Pot Strategies' which includes reference to selective activation techniques and the author's 'Optimer' programme.[22] Further relevant reviews of a general nature in the area have been published: the numbers of references cited in each are as follows 65,[23] 40,[24] 21,[25] 15[26] and 8.[27]

Two papers have dealt with the use of enzymes to make chitooligosaccharides: a review covered the use of chitinase and glycanase for this purpose,[28] and terminating β-(1→3) and β-(1→4) galactosylated chitooligosaccharides were made by transgalactosylation from p-nitrophenyl β-D-galactopyranoside to chitobiose and chitotriose using bovine testes β-galactosidase.[29] Chitinase catalyses the transfer of GlcNAc and LacNAc from their oxazoline derivatives.[30] A 'glycosynthase' from *Humicola insolens* transfers lactose from its α-glycosyl fluoride in β-(1→4) manner to a range of mono- and disaccharide glycosides including β-glucosides, β-mannosides, β-cellobiosides and various halogenated derivatives.[31]

Other reviews or articles of general relevance to have appeared cover the oligosaccharide antibiotics of *Ziracia* which are of the orthosomycin class and include everninomycin;[32] water-soluble, lectin-binding hybrids of chitosan and sialic acid, made by the reductive coupling of α-NeuNAcOC$_6$H$_4$CHO-(p) with

chitosan amino groups;[33] and the copper-catalysed acetylenic coupling of Vas-
ellas' sugar-alkynes.[34]

A different matter of general significance is the publication of a method for
eliminating interfering signals of some protecting groups (*e.g.* benzyl and acetal
methylene resonances) from 2D NMR spectra of oligosaccharide derivatives.[35]

2   Trisaccharides

2.1   General. – Compounds in Sections 2.2–2.4 are treated according to their
non-reducing end sugars. Unless otherwise specified, the sugar name abbreviations (Glc *etc.*) imply the pyranosyl ring forms.

A library of 72 trisaccharides derived from Glc, Gal and Man have been made
in solution by one-pot procedures using glycosyl bromides, phenylthio glycosides as donors and 2-bromoethyl glycosides as acceptors.[36]

2.2   Linear Homotrisaccharides. – Maltose, cellobiose and lactose as *O*-pro-
tected β-thioglycosides have been tethered *via* their O-2 to O-3 of benzyl 6-*O*-
benzyl-2-*O*-benzoyl-α-D-glucopyranoside. Cyclization induced by NIS gave α-
linked trisaccharide products in about 60% yield. This approach has therefore
given access to α-D-Glc-(1→4)-α-D-Glc-(1→4)-D-Glc, β-D-Glc-(1→4)-α-D-Glc-
(1→4)-D-Glc and β-D-Gal-(1→4)-α-D-Glc-(1→4)-D-Glc.[37]

The β-(1→6), β-(1→6) linked trimers (and tetramers) of Glc and Gal and the
α-(1→6), α-(1→6) analogues of Man have been made by use of the 1,2-orthoester
approach.[38] The 4-pentenyl β-glycoside of the β-(1→6) linked glucotriose was
prepared *via* the resin-bound **2** and cleaved from the resin by elimination of
bromine and application of Grubbs' catalyst as an alkene cleaving agent.[39] By

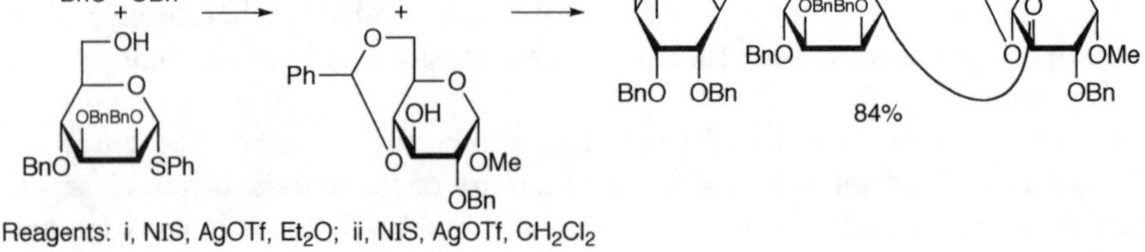

Reagents: i, NIS, AgOTf, Et₂O; ii, NIS, AgOTf, CH₂Cl₂

Scheme 1

use of disarmed donor/acceptor **3** 6-*O*-α-glycosylation was effected with tetra-*O*-benzyl-β-D-glucosyl fluoride and the product was then activated as a donor to allow the synthesis of α-D-Glc-(1→6)-β-D-Glc-(1→6)-D-Glc in high yield,[40] and the β-(1→6), β-(1→6) isomer was also made with high efficiency using an analogous approach.[41] [See ref 59 and Scheme 1 for a similar approach which depends on solvent change for preferential activation of thioglycoside donors]. Enzymic glucosylation of gentiobiose also gave the β-(1→6)-linked glucotriose, in 13% yield, and likewise afforded β-D-Glc-(1→6)-β-D-Glc-(1→4)-D-Glc and the β-(1→6), β-(1→6) isomers from cellobiose and laminaribiose, respectively. Sophorose, the β-(1→2) linked glucobiose did not act as an acceptor.[42]

By use of the 6-*O*-acetyl-tri-*O*-benzyl-mannosyl dimethylphosphonothioate as donor α-D-Man-(1→6)- α-D-Man-(1→2)-α-D-Man-OPr was made as the 6″-phosphate as a component of the glycosyl phosphatidylinositol anchor compound.[43] The galacturonic acid trimer α-D-GalA-(1→4)- α-D-GalA-(1→4)-α-D-GalA was made as the trimethyl ester by the use of thioglycoside coupling, as a fragment of pectin galacturonans.[44]

2.3 Linear Heterotrisaccharides. – As above (refs. 40, 41), the monohydroxy thioglycoside approach was used by the Mukaiyama group in the synthesis of a derivative of β-D-Glc-(1→4)-β-D-GlcNH₂-(1→6)-α-D-Glc-OMe.[45] The use of polystyrylboronic acid to bind the 4,6-diol of methyl 3-*O*-benzyl galactoside is a novel approach which allows a loading, release, loading sequence in oligosaccharide synthesis. It was applied in the preparation of a derivative of β-D-Glc-(1→3)-α-D-Gal-(1→2)-D-Gal,[46] and α-D-Glc-(1→4)-α-L-Rha-(1→3)-β-GlcNAc was made as a fragment of the O-specific polysaccharide of *Shigella flexneri*.[47] Several β-Glc-(1→3)-α-L-Rha-(1→6)-D-Glc and -Gal trisaccharides have been reported as components of flavonoids of Green Tea.[48]

Several modifications of α-D-Gal-(1→3)-β-D-Gal-(1→4)-D-Glc have been made with specific deoxygenation at each ring position of the non-reducing end unit, and with β-L-arabinopyranose in place of that moiety. All were prepared as glycosides with a C₉ spacer arm joining a chromophoric aglycon.[49] Enzymic α-galactosylation to lactose gave the β-(1→3′)-linked anomer.[50]

Globotriose, α-D-Gal-(1→4)-β-D-Gal-(1→4)-β-D-Glc, has been *C*-linked to a peptide and the peptide was also diglycosylated, the products being evaluated as ligands for a Shiga-like toxin binding unit.[51] The closely related α-D-Gal-(1→4)-β-D-Gal-(1→4)-D-GlcNAc, the human blood group P₁- antigenic determinant, has also been made,[52] as has β-D-Gal-(1→6)-β-D-Gal-(1→4)-D-Glc, a human milk trisaccharide.[53]

β-D-Man-(1→4)-β-D-GlcNAc-(1→4)-β-D-GlcNAc was described in a paper on glycopeptide work.[54]

Compounds having 2-amino-2-deoxy sugars at the non-reducing position to have been reported are α-D-GlcNAc-(1→4)-β-D-Gal-(1→4)-β-D-GalNAcOC₁₆H₃₃ made to form micelles corresponding to structures found in ovarian cyst fluids. A range of protecting groups had to be tried for suitable coupling with a disaccharide glycosyl bromide.[55] β-D-GlcNAc-(1→4)-β-D-GlcA-(1→3)-D-GlcNAc, a repeating unit of hyaluronan, and the alternative trisaccharide β-D-GlcA-(1→3)-β-D-GlcNAc-(1→4)-D-GlcA have been made as their methyl β-glycosides.[56]

In the field of 6-deoxyhexose-terminating trisaccharides β-L-Rha-(1→4)-β-D-Glc-(1→6)-β-D-Glc[57] and α-L-Rha-(1→2)-β-D-Gal-(1→2)-β-D-GlcA[58] have been isolated from plant products, and α-L-Rha-(1→6)-α-D-Man-(1→3)-D-Glc was synthesized by a notable, new approach. Use was again made of a monohydroxy thiomannoside which was first rhamnosylated with a thioglycoside and the dimer became the glycosylating agent simply by change of solvent in a one-pot procedure (Scheme 1).[59] This approach is similar to those used by Mukaiyama as noted above (refs. 40, 41).

Synthesis of α-L-Fuc-(1→2)-β-D-Gal-(1→2)-D-Xyl as its α- and β-glycosides has been reported,[60] and the highly modified trisaccharide **4** containing eight deoxy centres has been found as a component of the anthraquinone *C*-glycosidic vineomycin C.[61] 3,6-Dideoxy-D-*arabino*-hexose-β-(1→3)-β-D-GalNAc-(1→4)-β-D-GlcNAc, found in the *Trichinella spiralis* parasite, has been synthesized.[61a]

4

The chondroitin 6-sulfate trisaccharide β-D-GlcA-(1→3)-β-D-GalNAc-6-sul-fate-(1→4)-β-D-GlcAOMe has been made[62] as has the rhamnogalacturonan fragment α-D-GalA-(1→2)-α-L-Rha-(1→4)-D-GalA as its dimethyl ester.[63]

Previous syntheses of ganglioside GM₃, α-NeuNAc-(2→3)-β-D-Gal-(1→4)-β-D-GlcOCer, have been reviewed and new chemical and chemoenzymic approaches have been investigated.[64] Other workers, using glycosylating agents with phosphorus-containing leaving groups, have completed a synthesis.[65]

O-Acetylated NeuNAcSPh was the key donor in the syntheses of α-NeuNAc-(2→3)-β-D-Gal-(1→3)-α-D-GalNAcOBu (see also next paragraph) and α-NeuNAc-(2→3)-β-D-Gal-((1→4)-β-D-GlcNAcOMe and its 6′ sulfate,[66] and the unsulfated trimer was also produced glycosidically linked to ovalbumin.[67] In this work ¹³C labelling in the NeuNAc and Gal moieties allowed the assignment of all the protons in the glycoprotein by use of 2D HMQC and ID HSQC-TOCSY and the new sequences 1D and 2D HSQC-TOCSY-NOESY-TOCSY.

Chemoenzymic methods have been used to make sialyl-T antigen α-NeuNAc-(2→3)-β-D-Gal-(1→3)-α-D-GalNAc-OThr,[68,69] the former paper also describing the α-(2→6)-β-(1→3) linked isomer. Solid phase work led to the preparation of the *N*-terminal glycopentapeptide of human glycophorin AM which has three copies of the sialyl-T antigen glycosidically linked within the pentapeptide.[70]

Glycosyl 2-pyridyl sulfones *e.g.* **5** activated with samarium triflate allow selective activation of armed *O*-benzylated compounds in the presence of *O*-benzoylated analogues or thioglycosides. By their use α-D-Araf-(1→6)-D-Man-(1→6)-α-D-ManOMe has been prepared.[71]

2.4 Branched Homotrisaccharides. – Molecular modelling studies of highly branched α-1,3-glucan have involved investigations of relevant components including α-isomaltose, α-D-Glc-(1→6)-α-D-Glc-(1→3)-α-D-Glc and the branched α-D-Glc-(1→3)-[α-D-Glc-(1→6)]-α-D-Glc. A conclusion was that the (1→6)-linked side residues did not interfere significantly with the (1→3)-linked backbone stereochemistry.[72]

On treatment with NIS, TfOAg a mixture of the 2,6-tethered **6** and phenyl tetra-*O*-acetyl-1-thio-α-D-mannopyranoside underwent (1→6)-cyclization of **6** and slower 3-α-mannosylation of the product, the former step being favoured by its intramolecular character in this one-pot procedure. α-D-Man-(1→3)-[α-D-Man-(1→6)]-α-D-ManOMe was consequently made in 30% yield.[73] Alternatively this trisaccharide was made in 42% yield by direct selective mannosylation with the *O*-acetylated trichloroacetimidate (2 equiv) of *p*-nitrophenyl α-D-mannopyranoside. The product was converted to the *p*-acrylamide glycoside and copolymerized with 2-acrylamido-*N*,*N*-dimethylamine to give a product tested for lectin binding.[74]

The syntheses of α-L-Fuc-(1→3)-[α-L-Fuc-(1→4)]-α-L-FucOPr and the (1→2), (1→3) linked isomer have been reported.[75]

2.5 Branched Heterotrisaccharides. – Compounds in this section are categorized according to their reducing end sugars.

The following glucose-based compounds have been made: β-D-Gal-(1→4)-[α-L-Fuc-(1→3)]-D-Glc (a human milk compound, made by chemoenzymic methods)[53] and β-D-Man-(1→4)-[α-L-Fuc-(1→3)]-β-D-GlcO-Cer (a glycosphingolipid from a millipede; chemical methods).[76] The galactose derivatives β-D-Glc-(1→4)-[β-D-GlcNH₂-(1→3)]-β-D-Gal, Type 1A, GroupB *Streptococcal* capsular polysaccharide component, has been made on a polymer support,[77] and the *E. coli* O128 O-antigenic polysaccharide repeating unit β-D-GalNAc-(1→6)-[α-L-Fuc-(1→2)]-D-Gal has been synthesized.[78]

In the field of aminosugar-based branched compounds β-D-Gal-(1→4)-[α-L-Fuc-(1→3)]-D-GlcNAc, the Le^x trisaccharide, was synthesized and dimerized by 6,6-methylene acetal linking to give a compound which is a model for cooperative Ca²⁺ binding.[79] Synthesis of the following GalNAc derivatives have also been reported: β-D-GlcNAc-(1→6)-[β-D-GlcNAc-(1→3)]-D-GalNAc, β-D-GlcNAc-(1→6)-[β-D-Gal-(1→3)]-D-GalNAc,[80] and β-D-GalNAc-(1→6)-[β-D-GlcA-(1→3)]-β-D-GalNAcO(CH₂)₃NH₂ a *Schistosoma mansoni* antigen.[81]

α-D-Glc-(1→4)-[α-L-Rha-(1→3)]-L-Rha, a fragment of the O-specific polysaccharide of *Shigella flexneri* serotype 2a, has been made[47] and β-D-Glc-(1→2)-[β-D-Xyl-(1→3)]-<F4·d-Fuc has been prepared as its ruscogenin-1-yl (a triterpene) glycoside.[82]

2.6 Analogues of Trisaccharides and Compounds with Anomalous Linking. – While orthoester **7** preferentially reacts with equatorial hydroxyl groups on saturated 6-membered rings the *n*-pentenyl glycoside **8** predominantly reacts at axial sites. When *myo*-inositol derivative **9** was treated with a mixture of equivalent amounts of **7** and **8** only one product was formed – the di-α-D-mannosyl derivative with silylated substituent at the axial site and a 2-*O*-benzoylated mannosyl group at the equatorial one.[83] This provides yet another ingenious one-pot selective substitution procedure.

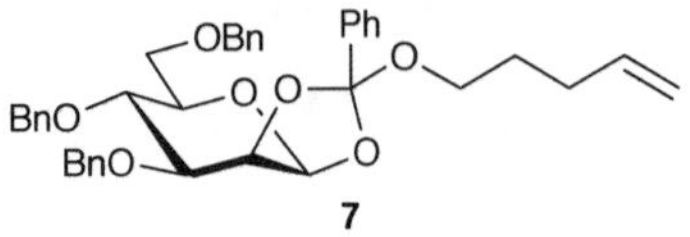

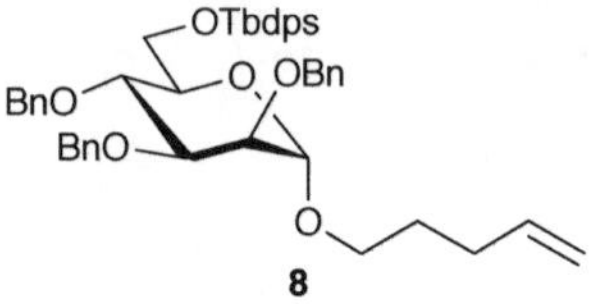

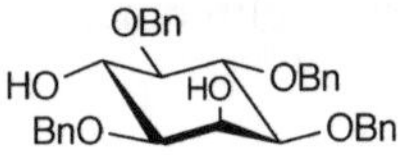

Analogue **10** of neamine with 5-amino-5-deoxy-β-D-ribofuranose attached by a flexible linker has been made as a potential inhibitor of HIV,[84] and the epimeric GM1 mimetics **11** have been described.[85] Rearrangement of the triunsaturated maltotriose derivative **12**, made from the 6,6′,6″-trideoxy-triiodo-maltotrioside derivative, by treatment with tri-isobutylaluminium, followed by DMSO, acetic anhydride oxidation of the triol formed, gives the tricarba-trisaccharide tricarbonyl derivative **13** in 33% yield.[86] This, importantly, adapts the known conversion of 6-deoxyhex-5-enosyl compounds into deoxyinososes in such a way that the glycosidic bonds are preserved.

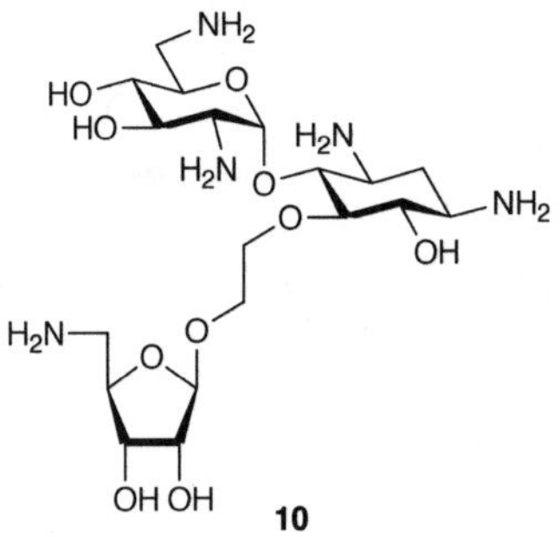

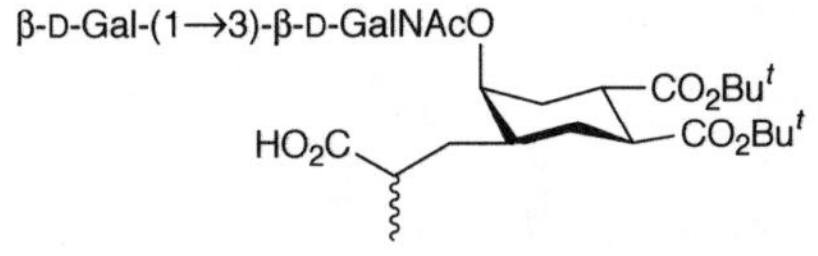

In the area of nitrogen-in-the-ring sugar analogues compound **14** has been made as a SiaLe^x mimetic.[87]

In the field of anomalously linked trisaccharides, the unusual GM3 analogue **15**, having an ether rather than a lactone linkage between the sialic acid and galactose moieties, has been made as a target for antibody-based cancer therapy,[88] and several analogues of SiaLe^x having ethylenedioxa linkages in place of

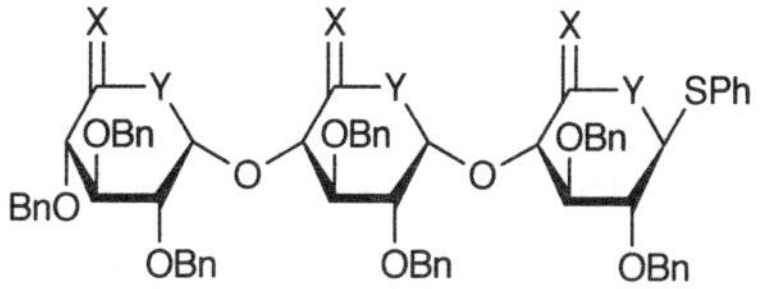

12 X = CH$_2$, Y = O → **13** X = O, Y = CH$_2$

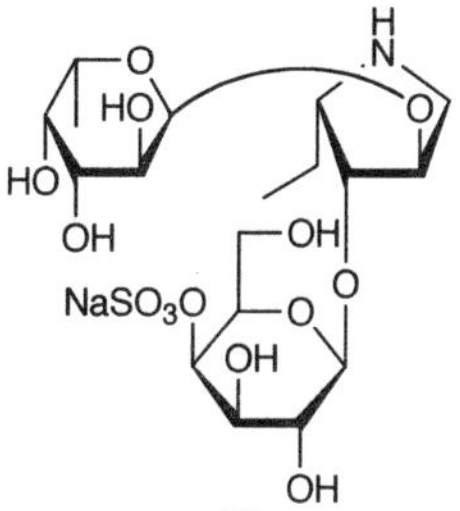

14

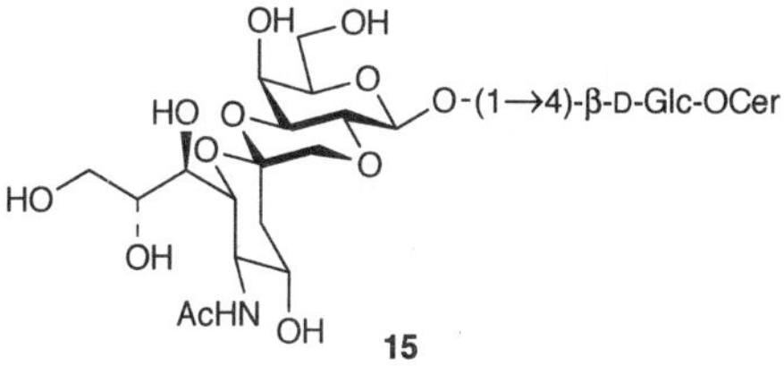

15

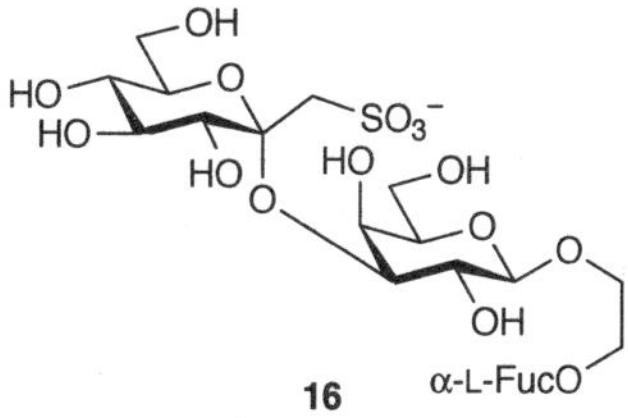

16

glycoside oxygen atoms have been reported. Compound **16** is an example.[89]

α-D-Man-(1-CH$_2$-3)-[α-D-Man-(1-CH$_2$-6)]-D-Man is a doubly *C*-linked trisaccharide made by simultaneous nucleophilic addition of a glycosyl SmI$_2$ reagent to a methyl 3,6-dideoxy mannoside derivative having formyl groups bonded at C-3 and C-6.[90]

3   Tetrasaccharides

Compounds of this set and higher oligosaccharides are classified according to whether they have linear or branched structures and then by the sugars at the reducing ends.

3.1   Linear Homotetrasaccharides. – Orthoester coupling was used to make a key fully protected allyl 2-*O*-acetyl-α-D-mannosyl-(1→2)-α-D-mannoside from which the glycosyl trichloroacetimidate donor and a 2′-unprotected acceptor were derived for coupling to give the α-(1→2)-linked mannotetrose. While this approach was suitable for making the analogous rhamnose tetramer it could not be applied successfully in the galactose or glucose series.[91]

3.2   Linear Heterotetrasaccharides. – Schmidt and co-workers have made β-D-Gal-(1→4)-β-D-GlcNAc-(1→3)-β-D-Gal-(1→4)-D-Glc by chemical methods as well as the lacto-*N*-neohexose and -neooctaose compounds having one and two further lactosamine units within the chains.[92] Other workers have made the tetramer by chemoenzymic methods and in the course of the work obtained a 2→3 linked NeuNAc derivative having the extra sugar at the non-reducing end.[93] The tetrasaccharide fragment β-D-Man-(1→4)-β-D-Glc-(1→4)-α-L-Rha-(1→3)-β-D-Glc *O*-linked to 5-aminopentanol, which is a fragment of the exopolysaccharide of *Arthrobacter* sp. CE-17, has been produced by effecting an intramolecular β-mannosylation of a malonyl-tethered disaccharide derivative in the key step.[94]

α-D-Gal-(1→2)-α-D-Man-(1→2)-α-D-Man-((1→6)-α-D-ManOMe was made, together with the lactose-terminating pentasaccharide analogue, for testing anti-metastatic properties.[95]

α-L-Rha-(1→3)-α-D-Man-(1→4)-α-D-Man-(1→3)-β-D-GlcNAc with 3-*O*-[(*R*)-1-methoxycarbonyl]ethyl at O-3 of the Rha and acetyl at O-2 of the Man moieties, was synthesized as its CH_2CH_2Tms glycoside. It is related to the repeating unit of the antigen from *Shigella dysenteriae*.[96]

The uronic acid-containing tetramer α-L-Rha-(1→4)-α-D-GalA-(1→2)-α-L-Rha-(1→4)-β-D-GalA was made from the analogue containing two galactose units which were oxidized using TEMPO, KBr, NaOCl. The product was used to model an anti-ulcer polysaccharide.[97]

A 2 + 2 strategy was employed in the chemical synthesis of α-D-Fuc-(1→4)-α-D-Fuc-(1→4)-α-D-Qui-(1→4)-D-Qui (Qui = 6-deoxy-D-glucose) which is a carbohydrate component of asterosaponin A from a starfish,[98] and the multi-deoxy tetrasaccharide 17 has been found as a component of the aristostatin antibiotics.[99]

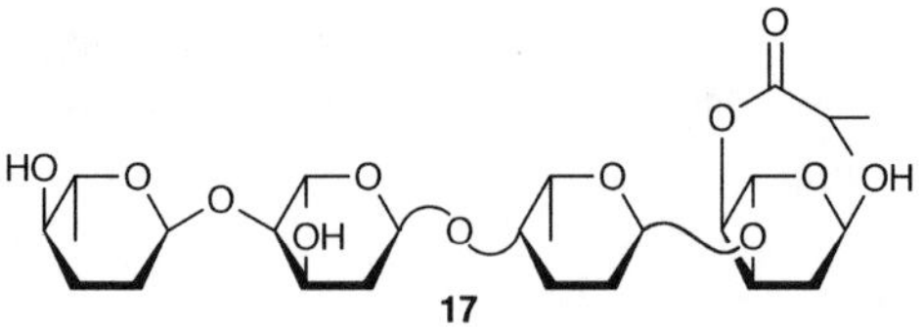

17

α-D-Man-(1→2)-α-D-Man-(1→5)-β-D-Ara*f*-(1→2)-α-D-Ara*f*-O(CH$_2$)$_8$CO$_2$Me, the tetrasaccharide cap of the lipoarabinomannan of *M. tuberculosis*, was made by a convergent method that used arabinofuranose 1,2,5-orthoesters to obtain both the donor and acceptor species employed to introduce the pentose moieties.[100]

3.3 Branched Homotetrasaccharides. – A series of α-D-arabinofuranoyl oligosaccharides which are fragments of the cell wall polysaccharides of *M. tuberculosis* to have been made include α-D-Ara*f*-(1→5)-[α-D-Ara*f*-(1→3)]-α-D-Ara*f*-(1→5)-D-Ara.[101]

3.4 Branched Heterotetrasaccharides. – A 'facile' synthesis has been reported of the repeating unit of the mannoglucan of *Microellobosporia grisea* α-D-Man-(1→3)-[α-D-Man-(1→6)]-β-D-Glc-(1→4)-D-Glc *i.e.* a dimannosylcellobiose.[102] Several linear and branched tetrasaccharides including β-D-Glc-(1→4)-α-L-Rha-(1→4)-[α-L-Rha-(1→2)]-β-D-Glc have been made as their diosgenin glyco-sides during synthesis of diogenyl saponins.[103] The stem bark of *Albizia julibrissin* has yielded related compounds with cytotoxic activity that included β-D-Glc-(1→3)-[α-L-Ara*f*-(1→4)]-α-L-Rha-(1→2)-β-D-Glc.[104]

A set of compounds terminating in mannose to have been synthesized are:

α-D-Man-(1→2)-α-D-Man-(1→6)-[β-D-Gal-(1→3)]-D-Man, which is related to the core structure of the GPI anchor of *Trypanosoma brucei*;[105] the closely related compound with the mannobiose and galactose moieties 1,2- and 1,4-linked to the reducing end Man, which is the immunologically important tetrasaccharide of the lipophosphoglycan of *L. donovani*;[106] β-D-Gal*f*-(1→3)-α-D-Man-(1→2)-[β-D-Gal*f*-(1→3)]-D-Man. In the last case phenyl tetra-*O*-acetyl-4-thio-1-selenogalac-tofuranoside was used to make an analogue with the Gal units having sulfur as the ring hetero atom.[107]

β-D-Gal-(1→6)-β-D-Gal-(1→6)-[α-L-Ara*f*-(1→2)]-β-D-Gal, an epitope of an arabinogalactan, has been made as its dodecyl glycoside.[108]

The SiaLe^x epitope α-NeuNAc-(2→3)-β-D-Gal-(1→4)-[α-L-Fuc-(1→3)]-D-GlcNAc has been glycosidically linked to a set of lipids to give products with good selectin-binding properties in a dynamic test system.[109] Formal syntheses have given the 6- and 6'-sulfate esters of this epitope,[110] and the corresponding glycal has been made by use of a fucosyl transferase.[110a] In work directed towards drug delivery to inflammatory lesions the isomer α-NeuNAc-(2→3)-[α-L-Fuc-(1→2)]-β-D-Gal-(1→4)-β-D-GlcNAc as a polyoxyethylene glycoside has been shown to be effective.[111] The Sia Le^A tetrasaccharide α-NeuNAc(2→3)-β-D-Gal-(1→3)-[α-L-Fuc-(1→4)]-β-D-GlcNAc has been made as the glycosyl azide and used in the preparation of Sia Le^A asparagine conjugates.[112,113]

Synthesis of calonyctin A2 (**18**), a macrolide glycolipid with plant growth-promoting activity, has been reported.[114]

18

3.5 Analogues of Tetrasaccharides and Compounds with Anomalous Linking. – Several cyclitols have been incorporated into pseudo-tetrasaccharides. For example, extensive syntheses of guanidino compounds have yielded guanidino-neomycin **19** and several other tetra- and tri-saccharide analogues.[115] The synthesis has also been reported of α-D-Man-(1→3)-α-D-Man-(1→4)-α-D-GlcNH₂O-*myo*-inositol, with phosphate esters on the inositol and at C-6 of the terminal mannose unit, which is an inositol phosphoglycan fragment of compounds found in *Leishmania* parasites.[116] A suit of compounds **20**, R = CH₂NR¹R² (24 examples) were made as E-selectin antagonists and compared with the known compound **20**, R = CH₂OH.[117]

A tetrasaccharide containing two Gal*f* having sulfur as the ring hetero-atom moieties has already been noted.[107]

Compounds with anomalous linking to have been made include the analogue of β-D-Glc-(1→6)-[β-D-Glc-(1→3)]-β-D-Glc-(1→3)-D-Glc with sulfur as the inter-unit linking atom at all three sites. A spacer group glycoside of the compound was linked to human serum albumin for immunostimulation testing.[118]

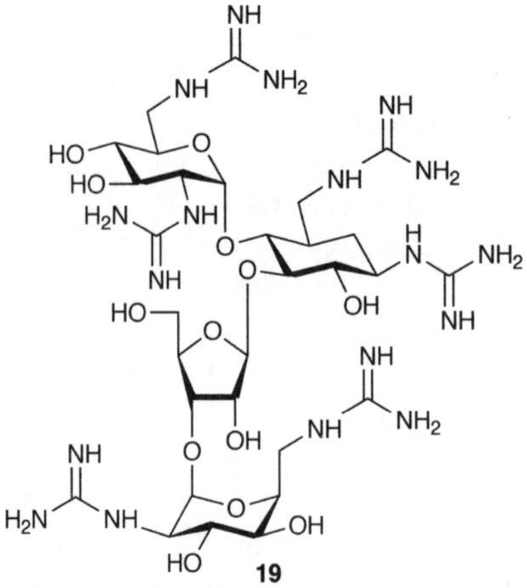

α-D-Man-(1-PO$_3$H-6)-β-D-Gal-(1→4)-α-D-Man-(1-PO$_3$H-6)$_n$-β-D-GalO(CH$_2$)$_8$CH=CH$_2$, (n = 1,2), further fragments of *Leishmania* lipophosphoglycan having phosphodiester linkages within the tetrasaccharide, have been made by MPEG soluble polymer techniques.[119]

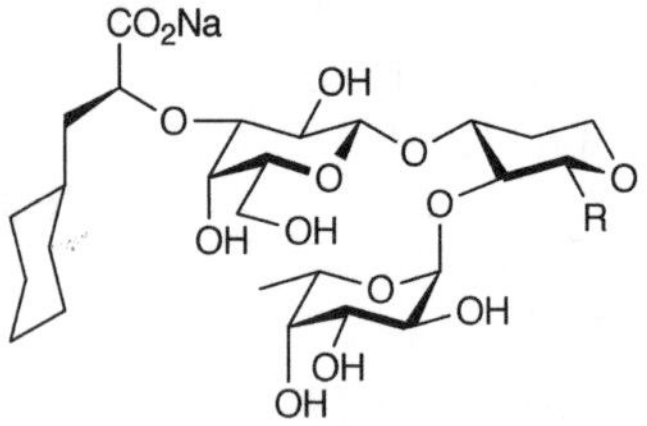

Two papers have dealt with analogues of calicheamicin γ$_1$[1] carbohydrate: compound **21**[120] and others containing more major structural changes.[121]

4   Pentasaccharides

4.1   Linear Homopentasaccharides. – In 'state of the art' work Ley's group has extended their sophisticated use of tuned glycosyl donors and acceptors to the one-pot synthesis of the α-1,2-linked D-mannopentamer, a GPI anchor compound.[122]

The α-(1→4)-linked pentamer of the daunosamine derivative 2,3,6-trideoxy-3-*N*-[(guanidino)acetamido]-L-*lyxo*-hexose **22** has been produced by a glycosyl sulfoxide method and shown to bind to the minor groove of DNA.[123]

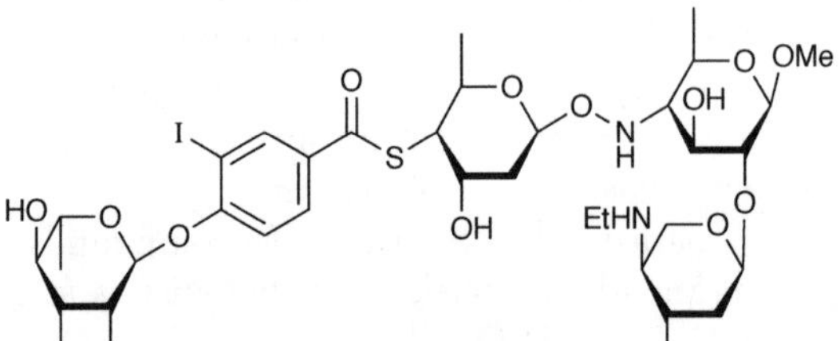

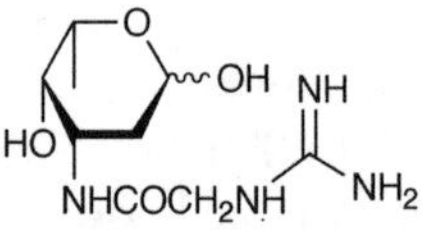

4.2   Linear Heteropentasaccharides. – As in recent years, of the compounds of this category reported in 2000 the great majority are terminally X substituted derivatives of the commonly occurring lacto-*N*-neotetraose *i.e.* X-β-D-Gal-(1→4)-β-D-GlcNAc-(1→3)-β-D-Gal-(1→4)-Glc. Three papers have reported the pentamer with X = NeuNAc (2→3) linked to the terminal galactose. One notes analogues with various acyl groups in place of acetyl in the NeuNAc moiety,[124] one deals with sulfates at O-6 of the GlcNAc or of the sialylated Gal which were tested as substrates of fucosyl transferase,[125] while the third is notable for the efficiency of the enzymic sialylating procedure.[93]

The pentamer with X = α-L-Fuc-(1→3), an antigen involved in the hyperacute rejection of xenotransplants, has been synthesized,[126] and a related pentamer, β-L-Fuc*f*-(1→4)-α-D-Gal-(1→4)-α-NeuNAc(2→3)-β-D-Gal-(1→4)-Glc has been isolated from the starfish *Acanthaster planci*.[127]

Lacto-*N*-neotetraose with X (see above) = β-D-GlcA3-SO$_3$H has been made as a 2-aminoethyl glycoside and converted to biotinylated conjugates.[128]

Sea cucumber has yielded a compound which is α-L-Fuc-(1→4)-α-Neu-(1→4)-α-NeuNAc-(2→3)-α-NeuNAc-(1→6)-β-D-GlcO-Cer but, significantly, with the central Neu moiety carrying a glycolic acid *N*-substituent which is bonded to the neighbouring sialic acid.[129]

Compounds terminating in sugars other than glucose to have been described are β-D-Gal-(1→4)-β-D-Glc-(1→2)-α-D-Man-(1→2)-α-D-Man-(1→6)-α-D-Man, which was made as a potential antimetastatic agent,[130] two unusual pentamers made from glucose, cymarose, oleandrose (2,6-dideoxy-3-*O*-Me-D-*ribo*- and L-*arabino*-hexose) and digitoxose (2,6-dideoxy-D-*ribo*-hexose) found in steroidal plant root compounds,[131] and chondroitin pentasaccharide **23** which has been used as the source of its O-4 and O-6 sulfates.[132]

23

4.3 Branched Homopentasaccharides.

– Mycobacterial cell walls contain a component of which β-D-Ara*f*-(1→2)-α-D-Ara*f*-(1→6)-[β-D-Ara*f*-(1→2)-α-D-Ara*f*-(1→3)]-D-Ara*f* is part. 1,2,5-Orthoesters of the sugar are key starting materials for the synthesis.[133] (*cf.* ref. 101).

4.4 Branched Heteropentasaccharides.

– Danishefsky has reported in detail on the synthesis of ganglioside GM1 α-Neu-(2→3)-[β-D-Gal-(1→3)-β-D-GalNAc-(1→4)]-β-D-Gal-(1→4)-D-Glc.[134] β-D-Gal-(1→4)-[α-L-Fuc-(1→3)]-β-D-GlcNAc-(1→3)-β-D-Gal-(1→4)-D-Glc with the terminal Gal moiety sulfated at O-3 and O-6 is a Le^x pentasaccharide made as a candidate ligand for L-selectin.[135]

β-D-Gal*f*-(1→3)-α-D-Man-(1→2)-[β-D-Gal*f*-(1→3)]-α-D-Man-(1→2)-D-Man has been made. It is the terminal section of the GPI of *T. cruzi*, the causative agent of Chagas disease.[136]

The D-Gal terminating β-D-Gal-(1→4)-[α-L-Fuc-(1→3)]-β-D-Glc-(1→3)-[β-D-Gal-(1→6)]-D-Gal occurs as a ceramide glycoside in the parasite *Spirometea erinacei*. It has now been made as the neutral glycoside of a C$_{30}$ fatty acid.[137]

In the area of pentamers terminating in 2-N-acetamido sugars α-D-Man-(1→6)-[α-D-Man-(1→3)]-X-D-Man-(1→4)-Y-D-GlcNAc-(1→4)-β-D-GlcNAc as their glycosylamines (X = Y = α or β) have been made and amide linked to a derivative of asparagine.[138]

Five reports have appeared on closely related pentasaccharides terminating in GalNAc. Two have described the synthesis of β-D-Gal-(1→4)-[α-L-Fuc-(1→3)]-β-D-GlcNAc-(1→6)-[β-D-Gal-(1→3)]-D-GalNAc with a carboxymethyl group at O-3 of the branching Gal unit,[139] and with sulfate groups at O-6 of this moiety or at O-6 of the GlcNAc.[140] Chemoenzymic methods were used to make the SiaLe^x-containing   α-NeuNAc-(2→3)-β-D-Gal-(1→4)-[α-L-Fuc-(1→3)]-β-D-GlcNAc-(1→6)-α-D-GalNAc O-linked to an octapeptide,[141] and the similar immunoreactive pentasaccharide part of the circulating anodic antigen of the parasite *Schistosoma mansoni* β-D-GalNAc-(1→6)-[β-D-GlcA-(1→3)]-β-D-GalNAc-(1→6)-[β-D-GlcA-(1→3)]-D-GalNAc has also been made.[81]

The D-alluronic acid-containing β-D-AllA-(1→3)-β-D-Gal-(1→4)-[α-L-Fuc-(1→2)]-β-D-Gal-(1→3)-D-GalNAc-ol has been isolated by reductive elimination of the oviducal mucin of *Rana temporaria*.[142]

Several branched pentamers with terminal 6-deoxyhexose units have been reported, some having been synthesized and some isolated from natural sources. β-D-Fuc-(1→2)-β-D-Qui-(1→4)-[3-*O*-Me-β-D-Qui]-β-D-Xyl-(1→4)-β-D-Qui has been isolated as a carbohydrate component of different triterpenoid starfish saponins,[143] and saponins from *Quillaja saponaria* have yielded β-D-Api-(1→3)-β-D-Xyl-(1→4)-[β-D-Glc-(1→3)]-α-L-Rha-(1→2)-D-Fuc.[144]

In connection with synthetic studies of fragments of the O-antigens of the polysaccharide of *Shigella flexneri* three pentasaccharides were made based on the tetramer β-D-GlcNAc-(1→2)-α-L-Rha-(1→2)-α-L-Rha-(1→3)-L-Rha with modifications as follows: α-D-Glc substituted at O-4 of the terminal Rha unit,[145] the same substituent at the O-3 of the central Rha unit,[146] and α-D-Glc attached at O-3 of the terminal Rha and also, at the non-reducing end, the disaccharide moiety inverted in α-L-Rha-(1→3)-β-D-GlcNAc-(1→2)-α-L-Rha-(1→2)-[α-D-Glc-(1→3)]-L-Rha.[147]

4.5   Analogues of Pentasaccharides and Compounds with Anomalous Linking. – The *myo*-inositolphosphoglycan **24** has been made in the course of work on GPI anchor compounds. Its conformation in solution and biological activities were determined.[148]

α-D-Man-(1→2)-α-D-Man-(1→6)-α-D-Man-(1→4)-α-D-GlcNH$_2$—O

24

An iterative procedure involving coupling of a β-*C*-formyl-D-galactoside and a galactoside 6-Wittig reagent has allowed the synthesis of the β (1→6)-CH$_2$-linked galactopentaose.[149]

5 Hexasaccharides

As is customary in these volumes, an abbreviated method is now used for representing higher saccharides. Sugars will be numbered as follows, and linkages will be indicated in the usual way:

1 D-Glc*p*	**2 D-Man*p***	**3 D-Gal*p***
4 D-Glc*p*NAc	**5 D-Gal*p*NAc**	**6 Neu*p*NAc**
7 L-Rha*p*	**8 L-Fuc*p***	**9 D-Xyl*p***
10 D-Glc*p*NH$_2$	**11 D-Glc*p*A**	**12 D-Qui (6-deoxy-D-glucopyranose)**
13 L-*Glycero*-D-*manno*-heptose	**14 L-Ara*f***	**15 Kdn**
	16 D-Gal*f*	

5.1 Linear Hexasaccharides. – Difficulties with using the globo-H breast tumour antigen **25** (R = allyl) (Vol 31, p. 73, ref. 161) have led to the 4-pentenyl analogue (**25**, R = pentenyl) which, as a glycosyl donor, has been linked to a protein to give a 'second generation antigen'.[150] Compound **26**, a 'masked form' of the hexasaccharide of landomycin A, an angucycline antibiotic, has been synthesized,[151] and oligomers up to the hexamer of the α-(1→4)-linked 2-deoxy-2-fluoro-D-glucose have been made by use of glycosyl chlorides derived from *p*-chlorophenyl thioglycosides.[152] Compound **27**, a structure found in *Haemophilus ducreyi* lipopolysaccharide has been prepared as its 2-(*p*-trifluoro-acetamidophenyl)ethyl glycoside.[153]

(α8)1→2(β3)1→3(β5)1→3(α3)1→4(β3)1→4(β1)—OR

25

26

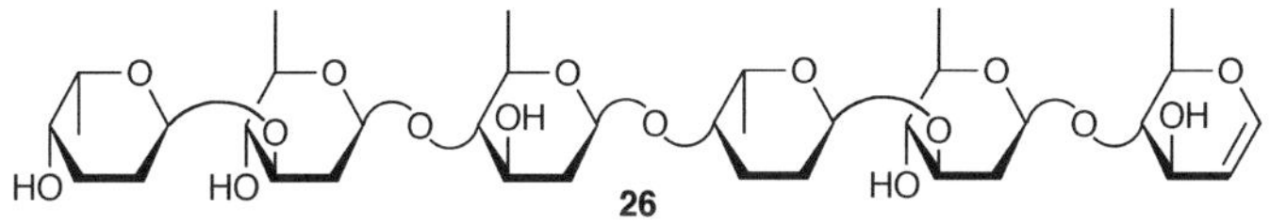

5.2 Branched Homohexasaccharides. – A one-pot synthesis of a 3- and 6-*O*-differentially protected glucoside gave access to a highly efficient synthesis of the β-(1→3) linked glucotetraose with β-(1→6) branches on units 1 and 3.[154]

5.3 Branched Heterohexasaccharides. – 3′,6′-Di-*O*-lactosaminyl-D-lactose has been made by a solid phase procedure using a ring-closing methathesis linking strategy.[155] Glucose terminating Le^b hexasaccharide compound **28** has been converted to human serum albumin conjugates by use of its 2-aminoethyl glycoside.[156] The galactose-terminating **29** has been identified as a component of a new plant saponin.[157]

Considerable interest is being taken in compounds of the category that terminate in amino-sugar residues. The nucleotide phosphosulfate **30** has been used in

(β3)1→4(β4)1→3(β3)1→4(α13)1→6(β1)1→4(13)

27

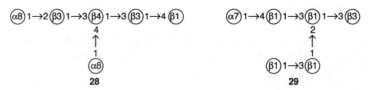

the synthesis of the sulfated SiaLe^x derivative **31** as well as the compounds less
one and plus one GlcNAc unit in the chain.[158] Compound **32** has been prepared
by a solid support procedure.[159]

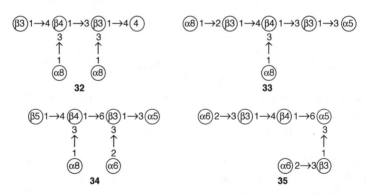

Compounds ending in GalNAc, however, have attracted most attention.
Notably Danishefsky's group have reported the synthesis of the *O*-serine glycoside of hexamer **33** and its incorporation with a disaccharide and monosaccharide glycoside of the amino-acid into a tetrapeptide containing multiple carbohydrate antigens for the purpose of producing an anticancer vaccine.[160] In related
work the more highly branched **34**[161] and **35**[162] were made, the former having a
sulfate ester group at O-6 of the galactose moiety and the latter having the
carboxylic acid group of the NeuNAc moieties lactonized to O-4 of the adjacent
galactose units.

Compound **36** with a sulfate group at O-6 of the GlcNAc moiety and closely
related hexamers were made and represent the chains present in respiratory
mucins.[163-165]

Starfish saponins continue to provide complex oligosaccharides; compound
37 is the latest example. It is notable that the fucose was present as the D-
enantiomer.[166]

5.4   Analogues of Hexasaccharides. – The D-*myo*-inositol glycoside of the pentasaccharide   β-D-GlcNAc-(1→6)-[α-D-Gal-(1→6)-α-D-Gal-(1→3)]-α-D-Man-

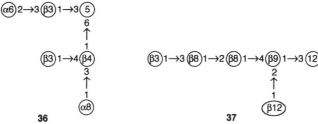

36   **37**

(1→4)-α-D-GlcNAc was made during work aimed at making inositolphospho-glycans.[167]

6   Heptasaccharides

The branched glucoheptaose **38**, a repeating unit of lentinan which is used for anti-tumour purposes in China and Japan, has been synthesized.[168]

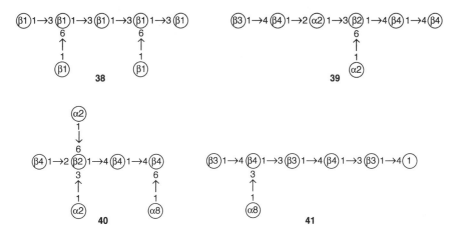

38   **39**

40   **41**

Syntheses of **39–41**, the first two being *N*-linked to asparagine, and the third as its 2-Tms-ethyl glycoside have been reported.[169–171] The concisely made glucosyl mannohexaose **42** has been linked to a cholesterol-containing lipophilic aglycon for targeting of liposomes to macrophages.[172]

Two reports have appeared on the synthesis of GPI anchor hexasaccharide-inositol compounds which are heptasaccharide mimics. The first product, **43**, is the anchor compound of *Trypanosoma brucei*,[173] while the second, **44**, is an analogous product of *Leishmania* parasites.[174]

7   Octasaccharides

An octosaccharide mimic containing an inositol group akin to the heptasacchar-ide mimic **44** has been synthesized.[174] The trichloroacetimidate method has been

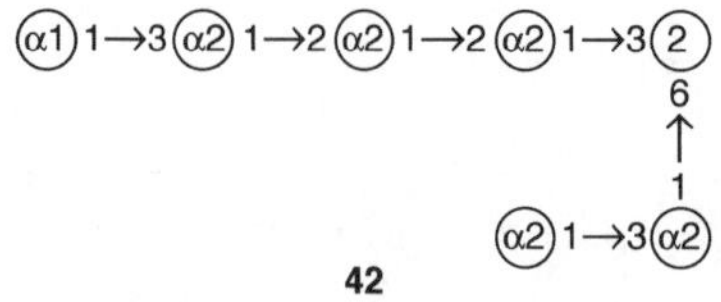

42

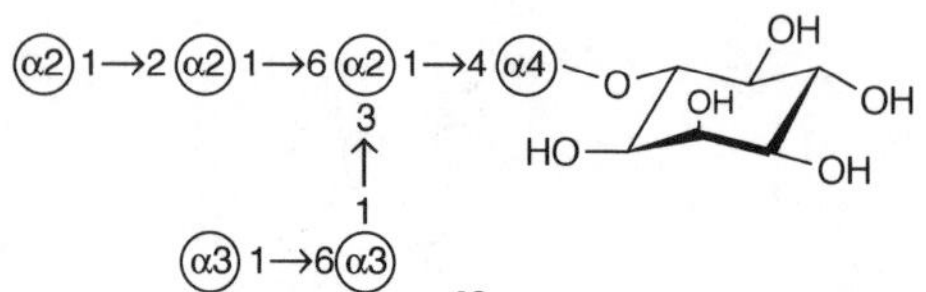

43

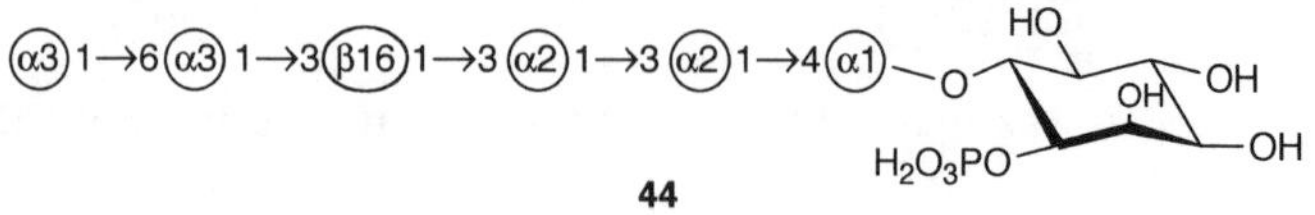

44

used in a concise preparation of the α-(1→5)-linked arabino-octaose as well as the tetramer and hexamer,[175] and the same group have used similar methods to make the β-(1→6)-linked gluco-octaose.[176] Lacto-N-neo-octaose comprising three 1,3-linked LacNAc units 1,3-bonded to lactose has been prepared together with the tetraose and hexaose of the series.[92]

After extensive NMR and mass spectrometric study the structure of the complex and novel octosaccharide **45** isolated from the fermentation broth of *Micromonospora carbonaceae* has been reported.[177]

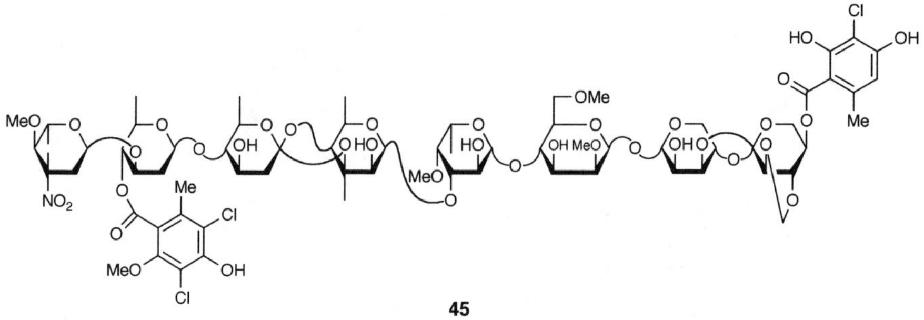

45

8   Nonasaccharides

Compounds of this set to have been prepared are **46**, a core fucosylated, biantennary N-glycan with a novel GlcNAc residue in the core region,[178] **47**, a Sia Le^x containing compound with and without a sulfate ester at O-6 of the GlcNAc residue,[179] and **48** made by chemoenzymic methods and linked to a fluorophore to provide means of visualization of cell-surface selectively bound SiaLe^x compounds.[180]

9   Higher Saccharides

The chemoenzymic synthesis of trimeric SiaLe^x decasaccharide **49** was achieved by enzymically sialylating and fucosylating the linear hexamer.[181] The analogue

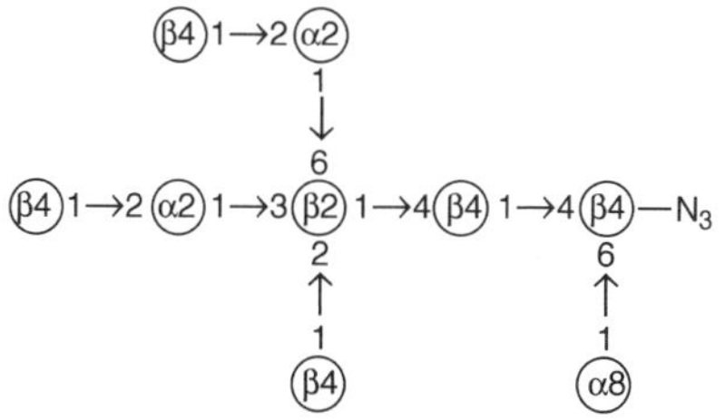

(β4)1→2(α2)

↓
1
6
(β4)1→2(α2)1→3(β2)1→4(β4)1→4(β4)—N₃
2   6
↑   ↑
1   1
(β4)   (α8)

46

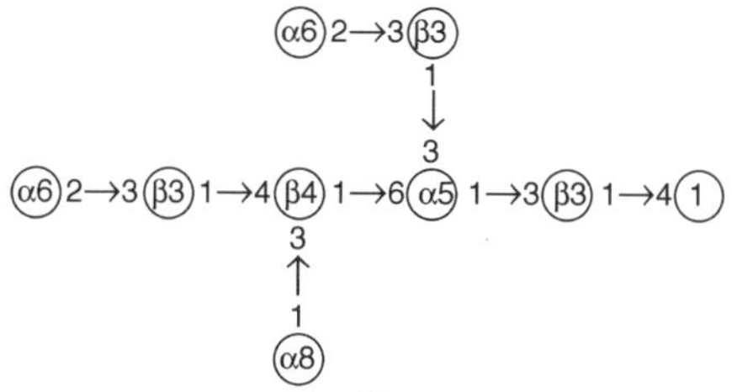

(α6)2→3(β3)
1
↓
3
(α6)2→3(β3)1→4(β4)1→6(α5)1→3(β3)1→4(1)
3
↑
1
(α8)

47

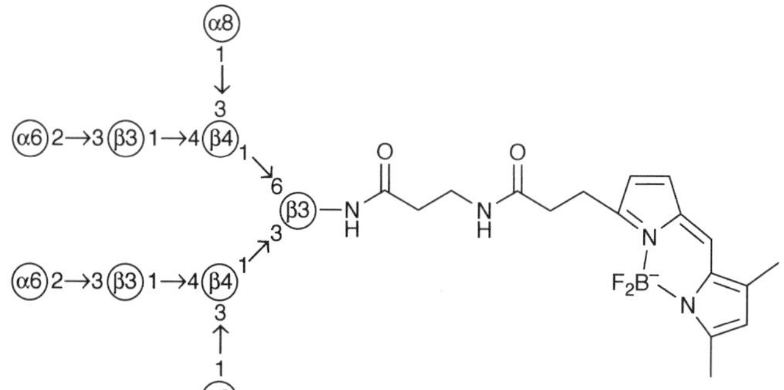

(α8)
1
↓
3
(α6)2→3(β3)1→4(β4)₁
                        ↘6
                     (β3)—N—
                        ↗3   H
(α6)2→3(β3)1→4(β4)₁
3
↑
1
(α8)

48

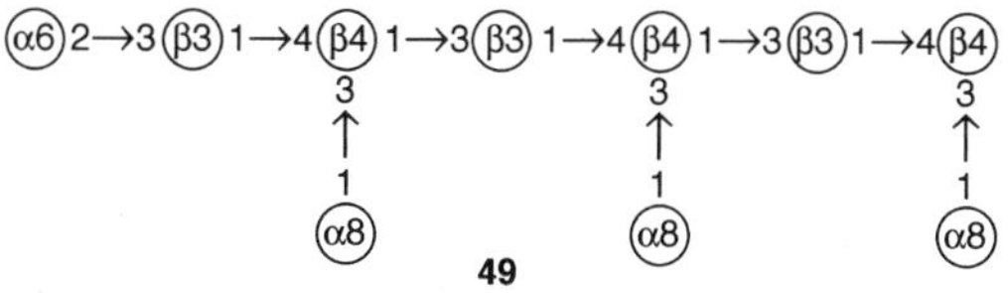

(α6)2→3(β3)1→4(β4)1→3(β3)1→4(β4)1→3(β3)1→4(β4)
3   3   3
↑   ↑   ↑
1   1   1
(α8)   (α8)   (α8)

49

β-L-Gal
1
↓
3
(α6)2→3(α3)1→4(β4)1→2(α2)₁
                            ↘6
                          (2)
                          ↗3
(α6)2→3(α3)1→4(β4)1→2(α2)₁
3
↑
1
β-L-Gal

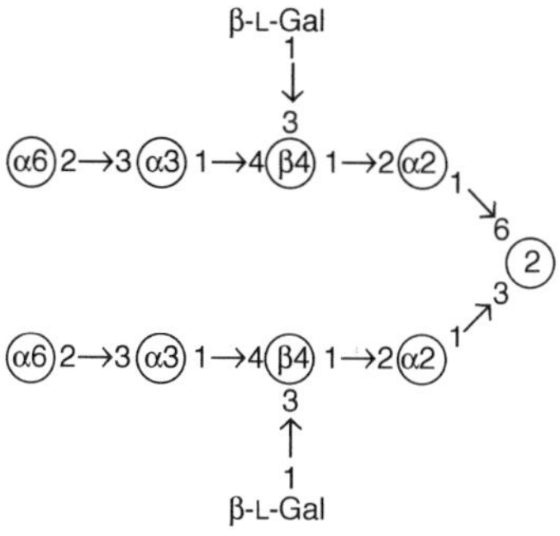

50

50 of dimeric SiaLeˣ was made by L-galactosylation with α-1,3-fucosyltransferase.[182]

The closely related **51** undecasaccharide was made as a prototype of the complex mammalian *N*-glycans,[183] as was the dodecamer having an additional β-GlcNAc moiety bonded to O-4 of the branching β-Man residue.[184]

$$(\alpha 6)2{\to}3\,(\alpha 3)\,1{\to}4\,(\beta 4)\,1{\to}2\,(\alpha 2)_1$$
$$\searrow 6$$
$$(\beta 2)1{\to}4\,(\beta 4)\,1{\to}4\,(\beta 4)$$
$$\nearrow 3$$
$$(\alpha 6)2{\to}3\,(\alpha 3)\,1{\to}4\,(\beta 4)\,1{\to}2\,(\alpha 2)^1$$

51

A tetramer of α-L-Fuc-(1→2)-6-*O*-Bn-α-D-Gal-(1→3)-α-D-GlcNAc-(1→3)-L-Fuc joined by α-(1→3) linkages between the fucosyl units gave a linear 24 unit product which was purified by reverse phase chromatography by use of a lipophilic ester derivative.[185]

10   Cyclodextrins

10.1 General Matters. – Oligomerization of carbamate **52** in the presence of sodium hydride gave only cyclic products which are cyclodextrin analogues.[186]

A procedure for expanding γ-cyclodextrins to δ-analogues involves acid catalysed cleavage with TmsSPh of one glycosidic bond of the permethyl derivative to give a phenylthio glycosidic group at one end and a silyl ether at O-4 of the non-reducing unit. The former group was then used to glycosylate a 4-hydroxy phenylthio glucoside and the resulting heptaose derivative was recyclized after desilylation.[187]

Hexa-6-*O*-Tbdms α-CD was converted to the hexa-2,3-anhydro-D-*manno* compound *via* a 2-sulfonylated intermediate and hence to the hexa-*altro*-CD.[188] A fructanotransferase catalyses the degradation of insulin into cycloinulo-saccharides.[189]

The AMBER force field programme has been modified to reproduce the observed properties of α-CD, β-CD and larger cyclomaltosaccharides.[190]

Hepta-2,6-di-*O*-methyl-β-CD enhanced solubility (by up to 100-fold) such that α-NeuNAc-(2→3)-β-D-Gal-(1→3)-α-D-GalNAc as a protected threonine glycoside was made enzymically from the corresponding α-D-GalNAc glycoside in 50% yield.[191]

10.2 Branched Cyclodextrins. – From the $6^A,6^N$-di-*O*-(dimethoxytrityl)-peracetates the corresponding set of $6^A,6^N$-di-*O*-α-D-Gal-γ -CD derivatives were made by trichloroacetimidate glycosylation following detritylation.[192]

By use of debranching enzyme acting in reverse the trisaccharide β-D-Gal-(1→4)-α-D-Glc-(1→4)-D-Glc was substituted into β-CD at O-6 and the three disubstituted products were also observed, the AC and AD isomers predominating.[193] Similar work with α-D-Man -(1→6)-α-D-Glc-(1→4)-α-D-Glc-(1→4)-D-Glc gave products with one and two tetrasaccharide branches introduced into β-CD.[194]

10.3 Cyclodextrin Ethers. – Octakis 6-*O*-Tbdms-γ -CD underwent selective 2-

O-benzylation, and from the product the 2-*O*-benzyl-3-*O*-methyl and 2-*O*-acetyl-3-*O*-methyl derivatives were obtained for use as stationary phases for capillary gas chromatography.[195] For complementary use in HPLC β-CD derivatives monofunctionalized with 6-aminohexyl and 7,8-epoxyoctyl groups at O-2, O-3 or O-6 were immobilized on functionalized silica gel.[196]

A novel way of generating selectively *O*-substituted CDs arises from the selectivity found on DIBALH-promoted debenzylation of perbenzylated derivatives. For example, the α-CD perether can be converted to the 6-monohydroxy compound or, in 82% yield, to the AD, 6,6′-dihydroxy product. Similar results were recorded for the β-CD and γ -CD compounds.[197]

Photoaddition of tetra-*O*-acetyl-1-thio-β-D-glucopyranose to the allyl groups of per 2-*O*-allyl, per 6-*O*-allyl or per 2,6-diallyl ethers of β-CD affords a simple and efficient way of attaching 1-thio glucose units to the cyclodextrin by way of propan-1,3-diyl linkages.[198] A related approach allows the introduction of further sugars into CDs by way of propargyl ether groups. Thus, for example, use of the galactosyl nitrile oxide afforded the *C*-glycosidic compound **53** when the 6-propargyl heptaether was employed.[199]

10.4 Cyclodextrin Esters. – CDs with mono-6-benzoate or monosubstituted 6-benzoates and related compounds with benzoylamino-deoxy groups in C-6 positions have been used for enantio-differentiating photoisomerization of (*Z*)-cyclooctene to the chiral (*E*)-isomer.[200]

Interest continues in the use of disulfonating agents for carrying out selective disubstitutions. Benzophenone 3,3′-disulfonyl imidazole reacts only with the O-2 hydroxyl groups on adjacent glucose units of γ-CD,[201] while the corresponding derivative of 1,4-dibenzoylbenzene reacts with O-2^A, O-2^C of β-CD to give 24% of the cyclic diester with 3% and 1% of the 2^A, 2^B and 2^A, 2^D derivatives respectively.[202]

Alternatively compound **54** reacted with mono-6-*O*-tosyl-β-CD to give mainly the cyclic diesters involving O-6 at the A,C related units furthest from the tosyl-substituted moiety. When these trisulfonates were treated with thiophenol the 6^A, 6^C, 6^E trithio product was obtained.[203] Similarly, **54** formed the 6^A, 6^C cyclic diester with α-CD.[204] By use of heptakis 2,6-di-Tbdms-β-CD the heptakis 3-*O*-methyl-2,6-disulfate was made as a possible chiral resolving agent for capillary electrophoresis.[205]

10.5 Amino Derivatives. – An improved procedure for making 6^A-alkylamino and 6^A-arylamino-6^A-deoxy derivatives of β-CD *via* the tosylate has been reported.[206] Mono-6^A-amino-6^A-deoxy-β- and γ-CD have been amide linked to peptides including substance P. Antibodies to CDs and to substance P recognized the corresponding portions of the peptidyl-CDs.[207]

As part of a study of the cleavage of DNA by ribose-containing biopolymers various derivatives of 5-amino-5-deoxyribose were *N*-bonded to C-6′of β-CD *via* the monotosylate to give model enzymes. They showed catalytic activity in hydrolysing phosphodiesters (nuclease activity) as well as phosphatase and ligase activities. The vicinal *cis*-diol of the ribofuranose was required for expression of these activities.[208]

6^A-*O*-Tosyl-β-CD on reaction with triethylenetetramine gave the 1:1 6-amino CD product in high yield and subsequently the disubstituted compound having two CD rings linked by the tetramine. This product was then used as a ligand with Cu^{2+} to give a coordination compound and to bind to fluorescent dyes and thereby enhance their fluorescence again.[209] In related work formation of copper(II) complex **55** quenched the fluorescence of the uncomplexed triamino compound, and since amino-acids added to the complex 'switched on' the fluorescence, and since the system was enantioselective with Pro, Phe and Trp, it could be used for determining the optical purity of samples of these amino acids. Studies were conducted with both diastereoisomers of **55**.[210] Other similar work has led to a derivative of 6-amino-6-deoxy-dodecamethyl-α-CD having *N*-hydroxy-cytosine linked by way of the primary amino groups by an *N*-propanoyl tether. These substituent groups bind Fe^{3+} ions to give complexes which show strong circular dichroism spectra.[211]

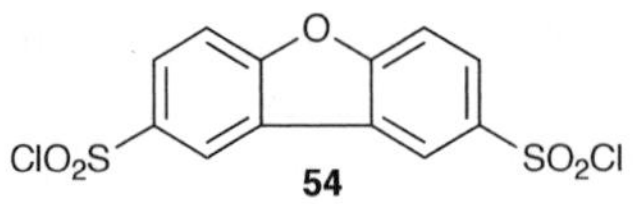

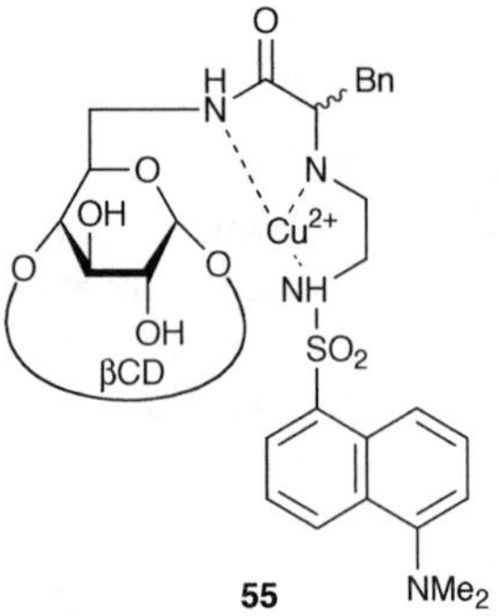

The aldehyde formed by oxidation of one of the primary hydroxyl groups of β-CD on reductive animation with sodium cyanoborohydride in the presence of the porphyrin with three *p*-sulfonophenyl and one *p*-aminophenyl substituent groups gave the 6-linked CD-porphyrin compound. This has a propensity to dimerize by inclusion of the aromatic macrocycle in the CD ring.[212]

10.6 Thio Derivatives. – The β-CD derivative having per 2,3-di-*O*-acetyl-6-thiocyanato-D-glucose components, by treatment with trimethyl(trifluoromethyl)silane and tetrabutylammonium fluoride, gave the corresponding heptatrifluoromethylthio compound which was deacetylated to give the product which, because of the effect of the CF$_3$S groups, showed amphiphilic behaviour at the air–water interface.[213] α-CD derivatives composed of 2,3-di-*O*-alkyl-6-thio-D-glucose (alkyl = pentyl, hexyl, heptyl) also form stable monolayers at the interface.[214]

6-Thioderivatives have been used for linking other sugar units to cyclodextrins. By use of the -S(CH$_2$)CONH(CH$_2$)$_3$O-sugar linkages, Gal, GlcNAc, Lac and LacNAc have been introduced, the key reaction involving heptadeoxy-heptaiodo-β-CD and the glycosides NaS(CH$_2$)$_2$CONH(CH$_2$)$_3$ O-sugar. The products showed enhanced affinity for lectins.[215] In related studies *O*-substituted 2-thioNeuNAc was linked to each of the primary positions of β-CD by the SCH$_2$CONH or SCH$_2$CONH(CH$_2$)$_6$NHCOCH$_2$S tethers.[216]

10.7 Oxidized Derivatives. – Oxidation of β-CD with TEMPO and NaClO$_4$ gave the heptacarboxylic acid which was amide-coupled with the 6-aminohexyl glycosides of α-D-Gal, α-D-Man and β-GlcNAc, and the products were identified as multi-valent conjugates which bind to lectins.[217]

References

1.   S.J. Danishefsky and J.R. Allen, *Angew. Chem., Int. Ed. Engl.*, 2000, **39**, 836.
2.   K.C. Nicolaou, R.M. Rodríguez, H.J. Mitchell, H. Suzuki, K.C. Fylaktakidou, O. Baudoin and F.L. van Delft, *Chem. Eur. J.*, 2000, **6**, 3095.
3.   K.C. Nicolaou, H.J. Mitchell, K.C. Fylaktakidou, R.M. Rodríguez and H. Suzuki, *Chem. Eur. J.*, 2000, **6**, 3116.
4.   K.C. Nicolaou, H.J. Mitchell, R.M. Rodríguez, K.C. Fylaktakidou, H. Suzuki and S.R. Conley, *Chem. Eur. J.*, 2000, **6**, 3149.
5.   K.C. Nicolaou, H.J. Mitchell, H. Suzuki, R.M. Rodríguez, O. Baudoin and K.C. Fylaktakidou, *Angew. Chem., Int. Ed. Engl.*, 2000, **39**, 3334, 3340, 3345.
6.   B.G. Davis, *J. Chem. Soc., Perkin Trans.*1, 2000, 2137.
7.   K.-H. Jung, M. Müller and R.R. Schmidt, *Chem. Rev.*, 2000, **100**, 4423.
8.   B. Fraser-Reid, R. Marsden, A.S. Campbell, C.S. Roberts and M.J. Robert, *Bioorg. Chem.: Carbohydr.*, 1999, **89**, 559 (*Chem. Abstr.*, 2000, **132**, 293 926).
9.   X.-S. Ye and C.-H. Wong, *J. Org. Chem.* 2000, **65**, 2410.
10. M.J. Sofia, *Annu. Rep. Comb. Chem. Mol. Diversity*, 1999, **2**, 41 (*Chem. Abstr.*, 2000, **132**, 208 007).
11. P.M. St. Hilaire and M. Meldal, *Angew. Chem. Int. Ed. Engl.*, 2000, **39**, 1162.
12. P.H. Seeberger and W.-C. Haase, *Chem. Rev.*, 2000, **100**, 4349.
13. H. Herzner, T. Reipen, M. Schultz and H. Kunz, *Chem. Rev.*, 2000, **100**, 4495.
14. A. Ishii, H. Hojo, A. Kobayashi, K. Nakamura, Y. Nakahara, Y. Ito and Y. Nakahara, *Tetrahedron*, 2000, **56**, 6235.
15. Y. Suhara and Y. Ichikawa, *Tennen Yuki Kagobutsu Toronkai Koen Yashishu*, 1999, **41**, 145 (*Chem. Abstr.*, 2000, **132**, 293 954).
16. G.-J. Boons and A.V. Demchenko, *Chem. Rev.*, 2000, **100**, 4539.
17. K.C. Nicolaou and N.J. Bockovich, *Biorg. Chem.: Carbohydr.*, 1999, **134**, 565 (*Chem. Abstr.*, 2000, **132**, 293 927).

18.   M.-C. Cheng, C.-H. Lin, H.-Y. Wang, H.-R. Lin and S.H. Wu, *Angew. Chem., Int. Ed. Engl.*, 2000, **39**, 772.

19.   P. Kosma, *Endotoxin Health Dis.*, 1999, 257 (*Chem. Abstr.*, 2000, **132**, 93 548).

20.   T. Ikami, H. Ishida and M. Kiso, *Methods Enzymol.*, 2000, **311**, 547 (*Chem. Abstr.*, 2000, **132**, 265 348).

21.   H. Miller-Podraza, *Chem. Rev.*, 2000, **100**, 4663.

22.   K.M. Koeller and C.-H. Wong, *Chem. Rev.*, 2000, **100**, 4465.

23.   M.M Palcic, *Curr. Opin. Biotechnol.*, 1999, **10**, 616 (*Chem. Abstr.*, 2000, **132**, 208 009).

24.   S.G. Whithers, *Spec. Publ., R. Soc. Chem.*, 1999, **246**, 62 (*Chem. Abstr.*, 2000, **132**, 222 714).

25.   O. Hindsgaul, *Essent. Glycobiol.*, 1999, 509 (*Chem. Abstr.*, 2000, **132**, 108 159).

26.   C. Bucke, J. Packwood, S. Suivasono and R.A. Rastall, *Methods Biotechnol.*, 1999, **10**, 213 (*Chem. Abstr.*, 2000, **132**, 108 148).

27.   D.H.G. Crout, P. Critchley, D. Muller, M. Scigelova, S. Singh and G. Vic, *Spec. Publ., R.. Soc. Chem.*, 1999, 15 (*Chem. Abstr.*, 2000, **132**, 222 713).

28.   S. Shoda and M. Fujita, *Kagaku to Seibutsu*, 2000, **38**, 309 (*Chem. Abstr.*, 2000, **132**, 347 799).

29.   U. Gambert, H.S. Conradt, M. Nimtz and J. Thiem, *J. Carbohydr. Chem.*, 2000, **19**, 621.

30.   S.-I. Shoda, T. Kiyosada, H. Mori and S. Kobayashi, *Heterocycles*, 2000, **52**, 599 (*Chem. Abstr.*, 2000, **132**, 279408).

31.   S. Fort, V. Boyer, L. Greffe, G.J. Davies, O. Moroz, L. Christiansen, M. Schülein, S. Cottaz and H. Driguez, *J. Am. Chem. Soc.*, 2000, **122**, 5429.

32.   A.K. Ganguly, *J. Antibiotics*, 2000, **53**, 1038.

33.   M. Sashiwa, Y. Makimura, Y. Shigemasa and R. Roy, *Chem. Commun.*, 2000, 909.

34.   P. Siemsen, R.C. Livingston and F. Diederich, *Angew. Chem., Int. Ed. Engl.*, 2000, **39**, 2633.

35.   K.E. Kövér, K. Fehér, L. Szilágyi, A. Borbás, P. Herczegh and A. Lipták, *Tetrahedron Lett.*, 2000, **41**, 393.

36.   T. Takahashi, M. Adachi, A. Matsuda and T. Doi, *Tetrahedron Lett.*, 2000, **41**, 2599.

37.   G. Lemanski, T. Lindenberg, H. Fakhrnabavi and T. Ziegler, *J. Carbohydr. Chem.*, 2000, **19**, 727.

38.   Y. Zhu and F. Kong, *J. Carbohydr. Chem.*, 2000, **19**, 837.

39.   L.G. Melean, W.-C. Haase and P.H. Seeberger, *Tetrahedron Lett.*, 2000, **41**, 4329.

40.   H. Jona, K. Takeuchi and T. Mukaiyama, *Chem. Lett.*, 2000, 1278.

41.   H. Jona, K. Takeuchi, T. Saitoh and T. Mukaiyama, *Chem. Lett.*, 2000, 1178.

42.   H. Kono, S. Kawano, T. Erata and M. Takai, *J. Carbohydr. Chem.*, 2000, **19**, 127.

43.   T. Yananoi, J. Inagaki, M. Mizuno, K. Yamashita and T. Inazu, *Heterocycles*, 2000, **52**, 921 (*Chem. Abstr.*, 2000, **132**, 279 417).

44.   S. Kramar, B. Nolting, A.-J. Ott and C. Vogel, *J. Carbohydr. Chem.*, 2000, **19**, 891.

45.   K. Takeuchi, T. Tamura and T. Mukaiyama, *Chem. Lett.*, 2000, 124.

46.   G. Belogi, T. Zhu and G.-J. Boons, *Tetrahedron Lett.*, 2000, **41**, 6965.

47.   L.A. Mulard, C. Costachel and P.J. Sansonetti, *J. Carbohydr. Chem.*, 2000, **19**, 849.

48.   S. Wada, P. He, I. Hashimoto, N. Watanabe and K. Sugiyama, *Biosci. Biotechnol. Biochem.*, 2000, **64**, 2262.

49.   K. Sujino, T. Uchiyama, O. Hindsgaul, N.O.L. Seto, W.W. Wakarchuk and M.M. Pacic, *J. Am. Chem. Soc.*, 2000, **122**, 1261.

50.   V. Perrin, B. Fenet, J.-P. Praly, F. Lecroix and C.D. Ta, *Carbohydr. Res.*, 2000, **325**, 202.

51.　J.J. Lundquist, S.D. Debenham and E.J. Toone, *J. Org. Chem.*, 2000, **65**, 8245.
52.　S. Koto, M. Hirooka, K. Yago, M. Komiya, T. Shimazu, K. Kato, T. Takehara, A. Ikefuji, A. Iwasa, S. Hagino, M.Sekiya, Y. Nakase, S. Zen, F. Tomonaga and S. Shimada, *Bull. Chem. Soc. Jpn.*, 2000, **73**, 173.
53.　B. La Ferla, L. Lay, L. Poletti, G. Russo and L. Panza, *J. Carbohydr. Chem.*, 2000, **19**, 331.
54.　Y. Ito, M. Gerz and Y. Nakahara, *Tetrahedron Lett.*, 2000, **41**, 1039.
55.　T. Buskas and P. Konradsson, *J. Carbohydr. Chem.*, 2000, **19**, 25.
56.　B.K.S. Yeung, D.C. Hill, M. Janicka and P.A. Petillo, *Org. Lett.*, 2000, **2**, 1279.
57.　C. Wegner, M. Hamburger, O. Kunert and E. Haslinger, *Helv. Chim. Acta*, 2000, **83**, 1454.
58.　T. Ikeda, J. Kinjo, T. Kaijmoto and T. Nohara, *Heterocycles*, 2000, **52**, 775 (*Chem. Abstr.*, 2000, **132**, 279 430).
59.　M. Lahmann and S. Oscarson, *Org. Lett.*, 2000, **2**, 3881.
60.　D.K. Watt, D.J. Brasch, D.S. Larsen, L.D. Melton and J. Simpson, *Carbohydr. Res.*, 2000, **325**, 300.
61.　K.A. Alvi, D.D. Baker, V. Stienecker, M. Hosken and B.G. Nair, *J. Antibiotics*, 2000, **53**, 496.
61a.　M. Nitz and D.R. Bundle, *J. Org. Chem.*, 2000, **65**, 3064.
62.　F. Bélot and J.-C. Jacquinet, *Carbohydr. Res.*, 2000, **325**, 93.
63.　B. Nolting, H. Boye and C. Vogel, *J. Carbohydr. Chem.*, 2000, **19**, 923.
64.　R.I. Duclos Jr., *Carbohydr. Res.*, 2000, **328**, 489.
65.　H. Sakamoto, S. Nakamura, T. Tsuda and S. Hashimoto, *Tetrahedron Lett.*, 2000, **41**, 7691.
66.　J. Xia, J.L. Alderfer, C.F. Piskorz, R.D. Locke and K.L. Matta, *Carbohydr. Res.*, 2000, **328**, 147.
67.　T. Mikazaki, H. Sato, T. Sakakibara and Y. Kajihara, *J. Am. Chem. Soc.*, 2000, **122**, 5678.
68.　D. Schmidt, B. Sauerbrei and J. Thiem, *J. Org. Chem.*, 2000, **65**, 8518.
69.　J. Satyanarayana, T.L. Gururaja, G.A. Naganagowda, S. Narasimhamurthy and M.J. Levine, *Carbohydr. Lett.*, 1999, **3**, 297 (*Chem. Abstr.*, 2000, **132**, 137 645).
70.　S. Ando, Y. Nakahara, Y.Ito, T. Ogawa and Y. Nakahara, *Carbohydr. Res.*, 2000, **329**, 773.
71.　G.X. Chang and T.L. Lowary, *Org. Lett.*, 2000, **2**, 1505.
72.　T.Yui, K. Goto, Y. Kawano and K. Ogawa, *Biosci. Biotechnol. Biochem.*, 2000, **64**, 52.
73.　S. Valverde, M. Garcia, A.M. Gómez and J.C. Lopez, *Chem. Commun.*, 2000, 813.
74.　H. Tanaka, Y. Nishida and K. Kobayashi, *J. Carbohydr. Chem.*, 2000, **19**, 413.
75.　E.A. Khatuntseva, N.E. Ustuzhanina, G.V. Zatonskii, A.S. Shashkov, A.I. Usov and N.E. Nifant'ev, *J. Carbohydr. Chem.*, 2000, **19**, 1151.
76.　N. Hada, I. Ohtsuka, M. Sugita and T. Takeda, *Tetrahedron Lett.*, 2000, **41**, 9065.
77.　S. Mehta and D.M. Whitfield, *Tetrahedron*, 2000, **56**, 6415.
78.　P. Sengupta, S. Basu and B.P. Chatterjee, *J. Carbohydr. Chem.*, 2000, **19**, 243.
79.　A. Geyer, C. Gege and R.R. Schmidt, *Angew. Chem., Int. Ed. Engl.*, 2000, **39**, 3246.
80.　G.V. Pazynina and N.V. Bovin, *Mendeleev. Commun.*, 2000, 132 (*Chem. Abstr.* 2000, **133**, 322 066).
81.　H.J. Vermeer, K.M. Halkes, J.A. van Kuik, J.P. Kamerling and J.F.G. Vliegenthart, *J. Chem. Soc., Perkin Trans.*1, 2000, 2249.
82.　M. Liu, B. Yu, X. Wu, Y. Hui and K.-P. Fung, *Carbohydr. Res.*, 2000, **329**, 745.
83.　G. Anilkumar, L.G. Nair and B. Fraser-Reid, *Org. Lett.*, 2000, **2**, 2587.

84.   N. Nishizono and V. Nair, *Nucleosides, Nucleotides, Nucleic Acids*, 2000, **19**, 283.
85.   A. Bernardi, L. Carrettoni, A.G. Ciponte, D. Monti and S. Sonnino, *Bioorg. Med. Chem. Lett.*, 2000, **10**, 2197.
86.   A.J. Pearce, R. Chevalier, J.-M. Mallet and P. Sinaÿ, *Eur. J. Org. Chem.*, 2000, 2203.
87.   E. Dechaux, P. Savy, S. Bouyain, C. Monneret and J.-C. Florent, *J. Carbohydr. Chem.*, 2000, **19**, 485.
88.   L.F. Tietze, H. Keim, C.O. Janssen, C. Tappertzhofen and J. Olschimke, *Chem. Eur. J.*, 2000, **6**, 2801.
89.   A. Borbás, G. Szabovik, Z. Antal, K. Fehér, M. Csávás, L. Szilágyi, P. Herczegh and A. Lipták, *Tetrahedron: Asymmetry*, 2000, **11**, 549.
90.   L.M. Mikkelsen, S.L. Krintel, J.J. Barbero and T. Skrydstrup, *Chem. Commun.*, 2000, 2319.
91.   Y. Zhu and F. Kong, *Synlett*, 2000, 1783.
92.   M.R.E. Aly, E.-S.I. Ibrahim, E.-S.H.E. El-Ashry and R.R. Schmidt, *Eur. J. Org. Chem.*, 2000, 319.
93.   F. Yan, W.W. Wakarchuk, M. Gilbert, J.C. Richards and D.M. Whitfield, *Carbohydr. Res.*, 2000, **328**, 3.
94.   G. Lemanski and T. Ziegler, *Eur. J. Org. Chem.*, 2000, 181.
95.   H. Li, Q. Li, M.-S.Cai and Z.-J. Li, *Carbohydr. Res.*, 2000, **328**, 611.
96.   I. Mukherjee, S.K. Das, A. Mukherjee and N. Roy, *Carbohydr. Res.*, 2000, **325**, 245.
97.   M. Maruyama, T. Takeda, N. Shimizu, N. Hada and H. Yamada, *Carbohydr. Res.*, 2000, **325**, 83.
98.   S. Koto, A. Kusunoki and M. Hirooka, *Bull. Chem. Soc. Jpn*, 2000, **73**, 967.
99.   Y. Igarashi, K. Takagi, Y. Kan, K. Fujii, K. Harada, T. Furumai and T. Oki, *J. Antibiotics*, 2000, **53**, 233.
100. T. Bamhaoud, S. Sanchez and J. Prandi, *Chem. Commun.*, 2000, 659.
101. F.W. D'Souza, J.D. Ayers, P.R. McCarren and T.L. Lowary, *J. Am. Chem. Soc.*, 2000, **122**, 1251.
102. Y. Zhu and F. Kong, *Carbohydr. Res.*, 2000, **329**, 199.
103. H. Yu, B. Yu, X. Wu, Y. Hui and X. Han, *J. Chem. Soc., Perkin Trans.1*, 2000, 1445.
104. K. Zou, Y. Zhao, G. Tu, J. Cui, Z. Jia and R. Zhang, *Carbohydr. Res.*, 2000, **324**, 182.
105. H. Ziegler, R. Dettmann and M. Duszenko, *Carbohydr. Res.*, 2000, **327**, 367.
106. M. Upreti, D. Ruhela and R.A. Vishwakarma, *Tetrahedron*, 2000, **56**, 6577.
107. K.D. Randell, B.D. Johnston, E.E. Lee and B.M. Pinto, *Tetrahedron: Asymmetry*, 2000, **11**, 207.
108. Y. Du, Q. Pan and F. Kong, *Carbohydr. Res.*, 2000, **323**, 28.
109. C. Gege, J. Vogel, G. Bendas, U. Rothe and R.R. Schmidt, *Chem. Eur. J.*, 2000, **6**, 111.
110. A.K. Misra, Y. Ding, J.B. Lowe and O. Hindsgaul, *Bioorg. Med. Chem. Lett.*, 2000, **10**, 1505.
110a. B. Ernst, B. Wagner, G. Boisch, A. Kaztopodis, T. Winkler and R. Ohrleim, *Can. J. Chem.*, 2000, **78**, 892.
111. M. Sakagami, K. Horie, K. Nakamoto, T. Kawaguchi and H. Hamana, *Chem. Pharm. Bull.*, 2000, **48**, 1256.
112. K. Peilstöcker and H. Kunz, *Synlett*, 2000, 820.
113. K. Peilstöcker and H. Kunz, *Synlett*, 2000, 823.
114. J.-i. Furukawa, S. Kobayashi, M. Nomizu, N. Nishi and N. Sakairi, *Tetrahedron Lett.*, 2000, **41**, 3453.
115. N.W. Luedtke, T.J. Baker, M. Goodman and Y. Tor, *J. Am. Chem. Soc.*, 2000, **122**, 12035.

116. K. Ruda, J. Lindberg, P.J. Garegg, S. Oscarson and P. Konradsson, *Tetrahedron*, 2000, **56**, 3969.
117. R. Bänteli, P. Herold, C. Bruns, J.T. Patton, J.L. Magnani and G. Thoma, *Helv. Chim. Acta*, 2000, **83**, 2893.
118. Y. Ding, S.S. Alkan, G. Baschang and J. Defaye, *Carbohydr. Res.*, 2000, **328**, 71.
119. A.J. Ross, I.A. Ivanova, A.P. Higson and A.V. Nikolaev, *Tetrahedron Lett.*, 2000, **41**, 2449.
120. K. Biswas, S. Pal, J.D. Carbek and D. Kahne, *J. Am. Chem. Soc.*, 2000, **122**, 8413.
121. D.L.J. Clive, Y. Tao, Y. Bo, Y.-Z. Hu, N. Selvakumar, S. Sun, S. Daigngeault and Y.-J. Wu, *Chem. Commun.* 2000, 1341.
122. D.K. Baeschlin, L.G. Green, M.G. Hahn, B. Hinzen, S.J. Ince and S.V. Ley, *Tetrahedron: Asymmetry*, 2000, **11**, 173.
123. H. Xuereb, M. Maletic, J. Gildersleeve, I. Pelczer and D. Kahne, *J. Am. Chem. Soc.*, 2000, **122**, 1883.
124. E. Tanahashi, K. Fukunaga, Y. Ozawa, T. Toyoda, H. Ishida and M. Kiso, *J. Carbohydr. Chem.*, 2000, **19**, 747.
125. K. Fukunaga, K. Shinoda, H. Ishida and M. Kiso, *Carbohydrate Res.*, 2000, **328**, 85.
126. C. Gege, W. Kinzy and R.R. Schmidt, *Carbohydr. Res.*, 2000, **328**, 459.
127. T. Miyamoto, A. Yamamoto, M. Wakabayashi, Y. Nagaregawa, M. Inagaki, R. Higuchi, M. Iha and K. Teruya, *Eur. J. Org. Chem.*, 2000, 2295.
128. A.V. Kornilov, A.A. Sherman, L.O. Kononov, A.S. Shashkov and N.E. Nifant'ev, *Carbohydr. Res.*, 2000, **329**, 717.
129. K. Yamada, Y. Harada, T. Miyamoto, R. Isobe and R. Higuchi, *Chem. Pharm. Bull.*, 2000, **48**, 157.
130. H. Li, Q. Li, M.-S. Cai and Z.-J. Li, *Carbohydr. Res.*, 2000, **328**, 611.
131. R.-S. Zhang, Y.-P. Ye, Y.-M. Shen and H.-L. Liang, *Tetrahedron*, 2000, **56**, 3875.
132. F. Bélot and J.-C. Jacquinet, *Carbohydr. Res.*, 2000, **326**, 88.
133. S. Sanchez, T. Bamhaoud and J. Prandi, *Tetrahedron Lett.*, 2000, **41**, 7447.
134. S.K. Bhattacharya and S.J. Danishefsky, *J. Org. Chem.*, 2000, **65**, 144.
135. A. Lubineau, J. Alais and R. Lemoine, *J. Carbohydr. Chem.*, 2000, **19**, 151.
136. K.D. Randell, B.D. Johnston, P.N. Brown and B.M. Pinto, *Carbohydr. Res.*, 2000, **325**, 253.
137. N. Hada, M. Kuroda and T. Takeda, *Chem. Pharm. Bull.*, 2000, **48**, 1160.
138. M. Takatani, T. Nakama, K. Kubo, S. Manabe, Y. Nakahara, Y. Ito and Y. Nakahara, *Glycoconjugate J.*, 2000, **17**, 361.
139. J. Xia, J. Alderfer and K.L. Matta, *Biorg. Med. Chem. Lett.*, 2000, **10**, 2485.
140. J. Xia, T. Srikrishnan, J.L. Alderfer, R.K. Jain, C.F. Piskorz and K.L. Matta, *Carbohydr. Res.*, 2000, **329**, 561.
141. K.M. Koeller, M.E.B. Smith, R.-F. Huang and C.-H. Wong, *J. Am. Chem. Soc.*, 2000, **122**, 4241.
142. E. Maes, D. Florea and G. Strecker, *Carbohydr. Lett.*, 2000, **3**, 431 (*Chem. Abstr.*, 2000, **133**, 135 525).
143. S. De Marino, M. Iorizzi, F. Zollo, C.D. Amsler, S.P. Greer and J.B. McClintock, *Eur. J. Org. Chem.*, 2000, 4093.
144. N.T. Nyberg, L. Kenne, B. Ronnberg and B.G. Sundquist, *Carbohydr. Res.*, 2000, **323**, 87.
145. C. Costachel, P.J. Sansonetti and L.A. Mulard, *J. Carbohydr. Chem.*, 2000, **19**, 1131.
146. L.A. Mulard and J. Ughetto-Monfrin, *J. Carbohydr. Chem.*, 2000, **19**, 503.
147. L.A. Mulard and J. Ughetto-Monfrin, *J. Carbohydr. Chem.*, 2000, **19**, 193.
148. M. Martín-Lomas, N. Khiar, S. García, J.-L. Koessler, P.M. Nieto and T.W.

Rademacker, *Chem. Eur. J.*, 2000, **6**,3608.

149. A. Dondoni, M. Mizuno and A. Marra, *Tetrahedron Lett.*, 2000, **41**, 6657.

150. J.R. Allen, J.G. Allen, X.-F. Zhang, L.J. Williams, A. Zatorski, G. Ragupathi, P.O. Livingston and S.J. Danishefsky, *Chem. Eur. J.*, 2000, **6**, 1366.

151. W.R. Roush and C.E. Bennett, *J. Am., Chem. Soc.*, 2000, **122**, 6124.

152. S. Sugiyama, W. Haque and J. Diakur, *Org. Lett.*, 2000, **2**, 3489.

153. C. Bernlind, S. Bennett and S. Oscarson, *Tetrahedron: Asymmetry*, 2000, **11**, 481.

154. Y.Du, M. Zhang and F. Kong, *Org. Lett.*, 2000, **2**, 3797.

155. L. Knerr and R.R. Schmidt, *Eur. J. Org. Chem.*, 2000, 2803.

156. A. Chernyak, S. Oscarson and D. Turek, *Carbohydr. Res.*, 2000, **329**, 309.

157. M. Itabashi, K. Segawa, Y. Ikeda, S. Kondo, H. Naganawa, T. Koyano and K. Umezawa, *Carbohydr. Res.*, 2000, **323**, 57.

158. M.D. Burkart, M. Izumi, E. Chapman, C.-H. Lin and C.-H. Wong, *J. Org. Chem.*, 2000, **65**, 5565.

159. T. Zhu and G.-J. Boons, *J. Am. Chem. Soc.*, 2000, **122**, 10222.

160. L.J. Williams, C.R. Harris, P.W. Glunz and S.J. Danishefsky, *Tetrahedron Lett.*, 2000, **41**, 9505.

161. W. Liao, R.D. Locke and K.L. Matta, *Chem. Commun.*, 2000, 369.

162. L. Singh, Y. Nakahara, Y. Ito and Y. Nakahara, *Carbohydr Res.*, 2000, **325**, 132.

163. J. Xia, J.L. Alderfer, C.F. Piskorz and K.L. Matta, *Chem. Eur. J.*, 2000, **6**, 3442.

164. J. Xia, C.F. Piskorz, J.L. Alderfer, R.D. Locke and K.L. Matta, *Tetrahedron Lett.*, 2000, **41**, 2773.

165. B.-G. Huang, R.K. Jain, R.D. Locke, J.L. Alderfer, W.A. Tabaczynski and K.L. Matta, *Tetrahedron Lett.*, 2000, **41**, 6279.

166. M. Sandvoss, L.H. Pham, K. Levsen, A. Preiss, C. Mügge and G. Wünsch, *Eur. J. Org. Chem.*, 2000, 1253.

167. M. Martin-Lomas, M. Flores-Mosquera and J.L. Chiara, *Eur. J. Org. Chem.*, 2000, 1547.

168. G. Yang and F. Kong, *Synlett*, 2000, 1423.

169. M.V. Chiesa and R.R. Schmidt, *Eur. J. Org. Chem.*, 2000, 3541.

170. Y. Ohnishi, H. Ando, T. Kawai, Y. Nakahara and Y. Ito, *Carbohydr. Res.*, 2000, **328**, 263.

171. Y. Zhang, B. Dausse, P. Sinaÿ, M. Afsahi, P. Berthault and H. Desvaux, *Carbohydr. Res.*, 2000, **324**, 231.

172. A. Düffels, L.G. Green, S.V. Ley and A.D. Miller, *Chem. Eur. J.*, 2000, **6**, 1416.

173. D.K. Baeschlin, A.R. Chaperon, L.G. Green, M.G. Hahn, S.J. Ince and S.V. Ley, *Chem. Eur. J.*, 2000, **6**, 172.

174. K. Ruda, J. Lindberg, P.J. Garegg, S. Oscarson and P. Konradsson, *J. Am. Chem. Soc.*, 2000, **122**, 11067.

175. Y. Du, Q. Pan and F. Kong, *Carbohydr. Res.*, 2000, **329**, 17.

176. Y. Zhu and F. Kong, *Synlett*, 2000, **663**.

177. M. Chu, R. Mierzwa, M. Patel, J. Jenkins, P. Das, B. Pramanik and T.-M. Chan, *Tetrahedron Lett.*, 2000, **41**,6689.

178. I. Prahl and C. Unverzagt, *Tetrahedron Lett.*, 2000, **41**, 10189.

179. N. Otsubo, H. Ishida and M. Kiso, *Tetrahedron Lett.*, 2000, **41**, 3879.

180. V. Wittmann, A.K. Datta, K.M. Koeller and C.-H. Wong, *Chem. Eur. J.*, 2000, **6**, 162.

181. K.M. Koeller and C.-H. Wong, *Chem. Eur. J.*, 2000, **6**, 1243.

182. A. Düffels, L.G. Green, R. Lenz, S.V. Ley, S.P. Vincent and C.-H. Wong, *Biorg. Med. Chem.*, 2000, **8**, 2519.

183. J. Seifert, M. Lergenmüller and Y. Ito, *Angew. Chem., Int. Ed. Engl.*, 2000, **39**, 531.
184. C. Unverzagt and J. Seifert, *Tetrahedron Lett.*, 2000, **41**, 4549.
185. V. Pozsgay, *Tetrahedron: Asymmetry*, 2000, **11**, 151.
186. P.Y. Chong and P.A. Petillo, *Org. Lett.*, 2000, **2**, 1093.
187. E. Bourgeaux and J.-C. Combret, *Tetrahedron: Asymmetry*, 2000, **11**, 4189.
188. S. Immel, K. Fujita, H.J. Lindner, Y. Nogami and F.W. Lichtenthaler, *Chem. Eur. J.*, 2000, **6**, 2327.
189. M. Kawamura, H. Nakai and T. Uchiyama, *Carbohydr. Res.*, 2000, **323**, 49.
190. F.A. Momany and J.L. Willett, *Carbohydr. Res.*, 2000, **326**, 194.
191. G. Dudziak, N. Bézay, T. Schwientek, H. Clausen, H. Kunz and A. Liese, *Tetrahedron*, 2000, **56**, 5865.
192. A. Ikuta, K. Koizumi and T. Tanimoto, *J. Carbohydr. Chem.*, 2000, **19**, 13.
193. S. Kitahatta, T. Tanimoto, A. Ikuta, K. Tanaka, K. Fujita, H. Hashimoto, H. Murakami, H. Nakano and K. Koizumi, *Biosci. Biotechnol. Biochem*, 2000, **64**, 1223.
194. S. Kitahata, T. Tanimoto, Y. Okada, A. Ikuta, K. Tanaka, H. Murakami, H. Nakano and K. Koizami, *Biosci. Biotechnol. Biochem.*, 2000, **64**, 2406.
195. G. Cravotto, G. Palmisano, L. Panza and S. Tagliapietra, *J. Carbohydr. Chem.*, 2000, **19**, 1235.
196. H. Dittmann, K. Scharwachter and W.A. Konig, *Carbohydr. Res.*, 2000, **324**, 75.
197. A.J. Pearce and P. Sinaÿ, *Angew. Chem., Int. Ed. Engl.*, 2000, **39**, 3610.
198. D.A. Fulton and J.F. Stoddart, *Org. Lett.*, 2000, **2**, 1113.
199. F.G. Calvo-Flores, J. Isac-Garcia, F. Hernández-Mateo, F. Pérez-Balderas, J.A. Calvo-Asín, E. Sanchéz-Vaquero and F. Santoyo-González, *Org. Lett.*, 2000, **2**, 2499.
200. Y. Inoue, T. Wada, N. Sugahara, K. Yamamoto, K. Kimura, L.-H. Tong, X.-M. Gao, Z.-J. Hou and Y. Liu, *J. Org. Chem.*, 2000, **65**, 8041.
201. K. Teranishi, M. Hisamatsu and T. Yamada, *Tetrahedron Lett.*, 2000, **41**, 933.
202. K. Teranishi, *Tetrahedron Lett.*, 2000, **41**, 7085.
203. M. Atsumi, M. Izumida, D.-Q. Yuan and K. Fujita, *Tetrahedron Lett.*, 2000, **41**, 8117.
204. K. Koga, D.-Q. Yan and K. Fujita, *Tetrahedron Lett.*, 2000, **41**, 6855.
205. D. Kirby Mayanard and G. Vigh, *Carbohydr. Res.*, 2000, **328**, 277.
206. R. Sebesta and M. Salisova, *Enantiomer*, 1999, **4**, 271 (*Chem. Abstr.*, 2000, **132**, 78 782).
207. C. Pean, C. Créminon, A. Wijkhuisen, J. Grassi, P. Guenot, P. Jéhan, J.-P. Dalbiez, B. Perly and F. Djedaïni-Pilard, *J. Chem. Soc., Perkin Trans, 2*, 2000, 853.
208. M.J. Han, K.S. Yoo, J.Y. Chang and T.-K. Ha, *Angew. Chem., Int. Ed. Engl.*, 2000, **39**, 347.
209. Y. Liu, C.-C. You, T. Wada and Y. Inoue, *Tetrahedron Lett.*, 2000, **41**, 6869.
210. S. Pagliari, R. Corradini, G. Galaverna, S. Sforza, A. Dossena and R. Marchelli, *Tetrahedron Lett.*, 2000, **41**, 3691.
211. T. Masuda, J.-C. Hayashi and S. Tamagaki, *J. Chem. Soc. Perkin Trans. 2*, 2000, 161.
212. T. Carofiglio, R. Fornasier, V. Lucchini, L. Simonato and U. Tonellato, *J. Org. Chem.*, 2000, **65**, 9013.
213. C.E. Granger, C.P. Félix, H.P. Parrot-Lopez and B.R. Langlois, *Tetrahedron Lett.*, 2000, **41**, 9257.
214. M.. Wazynska, A. Temeriusz, K. Chmurski, R. Bilewicz and J. Jurczak, *Tetrahedron Lett.*, 2000, **41**, 9119.

215. T. Furuike, S. Aiba and S.-I. Nishimura, *Tetrahedron*, 2000, **56**, 9909.
216. R. Roy, F. Hernández-Mateo and F. Santoyo-Gorzález, *J. Org. Chem.*, 2000, **65**, 8743.
217. M. Ichikawa, A.S. Woods, H. Mo, I.J. Goldstein and Y. Ichikawa, *Tetrahedron: Asymmetry*, 2000, **11**, 389.

5
Ethers and Anhydro-sugars

1 Ethers

1.1 Methyl Ethers. – All possible monomethyl ethers of 4-nitrophenyl α-D-gluco-, α-D-galacto- and α-D-mannopyranosides have been prepared by standard methods.[1] 3-*O*-Methyl-D-mannose[2] and all positional isomers of partially methylated and acetylated or benzoylated 1,5-anhydroribitol[3] and methyl 2-(acetylmethylamino)-2-deoxy-β-D-glucopyranoside[4] have been prepared as analytical standards. A convenient procedure for the preparation of 3-*O*-[^{11}C]-methyl-D-glucose has been described.[5] The 2'-, 3'- and 4'-*O*-methyl ethers of the disaccharide glycoside octyl α-L-Fuc*p*-(1→2)-β-D-Gal*p* have been synthesized and tested as acceptors for human blood group glycosyltransferases.[6]

1.2 Other Alkyl and Aryl Ethers. – Pilosidine **1**, a fused ether-glycoside, was isolated from rhizomes of *Curculigo pilosa*, an African plant.[7] The synthesis and applications of carbohydrate-derived amphiphilic perfluoroalkyl ethers have been reviewed.[8]

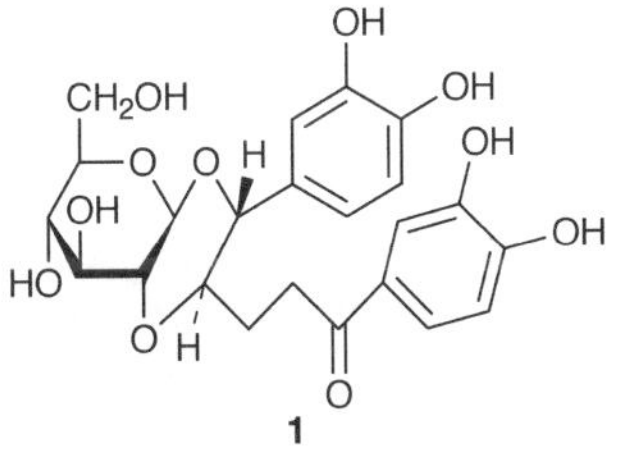

1

Mono-*O*-allyl derivatives of D-fructose have been prepared in order to test the specificity of the fructose transporter GLUT5.[9] Tris(triphenylphosphine)ruthenium(II) dichloride is an excellent catalyst for the isomerization of allyl to prop-1-enyl ethers in a de-*O*-allylation procedure. The catalyst avoids the problem of reduction of the alkene as a side reaction.[10] Another method uses Pd(0)-catalysed allyl transfer to *p*-toluenesulfinic acid for the cleavage of allyl ethers on a solid support.[11]

Carbohydrate Chemistry, Volume 34
© The Royal Society of Chemistry, 2003

The tritylation of alcohols under neutral conditions has been achieved using *p*-methoxybenzyl trityl ether in the presence of DDQ.[12] A selective cleavage of trityl ethers by CBr_4–MeOH has been attributed to the formation of traces of HBr.[13] Ceric ammonium nitrate on silica gel rapidly cleaves trityl and mono and di-methoxytrityl ethers[14] whereas $CeCl_3$–NaI in refluxing acetonitrile effects the selective cleavage of trityl and dimethoxytrityl ethers under neutral conditions.[15]

Reductive ring opening reactions of 4,6-*O*-benzylidene acetals of mono- and di-saccharide derivatives using Et_3SiH–TfOH or Et_3SiH–$PhBCl_2$ have afforded, with excellent selectivity, the primary and secondary benzyl ethers respectively.[16] Similarly, standard conditions ($LiAlH_4$–$AlCl_3$ or $NaCNBH_3$–HCl) have been utilized to convert 4,6-*O*-(2-naphthyl)methylene acetals of glycosides into the 4-*O*- and 6-*O*-(2-naphthyl)methyl ethers, respectively, with good selectivity,[17] while the 2-naphthylmethyl ether group has been touted as a versatile protecting group, stable to acidic conditions and readily removed with DDQ in CH_2Cl_2.[18] Benzyl ethers situated vicinally to alcohols may be selectively deprotected using either NIS or diacetoxyiodosobenzene-I_2 (Scheme 1). In some circumstances, however, the corresponding benzylidene acetal can be formed.[19] A chemoselective deprotection of benzyl ethers in the presence of Cbz groups uses $NaBrO_3$/$Na_2S_2O_4$, which is assumed to result in radical bromination.[20] While *p*-methoxybenzyl ethers may be cleaved with $SnCl_4$/PhSH at −78 °C, esters, benzyl and allyl ethers, isopropylidene acetals and Tbdps groups are stable under these conditions.[21]

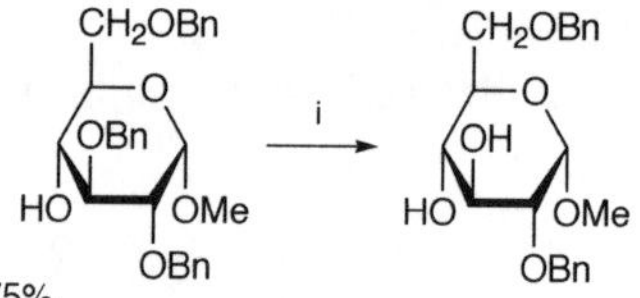

Reagents: i, NIS, MeCN, reflux, 75%

Scheme 1

Dimeric carbohydrate-based surfactants such as **2** have been prepared,[22] and 5-*O*-(2-iodoethyl)-D-glucose was synthesized as a substrate suitable for single photon emission computer tomography.[23] Some monosaccharide poly-*O*-(2-azidoethyl) ethers were synthesized and the azide moieties transformed into hydrophobic aminoalkyl groups.[24] Cyclic polyethers incorporating some carbohydrates have been synthesized and evaluated as facilitators of transmembrane sodium ion transport.[25] Butenolides (*e.g.* **3**) were obtained by reaction of the alkoxide **4** with difluorobutenolides such as RF.[26] 3-*O*-Substituted derivatives such as **5** have been obtained by [2 + 2] cycloadditions of chlorosulfonyl isocyanate and the corresponding vinyl ether.[27] The e.e.s of organic amine salts have been determined by FAB MS relying on differential complexation with crown ether-type D- or L-galactose derivatives.[28]

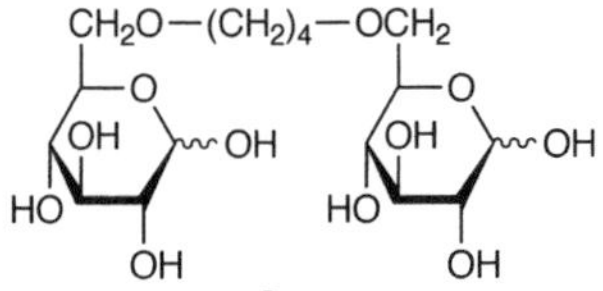

2

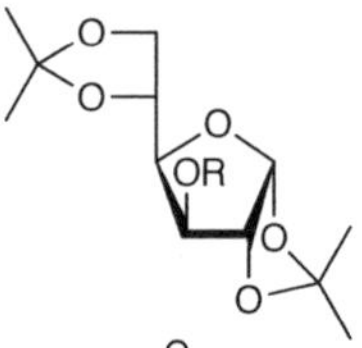

3 R =

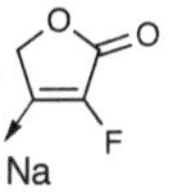

4 R = Na

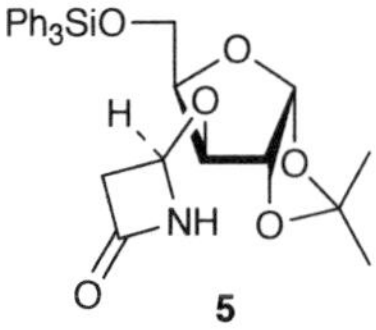

5

1.3   Silyl Ethers. – Primary Tbdms groups are selectively cleaved in the presence of secondary Tbdms moieties with BCl_3.[29]

2   Intramolecular Ethers (Anhydro-sugars)

2.1   Oxiranes. – The reaction of terminal diols with $TPP/imidazole/I_2$ under certain conditions has afforded epoxides.[30] The previously reported 3,4-anhydro-D-tagatopyranose derivative **6** has been found to equilibrate under basic conditions to give a 2:1 mixture of **7** and **6**.[31] Resin-bound (*via O*-4) 1,6:2,3-dianhydro-β-D-mannopyranose has been treated with various nucleophiles to give D-*gluco*- products of attack at C-2 with opening of the epoxide.[32]

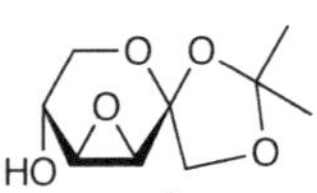

6 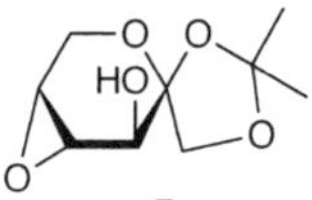   **7**

2.2   Other Anhydrides. – The production and physiological properties of difructose dianhydrides have been reviewed.[33] Treatment of some anhydroalditol derivatives (*e.g.* **8**) with diacetoxyiodosobenzene led to oxygen radical formation from the hydroxy group, followed by intramolecular hydrogen atom abstraction and subsequent anhydro-sugar formation to give **9** in good yields.[34]

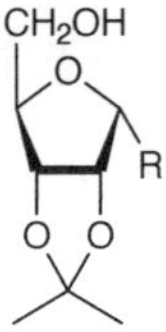

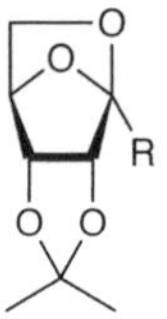

8 R = CH$_2$OH, (CH$_2$)$_2$OH **9**

References

1.   W. Hakamata, T. Nishio, R. Sato, T. Mochizuki, K. Tsuchiya, M. Yasuda and T. Oku, *J. Carbohydr. Chem.*, 2000, **19**, 359.
2.   X.-X. Tian, A. Li, I.V. Farrugia, X. Mo, D. Crich and M.J. Groves, *Carbohydr. Res.*, 2000, **324**, 38.
3.   C.M. El Hilali, T.A. Lamosse and G.R. Gray, *Carbohydr. Res.*, 2000, **329**, 189.

4.   L.E. Elvebak II, S.O. Smith and G.R. Gray, *Carbohydr. Res.*, 2000, **329**, 799.
5.   P. Mading, H. Kasper, J. Zessin, M. Gnauck, F. Fuchtner, P. Brust and J. Steinbach, *Wiss.-Tech. Ber. – Forschungszent. Rossendorf*, 1999, 141 (*Chem. Abstr.*, 2000, **132**, 265 352).
6.   A. Mukherjee, M.M. Palcic and O. Hindsgaul, *Carbohydr. Res.*, 2000, **326**, 1.
7.   G. Palazzino, C. Galeffi, E. Federici, F. Delle Monache, M.F. Cometa and M. Palmery, *Tetrahedron*, 2000, **55**, 411.
8.   J.G. Riess and J. Greiner, *Carbohydr. Res.*, 2000, **327**, 147.
9.   A. Tatibouet, J. Yang, C. Morin and G.D. Holman, *Bioorg. Med. Chem.*, 2000, **8**, 1825.
10. Y.-J. Hu, R. Dominique, S.K. Das and R. Roy, *Can. J. Chem.*, 2000, **78**, 838.
11. T. Opatz and H. Kunz, *Tetrahedron Lett.*, 2000, **41**, 10185.
12. G.V.M. Sharma, A.K. Mahalingham and T. Rajendra Prasad, *Synlett*, 2000, 1479.
13. J.S. Yadav and B.V.S. Reddy, *Carbohydr. Res.*, 2000, **329**, 885.
14. J.R. Hwu, M.L. Jain, F.-Y. Tsai, A. Balakumar and G.H. Hakimelahi, *J. Org. Chem.*, 2000, **65**, 5077.
15. J.S. Yadav and B.V.S. Reddy, *Synlett*, 2000, 1275.
16. M. Sakagami and H. Hamana, *Tetrahedron Lett.*, 2000, **41**, 5547.
17. A. Lipták, A. Borbás, L. Jánossy and L. Szilágyi, *Tetrahedron Lett.*, 2000, **41**, 4949.
18. J. Xia, S.A. Abbas, R.D. Locke, C.F. Piskorz, J.L. Alderfer and K.L. Matta, *Tetrahedron Lett.*, 2000, **41**, 169.
19. J. Madsen, C. Viuf and M. Bols, *Chem. Eur. J.*, 2000, **6**, 1140.
20. M. Adinolfi, L. Guariniello, A. Iadonisi and L. Mangoni, *Synlett*, 2000, 1277.
21. W. Yu, M. Su, X. Gao, Z. Yang and Z. Jin, *Tetrahedron Lett.*, 2000, **41**, 4015.
22. M.J.L. Castro, J. Kovenski and A.F. Cirelli, *J. Carbohydr. Chem.*, 2000, **19**, 1175.
23. C. Morin and L. Ogier, *J. Carbohydr. Chem.*, 2000, **19**, 111.
24. M. Bessodes, C. Dubertret, G. Jaslin and D. Scherman, *Bioorg. Med. Chem Lett.*, 2000, **10**, 1393.
25. C.G. Espinola, R. Pérez and J.D. Martín, *Org. Lett.*, 2000, **2**, 3161.
26. O. Paleta, A. Pelter, J. Kebrle, Z. Duda and J. Hajduch, *Tetrahedron*, 2000, **56**, 3197.
27. Z. Kaluza, B. Furman, P. Krajewski and M. Chmielewski, *Tetrahedron*, 2000, **56**, 5553.
28. M. Shizuma, H. Imamura, Y. Takai, H. Yamada, T. Takeda, S. Takahashi and M. Sawada, *Chem. Lett.*, 2000, 1292.
29. Y.-Y. Yang, W.-B. Yang, C.-F. Teo and C.-H. Lin, *Synlett*, 2000, 1634.
30. H.B. Merayala, P.M. Goud, R.R. Gadikota and K.R. Reddy, *J. Carbohydr. Chem.*, 2000, **19**, 1211.
31. M. Ataie, J.G. Buchanan, A.R. Edgar, R.G. Kinsman, M. Lyssikatou, M.F. Mahou and P.M. Welch, *Carbohydr. Res.*, 2000, **323**, 36.
32. S. Wendeborn, A. de Mesmaeker, W.K. Brill and S. Berteina, *Acc. Chem. Res.*, 2000, **33**, 215.
33. K. Saito and F. Tomita, *Biosci. Biotechnol. Biochem.*, 2000, **64**, 1321.
34. C.G. Francisco, A.J. Herrera and E. Suárez, *Tetrahedron Lett.*, 2000, **41**, 7869.

6
Acetals

1 Synthesis

The pyridinium *p*-toluenesulfonate-promoted reaction of D-galactose with cyclohexanone in DMF has been studied in detail to identify conditions for optimal yields of the furanose 1,2:5,6-di-acetal **1**. (See Chapter 3 for use of this derivative in the synthesis of the disaccharide epitope α-D-Gal*p*(1→3)-D-Gal.)[1] A new method for the preparation of benzylidene acetals involves exposure of a benzyl ether with a neighbouring free hydroxyl group to NIS (*e.g.* **2**→**3**).[2] The 4,6-*O*-(2-naphthyl)methylene acetal **4** was obtained by conventional, acid-catalysed reaction of methyl α-D-glucopyranoside with 2-naphthaldehyde or its dimethyl acetal. Reductive cleavage to give the 4- or 6-*O*-2-naphthylmethyl ethers was effected by applying the standard procedures for these transformations (see Chapter 5). The diastereomeric dioxolane type acetals **5** were obtained in a similar way from methyl α-L-rhamnopyranoside.[3]

1

2 R¹ = Bn, R² = H

3 R¹,R² = ≻~Ph

4

5

6

Carbohydrate Chemistry, Volume 34

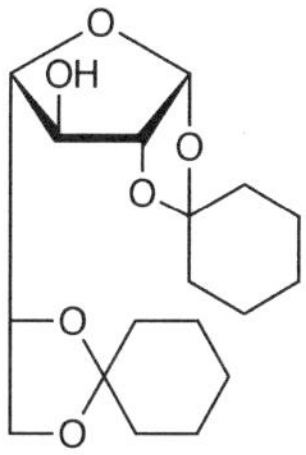

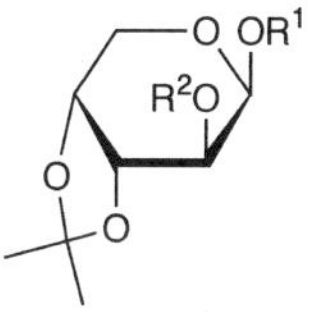

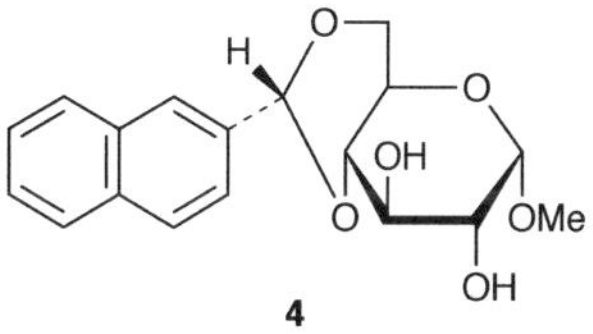

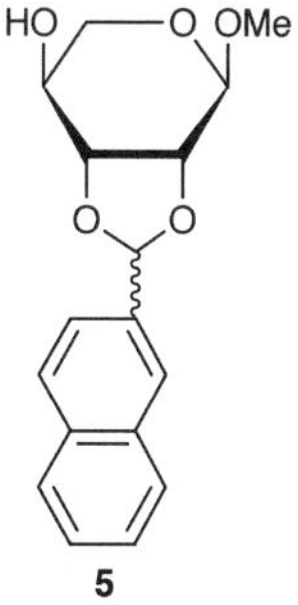

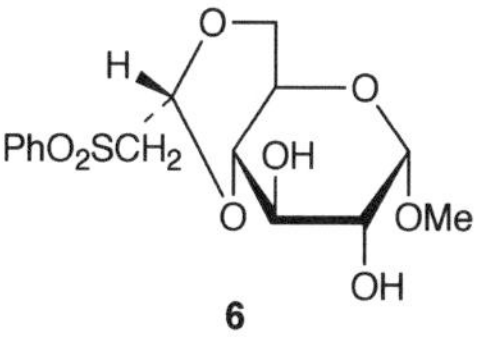

4,6-*O*-Phenylsulfonylethylidene acetals, such as methyl α-D-glucopyranoside derivative **6**, were formed in high yields by exposure of 4,6-diols to 1,2-(diphenyl-sulfonyl)ethene and sodium hydride. They resist acid hydrolysis but are cleaved by LAH.[4] Lanthanide [(Yb(III), Er (III)] cation exchange resins catalysed the acetalation of unprotected sucrose by α,β-unsaturated and aromatic aldehydes in a one-pot, two-step process, as indicated in Scheme 1.[5]

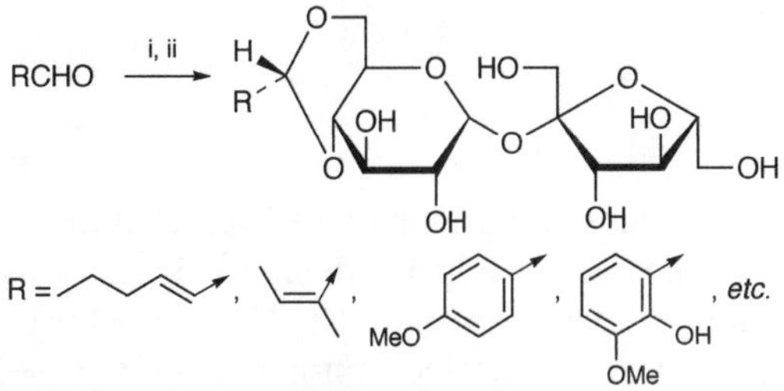

Reagents: i, HC(OMe)$_3$, Yb(III)-exchanged amberlyst; ii, sucrose

Scheme 1

When 3-hydroxy-2-*O*-thiobenzoate **7** was irradiated in dichloromethane in the presence of triethylamine, the monothioacetals **8** were obtained in near quantitative yield. The mechanism proposed for this reaction, involving photo-induced electron-transfer and incorporation of the solvent, is outlined in Scheme 2.[6]

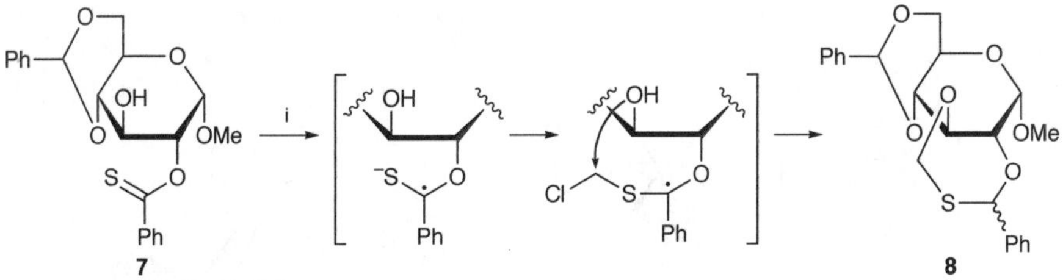

Reagents: i, CH$_2$Cl$_2$, Et$_3$N, *hv*

Scheme 2

Spirophostins **9**, which have been prepared as conformationally restricted analogues of adenophostins, and some *spiro*-acetals obtained from enol esters are covered in Chapter 2.

2   Hydrolysis and Reductive Ring-opening

Studies on the selective deprotection of lactose derivative **10** by acid hydrolysis showed the stabilities of the acetals to be dimethyl acetal ≈ 2,3- > 3′,4′- > 5,6-*O*-isopropylidene.[7]

HCl in dichloromethane or acetone has been used to cleave isopropylidene and benzylidene acetals in the presence of other acid-sensitive groups.[8] Iso-propylidene-protected sugar phosphates and thiophosphates, such as **11**, were deacetalated by heating in water without added acid, the aq. solutions having a

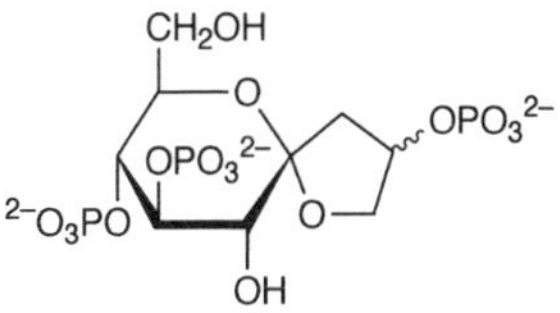

9 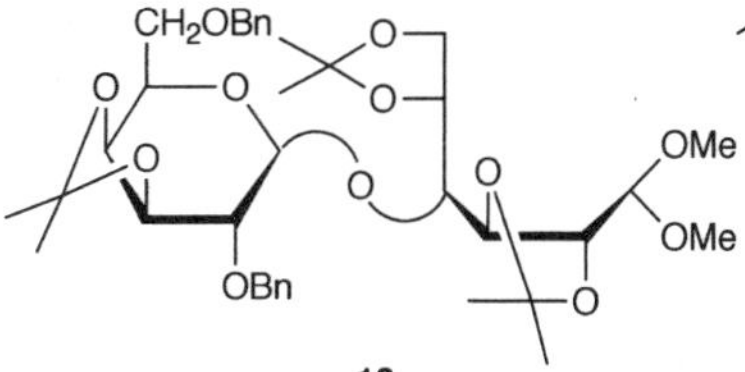   **10** 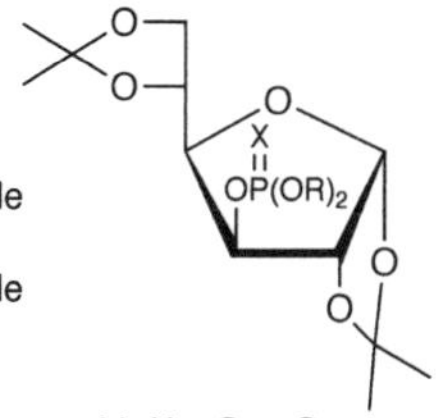   **11** X = O or S;
R = Et or Ph

pH of 3–4.[9] Selective hydrolysis of terminal isopropylidene and benzylidene acetals in the presence of a variety of other protecting groups has been achieved by use of $Zn(NO_3)_2 \cdot 6H_2O$ in acetonitrile (*e.g.* **12→13**). Bn, Ac, Bz, Tbdms, Ts, but not THP and Tr were stable towards these conditions.[10] In the regioselective cleavage of dioxolanes with Me_3SiCH_2MgCl under chelation control, acetals which are *trans*-fused to inositol rings reacted faster then *cis*-fused ones (*e.g.* **14→15**).[11] A variety of terminal acetals and ketals have been cleaved regioselectively by solvent-free peroxymonosulfate on alumina under microwave irradiation.[12] Acetal hydrolysis in the presence of silyl groups by use of BCl_3 is referred to in Chapter 2.

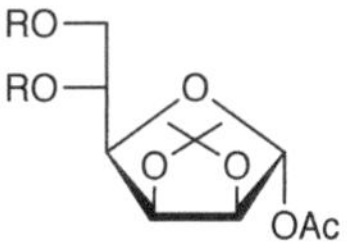

12 R,R = ⤬ Me, Me
13 R = H

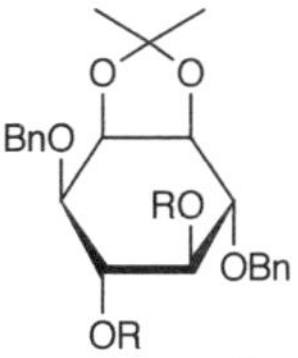

14 R,R = ⤬ Me, Me
15 R = H

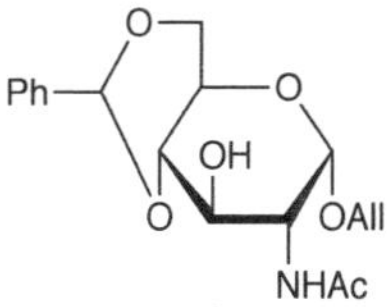

16

Et$_3$SiH in the presence of TfOH or PhBCl$_2$ cleaved the acetal rings of 4,6-*O*-benzylidene-protected mono- and di-saccharides reductively to give 6-*O*-and 4-*O*-benzyl ethers, respectively, with excellent selectivity and in high yields.[13] Opening of the benzylidene ring in compound **16** with NaCNBH$_3$-TmsCl gave the 6-benzyl ether as the main product rather than the expected 4-*O*-Bn isomer. The latter was obtained, however, in 90% yield by use of BH$_3$ · NMe$_3$-Me$_2$BBr at − 78 °C.[14] The reductive cleavage of 4,6-*O*-(2-naphthyl)methylene acetals to give the 4- or 6-*O*-2-naphthylmethyl ethers is referred to above.

3   Other Reactions and Properties

Contra-thermodynamic *trans*/*cis*-epimerization of vicinal diols on furanose rings relied on isomerization of the *trans*-isopropylidene derivatives to the more stable *cis* isomers on treatment with (Bu^{t}O)$_3$SiSH in the presence of a radical initiator.[15] 1,2:5,6-Di-*O*-isopropylidene-α-D-glucose (**17**) has been converted to 1,2:3,5-di-*O*-isopropylidene-β-L-idose (**18**) in a four step reaction sequence in which acetal migration was followed by elimination-borohydration, resulting in epimerization at C-5 (Scheme 3).[16]

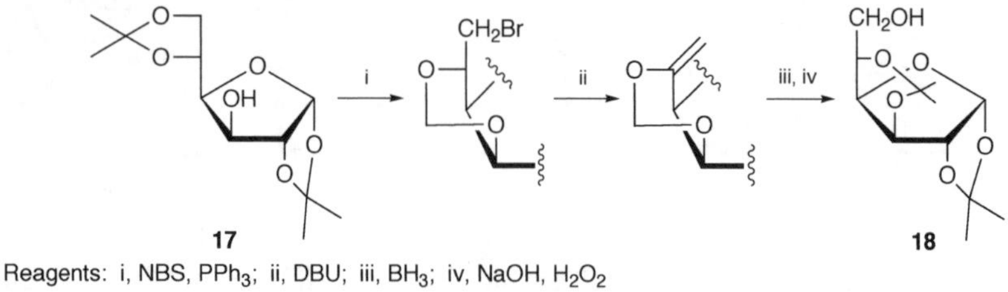
17
18
i
ii
iii, iv
CH2Br
CH2OH
OH
Reagents: i, NBS, PPh3; ii, DBU; iii, BH3; iv, NaOH, H2O2

Scheme 3

D-Arabino-, D-manno-, D-galacto- and L-fucopyranosyl azides were epimerized at C-3 by the method of Miethchen (see Vol. 33, Chapter 2, Ref. 14, Vol. 32, Chapter 6, Ref. 8; Vol. 30, p. 99, Refs. 17–19) to give D-lyxo-, D-altro-, D-gulo- and L-gulo-configurated products, respectively, in high yields (*e.g.* **19**→**20**).[17]

Methyl 4,6-*O*-benzylidene-α-D-mannopyranoside and methyl 4,6-*O*-*p*-nitrobenzylidene-α-D-galactopyranoside, which are both scarcely soluble in apolar solvents, act as 'supergelators' when forced into solution by heating in a sealed tube above the normal boiling points of, for example, hexane or carbon tetrachloride.[18]

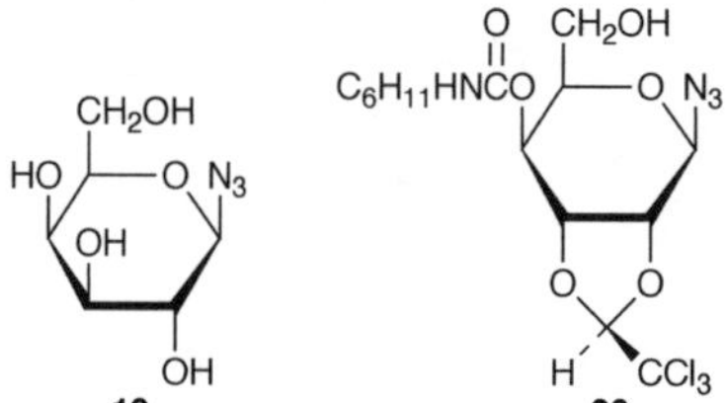
CH2OH
HO
N3
OH
OH
C6H11HNCO
CCl3
H

19 **20**

References

1. I. Sakamoto and H. Ohrui, *Biosci. Biotechnol. Biochem.*, 2000, **64**, 1974.
2. J. Madsen, C. Viuf and M. Bols, *Chem. Eur. J.*, 2000, **6**, 1140.
3. A. Lipták, A. Borbás, L. Jánossy and L. Szilágyi, *Tetrahedron Lett.*, 2000, **41**, 4949.
4. F. Chéry, P. Rollin, D. De Lucchi and S. Cossu, *Tetrahedron Lett.*, 2000, **41**, 2357.
5. S. Porwanski, P. Salanski, G. Descotes, A. Bouchu and Y. Queneau, *Synthesis*, 2000, 525.
6. A.L. Schreiber, M.A. Fashing and C.J. Abelt, *J. Chem. Soc., Perkin Trans. 2*, 2000, 953.
7. G. Catelani, F. D'Andrea and L. Puccioni, *Carbohydr. Res.*, 2000, **324**, 204.
8. A.G. Cioletti, R.J. Alves, J.D. de Filho, J.G. Chaves and M.A.F. Prado, *Synth. Commun.*, 2000, **30**, 2019.
9. D. Postel, G. Ronco and P. Villa, *J. Carbohydr. Chem.*, 2000, **19**, 171.
10. S. Vijayasaradhi, J. Singh and I.S. Aidhen, *Synlett*, 2000, 110.
11. C.-C. Chiang, Y.-H. Chen, Y.-T. Hsieh and T.-Y. Luh, *J. Org. Chem.*, 2000, **65**, 4694.
12. D. Subhas Bose, B. Jayalakshmi and A. Venkat Narsaiah, *Synthesis*, 2000, 67.
13. M. Sakagami and H. Hamana, *Tetrahedron Lett.*, 2000, **41**, 5547.

14. M. Ghosh, R.G. Dulina, R. Kakarla and M.J. Sofia, *J. Org. Chem.*, 2000, **65**, 8387.
15. H.-S. Dang and B.P. Roberts, *Tetrahedron Lett.*, 2000, **41**, 8595.
16. S.C. Hung, R. Puranik and F.-C. Chi, *Tetrahedron Lett.*, 2000, **41**, 77.
17. C. Hager, R. Miethchen and H. Reinke, *Synthesis*, 2000, 226.
18. R. Luboradzki, O. Gronwald, A. Ikeda and S. Shinkai, *Chem. Lett.*, 2000, 1148.

7
Esters

1 Carboxylic and Thiocarboxylic Esters

1.1 Synthesis. – A report on perspectives on alkyl carbonates in organic synthesis has appeared. This contains applications of alkyl carbonate protecting groups in carbohydrate chemistry, carbonate exchange reactions and subsequent modifications as applied to carbohydrate derivatives and promotion of glycosylations.[1]

1.1.1 Chemical Acylation and Deacylation. Several methods for exhaustive acylations have been reported. Acetylation of D-glucose to give penta-*O*-acetyl-D-glucofuranose was accomplished using a two-step reaction involving treatment with boric acid in AcOH followed by an acetic anhydride/sulfuric acid mixture. The reaction occurs in 93% yield and a 1:1.8 ratio of the α- to β-isomers is obtained. Importantly, crystalline 1,2,3,5,6-penta-*O*-propanoyl-β-D-glucose is readily available in 58% yield by this method.[2] Sucrose octaacetate was synthesized in 91% from sucrose in acetic anhydride using a SO_4^{2-}/TiO_2 solid catalyst.[3] The kinetics of the acetolysis and accompanying anomerization of methyl tetra-*O*-acetyl-α- and β-D-mannopyranosides with H_2SO_4 in $Ac_2O/HOAc$ mixtures have also been reported.[4] Finally, an improved method for the esterification of sugar acetals with dicyclohexylcarbodiimide and dimethylaminopyridine has been reported.[5]

New routes to glycosyl carbamates and esters have been described. Reaction of **1** with the amino-sugar **2** gives the carbamate **3** in 92% yield (Scheme 1).[6] Four new tetraacetyl glucosyl esters of 5-aryl-2-furoic acids **4** (R = 2-NO$_2$, 3-NO$_2$, 4-NO$_2$, H) were also prepared by glycosylation of the acids under phase transfer catalysis [$CH_3(CH_2)_{15}Me_3NBr$].[7]

Numerous regioselective acylations have also been reported. Acetylation of a spacered 4,6-*O*-benzylidene-β-D-galactopyranoside with acetylimidazole gives the 3-*O*-acetyl derivative in 97% yield. Alternatively, a one-pot procedure involving chloroacetylation, acetylation and then dechloroacetylation gave the 2-*O*-acetyl isomer in 85% yield.[8] Regioselectivity for the introduction of 2-*O*-alkoxycarbonyl groups in methyl 4,6-*O*-benzylidene-β-D-glucopyranoside using

Carbohydrate Chemistry, Volume 34
© The Royal Society of Chemistry, 2003

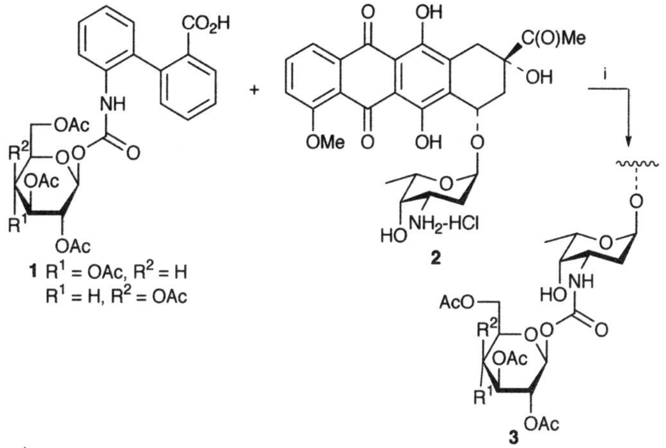

Reagents: i, Pr$^i_2$NEt, Benzotriazol-1-yloxy-tris(dimethylamino)phosphonium hexafluorophosphate

Scheme 1

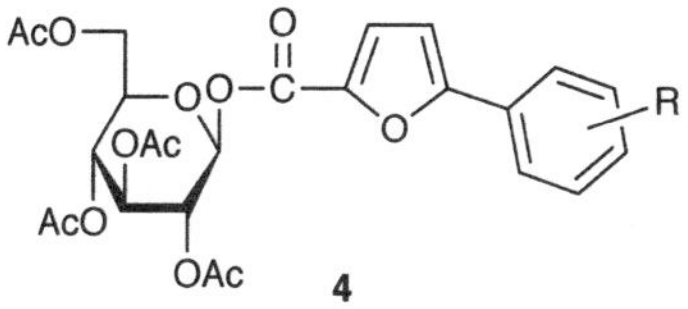

tetramethylethylenediamine (TMEDA) was also obtained. Yields of benzyloxy- and allyloxy-carbonyl esters were 65 and 67% respectively.[9] Tin-mediated benzoylation of methyl 4,6-O-benzylidene-α-D-glucopyranoside gives the 2-O-benzoate. The intermediates in the process do not equilibrate under the esterification condititions; therefore the 2-O-benzoate is formed under kinetic and not thermodynamic control.[10] Tin-mediated regioselective acylations on the solid phase have been accomplished using methyl α-D-glucopyranoside, methyl α-D-mannopyranoside and methyl β-D-galactopyranoside bound *via* O-6 to a copolystyrene-DVB resin through a trityl ether linker. Treatment with Bu$_2$SnO then benzoyl chloride gave 2-O-benzoyl derivatives for the glucoside and 3-O-benzoyl derivatives for the galactoside and mannoside with high regioselectivity.[11] Pivaloylation of the same three pyranosides, in solution and by the dibutyltin oxide method, furnished mixtures of three products in each case. However, direct pivaloylation of the mannopyranoside provided the 3,6-di-O-pivaloyl derivative as the sole product.[12] 3-Acryloyl-D-glucose has been synthesized by acylation of the corresponding diisopropylidene derivative followed by hydrolysis with Cationite.[13] Syntheses of the caffeoylglucosides **5** (both α- and β-OMe) have been accomplished and their HIV inhibition properties reported.[14]

Preparation of the Fmoc protected **6** was accomplished and used in oligosaccharide synthesis. After formation of the trichloroacetimidate donor and glycosylation, the Fmoc group was quantitatively cleaved using mild basic conditions.[15] The synthesis of conandroside **7** was accomplished using conventional methods, with the key intermediate being 2-(3,4-di-O-benzyloxyphenyl)ethyl 2,6-di-O-acetyl-4-O-(E,Z)-3,4-O-benzylcaffeoyl)-β-D-glucopyranoside.[16] 6-*O*-Amino acid esters of D-glucose and methyl α-D-glucopyranoside were prepared using pentachlorophenyl esters of *N*-Boc-L-phenylalanine and *N,O*-di-Boc-tyrosine. The products were subsequently subjected to degradation in Py–HOAc, which was accelerated by the presence of the NH$_2$ groups.[17] Dimeric carbohydrate-based surfactants were synthesized and were found to perform better than the corresponding monomers (Scheme 2).[18]

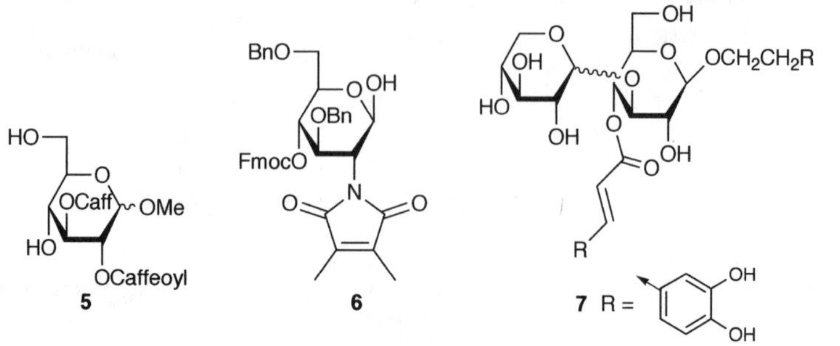

5   **6**   **7** R =

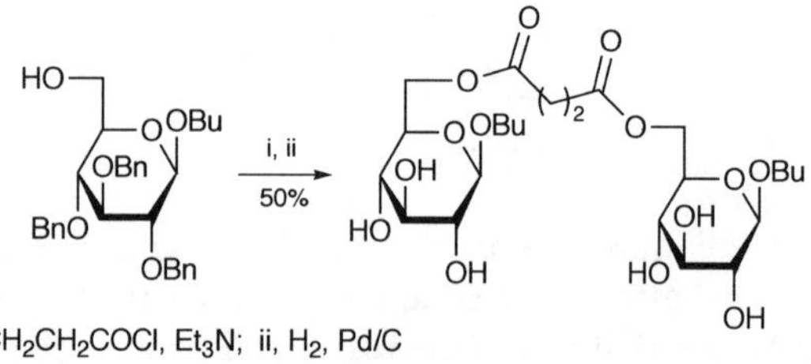

Reagents: i, Cl(O)CCH$_2$CH$_2$COCl, Et$_3$N; ii, H$_2$, Pd/C

Scheme 2

Acylations of a variety of other types of carbohydrates, including inositols, sucrose and neuraminic acid derivatives, have been reported. For example, two protected gallotannin precursors, with the phenolic unit specifically at 1'- or 6'-positions of the fructose unit of sucrose have been prepared. Galloylation was carried out with (MeO)$_3$galloyl chloride and TMEDA in dichloromethane.[19] Sucrose 2,3,4,6,3',4',6'-heptabenzoate was prepared by selective desilylation/benzoylation of 2,3,4,3',4'-pentabenzoyl-6,1',6'-tri(*tert*-butyldiphenylsilyl)sucrose at the 6,6'-O positions with aqueous HF, then desilylation at 1' with TBAF.[20] Synthesis of *N*-alkylaminoacyl sucrose derivative **8** was accomplished by treatment of sucrose with maleic anhydride and NaOAc in DMSO followed by the fatty amine.[21] Analogues of 4-guanidino-Neu5Ac2en (Zanamivir, a potent inhibitor of influenza virus) were prepared, and two were found to be as active as the parent compound. These were the 7-*O*-(6-aminohexyl)carbamoyl and 7-*O*-(heptyl)carbamoyl esters.[22] The precursor of an inositolphosphoglycan was prepared and used a boron–tin exchange reaction to achieve the key selective acylation (Scheme 3). This overcomes the insolubility problem of *myo*-inositol in most organic solvents.[23] Synthesis of 3-*O*-benzyl-β-L-arabinofuranose 1,2,5-orthopivaloate from l-arabinose was accomplished following the protocol used to make the galactose analogue (Vol. 32, p. 100, ref. 54).[24] Positional isomers of partially methylated and acetylated or benzoylated 1,5-anhydroribitol were obtained in pure form from 1,5-anhydroribitol by sequential methylation and benzoylation. Debenzoylation and acetylation yielded the desired acetates.[25]

Deacylations of acylated sugars that have been described include the use of diazabicyclo[2.2.2]octane (DABCO) as a highly selective dechloroacetylation reagent. This gives complete and selective cleavage of the chloroacetyl group in the presence of other ester moieties such as benzoyl and acetyl groups at primary or secondary positions.[26] The full deprotection of the disaccharide **9**, including removal of the Fmoc carbamate, occurred with the use of hydrazine in methanol.[27]

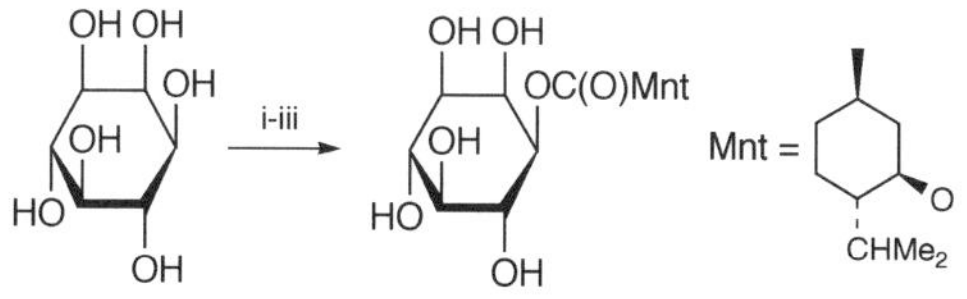

Reagents: i, Et₃B–BuCO₂BEt₂, hexane; ii, Bu₂Sn(acac)₂; iii, MntCOCl, NMI

Scheme 3

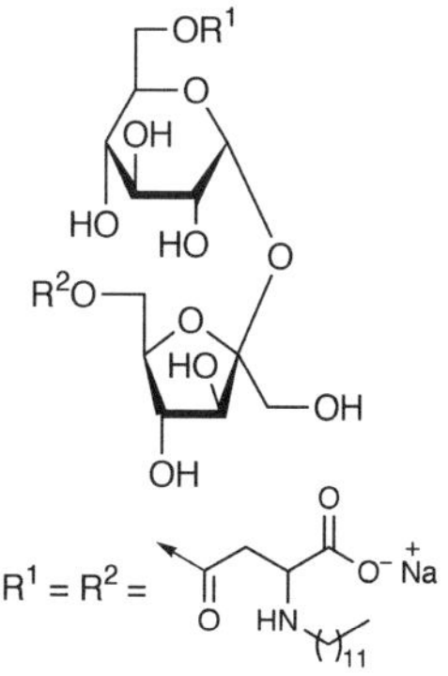

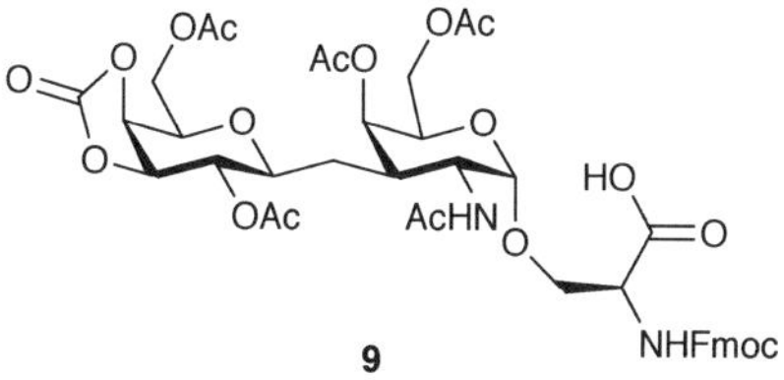

1.1.2 Enzymic Acylation and Deacylation. By use of a protease (at 100 g l⁻¹) from *Bacillus licheniformis*, enzymic acryloylation of sucrose (1 M) with vinyl acrylate (4 M) was carried out in anhydrous pyridine to give sucrose acrylate esters with greater than 90% removal of the substrate. After five days, the ester products consisted of 70% sucrose 1'-acrylate and 30% of the 6',1'-diacrylate and 6,1'-diacrylate.[28] Fatty acid esters of ascorbic acid were synthesized in a solid-phase system catalysed by immobilized lipase B from *Candida antarctica*. The highest reaction rates and yields were obtained using fatty acid vinyl esters and produced 6-*O*-palmitoyl-L-ascorbic acid in 91% yield.[29] The conversion of 3-*O*-benzylmorphine to 3-*O*-benzylmorphine-6-glucuronide was achieved using enzymatic and biochemical synthesis.[30] The partial *N*-acetylation of chitosan tetramer has been achieved using a reverse hydrolysis reaction of a chitin deacetylase thus giving β-D-GlcNAc-(1→4)-β-D-GlcNAc-(1→4)-β-D-GlcNAc-(1→4)-D-GlcNH₂.[31] 1-*O*-*p*-Hydroxybenzoyl β-galactose (*p*HB-Gal) and *p*-nitrophenyl β-galactoside (*p*NP-Gal) were chemically synthesized in order to compare the hydrolytic activities of β-galactosidases. The enzymes from *Penicillium multicolor*, *Escherichia coli* and *Aspergillus oryzae* all hydrolysed the substrates at nearly equal rates, while enzymes from *Bacillus circulans*, *Saccharomyces fragilis* and bovine liver showed much lower activities. Inhibition analysis, pH–activity profiles and kinetic properties of the reactions suggested that β-galactosidase had only one active site for hydrolysis of both the galactosyl ester and the galactoside.[32]

1.2 Natural Products. – Two compounds, isolated from the stembark of *Juglans mandshurica*, were tested for activity against the reverse transcriptase (RT) of HIV-1, since this has both RT and RNase-H activity in different domains of the enzyme. The glucose esters 1,2,6-tri- and 1,2,3,6-tetra-*O*-galloyl-D-glucopyranose had IC$_{50}$ values of 67 and 40 nM against RT, while the latter had an IC$_{50}$ value of 39 (181)M against Rnase-H.[33] Isolated from the wood of *Nyssa sylvatica*, 7-caffeoxylsedoheptulose was shown to adopt the α,β-furanose form in a DMSO solution.[34] The major anthocyanin pigments of blue *Agapanthus* flowers were found to contain succinoyldiester linked delphinidin diglycosides and triglycosides **10**.[35] Similarly, the two novel copigments **11** from *Allium schoenoprasum* have similar structural characteristics.[36] Phenylpropanoid esters of rhamnose, such as **12**, were isolated from *Scrophularia buergeriana* and found to have neuroprotective effects (prevent the death of contical neuron cell cultures that are treated with the neurodegenerative agent glutanate).[37] Five new hydrolysable

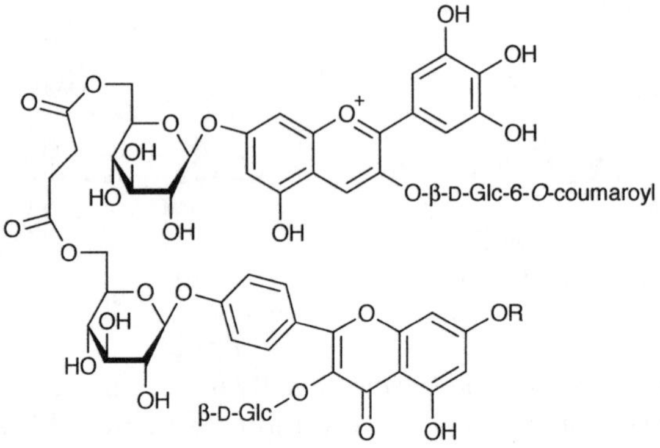

10 R = β-D-Xyl or β-D-Glc

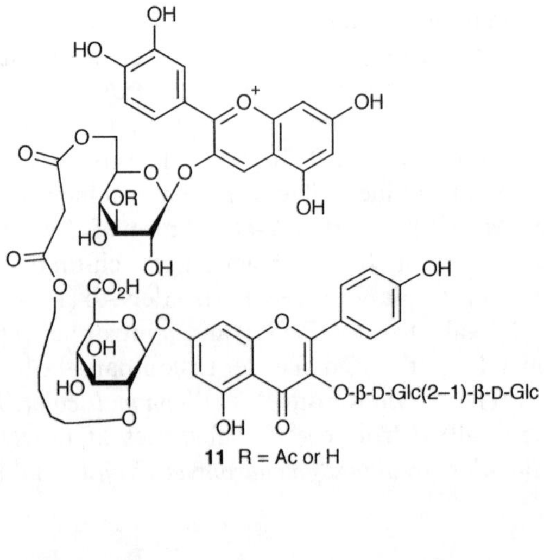

11 R = Ac or H

tannin metabolites have been isolated from *Pelargonium reniforme*. These pelargonins have the 2,4-diester moieties as in **13**.[38] *E*- and *Z*-Venusol A **14** were isolated from the leaves of *Umbilicus pendulinus*; however, only the *E*-isomer is present naturally, and the *Z*-isomer forms on exposure to light.[39] Metabolites from five different strains of *Streptomyces* were detected and found to have the novel acyl rhamnopyranose structures **15**.[40]

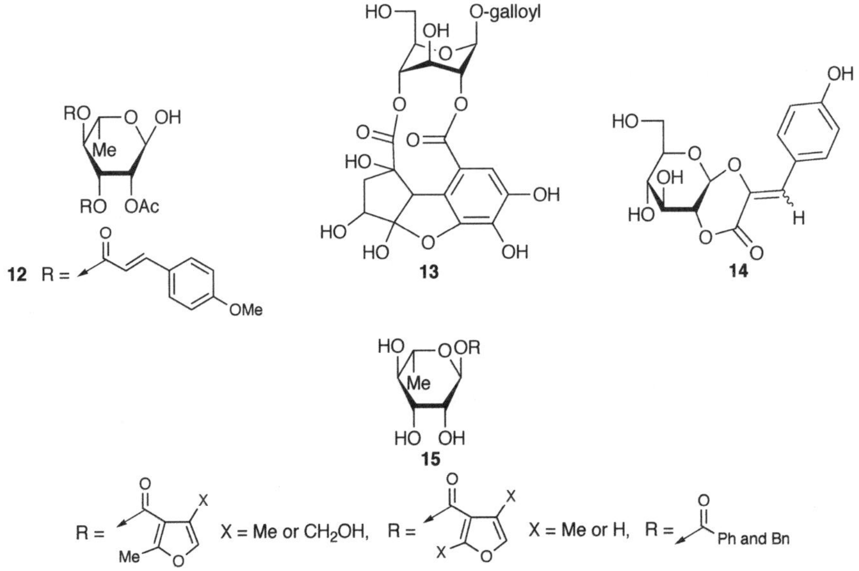

2 Phosphates and Related Esters

Hanessian's 'remote activation concept', the stereocontrolled glycosylation reaction using methoxypyridyl (MOP) *O*-unprotected glycosyl donors to give glycosyl 1-phosphates, has been reviewed.[41] The 1-triphosphates of D-glucuronic acid, D-glucose-6-phosphate and D-xylose were obtained in 43, 32 and 42% yields, respectively, by incubation with sodium cyclo-triphosphate at pH 12. However, replacement of the hydroxyl moiety at C-2, as in 2-amino-2-deoxy-β-D-glucopyranose and 2-deoxy-D-glucose, results in low yields of only 14 and 13%.[42] The reaction of D-glucose with cyclotriphosphate has been reported to give β-D-glucopyranosyl 1-triphosphate. This has been extended to D-ribose, D-lyxose, D-mannose and D-arabinose, which gave poor yields of the corresponding pyranose 1-triphosphates. However, D-galactose, D-xylose and D-allose gave yields of around 30% for the β-D-pyranose 1-triphosphates.[43] Four monodeoxy α-D-mannose-1-phosphate derivatives were prepared and evaluated as substrates for a recombinant GDP-mannose pyrophosphorylase. The derivatives contained deoxygenation at C-2, C-3, C-4 or C-6 and all could be used successfully.[44] Tetrabutylammonium benzyl dihydrophytylphosphate was coupled to *S*-phenyl 2,3-di-*O*-benzyl-4,6-*O*-benzylidene-1-thio-α-D-mannopyranoside *S*-oxide with triflic anhydride as a catalyst to give the corresponding β-mannosyl phosphate in 56% yield. Treatment with sodium in liquid ammonia afforded the unprotected mannosyl phosphoisoprenoid **16**.[45] The syntheses of partial structures of the glycosyl phosphatidylinositol anchor were accomplished using donor **17** and the dimethyl phosphonothioate method.[46] Reactions of 2,3,4,6-tetra-*O*-acetyl-α-D-glucopyranosyl bromide with salts of citronellyl, dolichyl and citronellyl benzyl phosphates gave the corresponding β-D-glucopyranosyl benzylphosphates. The reaction yields were found to be dependent on the structures of the phosphate reagents and the reaction conditions.[47] Glycosyl boranophosphate diesters, such as **19**, have been reported as stable analogues of glycosyl phosphates. These have been prepared in a one-pot procedure from the *H*-phosphonate **18** (Scheme 4).[48]

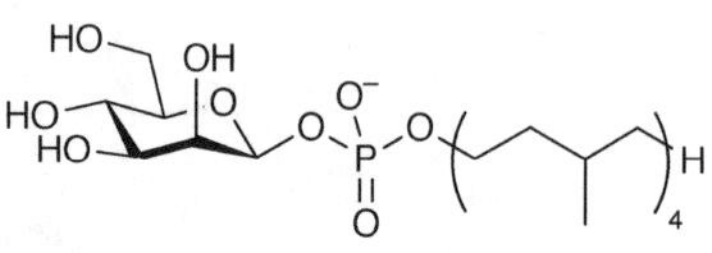

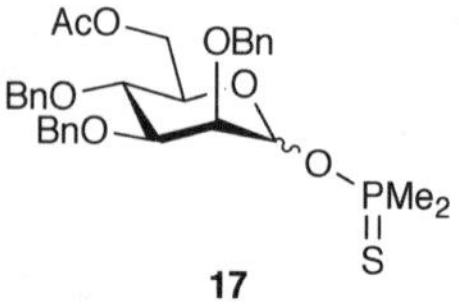

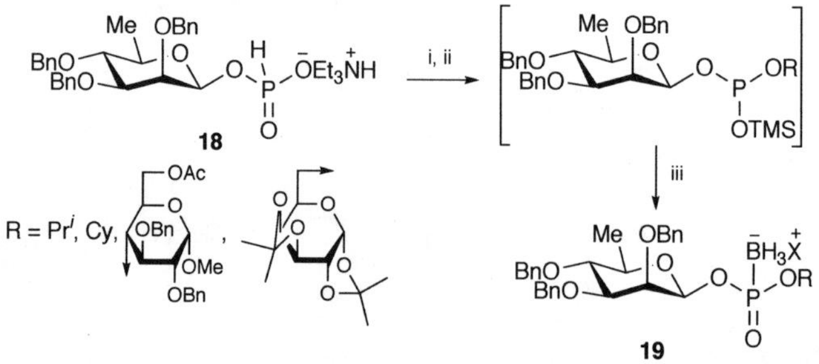

Reagents: i, ROH, PivCl, Py, THF; ii, *N,O*-bis(tms)-trifluoroacetamide; iii, BH₃–amine complex

Scheme 4

Examples of the synthesis of phosphite and phosphate esters of sugars are included in a review on trivalent phosphorous acids (Scheme 5).[49] The synthesis of an oxyphosphorane has also been reported (Scheme 6).[50] A strategy for regioselective phosphorylation of aldoses in aqueous solution involves reversible attachment of activated diamidophosphate (DAP). The carbonyl addition prod-

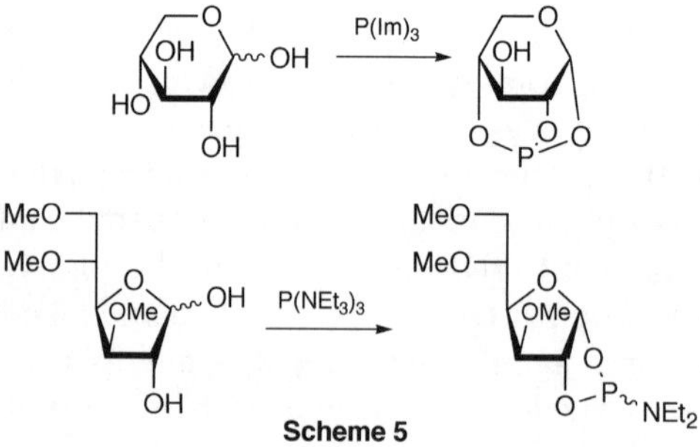

Scheme 5

uct subsequently undergoes irreversible intramolecular delivery of the phosphate group to the proximal site (Scheme 7).[51]

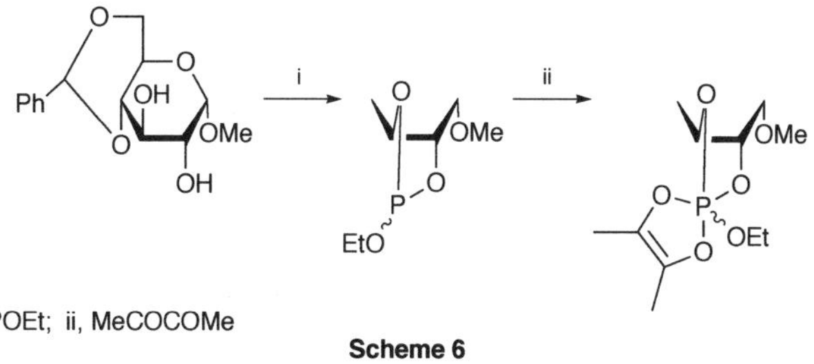

Reagent: i, Cl₂POEt; ii, MeCOCOMe

Scheme 6

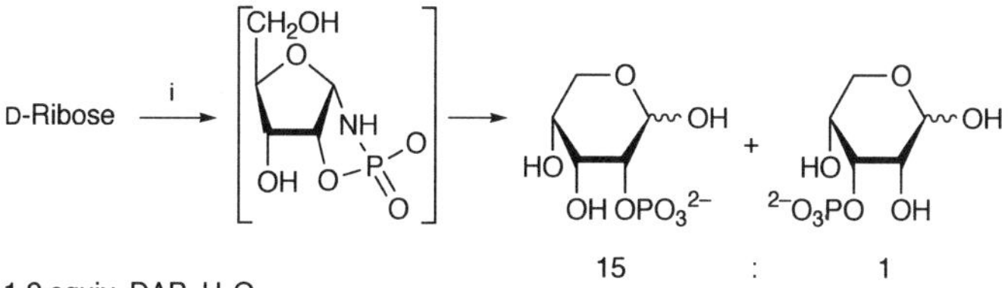

Reagent: i, 1.2 equiv. DAP, H₂O

Scheme 7

The synthesis of lipid A-type analogues **20** and **21** by standard methods has been described. Analogue **20** was found to be an LPS agonist, while **21** is an LPS antagonist.[52] Nucleoglycoconjugates such as **22** have been synthesized using phosphoramidite chemistry then coupled on a solid support to DNA oligonucleotides. Cleavage from the resin and deprotection of the conjugates gave stable duplexes with complementary DNA.[53]

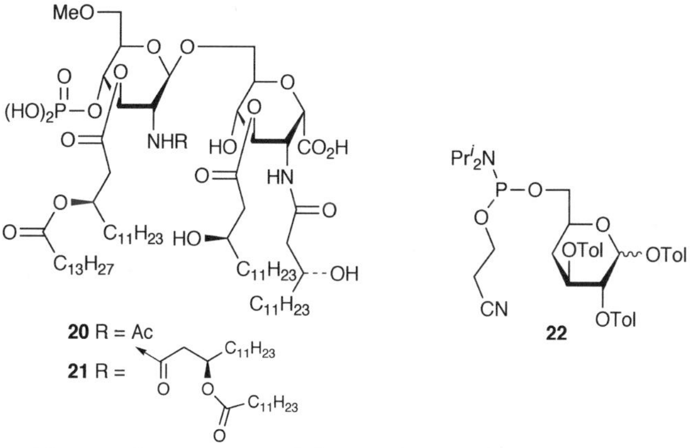

The tritiated benzophenone-containing derivative of glucose 6-phosphate **23** was made as a photoactivatible affinity label for proteins.[54] Phosphonate **24** was prepared to act as a hapten to elicit monoclonal antibodies for transesterification reactions between **25** (R = H) and **26**; however, the antibodies failed to catalyse the reaction.[55] D-Galactose 6-*O*-phosphate and *O*-thionophosphate monoesters **27** have been prepared and then deacetalated by heating in water without added acid.[56]

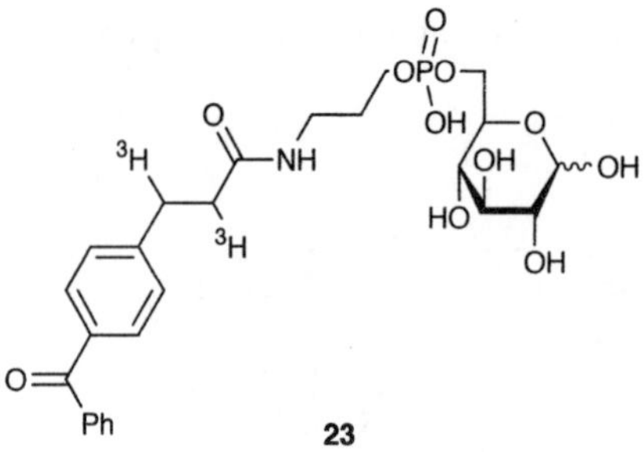

23

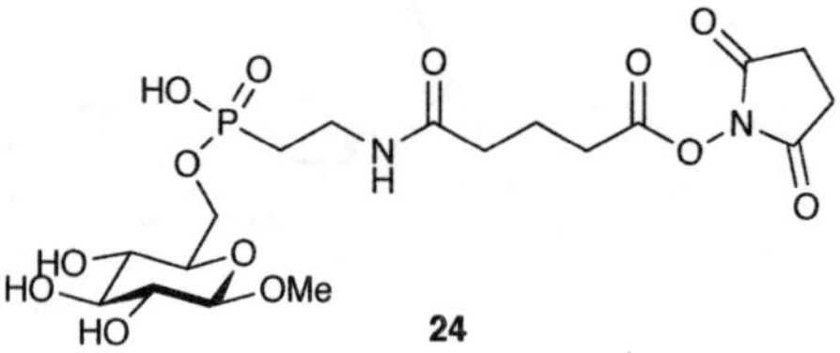

24

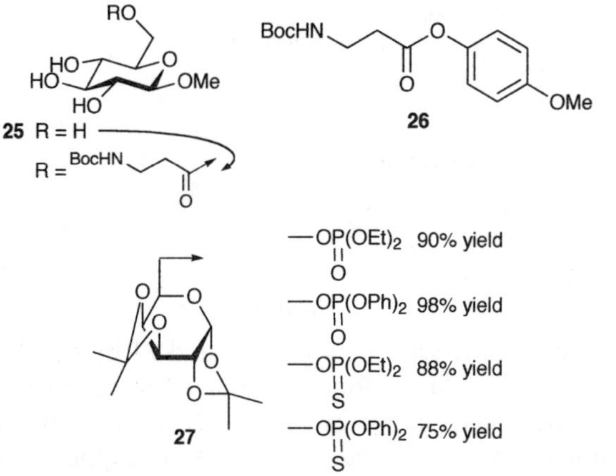

25 R = H

R = BocHN

26

27

—OP(OEt)$_2$ 90% yield
       ‖
       O

—OP(OPh)$_2$ 98% yield
       ‖
       O

—OP(OEt)$_2$ 88% yield
       ‖
       S

—OP(OPh)$_2$ 75% yield
       ‖
       S

3   Sulfates and Related Esters

Cyclic sulfites and sulfates in organic synthesis have been reviewed in a *Tetrahedron* report. The reactions of several carbohydrate-based sulfite and sulfate derivatives with nitrogen, carbon and sulfur nucleophiles are reported.[57] In the area of natural products, five sulfated saikosaponins containing the trisaccharide unit **28** have been isolated as minor constituents from *Bupleurum rigidum*.[58] The leaf-movement hormones, sulfated phenolic glycosides **29**, have been shown to be plant specific, rather than being common to all plants.[59]

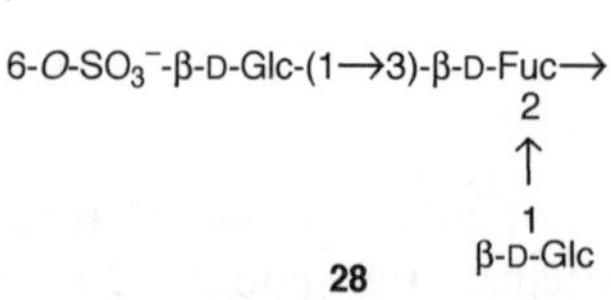

6-O-SO$_3^-$-β-D-Glc-(1→3)-β-D-Fuc→
                                       2
                                       ↑
                                       1
                                       β-D-Glc

28

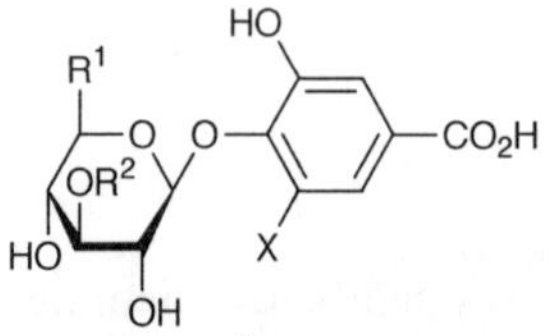

29 R^1 = CH$_2$OSO$_3$H or CO$_2$H, R^2 = H or SO$_3$H, X = H or OH

The sodium salts (**30**) of heptakis (2-O-methyl-3,6-di-O-sulfo)cyclomaltoheptaose have been made for use as possible chiral resolving agents for capillary electrophoresis.[60] The 3- and 6-monosulfated and the 3,6-disulfato-β-D-Gal-(1→4)-[α-L-Fuc-(1→3)]-β-D-GlcNAc-(1→3)-β-D-Gal-(1→4)-D-Glc have been constructed as candidate ligands for binding to human L-Selectin.[61] Also in the area of SiaLe^x chemistry, analogue **31**, in which the sulfo group was introduced selectively at the 4 position of the galactose moiety *via* a dibutylstannane intermediate, has been prepared as indicated.[62] Sulfated sialyl-α-(2→3)-neolactotetraose derivatives have been prepared with sulfation at either O-6 of the GlcNAc residue, at O-6 of the terminal Gal residue or at both positions. Their acceptor specificity for an α-(1→3)-fucosyl transferase in the biosynthesis of L-Selectin ligand was tested and only the GlcNAc-sulfated compound was recognized.[63] The sulfated GalNAc peptide analogue **32** was tested as an amphiphile and an anti-HIV agent; little activity was found.[64]

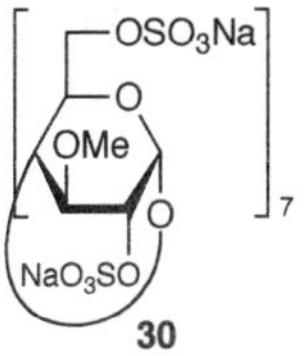

30

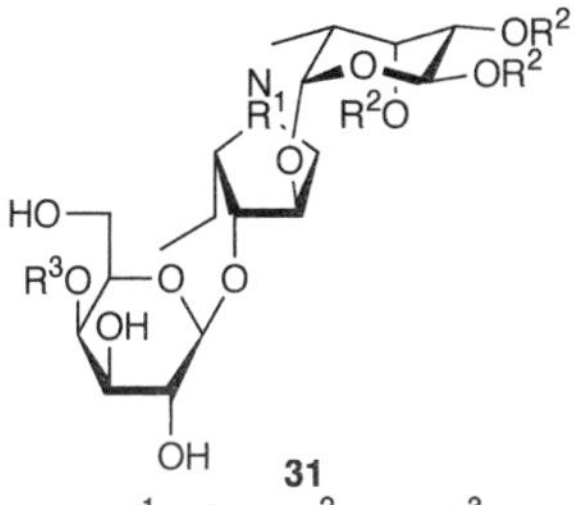

31

R^1 = Cbz, R^2 = Bn, R^3 = H
R^1 = R^2 = H, R^3 = SO_3Na

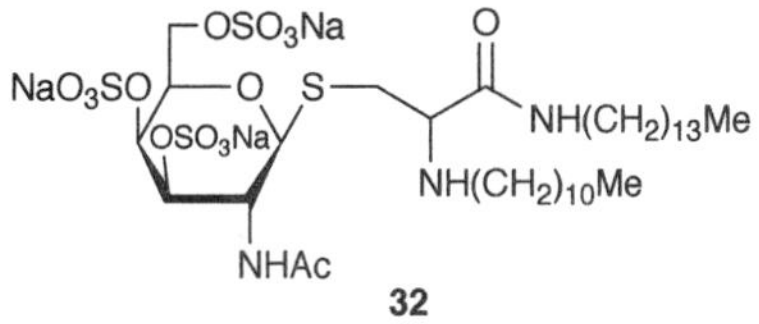

32

Several examples of ring-opening reactions of cyclic sulfates have been reported (Scheme 8). The opening of β-D-galactoside ester **33** with a benzoate nucleophile resulted in the formation of **34** in 97% yield,[65] and **35** was opened with the nitrogen bases Et_3N (shown), Py, DBU, $NaNH_2$ or $LiNPr^i_2$ to produce the 5-sulfates, *e.g.* **36**, in 80–90% yields.[66] Formation of the 3,4-epoxide by treatment of cyclic sulfate **37** with NaOH and MeOH in 86% yield has also been reported.[67] Heterocyclization of alditols *via* bis cyclic sulfates using Na_2S has been also accomplished (Scheme 9). Several examples of the formation of compounds with various ring sizes in 55–80% yields were reported.[68]

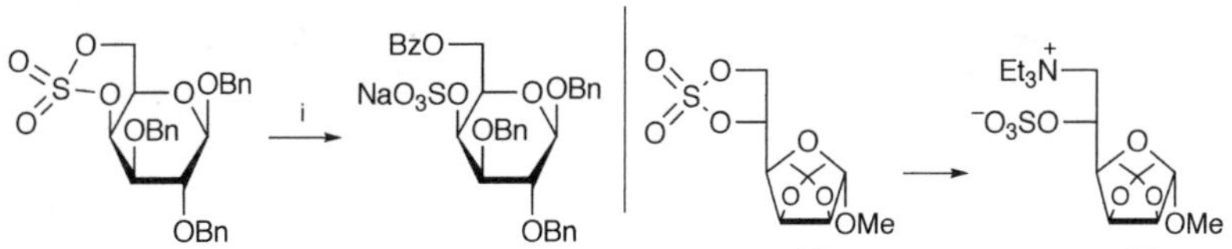

33 **34** **35** **36**

Reagents: i, Bu_4NOBz, DMF

Scheme 8

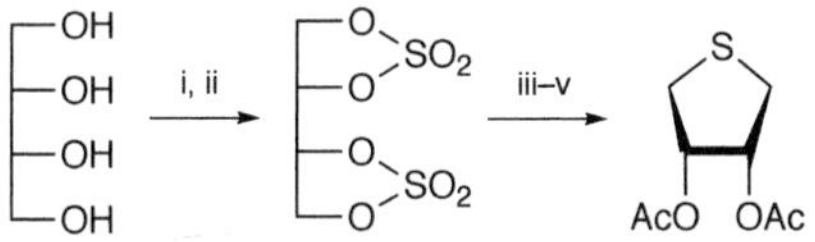

Reagents: i, $SOIm_2$; ii, $NaIO_4$, $RuCl_3$; iii, Na_2S; iv, H_3O^+; v, Ac_2O, Py

Scheme 9

4   Sulfonate Esters

Selective Bu$_2$SnO-catalysed monosulfonylation of methyl α-D-xylopyranoside gave 70% of the substitution at O-2, while similar sulfonylation of methyl β-D-xylopyranoside gave 90% of the esters at O-4.[69] TSAO-T **38** and TSAOm^3-T **39**, which have internal sulfonate ester moieties, are potent, selective HIV-I reverse transcriptase inhibitors. New analogues were made by the conversion of the alcohol (**40**, R^2 = OH) to esters, carbamates and thiocarbamates (R^2 = OCOMe, OCONMe$_2$, OCSNMe$_2$) and ethers (R^2 = OMe, OBn, OAll, OBu^i). Amines where R^2 = NMe, NEt$_2$, N(CH$_2$)$_5$ were also made. The products have moderate activity when R^2 is an ether or cyclic amine.[70] Chondroitin pentasaccharide **41** was di-*O*-sulfonated at O-6 or alternatively could be benzoylated at O-6 then sulfonated at O-4 of the amino sugar moiety.[71] A one-pot selective tosylation of adenosine using catalytic amounts of organotin mediators and triethyl- or tributyl-amines has been reported. A 90% yield of the 2-*O*-tosylate was achieved, and the methodology was subsequently applied to methyl D-glucopyranoside, D-galactopyranoside, and D-mannopyranoside to give 3,6-diesters in 80–90% yields.[72]

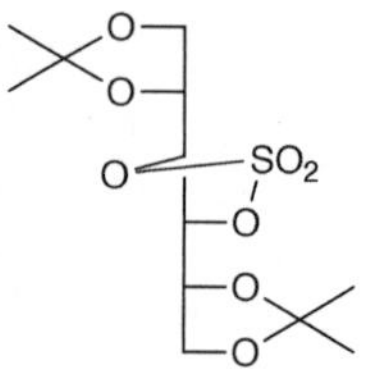

37

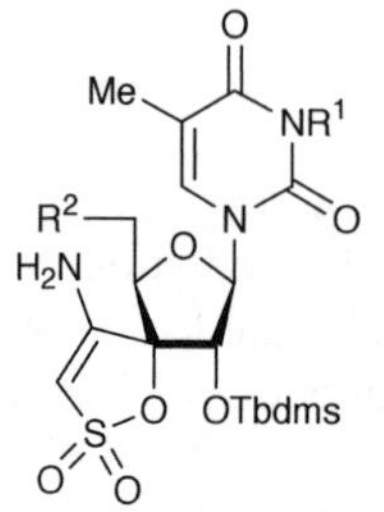

38 R^1 = H, R^2 = TbdmsO
39 R^1 = Me, R^2 = TbdmsO
40 R^1 = Me, R^2 = OH

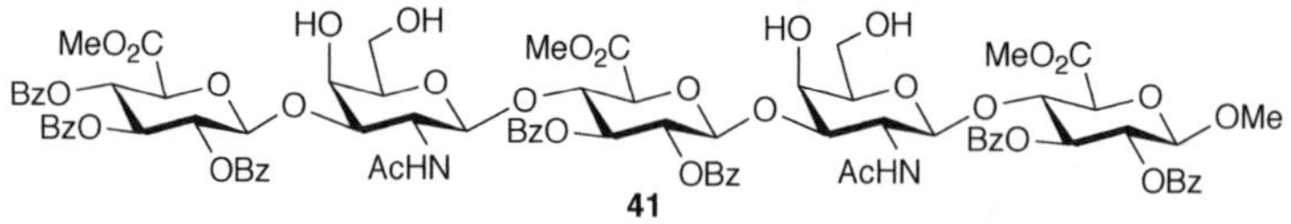

41

Regiospecific bifunctionalization onto the A, B C-2 hydroxyl moieties of α- and β-cyclodextrins has been achieved using benzophenone 3,3-disulfonylimidazole and molecular sieves (Scheme 10). Diepoxidation and amination products were subsequently formed from the sulfonate **42**.[73]

5   Other Esters

Polymer-bound protected sugar derivatives such as **43**, which are soluble in organic solvents, have been synthesized using xanthate transfer technology.[74] The synthesis of thioacetal **44** was achieved under photochemical conditions,

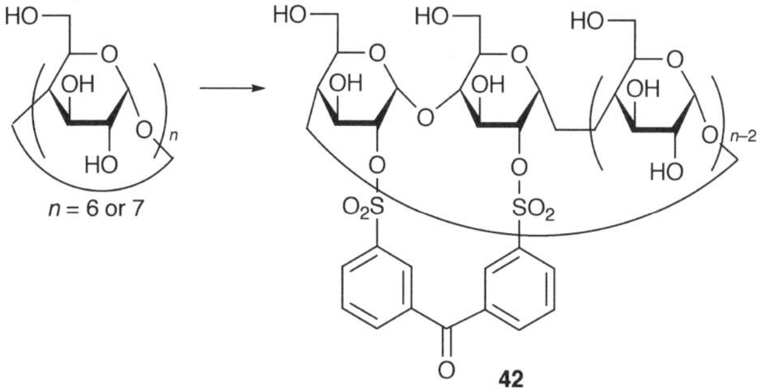

Scheme 10

which led to incorporation of a methylene moiety from the solvent dichloromethane (Scheme 11).[75] Ferrocene-containing substituted sugar esters have been made and tested for bioactivity (Scheme 12). Coupling of **45** and **46** under three different sets of conditions, followed by deprotection, gave the esters **47**, in which $R^1 = H$, $R^2 =$ ferrocene showed antimalarial activity.[76] A molecular imprinted surface was prepared resulting in the formation of **49**. Thus, diamide polymer **48**, with two boronic acid groups, was condensed with 4-nitrophenyl α-D-mannopyranoside to give **49**.[77] Ortho esters of the type **50** have been prepared from aldono lactones (Scheme 13, see also Chapter 3). Other analogues reported include those derived from D-galactono, D-mannono, 6-deoxy-D-galactono, 6-deoxy-L-galactono and 6-deoxy-L-mannono-lactones. These all give the *R*-configuration at the ortho ester site, the first three with O-6 α and the last two with O-4 α.[78]

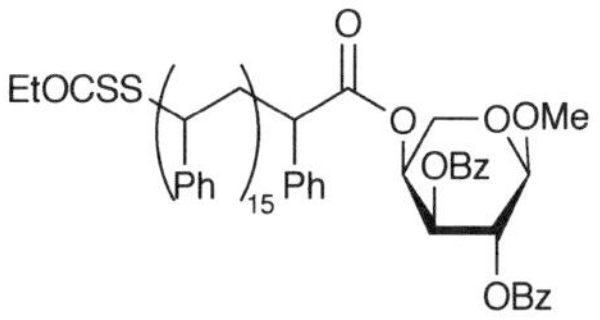

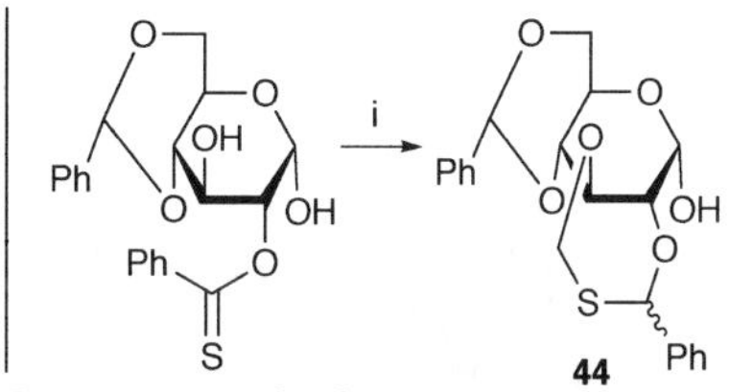

Reagents: i, *hv*, CH$_2$Cl$_2$, Et$_3$N

Scheme 11

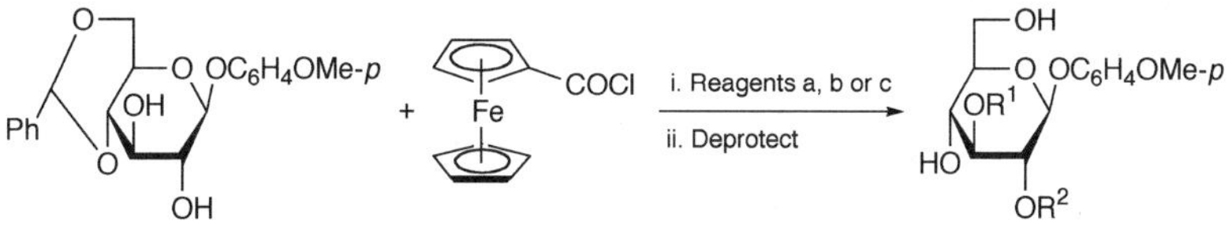

Reagents: a. DMAP (2.2 equiv.), CH$_2$Cl$_2$, 14 h
b. NaH, PhMe, 30 min
c. DMAP (0.2 equiv.), Py, CH$_2$Cl$_2$, 12 h

$R^1 = R^2 =$ Ferrocenecarbonyl
$R^1 = H$, $R^2 =$ Ferrocenecarbonyl
$R^1 =$ Ferrocenecarbonyl, $R^2 = H$

Scheme 12

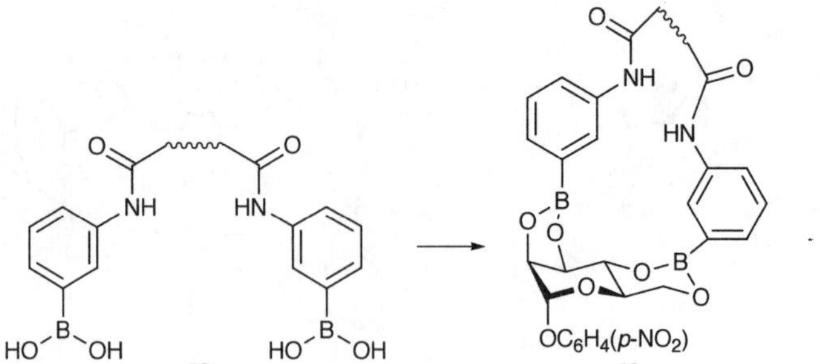

48   49

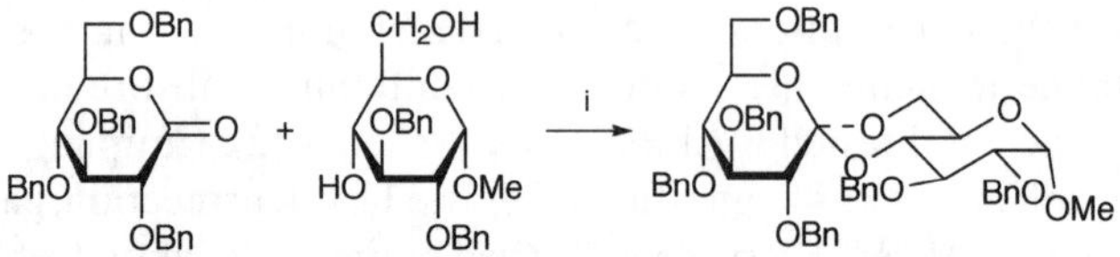

Reagents: i, TMSOMe, TMSOTf

50

Scheme 13

References

1. J.P. Parrish, R.N. Salvatore and K.W. Jung, *Tetrahedron*, 2000, **56**, 8207.
2. R.H. Furneaux, P.M. Rendle and I.M. Sims, *J. Chem. Soc., Perkin Trans.* 1, 2000, 2011.
3. J. Zhao, *Huaxue Shijie*, 1999, **40**, 586 (*Chem. Abstr.*, 2000, **132**, 122 870).
4. J. Kaczmarek, Z. Kaczynski, Z. Trumakaj, J. Szafranek, M. Bogalecka and H. Lonnberg, *Carbohydr. Res.*, 2000, **325**, 16.
5. C.C. Perrone, P.R.R. Costa and V.F. Ferreira, *Org. Prep. Proced. Int.*, 1999, **31**, 689 (*Chem. Abstr.*, 2000, **132**, 265 365).
6. R.G.G. Leenders and H.W.Q. Scheeren, *Tetrahedron Lett.*, 2000, **41**, 9173.
7. M.-G. Liu, K.-F. Tong, G.-F. Luo, Q.-B. Liao and S.-L. Fu, *Youji Huaxue*, 2000, **20**, 198 (*Chem. Abstr.*, 2000, **132**, 322 044).
8. I.V. Mikhura, A.A. Formanovski and N.V. Bovin, *Carbohydr. Lett.*, 1999, **3**, 305 (*Chem. Abstr.*, 2000, **132**, 137 622).
9. M. Adinolfi, G. Barone, L. Guariniello and A. Iadonisi, *Tetrahedron Lett.*, 2000, **41**, 9305.
10. M.W. Bredenkamp and H.S.C. Spies, *Tetrahedron Lett.*, 2000, **41**, 543.
11. F. Peri, L. Cipolla and F. Nicotra, *Tetrahedron Lett.*, 2000, **41**, 8587.
12. A.K.M.S. Kabir, *J. Bangladesh Chem. Soc.*, 1997, **10**, 75 (*Chem. Abstr.*, 2000, **132**, 137 620).
13. A.I. Slivkin and V.L. Lopenko *Izv. Vyssh. Uchebn. Zaved., Khim. Khim. Tekhnol.*, 1999, **42**, 112 (*Chem. Abstr.*, 2000, **132**, 152 032).
14. S.N. Kim, J.Y. Lee, H.J. Kim, C.-G. Shin, H. Park and Y.S. Lee, *Bioorg. Med. Chem. Lett.*, 2000, **10**, 1879.
15. F. Roussel, L. Knerr, M. Grathwohl and R.R. Schmidt, *Org. Lett.*, 2000, **2**, 3043.
16. T. Kawada, R. Asano, K. Makiuo and T. Sakuno, *Eur. J. Org. Chem.*, 2000, 2723.
17. I. Jeric, L. Simicic, M. Stipetic and S. Horuat, *Glycoconjugate J.*, 2000, **17**, 273.
18. M.J.L. Castro, J. Kovenski and A.F. Cirelli, *J. Carbohydr. Chem.*, 2000, **19**, 1175.

19. M.T. Barros, C.D. Maycock, F. Sineriz and C. Thomassigny, *Tetrahedron*, 2000, **56**, 6511.
20. M.T. Barros, C.D. Maycock and C. Thomassigny, *Carbohydr. Res.*, 2000, **328**, 417.
21. G. Kretzschmar, *J. Carbohydr. Chem.*, 2000, **19**, 879.
22. D.M. Andrews, P.C. Cherry, D.C. Humber, P.S. Jones, S.P. Keeling, P.T. Martin, CD Shaw and S. Swanson, *Eur. J. Med. Chem.*, 1999, **34**, 563 (*Chem. Abstr.*, 2000, **132**, 23 155).
23. M. Martin-Lomas, M. Flores-Mosquera and J.L. Chiara, *Eur. J. Org. Chem.*, 2000, 1547.
24. M. Hori and F. Nakatsubo, *J. Carbohydr. Chem.*, 2000, **19**, 399.
25. C.M. El Hilali, T.A. Lamosse and G.R. Gray, *Carbohydr. Res.*, 2000, **329**, 189.
26. D.J. Lefeber, J.P. Kamerling and J.F.G. Vliegenthart, *Org. Lett.*, 2000, **2**, 701.
27. L.J. Williams, C.R. Harris, P.W. Glunz and S.J. Danishefsky, *Tetrahedron Lett.*, 2000, **41**, 9505.
28. H.G. Park and H.N. Chang, *Biotechnol. Lett.*, 2000, **22**, 39 (*Chem. Abstr.*, 2000, **132**, 265 380).
29. Y. Yan, U.T. Bornscheuer and R.D. Schmid, *Biotechnol. Lett.*, 1999, **21**, 1051 (*Chem. Abstr.*, 2000, **132**, 251 346).
30. G.N. Jenkins, A.V. Stachulski, F. Sheinmann and N.J. Turner, *Tetrahedron Asymmetry*, 2000, **11**, 413.
31. K. Tokuyasu, H. Ono, M. Mitsutomi, K. Hayashi and Y. Mori, *Carbohydr. Res.*, 2000, **325**, 211.
32. T. Kiso, H. Nakano, H. Nakajima, T. Terai, K. Okamoto and S. Kitahata, *Biosci. Biotechnol. Biochem.*, 2000, **64**, 1702.
33. B.-S. Min, N. Nakamura, H. Miyashiro, Y.-H. Kim and M. Hattori, *Chem. Pharm. Bull.*, 2000, **48**, 194.
34. X.-C. Li, H.N. El Shohly and A.M. Clark, *Phytochemistry*, 2000, **53**, 1033.
35. S.J. Bloor and R. Falshaw, *Phytochemistry*, 2000, **53**, 575.
36. T. Fossen, R. Slimestad, D.O. Oustedal and O.M. Andersen, *Phytochemistry*, 2000, **54**, 317.
37. S.R. Kim and Y.C. Kim, *Phytochemistry*, 2000, **54**, 503.
38. K.P. Latté and H. Kolodziej, *Phytochemistry*, 2000, **54**, 701.
39. L. Viornery, C. Saliba, J.B. Daskiewicz, C. Bayet, G. Comte, B. Fenet, G. Butierrez and D. Barron, *Chem. Pharm. Bull.*, 2000, **48**, 1768.
40. S. Grond, H.-J. Langer, P. Henne, I. Sattler, R. Thiericke, S. Grabley, H. Zähner and A. Zeeck, *Eur. J. Org. Chem.*, 2000, 929.
41. S. Hanessian and B. Lou, *Chem. Rev.*, 2000, **100**, 4443.
42. H. Inoue, M. Watanabe, H. Nakayuama and M. Tsuhako, *Chem. Pharm. Bull.*, 2000, **48**, 802.
43. H. Inoue, H. Nakayama and M. Tsuhako, *Carbohydr. Res.*, 2000, **324**, 10.
44. G.M. Watt, S.L. Flitsch, S. Fey, L. Ellinq and U. Kragl, *Tetrahedron: Asymmetry*, 2000, **11**, 621.
45. D. Crich and V. Dudkin, *Org. Lett.*, 2000, **2**, 3941.
46. T. Yananoi, J. Inagaki, M. Mizuno, K. Yamashita and T. Inazu, *Heterocycles*, 2000, **52**, 921 (*Chem. Abstr.*, 2000, **132**, 279 417).
47. S.D. Maltsev, L.L. Danilov and V.N. Shibaev, *Bioorg. Khim.*, 1999, **25**, 299 (*Chem. Abstr.*, 2000, **132**, 78 769).
48. D. Prosperi, L. Panza, L. Poletti and L. Lay, *Tetrahedron*, 2000, **56**, 4811.
49. E.E. Nifantiev, M.K. Grachev and S. Yu. Burmistrov, *Chem. Rev.*, 2000, **100**, 3755.
50. J.-J. Hu, Y. Ju, Y.-F. Zhao and Y.-X. Cul, *Synth. Commun.*, 2000, **30**, 4087.

51.   R. Krishnamurthy, S. Euntha and A. Eschenmoser, *Angew. Chem. Int. Ed. Engl.*, 2000, **39**, 2281.
52.   T. Mochizuki, Y. Iwano, M. Shiozaki, S. Kurakata, S. Nanai and M. Nishijima, *Carbohydr. Res.*, 2000, **324**, 225.
53.   T.L. Sheppard, C.-H. Wong and G.F. Joyce, *Angew. Chem. Int. Ed. Engl.*, 2000, **39**, 3660.
54.   J. Peng, P.Y. Chen, R.B. Marchase and G.D. Preswich, *Bioorg. Med. Chem. Lett.*, 2000, **10**, 535.
55.   T. Lintunen and J.T. Yli-Kauhaluoma, *Bioorg. Med. Chem. Lett.*, 2000, **10**, 1749.
56.   D. Postel, G. Ronco and P. Villa, *J. Carbohydr. Chem.*, 2000, **19**, 171.
57.   H.-S. Byun, L. He and R. Bittman, *Tetrahedron*, 2000, **56**, 7051.
58.   S. Sánchez-Contreras, A.M. Diaz-Lanza, C. Bartolomé and M. Bernabé, *Phytochemistry*, 2000, **54**, 783.
59.   M. Ueda, H. Shigemori, N. Sata and S. Yamamura, *Phytochemistry*, 2000, **53**, 39.
60.   D. Kirby Maynard and G. Vigh, *Carbohydr. Res.*, 2000, **328**, 277.
61.   A. Lubineau, J. Alais and R. Lemoine, *J. Carbohydr. Chem.*, 2000, **19**, 151.
62.   E. Dechaux, P. Savy, S. Bouyain, C. Monneret and J.-C. Florent, *J. Carbohydr. Chem.*, 2000, **19**, 485.
63.   K. Fukunaga, K. Shinoda, H. Ishida and M. Kiso, *Carbohydr. Res.*, 2000, **328**, 85.
64.   B. Faroux-Corlay, L. Clary, C. Gadras, D. Hammache, J. Greiner, C. Santaella, A.-M. Aubertin, P. Vierling and J. Fatini, *Carbohydr. Res.*, 2000, **327**, 223.
65.   F. Dagron and A. Lubineau, *J. Carbohydr. Chem.*, 2000, **19**, 311.
66.   T. Gourlain, A. Wadouachi and D. Beaupère, *Tetrahedron Lett.*, 2000, **41**, 659.
67.   D.O. Jang, Y.H. Joo and D.H. Cho, *Synth. Commun.*, 2000, **30**, 448.
68.   V. Glaçon, M. Benazza, D. Beauperè and G. Demailly, *Tetrahedron Lett.*, 2000. **41**, 5053.
69.   M.J. Martinelli, R. Vaidyanathan and V.V. Khau, *Tetrahedron Lett.*, 2000, **41**, 3773.
70.   A. San-Félix, C. Chamorro, M.J. Pérez-Pérez, S. Velázquez, E. De Clercq, J. Balzarini and M.J. Camarasa, *J. Carbohydr. Chem.*, 2000, **19**, 635.
71.   F. Bélot and J.-C. Jacquinet, *Carbohydr. Res.*, 2000, **326**, 88.
72.   M. Kawana, M. Tsujimoto and S. Takahashi, *J. Carbohydr. Chem.*, 2000, **19**, 67.
73.   K. Teranismi, *Chem. Commun.*, 2000, 1255.
74.   B. Quiclet-Sire, A. Wilczewska and S.Z. Zard, *Tetrahedron Lett.*, 2000, **41**, 5673.
75.   A.L. Schreiber, M.A. Fashing and C.J. Abelt, *J. Chem. Soc., Perkin Trans. 2*, 2000, 953.
76.   T. Itoh, S. Shirakami, N. Ismida, Y. Yamashita, T. Yosmida, H.-S. Kim and Y. Wataya, *Bioorg. Med. Chem. Lett.*, 2000, **10**, 1657.
77.   T. Miyahara and K. Karihara, *Chem. Lett.*, 2000, 1356.
78.   H. Ohtake, N. Ichiba, M. Shiro and S. Ikegami, *J. Org. Chem.*, 2000, **65**, 8164.

8
Halogeno-sugars

1 Fluoro-sugars

A special issue of *Carbohydrate Research* devoted to fluoro-sugars has featured reviews on methods for the synthesis of glycosyl fluorides,[1] uses of glycosyl fluorides in glycosidation reactions[2] and in enzymatic reactions that catalyse glycosyl transfer,[3] advances in the synthesis of deoxyfluoro-sugars (focusing on the use of DAST, XeF$_2$ and electrophilic fluorinating reagents),[4] the synthesis and biological activity of fluorinated nucleosides,[5] the preparation of ^{18}F-labelled sugars,[6] and fluorinated branched sugars,[7] the use of fluorinated carbohydrates in liquid crystal compounds[8] and the NMR spectra of fluorinated carbohydrate derivatives.[9] Another review has described different methodologies for the preparation of selectively fluorinated carbohydrates.[10]

Some rearrangement reactions have been encountered during the treatment of 3-deoxy-3-*C*-methyl-3-nitro-hexopyranosides with DAST as exemplified in Scheme 1.[11] 4-Deoxy-4-fluoro analogues of GlcNAc and GalNAc have been prepared from appropriately protected GlcNAc derivatives using standard methods.[12] A range of deoxyfluoro-D-fructose derivatives has been prepared by an enzyme-catalysed isomerization of the corresponding D-*gluco*-substrates,[13] and a number of fluorinated derivatives of the monosaccharide determinants of *Vibrio cholerae* have been synthesized.[14]

Five to six degradation products of 6-deoxy-6-fluoro-L-ascorbic acid have been observed by ^{19}F NMR in a study of the feasibility of using this fluorinated derivative to identify the degradation products of L-ascorbic acid.[15] Dihydroxylation of the α-fluoroimide **1** has afforded a mixture of the corresponding D-*xylo*- and L-*lyxo*- derivatives from which 2-deoxy-2-fluoro-D-xylose and -L-lyxose were prepared.[16] Oligosaccharides have been assembled that are composed of up to six 2-deoxy-2-fluoro-sugar units.[17] A report on nucleophilic trifluoromethylation reactions employing TmsCF$_3$ additions to carbonyl compounds contains examples with carbohydrate ulose derivatives.[18]

2 Chloro-, Bromo- and Iodo-sugars.

– The preparation of glycosyl halides and their use in the solid-phase synthesis of oligosaccharides has been reviewed.[19] A new method for the preparation of glycosyl chlorides involves the treatment of otherwise protected free sugars with triphosgene in the presence of pyridine.[20]

Carbohydrate Chemistry, Volume 34

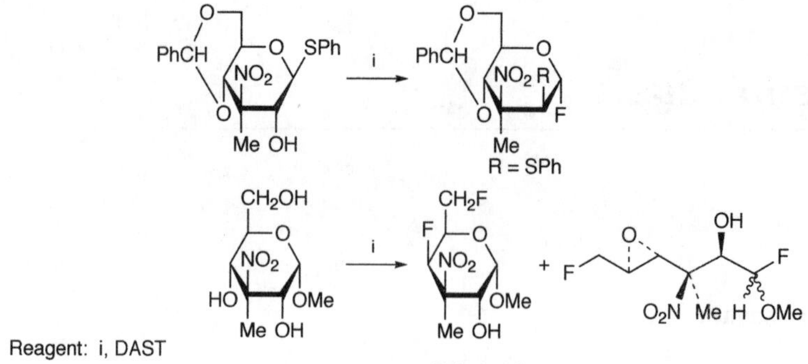

Reagent: i, DAST

Scheme 1

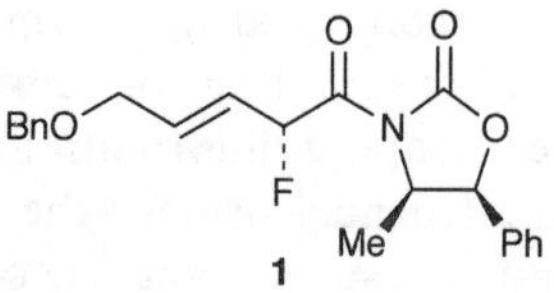

1

The varied uses of iodine (and hypervalent iodine) in carbohydrate chemistry have been reviewed,[21] and anhydrous hydrogen iodide (generated from I_2 and thioacetic acid) has been employed for the efficient preparation of glycosyl iodides from glycosyl acetates, and vicinal iodohydrins from epoxides.[22] The 6-deoxy-6-iodo-galactoside **2** has been prepared as an analogue of an artificial leaf-opening substance,[23] while daunomycinone deoxyhalo-glycosides with sugar moieties of increased lipophilicity have been synthesized and shown to be active against drug-resistant cancer cell lines.[24] The branched-chain and chain-extended iodinated glucose analogues **3–5** have been prepared.[25] The use of ω-iodo-α,β-unsaturated compounds prepared from 2-deoxy-D-ribose for the study of samarium(II) iodide-mediated reductive carbocyclizations is covered in Chapter 22, and similar cobalt-catalysed radical cyclizations are noted in Chapter 18.

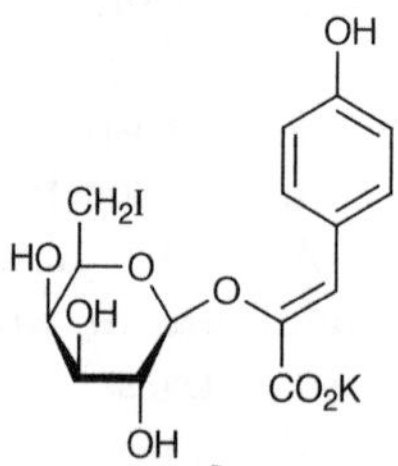

2

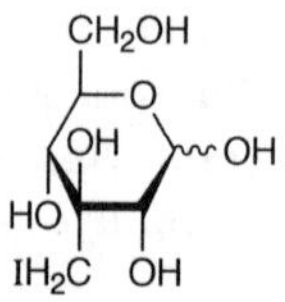

3

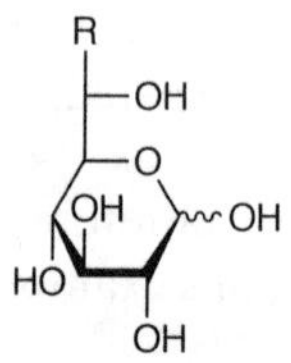

4 R = CH₂I
5 R = C₆H₄I

References

1. M. Yokoyama, *Carbohydr. Res.*, 2000, **327**, 5.
2. K. Toshima, *Carbohydr. Res.*, 2000, **327**, 15.
3. S.J. Williams and S.G. Withers, *Carbohydr. Res.*, 2000, **327**, 27.
4. K. Dax, M. Albert, J. Ortner and B.J. Paul, *Carbohydr. Res.*, 2000, **327**, 47.
5. K.W. Pankiewicz, *Carbohydr. Res.*, 2000, **327**, 87.
6. B. Beuthien-Baumann, K. Hamacher, F. Oberdorfer and J. Steinbach, *Carbohydr. Res.*, 2000, **327**, 107.
7. R. Plantier-Royon and C. Portella, *Carbohydr. Res.*, 2000, **327**, 119.
8. R. Miethchen and M. Hein, *Carbohydr. Res.*, 2000, **327**, 169.
9. M. Michalik, M. Hein and M. Frank, *Carbohydr. Res.*, 2000, **327**, 185.
10. B. Novo and G. Resnati, *Enantiocontrolled Synth. Fluoro-Org. Compd.*, 1999, 349 (*Chem. Abstr.*, 2000, **132**, 166 394).
11. P. Borrachero, F. Gabrera-Escribano, A.T. Carmona and M. Gómez-Guillén, *Tetrahedron: Asymmetry*, 2000, **11**, 2927.
12. A. Berkin, W.A. Szarek and R. Kisilevsky, *Carbohydr. Res.*, 2000, **326**, 250.
13. P. Hadwiger, P. Mayr, B. Nidetzky, A.E. Stütz and A. Tauss, *Tetrahedron: Asymmetry*, 2000, **11**, 607.
14. A.C.H. Chang, D. Horton and P. Kováč, *Tetrahedron: Asymmetry*, 2000, **11**, 595.
15. J. Madaj, Y. Nishikawa, V.P. Reddy, P. Rinaldi, T. Kurata and V.M. Monnier, *Carbohydr. Res.*, 2000, **329**, 477.
16. F.A. Davis, H. Qi and G. Sundarababu, *Tetrahedron*, 2000, **56**, 5303.
17. S. Sugiyama, W. Haque and J. Diakur, *Org. Lett.*, 2000, **2**, 3489.
18. R.P. Singh and J.M. Shreeve, *Tetrahedron*, 2000, **56**, 7613.
19. B. Fraser-Reid, R. Madsen, A.S. Campbell, C.S. Roberts and J.R. Merritt, *Bioorg. Chem.: Carbohydr.*, 1999, 89 (*Chem. Abstr.*, 2000, **132**, 293 926).
20. R.M. Cicchillo and P. Norris, *Carbohydr. Res.*, 2000, **328**, 431.
21. A.R. Vaino and W.A. Szarek, *Adv. Carbohydr. Chem. Biochem.*, 2000, **56**, 9.
22. S.M. Chervin, P. Abada and M. Koreeda, *Org. Lett.*, 2000, **2**, 369.
23. M. Ueda, Y. Sawai, Y. Wada and S. Yamamura, *Tetrahedron*, 2000, **56**, 5123.
24. N. Aligiannis, N. Pouli, P. Marakos, S. Mitaku, A.-L. Skaltsounis, S. Leonce, A. Pierce and G. Atassi, *Chem. Pharm. Bull.*, 2000, **48**, 150.
25. C. Morin and L. Ogier, *Tetrahedron: Asymmetry*, 2000, **11**, 629.

9
Amino-sugars

1 Natural Products

The biosynthesis of the macrolactam polyketide glycoside anti-tumour antibiotic vicenistatin L (**1**), including its vicenisamine moiety, has been studied using heavy atom-labelled precursors.[1] The bicyclic branched-chain diaminohepturonic acid derivative **2**, a tyrosyl tRNA synthetase inhibitor of unknown absolute stereochemistry, has been isolated from a *Microsmonospora* species fermentation medium.[2]

2 Syntheses

Syntheses covered in this section are grouped according to the method used for introducing the amino-functionality.

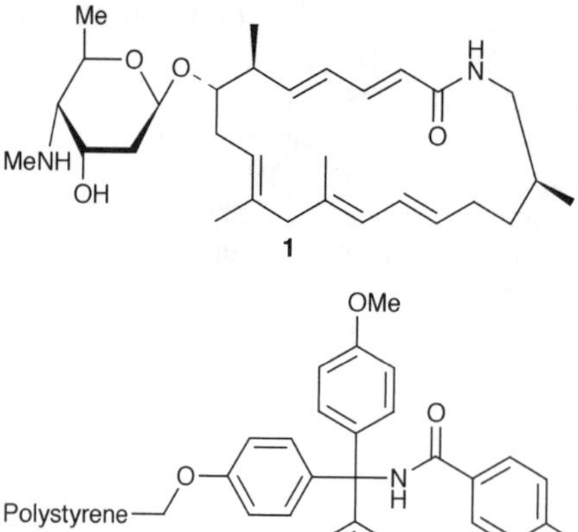

1

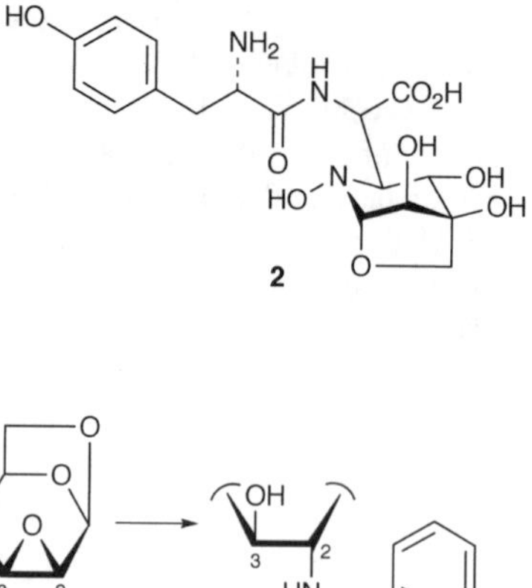

2

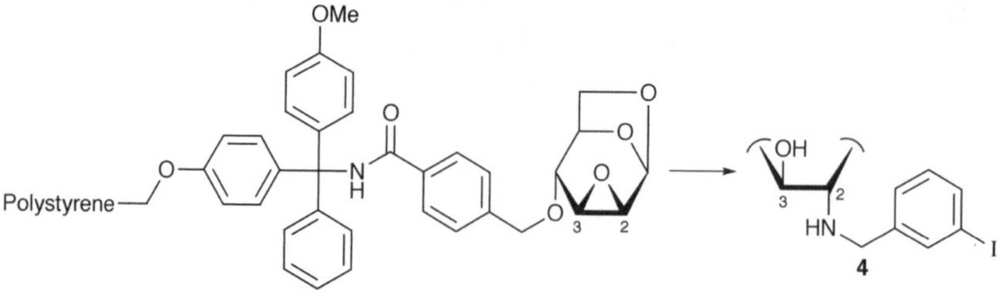

3 **4**

Scheme 1

2.1 By Epoxide Ring Opening. – A review of work at Novartis on the synthesis of complex molecules on solid supports included the nucleophilic ring opening of the immobilized levoglucosan-derived epoxide **3** at C-2 with various nucleophiles, *e.g.* to give **4** (Scheme 1).[3] Amphiphilic glucos-6-yl benzodiazepines, *e.g.*

Carbohydrate Chemistry, Volume 34

5, have been synthesized by alkylation at N-1 of the benzodiazepine, *e.g.* with a 5,6-anhydro-D-glucofuranose derivative as electrophile.[4] The pyrimidine-substituted L-ascorbic acid derivative **6** of has been obtained by ring opening addition to a 5,6-epoxide. A related 4,5-unsaturated analogue was also reported.[5]

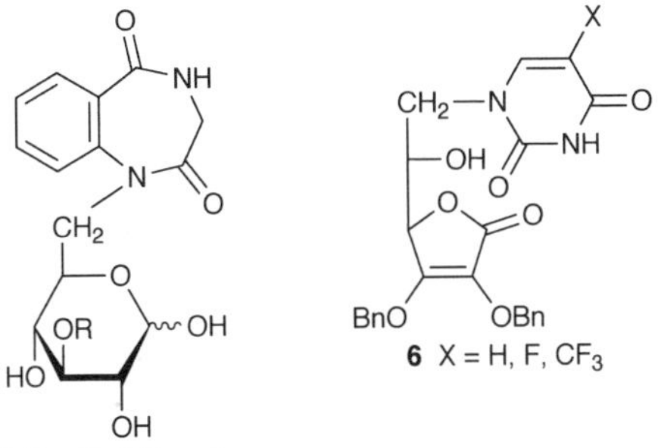

5 R = H, C$_8$H$_{17}$ *etc.*

6 X = H, F, CF$_3$

2.2 By Nucleophilic Displacement. – The Amadori compound **7** was synthesized by reaction of 2,3:4,5-di-*O*-isopropylidene-D-fructopyranose 1-triflate with the requisite primary amine, followed by acid-catalysed deprotection.[6] Syntheses of 2-azido-2-deoxy-α-D-glucopyranosides of *myo*-inositol, involving either inversion at C-2 of the corresponding α-mannoside 2-triflate or direct glycosylation with a pentenyl donor, are further detailed in Chapter 10.[7]

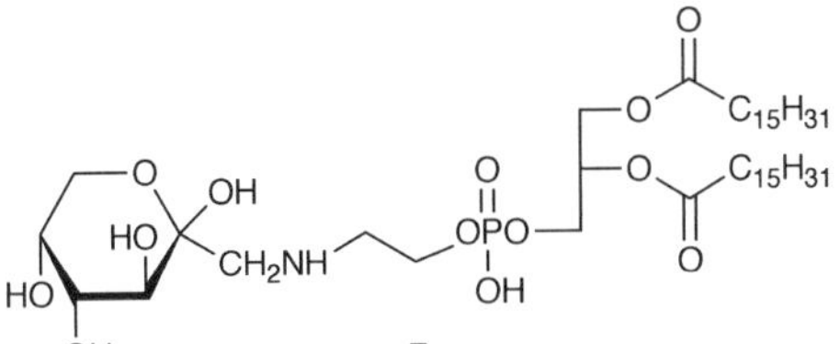

7

The butanediol-linked dimer **9** of a conformationally rigid bridged aminodeoxysugar is a potential RNA-binder. It was synthesized from D-mannose *via* the 3-acetamido-sugar **8** (Scheme 2), azide displacement with inversion being used for introduction of the amino-group.[8] The 3-amino-3,4,6-trideoxyhexoside **10** was synthesized in 10 steps from methyl 4,6-*O*-benzylidene-α-D-glucopyranoside. This involved 4,6-dichlorination then dehalogenation, inversion at C-3 by oxidation-reduction, then azide displacement with inversion to introduce the 3-amino-group. Amino-sugar nucleotide **10** was the predicted substrate for a methyltransferase involved in the synthesis of the desosamine residue in various macrolide antibiotics, and was indeed converted to **11** by this enzyme.[9]

The 5-amino-sugar **12** was synthesized from D-arabinose using conventional chemistry, including azide displacement of a tosylate group, and coupled with a 6-thiosugar to give the disaccharide analogue **14** (Scheme 3). While **14** was stable at low pH, it hydrolysed rapidly at pH >5. Unprotected 5-amino-5-deoxy-D-arabinose (**13**) gave the 2-keto-derivative **15** under mild acid conditions *via* an

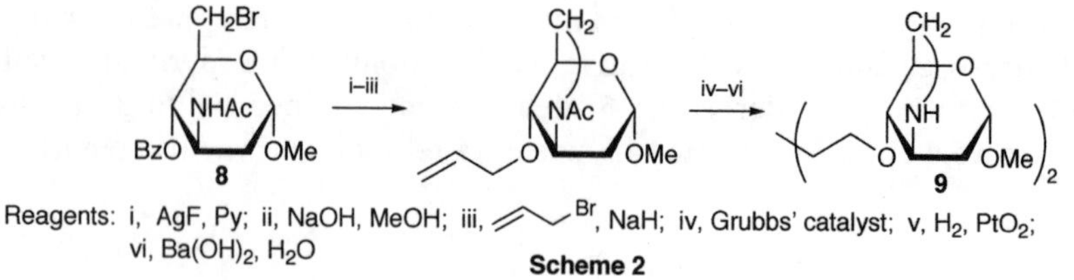

Reagents: i, AgF, Py; ii, NaOH, MeOH; iii, ⟋⟍Br, NaH; iv, Grubbs' catalyst; v, H₂, PtO₂; vi, Ba(OH)₂, H₂O

Scheme 2

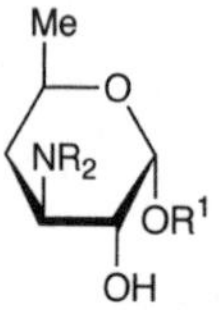

R¹ = thymidine-5′-(dihydrogenphosphate)

10 R = H
11 R = Me

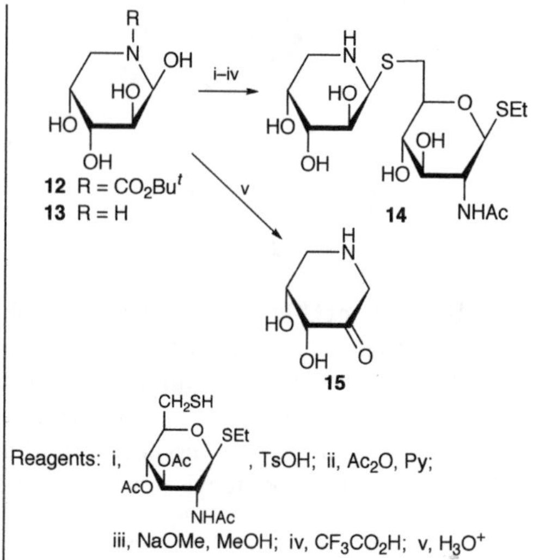

12 R = CO₂Buᵗ
13 R = H

14

15

Reagents: i, [structure], TsOH; ii, Ac₂O, Py;
iii, NaOMe, MeOH; iv, CF₃CO₂H; v, H₃O⁺

Scheme 3

imine then an enamine. The 2,3,4-tri-*O*-benzyl ether derivative of **12** was used in acid-catalysed syntheses of various *O*- and *S*-glycosides of the 5-aminodeoxy-sugar.[10] The photoaffinity derivatives **16** and **17** related to adenophostin, were synthesized by acylation of corresponding amino-derivatives and shown to be biologically effective agonists of D-*myo*-inositol 1,4,5-triphosphate receptor. The 5-amino-group in **16** was introduced by reaction of a 5-hydroxy-derivative under Mitsunobu conditions [Ph₃P, DEAD, (PhO)₂PN₃].[11] Syntheses of 5′-deoxy-5′-dialkylamino-derivatives of a spiro-cyclic nucleoside inhibitor of HIV-1 reverse transcriptase are covered in Chapter 19. The disaccharide scaffold **18**, based on tunicamycin and required for further elaboration into bioactive compounds, was made by glycosylation of a nucleoside derivative followed by introduction of a 6-azido group by displacement of a 6-tosylate group.[12]

Sugar analogues of anti-cancer agents cisplatin and carboplatin (*e.g.* **19**) were synthesized by displacement of a primary iodide or tosylate group with ethylenediamine, followed by complex formation with platinum(II).[13] The 6-amino-2,5-anhydro-D-mannonic acid derivative (**20**), synthesized *via* ring closure of a 1,2-epimino-D-glucitol derivative, was oligomerized into peptidomimetics up to an octamer.[14] Syntheses of hybrid peptidomimetics containing both amino acid and 6-amino-2,5-anhydro-hexonic or 2,5-anhydro-aldaric acid units are covered in Chapters 16. 1,6-Lactam **21** and an analogue were synthesized by the route outlined in Scheme 4.[15]

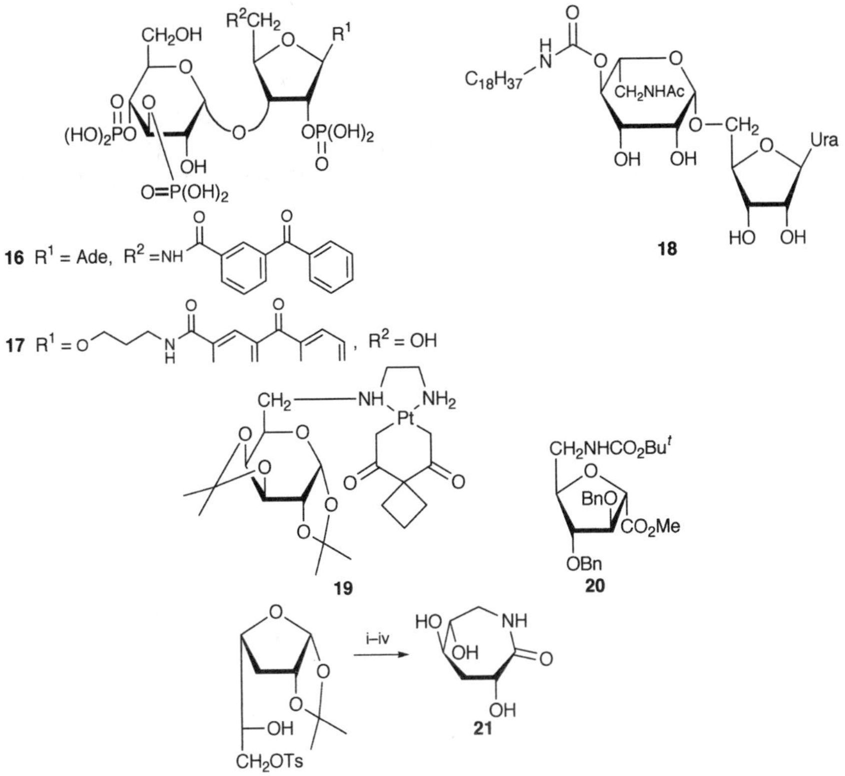

Reagents: i, NaN₃; ii, H₃O⁺; iii, Br₂, BaCO₃, H₂O; iv, H₂, Pd/C

Scheme 4

2.3 By Amadori Reaction. – *N*-Arylsulfonyl-derivatives of 1- and 6-aminodeoxy-D-fructose have been synthesized and shown to have good affinities for the glucose transporter of the protozoan parasite *Trypanosoma brucei*. 1-Amino-1-deoxy-D-fructose was prepared by Amadori reaction of D-glucose with toluidine, then hydrogenoltyic removal of the tolyl group (with H₂, PdO/BaSO₄). 6-Amino-6-deoxy-D-fructose was prepared by aldolase-catalysed chain-extension of 3-azido-3-deoxy-D-glyceraldehyde.[16]

2.4 From Azido-sugars. – Samarium diiodide in water has been used to reduce azido-sugars to their corresponding amines, and the method has been applied in the synthesis of chito-oligosaccharides.[17] A ferric chloride–zinc powder combination effected similar transformations in two isopropylidene-protected azido-sugar derivatives.[18] All eight D-hexono-1,5-lactams have been made in connection with glycosidase inhibition studies, 5-azido-5-deoxy-hexono-1,4-lactone derivatives being the precursors.[19] 'Aminoketalic' castanospermine analogues **22** (Scheme 5) and their β-L-*ido*-analogues were separately synthesized and shown to be present exclusively as α-anomers in aqueous solution and to be only weak glycosidase inhibitors.[20] Sugar urea and thiourea derivatives such as **23** were synthesized by Ph₃P-induced condensation of the corresponding protected 6-azido-sugar with various isothiocyanates followed by either hydrolysis (HOAc, H₂O) or thiolysis (H₂S, silica gel), respectively, then deprotection.[21] 2-, 3- and 4-Deoxy-isothiocyanato-sugar derivatives were obtained from the corresponding azides by reduction (H₂, P/C) then reaction with thiophosgene.[22] Branched glycopeptidomimetics such as **24** have been synthesized from 6-azido-glucosides.[23]

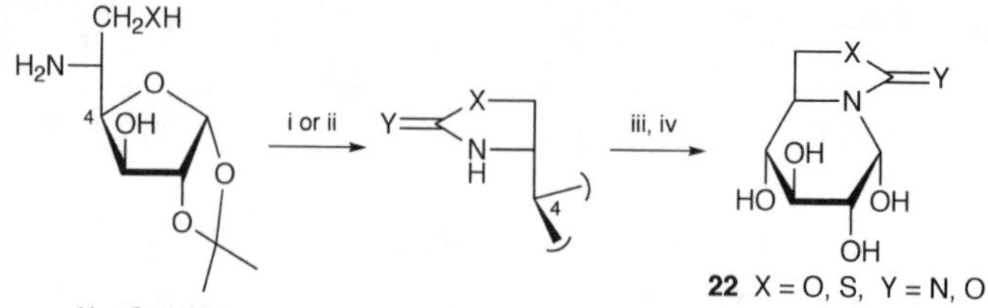

X = O or NH

Reagents: i, CS₂, DCC; ii, triphosgene, Pr$^i_2$NEt; iii, H₃O⁺; iv, Resin(OH⁻)

Scheme 5

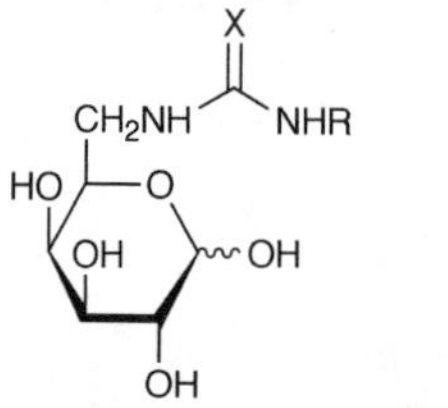

23 X = O or S, R = Me, β-D-Glc*p*

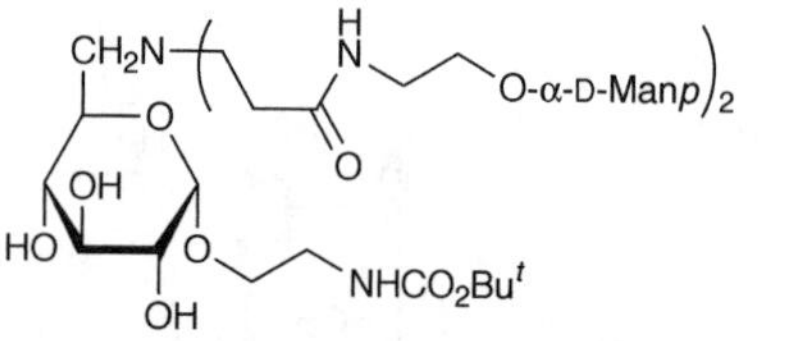

24

2.5 From Nitro-sugars. – The 2-acetamido-2-deoxy-β-D-galactopyranosylamine **25** was obtained by addition of benzimidazole to a 2-nitro-galactal derivative (see Chapter 10).[24]

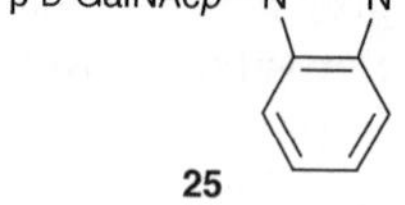

β-D-GalNAc*p*—N

25

2.6 From Unsaturated Sugars. – *cis*-1,2-Aminoalcohols were obtained in high yields (60–83%) by reactions of a variety of sugar allylic alcohols with arylisocyanates and either iodoxybenzoic acid or 1-hydroxy-1,2-benziodoxol-3(1*H*)-one 1-oxide ('IBX'), the reaction involving intramolecular delivery of the amino-function. Applied to D-glucal or D-galactal derivatives, *e.g.* **26**, reaction conditions determined whether 2-amino-1,5-anhydro-2-deoxyhexitols, *e.g.* **27**, or 2-amino-2-deoxy-sugars, *e.g.* **28**, were formed (Scheme 6). In the same way, 4,6-di-O-acetyl-2,3-dideoxy-D-*erythro*-hex-2-enopyranose yielded the 2,3-dideoxy-2-amino-sugar **29**, after cleavage of the urethane moiety (i, CAN; ii, NaOH).[25] An iodonium ion catalysed conversion of a glycal into an 2-amino-1-thioglycoside has been employed in the synthesis of a high mannose N-linked glycopeptide (see Chapter 10).[26] A new sulfonium reagent **30**, prepared by reaction of thianthrene-5-oxide with triflic anhydride, has been employed in a one-pot acetamidoglycosylation of glycal derivatives that involves an acid-catalysed 1,2-oxazoline ring opening by the nucleophile (ROH) (*e.g.* Scheme 7).[27]

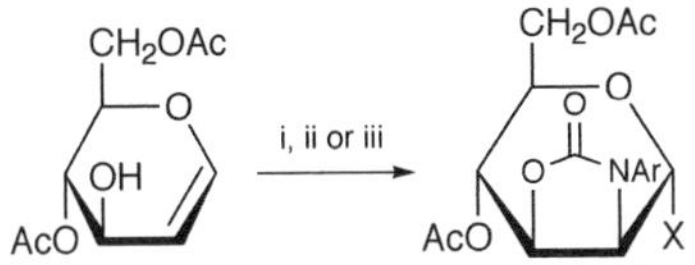

Reagents: i, 4-MeOC$_6$H$_4$NCO, DBU; ii, IBX, anhydrous (for X = H); iii, IBX, H$_2$O (for X = OH)

Scheme 6

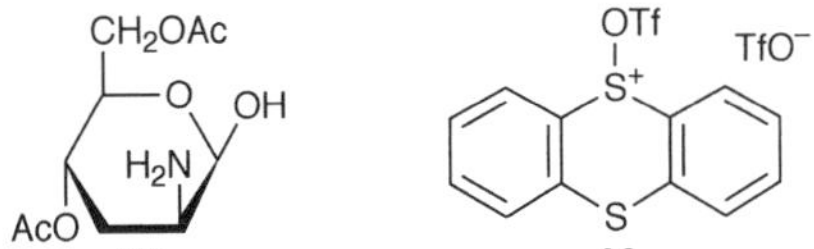

R = e.g. Bn, Pr^i,

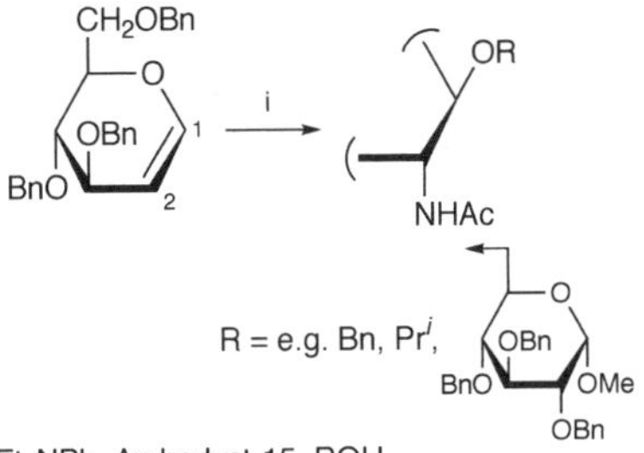

Reagents: i, **30**, AcNHSiMe$_3$, Et$_2$NPh, Amberlyst-15, ROH

Scheme 7

Attempted intramolecular glycosylation with **31** gave mostly the 3-deoxy-3-pyrimidinyl-D-glucal **32** and only a little of the isomeric 2,3-unsaturated nucleoside (Scheme 8).[28] In a long synthesis of methyl α-L-vancosaminide (**34**) from di-*O*-acetyl-L-rhamnal, the key step involved intramolecular amino-mercuration of the unsaturated branch-chain sugar **33** (Scheme 9).[29] New routes to precursors of lincosamine (6-amino-6,8-dideoxy-D-*erythro*-D-*galacto*-octose) involved intramolecular rearrangement of the allylic trichloroacetimidate **35** or the isothiocyanate **36** (Scheme 10).[30]

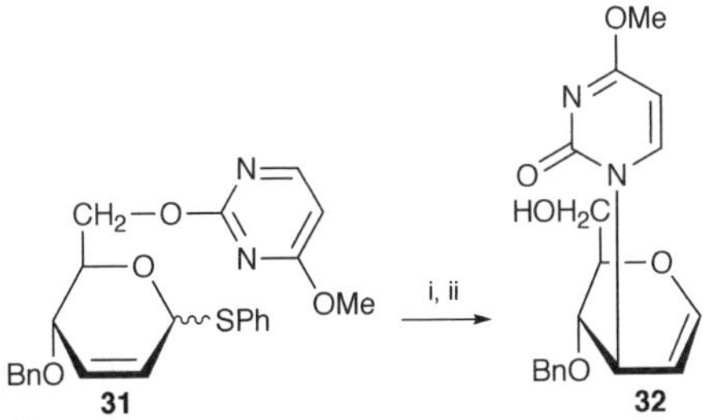

Reagents: i, Me$_2$S–SMe·BF$_4$; ii, NaOH, H$_2$O

Scheme 8

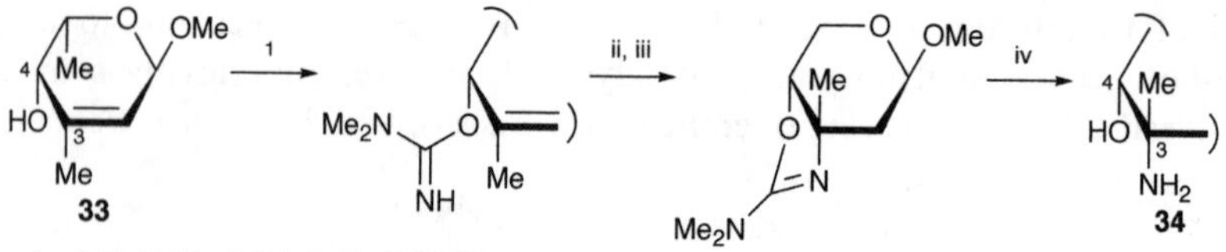

Reagents: i, NaH, Me$_2$NCN; ii, Hg(OCOCF$_3$)$_2$; iii, NaBH$_4$; iv, Ba(OH)$_2$

Scheme 9

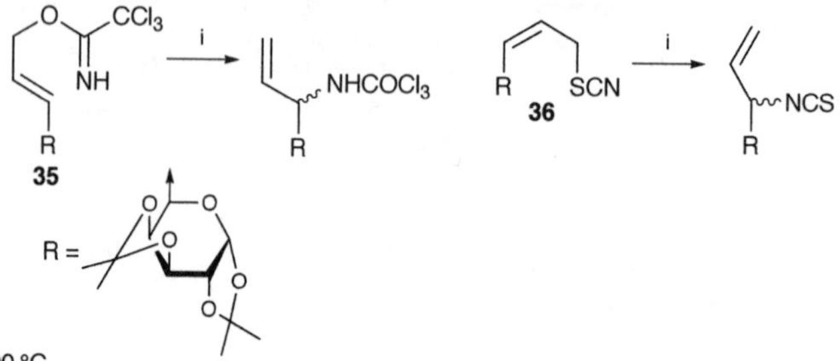

Reagent: i, 200 °C

Scheme 10

Syntheses of amino- and azido-deoxy-sugars by conjugate addition of *N*-nucleophiles to unsaturated sugars has been included in a review.[31] The anthracycline antibiotic acosaminyl-ε-iso-rhodomycinone (**38**) has been constructed using the L-*arabino*-azide **37** obtained as the major product from L-rhamnal diacetate on a 100-gram scale (Scheme 11). The daunosaminyl (L-*lyxo*), ristosaminyl (L-*ribo*) and *epi*-daunosaminyl (L-*xylo*) analogues were also reported.[32] Addition of primary or secondary amines to the vinylsulfone **39** gave 2-amino-2-deoxy-β-glucosides, *e.g.* **40** (Scheme 12). Addition of primary amines to the corresponding α-anomer of **39** similarly gave α-glucosides, whereas addition of secondary amines gave α-glucosides and α-mannosides as major and minor products, respectively.[33] The 2,3-aziridino-1,4-pentonolactone **43** has been synthesized from D-ribono-1,4-lactone (**41**) *via* the vinylic triflate **42** (Scheme 13).[34]

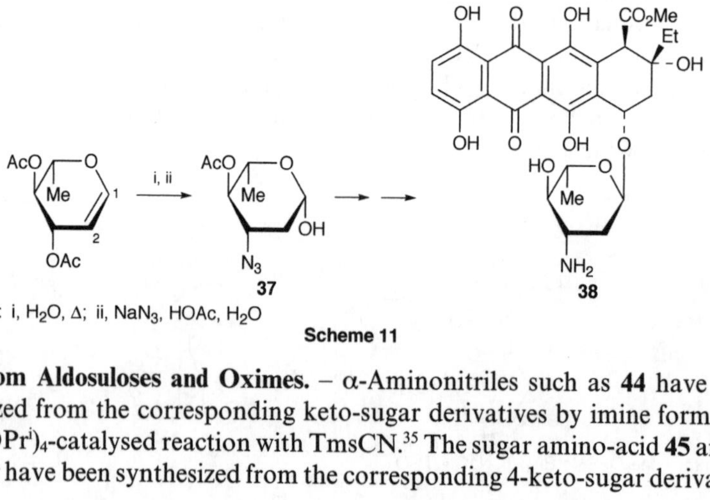

Reagents: i, H$_2$O, Δ; ii, NaN$_3$, HOAc, H$_2$O

Scheme 11

2.7 From Aldosuloses and Oximes. – α-Aminonitriles such as **44** have been synthesized from the corresponding keto-sugar derivatives by imine formation and Ti(OPr^i)$_4$-catalysed reaction with TmsCN.[35] The sugar amino-acid **45** and its 5-epimer have been synthesized from the corresponding 4-keto-sugar derivatives

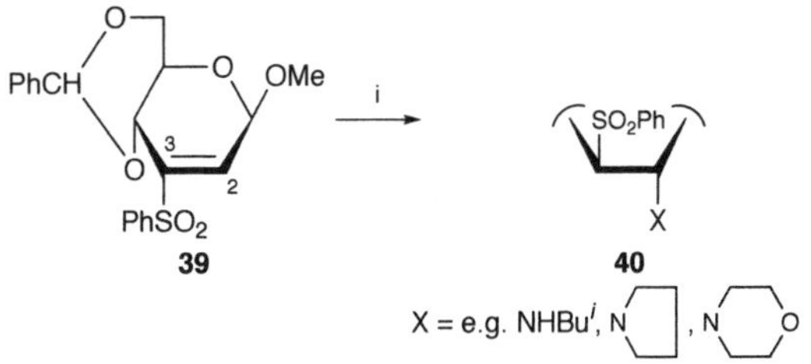

Reagent: i, 100 °C

Scheme 12

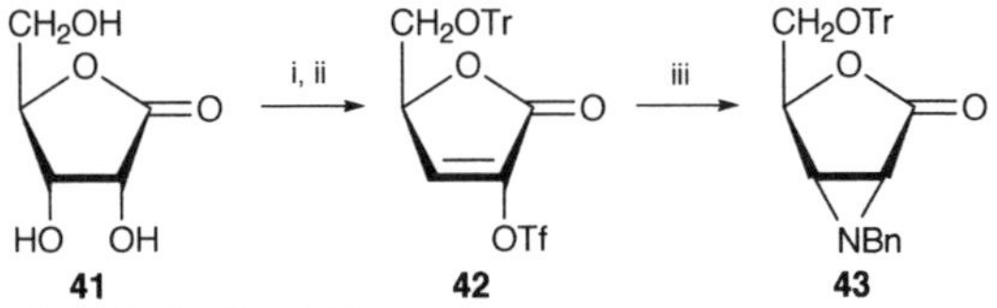

Reagents: i, TrCl, Py; ii, Tf$_2$O, Py; iii, BnNH$_2$

Scheme 13

by conversion to α-aminonitriles [KCN, (NH$_4$)$_2$CO$_3$] then formation and hydrolysis of *spiro*-hydantoins [i, (NH$_4$)$_2$CO$_3$, EtOH, H$_2$O, reflux; ii, H$_3$O$^+$; iii, OH$^-$].[36] 4-*O*-(2-Acetamido-2-deoxy-β-D-talopyranosyl)-D-glucose was synthesized from a known lactose derivative with a single unprotected 2-hydroxy-group (*cf.* Vol. 31, Ch. 6, Ref. 6) by an oxidation – oxime formation – LiAlH$_4$ reduction – acetylation sequence that resulted predominantly in inversion at C-2.[37] Reactions of the 2-oximinoglycosyl donor **46** with sugar alcohol acceptors catalysed by NIS–TfOH, followed by LiAlH$_4$ reduction–acetylation, gave predominantly disaccharides with an α-D-GlcNAc moiety when the glycosylation reaction was conducted in dichloromethane as solvent, but with a β-D-ManNAc moiety when acetonitrile was used.[38]

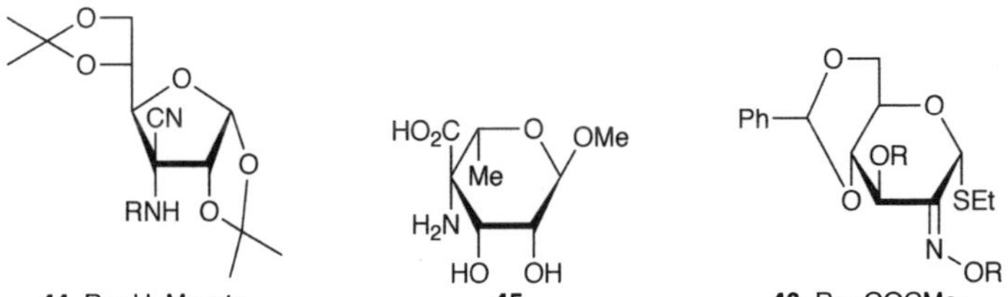

2.8   From Chiral Non-carbohydrates. – The 3-amino-3,6-dideoxy-hexose derivative **50** was the major product obtained from aldol condensation of the tricarbonyliron–diene complex **47** with the (*R*)-lactaldehyde derivative **48** (Scheme 14). Adduct **49** was the major (45%) of four isomers formed. If decomplexation (step iv) preceded reduction (step ii), then the 2-epimer of **50** predominated.[39] *N*-Acetyl-L-daunosamine and its 5-*C*-isobutyl analogue (**52**) were synthesized from the corresponding cyanohydrin derivatives **51** by chain extension (Scheme 15).[40] Aziridination of chiral *trans*-α,β-unsaturated esters such as **53** gave mainly two isomeric 2,3-epimines, *e.g.* **54** and **55** in the ratio 67:33 (Scheme

16). The *cis*-isomeric alkenes gave all four possible isomers.[41] The galantinic acid derivative **56** was obtained from an L-serinal derivative in seven steps including two chain-extension reactions.[42]

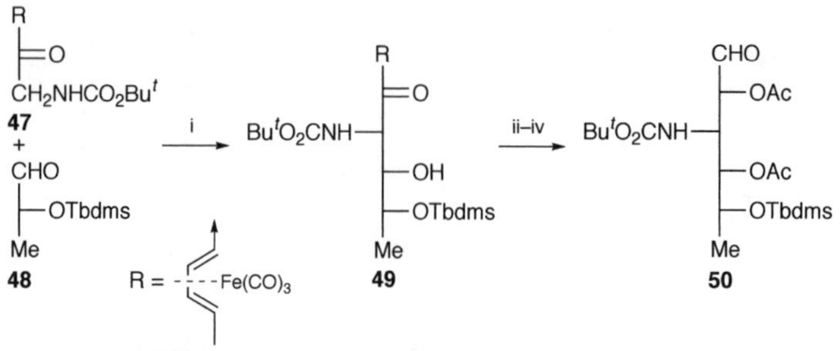

Reagents: i, $Sn(OTf)_2$, HNNEt; ii, $BH_3 \cdot SMe_2$; iii, Ac_2O, Et_3N; iv, $(NH_4)_2Ce(NO_3)_6$; v, O_3

Scheme 14

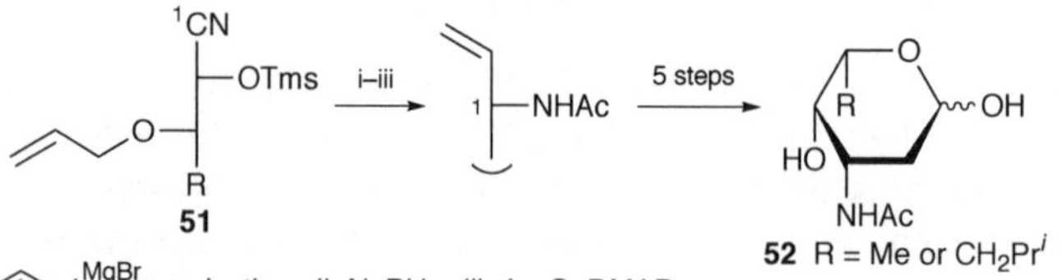

Reagents: i, MgBr, sonication; ii, $NaBH_4$; iii, Ac_2O, DMAP

Scheme 15

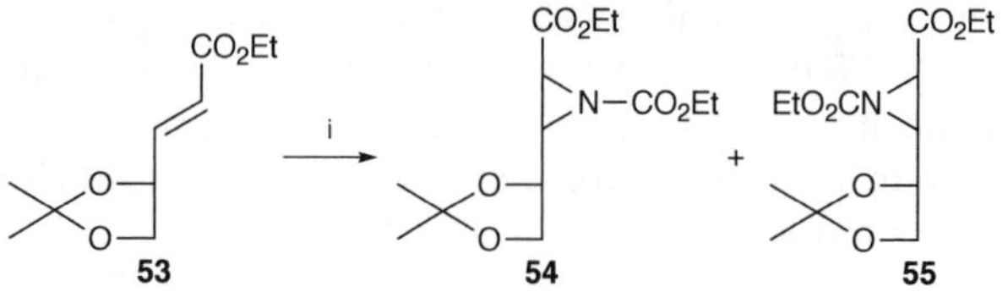

Reagents: i, $4\text{-}O_2N\text{-}C_6H_4SO_2NHCO_2Et$, CaO (neat)

Scheme 16

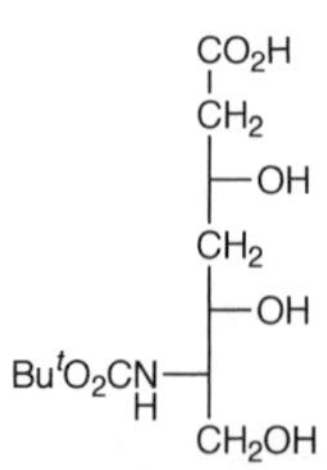

2.9 From Achiral Non-carbohydrates. – Epoxide **57**, available from methyl sorbate by a chemoenzymatic synthesis (Ono *et al.*, *Tetrahedron: Asymmetry*, 1996, **7**, 2595), was used in the construction of a set of isomeric 3-amino-2,3,6-trideoxyhexoses. Addition of benzylamine to the α,β-unsaturated acid moiety before or after opening of the epoxide ring led to the D-acosamine derivative **58** or the lactone **59**, respectively (Scheme 17).[43] Lactone **59** has previously been converted to L-daunosamine and D-acosamine derivatives (Vol. 14, p. 72, ref. 14).

In a model study towards the synthesis of 4-amino-2,3,6-trideoxyhexoses, the α,β-unsaturated ester **60** and related derivatives proved excellent substrates for enantio- and regio-selective Sharpless asymmetric aminohydroxylation (Scheme 18).[44]

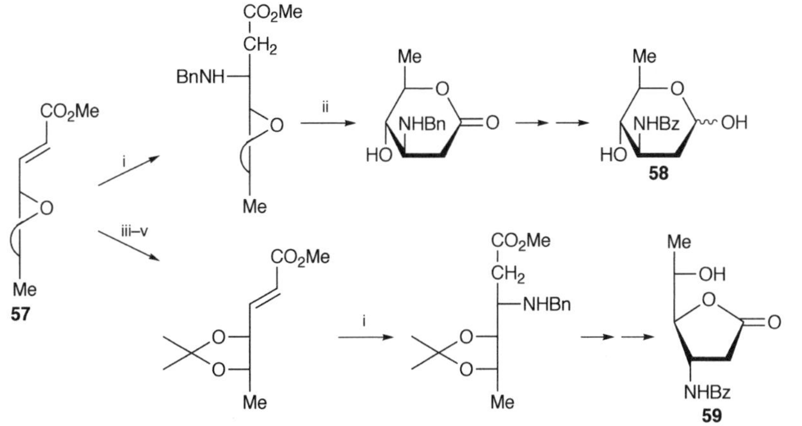

Reagents: i, BnNH₂; ii, TfOH; iii, BnOH, BF·OEt₂; iv, AlCl₃, *m*-xylene; v, Me₂C(OMe)₂, TsOH, Me₂CO

Scheme 17

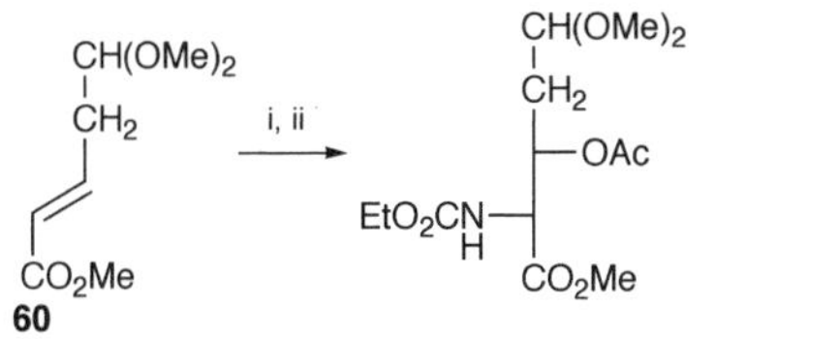

Reagents: i, K₂OsO₂(OH)₄, (DHQ)₂AQN, EtOCONH₂, BuᵗOCl;
ii, Ac₂O, Et₃N, DMAP

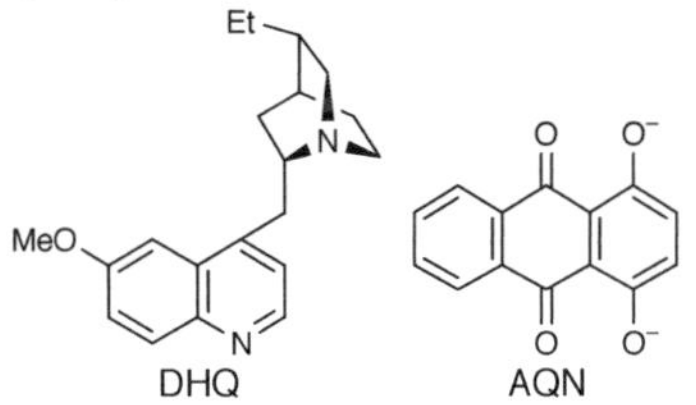

Scheme 18

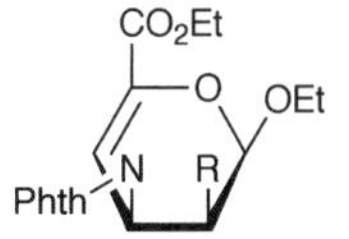

Hetero-Diels–Alder reactions have been used in the construction of amino-sugars, asymmetry being induced by the use of either a chiral Lewis acid catalyst or a chiral auxiliary. Thus, the 3-amino-hexuronic acid derivatives **61** were obtained with high *de* and *ee* using a copper(II) triflate catalyst with a chiral bisoxazoline ligand.[45,46] 4-Amino-4-deoxy-D-erythrose and 4-amino-4,5-dideoxy-L-lyxose (**63**), the latter a potent inhibitor of α-L-fucosidase, were obtained from the D-pyroglutamic acid-containing diene **62** (Scheme 19).[47]

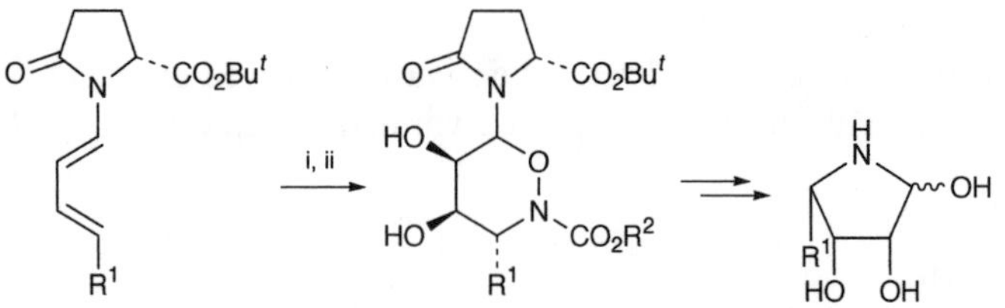

62 R^1 = H or Me   **63** R^1 = H or Me

Reagents: i, R^{1}O$_2$C—N=O (R^2 = Bu^t when R^1 = H; R^2 = Bn when R^1 = Me; ii, OsO$_4$, NMNO

Scheme 19

3 Reactions and Derivatives

3.1 Interconversion Reactions. – 2-Acetamido-2-deoxy-D-galactopyranoside derivatives have been prepared from the corresponding 2-acetamido-2-deoxy-D-glucopyranosides using either an oxidation–reduction sequence at C-4 or a nucleophilic displacement of a 4-triflate group.[48] The 2-amino-2-deoxy-D-allopyranoside **64** and its D-*galacto*-isomer were obtained in a ratio of 20:1 from the corresponding 3,4-ene by a *syn*-selective dihydroxylation with catalytic OsO$_4$ and quinuclidine-*N*-oxide as reoxidant.[49] 3,4-Dihydroxy-L-glutamic acid has been synthesized from a 2,3-epimino-derivative of D-lyxaro-1,4-lactone.[50]

The anomers **65**, containing the intercalator and carbohydrate moieties of the endiyne antibiotic neocarzinostatin, were synthesized from 2-amino-2-deoxy-D-galactose. They were shown to cleave DNA under photoirradiation (365 nm), but with a different site selectivity from that of neocarzinostatin.[51] A multi-gram synthesis of UDP-*N*-acetylmuramic acid was made possible by an improved synthesis of the 1-phosphate precursor.[52]

The nojirimycin α-trichloroacetimidate **66** was synthesized from a protected 5-amino-5-deoxy-D-glucofuranoside and used as a glycosyl donor to prepare the corresponding β-thioglycosides, including one with a glucosid-3-yl moiety as an aglycon. These could be *O*-deprotected.[53]

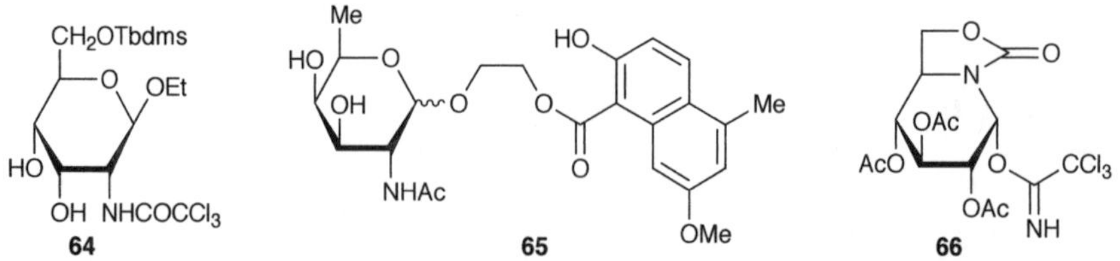

64   **65**   **66**

3.2 *N*-Acyl and *N*-Carbamoyl Derivatives. – Selective *N*-deprotection of 2-methoxycarbonylamino-2-deoxy-D-glucopyranoside derivatives was achieved by treatment with MeSiCl$_3$ and Et$_3$N in dry THF at 60 °C and subsequent hydrolysis. Azido- and *N*-(2,2,2-trichloroethoxy)carbonyl groups were unaffected.[54] Applications of a new *N*-protecting group, the TDG (thiodiglycoloyl) group (see **67**), have been described. Thus, 2-deoxy-2-thiodiglycoloylamino-D-glucopyranoside tetraacetate was synthesized by reaction of 2-amino-2-deoxy-D-glucose with thiodiglycolic anhydride followed by acetylation (with Ac$_2$O, Py).

Conversion to the β-trichloroacetimidate **67** provided a good β-glycosyl donor. The TDG group could be cleaved by reaction with MeONa in MeOH followed by Bu₃SnH.[55] The *N*-(*p*-nitrobenzyloxycarbonyl)amino-derivative **68**, obtained in five-steps from 2-amino-2-deoxy-D-galactose, was shown to be a good glycosyl donor in the construction of a mucin core-2 analogue. The *N*-protecting group could be removed under mild conditions (with neutral Na₂S₂O₄).[56] The *N*-(2,2,2-trichloroethoxy)carbonyl, *N*-phthaloyl and *N*-acetoxyacetyl derivatives of 1,3,4,6-tetra-*O*-acetyl-2-amino-2-deoxy-α-D-glucopyranose have been prepared by *N*-acylation.[57] *p*-Nitrophenyl 2-deoxy-2-*p*-nitrobenzamido-β-D-glucopyranoside has been prepared in five step from 2-amino-2-deoxy-D-glucose.[58]

67 X = —N⟨⟩S, R¹ = OAc, R² = H

68 X = —NHCO₂CH₂—⟨⟩—NO₂, R¹ = H, R² = OAc

69

Amphiphilic 2-aminoacylamino-2-deoxy-β-D-glucopyranosylamines such as **69** have been synthesized and shown to have immune stimulatory or immunoadjuvant activities.[59] Syntheses of variously 2-*N*-acyl derivatives of glyceryl 2-amino-2-deoxy-β-D-glucopyranosides are covered in Chapter 3.

The set of variously *N*-acylated 2,3-unsaturated sialic acids **70** was synthesized to probe the activity of the sialidase of *Vibrio cholerae*, the causative agent of cholera.[60] Syntheses of a range of fluorinated analogues of the methyl glycoside of 4-(2,4-dihydroxybutryoyl)amino-2,6-dideoxy-D-mannopyranose, the monosaccharide determinant of *Vibrio cholerae* are covered in Chapter 8.

The bis-sugar substituted calix[4]arene **71** could be prepared, but the sugar moiety of the product could not be deprotected (another analogue was deprotected, see Chapter 16).[61] Radical cyclization of the 4-*O*-allyl-6-(*o*-iodobenzamido)-D-glucoside **72** yielded a macrocycle (in this case **73**) in better yield than in a model study in which the groups involved in the reaction were attached to 3-amino-1-propanol (40 *vs.* 14%).[62]

3.3 Urea, Thiourea, Isothiocyanato and Guanidino Derivatives. – The nitrosourea derivative **74** was the most potent antitumour compound in a set of related di- and tri-deoxy-sugars.[63] Various mono- and di-*N*- and/or *S*-acetylated regioisomers were isolated from acetylation of 5-deoxy-5-*N*-thioureido-D-xylofuranose and -D-glucofuranose derivatives.[64] Methyl 6,7-dideoxy-7-isothiocyanato-heptoses with the α-D-*gluco*, α-D-*manno*- or α-D-*galacto*-stereochemistry were synthesized from the corresponding 6-iodo-hexosides by reaction with cyanide, reduction and reaction with thiophosgene.[65] Various

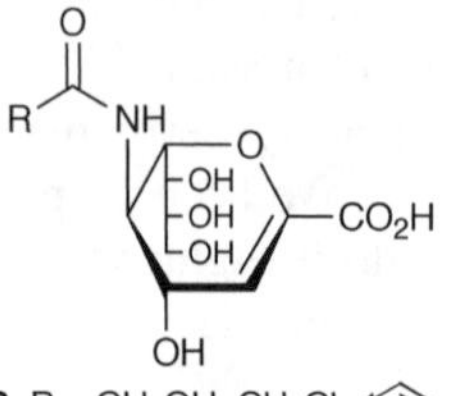

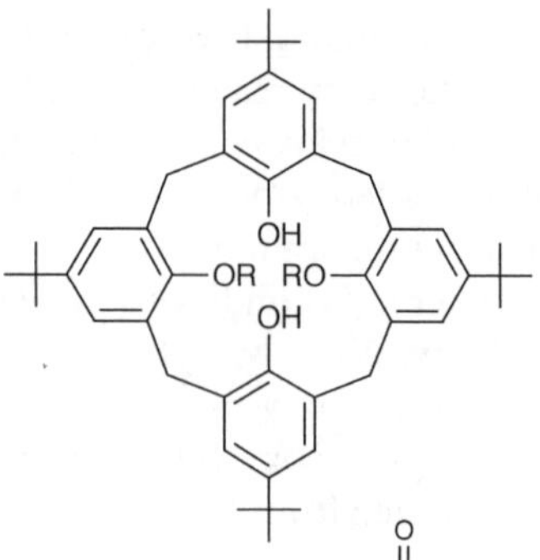

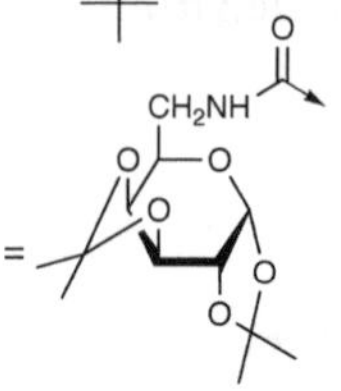

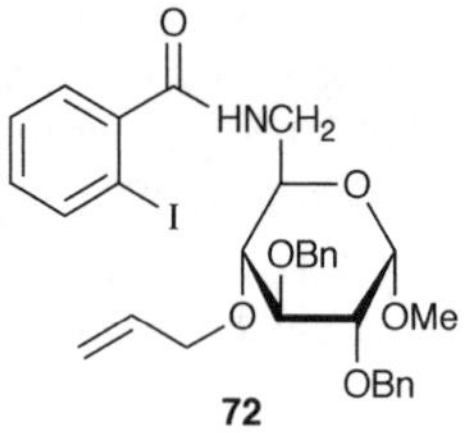

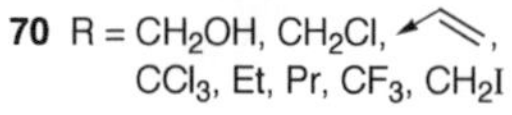

70 R = CH$_2$OH, CH$_2$Cl, ⌃⌃⌃,
CCl$_3$, Et, Pr, CF$_3$, CH$_2$I

71 R =

72

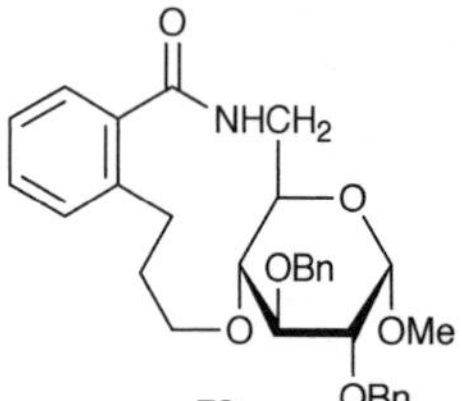

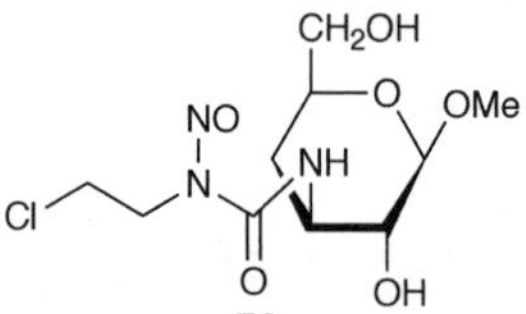

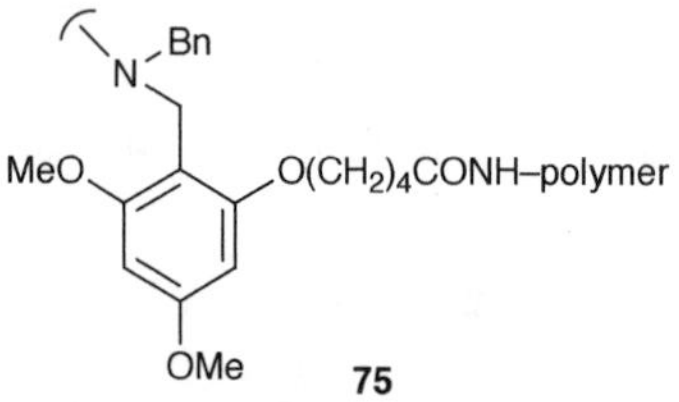

73

74

75

unprotected amino-sugars and aminoglycoside antibiotics were converted directly in the presence of water to the corresponding guanadino-analogues by use of
a new N-selective reagent, ^{t}BuO$_2$C-HNC($=$NTf)NHCO$_2$Bu^t.[66,67] The products
formed from tobramycin and neomycin had greatly increased anti-HIV activities.[66]

3.4 N-Alkyl, N-Alkenyl and N-Glycosyl Derivatives. – Diisopropyl azodicar-
boxylate effected selective N-debenzylation in the presence of O-benzyl groups in
a set of 2-, 3- and 4-benzylaminodeoxy derivatives of isomeric 1,6-anhydro-β-D-
hexopyranose derivatives.[68] The N-trisalkoxybenzyl group **75**, was used as a
linker at the 2-position of 2-amino-2-deoxy-D-glucose for solid phase glycosylations. It is stable to conc. trifluoroacetic acid and a range of Lewis acids, but
cleaved upon treatment with a 19:1 mixture of TFA–H$_2$O.[69] Copper(II) complexes formed with amino-sugar-derived β-oxoenamine ligands such as **76** contain elements found in certain metalloenzymes and were shown to catalyse
oxidation of a catechol to an o-quinone with molecular oxygen.[70] The orthogonally protected building block **77** has been used in the construction of
amide-linked glycodendrimers.[71]

3.5   Lipid A Analogues. – Of a set of eleven new N-acylated L-serine-containing
2-amino-2-deoxy-D-glucose derivatives synthesized as Lipid A analogues and
evaluated as synthetic immunoadjuvants, compound **78** had the greatest

mitogenicity.[72] The *Helicobacter pylori* Lipid A (**79**), along with an analogue **80**, was synthesized in 14 steps and 2.2% overall yield from 2-amino-2-deoxy-D-glucose.[73]

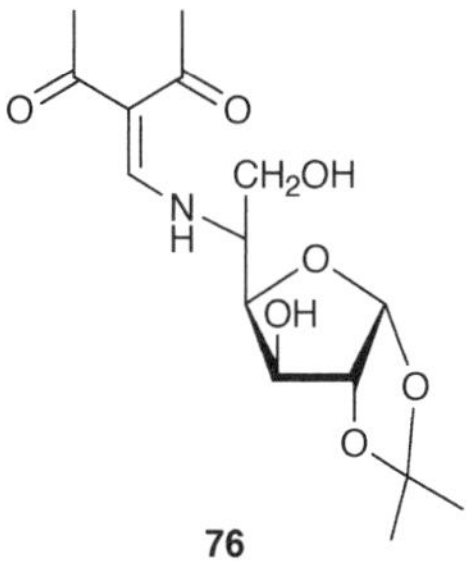

76

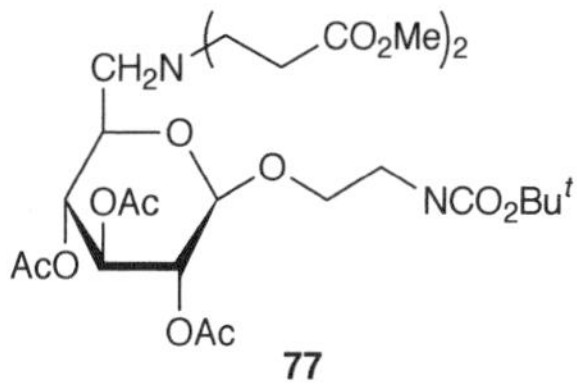

77

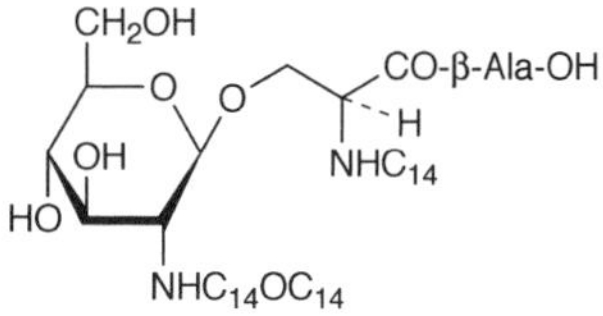

78

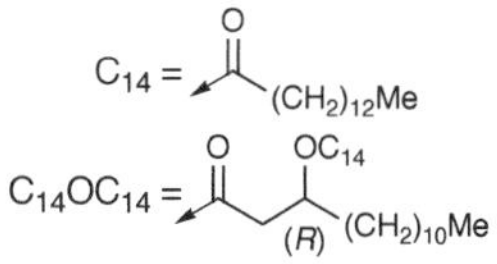

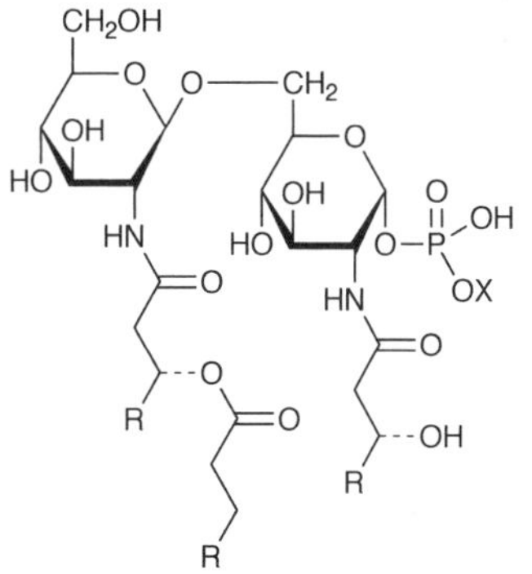

$$C_{14} = \begin{array}{c} O \\ \end{array} (CH_2)_{12}Me$$

$$C_{14}OC_{14} = \begin{array}{c} O \quad OC_{14} \\ \end{array} (R) (CH_2)_{10}Me$$

79 X = CH$_2$CH$_2$NH$_2$, R = —(CH$_2$)$_{14}$Me
80 X = H, R = —(CH$_2$)$_{14}$Me

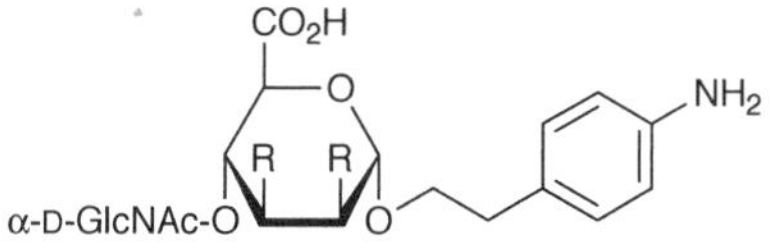

81 R = NHAc

α-D-GlcNAc-O

4   Diamino-sugars

The synthesis of the spacer-containing disaccharide fragment **81** of *Bordetella pertussis* lipopolysaccharide has been described. Ethyl 4,6-*O*-benzylidene-1-thio-β-D-glucopyranoside was subjected to 2-*O*-tosylation, 2,3-epoxide formation, azide ring opening, and formation and azide displacement of a 3-triflate to yield a 2,3-diazido-2,3-dideoxy-β-mannoside derivative. This then participated as a donor then an acceptor in two consecutive glycosylation reaction to yield **81** after protecting group manipulation.[74] A panel of 2,4- and 2,3-diacetamido-2,4,6- and 2,3,6-trideoxy-L-hexopyranosides has been prepared by epoxidation of the 2-acetamido-hex-3-enoside **82** followed by ring opening with azide to give separable mixtures of the azides **83** and **84** (Scheme 20). The stereochemistry at the carbon atom bearing the unprotected hydroxyl group in either compound could be inverted by a triflate displacement reaction.[75] Prop-2-yl D-*epi*-purpurosamine (**86**) was synthesized from the hex-3-enoside **85** by a double triflate displacement reaction with azide (Scheme 21).[76]

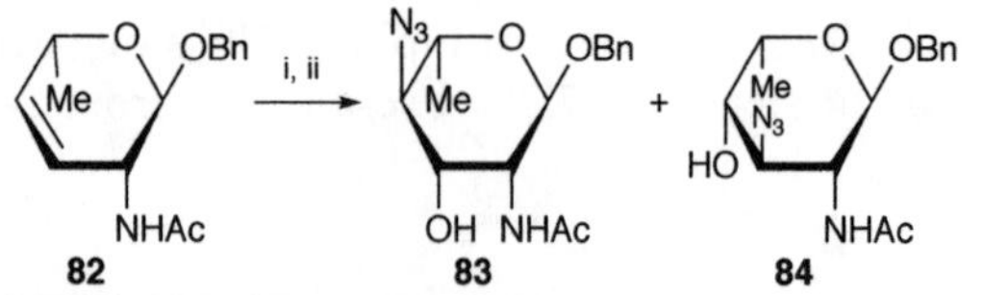

Reagents: i, MCPBA; ii, TMSN$_3$, BF$_3$·OEt$_2$ or NaN$_3$, NH$_4$Cl

Scheme 20

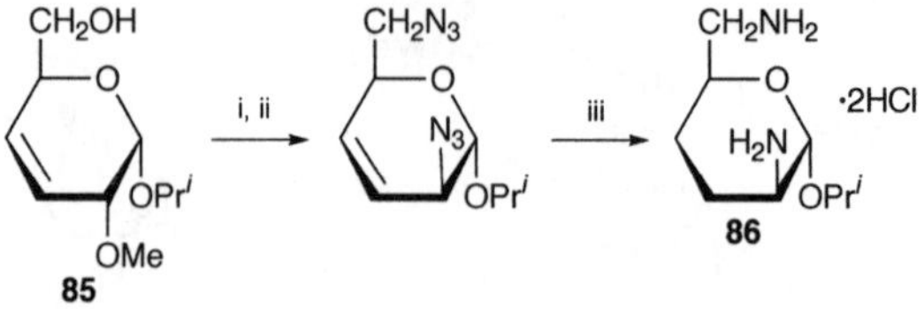

Reagents: i, Tf$_2$O; ii, NaN$_3$, DMF; iii, H$_2$, Pd/C

Scheme 21

Reductive amination was used to link the 2,4-diamino-sugar and di-fluorophosphonomethyl-iminoalditol moieties of the disaccharide mimic **87**.[77] The 2,5-diamino-2,3,5-trideoxy-hexosides **89** were obtained as the major stereoisomers by Michael additions to the chiral (2S)-hydroxymethyl-dihyd-ropyridine **88** (derived from D-glucal in 6-steps) followed by reduction (Scheme 22).[78]

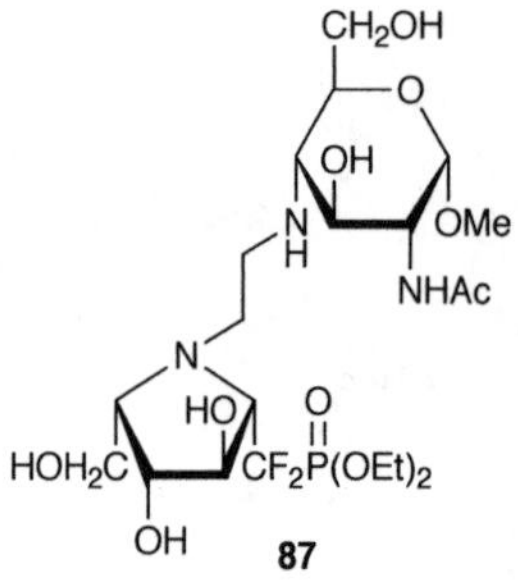

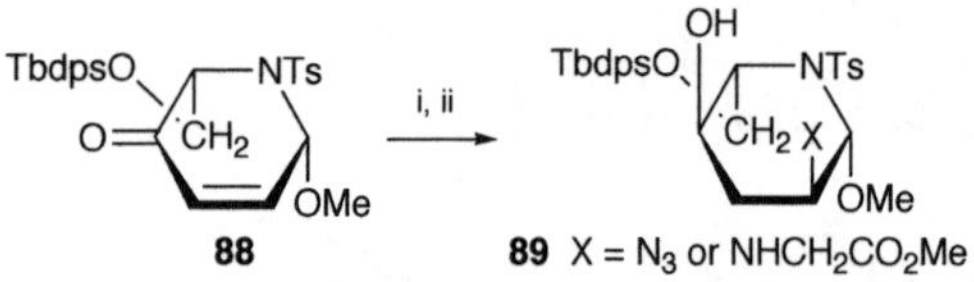

Reagents: i, NaN$_3$ or H$_2$NCH$_2$CO$_2$Me; ii, NaBH$_4$, pH 8

Scheme 22

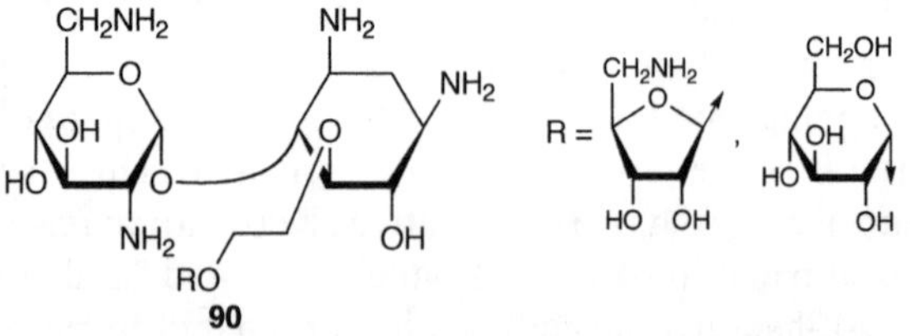

A review of strategies for the design of drugs targeting RNA and RNA-protein complexes included RNA-binding aminoglycoside antibiotics such as neomycin B, tobramycin and simplified semisynthetic mimetics of them.[79] Simplified analogues **90** have been obtained by substitution on neamine, itself obtained by degradation of Neomycin B.[80] While these showed little anti-HIV activity, coupling of the aromatic hydrocarbon pyrene group to neamine by acylation at N-6' gave a compound that bound at sub-micromolar levels to RNA from HIV.[81]

References

1. M. Otsuka, M. Fujita, Y. Matsushima, T. Eguchi, K. Shindo and K. Kakinuma, *Tetrahedron*, 2000, **56**, 8281.
2. C.S.V. Houge-Frydrych, S.A. Readshaw and D.J. Bell, *J. Antibiotics*, 2000, **53**, 351.
3. S. Wendeborn, A. de Mesmaeker, W.K.-D. Brill and S. Berteina, *Acc. Chem. Res.*, 2000, **33**, 215.
4. D. Bouhlal, P. Godé, G. Goethals, M. Massoui, P. Villa and P. Martin, *Carbohydr. Res.*, 2000, **329**, 207.
5. S. Raic-Malic, D. Svedruzic, T. Gazivoda, A. Marunovic, A. Hergold-Brundic, A. Nagl, J. Balzarini, E. De Clercq and M. Mintas, *J. Med. Chem.*, 2000, **43**, 4806.
6. C.M. Utzmann and M.D. Lederer, *Carbohydr. Res.*, 2000, **325**, 157.
7. G. Anilkumar, L.G. Nair, L. Olsson, J.K. Daniels and B. Fraser-Reid, *Tetrahedron Lett.*, 2000, **41**, 7605.
8. A. Kirschning and G.-W. Chen, *Synthesis*, 2000, 1133.
9. C.-W. Chang, L. Zhao, H. Yamase and H.-W. Liu, *Angew. Chem. Int. Ed. Engl.*, 2000, **39**, 2160.
10. K. Suzuki and H. Hashimoto, *Carbohydr. Res.*, 2000, **323**, 14.
11. M. de Kort, J. Luijendijk, G.A. van der Marel and J.H. van Boom, *Eur. J. Org. Chem.*, 2000, 3085.
12. D.J. Silva and M.J. Sofia, *Tetrahedron Lett.*, 2001, **41**, 855.
13. M. Vieira De Almeida, E. Teixeira Cesar, A.P. Soares Fontes and E. de Castro Antunes Felicio, *J. Carbohydr. Chem.*, 2000, **19**, 323.
14. T.K. Chakraborty, S. Jayapraksah, P. Srinivasu, M.G. Chary, P.V. Diwan, R. Nagaraj, A.R. Sankar and A.C. Kunwar, *Tetrahedron Lett.*, 2000, **41**, 8167.
15. M. Hamernikova, S. Pakhomova, J. Havlicek, H. Votavova and K. Kefurt, *Carbohydr. Res.*, 2000, **325**, 56.
16. L. Azema, F. Bringaud, C. Blonski and J. Périé, *Bioorg. Med. Chem.*, 2000, **8**, 717.
17. F.-Y. Dupradeau, F. Gaillard and J.-M. Beau, *Carbohydr. Lett.*, 1999, **3**, 317 (*Chem. Abstr.*, 2000, **132**, 166 414).
18. D. Pathak, D.D. Laskar, D. Prajapati and J.S. Sandhu, *Chem. Lett.*, 2000, 816.
19. Y. Nishimura, H. Adachi, T. Satoh, E. Shitara, H. Nakamura, F. Kojima and T. Takeuchi, *J. Org. Chem.*, 2000, **65**, 4871.
20. V.M. Díaz Pérez, M.I. García Moreno, C. Ortiz Mellet, J. Fuentes, J.C. Díaz Arribas, F. Javier Canada and J. M. García Fernández, *J. Org. Chem.*, 2000, **65**, 136.
21. V.M. Diaz Pérez, C. Ortiz Mellet, J. Fuentes and J.M. García Fernández, *Carbohydr. Res.*, 2000, **326**, 161.
22. J. Fuentes, D. Olano, C. Gasch and M.A. Pradera, *Tetrahedron: Asymmetry*, 2000, **11**, 2471.
23. C. Kieburg, K. Sadalapure and T.K. Lindhorst, *Eur. J. Org. Chem.*, 2000, 2035.
24. G.A. Winterfeld, J. Das and R.R. Schmidt, *Eur. J. Org. Chem.*, 2000, 3047.
25. K.C. Nicolaou, P.S. Baran, Y.-L. Zhong and J.A. Vega, *Angew. Chem. Int. Ed. Engl.*, 2000, **39**, 2525.
26. Z.-G. Wang, X. Zhang, D. Live and S.J. Danishefsky, *Angew. Chem. Int. Ed. Engl.*, 2000, **39**, 3652.

27.   V. Di Bussolo, J. Liu, L.G. Huffman, Jr. and D.Y. Gin, *Angew. Chem. Int. Ed. Engl.*, 2000, **39** 204.
28.   H. Sugimura, *Nucleosides, Nucleotides, Nucleic Acids*, 2000, **19**, 629.
29.   G.R. Smith and R.M. Giuliano, *Carbohydr. Res.*, 2000, **323**, 208.
30.   J. Gonda, E. Zavacka, M. Budesinsky, I. Cisarova and J. Podlaha, *Tetrahedron Lett.*, 2000, **41**, 525.
31.   B. Becker, *J. Carbohydr. Chem.*, 2000, **19**, 253.
32.   B. Renneberg, Y.-M. Li, H. Laatsch and H.-H. Fiebig, *Carbohydr. Res.*, 2000, **329**, 861.
33.   B. Ravindran, K. Sakthivel, C.G. Suresh and T. Pathak, *J. Org. Chem.*, 2000, **65**, 2637.
34.   C. de Saint-Fuscien, A. Terrade, P. Dauban and R.H. Dodd, *Tetrahedron Lett.*, 2000, **41**, 6393.
35.   D. Postel, A.N.V. Nhien, M. Pillon, P. Villa and G. Ronco, *Tetrahedron Lett.*, 2000, **41**, 6403.
36.   M. Koos, B. Steiner, V. Langer, D. Gyepesova and M. Durik, *Carbohydr. Res.*, 2000, **328**, 115.
37.   P.L. Barili, G. Berti, G. Catelani, F. D'Andrea and L. Puccioni, *J. Carbohydr. Chem.*, 2000, **19**, 79.
38.   S.C. Ennis, J.J. Gridley, H.M.I. Osborn and D.G. Spackman, *Synlett*, 2000, 1593.
39.   M. Franck-Neumann, L. Miesch-Gross and C. Gateau, *Eur. J. Org. Chem.*, 2000, 3693.
40.   F. Effenberger and J. Roos, *Tetrahedron: Asymmetry*, 2000, **11**, 1085.
41.   A. Fazio, M.A. Loretto, P.A. Tardella and D. Tofani, *Tetrahedron*, 2000, **56**, 4515.
42.   S.-I. Kiyooka, K. Goh, Y. Nakamura, H. Takesue and M.A. Hena, *Tetrahedron Lett.*, 2000, **41**, 6599.
43.   C. Saotome, M. Ono and H. Akita, *Tetrahedron: Asymmetry*, 2000, **11**, 4137.
44.   R.M. Davey, M.A. Brimble and M.D. McLeod, *Tetrahedron Lett.*, 2000, **41**, 5141.
45.   A. Zhuang, J. Thornmauge and K.A. Jørgensen, *Chem. Commun.*, 2000, 459.
46.   K.A. Jørgensen, *Angew. Chem. Int. Ed. Engl.*, 2000, **39**, 3559.
47.   M. Joubert, A. Defoin, C. Tarnus and J. Streith, *Synlett*, 2000, 1366.
48.   C.J. Moule, P.R. Clements, J.J. Hopwood and G.T. Crisp, *Synth. Commun.*, 2000, **30**, 1489.
49.   K. Blades, T.J. Donohoe, J.J.G. Winter and G. Stemp, *Tetrahedron Lett.*, 2000, **41**, 4701.
50.   P. Dauban, C. de Saint-Fuscien, F. Archer, L. Prézeau, I. Brabet, J.-P. Pin and R.H. Dodd, *Bioorg. Med. Chem. Lett.*, 2000, **10**, 129.
51.   K. Toshima, S. Takai, Y. Maeda, R. Takano and S. Matsumura, *Angew. Chem. Int. Ed. Engl.*, 2000, **39**, 3656.
52.   C. Dini, N. Drochon, P. Ferrari and J. Aszodi, *Bioorg. Med. Chem. Lett.*, 2000, **10**, 143.
53.   T. Fuchss and R.R. Schmidt, *J. Carbohydr. Chem.*, 2000, **19**, 677.
54.   B.K.S. Yeung, S.L. Adamski-Werner, J.B. Bernard, G. Poulenat and P.A. Petillo, *Org. Lett.*, 2000, **2**, 3135.
55.   J.C. Castro-Palomino and R.R. Schmidt, *Tetrahedron Lett.*, 2000, **41**, 629.
56.   W.L. Conrad, F. Piskorz, R.D. Locke and K.L. Matta, *Bioorg. Med. Chem. Lett.*, 2000, **10**, 793.
57.   E.V. Shipova, N.E. Byramova and N.V. Bovin, *Bioorg. Khim.*, 1999, **24**, 634 (*Chem. Abstr.*, 2000, **132**, 137 636).
58.   Y. Zhang and P. Zhang, *Guangdong Yaoxueyuan Xuebao*, 1999, **15**, 252 (*Chem.*

Abstr., 2000, **132**, 334 677).

59.   O. Lockhoff and Y. Hayauchi, J. Carbohydr. Chem., 2000, **19**, 603.

60.   J.C. Wilson, R.J. Thomson, J.C. Dyason, P. Florio, K.J. Quelch, S. Abo and M. von Itzstein, Tetrahedron: Asymmetry, 2000, **11**, 53.

61.   J. Budka, M. Tkadlecová, P. Lhoták and I. Stibor, Tetrahedron, 2000, **56**, 1883.

62.   M.A.F. Prado, R.J. Alves, J.D. Souza Filho, R.B. Alves, M.T.C. Pedrosa, R.F. Prado and A.A.G. Faraco, J. Chem. Soc., Perkin Trans. 1, 2000, 1853.

63.   C. Monneret, S. Risse, P. Ardouin and A. Gouyette, Eur. J. Med. Chem., 2000, **35**, 137 (Chem. Abstr., 2000, **132**, 347 823).

64.   M.I. Garcia-Moreno, J.M. Benito, D.O. Mellet and J.M.G. Fernandez, Tetrahedron: Asymmetry, 2000, **11**, 1331.

65.   J.M. Benito, C.O. Mellet and J.M.G. Fernandez, Carbohydr. Res., 200, **323**, 218.

66.   T.J. Baker, N.W. Luedtke, Y. Tor and M. Goodman, J. Org. Chem., 2000, **65**, 9054.

67.   N.W. Luedtke, T.J. Baker, M. Goodman and Y. Tor, J. Am. Chem. Soc., 2000, **122**, 12035.

68.   J. Kroutil, T. Trnka and M. Cerny, Org. Lett., 2000, **2**, 1681.

69.   J.F. Tolborg and K.J. Jensen, Chem. Commun., 2000, 147.

70.   R. Wegner, M. Gottschaldt, H. Gorls, E.-G. Jager and D. Klemm, Angew. Chem. Int. Ed. Engl., 2000, **39**, 595.

71.   K. Sadalapure and T.K. Lindhorst, Angew. Chem. Int. Ed. Engl., 2000, **39**, 2010.

72.   K. Ikeda, K. Miyajima, T. Shimizu and K. Achiwa, Chem. Pharm. Bull., 2000, **48**, 32.

73.   Y. Sakai, M. Oikawa, H. Yoshizaki, T. Ogawa, Y. Suda, K. Fukase and S. Kusumoto, Tetrahedron Lett., 2000, **41**, 6843.

74.   M. Nilsson and T. Norberg, Carbohydr. Res., 2000, **327**, 261.

75.   A. Banaszek and B. Janisz, Tetrahedron: Asymmetry, 2000, **11**, 4693.

76.   C.A. Iriarte Capaccio and O. Varela, Tetrahedron: Asymmetry, 2000, **11**, 4945.

77.   I. Gautier-Lefebvre, J.-B. Behr, G. Guillerm and N.S. Ryder, Bioorg. Med. Chem. Lett., 2000, **10**, 1483.

78.   S.D. Koulocheri, P. Magiatis, A.-L. Skaltsounis and S.A. Haroutounian, Tetrahedron, 2000, **56**, 6135.

79.   T. Hermann, Angew. Chem. Int. Ed. Engl., 2000, **39**, 1890.

80.   N. Nishizono and V. Nair, Nucleosides, Nucleotides, Nucleic Acids, 2000, **19**, 283.

81.   K. Hamasaki, M.-C. Woo and A. Ueno, Tetrahedron Lett., 2000, **41**, 8327.

10

Miscellaneous Nitrogen-containing Derivatives

1 Glycosylamines and Related Glycosyl-*N*-bonded Compounds

1.1 Glycosylamines and Glycosylammonium Salts. – X-Ray crystallography has shown that *N*-phenyl-, *N*-4-chlorophenyl- and *N*-carboxamido-D-mannosylamine are in the β-pyranosyl form in the solid state, whereas mannose oxime is in the open chain form.[1] The conclusion from an examination of the conformations of various 2,3,4,6-tetra-*O*-acetyl-*N*-aryl-D-glucopyanosylamines in neutral and protonated form was that there was no evidence to support the existence of a reverse anomeric effect.[2]

A 27-step synthesis of spicamycin amino-nucleoside (**1**) from *myo*-inositol featured a palladium-catalysed coupling of 5-chloropurine and with a glycosylamine derivative.[3] Full details have been published on the products obtained from reactions of 5,6-diamino-2-methoxy-3-methylpyrimidi *N*-4(3H)-one with xylose and with other pentoses in alcohol solutions (*cf.* Vol. 27, p. 127, ref. 8).[4]

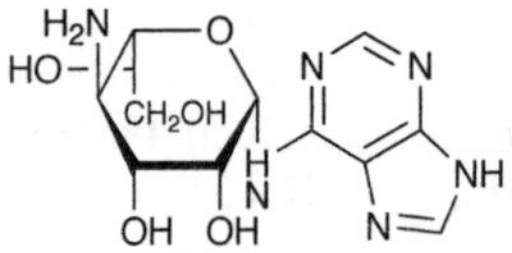

1

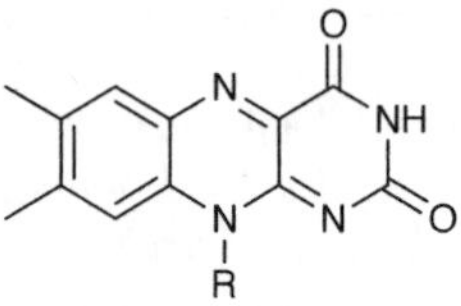

2 R = β-D-Rib*p*
or β-D-Rib*f*-Ac₃

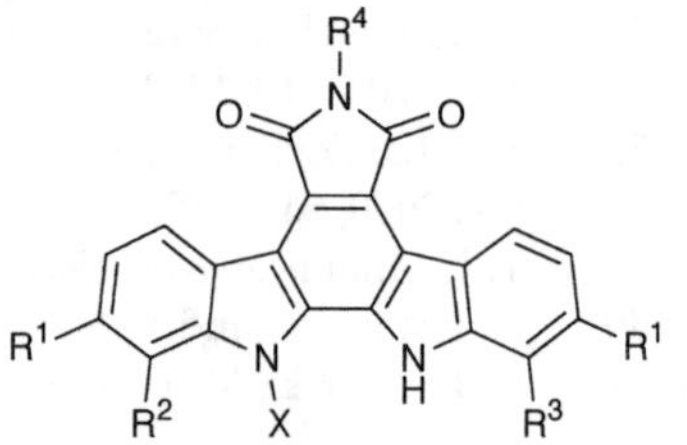

3 X = β-D-Rib*f*
R¹ = OBn, R² = R³ = H,
R⁴ = NHCH(CH₂OH)₂

4 X =

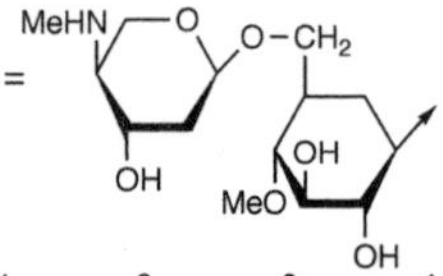

R¹ = H, R² = Cl, R³ = H, R⁴ = Me

The flavin ribosides **2**, required asprobes of the pairing properties of flavin-containing oligonucleotides, were synthesized by elaboration of the corresponding *N*-(2-nitro-4,5-dimethylphenyl)-riboside triacetates. The *O*-deprotected

Carbohydrate Chemistry, Volume 34
© The Royal Society of Chemistry, 2003

furanosyl derivative was too unstable to be isolated.[5] *N*-Glycosylations of imidazo[4,5-d]phenazine, its 2-methyl- and 2-trifluoromethyl-derivatives, and 1,2,3-triazolo[4,5-d]phenazine have been reported.[6] Further rebeccamycin analogues have been synthesized. The β-D-ribofuranosyl derivative **3** was identified as the most potent topoisomerase I inhibitor of a set of sugar variants, the synthesis of which involved base-catalysed reaction of an aglycon (in which $R^4 =$ Me) with a perbenzylated aldopyranosyl chloride. These reactions gave predominantly β-anomersexcept in the case of the D-allose derivative.[7] Analogue **4**, the antibiotic AT 2433-A1, was obtained by a direct, acid-catalysed condensation of a free disaccharide with a tetrahydro-precursor of the aglycon, followed by aromatization. Syntheses of a variety of model compounds and analogues were also reported.[8] Michael additions of heterocyclic bases to the 2-nitro-D-galactal **5** yielded 2-deoxy-2-nitro-β-D-galactosylamine derivatives **6**, one being converted to the 2-acetamido-2-deoxy-β-D-galactosylamine derivative **7** (Scheme 1).[9]

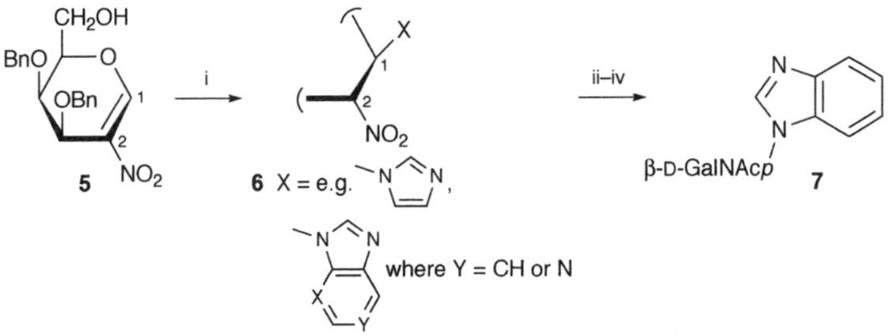

Reagents: i, XH, base, THF; ii, H_2, Raney Ni; iii, Ac_2O, py; iv, H_2, Pd(OH)$_2$

Scheme 1

Activation of Schiff bases by *N*-glycosylation has been shown to induce asymmetric Mannich reactions with *O*-silyl ketene acetals to give β-amino acid esters in good yields, as exemplified in Scheme 2. The predominant diastereomer was favoured by a ratio of 8:1 to 5:4, depending on the reactants. The *N*-glycosyl linkage could be cleaved with HCl in MeOH.[10] A review of asymmetric hetero-Diels–Alder reactions involving imines included the use of *N*-glycosyl-imines.[11]

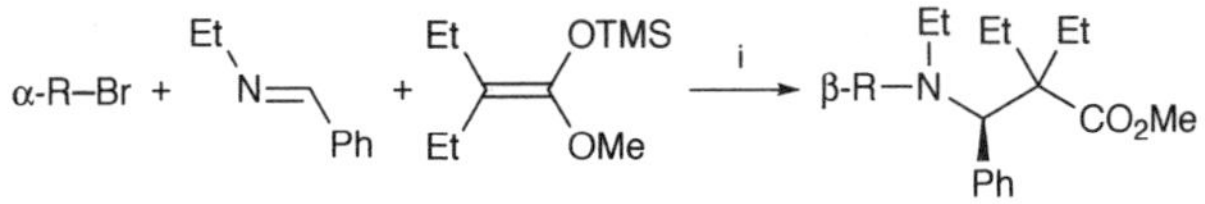

Reagents: i, AgOTf, 2,6-lutidine

Scheme 2

Several 4-thio-D-glucofuranosylamine disaccharide derivatives such as **8** (α:β, 1:3) were obtained by direct condensation of 4-thio-D-glucose with an amino-sugar.[12] A review of the application of palladium π-allyl complexes of unsaturated carbohydrates in fuctionalization at the anomeric centre included the use of *N*-nucleophiles to provide unsaturated *N*-glycosyl derivatives.[13] Cyclic 1,2-*N,O*-phosphate derivatives of aldoses feature as intermediates in selective 2-*O*-phosphorylation reactions (see Chapter 7).

Glycosylamine derivatives **10** were prepared from the β-glycosyl bromide derivative **9** of methyl D-*arabino*-hex-2-ulosonate, either by Hg(CN)$_2$-catalysed condensation with the free amine, or by preparation, reduction, anomerisation and acylation of the corresponding α-glycosyl azide.[14] 'Fused glycosyl-glycines' such as **11**, its sialic acid analogue and 5-*epi*-hydantocidin (**12**) have been prepared for use as synthons for artificial glycopeptides.[15] 4-*C*-Substituted α-L-lyxofuranose nucleoside analogues such as **13** were obtained from the corresponding 4-*C*-bromo-D-ribose derivative by displacement with inversion.[16] Four bicyclic glycosylamine derivatives, *e.g.* **14**, were obtained by use of intramolecular hydrogen abstraction reactions (Scheme 3).[17]

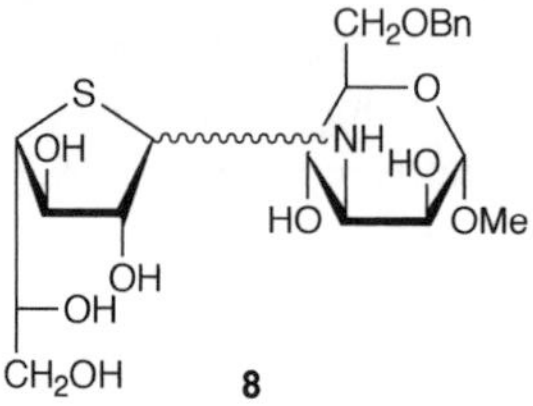

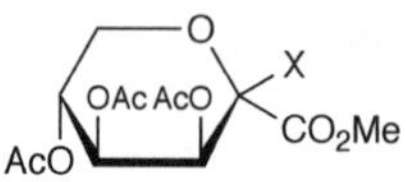

Reagents: i, PhI(OAc)$_2$ or PhIO, I$_2$

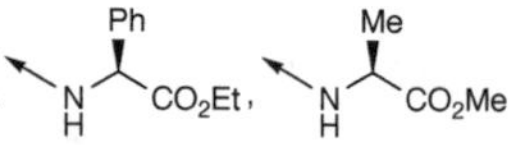

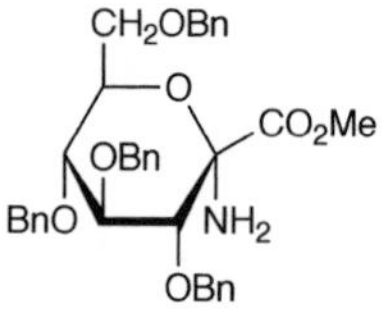

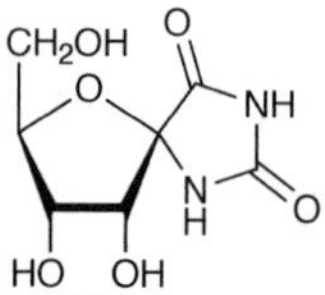

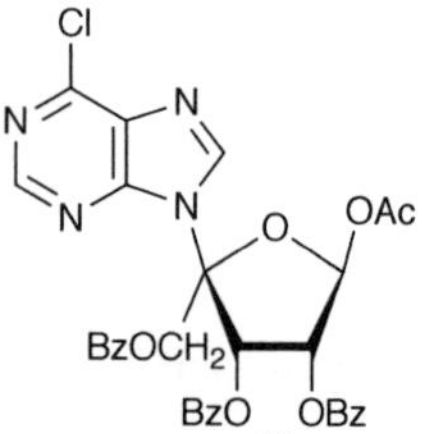

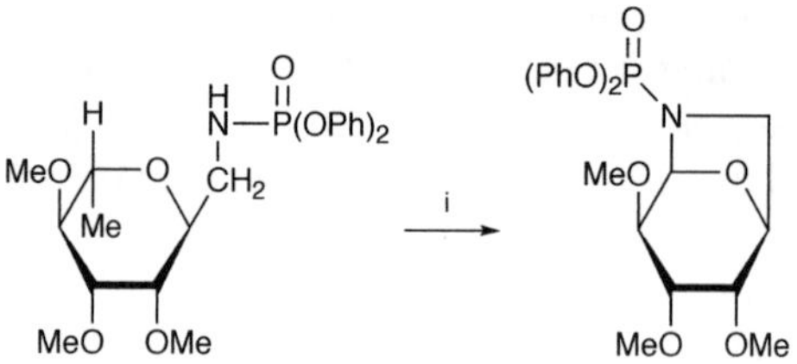

Scheme 3

The preparation, X-ray crystal structure and *ab initio* calculation of the charge distribution of *N*-(2-acetamido-3,4,6-tri-*O*-acetyl-2-deoxy-β-D-glucopyranosyl)-pyridinium chloride have been reported.[18] The anomeric *N*-(2,3,4,6-tetra-*O*-acetyl-D-glucopyranosyl)-4-dodecylpyridinium bromides were prepared as novel cationic surfactants.[19] The panel of five substituted-pyridinium salts **15** were synthesized from the peracetylated glycosyl chloride of β-D-*N*-acetyl-neuraminic acid and used to probe the mechanisms of enzymatic and non-enzymatic hydrolysis of neuraminides.[20] Detailed analysis of the solvolysis of *N*-(2-deoxy-α- and β-D-glucopyranosyl)-4-bromoisoquinolinium tetrafluoroborate in aqueous alcoholic solvents indicated S$_N$2-like transition states.

D-Glucal and 1,6-anhydro-2-deoxy-β-D-glucopyranose were found amongst the reaction products.[21] The photohydration of the aglycon of N-(α-D-glucopyranosyl)-pyridinium chloride is covered in Chapter 22.

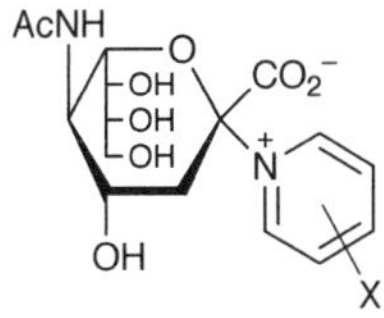

15 X = H, 3-OMe, 3-Me, 4-Me, 3,4-Me₂   **16** X = Y = H or X = OH, Y = H or X = OH, Y = NH₂

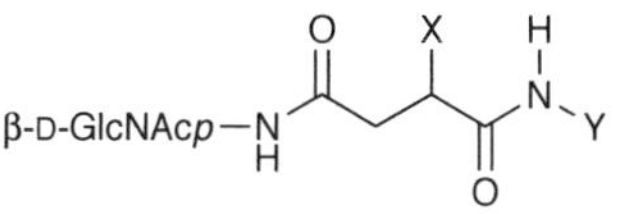

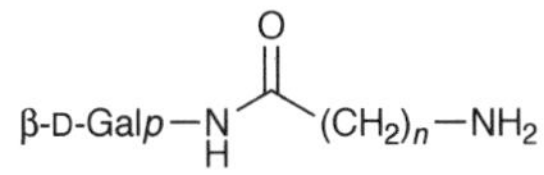

17 n = 3, 5 or 10

18 R = 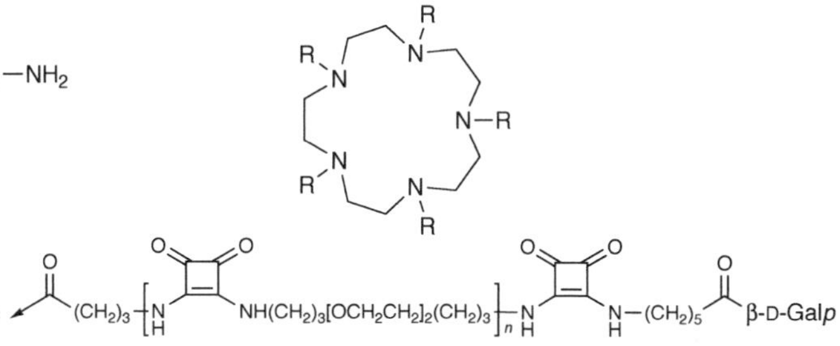   Where n = 1, 2, 3 or 4

1.2 Glycosylamides Including N-Glycopeptides. – Glycosylamides were prepared directly in high yield by reaction of glycosyl azides with acyl chlorides and Ph₃P. β-Azides gave β-products, while α-azides gave α,β-mixtures.[22]

The three N^4-(2-acetamido-2-deoxy-β-D-glucopyranosyl)-L-asparagine analogues **16** were synthesized by acylation of the corresponding glycosylamine.[23] Amphiphilic 2-acetamido-2-deoxy-β-D-glucopyranosylamide derivatives bearing an amino acid moiety at C-2 and two long chain fatty acyl moieties on the aglycon have been prepared (see Chapter 9, ref 59 for an example).[24]

Polymers bearing N-linked sugar units have been prepared. In studies directed towards specific drug delivery methods, microspheres have been made from the polymers obtained by condensation of polylactic acid with the galactosylamide-amines **17**, synthesized from tetra-O-acetyl-β-D-galactopyranosyl isothiocyanate. These microspheres were recognized by galactose-specific lectins.[25] The pentavalent β-D-galactosylamide dendrimers **18** were prepared by a modular approach using squaric acid coupling reactions. The largest (n = 4) had the highest affinity for the heat-labile enterotoxin of *Escherichia coli*.[26] The Ugi reaction (which involves one-pot condensation of an amine, an isonitrile, a carbonyl compound and a carboxylic acid) has been used to append 1–10 N-glycosylamide units to bovine serum albumin (BSA) or horseradish peroxidase. For example, condensation of BSA with β-D-glucopyranosyl isocyanide, acetone and Rhodamine B (a carboxylic acid) gave the adduct **19**.[27]

An extensive review (261 references) has been published on the synthesis of glycopeptides containing carbohydrate and peptide recognition motifs.[28] N-Acylation with fluoride **20** wasused to prepare glycosylamino acid derivatives such as **21**, the amino acid moiety then being elaborated into a pentapeptide.[29]

Danishefsky's glycal iodosulfonimidation-ethanethiolate rearrangement method has been applied to the total synthesis of the high-mannose *N*-linked glycopeptide **22** (Scheme 4).[30] *N*-Glycosylated α-aminooxyacids **23** were obtained by condensation of the glycosylamine with an L-malic acid derivative.[31] In a new synthesis of neoglycopeptides, the amine **25** was synthesized from *N*-allyl-β-maltosylamine **24** (Scheme 5) and coupled to a dipeptide (benzyloxycarbonyl-glutaminyl-glycine) through the acyl transferase action of transglutaminase. The method was applicable to amines varying in the sugar or spacer moiety.[32]

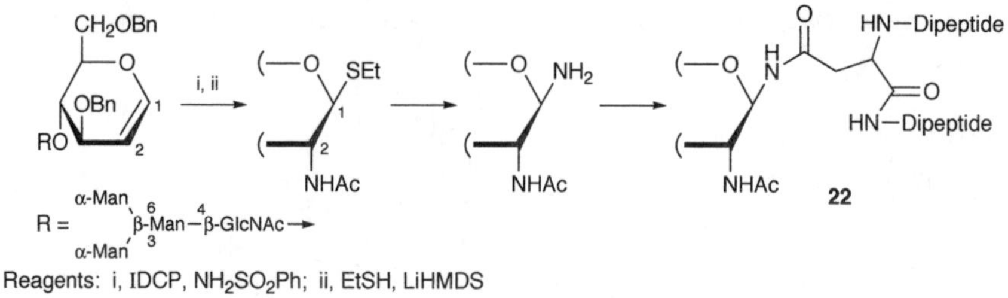

Reagents: i, IDCP, NH$_2$SO$_2$Ph; ii, EtSH, LiHMDS

Scheme 4

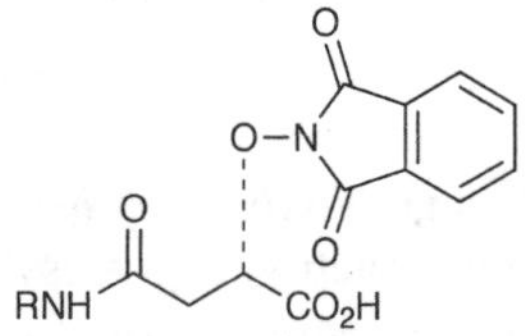

23 R = β-D-Glc(Ac)$_4$, β-D-GlcNHAc(Ac)$_3$, β-lactosyl(Ac)$_7$

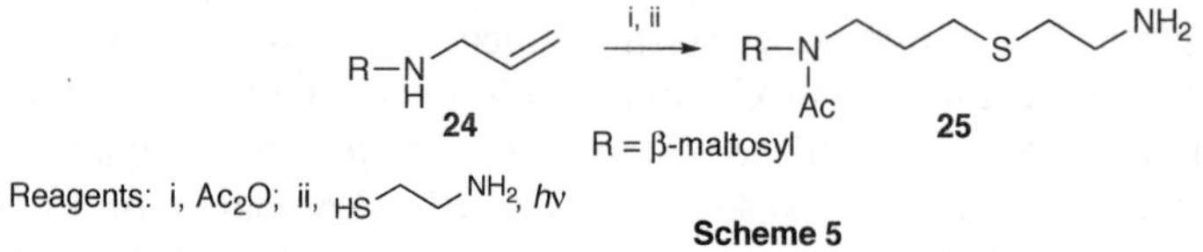

R = β-maltosyl

Reagents: i, Ac$_2$O; ii, HS$\frown$NH$_2$, hv

Scheme 5

Solid phase glycopeptide syntheses continue to be reported. A modified Staudinger reaction was used to synthesize the polymer-bound β-glucuronosylamide **26** from the corresponding β-azide. The leucine moiety was then converted into a leucine-enkephaline pentapeptide motif.[33] The polymer-bound β-galactosylamine **27**, obtained from the corresponding β-azide, was used as an asymmetric ammonia equivalent in stereoselective Ugi reactions, an example of one combination from a set of five aldehydes and three isocyanates being shown in Scheme 6. Diastereomeric ratios ranging of 10:1 to 15:1 were achieved. A set of D-amino-acid derivatives was obtained following removal of product from the resin and cleavage of the glycosidic link (CF$_3$CO$_2$H, then HCl, MeOH).[34] A range of β-glycosylamines were polymer-bound and coupled to asparagine and glutamine derivatives to give products such **28** and **29**, respectively. Following cleavage from the resin (CF$_3$CO$_2$H), these were used in the construction of glycopeptide libraries.[35]

The synthesis and glycosidase inhibitory properties of several nitrogen-in-the-

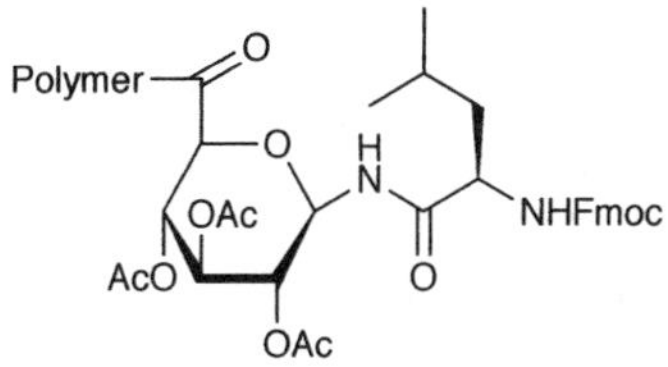

Polymer

OAc
AcO
OAc

NHFmoc

26

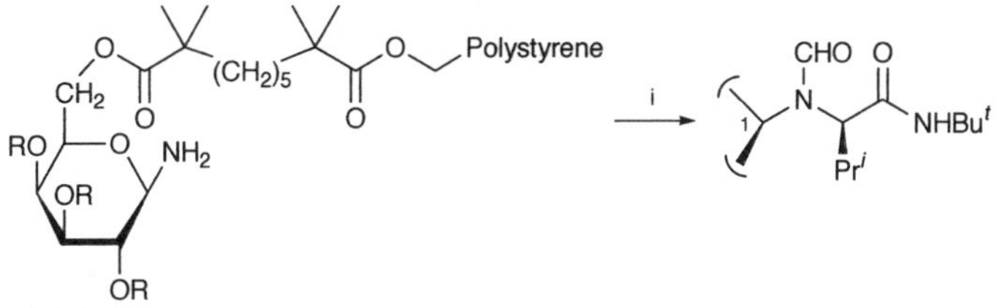

CH₂

(CH₂)₅

O—Polystyrene

RO
OR
OR
NH₂

27 R = Piv

i

CHO O

N
Pr^i
NHBu^t

Reagents: i, HCO₂H, Pr^{i}CHO, Bu^{t}N≡C, ZnCl₂

Scheme 6

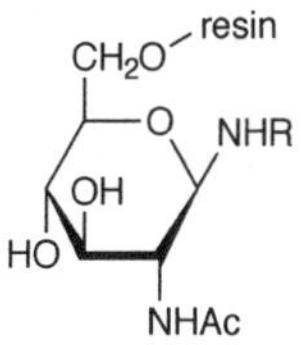

CH₂O—resin

OH
HO
NHR
NHAc

28 R =

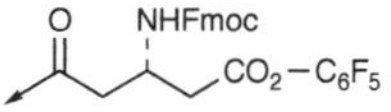

O NHFmoc

CO₂–C₆F₅

29 R =

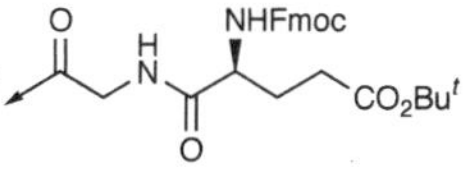

O

NHFmoc

N
H

CO₂Bu^t

ring glycosylamines have been reported. The 5-amino-5-deoxy-4-C-methyl-D-ribopyranosylamines **31** were obtained from the known lactam **30** (Scheme 7) and shown to be potent α-L-fucosidase inhibitors, the trichloro-derivative having a K_i value of 9×10^{-9} M.[36] In a similar way, the potent β-glucuronidase inhibitor **32** ($K_i = 6.5 \times 10^{-8}$ M) and three isomers were obtained in 12 steps from 5,6-O-isopropylidene-L-galactono-1,4-lactone.[37] Siastatin B (**33**), a sialidase inhibitor, has been synthesized in 15 steps from a resolved 4,5-dehydropiperidine-3-carboxylate. Two analogues, the N-trifluorocetate and 5-C-benzyl derivatives, were also reported.[38] Natural siastatin B was converted into compounds **34**, which are mimics of known sialidase inhibitors.[39]

1.3 N-Glycosyl-carbamates, -ureas, -isothiocyanates, -thioureas and Related Compounds. – N-Glucosyl cyclic carbamates such as **35** were obtained from silver ion-mediated desulfurization – condensation of peracetylated β-D-glucopyranosyl isothiocyanate with α-hydroxy-acids. When applied to aliphatic β-hydroxy-acids, the initially formed cyclic carbamates underwent decarboxylation. 3-Hydroxybutanoic acid thus gave **36**, while salicylic acid gave **37**.[40] The cyclic carbamates **38**, prepared from D-glucose, were employed as chiral auxiliaries. Thus N-acylation, conversion to a boron enolate and condensation with an aldehyde gave diastereomeric mixtures enriched in *syn*-products such as **39**.[41]

N-Glycosyl-urea derivatives such as **40**, required as building blocks for glycopeptide mimics, were synthesized from the corresponding glycosyl isocyanides, these being available from the corresponding α- or β-glucosyl azides (Scheme 8). While the α-azide gave a 4:1 mixture of α- and β-isocyanides, no epimerization occurred in the conversion of the β-azide, nor in the formation of the *N*-glycosyl-ureas.[42] The 5′,6-cyclo-dihydro-2′-deoxyuridine **41**, a major product of gamma-irradiation of oxygen-free aqueous solutions of 2′-deoxycytidine, has been synthesized and incorporated into DNA oligomers.[43]

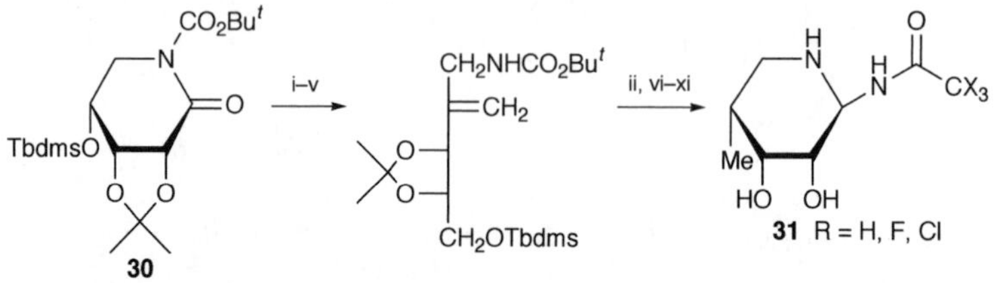

Reagents: i, NaBH₄; ii, Bu₄NF; iii, Tbdms–Cl, Im, DMF; iv, Dess-Martin oxidation; v, Ph₃P=CH₂; vi, Swern oxidation; vii, PhthNH, Ph₃P, DEAD; viii, H₂, Pd/C; ix, NH₂NH₂, MeOH; x, Ac₂O, (CF₃CO)₂ or Cl₃CCOCl; xi, HCl

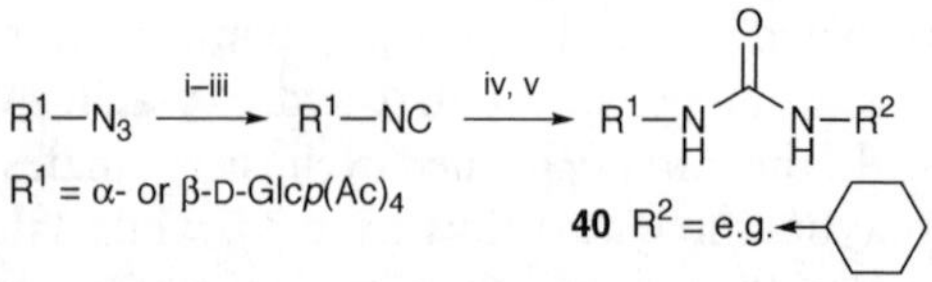

Scheme 7

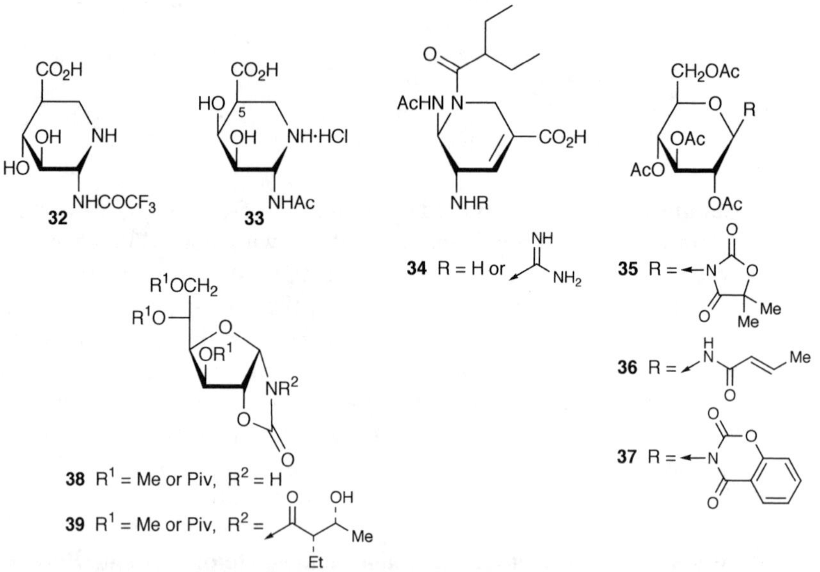

Reagents: i, H₂, Pd/C; ii, AcOCHO; iii, CBr₄, Ph₃P, Et₃N; iv, C₅H₅N → O, I₂; v, R²–NH₂

Scheme 8

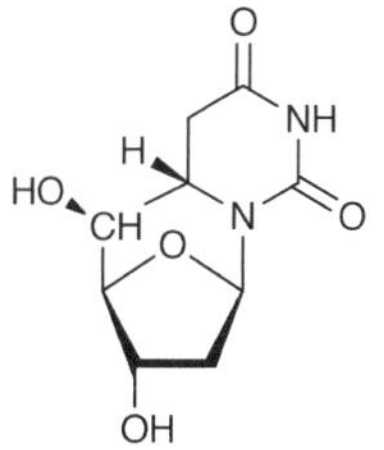

41 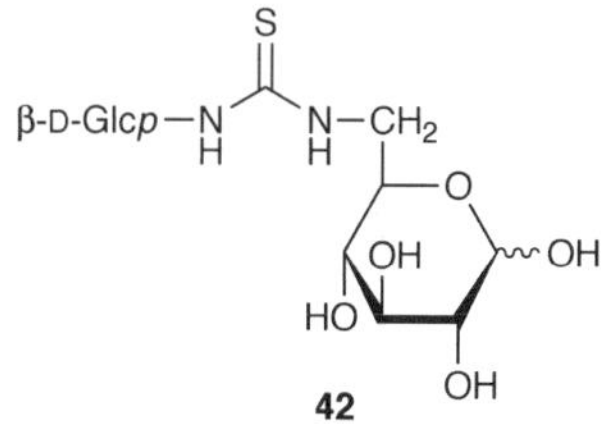   **42**

The synthesis of β-D-glucopyranosyl-thioureas has been reported.[44] *N'*-Glucosylated sugar urea and thiourea derivatives such as **42** were synthesized by Ph₃P-induced condensation of the corresponding protected 6-azido-sugar with tetra-*O*-acetyl-β-D-glucopyranosyl isothiocyanate followed by either hydrolysis (HOAc, H₂O) or thiolysis (H₂S, silica gel), respectively, then deprotection.[45] Acetylation of such doubly substituted thiourea derivatives led to various regioisomeric *N*- and *S*-acetylated derivatives.[46]

Imidazoline-2-thione derivatives such as **43** were synthesized by condensation of 1-amino-1-deoxy-D-fructose (or its *N*-alkylated derivatives) with glycosyl isothiocyanates and converted to dehydration products or spiro-cyclic derivatives such as **44** and **45**, respectively, under different acidic conditions.[47] The spiro-thiohydantoin analogues **47** and **48** were synthesized from the photobromination product **46** (Scheme 9), the latter being able to interfere with the formation of glucose from glycogen in rats.[48,49]

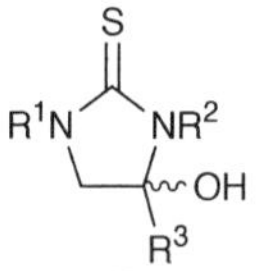

43 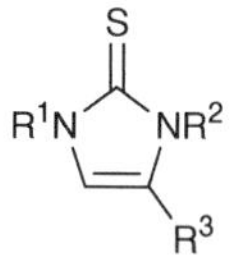   **44** 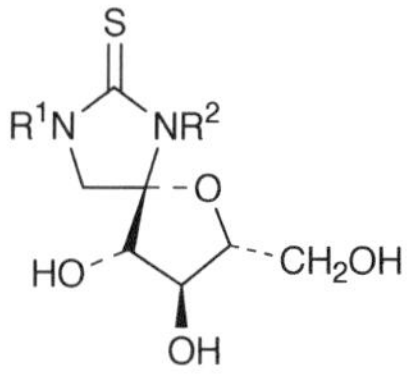   **45**

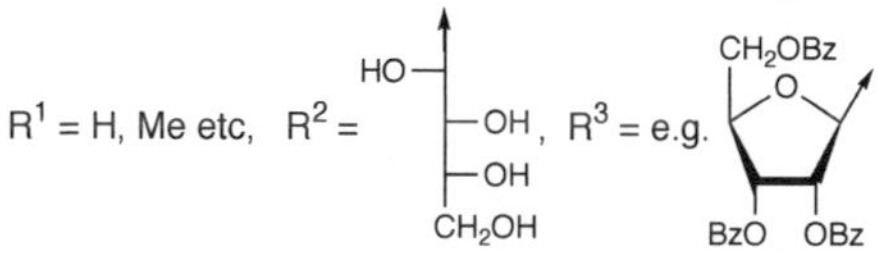

R^1 = H, Me etc, R^2 =   , R^3 = e.g.

2 Azido-sugars

2.1 Glycosyl Azides and Triazoles. – Peracetylated 1,2-*trans*-glycosyl azides have been synthesized from the corresponding 1-*O*-acetates using NaN₃ and BF₃.OEt₂,[50] or the corresponding 1-bromides or chlorides using NaN₃ in aqueous acetone.[51] The synthesis of 1,2-*trans*-glycofuranosyl azides from the corresponding peresters with TmsN₃ and various Lewis acids has been studied. Variable amounts of the 1,2-*cis*-isomers were produced, depending upon the Lewis acid. The best 1,2-*trans*-selectivity was achieved with TmsN₃ and catalytic SnCl₄. Poorer selectivity was observed with the corresponding 1,2-*cis*-glycosyl

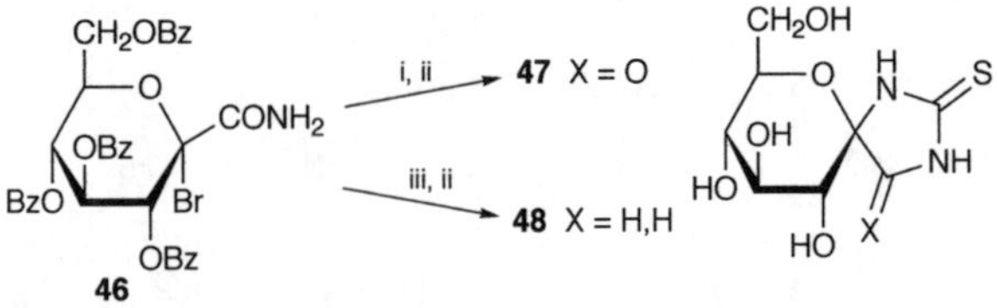

Reagents: i, NH_4SCN; ii, NaOMe; iii, NH_4SCN, S_8

Scheme 9

peresters.[52] L-Arabino-, D-ribo- and D-xylo-furanosyl azides were prepared by treatment of reducing pentose derivatives with $(PhO)_2P(O)Cl$ or TsCl and NaN_3 under phase-transfer catalysis conditions.[53] The *O*-protected α-D-ribo-, β-D-arabino- and 2-deoxy-2-fluoro-β-D-arabino-furanosyl azides, all with a 1,2-*cis*-stereochemistry, were the predominant products from reaction of the corresponding 1,2-*trans*-glycofuranosyl halides with NaN_3 under phase-transfer catalysis conditions. 3,5-Di-*O*-(4-methylbenzoyl)-2-deoxy-β-D-*erythro*-pentofuranosyl azide was obtained as the major isomer from reaction of the corresponding α-chloride with CsN_3 in DMSO.[54]

Reaction of β-D-galactopyranosyl azide with chloral and dicyclohexylcarbodiimide in 1,2-dichloroethane gave the D-gulopyranosyl product **49** as result of epimerization at C-3, in a one-pot acetalation – epimerization procedure that is generally applicable to cyclic polyols with three contiguous hydroxyls in a *cis–trans* sequence. Thus, D-arabino-, L-fuco-, and D-manno-pyranosyl azides gave D-lyxo-, L-gulo- and D-altro-products, the reaction being independent of the presence of the azido-function.[55] Regioisomeric pairs of *N*-glycosyl-1,2,3-triazoles such as **50** were obtained by dipolar cycloaddition of the corresponding azide with 3,3,3-trifluoropropynylbenzene.[56]

2.2 Other Azides and Triazoles. – 2-Azido-2-deoxy-α-D-glucopyranosyl inositol derivatives have been synthesized by two approaches: (a) the use of the conformationally constrained 5-pentenyl 2-azido-3-*O*-benzyl-4,6-*O*-benzylidene-2-deoxy-β-D-glucoside as a glycosyl donor which has a non-participating group at C-2; and (b) the use of a 1,2-*O*-(5-pentenyl orthobenzoate)-β-D-mannosyl donor, followed by azidation with inversion at C-2 through displacement of a 2-triflate with $TmsN_3$ and Bu_4NF.[57] In a similar way, 2,3-diazido-2,3-dideoxy-β-D-mannopyranosides were synthesized either directly using the corresponding α-mannosyl bromide as donor, or indirectly using a 3-azido-3-deoxy-

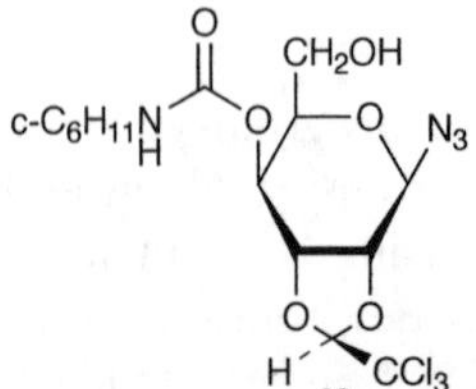

49

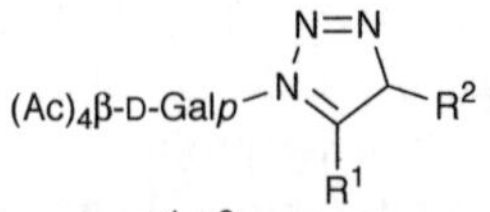

50 $R^1,R^2 = CF_3$, Ph

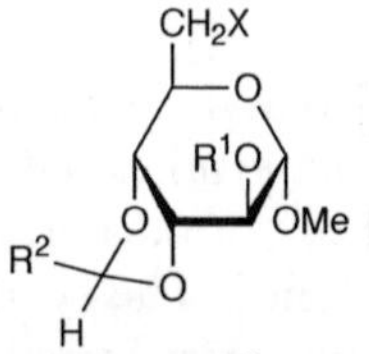

51 $X = N_3$, $R^1 = Ac$, $R^2 = Me$

52 X = OH, R^1 = C(O)NH(c-C_6H_{11}), $R^2 = CCl_3$

D-glucopyranosyl donor, and subsequent displacement of a 2-triflate with azide ion.[58] The crystalline methyl 3-azido-2,3-dideoxy-α-D-glucopyranoside has been synthesized from D-glucal triacetate by hydration, Michael addition of azide, methanolysis, chromatographic fractionation and deacetylation.[59]

6-Azido-sugar derivatives such as **51** were synthesized conventionally from the known trichloroethylidene carbamate derivatives such as **52**, prepared by a one-pot acetalation-epimerization procedure from methyl α-D-mannopyranoside (see Section 2.1).[60] The 2,6-anhydro-5-azido-2-*C*-hydroxymethyl-L-alloheptose derivative **53** has been synthesized from a racemic norbornenone derivative.[61] 6-Azido-6-deoxy-, and 5,6-diazido-, 5-azido-6-fluoro- and 6-azido-5-fluoro-5,6-dideoxy-D-fructoses **54** have been prepared by polyol dehydrogenase-catalysed isomerization of the corresponding D-glucose derivatives.[62]

Fleet and co-workers have prepared tetrahydrofuran azido-acids such as **55** and **56** and demonstrated their use as building blocks for peptidomimetics.[63,64] The synthesis of 1,4-anhydro-3,6-diazido-4-thio-alditol derivatives is detailed in Chapter 11.[65]

The thermal cycloaddition of 3-azido-3-deoxy-1,2:5,6-di-*O*-isopropylidene-α-D-glucofuranose to C$_{60}$ has been described.[66]

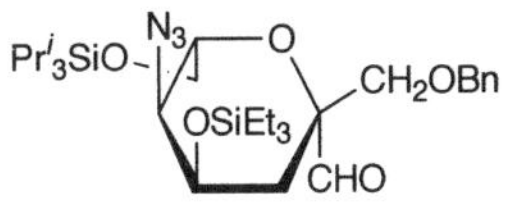

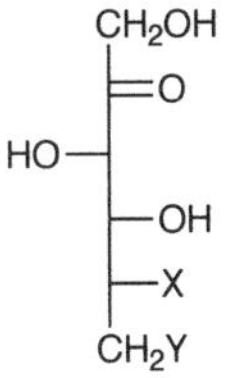

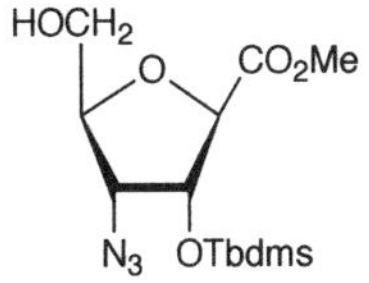

3   **Nitro- and Nitroso-sugars**

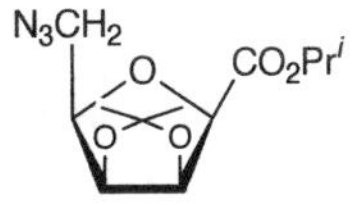

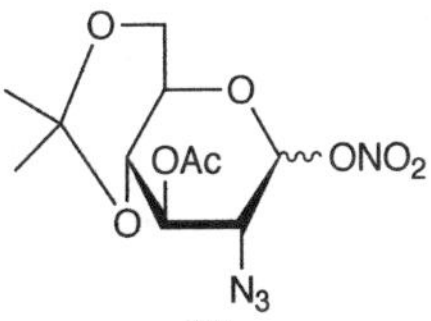

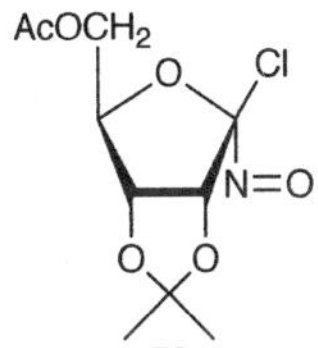

Reviews have covered diastereoselective nitro-aldol reactions for the synthesis of amino-trideoxyhexoses,[67] and anionic conjugate addition reactions of unsaturated nitro-sugars.[68] Azidonitration of conformationally constrained D-glucal derivatives yielded predominantly the D-*gluco*-isomers such as **57**.[69] The total synthesis of the antibiotic everninomicin, which contains the nitro-sugar evernitrose, is covered in Chapter 4.

The α-chloronitroso-sugar **58** was synthesized by reaction of the corresponding oximino-lactone with *tert*-butyl hypochlorite and used as a dienophile in asymmetric hetero-Diels–Alder reactions. Adducts were produced as hydrochloride salts in 93–99% *ee* with simultaneous solvolytic removal of the auxiliary.[70]

The synthesis of 1,4:3,6-dianhydro-D-glucitol 5-nitrate from the 2,5-dinitrate is covered in Chapter 18.

4   Oximes, Hydroxylamines and Nitrones

The synthesis, properties and applications of hydoxyimino derivatives of pyranoses and pyranosides have been reviewed (67 references).[71] The 2-nitroso-α-D-xylopyranosyl chloride derivative **59** has been converted to a 1:1 mixture of N-glycosyl-pyrazole oxime α-**60** and its β-anomer. The β-anomer of **60** could be 3-deoxygenated (on treatment NaBH₄) or converted to a mixture of the N-β-D-arabino- and xylo-pyranosyl-pyrazoles (by reaction with CH_3CHO then NaBH₄). The α-anomer gave mostly the 3-azide **61** on reaction with sodium azide (Scheme 10).[72]   Tri-O-acetyl-2-deoxy-2-nitroso-α-D-galactopyranosyl chloride was similarly converted to anomeric mixtures of the thioglycoside 2-oximes **62**.[73]

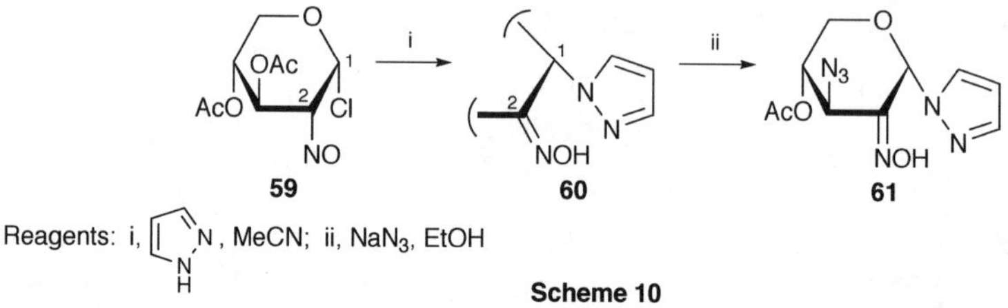

Reagents: i, (pyrazole), MeCN; ii, NaN_3, EtOH

Scheme 10

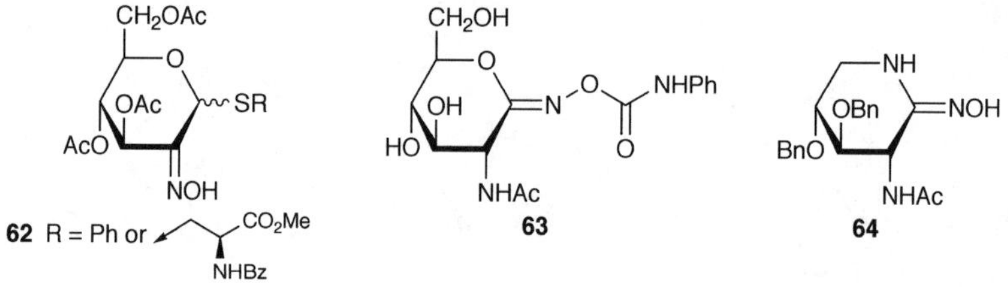

62 R = Ph or (CO₂Me / NHBz)   **63**   **64**

An improved synthesis of the oximino-lactone derivative **63** ('PUGNAc') in five steps from 2-acetamido-2-deoxy-α-D-glucopyranose tetraacetate has been described, the critical step being a low temperature oxidation of an acyclic aldose oxime intermediate to a cyclic oximino-lactone with NCS and DBU.[74] The D-xylono-1,5-lacatam oxime derivative **64** has been synthesized from an L-serinal derivative as a potential N-acetylglucosaminyl transferase inhibitor.[75]

Reduction of 1-nitro-1-alkene derivatives such as **65** provided 2-deoxy-aldose oximes and their elimination products such as **66** and **67**, respectively (Scheme 11). The oxime **66** could be converted into the corresponding free 2-deoxy-sugar, 2-deoxy-aldononitrile and 2-deoxy-aldonic acid.[76] Radical cyclization of 5-keto-aldose aldoximes to give aminocyclopentitol derivatives is covered in Chapter 18.

Glycosylamines produced by reaction of unprotected sugars (*e.g.* LacNAc, Le^x trisaccharide and sialyl-Le^x tetrasaccharide) with the photoreactive, carbene-

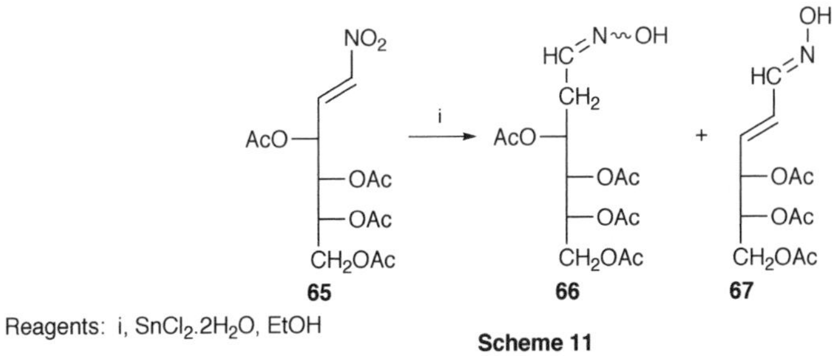

Reagents: i, SnCl$_2$.2H$_2$O, EtOH

Scheme 11

generating *O*-substituted hydroxylamine **68** enable chemiluminescence detection of lectins by photoaffinity labelling.[77] Trichostatin D (**69**) has been isolated as a new inducer of phenotypic reversion in transformed cells, the β-anomer having been reported previously (Vol. 12, p. 153–4).[78] The glucuronide **70** of Trocade™, a collagenase inhibitor in Phase III clinical trials, has been obtained by *N*-acylation of an *O*-glucuronosyl-hydroxylamine derivative, synthesized by glucuronosylation of *N*-hydroxy-phthalimide.[79]

A review on nucleophilic additions to glycines has included a section on *N*-glycosyl nitrones, *e.g.* the use of an *N*-mannofuranosyl nitrone in the synthesis of chiral amines.[80] The chiral nitrone **72** has been synthesized from L-erythrulose (i, NH$_2$OH; ii, Me$_2$CO, H$^+$), oxidation of the initial adduct **71** appearing to occur spontaneously. It was used in 1,3-dipolar cycloaddition reactions with ethyl acrylate and ethyl propiolate to give isoxazolidines *e.g.* **73**.[81] 3-Glycosyl-5-substituted isoxazoles **75** were obtained from 1,3-dipolar cycloaddition reactions of the nitrile oxide generated *in situ* from the oxime **74** in the presence of chloramine-T.[82]

5   Nitriles, Tetrazoles and Related Compounds

The photo- and thermo-induced ring expansions of 1-cyano-glycosyl azides have been studied. Under both conditions, both anomers of the D-galactopyranosyl derivative gave **76**.[83] The 6-cyano-D-galactose derivative **77** reacted with sodium azide to give the tetrazole **78**, which formed isomeric dimers **79** and **80** in DMF at 110 °C, and the derivative **81** upon treatment with acetic anhydride (Scheme 12).[84]

6   Hydrazines, Hydrazones and Related Compounds

The preparation and amphiphilic properties of a new class of surfactants, *N*-(fatty acyl)-*N*′-β-D-glucopyranosyl or β-maltosyl-hydrazines, has been described.[85] 1-Glycosyl-3-methylpyrazol-5-ones such as **82**, formed by acid-catalysed reaction of aldose hydrazones with 2,2,2-trifluoroethyl acetoacetate, were sensitive to oxidation. Coupling with a diazonium salt prevented degradation and provided highly coloured adducts such as **83**, suitable as derivatives for

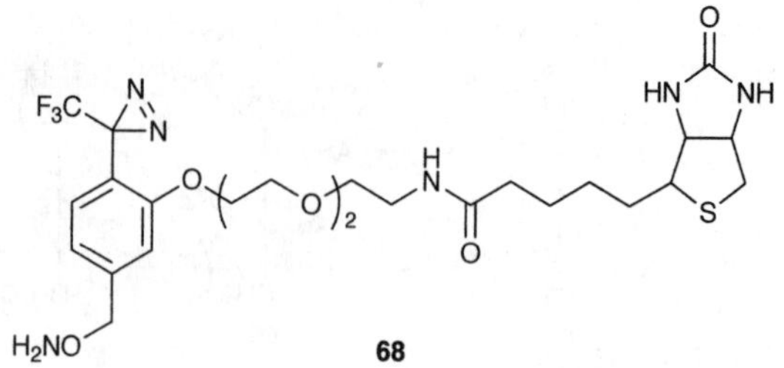

68

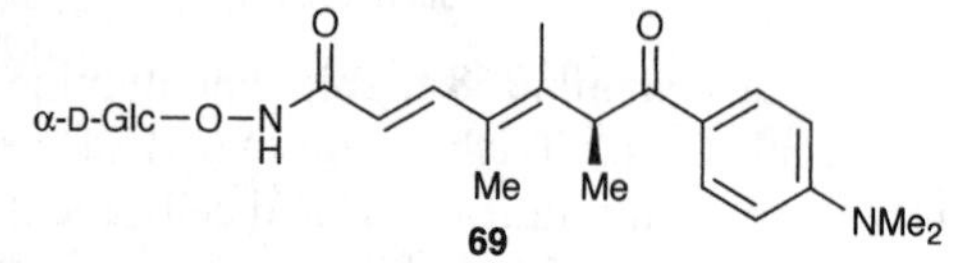

69

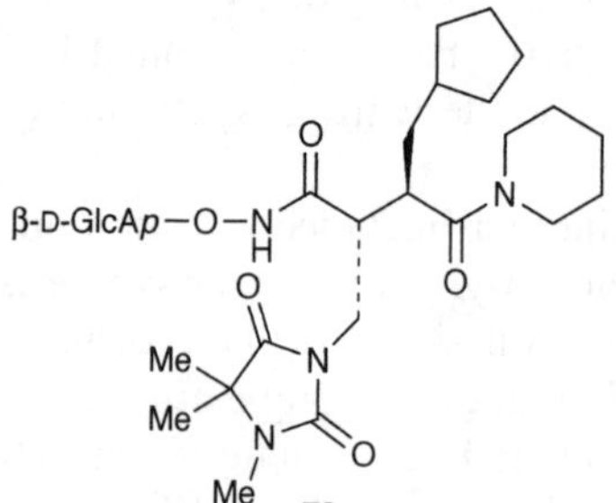

70 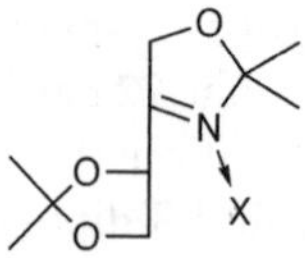   **71** X = :
          72 X = O

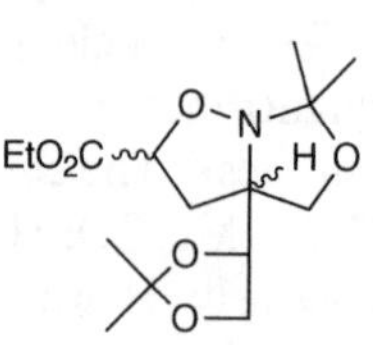

73

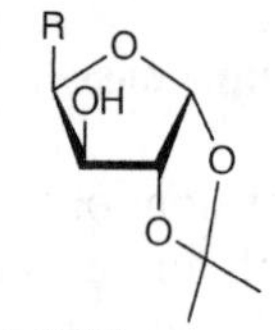

74 R = CH=NOH
75 R = ⟨N,O⟩, X = e.g. Ph, CH₂OH

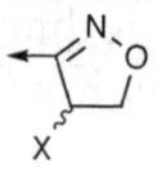

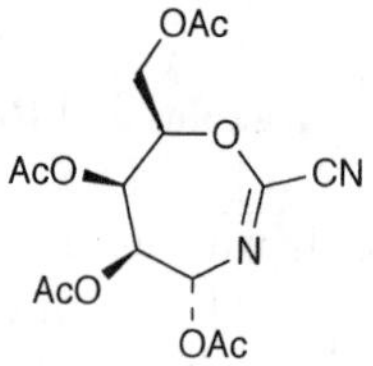

76

chromatographic analysis. Pyrazolone regioisomers with the sugar bonded to N-2 were also encountered.[86] The formation of pentose and hexose hydrazones on reaction with 2-hydrazino-4-oxo-6-phenyl-pyrimidine and their cyclization to the weakly antimicrobial derivatives **84**, has been described.[87] Condensation of benzamide hydrazone hydroiodide with D-glucose gave the β-pyranosyl hydrazine **85**, which formed the triazole **86** on acetylation. When applied to D-galactose and D-arabinose, this condensation-acetylation sequence provided the corresponding α-galactofuranosyl triazole derivative and the acyclic derivative **87**, respectively.[88]

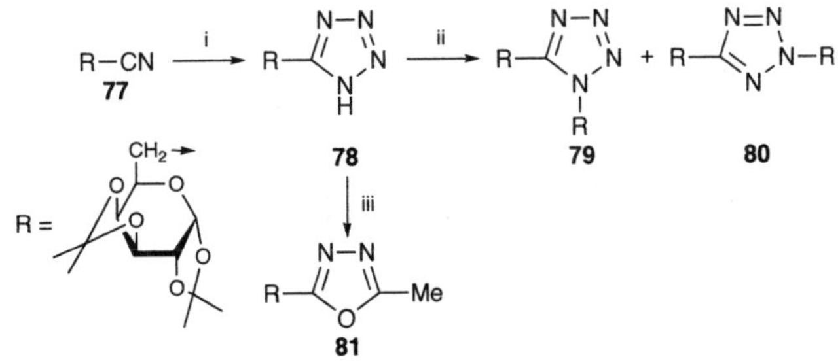

Reagents: i, NaN_3; ii, DMF, 110° C; iii, Ac_2O

Scheme 12

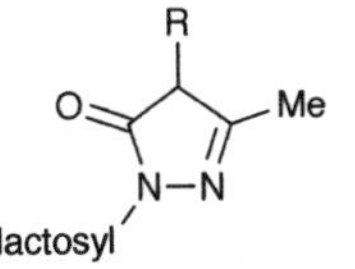

82 R = H

83 R =

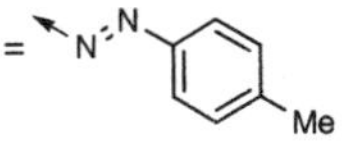

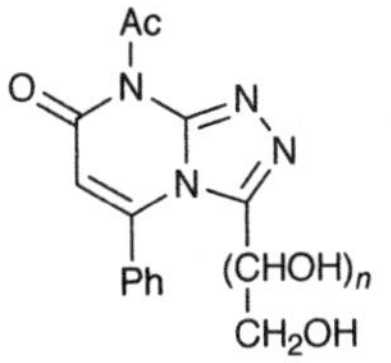

84 n = 3, 4

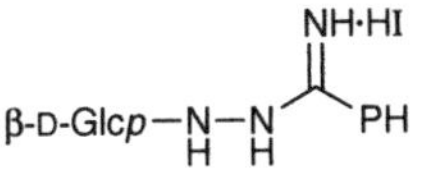

85

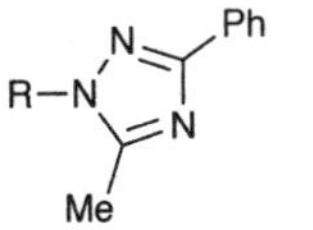

86 R = β-D-Glc$p$(Ac)$_4$

87 R =

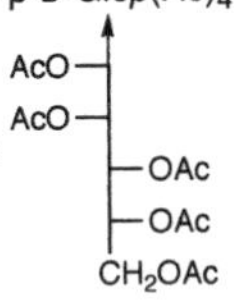

The thiosemicarbazone **88** gave the 1,3,4-thiadiazole **89** with $FeCl_3$ or the dihydrothiadiazole **90** on acetylation.[89] Similar reactions have been reported for D-galactosyl, D-glucosyl and D-mannosyl thiosemicarbazones.[90] Tetradentate metal ion (Cu^{2+}, Pt^{2+} and Pd^{2+}) complexes of the 1,2-bis(thiosemicarbazone) **91** have been prepared.[91] Triazenes such as **92** were obtained by reactions of β-D-glycopyranosylamines with a diazonium salt.[92]

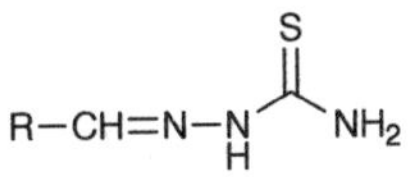

88

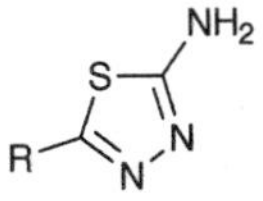

89

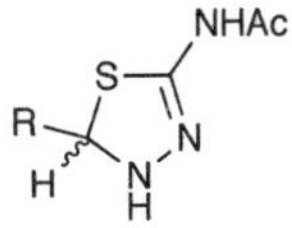

90

R =

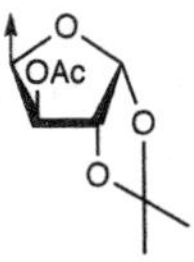

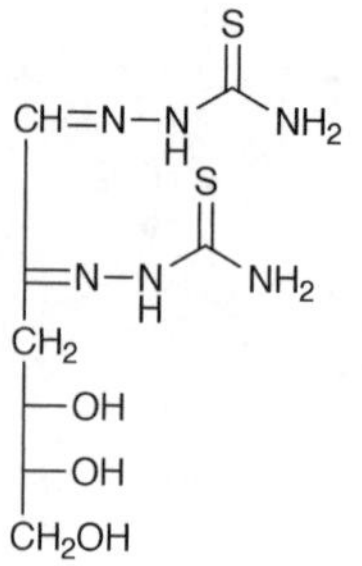

91 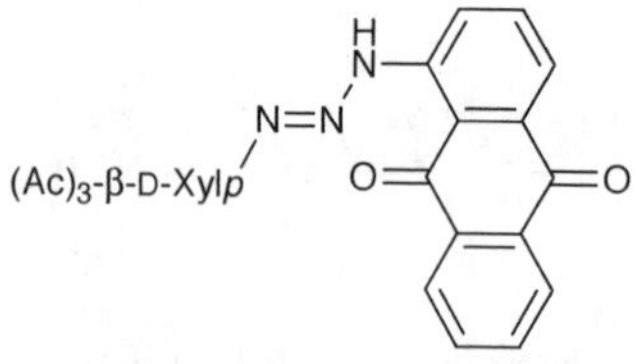　　　　　　　　　　　　　　　　　　　　92

The syntheses of the potent glucosidase inhibitor 1-azafagomine (**93**) and the related D-*lyxo*-isomer **96** are covered in Chapter 18.[93,94] Oxidation of **93** provided the imines **94** and **95** (Scheme 13), but these were very poor glucosidase inhibitors, suggesting that a half-chair conformation is less important than appropriate protonation for a transition state analogue.[95] The synthesis of the related dihydro- and tetrahydro-pyridazinones **98** and **99** from tri-*O*-acetyl-D-glucal (**97**) is shown in Scheme 14. While the D-*arabino*-configured tetrahydro-pyridazinone **99** (R = H) was a micromolar inhibitor of several glycosidases, *N*-substituted analogues and the dihydro-analogues **98** are weak inhibitors.[96]

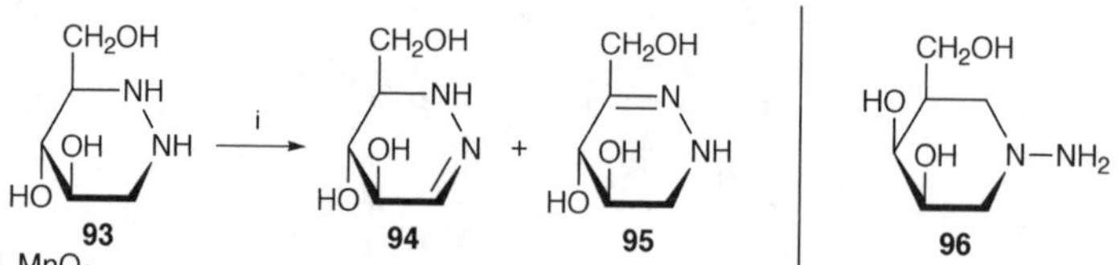

Reagent: i, MnO$_2$

Scheme 13

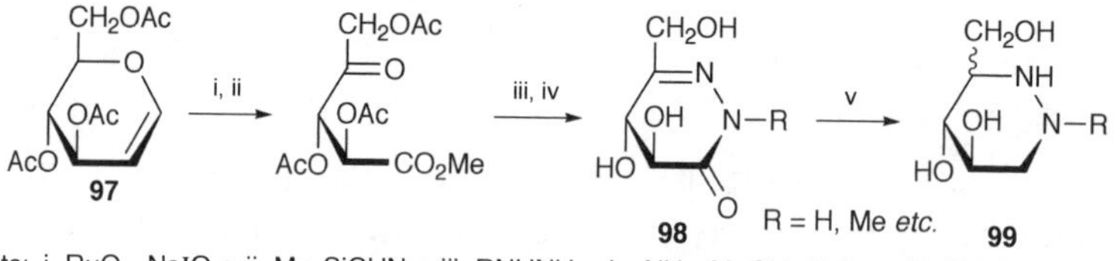

Reagents: i, RuO$_3$, NaIO$_4$; ii, Me$_3$SiCHN$_2$; iii, RNHNH$_2$; iv, NH$_3$, MeOH, H$_2$O; v, NaBH$_3$CN, AcOH

Scheme 14

7　Other Heterocycles

The fused oxazole derivative **100** was synthesized from 3-azido-3-deoxy-1,2:5,6-di-*O*-isopropylidene-α-D-allofuranose, the key final step being the reaction of methyl 3-amino-3-deoxy-5,6-O-isopropylidene-β-D-alloside with DMF dimethyl acetal.[97] The annelated pyranoside **101** was obtained by cyclization of a branched chain hexosulose derivative,[98] and the fused triazole-piperidinoses such as **102** were made by a radical cyclization of a 3-pyrazolo-6-iodo-sugar derivative.[99] The spiro-isoxazoline **103** and related isomers have been synthesized by dipolar cycloadditions of mesitonitrile oxide to 2-deoxy-2-*C*-methylene-pentonic acids, themselves available in five steps from D-mannitol.[100] Intramolecular dipolar cycloaddition reactions of an allyl group in a C-3 substituent and a nitrileoxide moiety at C-5 led to 10–12-membered nitrogen and oxygen heterocycles, such as **104**, fused to an isoxazoline.[101] β-Lactams such as **105** have been synthesized by, for example, [2+2]cycloaddition of chlorosulfonyl isocyanate to the corresponding 3-*O*-vinyl ether, and converted into the polycyclic derivatives such as **106** by intramolecular alkylation.[102]

100

101

102

103

104 X = O or NTs

105

106

107 R = H, Cl, Me

108

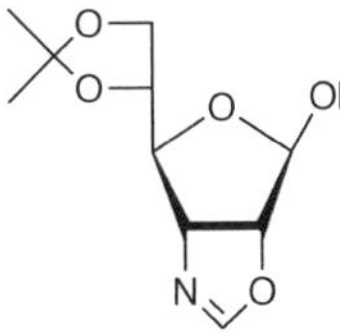

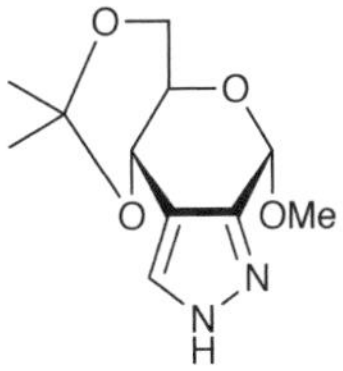

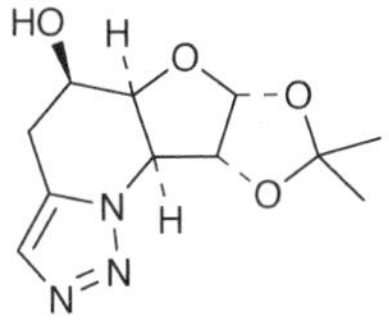

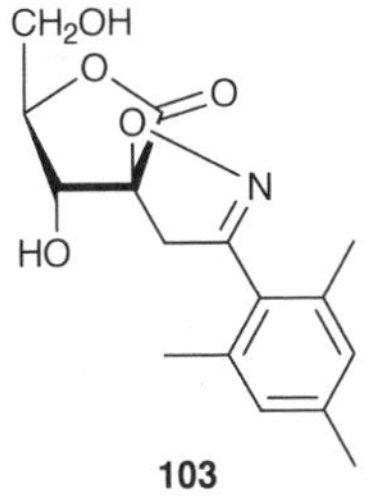

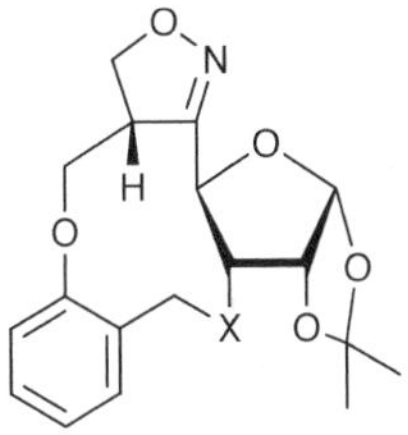

Condensation of D-glucose with 1,2-diaminobenzene or its 3-substituted-derivatives, then with hydrazine, provided the pyrazoloquinoxalines **107**.[103] Triazolopyrimidopurines such as **108** (and a regioisomer) were obtained by cycloaddition of an acetylene-substituted heterocycle with a 1-azido-1-deoxyalditol derivatives.[104]

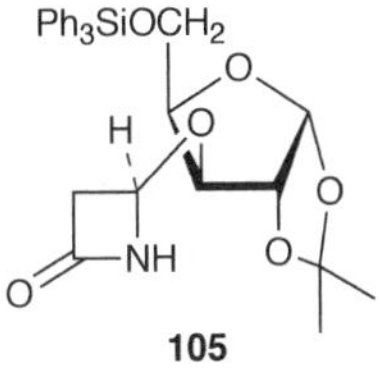

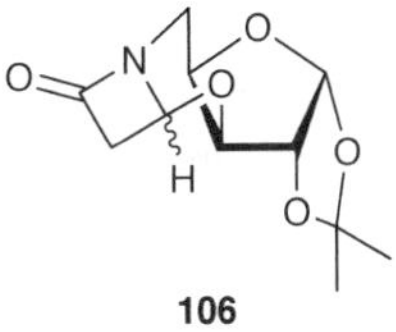

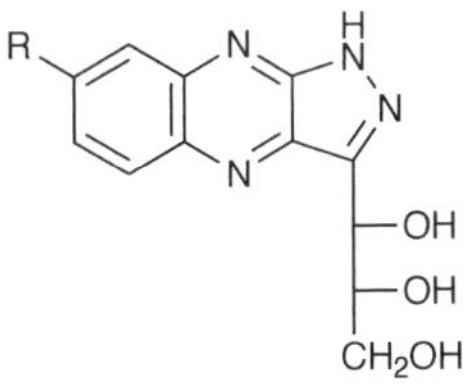

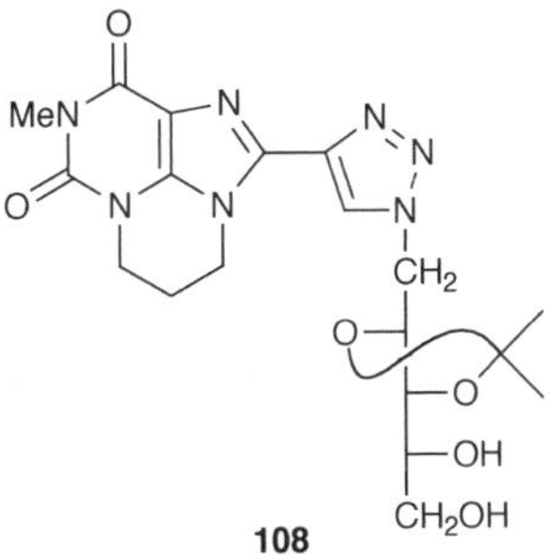

The thiazoline derivative **109** was obtained as a single isomer by condensation of the corresponding 1,6-D-galactodialdehyde diacetal with 2-chloro-2-methylpropanal, ammonia and sodium sulfide.[105] The multi-step synthesis of 3-trialkylsilypyrazole derivatives such as **110** from 5,6-dihydroxy-sugar derivatives has been described.[106]

Fused triazoles such as the D-arabinose-derived **112** were synthesized by intramolecular condensation of the 7-azide produced by displacement of the 7-tosylate from unsaturated ester **111** (Scheme 15).[107] Triazole **114** was synthesized by intramolecular dipolar cycloaddition of the azido-alkyne **113**, derived by

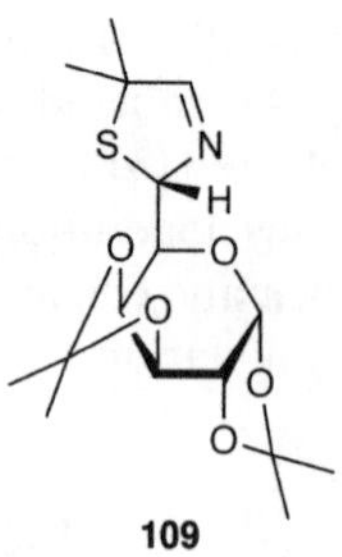

109 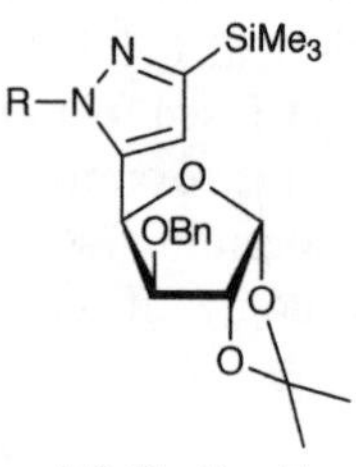 **110** R = H or Me

addition of ethynyl dichlorocerium(III) to a 5-azido-5-deoxy-D-glucofuranose derivative (Scheme 16), and was shown to be a poor glycosidase inhibitor. The *C*-disaccharide mimic **115** was obtained similarly through addition to **113** of a cerium(III)-5-*C*-ethynyl-glycoside derivative [108] Vasella and co-workers have provided a detailed analysis of the glycosidase inhibitory properties of a range of such heterocycle-fused aza-sugars. Tetrahydroimidazopyridines with hydrophobic, flexible substituents at C-2, syntheses of which were described, were very potent β-glucosidase inhibitors; thus **116** had K_i values of 1.2 and 0.11 nM against the enzymes from almonds and *Caldocellum saccharolyticum*, respectively.[109] Inhibition of these β-glucosidases by a set of 2-amino-2-deoxy-D-glucose mimics bearing a ring nitrogen atom and a fused unsaturated five membered ring containing a further zero, one or two nitrogen atoms has been reported. The inhibitors are bound as their ammonium salts, and the best inhibitors, such as

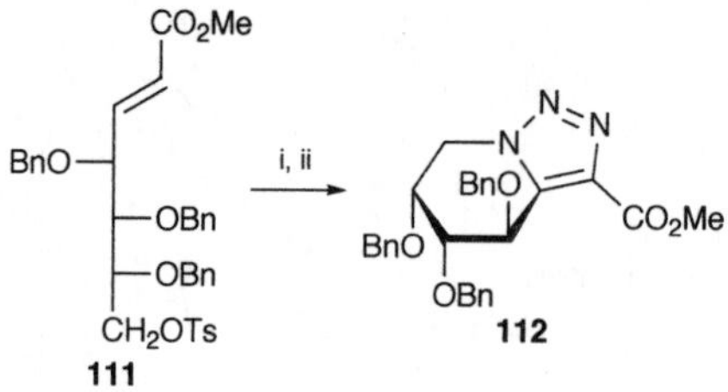

111 **112**

Reagents: i, NaN₃, DMF; ii, DBU

Scheme 15

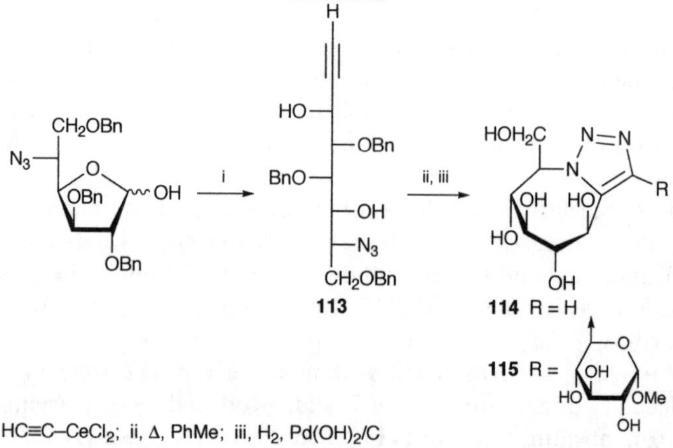

113 **114** R = H

115 R =

Reagents: i, HC≡C–CeCl₂; ii, Δ, PhMe; iii, H₂, Pd(OH)₂/C

Scheme 16

117 and **118**, lacked a heteroatom bonded *exo* to C-1.[110] The 2-acetamido-2-deoxy-D-glucose mimics **119** and **120** were potent inhibitors of bovine kidney *N*-acetylglucosaminidase (K_i values of 4 μM and 34 nM, respectively).[111] The isoquinuclidine derivative **121**, synthesized from 3-hydroxy-2-pyridone, mimics a mannopyranoside boat conformation and was an inhibitor of snail β-mannosidase (K_i = 20 μM).[112] The synthesis of N^5,N^6-bridged 6-amino-1,6-dideoxynojirimycins is covered in Chapter 18.

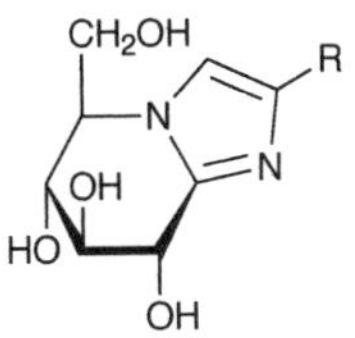

116 R = CH₂CH₂ **117** X = Y = N **118** X = CH, Y = CCO₂Me **119** 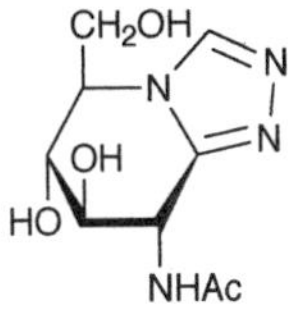 **120**

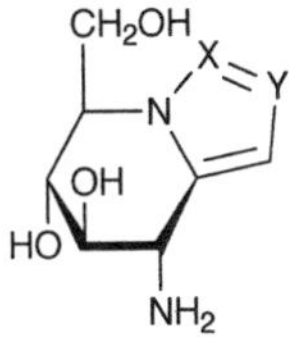

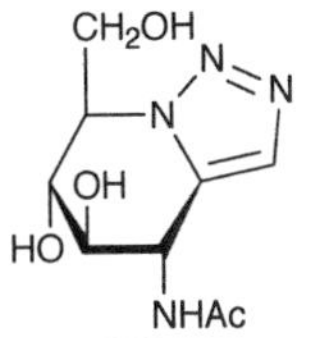

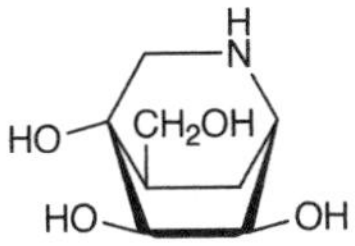

121

References

1. W.H. Ojala, J.M. Ostman and C.R. Ojala, *Carbohydr. Res.*, 2000, **326**, 104.
2. K.D. Randell, B.D. Johnston, D.F. Green and B.M. Pinto, *J. Org. Chem.*, 2000, **65**, 220.
3. T. Suzuki, S. Tanaka, I. Yamada, Y. Koashi, K Yamada and N. Chida, *Org. Lett.*, 2000, **2**, 1137.
4. M.D. López, M.L. Quijano, A. Sánchez and M. Nogueras, *J. Heterocycl. Chem.*, 2000, **37**, 1511.
5. A. Schwögler, V. Gramlich and T. Carell, *Helv. Chim. Acta*, 2000, **83**, 2452.
6. V.L. Makytruk, A.S. Shalamay and I.V. Kondratyuk, *Biopolim. Kletka*, 1997, **13**, 453 (*Chem. Abstr.*, 2000, **132**, 122 844).
7. M. Ohkubo, T. Nishimura, M. Kawamoto, M. Nakano, T. Honma, T. Yoshinari, H. Arakawa, M. Suda, H. Morishima and S. Nishimura, *Bioorg. Med. Chem. Lett.*, 2000, **10**, 419.
8. J.D. Chisholm and D.L. Van Vranken, *J. Org. Chem.*, 2000, **65**, 7541.
9. G.A. Winterfeld, J. Das and R.R. Schmidt, *Eur. J. Org. Chem.*, 2000, 3047.
10. P. Allef and H. Kunz, *Tetrahedron: Asymmetry*, 2000, **11**, 375.
11. K.A. Jørgensen, *Angew. Chem. Int. Ed. Engl.*, 2000, **39**, 3559.
12. K.D. Randell, B.D. Johnston and B.M. Pinto, *Carbohydr. Res.*, 2000, **326**, 145.
13. D. Sinou and C. Moineau, *Recent Res. Dev. Org. Chem.*, 1999, **3**, 1 (*Chem. Abstr.*, 2000, **132**, 237 247).
14. J. Andersch, L. Hennig and H. Wilde, *Carbohydr. Res.*, 2000, **329**, 693.
15. A. Dondoni and A. Marra, *Chem. Rev.*, 2000, **100**, 4395.
16. J.-C. Luo, G.-R. Chen, K.-C. Chen and Y.-Y. Xie, *Huadong Ligong Daxue Xuebao*, 1999, **25**, 538 (*Chem. Abstr.*, 2000, **132**, 108 219).

17. C.G. Francisco, A.J. Herrera and E. Suárez, *Tetrahedron: Asymmetry*, 2000, **11**, 3879.
18. B. Dmochowsak, D. Bednarczyk, A. Nowacki, A. Konitz, W. Wojnowski and A. Wisniewski, *Carbohydr. Res.*, 2000, **329**, 703.
19. G. Viscardi, P. Quagliotto, C. Barolo, P. Savarino, E. Barni and E. Fisicaro, *J. Org. Chem.*, 2000, **65**, 8197.
20. D.T.H. Chou, J.N. Watson, A.A. Scholte, T.J. Borgford and A.J. Bennet, *J. Am. Chem. Soc.*, 2000, **122**, 8357.
21. J. Zhu and A.J. Bennett, *J. Org. Chem.*, 2000, **65**, 4423.
22. P. Boullanger, V. Maunier and D. Lafont, *Carbohydr. Res.*, 2000, **324**, 97.
23. D.H. Huang and J.M. Risley, *Carbohydr. Res.*, 2000, **329**, 487.
24. O. Lockhoff and Y. Hayauchi, *J. Carbohydr. Chem.*, 2000, **19**, 603.
25. R. Kassab, B. Fenet, H. Fessi and H. Parrot-Lopez, *Tetrahedron Lett.*, 2000, **41**, 877.
26. E. Fan, Z. Zhang, W.E. Minke, Z. Hou, C.L.M.J. Verlinde and W.G.J. Hol, *J. Am. Chem. Soc.*, 2000, **122**, 2663.
27. T. Ziegler, S. Gerling and M. Lang, *Angew. Chem. Int. Ed. Engl.*, 2000, **39**, 2109.
28. H. Herzner, T. Reipen, M. Schultz and H. Kunz, *Chem. Rev.*, 2000, **100**, 4495.
29. Y. Ito, M. Gerz and Y. Nakahara, *Tetrahedron Lett.*, 2000, **41**, 1039.
30. Z.-G. Wang, X. Zhang, D. Live and S.J. Danishefsky, *Angew. Chem. Int. Ed. Engl.*, 2000, **39**, 3652.
31. I. Shin and J. Lee, *Synlett.*, 2000, 1297.
32. D. Ramos, P. Rollin and W. Klaffke, *Angew. Chem. Int. Ed. Engl.*, 2000, **39**, 396.
33. J.P. Malkinson, R.A. Falconer and I. Toth, *J. Org. Chem.*, 2000, **65**, 5249.
34. K. Oertel, G. Zech and H. Kunz, *Angew. Chem. Int. Ed. Engl.*, 2000, **39**, 1431.
35. L. Jobron and G. Hummel, *Angew. Chem. Int. Ed. Engl.*, 2000, **39**, 1621.
36. E. Shitara, Y. Nishimura, F. Kojima and T. Takeuchi, *Bioorg. Med. Chem.*, 2000, **8**, 343.
37. Y. Nishimura, E. Shitara, H. Adachi, M. Toyoshima, M. Nakajima, Y. Okami and T. Takeuchi, *J. Org. Chem.*, 2000, **65**, 2.
38. S. Knapp and D. Zhao, *Org. Lett.*, 2000, **2**, 4037.
39. E. Shitara, Y. Nishimura, K. Nerome, Y. Hiramoto and T. Takeuchi, *Org. Lett.*, 2000, **2**, 3837.
40. Y. Gama, I. Shibuya and M. Shimizu, *J. Carbohydr. Chem.*, 2000, **19**, 119.
41. M. Störer, A. Lützen and P. Köll, *Tetrahedron: Asymmetry*, 2000, **11**, 371.
42. Y. Ichikawa, T. Nishiyama and M. Isobe, *Synlett*, 2000, 1253.
43. E. Muller, D. Gasparutto, M. Jaquinod, A. Romieu and J. Cadet, *Tetrahedron*, 2000, **56**, 8689.
44. S.P. Deshmukh, *Orient. J. Chem.*, 2000, **16**, 143 (*Chem. Abstr.*, 2000, **133**, 177 388).
45. V.M. Diaz Pérez, C. Ortiz Mellet, J. Fuentes, J.M. García Fernández, *Carbohydr. Res.*, 2000, **326**, 161.
46. M.I. Garcia-Moreno, J.M. Benito, C.O. Mellet and J.M.G. Fernández, *Tetrahedron: Asymmetry*, 2000, **11**, 1331.
47. C. Gasch, M.A. Pradera, B.A.B. Salameh, J.L. Molina and J. Fuentes, *Tetrahedron: Asymmetry*, 2000, **11**, 435.
48. L. Somsák, V. Nagy, T. Docsa, B. Tóth and P. Gergely, *Tetrahedron: Asymmetry*, 2000, **11**, 405.
49. L. Somsák and V. Nagy, *Tetrahedron: Asymmetry*, 2000, **11**, 1719.
50. F.M. Ibatullin and K.A. Shabalin, *Carbohydr. Lett.*, 2000, **3**, 427 (*Chem. Abstr.*, 2000, **133**, 135 514).
51. F.M. Ibatullin and K.A. Shabalin, *Synth. Commun.*, 2000, **30** 2819.

52.   A. Stimac and J. Kobe, *Carbohydr. Res.*, 2000, **324**, 149.
53.   J. Bogusiak, *Pol. J. Chem.*, 2000, **74**, 503 (*Chem. Abstr.*, 2000, **133**, 43 716).
54.   A. Stimac and J. Kobe, *Carbohydr. Res.*, 2000, **329**, 317.
55.   C. Hager, R. Miethchen and H. Reinke, *Synthesis*, 2000, 226.
56.   C. Hager, R. Miethchen and H. Reinke, *J. Prakt. Chem.*, 2000, **342**, 414 (*Chem. Abstr.*, 2000, **133**, 120 549).
57.   G. Anilkumar, L.G. Nair, L. Olsson, J.K. Daniels and B. Fraser-Reid, *Tetrahedron Lett.*, 2000, **41**, 7605.
58.   Z. Szurmai, J. Rákó, K. Ágoston, A. Danan and D. Charon, *Org. Lett.*, 2000, **2**, 1839.
59.   A. Dabrowska, P. Dokurno, A. Konitz and Z. Smiatacz, *Carbohydr. Res.*, 2000, **323**, 230.
60.   C. Hager, R. Miethchen and H. Reinke, *J. Carbohydr. Chem.*, 2000, **19**, 997.
61.   F. Carrel and P. Vogel, *Tetrahedron: Asymmetry*, 200, **11**, 4661.
62.   P. Hadwiger, P. Mayr, B. Nidetzky, A.E. Stütz and A. Tauss, *Tetrahedron: Asymmetry*, 2000, **11**, 607.
63.   D.E.A. Brittain, M.P. Watterson, T.D.W. Claridge, M.D. Smith and G.W.J. Fleet, *J. Chem. Soc., Perkin Trans. 1*, 2000, 3655.
64.   N.L. Hungerford and G.W.J. Fleet, *J. Chem. Soc., Perkin Trans. 1*, 2000, 3680.
65.   A. Arcelli, V. Ceré, F. Peri, S. Pollicino and P. Sabatino, *Tetrahedron: Asymmetry*, 2000, **11**, 1389.
66.   J. Marco-Contelles, N. Jagerovic and C. Alhambra, *J. Chem. Res. (S)*., 1999, 680 (*Chem. Abstr.*, 2000, **132**, 122 830).
67.   V. Jager, R. Ohrlein, V. Wehner, P. Poggendore, B. Steuer, J. Raczko, H. Griesser, F.-M. Kiess and A. Menzel, *Enantiomer*, 1999, **4**, 205 (*Chem. Abstr.*, 2000, **132**, 64 452).
68.   B. Becker, *J. Carbohydr. Chem.*, 2000, **19**, 253.
69.   P.H. Seeberger, S. Roehrig, P. Schell, Y. Wang and W.J. Christ, *Carbohydr. Res.*, 2000, **328**, 61.
70.   A. Defoin, M. Joubert, J.-M. Heuchel, C. Strehler and J. Streith, *Synthesis*, 2000, 1719.
71.   B. Liberek and Z. Smiatacz, *Waid. Chem.*, 1999, **53**, 461(*Chem. Abstr.*, 2000, **132**, 137 617).
72.   B. Liberek and Z. Smiatacz, *J. Carbohydr. Chem.*, 2000, **19**, 1259.
73.   B. Liberek, A. Konitz, R. Frankowski and Z. Smiatacz, *Carbohydr. Res.*, 2000, **326**, 151.
74.   H. Mohan and A. Vasella, *Helv. Chem. Acta*, 2000, **83**, 114.
75.   L. Devel, A. Vidal-Cros and A. Thellend, *Tetrahedron Lett.*, 2000, **41**, 299.
76.   M. Koos, *Tetrahedron Lett.*, 2001, **41**, 5403.
77.   Y. Hayakawa, M. Nakai, K. Furihata, K. Shinya and H. Seto, *J. Antibiotics*, 2000, **53**, 179.
78.   Y. Hatanaka, U. Kempin and P. Jong-Jip, *J. Org. Chem.*, 2000, **65**, 5639.
79.   M.B. Mitchell and J.W.A. Whitcombe, *Tetrahedron Lett.*, 2000, **41**, 8829.
80.   M. Lombardo and C. Trombini, *Synthesis*, 2000, 759.
81.   M. Carda, R. Portoles, J. Murga, S. Uriel, J.A. Marco, L.R. Domingo, R.J. Zaragoza and H. Roper, *J. Org. Chem.*, 2000, **65**, 7000.
82.   M.L. Fascio, V.J. Montesano and N.B. D'Accorso, *J. Carbohydr. Chem.*, 2000, **19**, 393.
83.   J.-P. Praly, C. Di Stefano and L. Somsák, *Tetrahedron: Asymmetry*, 2000, **11**, 533.
84.   A.H. Klier, R.J. Alves, M. Auxiliadora, F. Prado, J.D. de S. Filho and N.B. D'Accorso, *Synth. Commun.*, 2000, **30**, 4361.

85.   J. Augé and N. Lubin-Germain, *J. Carbohydr. Chem.*, 2000, **19**, 379.
86.   W.C. Kett, M. Batley and J.W. Redmond, *Carbohydr. Res.*, 2000, **329**, 169.
87.   M.A.E. Shaban, A.Z. Nasr and A.E.A. Morgaan, *Pharmazie*, 2000, **55**, 87 (*Chem. Abstr.*, 2000, **132**, 308 580)
88.   E.S.H. El Ashry, L.F. Awad and M. Winkler, *J. Chem. Soc.*, *Perkin Trans. 1*, 2000, 829.
89.   M.A. Martins Alho and N.B. D'Accorso, *J. Heterocycl. Chem.*, 2000, **37**, 811.
90.   M.A. Martins Alho and N.B. D'Accorso, *Carbohydr. Res.*, 2000, **328**, 481.
91.   D. Horton and O. Varela, *Carbohydr. Res.*, 2000, **328**, 425.
92.   M. Weng and J.C. Jochims, *J. Prakt. Chem.*, 2000, **342**, 530 (*Chem. Abstr.*, 2000, **133**, 296 605).
93.   B.V. Ernholt, I.B. Thomsen, A. Lohse, I.W. Plesner, K.B. Jensen, R.G. Hazell, X. Liang, A. Jakobsen and M. Bols, *Chem. Eur. J.*, 2000, **6**, 278.
94.   H. Søhoel, X. Liang and M. Bols, *Synlett*, 2000, 347.
95.   S.U. Hansen and M. Bols, *J. Chem. Soc.*, *Perkin Trans. 2*, 2000, 665
96.   C.V. Ramana and A. Vasella, *Helv. Chim. Acta*, 2000, **83**, 1599.
97.   J. Molina and C. Simons, *J. Carbohydr. Chem.*, 2000, **19**, 991.
98.   B. Kuhla, K. Peseke, G. Thiele and M. Michalik, *J. Prakt. Chem.*, 2000, **342**, 240 (*Chem. Abstr.*, 2000, **133**, 74 195).
99.   J. Marco-Contelles and M. Rodríguez-Fernández, *Tetrahedron Lett.*, 2000, **41**, 381.
100. P. Micuch, L. Fisera, M.K. Cyranski, T.M. Krygowski and J. Krajcik, *Tetrahedron*, 2000, **56**, 5465.
101. S. Majundar, R. Mukhopadhyay and A. Bhattacharjya, *Tetrahedron*, 2000, **56**, 8945.
102. Z. Kaluza, B. Furman, P. Krajewski and M. Chmielewski, *Tetrahedron*, 2000, **56**, 5553.
103. M.A.E. Sallam, F.F. Louis, T. Anthonsen and E.G.K. Quartey, *Carbohydr. Res.*, 2000, **324**, 1.
104. O. Šimo, A. Rybár and J. Alföldi, *J. Heterocycl. Chem.*, 2000, **37**, 1033.
105. I. Schlemminger, H.-H. Janknecht, W. Maison, W. Saak and J. Martens, *Tetrahedron Lett.*, 2000, **41**, 7289.
106. S. Gérard, R. Plantier-Royon, J.-M. Nuzillard and C. Portella, *Tetrahedron Lett.*, 2000, **41**, 9791.
107. T. Flessner and C.-H. Wong, *Tetrahedron Lett.*, 2000, **41**, 7805.
108. K. Tezuka, P. Compain and O.R. Martin, *Synlett*, 2000, 1837.
109. N. Panday, Y. Canac and A. Vasella, *Helv. Chim. Acta*, 2000, **83**, 58.
110. N. Panday, M. Meyyappan and A. Vasella, *Helv. Chim. Acta*, 2000, **83**, 513.
111. N. Panday and A. Vasella, *Helv. Chim. Acta*, 2000, **83**, 1205.
112. E. Lorthiois, M. Meyyappan and A. Vasella, *Chem. Commun.*, 2000, 1829.

11

Thio-, Seleno- and Telluro-sugars

1 Thiosugars

1.1 Monosaccharides. – *1.1.1 Acyclic Compounds.* A new route to 2-deoxypentose dialkyl dithioacetals **2** involved opening of the L-ascorbic acid-derived epoxide **1** with lithiated formaldehyde dialkyl dithioacetals. Standard protecting group manipulations converted **2** (R = Bu^t) to the 4-mesylate **3**.[1]

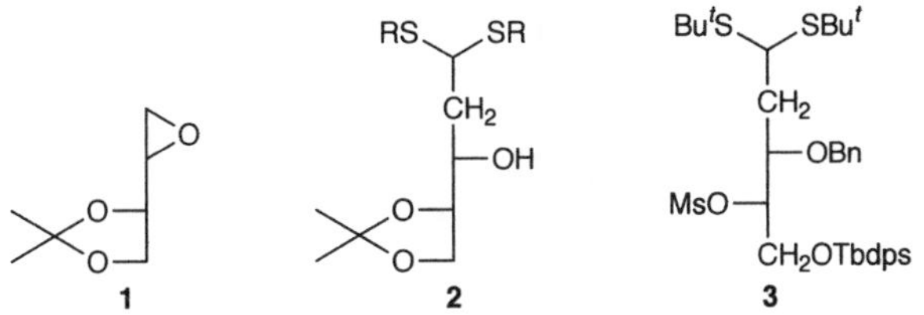

1.1.2 Compounds with Sulfur as the Ring Heteroatom. Thietanes **6** and **7** were formed preferentially on de-S-acetylation of the 6-S-acetyl-6-thio-polymesylates **4** and **5**, respectively. The latter transformation proceeded presumably by way of a 3,4-epoxide.[2]

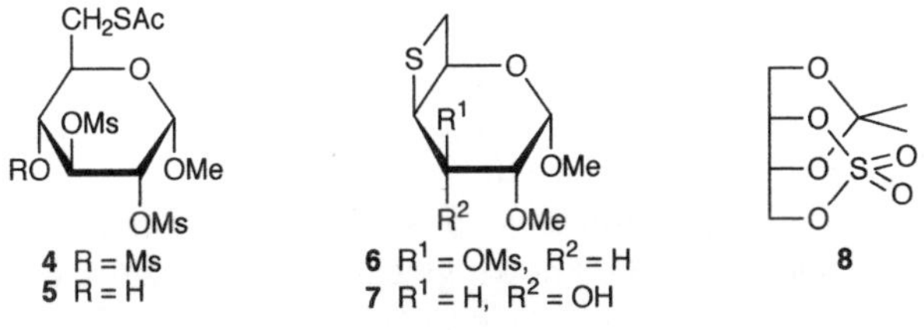

The natural α-glucosidase inhibitor salacinol (**9**) (see Vol. 31, p. 163, ref. 29) and its diastereomer **10** have been synthesized by opening of the cyclic sulfates **8** and its 1,3-cyclic sulfate equivalent, respectively, with 1,4-anhydro-4-thio-D-arabinitol.[3] On treatment with NaN₃ in DMSO the thiepane **11**, readily available from D-glucitol, underwent direct S$_N$2 substitution of one of the mesylate groups by azide, while the other mesylate group was displaced transannularly by sulfur. Attack of N₃⁻ on the the resulting bicyclic sulfonium salts afforded a 5:1 mixture of 1,4-anhydro-4-thio-hexitols **12** and **13**.[4]

An improved protocol for the known ring-closure of 2-deoxy-4-O-sulfonyl-

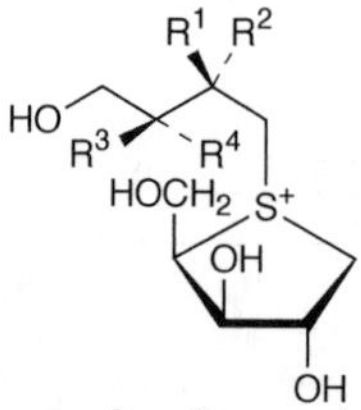

9 R^1 = OH, R^2 = R^3 = H, R^4 = OSO_3^-
10 R^1 = R^4 = H, R^2 = OH, R^3 = OSO_3^-

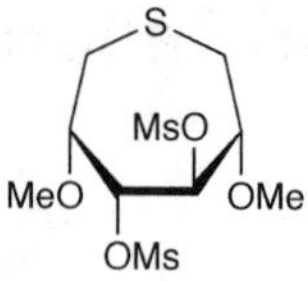

11

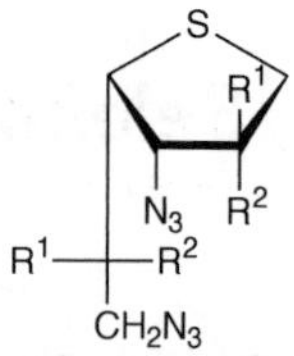

12 R^1 = OMe, R^2 = H
13 R^1 = H, R^2 = OMe

pentose dialkyl dithioacetals to furnish 1,4-dithiofuranose derivatives has been applied to the synthesis of 4′-thionucleoside precursors (*e.g.* 3→14). The preparation of 4′-thionucleosides from 15 is covered in Chapter 19.[1]

Introduction of a sulfur substituent at the 5-position of hex-1-enitol 15 with retention of configuration to give dithiocarbamate 16 was achieved by carrying out two consecutive Mitsunobu reactions, the second one employing $Zn(SCSNMe_2)_2$ as nucleophile. Following de-*S*-protection, NIS-promoted cyclization gave the hetero-*exo*-glycal 17.[5]

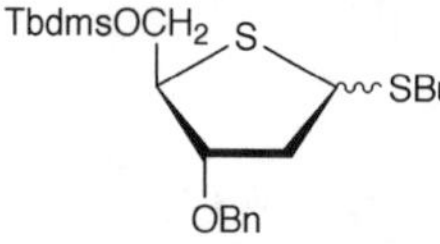

14

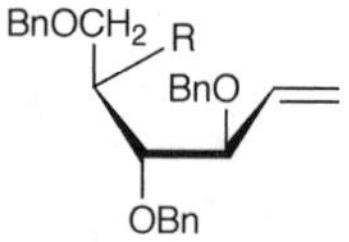

15 R = OH
16 R = SC(S)NMe₂

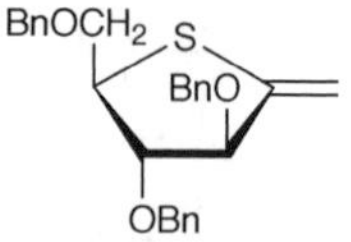

17

Glycosylation of 4-substituted-benzenethiols with 1,2,5-tri-*O*-acetyl-3,6-an-hydro-3-thio-α,β-D-glucofuranose (19), prepared from D-glucose *via* the 6-thioacetate-3-mesylate 18, then deprotection, furnished the new 3,6-anhydro-1,3-dithio-D-glucofuranoside analogues 20 of the antithrombotic agent beciparcil (see Vol. 33, Chapter 11, refs. 5–8).[6] Bicyclic beciparcil analogues 22 with 2,5-anhydro-1,6-dithio-α-D-*gluco*-septanoside frameworks were elaborated from diepoxide 21 in low overall yields by the lengthy reaction sequence outlined in Scheme 1,[7] which was also adapted to furnish the 3-azido-3-deoxy analogues of compounds 22. The 2-azido-2-deoxy-1,5-dithio-β-D-arabinopyranoside analogues 24 were obtained from the known glycal 23, using azidonitration as the key-step.[9]

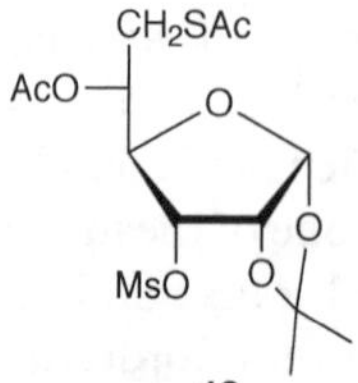

18

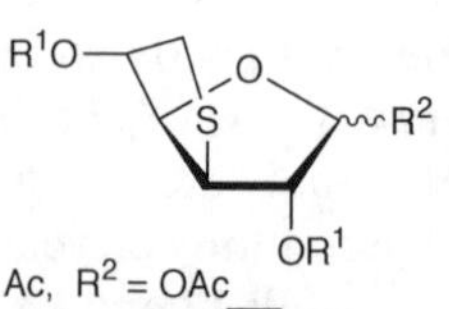

19 R^1 = Ac, R^2 = OAc
20 R^1 = H, R^2 = S—⟨ ⟩—X ; X = CN or NO_2

5-Thio-L-altrose (27) has been prepared from D-galactose by standard reactions *via* epoxide 25 and episulfide 26.[10] 2,6-Anhydro-2-thio-D-altrose (28),

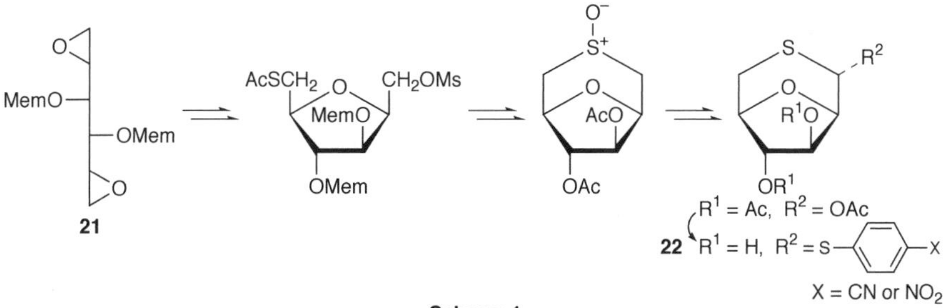

Scheme 1

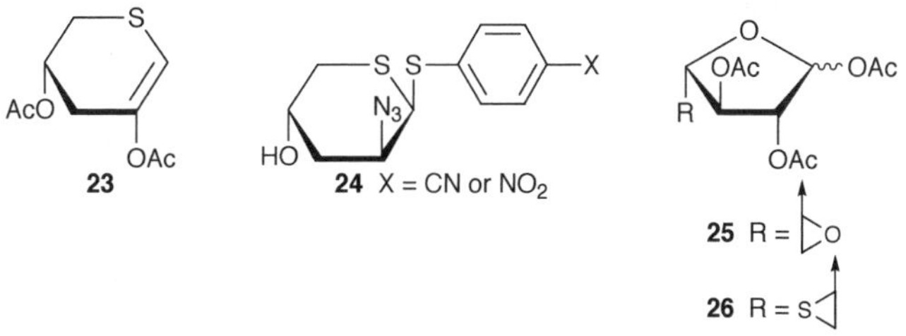

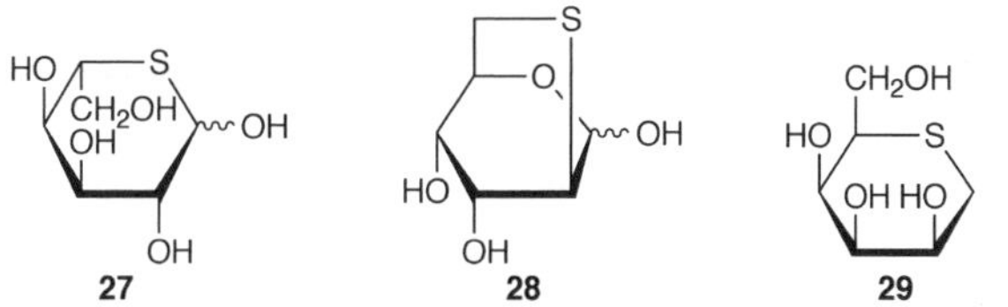

$R^1 = Ac,\ R^2 = OAc$

$22\ \begin{cases} R^1 = H,\ R^2 = S\text{—}\bigcirc\text{—}X \end{cases}$

$X = CN\ or\ NO_2$

23

24 X = CN or NO_2

25 R =

26 R =

27

28

29

formed on acid hydrolysis of its known methyl α-pyranoside, furnished 1,5-anhydro-5-thio-talopyranose (1-deoxythio-*talo*-nojirimycin **29**) on reduction with NaBH₄.[11] The sulfonium ion analogue **34** of castanospermine has been synthesized from ethyl 1,5-dithio-α-D-glucopyranoside by way of dialdose derivative **30** and its Wittig-extension product **31**, which was desulfurized at the anomeric position and reduced to alcohol **32**, then converted to bromide **33**. Cyclization took place on exposure of the latter compound to AgClO₄.[12]

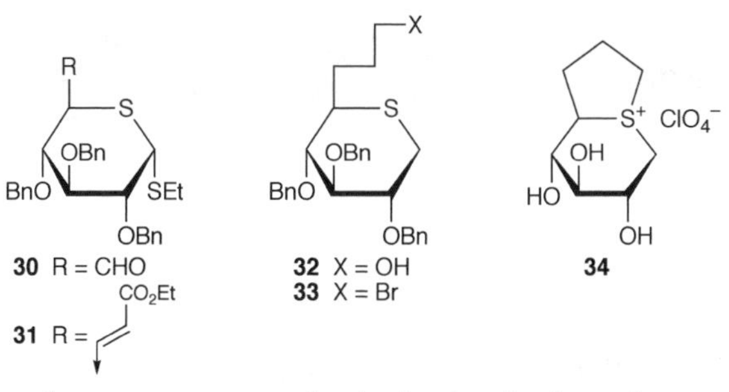

30 R = CHO

31 R =

32 X = OH
33 X = Br

34

A study on the consequences of substitution in the aglycon on the configurational equilibria of neutral or protonated *N*-aryl 5-thioglucopyranosylamines **35** concluded that there is no evidence in support of a generalized reverse anomeric effect.[13]

2,3,4-Tri-*O*-benzoyl-6-thiolevoglucosan was oxidized to *exo*-sulfoxide **36** and

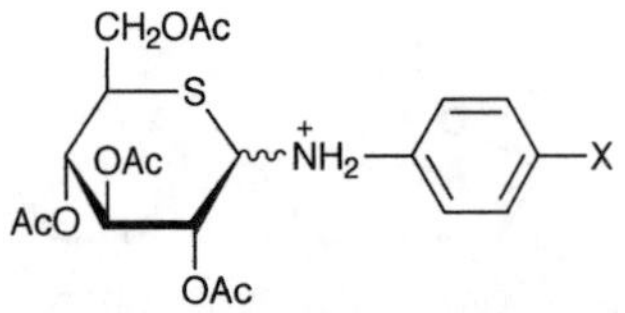

35 X = H, OMe, CF₃, NO₂

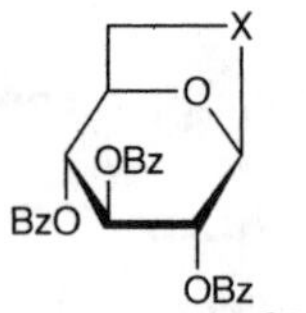

36 X = +S''^O⁻
36a X = SO₂

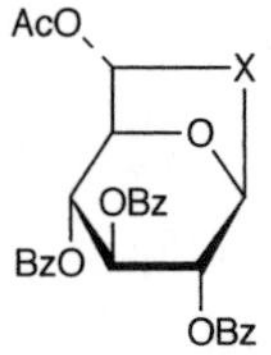

37 X = S
37a X = SO₂

sulfone **36a** by use of MCPBA and dimethyldioxirane, respectively; Pummerer rearrangement of compound **36** (NaOAc, Ac₂O) furnished the α-acetoxysulfide **37**, which gave sulfone **37a** on renewed dimethyldioxirane oxidation. All four compounds failed as glycosyl donors under various activating conditions.[14] 1,4-Anhydro-6-thiosugar septanoside **39** was formed on acid hydrolysis of 1,2-O:5,6-O/S-diisopropylidene-β-L-idofuranose **38** and subsequent acetylation.[15]

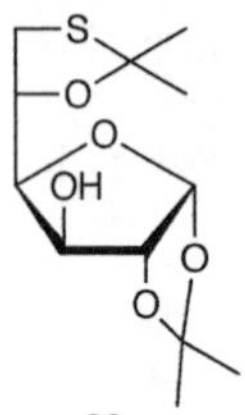

38 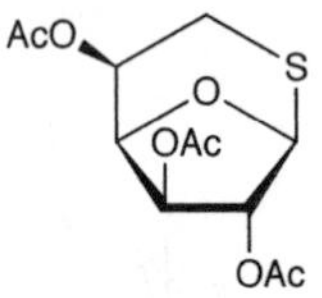   **39** 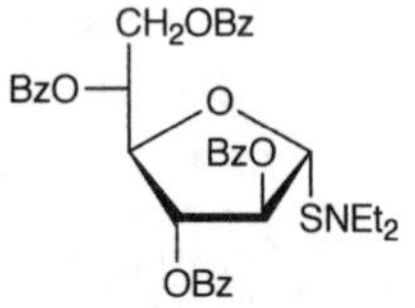   **40**

1.1.3 Other Monosaccharide Thiosugars.

N,N-Dialkyl-S-glycosylsulfenamides, such as **40**, were obtained in a one-pot procedure by treatment of a 1-thiosugar S-acetate with secondary amines in the presence of NBS or BrCH(CO₂Et)₂. The reactions are presumed to proceed by way of sulfenyl bromides.[16]

The hexopyranosyl methanethiosulfonates **41** have been generated by reaction of the corresponding glycosyl chlorides with NaSSO₂CH₃, for use as protein glycosylation reagents. Related glycomethanesulfonates are referred to in Chapter 3.[17] (Thioglycosides of nojirimycin are referred to in Chapter 18).

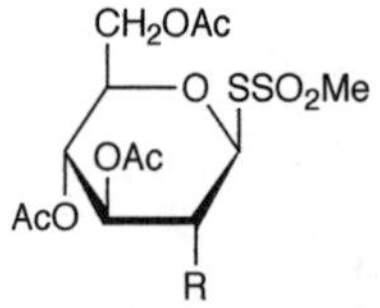

41 R = OAc or NHAc 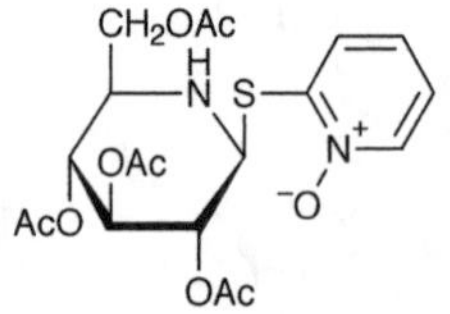   **42** 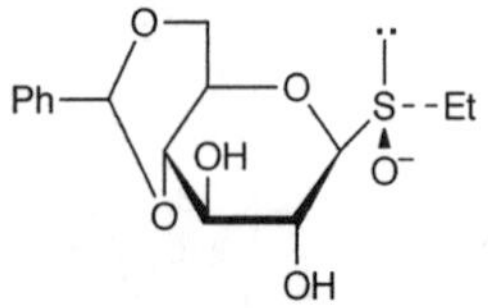   **43**

2-Thioglycosyl pyridine N-oxides, *e.g.* compound **42**, have been prepared as 'latent' glycosyl donors from peracetylated 1-thiosugars and 2-bromopyridine-N-oxide.[18] MCPBA oxidation of ethyl 1-thio-β-glycopyranosides gave the corresponding S_s-sulfoxides, for example **43**, as the major products. The preferred conformations of the minor R_s-epimers under the influence of the *exo*-anomeric effect are discussed in Chapter 21.[19]

1-Thioglycosides, in general, are covered in Chapter 3.

Unprotected D-mannitol was transformed into 2,5-anhydro-1,6-dithio-D-glucitol **46** in one step by use of the thio-Mitsunobu conditions (PPh₃-DEAD-AcSH-pyridine). The reaction is believed to proceed *via* epoxide **44** or phospholane intermediate **45**.[20] The thio-Mitsunobu reaction has also been applied to the synthesis of 5′-thioadenosine **48** from commercial 2′,3′-*O*-isopropylideneadenosine *via* the thioacetate **47**. This intermediate was, in addition, converted to *S*-adenosyl-L-homocystein analogues **49**.[21] 8-Aminooctyl 5-*S*-coniferyl-5-thio-α-L-arabinofuranoside (**50**), synthesized by standard procedures, was attached to Sepharose as an affinity ligand for feruloyl esterase of *Aspergillus niger*.[23] Isopropylidene-protected sugar-*S*-dithiophosphates, such as **51**, were obtained by exposure of the corresponding iodo-deoxy-sugars to (PhO)₂P(S)Cl. Their deacetalization by heating in water without added acid is referred to in Chapter 6, and related *O*-thionophosphates are covered in Chapter 7.[23]

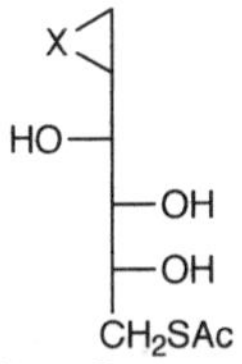

44 X = —O—
45 X = —O—PPh₃—O—

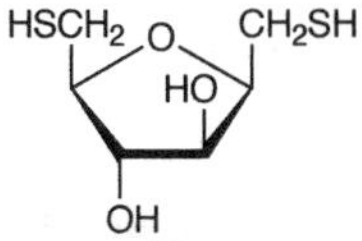

46

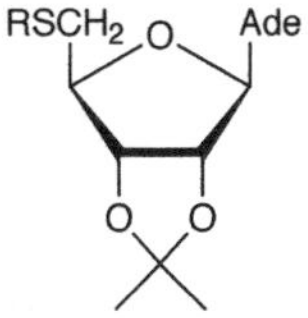

47 R = Ac
48 R = H
49 R = (CH₂)₃NH₂ or (CH₂)₃CO₂H

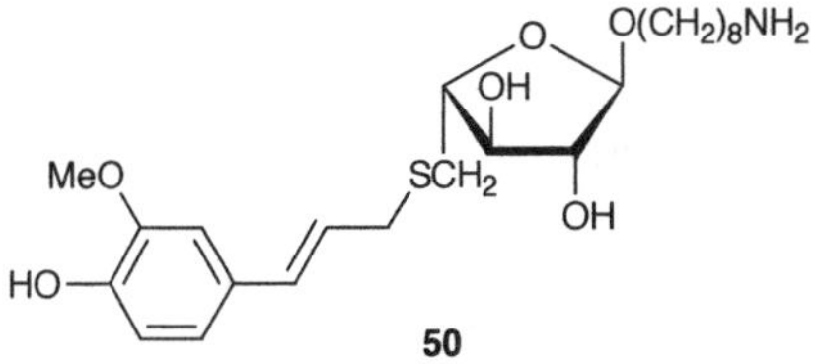

50

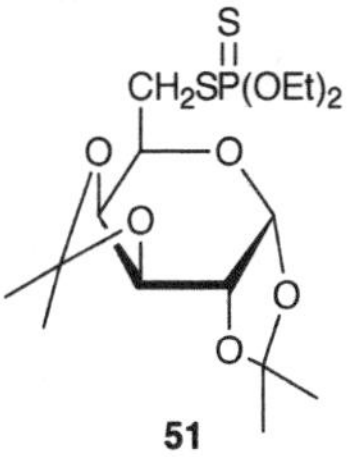

51

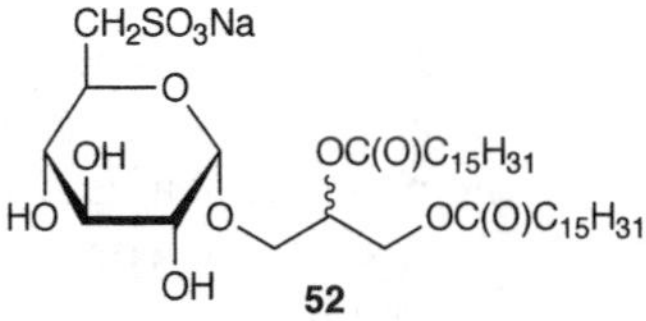

52

The synthesis of sulfoquinovosyl diacylglycerols **52**, potent inhibitors of DNA polymerases, by standard methods, starting from 6-*S*-acetyl-2,3,4-tri-*O*-benzyl-6-thio-α-D-glucopyranosyl trichloroacetimidate, has been reported.[24]

The BF₃-catalysed reactions of tri-*O*-acetyl-D-glucal with thiophenol have been reinvestigated. Whereas under anhydrous conditions the expected 2,3-unsaturated phenyl 1-thioglycopyranosides **53** were formed, an isomeric mixture of phenyl 2-deoxy-3-phenylthio-1-thiohexopyranosides **54** was obtained in almost quantitative yield when traces of water were present. It is proposed that these products arise from an intermediate enal.[25] The addition products of TolSCl to tri-*O*-benzyl-D-galactal formed diastereomeric episulfonium ions on exposure to SnCl₄. These were opened by *C*-nucleophiles (TmsCN, AllTms, *etc.*)

at the anomeric carbon atoms to give mainly the 2-*S*-(*p*-tolyl)-2-thio-β-D-galac-topyranose derivative **55**, accompanied by minor quantities of the α-D-talopyranose analogue.[26]

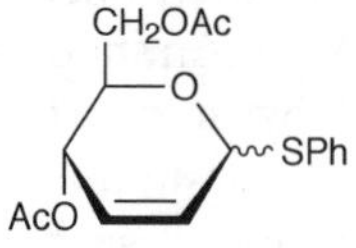

53 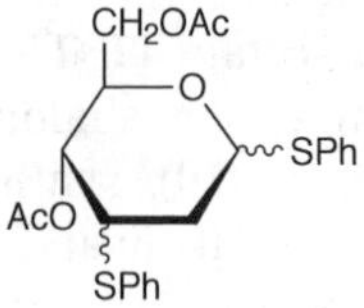   **54** 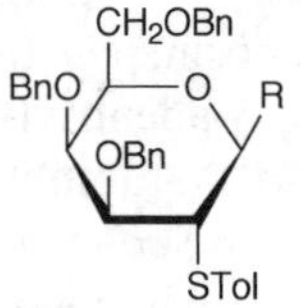   **55** R = CN, All, *etc.*

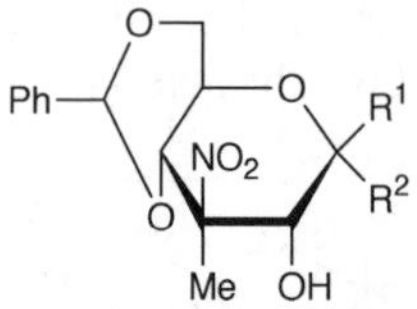

56 R^1 = SPh, R^2 = H
57 R^1 = H, R^2 = SPh
58 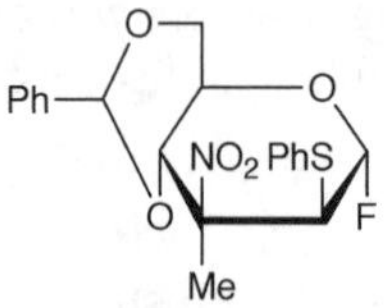
59 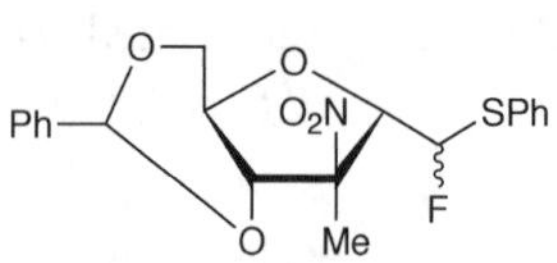

1,2-*Trans*- and 1,2-*cis*-phenyl 1-thioglycosides with a free hydroxyl group at the 2-positions underwent 1,2-*S*-migration or ring-contraction, respectively, on exposure to DAST (*e.g.* **56**→**58**; **57**→**59**).[27]

A review of the work at Novartis on the synthesis of complex compounds on solid supports includes the preparation and use of solid-phase bound thiosugars derived from levoglucosan.[28]

A study of the hydrolysis of the *cis*-methyl ester of 3'-deoxy-3'-thiothymidine 3'-*S*,5'-*O*-cyclic(phosphorothiolate) is referred to in Chapter 7.

1.2   Di- and Oligo-saccharides. – Sulfur-containing analogues, namely 4,6-thietane **61** and 4,6-dithiolane **62**, of the tumor-associated T-antigen β-D-Gal-(1→3)-α-D-GalNAc have been prepared by exposure of 6-thiocyanate **60** to MeO⁻/MeOH and KSCN followed by MeO⁻/MeOH, respectively.[29] A number of *S*-linked disaccharides have been prepared by direct alkylation of glycosyl thiolates with sugar triflates in the presence of a base, the very hindered base **63** being particularly effective.[30,31] The previously reported Pd(0)-catalysed route to O-linked disaccharides involving coupling of an unsaturated sugar carbonate with a protected free sugar (see Vol. 32, p. 30. ref. 176) has now been extended to the synthesis of thiodisaccharides. An example is shown in Scheme 2.[32]

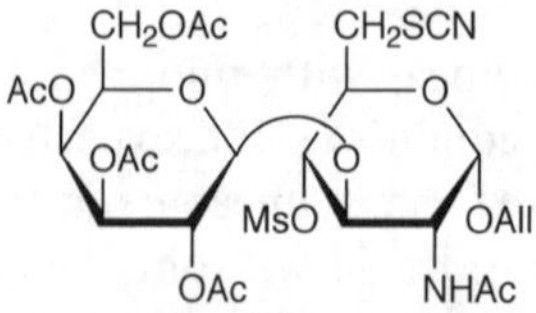

60

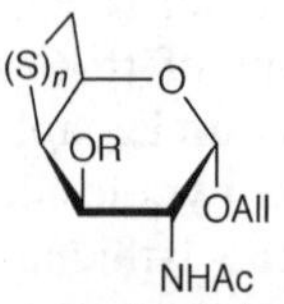

R = β-D-Gal*p*
61 *n* = 1
62 *n* = 2

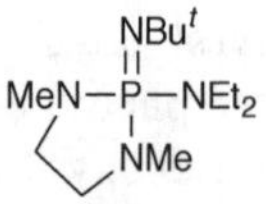

63

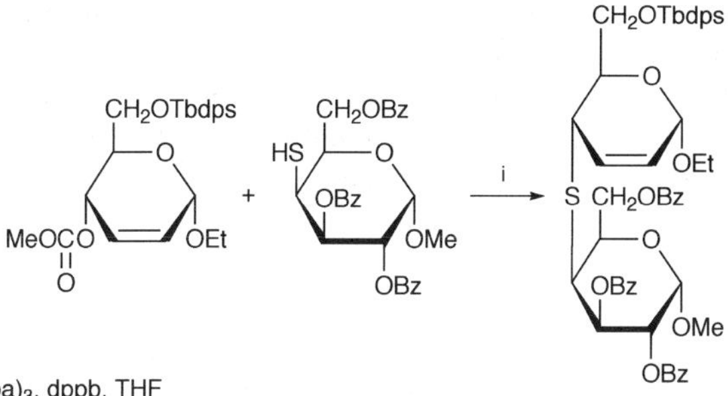

Reagents: i, Pd$_2$(dba)$_3$, dppb, THF

Scheme 2

Per-*O*-acetylated 1-thiosugars added stereoselectively to the levoglucosan-derived nitroalkene **64** (see Chapter 14 for synthesis) to furnish thiodisaccharides, such as **65**, after reduction of the nitro group, *N*-acetylation and de-*O*-acetylation.[33]

Sulfur-linked pseudodisaccharides, *e.g.*, compound **66**, with an iminosugar residue as the non-reducing moiety, were obtained by acid-induced displacement of 1-OH of L-*arabino*-nojirimycin by 6-thiosugar derivatives.[34] A new class of disaccharides having nitrogen in the interglycosidic linkage and 4-thio-Gal*f* as the non-reducing-end moiety is covered in Chapter 10.

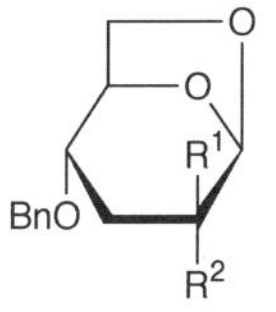

64 R^1,R^2 = ⟋⟍NO$_2$
65 R^1 = ⟋NHAc, R^2 = β-D-Glc*p*S
66 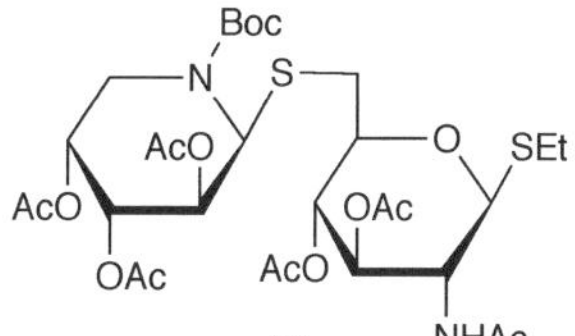

The all-sulfur-linked, tetrasaccharide **67** was prepared by standard methods from the β-peracetate (see Vol. 30, p. 68, ref. 100) and conjugated to human serum albumin for immunostimulation tests.[35] The synthesis of a neomycin-acridine conjugate, the two moieties being linked by an aminoethylthio-bridge from C-5 of the ribofuranose residue is referred to in Chapter 18, and the conformational behaviour of methyl 4-thio-α-lactoside is covered in Chapter 21.

$$\beta\text{-D-Glc}p\text{-}(1\rightarrow 3S)\text{-}\beta\text{-D-Glc}p\text{-}(1\rightarrow 3S)\text{-}\beta\text{-D-Glc}pO(CH_2)_8NH_2$$
$$6S$$
$$\uparrow$$
$$1$$
$$\beta\text{-D-Glc}p \qquad \textbf{67}$$

2 Seleno- and Telluro-sugars

Opening of epoxide **68** with sodium hydrogen selenide and concomitant mesylate displacement afforded methyl 2,5-anhydro-2-seleno-α-D-arabinofuranoside (**69**) and its 3,5-anhydro-3-seleno-α-D-xylofuranoside analogue **70** in equal proportions. The former was used in the synthesis of novel nucleoside analogues (see Chapter 19). The corresponding telluro compounds were similarly prepared by use of sodium hydrogen telluride.[36] The 5-seleno-D-ribopyranose derivative **71** was formed on SmI$_2$-mediated radical cyclization of 2,3,4-tri-*O*-benzyl-5-*Se*-benzyl-5-seleno-D-ribose.[37]

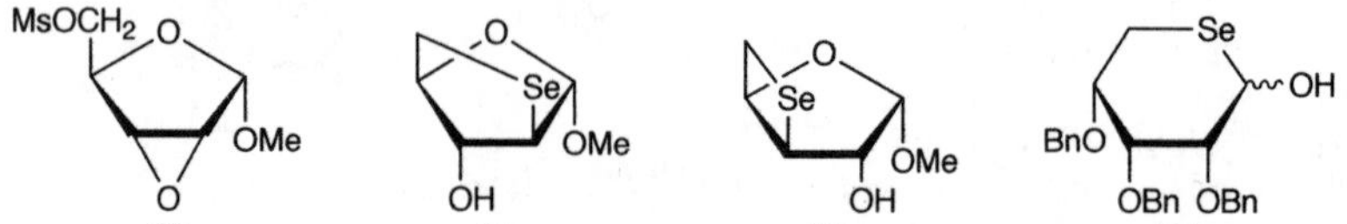

DAST-induced 1,2-selenium migration allowed the stereocontrolled construction of 2-deoxyglycosides, 2-deoxy-1-orthoesters and 2,3-unsaturated 1-orthoesters as illustrated in Scheme 3.[38] Practical application of these reactions is included in two vast papers describing the total synthesis of everninomycin.[39,40]

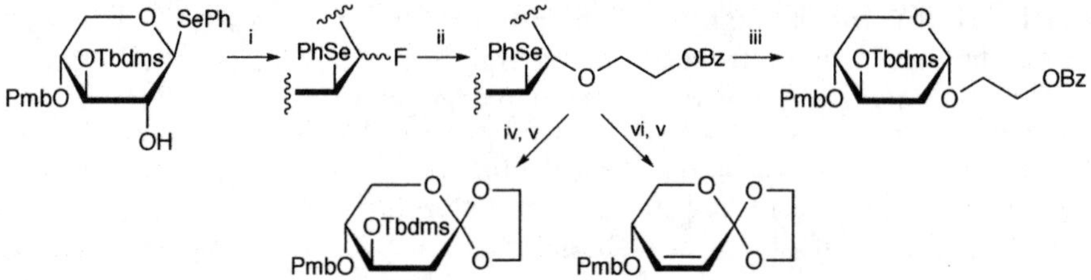

Reagents: i, DAST; ii, HO⌒⌒OBz, SnCl$_2$; iii, Bu$_3$SnH, AIBN; iv, MeO⁻, MeOH; v, NaIO$_4$, then Δ; vi, Bu$_4$NF

Scheme 3

References

1. J.A. Miller, A.W. Pugh, G.M. Ullah and C. Gutteridge, *Tetrahedron Lett.*, 2000, **41**, 10099.
2. G. Adiwidjaja, J.-S. Brunck, K. Polchow and J. Voss, *Carbohydr. Res.*, 2000, **325**, 237.
3. H. Yuasa, J. Takada and H. Hashimoto, *Tetrahedron Lett.*, 2000, **41**, 6615.
4. A. Arcelli, V. Cerè, F. Peri, S. Pollicino and P. Sabatino, *Tetrahedron: Asymm.*, 2000, **11**, 1389.
5. A. Tatibouët, P. Rollin and O.R. Martin, *J. Carbohydr. Chem.*, 2000, **19**, 641.
6. E. Bózo, S. Boros and J. Kuszmann, *Carbohydr. Res.*, 2000, **329**, 525.
7. E. Bózo, A. Medgyes, S. Boros and J. Kuszmann, *Carbohydr. Res.*, 2000, **329**, 25.
8. E. Bózo, S. Boros, L. Párkányi and J. Kuszmann, *Carbohydr. Res.*, 2000, **329**, 269.
9. E. Bózo and J. Kuszmann, *Carbohydr. Res.*, 2000, **325**, 143.
10. N.A. Hughes, *Carbohydr. Res.*, 2000, **326**, 323.
11. I. Izquierdo Cubero, M.T. Plaza-López-Espinosa, R. Asenjo Asenjo and A. Ramírez Fernández, *Carbohydr. Lett.*, 1999, **3**, 323 (*Chem. Abstr.*, 2000, **132**, 166 405).
12. L. Swansson, B.D. Johnston, J.-H. Gu, B. Patrick and B.M. Pinto, *J. Am. Chem. Soc.*, 2000, **122**, 10769.

13. K.D. Randell, B.D. Johnston, D.F. Green and B.M. Pinto, *J. Org. Chem.*, 2000, **65**, 220.
14. B.W. Skelton, R.V. Stick, D.M.G. Tilbrook, A.H. White and S.J. Williams, *Aust. J. Chem.*, 2000, **53**, 389.
15. N.A. Hughes and N.D. Todhunter, *Carbohydr. Res.*, 2000, **326**, 81.
16. D.J. Owen and M. von Itzstein, *Carbohydr. Res.*, 2000, **328**, 287.
17. B.G. Davis, M.A.T. Maugham, M.P. Green, A. Ullman and J.B. Jones, *Tetrahedron: Asymm.*, 2000, **11**, 245.
18. G. Pastuch and W. Szeja, *Pol. J. Chem.* 2000, **74**, 227 (*Chem. Abstr.*, 2000, **132**, 265 369).
19. N. Khiar, *Tetrahedron Lett.*, 2000, **41**, 9059.
20. O. Schulze, S. Bruns, J. Voss and G. Adiwidjaja, *Carbohydr. Res.*, 2000, **329**, 781.
21. M. Pignot, G. Pljevaljcic and E. Weinhold, *Eur. J. Org. Chem.*, 2000, 549.
22. Y. Zhao and M.L. Sinnott, *Bioorg. Med. Chem.*, 2000, **8**, 917.
23. D. Postel, G. Ronco and P. Villa, *J. Carbohydr. Chem.*, 2000, **19**, 171.
24. S. Hanashima, Y. Mizushima, T. Yamazaki, K. Ohita, S. Takahashi, H. Koshino, H. Sahara, K. Sakaguchi and F. Sugawara, *Tetrahedron Lett.*, 2000, **41**, 4403.
25. R. Blattner, R.J. Ferrier and R.H. Furneaux, *Tetrahedron: Asymm.*, 2000, **11**, 379.
26. M. Han, I.P. Smoliakova and L.N. Koikov, *Carbohydr. Res.*, 2000, **323**, 202.
27. P. Borrachero, F. Cabrera-Escribano, A.T. Carmona and M Gómez-Guillén, *Tetrahedron: Asymm.*, 2000, **11**, 2927.
28. S. Wendeborn, A. de Mesmaeker, W.K.-D. Brill and S. Berteina, *Acc. Chem. Res.*, 2000, **33**, 215.
29. H. Streicher, W. Schmid, I. Wenzl, C. Fiedler, H. Kählig and F.M. Unger, *Bioorg. Med. Chem. Lett.*, 2000, **10**, 1369.
30. O. Pàmies, G. Net, A. Ruiz, C. Claver and S. Woodward, *Tetrahedron: Asymm.*, 2000, **11**, 871.
31. W. Xu, S.A. Springfield and J.T. Koh, *Carbohydr. Res.*, 2000, **325**, 169.
32. A. Zawisza, B. Kryczka, P. Lhoste, S. Porwanski and D. Sinou, *J. Carbohydr. Chem.*, 2000, **19**, 795.
33. J.Z. Witczak, R. Chhabra and D. Boryczewski, *J. Carbohydr. Chem.*, 2000, **19**, 543.
34. K. Suzuki and H. Hashimoto, *Carbohydr. Res.*, 2000, **323**, 14.
35. Y. Ding, S.S. Alkan, G. Baschang and J. Defaye, *Carbohydr. Res.*, 2000, **328**, 71.
36. G. Adiwidjaja, O. Schulze, J. Voss and J. Wirsching, *Carbohydr. Res.*, 2000, **325**, 107.
37. M.A. Lucas, O.T.K. Nguyen, C.H. Schiesser and S.-L. Zheng, *Tetrahedron*, 2000, **56**, 3995.
38. K.C. Nicolaou, H.J. Mitchell, K.C. Fylaktakidou, H. Suzuki and R.M. Rodríguez, *Angew. Chem., Int. Ed. Engl.*, 2000, **39**, 1089.
39. K.C. Nicolaou, H.J. Mitchell, K.C. Fylaktakidou, .M. Rodríguez, and H. Suzuki, *Chem. Eur. J.*, 2000, **6**, 3116.
40. K.C. Nicolaou, K.C. Fylaktakidou, H.J. Mitchell, F.L. van Delft, R.M. Rodríguez, S.R. Conley and Z. Jin, *Chem. Eur. J.*, 2000, **6**, 3166.

12
Deoxy-sugars

The synthesis of 2-deoxyglycosides over 10 years to 1999 has been reviewed[1] as have the catalytic mechanisms of several characterized enzymes involved with reductive C–O bond cleavage.[2] A cardiac glycoside which contained a 6-deoxy-D-gulosyl moiety was isolated from *Erysimum cheiranthoide*.[3] The two *C*-glycosides **1** and **2** were isolated as artefacts from the reaction of the *O*-glycoside mithramycin under acidic conditions, and they have promise as anti-tumour agents.[4] A steroidal pentasaccharide with deoxysugar components is mentioned in Chapter 4.

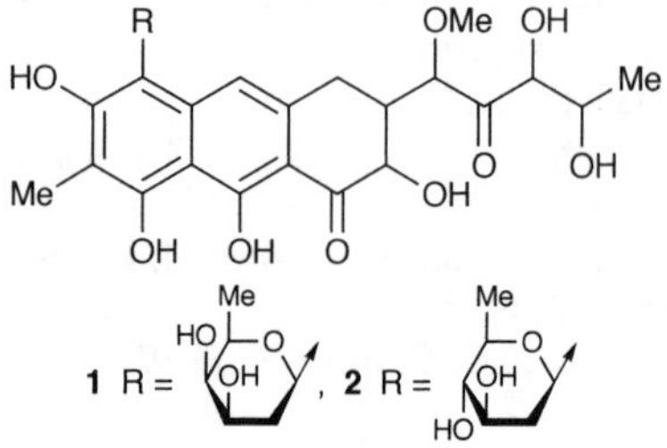

A mild one-pot procedure for the synthesis of 2-deoxyaldoses from glycals involves treatment with NIS in aqueous MeCN followed by reduction with $Na_2S_2O_4$.[5] The stereocontrolled construction of 2-deoxy-glycosides and glycosylidene orthoesters in solution and on solid support was facilitated by a 1,2-selenium migration (Scheme 1). The resulting 2-deoxy-2-selenoglycosyl fluorides were glycosylated, then subjected to a radical deselenation or oxidized at selenium followed by heat promoted *syn*-elimination and cyclization to give 2-deoxy-orthoesters.[6–8] The unsaturated 4-deoxy-L-sugar precursor **3** was synthesized from (*R*)-benzyl glycidyl ether by standard methods.[9] The 2-, 3-, 4- and 6-deoxy- analogues of 4-nitrophenyl α-D-galactopyranoside were synthesized and studied as potential substrates for α-galactosidases from several sources.[10] Similarly the 2'-, 3'- and 4'-deoxy- analogues of octyl α-L-fucopyranosyl-(1→2)-β-D-galactopyranoside were prepared and tested as acceptors for human blood group glycosyltransferases.[11] 4-Nitrophenyl β-D-fucofuranoside has been synthesized and shown not to be a substrate for the *exo* β-D-galactofuranosidase from *P. fellutanum*.[12] The 4-deoxy- analogues of GlcNAc and XylNAc have been

---

Carbohydrate Chemistry, Volume 34
© The Royal Society of Chemistry, 2003

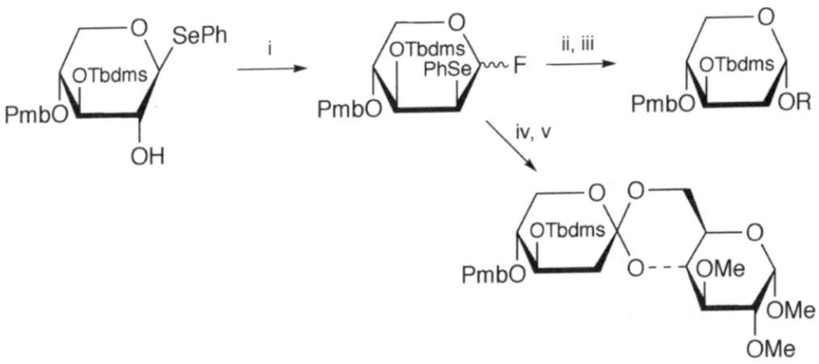

Reagents: i, DAST; ii, ROH, SnCl₂; iii, Bu₃SnH; iv, methyl 2-3-di-*O*-methyl-β-D-glucopyranoside, SnCl₂; v, NaIO₄, MeOH, heat

Scheme 1

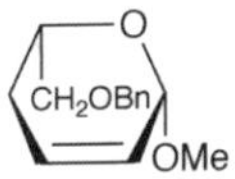

3

synthesized and evaluated for their ability to inhibit glycoconjugate biosynthesis.[13]

1,2-Di-*O*-acetyl-5-*O*-benzoyl-3-deoxy-L-*erythro*-pentofuranose, a precursor for the synthesis of β-L-nucleosides, has been prepared from L-xylose.[14] Some *O*-acetylated 1,2-dideoxy-1-nitrohex-1-enitols have been converted, using stannous chloride, into the oximes of the corresponding 2-deoxyhexoses.[15] An efficient synthesis of 2,3,6-trideoxy-4,5-*O*-isopropylidene-L-*threo*-hexose from diethyl L-tartrate *via* 1,4-dideoxy-1-iodo-2,3-*O*-isopropylidene-l-threitol has been achieved,[16] and syntheses of 3,6-dideoxy-D-*ribo*- and -*xylo*-hexopyranoses in forms suitable to act as glycosyl donors have been reported.[17] A short enzymatic synthesis of L-fucose has employed sequentially L-fuculose phosphate aldolase, alkaline phosphatase and L-fucose isomerase.[18] A chiral synthesis of 2-deoxy-L-ribose from the dihydroxyacetone derivative **4** has been outlined (Scheme 2).[19] Ring-opening reactions of galactopyranoside 3,4- and 4,6-cyclic sulfates, some of which generated deoxysugar products, are covered in Chapters 7 and 13.

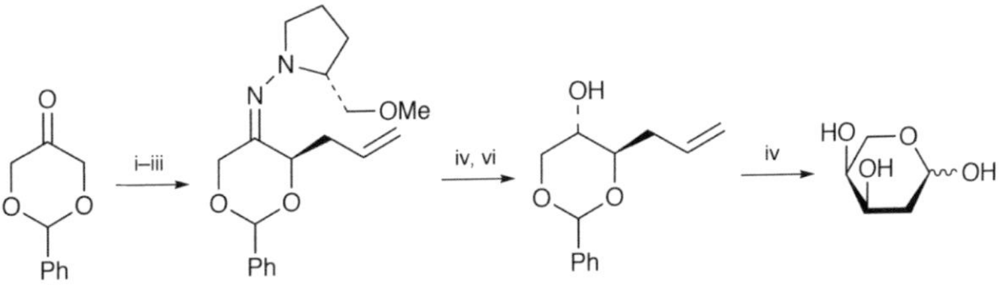

Reagents: i, ii, BuLi; iii, allyl bromide; iv, NH₄H₂PO₄; v, NaBH₄; vi, O₃, Me₂S, then TFA

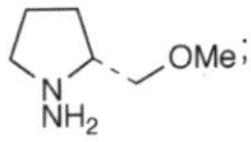

Scheme 2

References

1. C.H. Marzabadi and R.W. Franck, *Tetrahedron*, 2000, **56**, 8385.
2. D.A. Johnson and H.-W. Liu, *Biol.-Chem. Interface*, 1999, 351 (*Chem. Abstr.*, 2000, **132**, 100 160).
3. Z.-H. Lei, S. Yahara, T. Nahara, B.-S. Tai, J.-Z. Xiang and Y.-L. Ma, *Chem. Pharm. Bull.*, 2000, **48**, 290.
4. M. Tagashira, T. Kitahawa, N. Nozato, S. Isonishi, A. Okamoto, K. Ochiai and Y. Ohtake, *Chem. Pharm. Bull.*, 2000, **48**, 575.
5. V. Costantino, C. Imperatore, E. Faltorusso and A. Mangoni, *Tetrahedron Lett.*, 2000, **41**, 9177.
6. K.C. Nicolaou, H.J. Mitchell, K.C. Fylaktakidou, H. Suzuki and R.H. Rodríguez, *Angew. Chem. Int. Ed. Engl.*, 2000, **39**, 1089
7. K.C. Nicolaou, H.J. Mitchell, K.C. Fylaktakidou, F.L. van Delft, R.H. Rodríguez, S.R. Conley and Z. Jin, *Chem. Eur. J.*, 2000, **6**, 3166.
8. K.C. Nicolaou, H.J. Mitchell, H. Suzuki, R.H. Rodríguez, O. Baudoin and K.C. Fylaktakidou, *Angew. Chem. Int. Ed. Engl.*, 2000, **39**, 3334, 3340, 3345.
9. R. Caputo, A. Guarogna, G. Palumbo, S. Pedotella and F. Solla, *J. Carbohydr. Chem.*, 2000, **19**, 631.
10. W. Hakamata, T. Nishio and T. Oku, *Carbohydr. Res.*, 2000, **324**, 107.
11. A. Mukherjee, M.M. Palcic and O. Hindsgaul, *Carbohydr. Res.*, 2000, **326**, 1.
12. A. Chiocconi, C. Marino and R.M. De Lederkremer, *Carbohydr. Res.*, 2000, **323**, 7.
13. A. Berkin, M.A. Szarek, J. Plenkiewicz, W.A. Szarek and R. Kisilevsky, *Carbohydr. Res.*, 2000, **325**, 30.
14. C. Mathé, J.-L. Imbach and G. Gosselin, *Carbohydr. Res.*, 2000, **323**, 226.
15. M. Koos, *Tetrahedron Lett.*, 2000, **41**, 5403.
16. N. Satyamurthi and I.S. Aiden, *Carbohydr. Lett.*, 1999, **3**, 355 (*Chem. Abstr.*, 2000, **132**, 166 398).
17. H.N. Yu, P. Zhang, C.-C. Ling and D.R. Bundle, *Tetrahedron: Asymmetry*, 2000, **11**, 465.
18. W.-D. Fessner, C. Grosse, G. Jaeschke and O. Eyrisch, *Eur. J. Org. Chem.*, 2000, 125.
19. T. Ulven and P.J.H. Carlsen, *Synth. Commun.*, 2000, **30**, 2275.

13
Unsaturated Derivatives

1   General

The olefin metathesis reaction as applied to carbohydrate dienes has been reviewed,[1] as has the use of titanocene(III) and zirconocene(III) chlorides as mild reagents for generating 1-glycosyl radicals from glycosyl halides, and converting them into glycals.[2]

2   Pyranoid Derivatives

2.1   1,2-Unsaturated Compounds and Related Derivatives. – A new method for the production of glycals from glycosyl halides uses catalytic Cp_2TiCl_2 in the presence of Mn(0) and TmsCl (Scheme 1).[3] D-Glucal and D-galactal both afforded the furan diol **1** on exposure to $InCl_3$ in acetonitrile.[4] A new sulfonium reagent has been used for glycal activation in a one-pot acetamidoglycosylation (Scheme 2). The reaction proceeds *via* acid-mediated oxazoline ring opening by an alcohol.[5] An alternative approach to aminosugar derivatives from glycals is based on an *o*-iodoxybenzoic acid-mediated synthesis from allylic alcohols and aryl isocyanates (Scheme 3).[6]

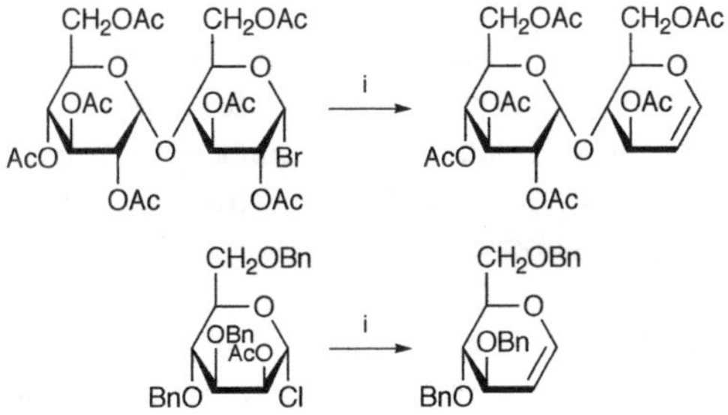

Reagents: i, Cp_2TiCl_2, Mn, TmsCl, THF, RT

Scheme 1

An extensive (re)investigation of the additions of IN_3, Br_2 and $Br_2/MeOH$ to tri-*O*-acetyl-D-glucal and tri-*O*-benzyl-D-galactal has been reported. The results are in keeping with previous knowledge.[7] The enolone **3** when treated with

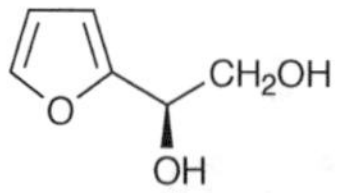

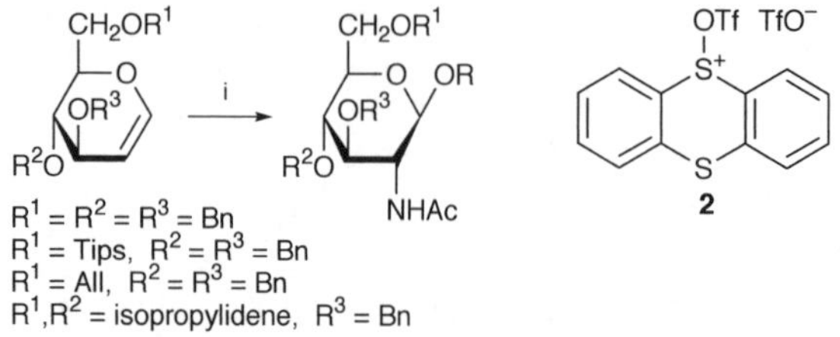

Reagents: i, 2, AcNHSiMe₃, Et₂NPh, then Amberlyst-15, ROH

Scheme 2

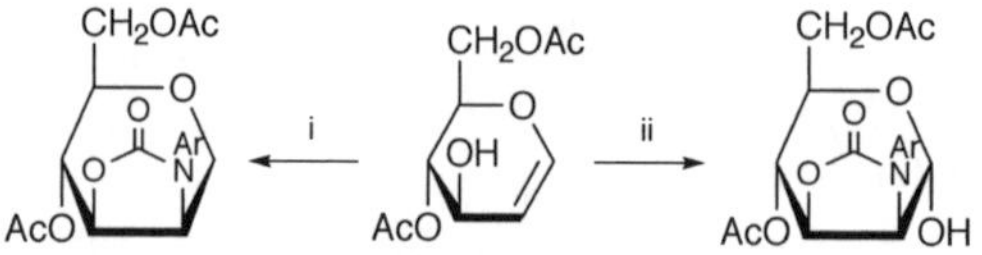

Reagents: i, ArNCO, DBU then o-iodoxybenzoic acid, anhydrous conditions; ii, as for i, wet conditions

Scheme 3

TmsCN afforded prdominantly the conjugate addition product **4**.[8] When the mannosyl acetylene **5**, or its α-anomer, was allowed to react with BuLi the glycal **6** was the major product. The same product, but in lower yield, was obtained from the corresponding glucosyl acetylenes in the same way.[9] In an extension of previous work, the 2,3-unsaturated glycoside **7** has afforded the branched glycal **8** by way of a Pd-catalysed Heck-type reaction.[10] Treatment of 2-C-formyl-substituted D-glucal derivatives with organocuprate reagents has produced 2-deoxy-2-C-formyl-C-glycosides with β-D-*manno* stereochemistry by 1,4-conjugate addition, whereas treatment with organocopper/BF₃.OEt₂ reagents generated 2,3-unsaturated-2-C-formyl-α-C-glycosides by an S_N2' reaction.[11] The relative rates of the base-catalysed rearrangements of **9** and **10** to **11** and **12**, respectively, have been compared and the pyranoid **9** was shown to be much more reactive than the carbocycle **10**.[12]

2.2   2,3-Unsaturated Compounds. – The Ferrier rearrangement of tri-O-acetyl-D-glucal or -galactal with diverse alcohols to the corresponding alkyl hex-2-enopyranosides can be catalysed by LiBF₄ in MeCN.[13] The effects of varying solvent, temperature and catalyst quantity have been studied for the same reaction catalysed by FeCl₃.[14] Additionally, Sc(OTf)₃ was found to be an efficient catalyst for this reaction with various alcohols and phenols; as usual the α-glycosides predominated.[15] O-Acetylated glycals react smoothly with alkyl hydroperoxides in the presence of BF₃.OEt₂ or SnCl₄ to give the corresponding alkylperoxy α-hex-2-enopyranosides.[16] The reaction of TmsCN with tri-O-acetyl-D-glucal to give the 2,3-unsaturated-hexopyranosyl nitrile proceeds without added catalyst at elevated temperatures.[17]

3 4 5 6

7 8 9 X = O 10 X = CH_2 11 X = O 12 X = CH_2

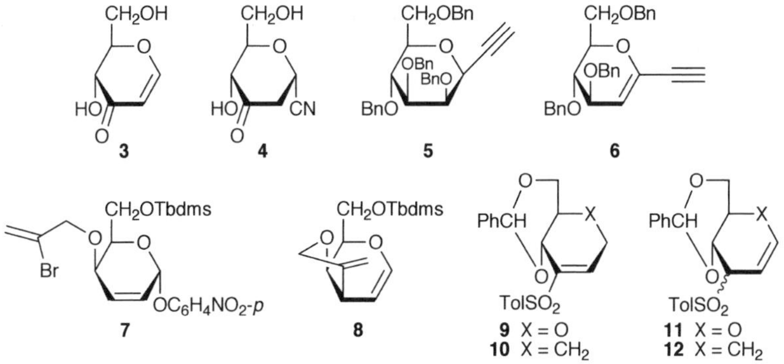

2.3 Other Unsaturated Compounds. – The dienals **13** (with D-*xylo*-, D-*lyxo*- and D-*arabino*-configurations) have been condensed with the phosphonate **14** to give the enone adducts **15**; the intramolecular Diels–Alder reactions of these adducts are mentioned in Chapter 22.[18] Treatment of the unsaturated lactone **16** with the Tebbe reagent has afforded **17**; its conversion into a cyclooctene derivative is covered in Chapter 18.[19] Exposure of the cyclic sulfate **18** to base has afforded a mixture of the *exo*- and *endo*-olefins **19** and **20**,[20] and a *syn*-selective dihydroxylation of the 3,4-unsaturated pyranoside **21** generated predominantly the D-*allo* product.[21] Synthesis of the branched-chain 5,6-ene **22** using standard methods, and its conversion into a cyclitol during the first total synthesis of (−)-tetracycline is mentioned in Chapter 22.

13 14 15 16 X = O 17 X = CH_2

18 19 20 21 22

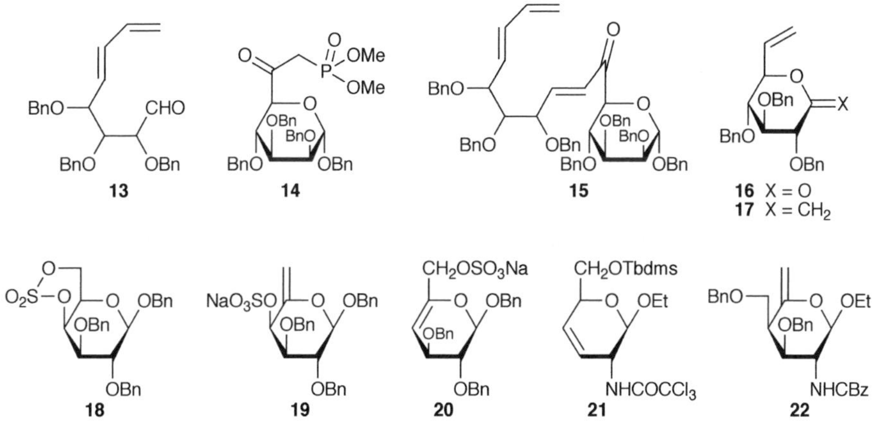

3 Furanoid Derivatives

Reaction of aldohexofuranose 5,6-diols with TPP/imidazole/I_2 has afforded the corresponding 5,6-unsaturated derivatives,[22] while elimination of HBr from the 6-bromo compound **23**, followed by hydroboration, has generated the product **24** with the L-*ido* configuration (Scheme 4).[23] Some sugar allyltin derivatives (*e.g.*

25) have been prepared and converted into 1,3-dienes (Scheme 5).[24] A Reformatsky reaction applied to nitrile **26** followed by a diastereoselective reduction, using either D- or L-tartaric acid in conjunction with NaBH$_4$, has allowed the synthesis of the L-2-deoxypentonolactones **27** and **28** respectively, from which the corresponding L-2-deoxypentoses are readily available.[25] The synthesis of isonucleosides with an exocyclic methylene group on the sugar moiety is covered in Chapter 19, as are C-5′ alkynyl-substituted adenosine analogues.

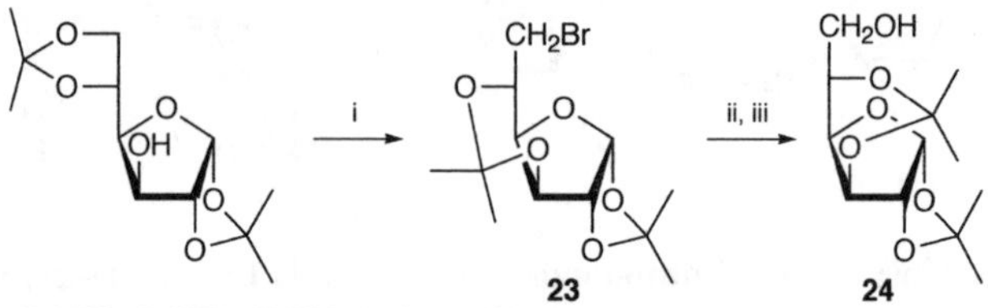

Reagents: i, Ph$_3$P, NBS; ii, DBU; iii, BH$_3$, H$_2$O$_2$, NaOH

Scheme 4

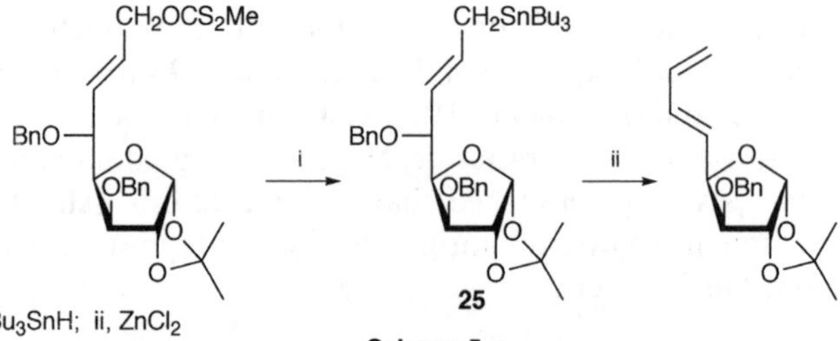

Reagents: i, Bu$_3$SnH; ii, ZnCl$_2$

Scheme 5

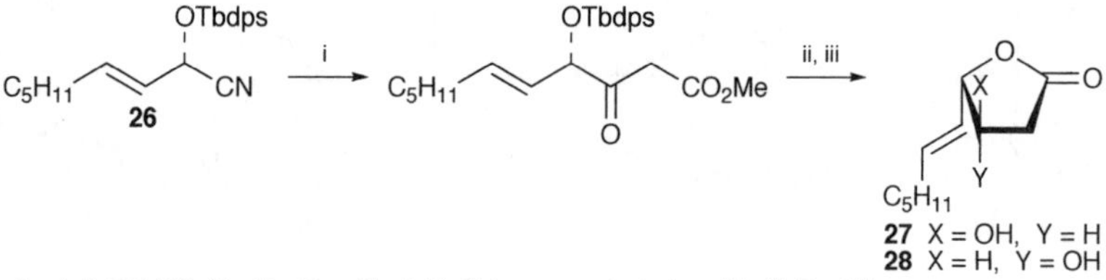

27 X = OH, Y = H
28 X = H, Y = OH

Reagents: i, BrCH$_2$CO$_2$Me, Zn, TmsCl; ii, NaBH$_4$, D- or L-tartaric acid; iii, Bu$_4$NF

Scheme 6

4   Acyclic Derivatives

Reduction of acyclic aldose 1-nitro-1-ene derivatives (*e.g.* **29**) with SnCl$_2$.2H$_2$O has provided an alternative route to 2-deoxy-aldose oximes (*e.g.* **30**). These may be converted into 2-deoxy-aldoses, -aldononitriles and -aldonic acids.[26] The zinc-mediated fragmentation of 6-deoxy-6-iodopyranosides to the corresponding acyclic 5,6-dideoxy-hex-5-enoses can be catalysed by vitamin B$_{12}$ affording higher yields under milder conditions.[27] In another study, enose products of this zinc-mediated reaction were subjected to Wittig reactions, and the resulting acyclic dienes were converted into cyclitols by olefin metathesis reactions.[28] A number of acyclic carbohydrate diene Horner–Wittig reagents have been prepared which, after reaction with aldehydes, generate trienes which undergo intramolecular Diels–Alder reactions (*e.g.* Scheme 7).[29]

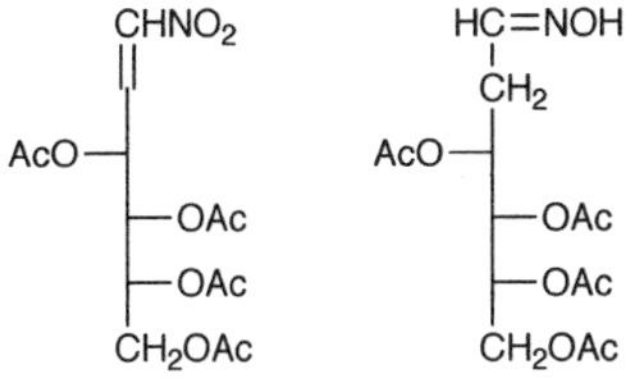

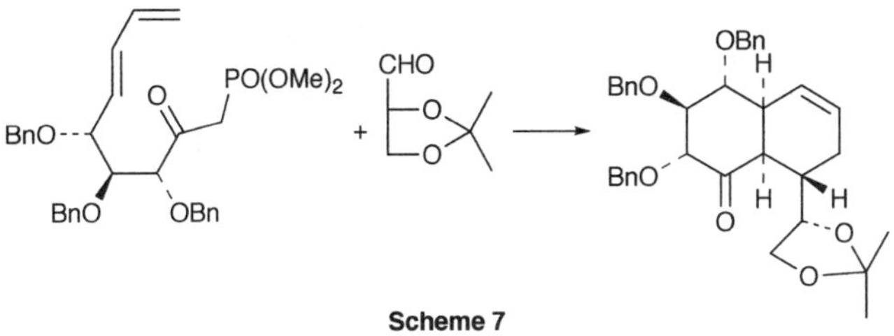

Scheme 7

References

1. M. Jorgensen, P. Hadwiger, R. Madsen, A.E. Stütz and T.M. Wrodnigg, *Curr. Org. Chem.*, 2000, **4**, 565.
2. R.P. Spencer and J. Schwartz, *Tetrahedron*, 2000, **56**, 2103.
3. T. Hansen, K. Daasberg and T. Skrydstrup, *Tetrahedron Lett.*, 2000, **41**, 8645.
4. B. Sobhana Babu and K.K. Balasubramanian, *J. Org. Chem.*, 2000, **65**, 4198.
5. V. Di Bussolo, J. Liu, L.G. Huffman, Jr. and D.Y. Gin, *Angew. Chem. Int. Ed. Engl.*, 2000, **39**, 204.
6. K.C. Nicolau, P.S. Baran, Y.-L. Zhong and J.A. Vega, *Angew. Chem. Int. Ed. Engl.*, 2000, **39**, 2525.
7. A. Boschi, C. Chiappe, A. De Rubertis and M.F. Ruasse, *J. Org. Chem.*, 2000, **65**, 8470.
8. M. Hayashi, H. Kawabata, S. Shimono and A. Kakehi, *Tetrahedron Lett.*, 2000, **41**, 2591.
9. T.M. Davis and T.L. Lowary, *Carbohydr. Res.*, 2000, **324**, 210.
10. K. Bedjeguelal, L. Joseph, V. Bolitt and D. Sinou, *J. Carbohydr. Chem.*, 2000, **19**, 221.
11. J. Cossy and H. Rakotoarisoa, *Synlett*, 2000, 734.
12. T. Sakokibora, M. Shindo, S. Narumi, C. Nagano and Y. Kajihara, *J. Carbohydr. Chem.*, 2000, **19**, 783.
13. B.S. Babu and K.K. Balasubramanian, *Synth. Commun.*, 1999, **29**, 4299 (*Chem. Abstr.*, 2000, **132**, 208 027).
14. C. Masson, J. Soto and M. Bessodes, *Synlett*, 2000, 1281.
15. J.S. Yadav, B.V. Subba Reddy, C.V.S.R. Murthy and G. Mahesh Kumar, *Synlett*, 2000, 1450.
16. G. Grynkiewicz and W. Priebe, *Pol. J. Chem.*, 1999, **73**, 1917 (*Chem. Abstr.*, 2000, **132**, 78 754).
17. M. Hayashi, H. Kawabata and K. Inoue, *Carbohydr. Res.*, 2000, **325**, 68.

18.   S. Jarosz and S. Skóra, *Tetrahedron: Asymmetry*, 2000, **11**, 1425.
19.   W. Wang, Y. Zhang, M. Sollogoub and P. Sinaÿ, *Angew. Chem. Int. Ed. Engl.*, 2000, **39**, 2466.
20.   F. Dagron and A. Lubineau, *J. Carbohydr. Chem.*, 2000, **19**, 311.
21.   K. Blades, T.J. Donohoe, J.J.G. Winter and G. Stemp, *Tetrahedron Lett.*, 2000, **41**, 4701.
22.   H.B. Merayala, P.M. Goud, R.R. Gadikota and K.R. Reddy, *J. Carbohydr. Chem.*, 2000, **19**, 1211.
23.   S.-C. Hung, R. Puranik and F.-C. Chi, *Tetrahedron Lett.*, 2000, **41**, 77.
24.   S. Jarosz, S. Skóra and K. Szewczyk, *Tetrahedron: Asymmetry*, 2000, **11**, 1997.
25.   D.V. Johnson, R. Fischer and H. Griengl, *Tetrahedron*, 2000, **56**, 9289.
26.   M. Koos, *Tetrahedron Lett.*, 2000, **41**, 5403.
27.   M. Kleban, U. Kautz, J. Greul, P. Hilgers, R. Kugler, H.-Q. Dong and V. Jäger, *Synthesis*, 2000, 1027.
28.   L. Hyldtoft and R. Madsen, *J. Am. Chem. Soc.*, 2000, **122**, 8444.
29.   S. Jarosz and S. Skóra, *Tetrahedron: Asymmetry*, 2000, **11**, 1433.

14

Branched-chain Sugars

1   Compounds with a C–$\overset{\overset{\text{R}}{|}}{\underset{\underset{\text{O}}{|}}{\text{C}}}$–C Branch-point

1.1 Branch at C-2 or C-3. – In new and versatile approaches to the synthesis of deuterium or tritium labelled branched tetritols, two separate syntheses of 2-C-methyl-D-erythritol (3) have been described starting from dimethyl fumarate (MeO$_2$CCH=CHCO$_2$Me, 1) and the xylofuranose derivative (5), respectively.[1-3] In the former, the unsaturated diester 1 was converted into 2, the five-carbon scaffold of 3, in one pot by ozonolysis of 1 followed by Wittig reaction with Ph$_3$P=C(Me)CHO and the subsequent borohydride reduction and benzyl protection of the generated alcohol. The ester 2 was then reduced with DIBAL-H and subjected to successive Sharpless epoxidation, Payne rearrangement (for stereoselective epoxide ring-opening) and hydrogenolysis to give 3. Conversion of 2 into 2-C-branched threitol 4 has also been described.[1] The latter method involved conversion of 5 into 6 by Swern oxidation followed by addition of MeMgCl and subsequent simple steps to convert 6 into the branched erythritol 3 as described in Scheme 1.[2] These methods also provide access to the corresponding isotopically enriched tetritols. Discussion on the use of some of these and other 2-C-branched erythritol derivatives in the elucidation of the 2-C-methyl-D-erythritol-4-phosphate pathway for isoprenoid biosynthesis has also been reported.[4-7] Molybdic acid-catalysed tautomerizations of D-xylulose and D-ribulose to yield 2-C-(hydroxymethyl)-D-erythrose and 2-C-(hydroxymethyl)-D-threose respectively have been reported.[8]

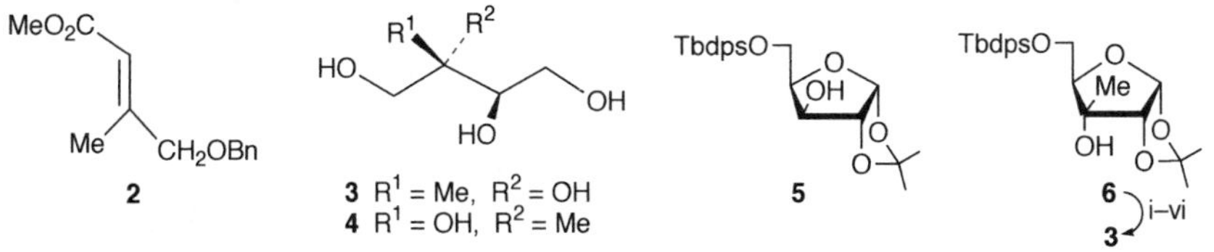

Reagents: i,Bu$^n_4$NF, THF; ii, NaH, BnBr, DMSO, 60 °C; iii, aq. TFA, –10 °C; iv, NaIO$_4$, aq. MeOH; v, NaBH$_4$, Pr^{i}OH; vi, H$_2$, Pd/C, EtOH

Scheme 1

Several 3-*C*-branched D-allofuranosyl (**7**) and 3-*C*-D-psicopyranosyl (**8**) derivatives have been obtained from 1,2;5,6-di-*O*-isopropylidene-α-D-*ribo*-hexofuranose-3-ulose and 1,2;4,5-di-*O*-isopropylidene-β-D-*erythro*-2-hexulopyranos-3-ulose respectively by treatment with lithium dianions of the respective carboxylic acids.[9] Synthesis of 3-*C*-methyl-rhamnose has been reported in connection with the total synthesis of eviminomicin 13,384-1, an orthosomicin class antibiotic (see also Section 2 and Chapter 4).[10]

R = CH_2CO_2H, $CH(Ph)CO_2H$, $CH_2CH{=}CHCO_2H$, *etc.*

Nucleophilic trifluoromethylation reactions of the 3-ulose derivative **9** with CF_3SiMe_3 and conversion of the products so obtained into nucleoside derivatives have been reported (see also Section 3.2).[11] Similar reactions of the hex-3-ulose derivative **10** have been reviewed (see also Vol. 32, Chapter 14, ref. 8 and *Nucleosides Nucleotides*, 1998, **17**, 2221).[12] Reaction of **10** with 2-lithiopyridine (obtained from 2-chloropyridine, Li and naphthalene) on the other hand has led to diastereoselective formation of the corresponding 3-(pyrid-2-yl)-derivative.[13] Addition of methylcerium (MeLi in the presence of $CeCl_3$) to the 2-deoxy-3-ulose derivatives **11** has been shown to result in the formation of a 5:1 mixture of *threo* and *erythro* configured products **12**.[14] Use of the alkyl cerium reagent is important as reaction with MeLi was found to lead to the formation of unsaturated products arising from β-elimination of MeOH (see also *J. Org. Chem.*, 1995, **60**, 202). Compound **12** was further converted in several steps into the methyl-branched sugar axenose (2,6-dideoxy-3-*C*-methyl-L-*xylo*-hexose) which occurs in disaccharides found in the antibiotics axenomycin, dutomycin and polyketomycin.[14] Syntheses of epoxides **14** (see also Section 5) and **15** from the corresponding ulose derivatives **13** and **10**, and their reductive ring opening to generate other functionalized branched-chain sugar derivatives, have been described.[15,16] Thus, treatment of **10**, for instance, with an equimolar or excess amount of Me_3SOI in the presence of Bu^{t}OK resulted in the formation of **15** in high yield. Compound **15** on reductive ring opening in the presence of Li and a catalytic quantity of 4,4'-di-*tert*-butylbiphenyl, resulted in the formation of the corresponding β-oxido organolithium derivative which upon reaction with various electrophiles (*e.g.* H_2O, Me_3SiCl, PhCHO, *etc.*) gave the desired branched-chain sugar derivatives (for structures see Vol. 32, Chapter 14, Scheme 6).[16] Application of this methodology to an oxetane derived from **10** has also been described.[16] Use of the 3-*C*-branched spiroepoxide **16** in the synthesis of the analogue **18** of the miharamycin sugar moiety has been documented. The synthesis was accomplished by treatment of **16** with LiCN followed by hydrolysis and spontaneous cyclization to afford the intermediate lactone **17** which upon stereoselective hydroxylation (MoO_5.py, HMPA), reduction ($LiAlH_4$) and cyclization (DEAD–Ph_3P) afforded **18**.[17]

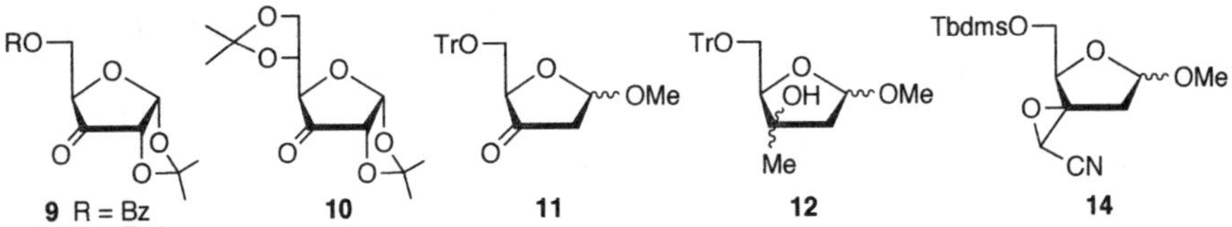

9 R = Bz
13 R = Tbdms

10 **11** **12** **14**

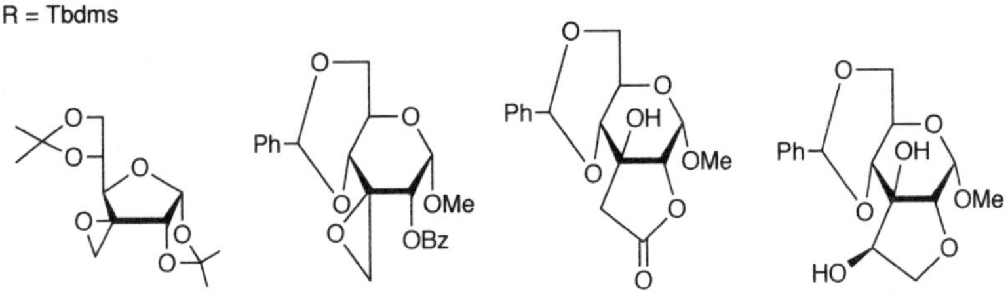

15 **16** **17** **18**

Application of ring-closing metathesis (RCM) to the preparation of enantiomerically pure annulated carbohydrates has been exemplified in the conversion of the branched-chain ald-2-uloside **19** into a series of tricyclic compounds **21** *via* **20** (see also Vol. 33, Chapter 14, structures **5** and **6** and ref. 5).[18] The alcohol **20** was obtained by the addition of the corresponding alkenylmagnesium chloride [CH$_2$=CH(CH$_2$)$_n$MgCl, n = 0, 1, 2 or 3] to **19** and the *cis*:*trans* ratio of **20** obtained was found to depend upon the chain length of the alkenyl residue being added. The RCM was conducted in the presence of the Ru complex [Cl$_2$(C$_6$H$_{11}$)$_3$P]$_2$RuCHPh in benzene. From the addition product obtained from vinylmagnesium chloride and **19**, however, only the *cis*-addition product underwent the RCM reaction. The assignment of the structure of **21** (n = 0; obtained from *cis*-**20**, n = 0) was supported by X-ray crystallography (see also Chapter 22).[18]

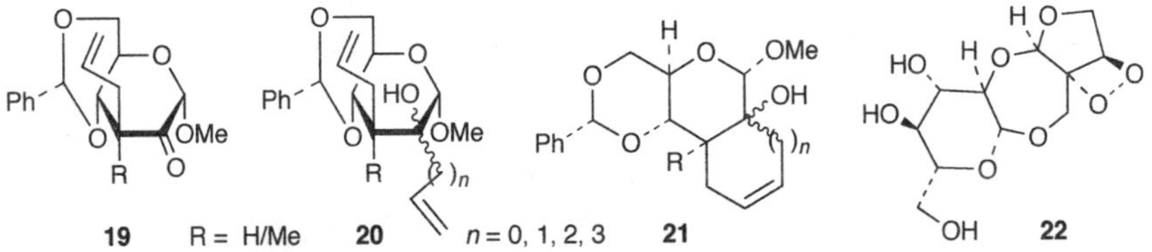

19 R = H/Me **20** n = 0, 1, 2, 3 **21** **22**

A novel disaccharide derivative, amabiose (**22**), has been isolated from rhizomes of *Hemsleya amabilis*. Although the structure of this compound has been reportedly elucidated by chemical and instrumental methods, discrepancies were noted in the structures depicted and the data described in the text by the authors. Structure **22** shown here is based on the text data (see also Chapters 3 and 21).[19] The biosynthetic pathways for the macrolactam polyketide glycoside anti-tumour antibiotic vicenistatins 1 and 2 (**23**, R = A, D-vicenisamine residue, and R = B, branched-chain sugar D-mycarose residue respectively) have been studied by feeding experiments with [^{13}C]- and [1,2-^{13}C$_2$]acetate, [1-^{13}C]propionate, DL-[2,3,3-^{2}H$_3$]glutamate, D-[6,6-^{2}H$_2$]glucose, L-[^{15}N]glutamate and L-[CH$_3$-^{13}C] methionine. The sugar units appeared to be synthesized by diverse modification of functional groups of a common intermediate (see also Chapters 3 and 9).[20]

Branched-chain sugar spirolactones **26** and **27**, potential inhibitors of ATPase, have been derived in very high yields from isoxazolines **24** and **25**, respectively by treatment with aqueous acetic acid.[21] The structure of **26** was confirmed by X-ray crystallography (see also Chapters 10 and 21).

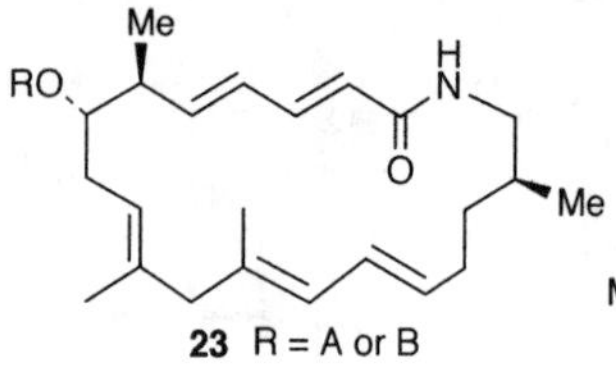

1.2   Branch at C-4. – The C-4-branched caryophyllose, a dodecose isolated from the phytopathogenic bacterium *Pseudomonas caryophylli*, and its methyl glycoside **31** have been prepared by two different routes.[22,23] While the first relied on the SmI$_2$-promoted coupling of acid chloride **28** with the cyclic ketone **29**, the second utilized reaction of lithiated dithioacetal **30** with **29** (Scheme 2). Synthesis of 4'-α-*C*-(2-aminoethyl)thymidine (**32**)-containing oligodeoxynucleotides (ODNs) having lipophilic groups at the terminal of the aminoethyl linker of **32** have been reported.[24] Thermal stabilities of the duplexes formed from these ODNs and RNA/DNA structures have also been discussed (see also Chapter 19).

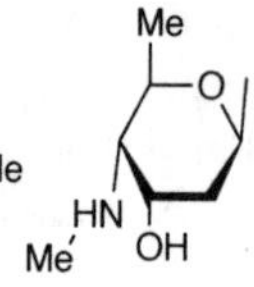

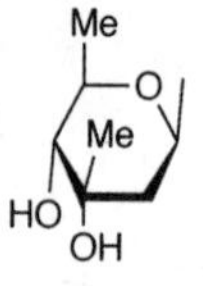

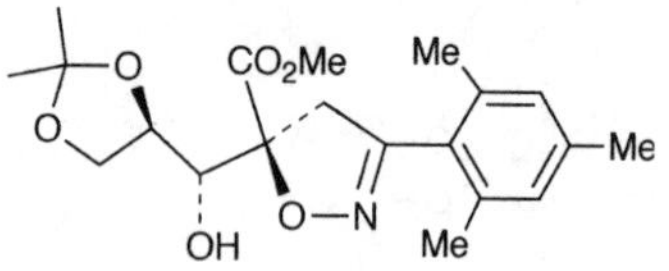

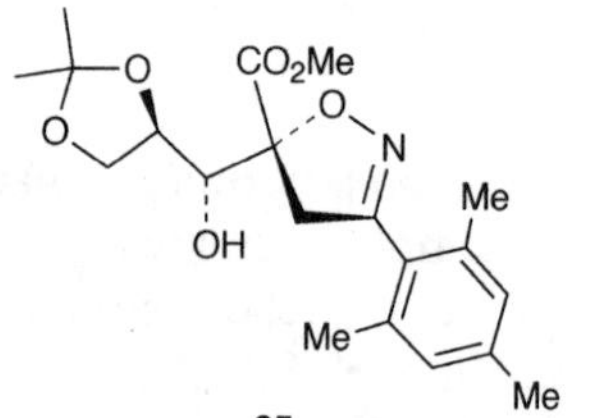

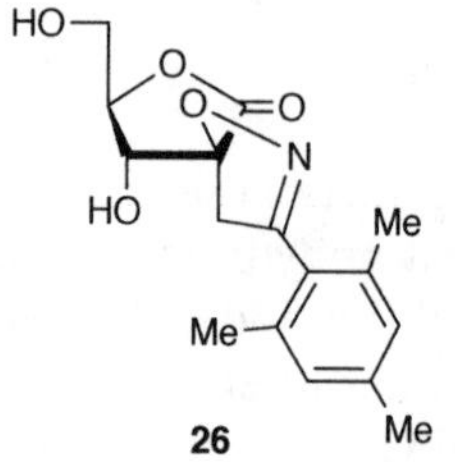

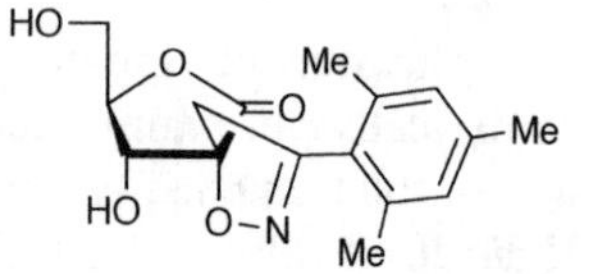

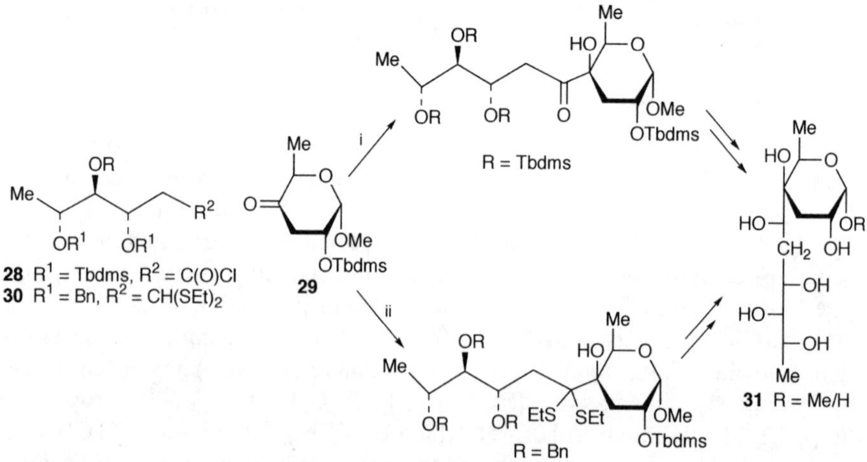

Reagents: i, SmI$_2$, THP, rt, 10 min; ii, Bu^{n}Li, Bu^{t}ONa, hexane, 0 °C, 1h, then **26**, THF, –78 °C, 10 min
Scheme 2

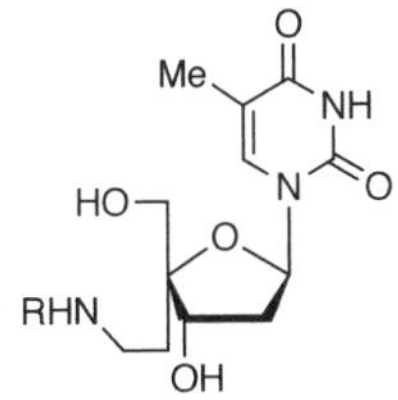

32 R = palmitoyl/oleoyl/cholesteryloxycarbonyl

2 Compounds with a C–C–C Branch-point (X = N or S)

α,β-Dihydroxy-γ-lactones (*e.g.* D-ribonolactone) have been converted into 2,3-aziridino-γ-lactones *via* Michael-type addition of benzylamine to a key 3-deoxy-2-*O*-triflyl-pent-2-enonolactone followed by *in situ* cyclization without affecting the lactone ring (Scheme 3; see also Chapter 9).[25] Stereoselective formation of α-aminonitriles (*e.g.* **33**) from various ulose derivatives (*e.g.* from **10**) and amines using Ti(IV) isopropoxide as a Lewis acid followed by TMSCN has been noted (Scheme 4).[26] The nitriles so obtained have been further transformed into other useful compounds (see also Chapter 14). A multi-step synthesis of methyl α-L-vancosaminide (**34**) from di-*O*-acetyl-L-rhamnal has been reported (see also Chapter 14).[27] On the other hand, glycosyl fluoride derivative **35** of evernitrose, a branched-chain sugar-constituent of everninomicin, has been used as a glycosyl donor in an everninomicin assembly (see also Section 1.1).[10,28]

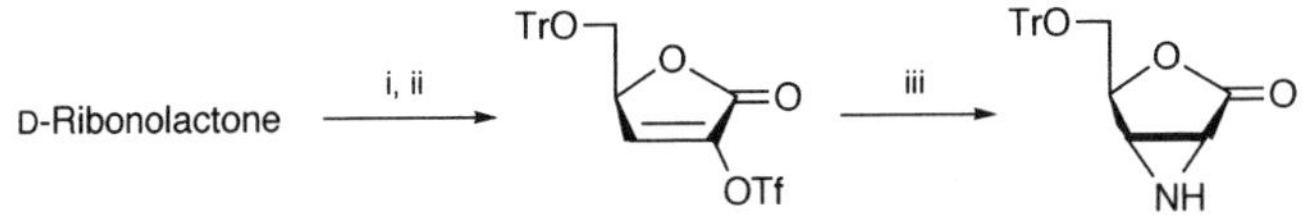

Reagents: i, TrCl, pyridine, 65 °C, 20 h; ii, Tf$_2$O, pyridine, CH$_2$Cl$_2$, –78 °C to –25 °C, 3 h; iii, BnH$_2$, DMF, –20 °C, 15 min **Scheme 3**

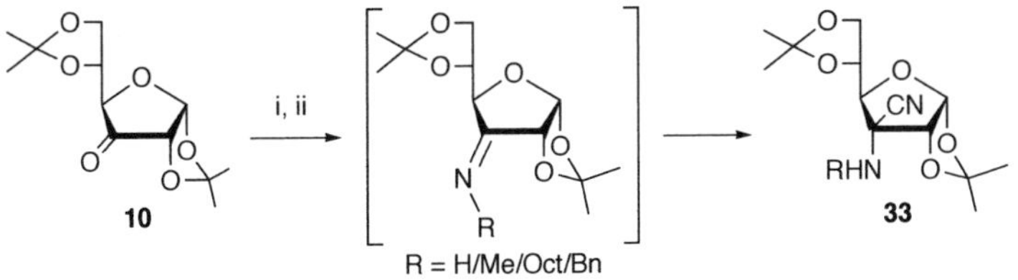

R = H/Me/Oct/Bn

Reagents: i, BnNH$_2$, molecular sieves, Pr^{i}OH; ii, RNH$_2$, Ti(OPr^i)$_4$, MeOH, then TMSCN
Scheme 4

Michael addition of *S*-nuclephiles (*e.g.* the thiolate from **36**) to sugar-based nitroalkenes (*e.g.* **37**) has been found to proceed stereoselectively and has thus been used to prepare thiodisaccharides (*e.g.* **38** from **36** and **37** after NaBH$_4$ reduction of the nitro group, acetylation of the amine so obtained and subsequent opening of the anhydro ring with acidic MeOH).[29]

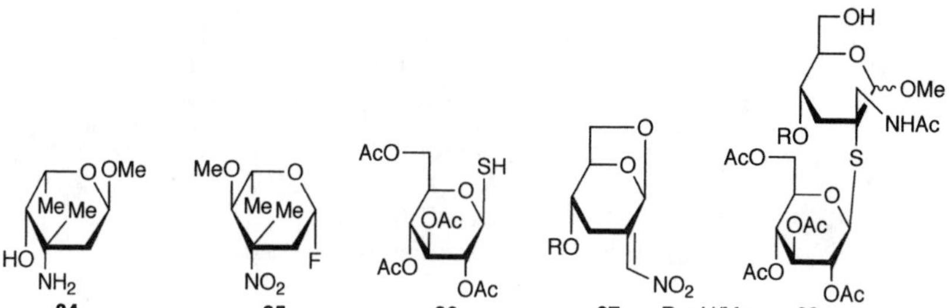

34   35   36   37   R = H/Me   38

2   **Compounds with a C–C–C Branch-point**

The branch-point carbon bears R above and H below.

3.1   Branch at C-2. – Conformationally restricted D- and L-2,3-dideoxy-2,3-*endo*-methylenepentofuranoses (**39**, made from 1,2;5,6-di-*O*-isopropylidene-D-mannitol in nine steps as its silylated derivative, and **40**, from L-glucono-γ-lactone in 10 steps) were synthesized and, after coupling with pyrimidine and purine bases, were tested as substrates for adenosine deaminase. Considerable difference in the activities of the D- and L-enantiomers was observed (see also Chapter 19).[30] Pt-catalysed ring opening of 1,2-cyclopropanated sugar derivatives has been extended. A series of 1,2-cyclopropanoid pyranoid derivatives, including substituted cyclopropanoid derivatives, have now been used as substrates in conjunction with a number of *O*-nucleophiles, including sugar alcohols (*e.g.* H₂O, MeOH, AllOH, Bu'OH, PhOH, methyl 2,3,4-tri-*O*-benzyl-α-D-glucopyranoside, *etc.*), for the ring opening reaction. Yields and the stereochemical outcomes of the reactions have been found to depend upon the nature of the cyclopropanoid substrate as well as the nucleophile used for the reaction (see Vol. 32, Chapter 14, pp. 181–182, reference 43; see also Chapter 3).[31]

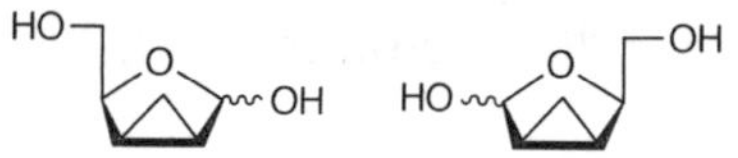

39   40

The evolution of an asymmetric synthesis of the differentially 2,6-anhydro-octitol ring of the altohyrtins has been described. An intramolecular epoxide ring opening was the key step in the synthesis, the epoxide itself being obtained *via* substrate-controlled epoxidation of the corresponding alkene (Scheme 5; see also Chapter 2).[32]

The use of 1,1,2,2-tetraphenyldisilane (**41**), a recently introduced stable radical-generating reagent, has been extended to radical cyclization reactions of carbohydrate derivatives. Thus a number of allyl 2-bromo (or phenylseleno)-2-deoxy-hexosides (*e.g.* **42**) underwent stereoselective 5-*exo* radical cyclization in the presence of **41** to give the corresponding *cis*-fused 5-*exo* cyclized bicyclic sugars (*e.g.* **43** from **42**).[33] This reagent resulted in improved yields and better stereocontrol, besides being more environmentally friendly, than the conventional

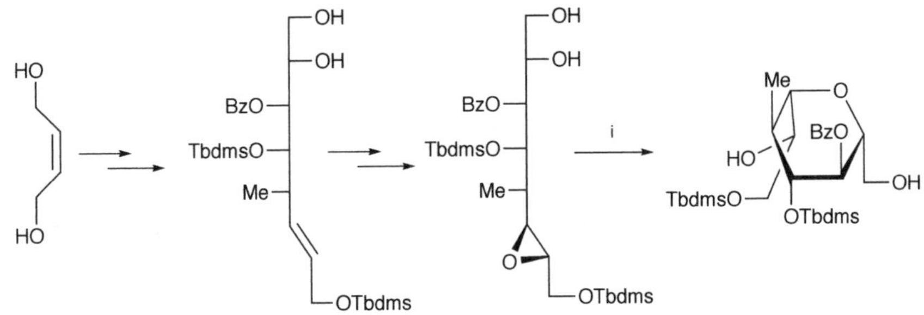

Reagents: i, Camphorsulfonic acid, CH_2Cl_2

Scheme 5

reagents, *e.g.* Sn-based reagents. Bu_3SnH-mediated intramolecular radical for-myl-transfer reactions have, however, been successfully used in the preparation of certain C-2 formyl carbohydrates (*e.g.* see Scheme 6).[34]

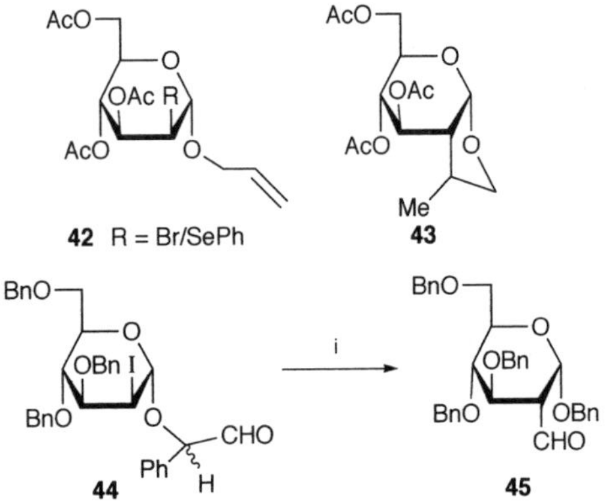

42 R = Br/SePh

43

44

45

Reagents: i, Bu_3SnH, AIBN, PhH, 84%

Scheme 6

Michael additions of dimethyl malonate and dibenzoylmethane to conforma-tionally flexible 3-*C*-nitro-2-enopyranosides **46** and **47** have been found to take place exclusively *trans* to the aglycon generating **48** and **49** respectively. Thus addition of the nucleophile to the α-glycoside **46** took place with the Michael acceptor in the $^0H_5$ conformation and the β-glycoside **47** reacted in the $^5H_0$ conformation. Addition of 2,4-pentanedione and ethyl acetoacetate, on the other hand, was accompanied by further intramolecular cyclization reaction to yield bicyclic products **50** and **51** from **46** and **47** respectively.[35]

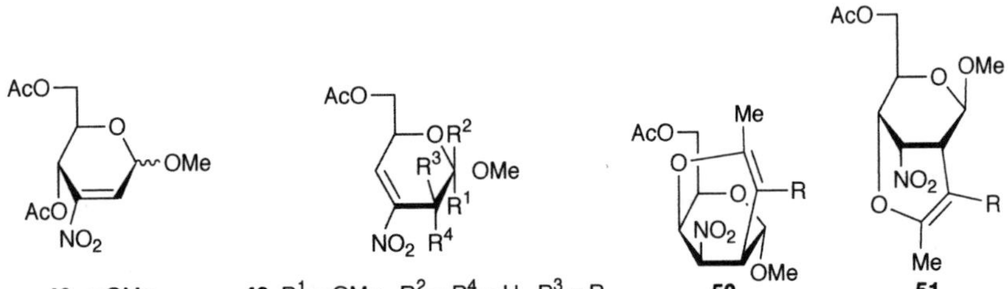

46 α-OMe
47 β-OMe

48 R^1 = OMe, R^2 = R^4 = H, R^3 = R
49 R^1 = R^3 = H, R^2 = OMe, R^4 = R
R = $CH(CO_2Me)_2/CH(COPh)_2$

50

51

R = $COMe/CO_2Et$

3.2 Branch at C-3. – The synthetic methods for preparing *C*-difluoromethyl-ene- and *C*-trifluoromethyl-containing carbohydrates as well as addition of C-nucleophiles to unsaturated uloses (in particular reactions of levogluco-senone), sugar lactones and nitrosugars for obtaining branched-chain carbohy-drates have been reviewed.[36,37] Similarly, cathodically initiated Michael addition to levoglucosenone has now been extended to the reactions of a series of 2-substituted (substituent groups being, 2-thienyl, 2-furyl-, *N*-methyl-2-pyrryl-, phenyl, 9-anthryl and ferrocenyl) 1,3-dinitropropanes to give tricyclic products containing a fused cyclohexane system. Although most of the structures were assigned from their NMR spectral data, the ferrocene-derived compound **52** was unambiguously characterized by X-ray crystallography[38] (see also http://www.electrosynthesis.com/news/w6content.html).

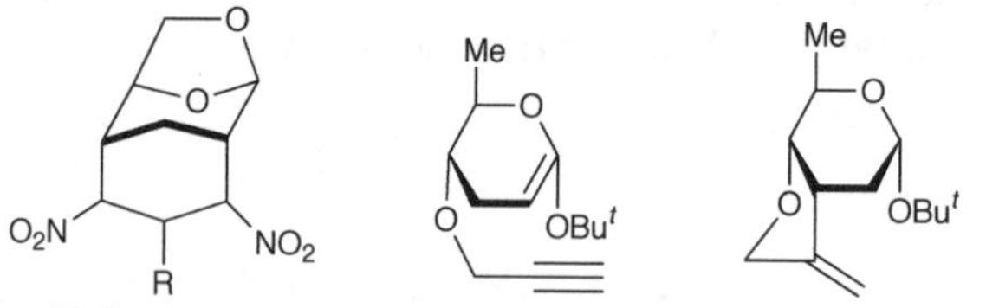

52 R = ferrocenyl   **53**   **54**

The 3-*C*-branched-chain sugar derivative **54** has been described during the total synthesis of the antifungal antibiotic sordarin analogue GM222712. It was obtained by a Sn-mediated radical cyclization of the 2,3-unsaturated α-glycoside **53**, derived from tri-*O*-acetyl-D-glucal.[39,40] A 3-*C*-homologated bicyclic glucoside derivative has also been similarly prepared by a Bu$^n_3$SnH-mediated radical process.[41] Synthesis of protected 3-aminoethyl-3-deoxy-1,2-*O*-isopropylidene-α-D-*ribo*-hexofuranuronic acid has been reported in a symposium on the chemis-try of nucleic acid components (see also Chapter 16). The C–C bond at the branch point at C-3 was presumably achieved by a Henry reaction.[42]

By use of the protected 3-xylulose-derived dithiane **55** as the key intermediate 3'-deoxy-3-*C*-formyl-5-methyluridine **56** has been synthesized and has been applied to the preparation of alternating methylene(methylimino)-linked phos-phodiester backbone oligonucleotides with 2'-OH and 2'-OMe groups (see also Chapter 19).[43] Diastereoselective addition of dialkylphosphonodithioyl radicals (**57**) to carbohydrate-based alkenes (*e.g.* **58**) on the other hand has afforded the corresponding 3-phosphonodithiomethyl-3-deoxyfuranose derivatives (*e.g.* **59**

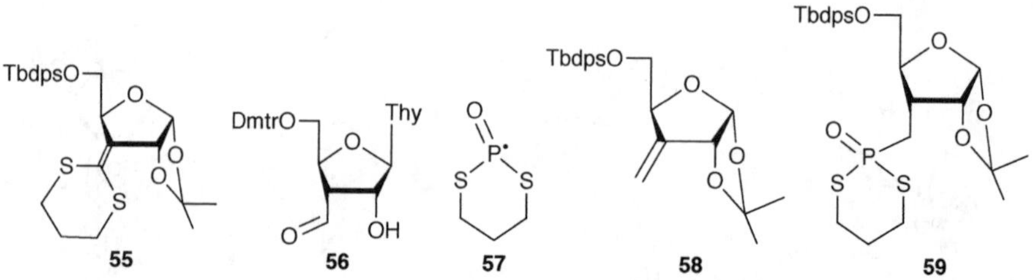

55   **56**   **57**   **58**   **59**

from **58**) suitable for conversion into interesting nucleoside/nucleotide ana-logues.[44] The branched-chain alcohol **60**, obtained from **10** by literature methods, has been converted into the dibromo-derivative **62** through the alkene intermediate **61**, and further into the cyclic phosphinic acid **63** by a double Arbuzov-type coupling of the dibromide with bistrimethylsilyl phosphonite (Scheme 7). Compound **63** was then transformed into a cyclic AMP analogue (see also Chapters 17 and 19).[45]

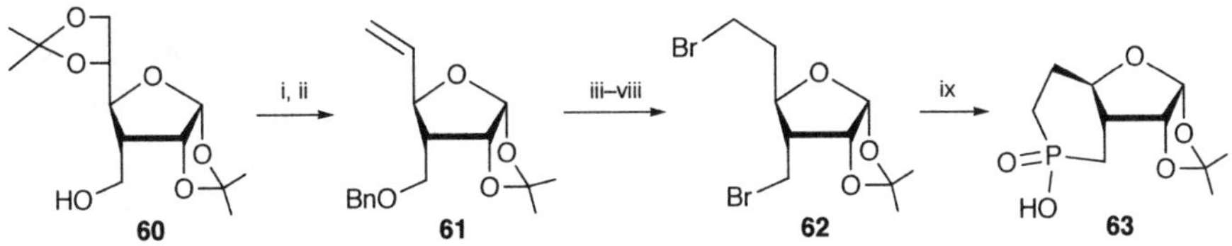

Reagents: i, BnBr, Bu₄NI, 50% aq. NaOH; ii, 75% aq. AcOH, 40 °C; iii, MsCl, Et₃N, CH₂Cl₂; iv, NaI, dimethoxyethane, reflux; v, BH₃.SMe₂. THF; vi, NaOH–H₂O₂; vii, H₂ (60 psi), Pd–C, EtOAc; viii, PPh₃, NBS, pyridine, DMF; ix, (Me₃SiO)₂PH, mesitylene, reflux

Scheme 7

3.3   Branch at C-4 or C-5. – A 4-*C*-benzyloxymethyl-branched hexosamine derivative prepared from the corresponding 4-*C-exo*-methylene compound (lit-erature methods) has been reported.[46] A multi-step synthesis of methylene-bridged disaccharide derivatives **69** and **70** has been described. The known *allo*-configured 3-deoxy-3-(formylmethyl)-furanosyl derivative **64** was subjected to a Wittig chain-extension with **65** to generate the enone **66** which, in the presence of Lewis acid, underwent facile cycloaddition reaction. Use of Eu(fod)₃ (fod, 1,1,1,2,2,3,3-heptafluoro-7,7-dimethyl-4,6-octanedione) as the Lewis acid was found to result in high *cis*-selectivity (*cis:trans*, 99:1) for the cycloaddition. The *cis*-cycloadducts (**67** and **68**) thus obtained were then converted into the *C*-linked-disaccharides with the newly formed saccharide residues having corre-spondingly the D- and L-configurations.[47]

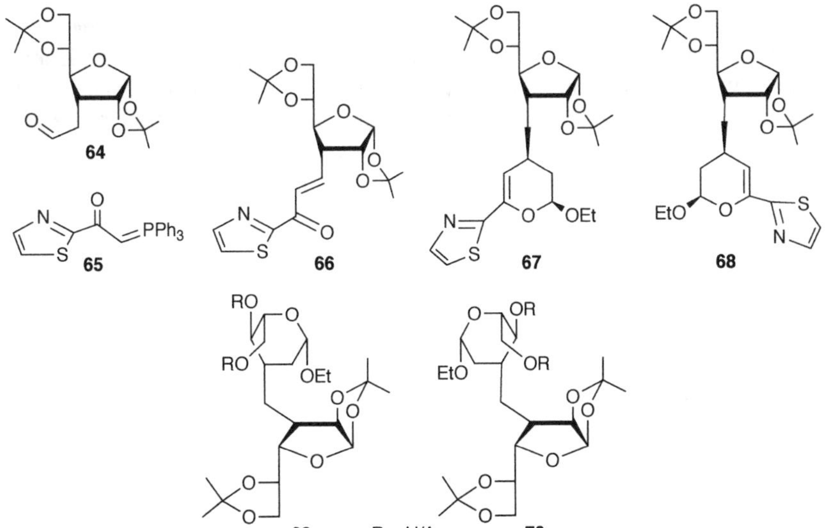

$$\overset{\text{R}}{\underset{\overset{|}{\text{R}}}{\text{4}\quad\textbf{Compounds with a C–}\overset{|}{\text{C}}\text{–C Branch-point}}}$$

[3,3]-Sigmatropic Claisen rearrangement of D-glyceraldehyde-derived (E)(S)-5-benzyloxy-1-t-butyldimethylsilanoxy-4-methylpent-3-en-2-ol has been exploited as a key step in its conversion into the tetrofuranosyl derivative (**71**) bearing a hydroxymethyl and a methyl group as 3-C-geminal substituents.[48] The latter was then transformed into 3′-C-methyl-apionucleosides for testing against HIV and hepatitis B viruses in human cell lines (see also Chapter 19). New syntheses, involving multi-step seqences, of (+)-L-noviose and some of its analogues and their inhibitory activities on DNA gyrase B have been documented.[49–53]

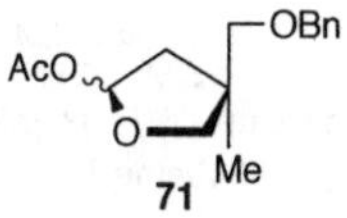

71

$$\overset{\text{R}}{\underset{}{\text{5}\quad\textbf{Compounds with a C–}\overset{\text{R}}{\overset{||}{\text{C}}}\text{–C or C–}\overset{\overset{|}{\text{R}}}{\text{C}}\text{–C Branch-point}}}$$

Reactions of 2-C-formyl-tri-O-benzyl-D-glucal (as well as other glucal derivatives) with the ylides derived from $MePh^+Ph_3Br^-$ and $CH_2{=}CHCH_2P^+Ph_3Br^-$ to give the corresponding C-2-vinyl and C-2-butadienyl derivatives have been reported. Also, the new glycals generated underwent electrophilic conjugate addition with alcohols, including sugar-derived alcohols, in the presence of $Ph_3P^+HBr^-$ (to give the corresponding O-glycosides, *e.g.* **72**) or with TsOH or benzotriazole in the absence of a catalyst (to give the corresponding C-1-linked sulfonyl triazolyl derivative, *e.g.* **73**; see also Chapter 3).[54] 2-C-Acetoxymethyl glucal derivatives have been used similarly as starting material for preparing O- and C-glycosides (see also Chapter 3).[55] C-Iminodisaccharides on the other hand have been prepared by a cross-aldol approach (*e.g.* **76** from **74** and **75**, Scheme 8).[56] Addition of N_2H_4 to the 2-ulose derivative **77** on the other hand has led to the corresponding branched-chain heterocyclic annulated pyranoside **78** (see also Chapter 9).[57,58] Isonucleosides bearing an exocyclic methylene group at the 3′-position, prepared from the corresponding nucleoside bases and the sugar

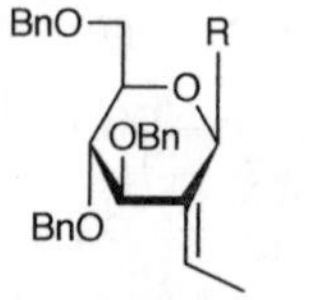

72 R = Et, A, *etc.*
73 R = Ts, B, C, *etc.*

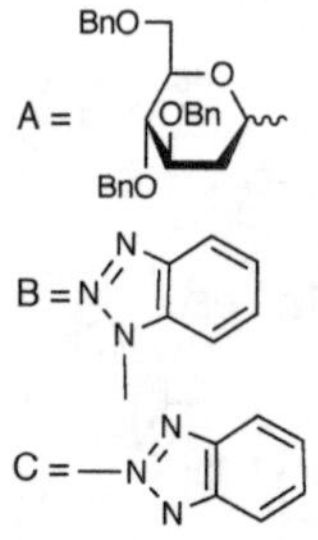

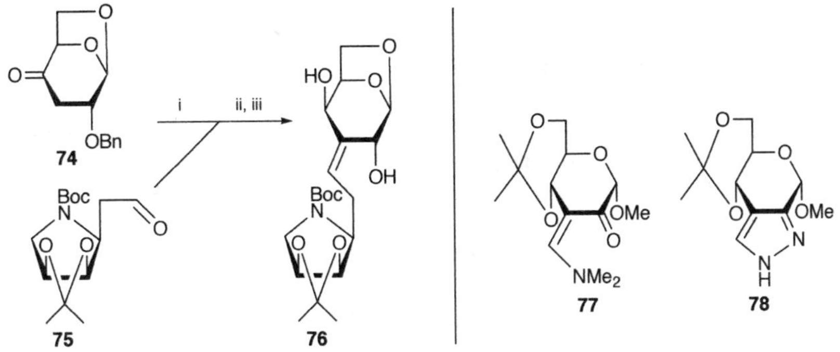

Reagents: i, BuLi, $(Me_3Si)_2NH$; ii, LAH; iii, $Pd(OH)_2/C$, H_2

Scheme 8

derivatives, which in turn were derived from the corresponding 3-ulose derivatives, under Mitsunobu conditions, were shown to be inactive as anti-HIV agents (see also Chapters 13 and 19).[59] Branched-chain sugar derivatives with exocyclic methylene groups at the C-4/C-5 positions, also obtained from the corresponding ulose derivatives, have been described as intermediates in the first total synthesis of ($-$)-tetracyclin.[46] Branched-chain sugars possessing exocyclic difluoromethylene groups have likewise been described in a review on the diastereoselective introduction of fluorinated methyl groups onto carbohydrates.[60]

References

1. A. Fontana, R. Messina, A. Spinella and G. Cimino, *Tetrahedron Lett.*, 2000, **41**, 7559.
2. J.-F. Hoeffler, C. Grosdemange-Billiard and M. Rohmer, *Tetrahedron Lett.*, 2000, **41**, 4885.
3. J.-F. Hoeffler, C. Pale-Grosdemange and M. Rohmer, *Tetrahedron*, 2000, **56**, 1485.
4. L. Chron, J.-F. Hoeffler, C. Pale-Grosdemange, L.M. Lois, N. Campos, A. Boronat and M. Rohmer, *Biochem. J.*, 2000, **346**, 737.
5. T. Kuzuyama, M. Takagi, K. Kaneda, T. Dairi and H. Seto, *Tetrahedron Lett.*, 2000, **41**, 703.
6. T. Kuzuyama, M. Takagi, K. Kaneda, H. Watanabe, T. Dairi and H. Seto, *Tetrahedron Lett.*, 2000, **41**, 2925.
7. M. Takagi, T. Kuzuyama, K. Kaneda, H. Watanabe, T. Dairi and H. Seto, *Tetrahedron Lett.*, 2000, **41**, 3395.
8. Z. Hricoviniova, M. Hricoviniova and L. Petrus, *J. Carbohydr. Chem.*, 2000, **19**, 827.
9. Z. Gonzalez and A. Gonzalez, *Carbohydr. Res.*, 2000, **329**, 901.
10. K.C. Nicolaou, H.J. Mitchell, H. Suzuki, R.M. Rodriguez, O. Boudoin and K.C. Fylaktakidou, *Angew. Chem. Int. Ed. Engl.*, 2000, **139**, 3334.
11. P.K. Sharma and V. Nair, *Nucleosides, Nucleotides, Nucleic Acids*, 2000, **19**, 757.
12. R.P. Singh and J.M. Shreeve, *Tetrahedron.*, 2000, **56**, 7613.
13. I. Gomez, E. Alonso, D.J. Ramon and M. Yus, *Tetrahedron*, 2000, **56**, 4043.
14. G.R. Smith, F.J. Villani Jr., L. Failli and R.M. Giuliano, *Tetrahedron: Asymmetry*,

2000, **11**, 139.

15.   Y. Lu and G. Just, *Tetrahedron Lett.*, 2000, **41**, 9223.

16.   T. Soler, A. Bachki, L.R. Falvello, F. Foubelo and M. Yus, *Tetrahedron: Asymmetry*, 2000, **11**, 493.

17.   A. Rauter, M. Ferreira, C. Berges, T. Duarte, F. Piedade, M. Silva and H. Santos, *Carbohydr. Res.*, 2000, **325**, 1.

18.   D.J. Holt, W.D. Barker, P.R. Jenkins, J. Panda and S. Ghosh, *J. Org. Chem.*, 2000, **65**, 482.

19.   Y.-K. Yang, M.-H. Chiu, C.-W. Gao, R.-L. Nie, Y. Lu and Q.-T. Zheng, *Tetrahedron*, 2000, **56**, 7433.

20.   M. Otsuka, M. Fujita, Y. Matsushima, T. Eguchi, K. Shindo and K. Kakinuma, *Tetrahedron*, 2000, **56**, 8281.

21.   P. Micuch, L. Fisera, M.K. Cyranski, T.M. Krygowski and J. Krajcik, *Tetrahedron*, 2000, **56**, 5465.

22.   J. Prandi and G. Couturier, *Tetrahedron Lett.*, 2000, **41**, 49.

23.   M. Adinolfi, G. Barone, P. Festa, L. Guariniello and A. Iadonisi, *Tetrahedron Lett.*, 2000, **41**, 4981.

24.   Y. Ueno, K. Tomino, I. Sugimoto and A. Matsuda, *Tetrahedron*, 2000, **56**, 7903.

25.   C. de Saint-Fuscien, A. Terrade, P. Dauban and R.H. Dodd, *Tetrahedron Lett.*, 2000, **41**, 6393.

26.   D. Postel, A.N.V. Nhien, M. Pillon, P. Villa and G. Ronco, *Tetrahedron Lett.*, 2000, **41**, 6403.

27.   G.R. Smith and R.M. Giuliano, *Carbohydr. Res.*, 2000, **323**, 208.

28.   K.C. Nicolaou, R.M. Rodriguez, H.J. Mitchell, H. Suzuki, K.C. Fylaktakidou, O. Boudoin and F.L. von Delft, *Chem. Eur. J.*, 2000, **6**, 3095.

29.   J.W. Witczak, R. Chhabra and D. Boryczewski, *J. Carbohydr. Chem.*, 2000, **19**, 543.

30.   B.K. Chun, S. Olgen, J.H. Hong, M.G. Newton and C.K. Chu, *J. Org. Chem.*, 2000, **65**, 685.

31.   J. Beyer, P.R. Skaanderup and R. Madsen, *J. Am. Chem. Soc.*, 2000, **122**, 9575.

32.   J.C. Anderson, B.P. McDermott and E.J. Griffin, *Tetrahedron.*, 2000, **56**, 8747.

33.   O. Yamazaki, K. Yamaguchi, M. Yokoyama and H. Togo, *J. Org. Chem.*, 2000, **65**, 5440.

34.   S.W.T. Choe and M.E. Jung, *Carbohydr. Res.*, 2000, **329**, 731.

35.   T. Sakakibara, K. Tokuda, T. Hayakawa and A. Seta, *Carbohydr. Res.*, 2000, **327**, 489.

36.   R. Plantier-Royan and C. Portella, *Carbohydr. Res.*, 2000, **327**, 119.

37.   B. Booker, *J. Carbohydr. Chem.*, 2000, **19**, 253.

38.   Z.I. Niazimbetova, D.H. Evans, C.D. Incarvito and A.L. Rheingold, *J. Electrochem. Soc.*, 2000, **147**, 1868.

39.   J.M. Bueno, J.M. Coteron, J.L. Chiara, A. Fernandez-Mayoralas, J.M. Fiandor and N. Valle, *Tetrahedron Lett.*, 2000, **41**, 4379.

40.   J.M. Coteron, J.L. Chiara, A. Fernandez-Mayoralas, J.M. Fiandor and N. Valle, *Tetrahedron Lett.*, 2000, **41**, 4373.

41.   D.R. Kelly and M.R. Picton, *J. Chem. Soc.*, *Perkin Trans. 1*, 2000, 1559.

42.   M. Turks, D. Katkevica, H. Smits and E. Bizdena, *Collect. Symp. Ser.*, 2000, **2**, 268 (*Chem. Abstr.*, 2000, **132**, 334 703).

43.   M. Prhavc, G. Just, B. Bhat, P.D. Cook and M. Manoharan, *Tetrahedron Lett.*, 2000, **41**, 9967.

44.   C. Lopin, A. Gautier, G. Gouhier and S.R. Piettre, *Tetrahedron Lett.*, 2000, **41**, 10195.

45. A.C. Regan, N. Sciammetta and P.I. Tattersall, *Tetrahedron Lett.*, 2000, **41**, 8211.
46. K. Tatsuta, T. Yoshimoto, H. Gunji, Y. Okada and M. Takahashi, *Chem. Lett.*, 2000, 646.
47. K. Kniezo, M. Budesinski, P. Voitisek and M. Martinkova, *Enantiomer*, 1999, **4**, 351 (*Chem. Abstr.*, 2000, **132**, 64 453).
48. J.H. Hong, M.-Y. Gao, Y. Choi, Y.-C. Cheng, R.F. Schinazi and C.K. Chu, *Carbohydr. Res.*, 2000, **328**, 37.
49. M. Takeuchi, T. Taniguchi and K. Ogasawara, *Tetrahedron Lett.*, 2000, **41**, 2609.
50. A.-M. Periers, P. Laurin, Y. Benedetti, S. Lachaud, D. Ferroud, A. Iltis, J.-L. Haesslein, M. Klich, G. L'Hermite and B. Musicki, *Tetrahedron Lett.*, 2000, **41**, 867.
51. A.-M. Periers, P. Laurin, D. Ferroud, J.-L. Haesslein, M. Klich, C. Dupuis-Hamelin, P. Mauvais, P. Lassaigne, A. Bonnefoy and B. Musicki, *Bioorg. Med. Chem. Lett.*, 2000, **10**, 161.
52. B. Musicki, A.-M. Periers, P. Laurin, D. Ferroud, Y. Benedetti, S. Lachaud, F. Chatreaux, J.-L. Haesslein, A. Iltis, C. Pierre, J. Khider, N. Tessot, M. Airault, J. Demassey, C. Dupuis-Hamelin, P. Lassaigne, A. Bonnefoy, P. Vicat and M. Klich, *Bioorg. Med. Chem. Lett.*, 2000, **10**, 1695.
53. C. Peixoto, P. Laurin, M. Klich, C. Dupuis-Hamelin, P. Mauvais, P. Lassaigne, A. Bonnefoy and B. Musicki, *Tetrahedron Lett.*, 2000, **41**, 1741.
54. B.-A. Feit, I.K. Kelson, A. Gerull, S. Abramson and R.R. Schmidt, *J. Carbohydr. Chem.*, 2000, **19**, 661.
55. A. Gupta and Y.D. Vankar, *Tetrahedron*, 2000, **56**, 8525.
56. F. Cardona, I. Robina and P. Vogel, *J. Carbohydr. Chem.*, 2000, **19**, 555.
57. B. Kuhla, K. Peseke, G. Thiele and M. Michalik, *J. Prakt. Chem.*, 2000, **342**, 240 (*Chem. Abstr.*, 2000, **132**, 74 195d).
58. M.G. Andreu, K. Peseke, H. Reinke, J.Q. Suarez and M. Michalik, *J. Prakt. Chem.*, 2000, **342**, 389.
59. S. Bera and V. Nair, *Helv. Chim. Acta*, 2000, **83**, 1398.
60. T. Yamazaki, S. Hirakoa and T. Kitazume, *ACS Symp. Ser.*, 2000, **746**, 142 (*Chem. Abstr.*, 2000, **132**, 308 549).

15
Aldosuloses and Other Dicarbonyl Compounds

1   Aldosuloses

Epoxides derived from 2,3,4-tri-O-protected-6-deoxyhex-5-enopyranosides hydrolyse spontaneously to give novel protected hexos-5-ulose derivatives. The products adopt a 1,6-anhydropyranos-5-ulose bicyclic structure in solution.[1] Treatment of the 5,6-cyclic sulfate **1** with BuLi has afforded the 6-deoxyhex-5-ulose derivative **2** following abstraction of H-5.[2] Oxidation (RuO$_2$, NaIO$_4$) of partially O-protected glycosyl H-phosphonates (*e.g.* **3**) gives the corresponding aldulosyl phosphates (*e.g.* **4**). These were coupled with nucleoside monophosphates to give biosynthetically relevant keto derivatives of nucleoside diphosphate sugars.[3] An NBS-mediated hydrolysis of a 2′-deoxy-4′-phenylselenyladenosine derivative has generated the corresponding 2-deoxypentos-4-ulose.[4]

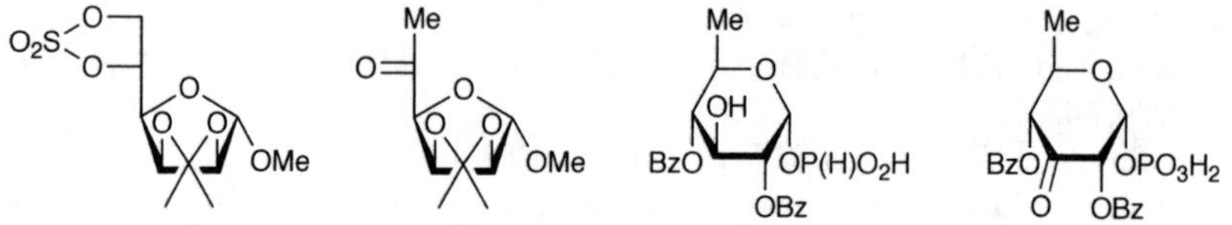

2   Other Dicarbonyl Compounds

The heptodialdose **5** was synthesized from cycloheptatriene by osmylation of its iron tricarbonyl complex and ozonolysis.[5] Addition of 1-ethoxyvinyl-lithium to the lactone **6** and subsequent acid hydrolysis has afforded 1-deoxy-D-*erythro*-hex-2,3-diulose (**7**).[6]

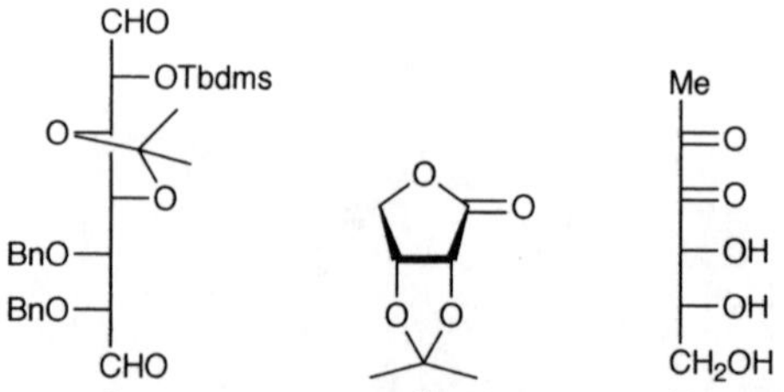

References

1.   P.M. Enright, K.M. O'Boyle and P.V. Murphy, *Org. Lett.*, 2000, **2**, 3929.
2.   T. Gourlain, A. Wadouachi and D. Beaupère, *Tetrahedron Lett.*, 2000, **41**, 659.
3.   A. Naundorf, S. Natsch and W. Klaffke, *Tetrahedron Lett.*, 2000, **41**, 189.
4.   M. Aso, K. Ryuo and H. Suemune, *Chem. Pharm. Bull.*, 2000, **48**, 1384.
5.   A.J. Pearson and S. Katiyar, *Tetrahedron*, 2000, **56**, 2297.
6.   M.A. Glomb and C. Pfahler, *Carbohydr. Res.*, 2000, **329**, 515.

16
Sugar Acids and Lactones

1   Aldonic Acids, Aldaric Acids, and Their Amides, Lactones and Lactams

2,3-Di-O-galloyl-4-O-(E)-caffeoyl-L-threonic acid and 2-O-galloyl-4-O-(E)-caffeoyl-L-threonic acid have been isolated from the leaves of *Cornus controversa*.[1]

The enantiomerically pure 2,3-unsaturated aldonic lactone **2** was prepared following Sharpless asymmetric dihydroxylation of 2-vinylfuran to give **1** (Scheme 1).[2] The chain-extended aldonolactones **4** and **5** were prepared from the β-keto ester **3** by use of a diastereoselective reduction with either D- or L-tartaric acid in conjuction with sodium borohydride (Scheme 2).[3] Reaction of Meldrum's acid (**5a**) with D-aldopentoses and hexoses gave 3,6-anhydro-2-deoxy-aldono-1,4-lactones. For example, D-ribose gave **6** in 41% yield. The isomer with inverted configuration at C-3 and C-4 was observed as a by-product of the reaction.[4] Bis-sugar substituted calix[4]arene **7** was prepared by an amide coupling reaction and found to be highly soluble in chloroform; it formed 1:1 complexes with octyl thioglycosides.[5] The thioamides **8** were prepared from gluconolactone by ring-opening with amines, then acetylation and thionation with Lawesson's reagent.[6] As shown in Scheme 3, 5-azido-5-deoxyhexono-γ-lactones have been used for the preparation of D-hexono-δ-lactams, which were used in glycosidase inhibition.[7]

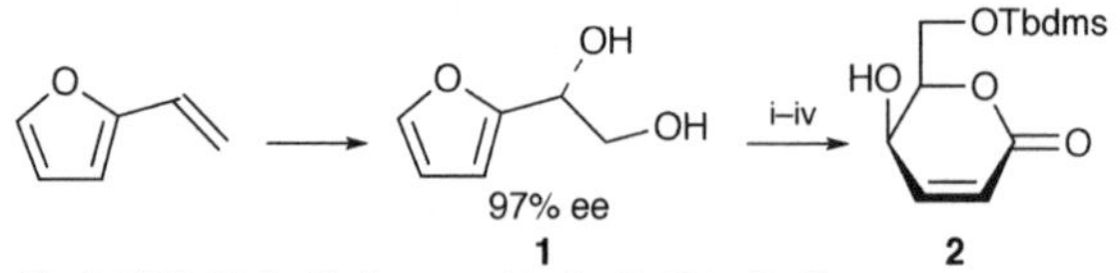

Reagents: i, TbdmsCl; ii, NBS, H$_2$O; iii, Jones oxid.; iv, NaBH$_4$, CeCl$_3$

Scheme 1

Numerous acyclic amide derivatives were reported this year. Bis-amides **9** were synthesized from L-mannaic acid and found to be potent HIV protease inhihibitors.[8] The diamide **11**, an isomer of a known HIV-1 protease inhibitor, was synthesized from L-sorbose *via* L-iditol **10** (Scheme 4). The 4-deoxy analogue of the product was also produced by a Barton deoxygenation.[9]

Both diastereomeric *N*-acetylneuraminic acid-derived nononic acids **12a** and

Carbohydrate Chemistry, Volume 34
© The Royal Society of Chemistry, 2003

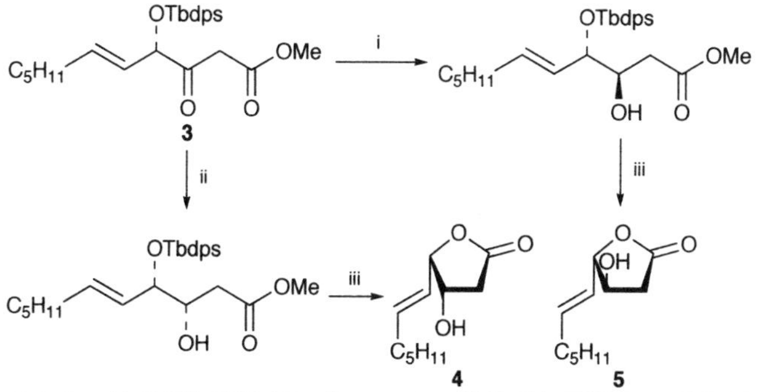

Reagents: i, D-tartaric acid–$NaBH_4$, THF, 93%; ii, L-tartaric acid–$NaBH_4$, THF, 89%; iii, TBAF, 85%

Scheme 2

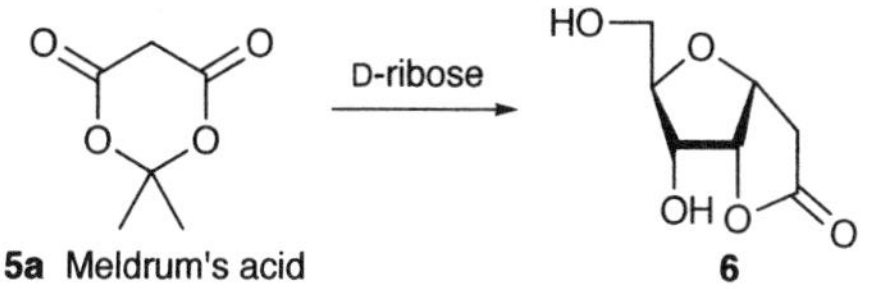

5a Meldrum's acid **6**

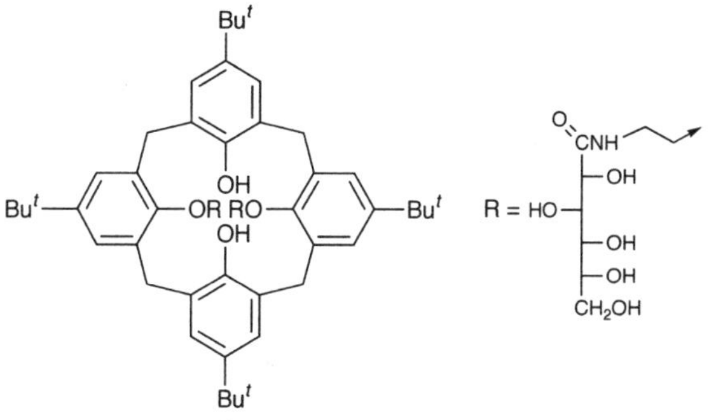

7

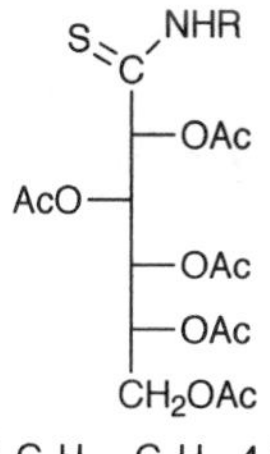

8 R = Me, Et, Pr, Pr^i, C_6H_{11}, C_6H_5, 4-$MeOC_6H_4$, 4-BrC_6H_4

12b have been prepared by reduction with sodium borohydride, acetylation, separation of the diastereomers and then deacetylation with NaOMe.[10] The preparation of 3-amino-3-deoxy-2,4,5,6-tetra-*O*-methyl-D-altronic acid hydrochloride **13** from methyl 3-azido-4,6-*O*-benzylidene-3-deoxy-α-D-mannopyranoside has been achieved in seven steps (1, TFA; 2, MeI/KOH; 3, HCl aq.; 4, thiolacetic acid/pyridine; 5, DMSO, Ac_2O; 6, MeI/KOH; 7, HCl).[11] The reac-

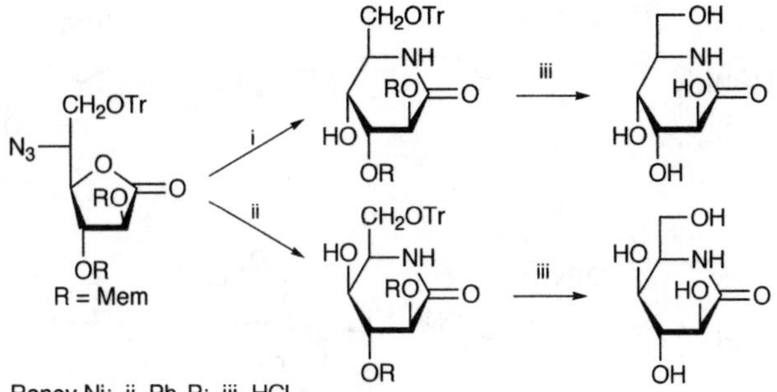

Reagents: i, H$_2$, Raney Ni; ii, Ph$_3$P; iii, HCl

Scheme 3

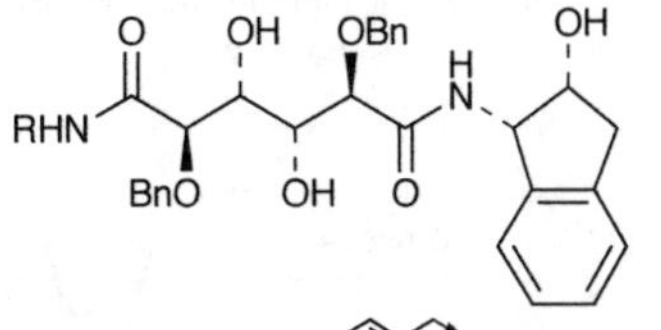

9 R = Me, Pr^i, Bn, [structure], HO[structure]

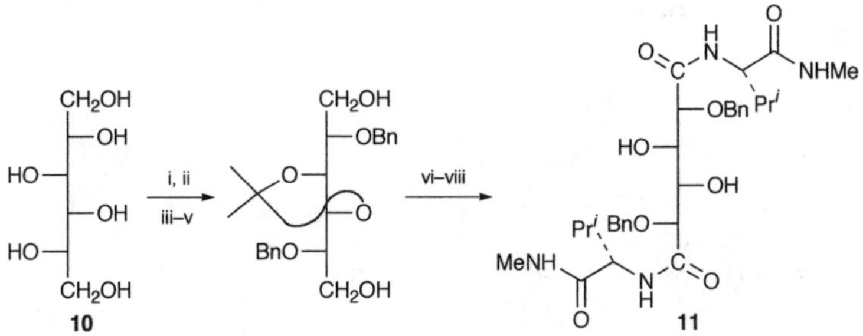

Reagents: i, Me$_2$C(OMe)$_2$, CSA, Me$_2$CO; ii, HOAc, H$_2$O; iii, DMTrCl, py; iv, BnBr, NaH; v, Cl$_2$CHCO$_2$H, CH$_2$Cl$_2$; vi, TEMPO, sodium hypochlorite aq.; vii, *N,N*-disuccinylimidyl carbonate, py; viii, L-valine methylamide

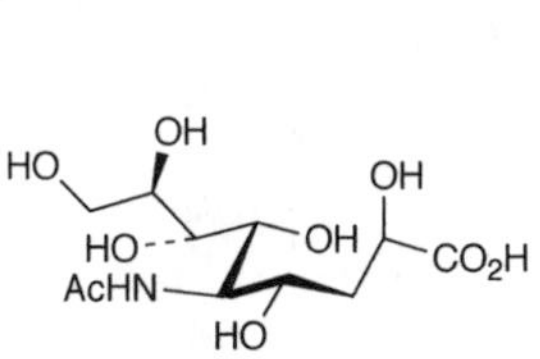

Scheme 4

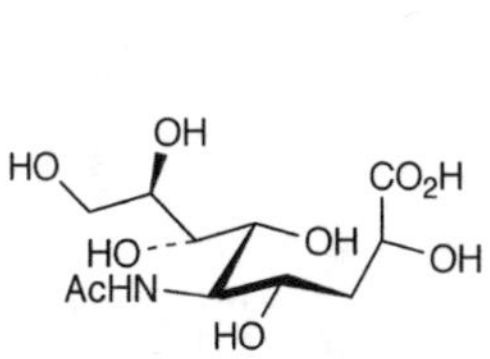

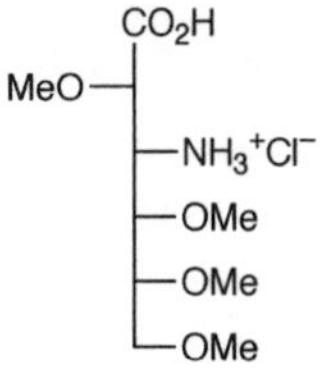

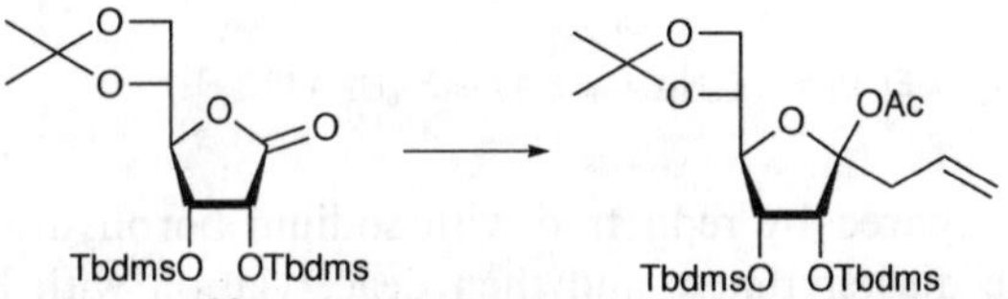

Scheme 5

tion of the 5,6-*O*-isopropylidene-2,3-*O*-Tbdms lactone **14** with an allyl Grignard reagent followed by acetylation was reported (Scheme 5). Subsequent ozonolysis or epoxidation with mCPBA was also performed.[12]

2 Ulosonic Acids

The ulosonic acid derivative **15** has been isolated from the marine sponge *Mycale mytilorum*; however, the stereochemistry has yet to be determined.[13]

A short review (11 pages, 14 references) on the synthesis and evaluation of mechanism-based inhibitors of Kdo8P synthase has appeared.[14] A review on the alkaline biocatalysis for the direct biosynthesis of *N*-acetyl-D-neuraminic acid (Neu5Ac) from *N*-acetyl-D-glucosamine has also been written.[15] The syntheses of furanose derivatives of 3-deoxy-D-*erythro*-2-hexulosonic acid and their 3-bromo and 3-deuterio analogues have been reported.[16] 3-Deoxyoctulosonic acid **16** has been prepared by use of *E. coli* heptulosonate synthase (Scheme 6).[17]

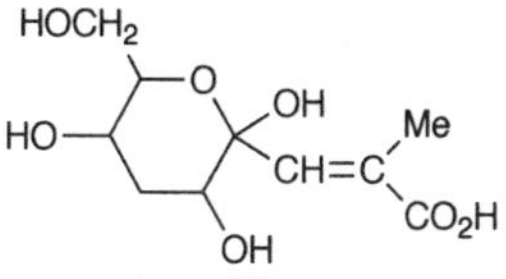

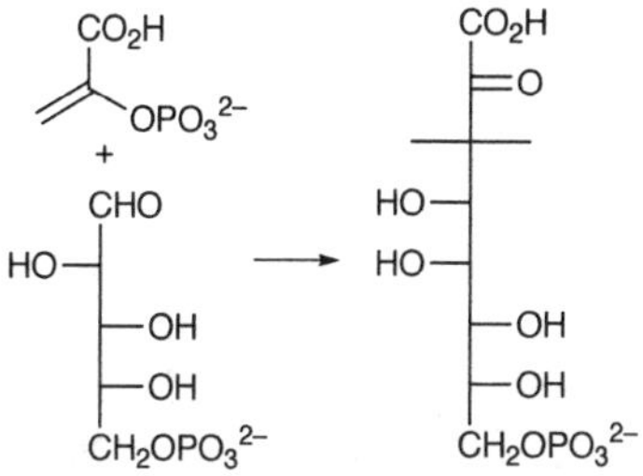

Scheme 6

A one- and two-dimensional NMR study of the acetic acid-induced lactonization of α-*N*-acetylneuraminyl-(2→3)-lactose showed that the major product **17** involved cyclization to O-2 of the galactose unit, while a minor lactone involved O-6 of this moiety.[18]

An enantioselective synthesis of **18** was achieved. This is a deoxy analogue of naturally occurring 3-deoxy-D-*glycero*-D-*galacto*-2-non-2-ulosonic acid (KDN).[19]

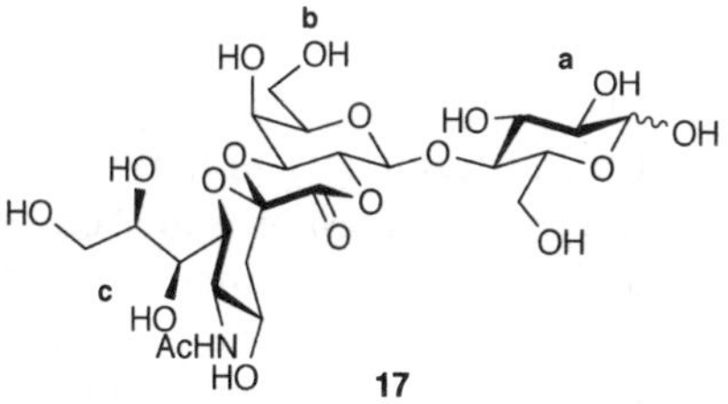

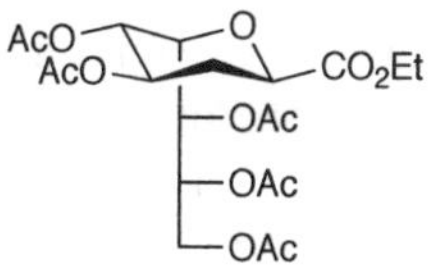

A new report of a practical synthesis of Kdn2en that also includes the synthesis of both 4-amino-4-deoxy analogues has been reported.[20] Mimetics of Kdn2en containing an aliphatic ether, thioether or acetamide in place of the natural C-6 glycerol side-chain have been synthesized from D-glucurono-6,3-lactone.[21] Analogues of 4-guanidino-Neu5Ac2en (Zanamivir, a potent inhibitor of influenza virus sialidases) were prepared with carbamate groups at the 7-position. Two analogues, 7-*O*-(6-aminohexyl)carbamoyl and 7-*O*-(heptyl)carbamoyl, were as active as the parent compound.[22] 2-Phenylthio- and 2-phenylseleno-glycosides of *N*-acetylneuraminic acid were transformed in 92–99% yields to the Neu5Ac2en derivatives **19** by treatment with dimethyl(thiomethyl)sulfonium triflate (DMTST) followed by DBU.[23] Other analogues, **20**, of Neu5Ac2en were prepared from the *N*-acetyl precursor by removal of the acetyl with $Ba(OH)_2$ and formation of the amides by standard DCC coupling. These were used to probe the sialidase of *Vibrio cholerae*, the causation agent of cholera.[24]

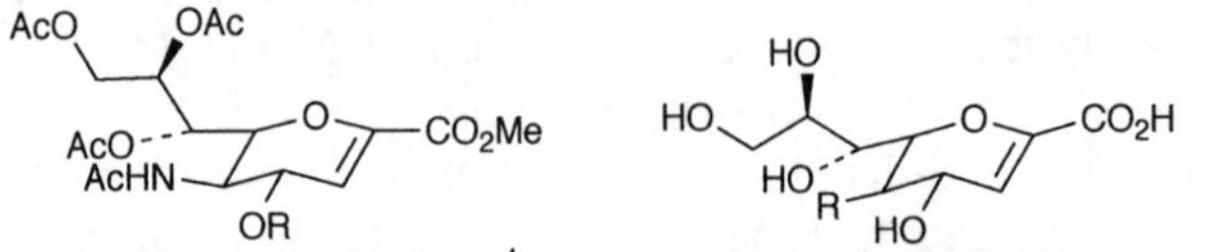

19 R = Ac, CH_2CN, CH_2COBu^t

20 R = $NHCOCH_2OH$
$NHCOCH_2Cl$
$NHCOCH=CH_2$
$NHCOCCl_3$
NHCOEt
NHCOPr
$NHCOCF_3$
$NHCOCH_2I$

Sialic acid and several analogues have been synthesized from D-glucose in 17 steps and 7% overall yield. A salenCo(II)-catalysed hetero-Diels–Alder reaction and oxidative azidation (CAN/NaN_3) of a silyl enol ether were the key steps.[25] The synthesis of KDN and other sialic acids using an indium-mediated allylation was reported (Scheme 7). The method enhanced the stereoselectivity by producing the *threo* product **21** only and simplified the purification through simple precipitation of the indium salt, thus avoiding chromatography.[26] Also in the area of total synthesis, a diasteroselective synthesis of Neu5Ac has been reported.[27]

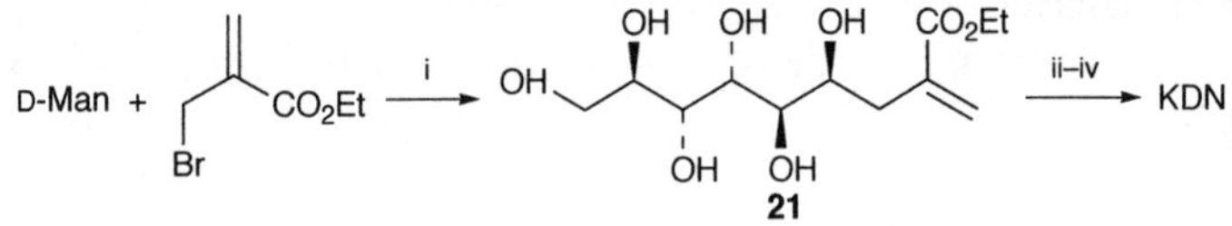

Reagents: i, In, $EtOH/HCl$; ii, O_3; iii, Me_2S, iv, OH^- (75% overall yield)

Scheme 7

The synthesis of C-9 modified *N*-acetylneuraminic acid derivatives was accomplished and the products were tested as substrates for Neu5Ac aldolase (*i.e.* the transformation of Neu5Ac back to *N*-acetylmannosamine). Analogues with OMe, OMOM and OAc at C-9 proved to be reasonable substrates, while that with $=CHCO_2H$ acted as an inhibitor.[28] Fluorinated and hydroxylated C-3 sialyl acids were tested for selective inhibition of neuraminidases and influenza hemagglutinin. They inhibited sialidase from *Clostridium perfringens*.[29] The synthesis of sialic acid glycosides of 6-nitro-5-deazaflavin derivatives was reported. However, their anti-tumour activities were similar or weaker than those of the non-sialylated flavins.[30] The synthesis of derivatives **22** was accomplished for use as molecular probes in elucidating the substrate specificity of human α-1,3-fucosyltransferase.[31] Cyclic Sialyl Le^x mimetics **23** were prepared and displayed excellent inhibition for binding with P-selectin. For example, when R = H the IC_{50} is 1000 better than that of SiaLe^x.[32] A chemoenzymatic deacetylation using lipase OF gave 4-*O*-acetyl-**24** from the 4,7,8,9-tetra-*O*-acetyl precursor in 79% yield.[33] Modifications of CMP-Neu5Ac **25** were prepared and tested with α-(2→6)-sialyltransferase from rat liver. Base modification is not tolerated but modifications at C-5, C-8 or C-9 are.[34] Dimerization of alkenyl sialosides **26** using Grubbs' catalyst gave a range of divalent sialoside derivatives (Scheme 8).[35]

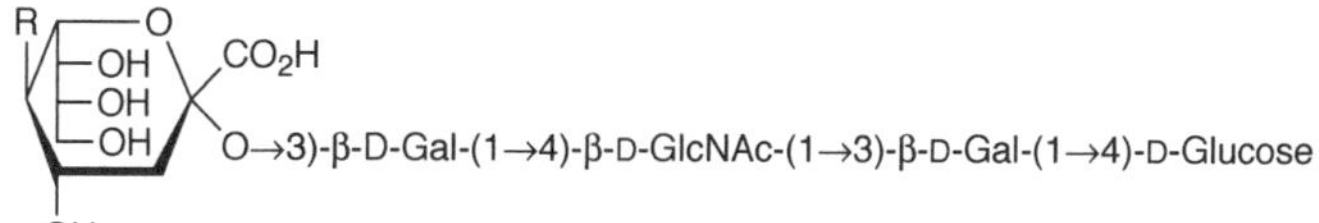

22 R = NHAc, NHCOCH$_2$OH, NHCO(CH$_2$)$_2$Me, OH

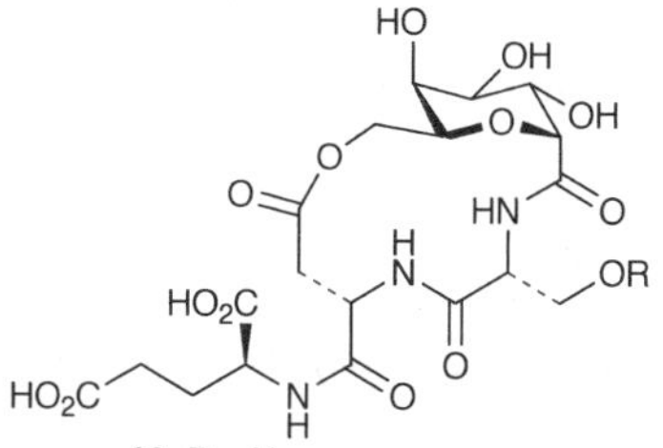

23 R = H
R = *p*-C$_6$H$_4$O(CH$_2$)$_6$Me

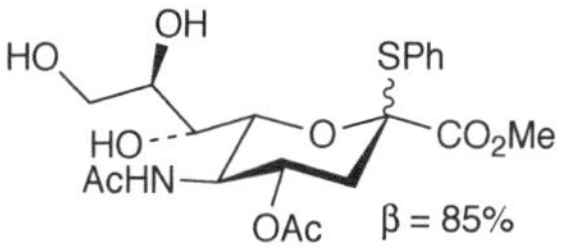

24 β = 85%

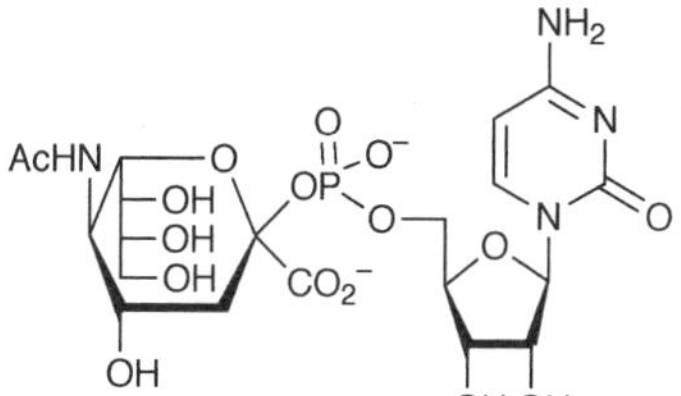

25

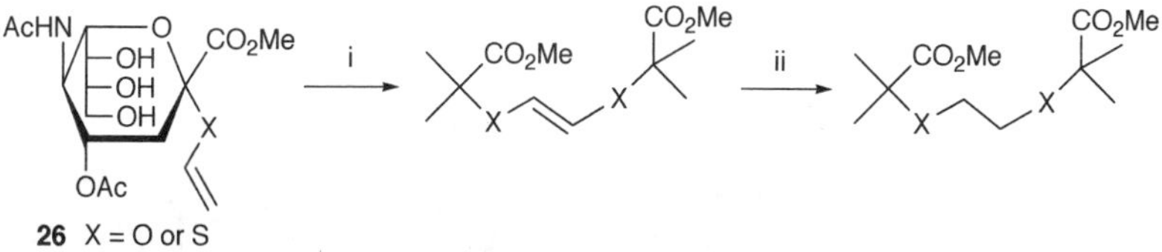

26 X = O or S

Reagents: i, Cl$_2$(PCy$_3$)$_2$Ru=CHPh; ii, H$_2$, Pd/C

Scheme 8

3 Uronic Acids

Numerous applications and variations of TEMPO oxidation for the formation of uronic acid derivatives have been reported. For example, oxidation of propyl α-L-Rha-(1→4)-α-D-Gal-(1→2)-α-L-Rha-(1→4)-β-D-Gal using TEMPO/KBr/NaOCl gave propyl α-L-Rha-(1→4)-α-D-GalA-(1→2)-α-L-Rha-(1→4)-β-D-GalA for use as a model, anti-ulcer compound.[36] The use of TEMPO and heterogeneous silver salts gave methyl α-D-glucopyranosiduronic acid in 90% from the corresponding pyranoside.[37] Similar TEMPO oxidations that use a platinum anode[38] or stoichiometric amounts of sodium hypochlorite and ultrasound[39] were reported, both methods giving methyl α-D-glucopyranosiduronic acid in over 90% yield. Also reported was the oxidation of sucrose using the TEMPO/ultrasound method to produce sucrose tricarboxylate.[40] A further use of a TEMPO/NaOCl oxidation applied to β-cyclodextrin gave the mono- and di-carboxylates.[41] Preparations of glucuronides have been achieved by oxidation of 6-phenylthio hexopyranosides prepared from 6-hydroxy-D-glucopyranosides (Scheme 9).[42]

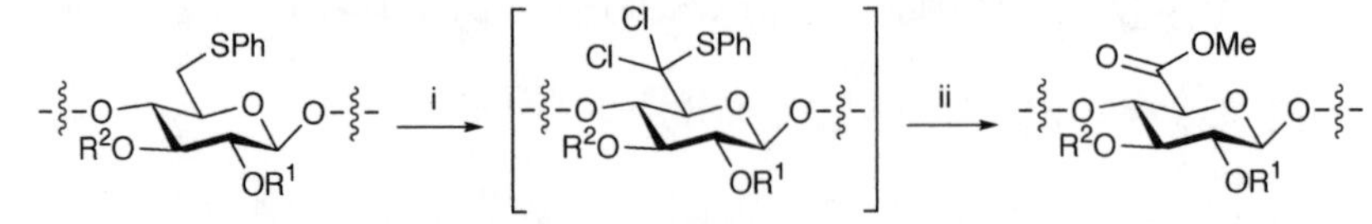

Reagents: i, SO$_2$Cl$_2$,Py; ii, HgCl$_2$ (10 equiv.), Py, MeOH/H$_2$O

Scheme 9

Uronic ester **27** was synthesized in 99% ee and good yield in a bisoxazoline Cu(OTf)$_2$ catalysed hetero Diels–Alder reaction (Scheme 10).[43] L-Iduronyl ester **28** (Scheme 11) has been used for combinatorial work on GAGs.[44] The syntheses of protected α-D-GalU-(1→4)-α-D-GalU-(1→4)-α-D-GalU derivatives, such as **29**,[45] and α-D-GalU-(1→2)-α-L-Rha-(1→4)-α-D-GalU are referred to in Chapter 4.[46] The synthesis of the C-1–C-6 fragment of macrolide antibiotics, carbonalide B, has been reported using a Wolff-rearrangement of α-diazoketones derived from furanuronic acids **30** (Scheme 12).[47]

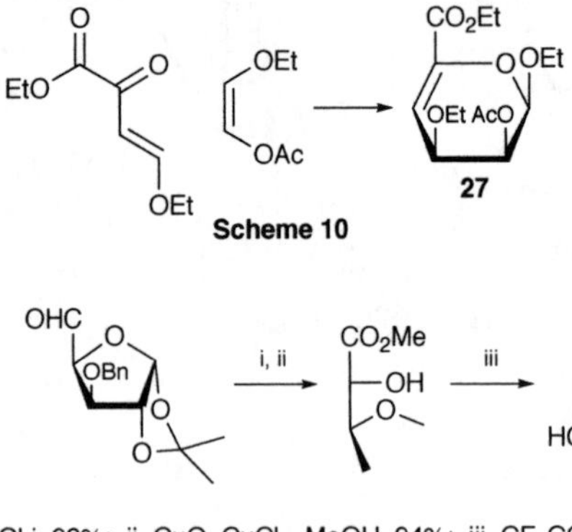

Scheme 10

Reagents: i, (PhS)$_3$CLi, 92%; ii, CuO, CuCl$_2$, MeOH, 94%; iii, CF$_3$CO$_2$H, 100%

Scheme 11

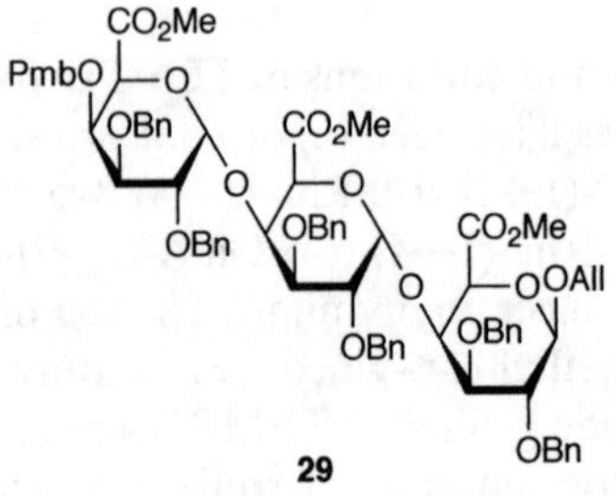

29

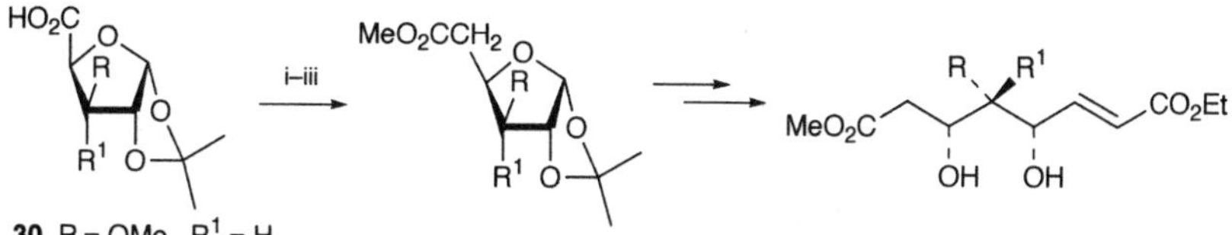

30 R = OMe, R^1 = H
R = OBn, R^1 = H
R = H, R^1 = OBn

Reagents: i, (COCl)$_2$; ii, CH$_2$N$_2$; iii, PhCOOAg, Et$_3$N, MeOH

Scheme 12

5   Ascorbic Acids

Reaction of ascorbic acid with *S*-nitrosothiols (RSNO) at low ascorbate concentration gives NO + RSSR.[48] Upon irradiation with UV light, 5,6-*O*-isopropylidene-L-ascorbic acid reduces quinones efficiently and rapidly to the corresponding hydroquinones rather than undergoing Paterno–Buchi oxetane formation.[49] Derivatives **31** were prepared from 5,6-di-*O*-acetyl-2,3-di-*O*-benzylascorbic acid by nucleophilic attack (with concurrent elimination of HOAc) of silylated bases. The products displayed modest anti-tumour activity.[50] Derivative **32** was also prepared and shown to possess stronger anti-oxidant activity than a combination of vitamins E and C.[51] Selective protection to give 3-ethers **33** of ascorbic acid has been reported. In one method, reaction of ascorbic acid with alkyl mesylates and NaHCO$_3$ in DMF gave 70–88% yields of **33**, where R = (CH$_2$)$_n$Me and *n* = 4–16.[52] Alternatively, treatment of ascorbic acid under Mitsunobu conditions (DEAD, Ph$_3$P then ROH) produced the ethers in 63–77% yields (R = Me, Pr, octyl, allyl and Bn).[53] Finally, the Michael adducts of acrolein and ascorbic acid have been prepared (Scheme 13) and shown to be mild inhibitors of Ser/Thr protein phosphatases.[54]

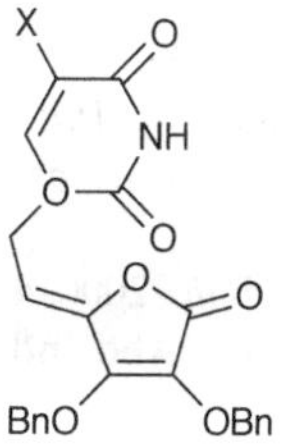

BnO OBn
31 X = H, F, CF$_3$

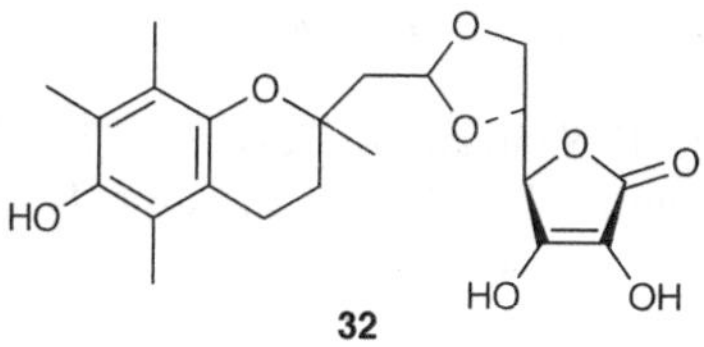

32

CH$_2$OH

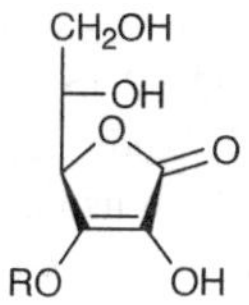

33

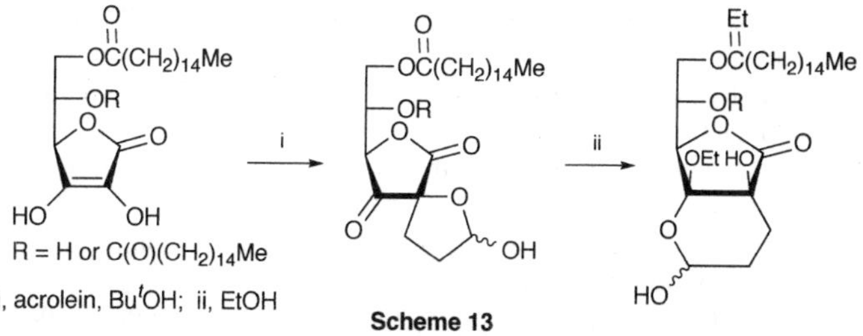

R = H or C(O)(CH$_2$)$_{14}$Me

Reagents: i, acrolein, Bu^{t}OH; ii, EtOH

Scheme 13

References

1. D. Lee, S.-J. Kang, S.-H. Lee, J. Ro, K. Lee and A.D. Kinghorn, *Phytochemistry*, 2000, **53**, 405.
2. J.M. Harris and G.A. O'Doherty, *Tetrahedron Lett.*, 2000, **41**, 183.
3. D.V. Johnson, R. Fischer and H. Griengl, *Tetrahedron*, 2000, **56**, 9289.
4. P. Köll, A. Wernicke, J. Kovacs and A. Lützen, *J. Carbohydr. Chem.*, 2000, **19**, 1019.
5. J. Budka, M. Tkadlecová, P. Lhoyák and I. Stibor, *Tetrahedron*, 2000, **56**, 1883.
6. M.J. Arévalo, M. Avalos, R. Babiaro, A. Cabanillas, P. Cintas, J.L. Jiménez and J.C. Palacios, *Tetrahedron: Asymmetry*, 2000, **11**, 1985.
7. Y. Nishimura, H. Adachi, T. Satoh, E. Shitara, H. Nakamura, F. Kojim and T. Takeuchi, *J. Org. Chem.*, 2000, **65**, 4871.
8. K. Oscarsson, B. Classon, I. Kvarnström, A. Hallberg and B. Samuelsson, *Can. J. Chem.*, 2000, **78**, 829.
9. J. Wachtmeister, A. Mühlman, B. Classon, I. Kvarnström, A. Hallberg and B. Samuelsson, *Tetrahedron*, 2000, **56**, 3219.
10. H.C. Doi, S.M. Marcuccio and W.R. Jackson, *Aust. J. Chem.*, 2000, **53**, 171.
11. M.G. García-Martín, M.V. de Paz Báñez and J.A. Galbis, *J. Carbohydr. Chem.*, 2000, **19**, 805.
12. W.-B. Yang, C.-H. Tsai and C.-H. Lin, *Tetrahedron Lett.*, 2000, **41**, 2569.
13. G.B.S. Reddy and N. Dhananjaya, *Bioorg. Med. Chem.*, 2000, **8**, 27.
14. T. Baasov and V. Belakhov, *Recent Res. Devel. Org. Chem.*, 1999, **3**, 195 (*Chem. Abstr.*, 2000, **132**, 237 249).
15. S. Blayer, J.M. Woodley, M.J. Dawson and M.D. Lilly, *Biotechnol. Bioeng.*, 1999, **66**, 131 (*Chem. Abstr.*, 2000, **132**, 152 026).
16. C. Di Nardo and O. Varela, *Carbohydr. Res.*, 2000, **328**, 605.
17. A.K. Sundaram and R.W. Woodard, *J. Org. Chem.*, 2000, **65**, 5891.
18. T. Nakamura, W.A. Bubb, T. Saito, I. Arai and T. Urashima, *Carbohydr. Res.*, 2000, **329**, 471.
19. X. Shen, Y.-L. Wu and Y. Wu, *Helv. Chim. Acta*, 2000, **83**, 943.
20. X.-L. Sun, N. Sato, T. Kai and K. Furuhata, *Carbohydr. Res.*, 2000, **323**, 1.
21. P. Florio, R.J. Thomson and M. von Itzstein, *Carbohydr. Res.*, 2000, **328**, 445.
22. D.M. Andrews, P.C. Cherry, D.C. Humber, P.S. Jones, S.P. Keeling, P.F. Martin, C.D. Shaw and S. Swanson, *Eur. J. Med. Chem.*, 1999, **34**, 563 (*Chem. Abstr.*, 2000, **132**, 23 155).
23. K. Ikeda, K. Konishi, K. Sano and K. Tanaka, *Chem. Pharm. Bull.*, 2000, **48**, 163.
24. J.C. Wilson, R.J. Thomson, J.C. Dyason, P. Florio, K.J. Quelch, S. Abo and M. von Itzstein, *Tetrahedron: Asymmetry*, 2000, **11**, 53.
25. L.-S. Li, Y.-L. Wu and Y. Wu, *Org. Lett.*, 2000, **2**, 891.
26. M. Warwel and W.-D. Fessner, *Synlett.*, 2000, 865.
27. S.H. Kang, H. Choi, J.S. Kim and J. Youn, *Chem. Commun.*, 2000, 227.
28. M.J. Kiefel, J.C. Wilson, S. Bennett, M. Gredley and M. von Itzstein, *Bioorg. Med. Chem.*, 2000, **8**, 657.
29. X.-L. Sun, Y. Kanie, C.-T. Guo, O. Kanie, Y. Suzuki and C.-H. Wong, *Eur. J. Org. Chem.*, 2000, 2643.
30. Y. Ikeuchi, M. Sumiya, T. Kawameto, N. Akimoto, Y. Mikata, M. Kishigami, S. Yano, T. Sasaki and F. Yoneda, *Bioorg. Med. Chem.*, 2000, **8**, 2027.
31. E. Tanahashi, K. Fukunaga, Y. Ozawa, T. Toyoda, H. Ishida and M. Kiso, *J. Carbohydr. Chem.*, 2000, **19**, 747.
32. C.-Y. Tsai, X. Huang and C.-H. Wong, *Tetrahedron Lett.*, 2000, **41**, 9499.

33. T.-L. Shih, M.-C. Cheng and S.-H. Wu, *Tetrahedron Lett.*, 2000, **41**, 7921.
34. G. Dufner, R. Schwörer, B. Müller and R.R. Schmidt, *Eur. J. Org. Chem.*, 2000, 1467.
35. Z. Gan and R. Roy, *Tetrahedron*, 2000, **56**, 1423.
36. M. Maruyama, T. Takeda, N. Shimizu, N. Hada and H. Yamada, *Carbohydr. Res.*, 2000, **325**, 83.
37. H. Kochkar, M. Morawietz and W.F. Holderich, *Stud. Surf. Sci. Catal.*, 2000, **130A**, 545 (*Chem. Abstr.*, **133**, 310 091).
38. K.D. Moeller, *Tetrahedron*, 2000, **56**, 9527.
39. S. Brochette-Lemoine, D. Joannard, G. Descotes, A. Bouchu and Y. Queneau, *J. Mol. Catal. A: Chem.*, 1999, **150**, 31 (*Chem. Abstr.*, 2000, **132**, 152 042).
40. S. Lemoine, C. Thomazeau, D. Joannard, S. Trombotto, G. Descotes, A. Bouchu and Y. Queneau, *Carbohydr. Res.*, 2000, **326**, 176.
41. C. Fraschini and M.R. Vignon, *Carbohydr. Res.*, 2000, **328**, 585.
42. B. Yu, X. Zhu and Y. Hui, *Org. Lett.*, 2000, **2**, 2539.
43. H. Audrain, J. Thorhauge, R.G. Hazell and K.A. Jorgensen, *J. Org. Chem.*, 2000, **65**, 4487.
44. A. Lubineau, O. Gavard, J. Alais and D. Bonnaffé, *Tetrahedron Lett.*, 2000, **41**, 307.
45. S. Kramer, B. Nolting, A.-J. Ott and C. Vogel, *J. Carbohydr. Chem.*, 2000, **19**, 891.
46. B. Nolting, H. Boye and C. Vogel, *J. Carbohydr. Chem.*, 2000, **19**, 923.
47. J.N. Tilekar, N.T. Patil and D.D. Dhavale, *Synthesis*, 2000, 395.
48. A.J. Holmes and D.L.H. Williams, *J. Chem. Soc., Perkin Trans. 2*, 2000, 1639.
49. M.G. Kulkarni and S.D. Kate, *J. Chem. Soc., Perkin Trans. 1*, 2000, 4242.
50. S. Raic-Malic, D. Svedruzic, T. Gazivoda, A. Marunovic, A. Hergold-Brundic, A. Nagl, J. Balzarini, E. De Clercq and M. Mintas, *J. Med. Chem.*, 2000, **43**, 4806.
51. S. Manfredini, S. Vertuani, B. Manfredi, G. Rossoni, G. Calviello and P. Palozza, *Bioorg. Med. Chem.*, 2000, **8**, 2791.
52. U. Beifuss, O. Kunz and G. Voss, *Tetrahedron*, 2000, **56**, 357.
53. H. Tahir and O. Hindsgaul, *J. Org. Chem.*, 2000, **65**, 911.
54. U.G. Witt, J.E. Schultz, M. Dolker and K. Eger, *Bioorg. Med. Chem.*, 2000, **8**, 807.

17
Inorganic Derivatives

1   Carbon-bonded Phosphorus Derivatives

A comprehensive review of carbohydrate complexes of platinum-group metals includes examples in which the sugar is C-linked to a phosphorus and thence to the metal.[1] A series of phosphonate analogues of glycosyl 1-phosphates was obtained by addition of $LiCH_2P(O)(OMe)_2$ to aldonolactones followed by Et_3SiH–TmsOTf reduction.[2] Some C-glycosyl phosphate and phosphonate analogues of N-acetylglucosamine 1-phosphate have been prepared using standard methods,[3] and difluoromethylphosphonate derivatives of iminoalditols are mentioned in Chapter 10. The α-fluorophosphonates 1 were prepared as substrate mimics for glucose 6-phosphate dehydrogenase by way of base-induced addition of diethyl phosphite to a 7-aldehydo-6-deoxyheptopyranose derivative and treatment of the products with DAST followed by deprotection.[4] Similarly the mannose 6-phosphate analogues 2 and 3 were prepared by addition of lithiated methyl diethylphosphonate to a 6-aldehydomannopyranose derivative.[5] The 'phostone' 4 has been prepared from L-fucose as a potential fucosidase inhibitor,[6] and the synthesis and use of some new sugar-derived phosphine ligands (e.g. 5 and 6) for asymmetric hydrogenation has been described.[7,8]

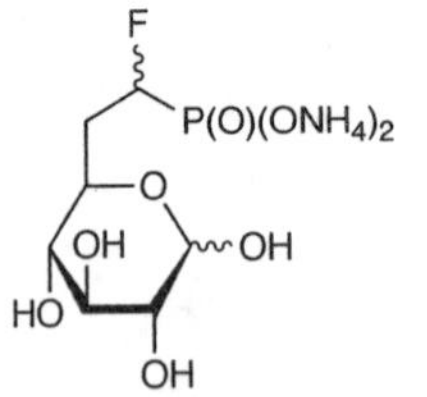

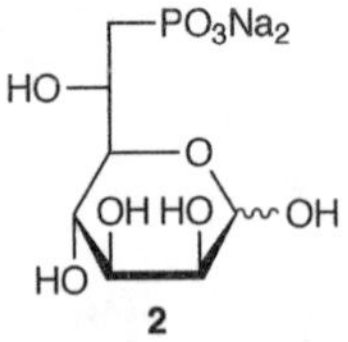

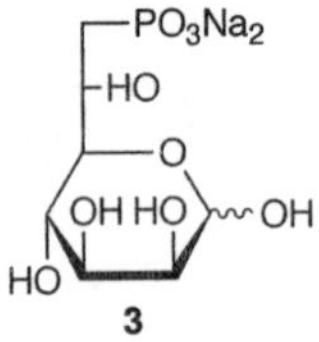

Some approaches to the synthesis of 2- and 3-phosphonomethyl derivatives of D-arabinose are covered in Chapter 14, and the generation of a dialkyl phosphonodithioyl radical from 7 and its addition to carbohydrate alkenes has allowed the synthesis of other C-phosphonomethyl-substituted sugar derivatives (Scheme 1).[9] The phosphinic acid analogue 8 of cyclic AMP has been prepared,[10] and the olefin metathesis reaction, with a newer Grubbs catalyst, has been used to synthesize some carbohydrate cyclic phosphonates, e.g. 9.[11]

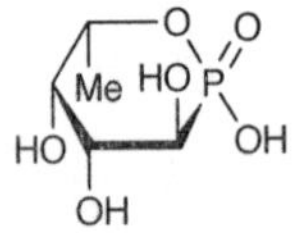

Carbohydrate Chemistry, Volume 34
© The Royal Society of Chemistry, 2003

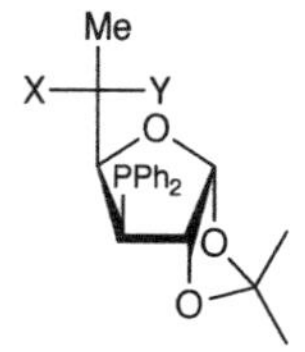

5 X = H, Y = PPh$_2$, or X = PPh$_2$, Y = H

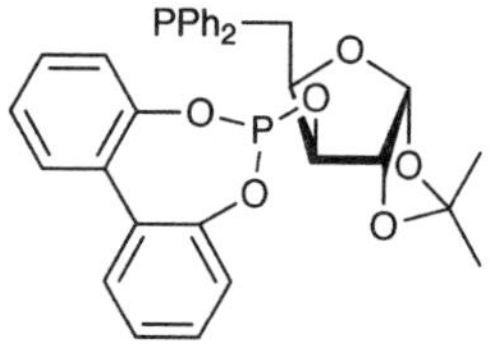

6

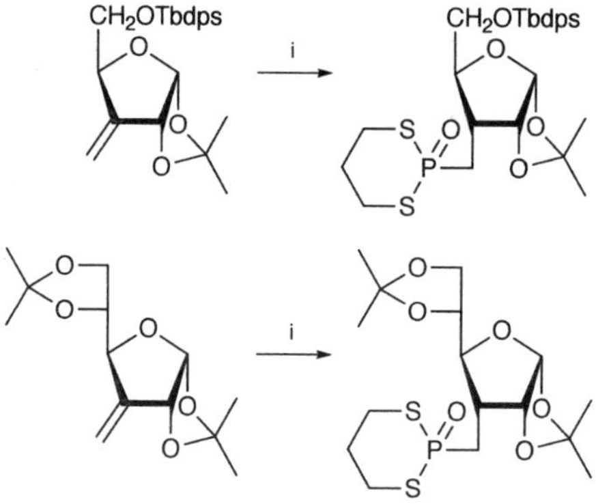

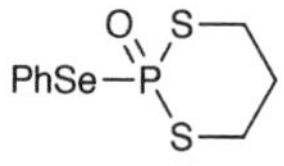

7

Reagents: i, (Tms)$_3$SiH, **7**, AIBN, toluene, reflux

Scheme 1

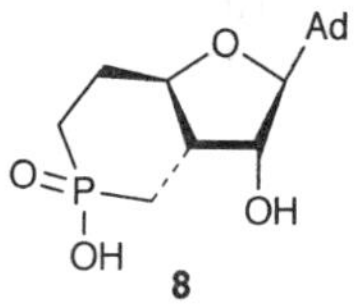

8

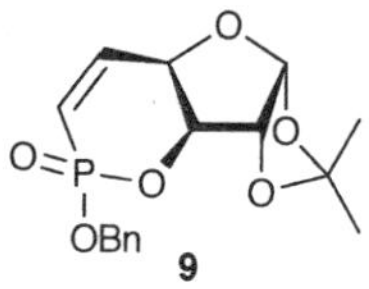

9

2   Other Carbon-bonded Derivatives

The arsenomethionine-based diastereomers **10** have been isolated from brown algae and separated.[12] The vinylstannane **11** has been converted into a chromium carbene enolate which was trapped with electrophiles to give chromium pyranosylidenes **12** with predominantly the D-*manno* configuration (Scheme 2).[13] A glycosylidene carbene derived from **13** inserts into a boron–alkyl bond of **14** to give glycosylborinates **15** with low stereoselectivity. These products are stable to base, acid and thermal conditions but are readily oxidized to yield the corresponding ketoses.[14] Some planar chiral ferrocene dicarboxylic acids have been resolved after esterification with a sugar diol.[15]

3   Oxygen-bonded Derivatives

O-Bonded carbohydrate complexes of platinum-group metals have been comprehensively reviewed,[1] as has the dibutyltin-mediated selective derivatization of compounds with three or more hydroxy groups.[16] An intermolecular migration

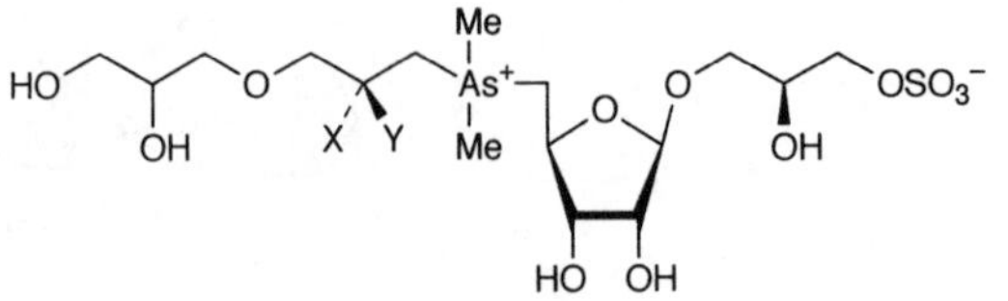

10 X, Y = H, OH

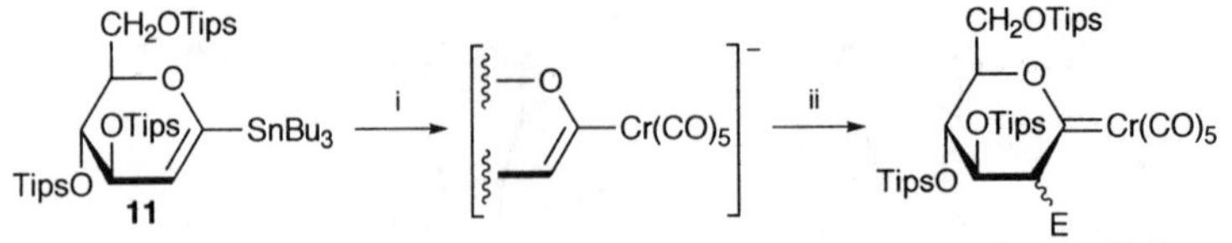

Reagents: i, BuLi, then Cr(CO)$_5$, THF; ii, ECl

Scheme 2

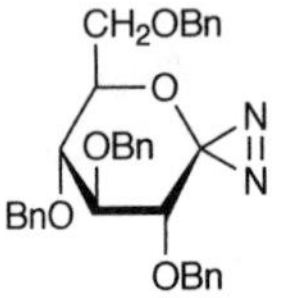

12 E = allyl, H, D

13 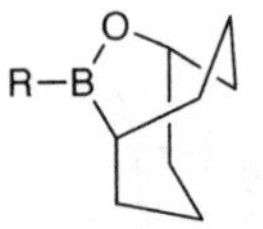   **14** 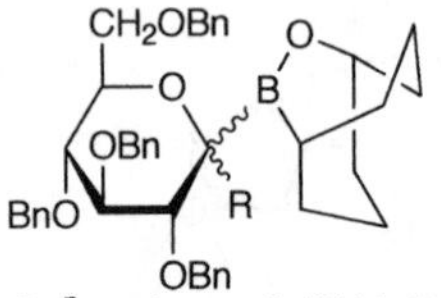   **15** R = Bu^n, cyclopentyl, (CH$_2$)$_2$C$_6$H$_4$Cl-*p*

of the dibutyltin groups has been observed with dibutylstannylene derivatives of pairs of sugar diols which contributes to the regioselectivity of substitution reactions.[17] The dimeric dibutylchlorostannyl derivative **16** has been character-ized as an intermediate that results from treating methyl 4,6-*O*-benzylidene-α-D-glucopyranoside with Bu$_2$SnO/BzCl.[18]

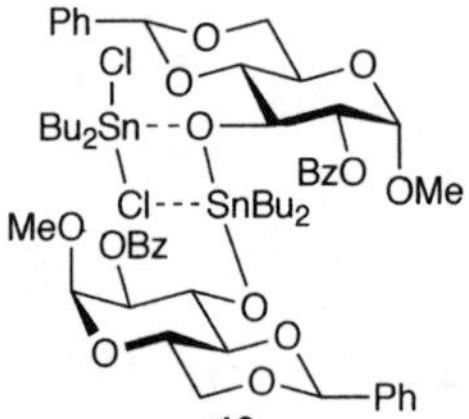

16

Complexes of D-glucose, D-mannose, D-galactose, D-arabinose, D-ribose, D-xylose, maltose and lactose with Pr(III) and Nd(III) have been synthesized and characterized. All monosaccharides formed nine-coordinate [M(sacch)$_3$]$^{3-}$ complexes.[19] The complexes MCl$_3$.*myo*-inositol.9H$_2$O (M = Pr, Nd or Sm) have been analysed by IR spectroscopy and the results suggest they have similar structures. That for M = Pr has been solved by X-ray crystallography.[20] Ten and seven complexes of D-gulonic acid with W(VI) and Mo(VI) respectively in aqueous solution have been identified. The nature of the complexes depended on pH and concentration.[21] A [Co(III)(phen)$_2$(sucrose)]$^{3+}$ complex forms with little apparent perturbation of the sucrose conformation. The sucrose acts as a biden-tate ligand possibly through O-2 (Glc) and O-1 (Fru).[22] Complexes of VO^{2+} and different saccharides were obtained as sodium salts from aqueous solutions and were characterized by spectroscopic methods.[23] A series of Ce(III) complexes with seven monosaccharides and two disaccharides have been synthesized and characterized.[24] The kinetics and relative rates of reduction of Cr(VI) by both α- and β-D-glucopyranose [to Cr(III) and D-gluconic acid] have been studied.[25]

High enantiomeric excesses and regioselectivities have been achieved in the Rh-catalysed hydroformylation of vinyl arenes using a chiral diphosphite derived from D-glucofuranose as a catalyst ligand.[26] Use of a D-glucose Schiff base ligand has allowed the synthesis of Cu(II) aggregates which assemble in a network containing hydrophobic helical channels.[27]

4   Nitrogen-bonded Derivatives

Tetradentate Cu, Pt and Pd complexes of the 1,2-bis(thiosemicarbazone) derived from 3-deoxy-D-*erythro*-hexos-2-ulose have been prepared,[28] and some sugar analogues of cisplatin which incorporate a mono-*N*-sugar-substituted ethylene diamine unit have been synthesized.[29]

References

1.   D. Steinborn and H. Junicke, *Chem. Rev.*, 2000, **100**, 4283.
2.   A. Dondoni, A. Marra and C. Pasti, *Tetrahedron: Asymmetry*, 2000, **11**, 305.
3.   O. Gaurat, J. Xie and J.-M. Valéry, *Tetrahedron Lett.*, 2000, **41**, 1187.
4.   D.B. Berkowitz, M. Bose, T.J. Pfannenstiel and T. Doukov, *J. Org. Chem.*, 2000, **65**, 4498.
5.   S. Vidal, C. Vidil, A. Morère, M. Garcia and J.-L. Montero, *Eur. J. Org. Chem.*, 2000, 3433.
6.   S. Hanessian and O. Rogel, *J. Org. Chem.*, 2000, **65**, 2667.
7.   M. Diéguez, O. Pàmies, A. Ruiz, S. Castillón and C. Claver, *Tetrahedron: Asymmetry*, 2000, **11**, 4701.
8.   O. Pàmies, M. Diéguez, G. Net, A. Ruiz and C. Claver, *Chem. Commun.*, 2000, 2383.
9.   C. Lopin, A. Gautier, G. Gautier and S.R. Piettre, *Tetrahedron Lett.*, 2000, **41**, 10195.
10.   A.C. Regan, N. Sciammetta and P.I. Tattersall, *Tetrahedron Lett.*, 2000, **41**, 8211.
11.   M.S.M. Timmer, H. Ovaa, D.V. Filippov, G.A. van der Marel and J.H. van Boom, *Tetrahedron Lett.*, 2000, **41**, 8635.
12.   J.S. Edmonds, *Bioorg. Med. Chem. Lett.*, 2000, **10**, 1105.
13.   C. Jäkel and K.H. Dötz, *Tetrahedron*, 2000, **56**, 2167.
14.   W. Wenger and A. Vasella, *Helv. Chim. Acta*, 2000, **83**, 1542.
15.   S. Shirakami and T. Itoh, *Tetrahedron: Asymmetry*, 2000, **11**, 2823.
16.   M.W. Bredenkamp, *S. Afr. J. Chem.*, 1999, **52**, 56 (*Chem. Abstr.*, 2000, **132**, 93 549).
17.   S. David and A. Malleron, *Carbohydr. Res.*, 2000, **329**, 215.
18.   M.W. Bredenkamp, H.S.C. Spies and M.J. van der Merwe, *Tetrahedron Lett.*, 2000, **41**, 547.
19.   A. Mukhopadhyay, E. Kolehmainen and C.P. Rao, *Carbohydr. Res.*, 2000, **328**, 103.
20.   L. Yang, Z. Wang, Y. Zhao, W. Tian, Y. Xu, S. Weng and J. Wu, *Carbohydr. Res.*, 2000, **329**, 847.
21.   M.L. Ramos, M.M. Caldeira and V.M.S. Gil, *Carbohydr. Res.*, 2000, **329**, 387.
22.   J. Parada, S. Bunel, C. Ibarra, G. Larrazabal, E. Moraga, N.D. Gillitt and C.A.

Bunton, *Carbohydr. Res.*, 2000, **329**, 195.

23. P.A.M. Williams, S.B. Etcheverry and E.J. Baran, *Carbohydr. Res.*, 2000, **329**, 41.
24. A. Mukhopadhyay, E. Kolehmainen and C.P. Rao, *Carbohydr. Res.*, 2000, **324**, 30.
25. S. Signorella, R. Lafarga, V. Daier and L.F. Sala, *Carbohydr. Res.*, 2000, **324**, 127.
26. M. Diéguez, O. Pàmies, A. Ruiz, S Castillón and C. Claver, *Chem. Commun.*, 2000, 1607.
27. A. Fragoso, M.L. Kahn, A. Castineiras, J.-P. Sutter, O. Kahn and R. Cao, *Chem. Commun.*, 2000, 1547.
28. D. Horton and O. Varela, *Carbohydr. Res.*, 2000, **328**, 425.
29. M. Vicira DeAlmeida, E. Teixeira Cesor, A.P. Soares Fontes and E. de Castro Antunes Felicio, *J. Carbohydr. Chem.*, 2000, **19**, 323.

18
Alditols and Cyclitols

---

1 Alditols and Derivatives

1.1 Alditols and Amino-alditols. – A preparation of the tetritol (C_3–C_6) fragment, **1**, of the macrolactone (+)-aspicilin, featured the desymmetrization of *meso*-erythritol through the formation and subsequent base-promoted diastereoselective acetal fission of a C_2-symmetric *bis*-sulfoxide intemediate, (Scheme 1, **2→3**).[1]

2-*C*-Methyl-D-erythritol 4-phosphate (**4**), a key isoprenoid precursor in the mevalonate-independent pathway leading to isopentenyl diphosphate, has been synthesized in eight steps from 1,2-*O*-isopropylidene-α-D-xylofuranose in such a way as to facilitate the incorporation of ^{14}C or ^{3}H radiolabels.[2] Syntheses of the non-phosphorylated derivative, **5**, and its L-threitol diastereomer, **6**, from D-glucose and D-galactose respectively, have also been reported.[3]

3β-(5'-D-Ribityl)cholestane (**7**), a putative biological precursor for fossil 3-alkylsteranes, has been synthesized from cholestanone by the stepwise and stereoselective construction and subsequent reductive opening of a 3β-(5'-deoxy-5'-yl-D-ribono-1,4-lactone) substituent.[4]

5-Deoxy-5-seleno-D-arabino-, D-ribo- and D-xylo-selenoformates **8**, **9** and **10** have been prepared from D-arabinose and known 2,3,4-tri-*O*-benzyl-D-ribose and -D-xylose diethyl dithioacetals, respectively.[5] The thermolysis of these compounds to afford 1,5-anhydro-5-deoxy-5-seleno-D-pentitols is covered in Chapter 11.

The efficient reduction of unprotected 5-bromo-5-deoxy-D-ribono-, -D-arabinono- and -D-xylono-1,4-lactones to the corresponding 1-bromo-1-deoxypentitols using sodium borohydride in ethanol has been reported. Displacement of bromide by azide, either prior, or subsequent to the reduction, leads to the 1-azido-1-deoxy-L-ribo-, -D-lyxo- and -L-xylo-pentitols which can be further reduced to the 1-amino-1-deoxypentitols by hydrogenation.[6]

The Cr(VI) oxidation of D-glucitol and -mannitol has been studied. Oxidation occurred selectively at the primary hydroxyl groups to afford the corresponding

---

Carbohydrate Chemistry, Volume 34
© The Royal Society of Chemistry, 2003

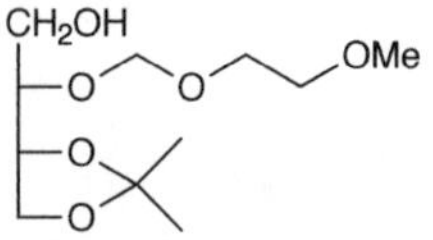

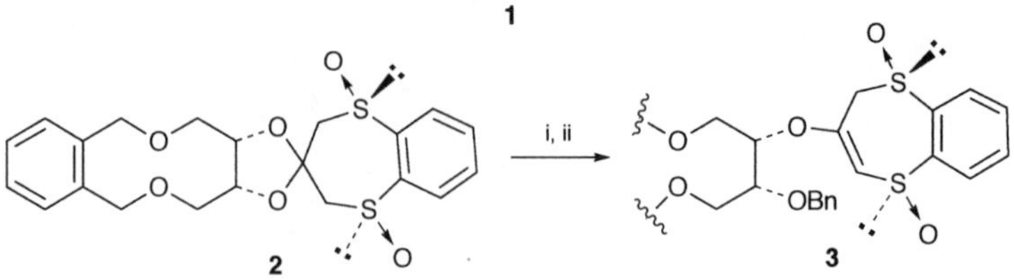

Reagents: i, LiHMDS, 18-crown-6; ii, BnBr

Scheme 1

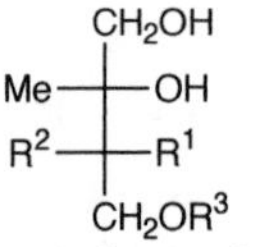

4 R^1 = OH, R^2 = H, R^3 = PO$_3$$^{2-}$
5 R^1 = OH, R^2 = H, R^3 = H
6 R^1 = H, R^2 = OH, R^3 = H

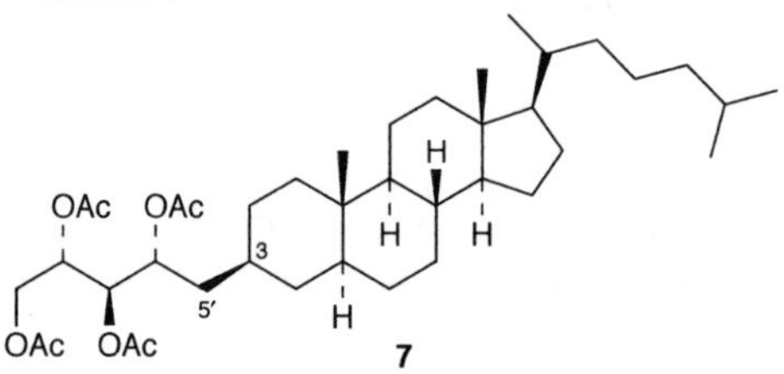

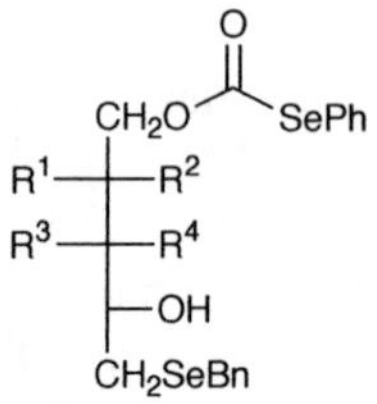

8 R^1 = OH, R^2 = H, R^3 = H, R^4 = OH
9 R^1 = H, R^2 = OH, R^3 = H, R^4 = OH
10 R^1 = H, R^2 = OH, R^3 = OH, R^4 = H

aldonic acids as the only products; a 5-membered Cr(V) chelate ring involving the primary hydroxyl is implicated.[7]

The hydrogenation of D-fructose to produce mixtures of D-glucitol and -mannitol has been carried out at atmospheric pressure on Ru/C catalysts and found to occur preferentially through the furanose form since interconversion between the furanose (α- and β-) tautomers is faster than interconversion between the corresponding pyranose forms.[8] In parallel studies using Pd or Pt catalysts the D-glucitol:D-mannitol selectivity could be modified somewhat by doping the catalyst with tin.[8] The single-step conversion of sucrose to D-glucitol and D-mannitol by simultaneous hydrolysis and hydrogenation at high pressure (5 Mpa, 130–140 °C) has also been reported. Yields in excess of 99% and a 4:1 D-glucitol:D-mannitol selectivity were claimed.[9]

The reaction of aldoses and their derivatives with dimethyl (diazomethyl)phosphonate, generated by *in-situ* methanolysis of dimethyl (1-diazo-2-oxopropyl)phosphonate, gives rise to 1,2-dideoxy-ald-1-ynitol derivatives in moderate to high yield (48–84%).[10] Eight examples, including the conversion of

2,3:5,6-di-*O*-isopropylidene-D-mannose to an acetylenic alditol in 84% yield (Scheme 2) are given, and from these results it appears that greater yields are obtained in those cases in which the hydroxyl groups are protected (**11** *vs.* **12**).

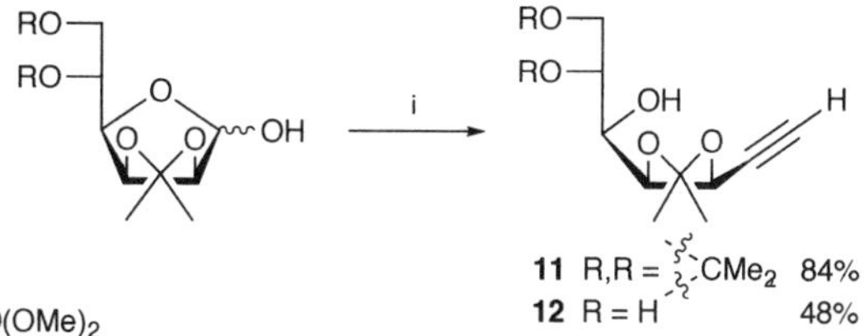

Reagents: i, N₂C⁻PO(OMe)₂

11 R,R = CMe₂ 84%
12 R = H 48%

Scheme 2

The regiospecific [3 + 2] 1,3-dipolar cycloaddition of 2-amino-1,3-thiazolium-4-olates (2-aminothioisomünchnones) to methyl propiolate to form pyridone derivatives has been investigated. In this study a dipolarophile, **13**, was tethered to 1-amino-1-deoxy-D-glucitol pentaacetate and formed pyridone **14** upon 1,3-dipolar cycloaddition to but-2-yndioic acid dimethyl ester.[11]

Glycosylated 'polyethylenimines' (PEIs) are typically synthesized by reductive amination of a di- or oligosaccharide in the presence of commercially available high molecular weight polyethylenimine using highly toxic sodium cyano-borohydride. The combination of titanium(IV) isopropoxide and sodium borohydride has been shown to be a safe and efficient substitute, and maltose and lactose have been converted to the corresponding glycosylated PEIs, **15** and **16**, in good yield by this means.[12]

Isomeric glucosylated amino-alditol derivatives, **17**, derived from reaction of 2-amino-2-deoxy-mannit-6-yl and -glucit-6-yl α-D-glucopyranosides with methyl methacrylate, were co-polymerized and the physical properties of the resulting polymers were examined in water.[13]

Sugar-substituted ligand **18** was prepared from 1-amino-1-deoxy-D-glucitol by reaction with salicylaldehyde in methanol. This Schiff base has been shown to form chiral tetranuclear aggregates with Cu(II) which organize themselves, *via* intermolecular H-bonding, into 3-D networks containing hydrophobic helical channels.[14] In the search for ee/de enhancement in organozinc addition reactions with aldehydes, chiral ligands **19** and **20** (the latter derived from the former) were prepared by standard methods from D-mannitol. Yields and ee's ranging from 62–97% and 43–92% respectively were recorded.[15]

A range of C_2-symmetric cyclic 1,6- and 2,5-diamino-hexitols have been prepared as potential glycosidase inhibitors. Reaction of guanidine with the known bis-epoxides, 1,2:5,6-dianhydro-3,4-*O*-isopropylidene-D-mannitol and -L-iditol, yields **21** and **22** respectively. The 2,5-diamino compounds **23** and **24** were derived from thiourea **25** by reaction with various amines, including methyl 6-amino-6-deoxy-α-D-glucopyranoside, in the presence of mercuric chloride.[16]

1.2   Anhydro-alditols. – The solvent-free thermal dehydration of tetritols on zeolites has been studied.[17] At temperatures in excess of 250 °C a mixture of products is obtained arising predominantly from intramolecular dehydra-

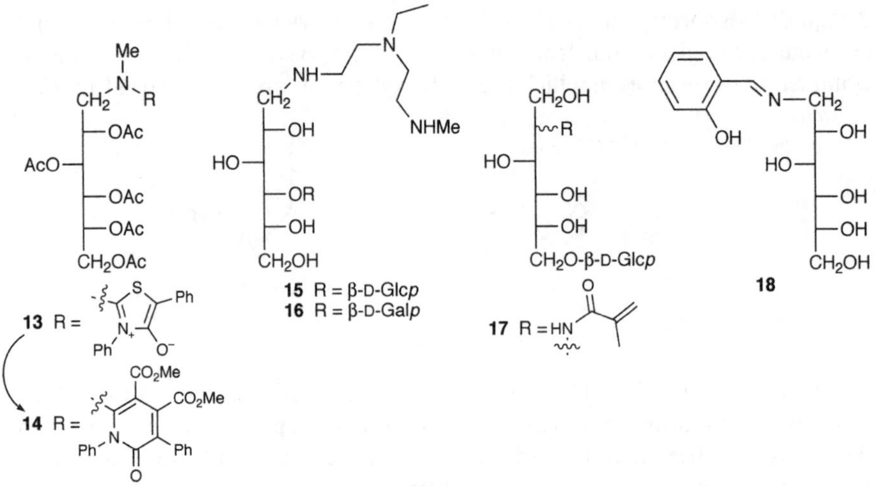

13 R =

14 R =

15 R = β-D-Glc*p*
16 R = β-D-Gal*p*

17 R = HN

18

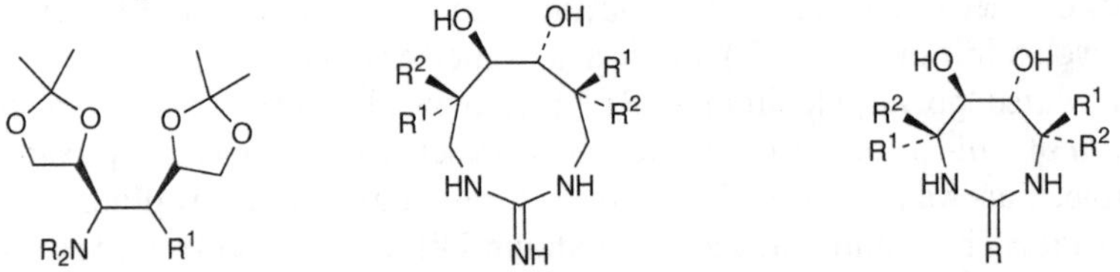

19 R^1 = OH, R = Bu^n, *n*-Oct or
      –(CH$_2$)$_n$–; *n* = 4,5,6
20 R^1 = SH, R = *n*-Oct

21 R^1 = H, R^2 = OH
22 R^1 = OH, R^2 = H

23 R^1 = CH$_2$OH, R^2 = H, R = NH
24 R^1 = H, R^2 = CH$_2$OH, R = NH or NBu^n
      or N-(6-deoxy-6-yl)-α-D-Glc*p*OMe
25 R^1 = H, R^2 = CH$_2$OH, R = S

tion/cyclization, but also from isomerization and intermolecular dehydration (dimerization). Heating erythritol, for example, on 3 Å molecular sieves to 290 °C under argon for six hours produced 1,4-anhydro-D-erythritol in 68% yield.

4,5-Anhydro-2-deoxy-pentitol **26** was synthesized in four steps from 5-benzyloxy-pent-2-en-1-ol as an intermediate in the synthesis of a γ-amino-β-hydroxy acid component of the cyclic depsipeptide, hapalosin.[18]

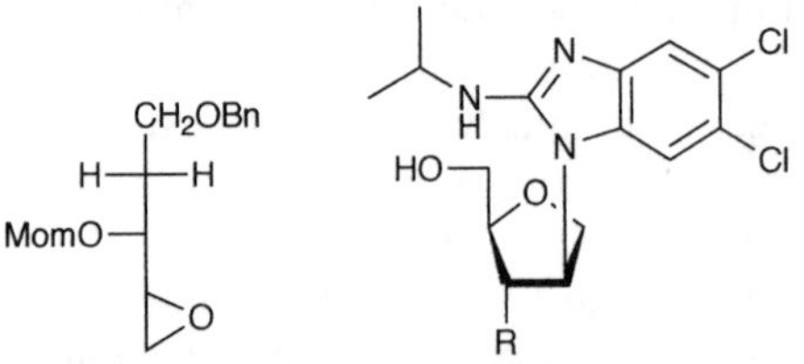

26

27 R = OH or H

1,4-Anhydro-2-*O*-*p*-toluenesulfonyl- and 1,4-anhydro-3-deoxy-2-*O*-*p*-toluenesulfonyl-D-ribitols, derived conventionally from D-ribose and D-lyxose derivatives respectively, have been coupled with substituted benzimidazoles by S$_N$ displacement of the tosyl groups to build the benzimidazole 2'-isonucleosides **27**.[19]

In a rapid synthesis of authentic standards for the reductive-cleavage method of polysaccharide structural analysis, the eight positional isomers of partially methylated and benzoylated (or acetylated) 1,5-anhydroribitol have been prepared simultaneously from 1,5-anhydroribitol by sequential methylation and benzoylation and isolated in pure form by HPLC. The acetates were obtained subsequently by debenzoylation of the pure isomers and acetylation.[20]

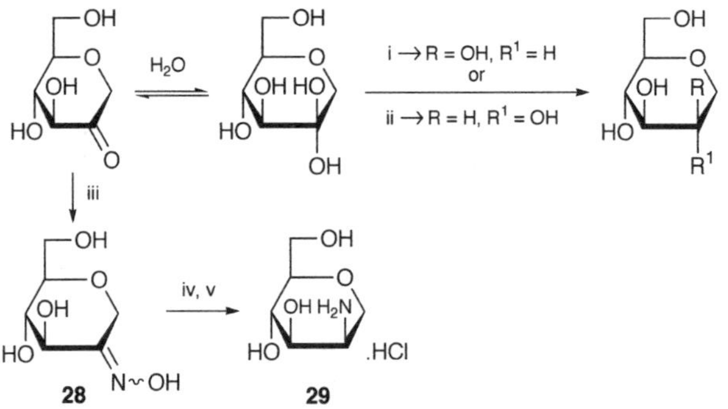

Reagents: i, H$_2$, Pd/C (58%); ii, NaBH$_4$ (60%); iii, HONH$_2$, EtOH (66%); iv, H$_2$, Pd/C, MeOH; v, HCl (72% 2 steps)

Scheme 3

1,5-Anhydro-D-glucitol and -D-mannitol can be obtained by stereoselective reduction of 1,5-anhydro-D-fructose as depicted in Scheme 3. Reaction with hydroxylamine in ethanol affords oxime **28** which undergoes stereoselective catalytic hydrogenation to produce 1,5-anhydro-2-amino-2-deoxy-D-mannitol (**29**), which in this case was isolated as its hydrochloride salt in high yield.[21] Alternatively, 1,5-anhydro-D-glucitol and -D-mannitol can be derived from the corresponding acetylated glycosyl bromides under nickel(II) catalysed radical conditions (Scheme 4).[22] Further, if this reaction is carried out in the presence of acrylate acceptors, C-glycosides are formed in good yield with high stereoselectivity (> 95% α-anomer, see Chapter 3).

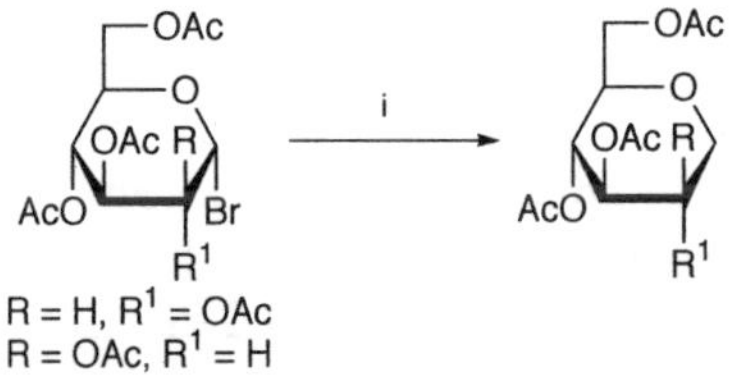

R = H, R^1 = OAc
R = OAc, R^1 = H

Reagents: i, Ni(tmc)$_2$(BF$_4$)$_2$ (20 mol%), Mn, PPh$_2$H, THF, Δ

Scheme 4

2,5-Anhydro-D-glucitol, 2,5-anhydro-D-mannitol and their 6-phosphate and 1,6-diphosphate derivatives have been prepared from protected D-mannose and D-glucose using methods developed with ^{2}H-labelling in mind.[23] 2,3,4,6-Tetra-*O*-benzyl-D-mannopyranose has been elaborated into both anhydro-alditols as shown in Scheme 5. Interestingly, the anhydro-alditols are formed by displacement of a 5-*O*-triflate by the benzyloxy substituent at C-3. ^{2}H-labelling at C-1/C-6 can be effected by substituting sodium borodeuteride in the reduction step. 2,5-Anhydro-D-glucitol was also synthesized from 2,3,4,6-tetra-*O*-benzyl-D-glucopyranose using analogous chemistry; to obtain the desired *gluco*-configuration in the product, the aldehydic precursor was epimerized using triethylamine in DMF prior to sodium borohydride reduction. If deuterium oxide is added to this epimerization mixture, ^{2}H incorporation occurs at C-5 by proton exchange.[23] A study of C_2-symmetric peptidomimetics as potential HIV inhibitors[24] describes the synthesis of peptide derivatives of 1,6-diamino-2,5-anhydro-1,6-dideoxy-D-iditol and -D-mannitol from 2,5-anhydro-3,4-di-O-benzyl-D-iditol[25] and -D-mannitol,[24] respectively.

Three methods for the synthesis of 1,4:3,6-dianhydro-D-glucitol 5-nitrate from the 2,5-dinitrate have been reported. Mono-denitration was effected enzymatically, by catalytic hydrogenation and by treatment with sodium borohydride and cobalt phthalocyanine.[26]

The di-O-alkylations of 1,4:3,6-dianhydro-D-glucitol, -mannitol and -iditol were reportedly facile under microwave irradiation in the presence of phase transfer catalysts. Yields in excess of 90% were routinely achieved after a reaction time of only a few minutes, and the method was found to be compatible with a wide range of alkylating agents.[27]

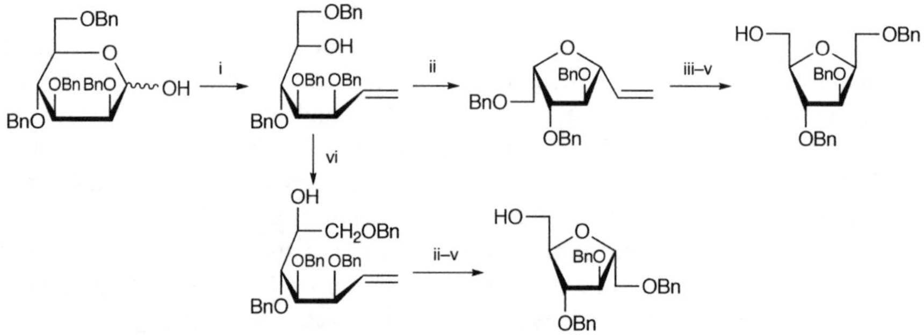

Reagents: i, Ph$_3$P=CH$_2$.HBr, BuLi; ii, Tf$_2$O; iii, OsO$_4$; iv, NaIO$_4$; v, NaBH$_4$; vi, PPh$_3$, DEAD, O$_2$NC$_6$H$_4$NO$_2$, Et$_3$N

Scheme 5

Whereas common methods for synthesizing glycosyl cyanides produce 1,2-*trans*-products, a new method based on the selective (radical) reduction of the nitro group of 2,6-anhydro-1-deoxy-1-nitro-hexitols and -heptitols (glycosyl nitromethanes) provides ready access to 1,2-*cis*-glycosyl cyanides; oximes are generated in 84–97% yield upon treatment with tributyltin hydride-ABCN (1,1'-azobis[cyclohexanecarbonitrile]) in refluxing benzene, and these are effectively dehydrated to afford the corresponding nitriles.[28] Numerous examples are reported including the use of such substrates as the peracetylated glycosyl nitromethanes with β-D-Gal*p*, β-D-Xyl*p*, β-D-Glc*p*NAc, α-D-Rib*f* and β-L-Rha*p* configurations.

Designed as a novel IP$_3$ receptor ligand, 3,7-anhydro-2-deoxy-D-*glycero*-D-*ido*-octitol 1,5,6-trisphosphate has been synthesized using a vinyl transfer strategy that hinges on susbtituent-induced conformational constraints (Scheme 6).[29] The use of bulky *tert*-butyldimethylsilyl protecting groups in the D-glucose substrate causes a conformational inversion allowing the successful radical cyclization of a temporary 2-O-vinylsilyl tether. Hydrogen peroxide oxidation

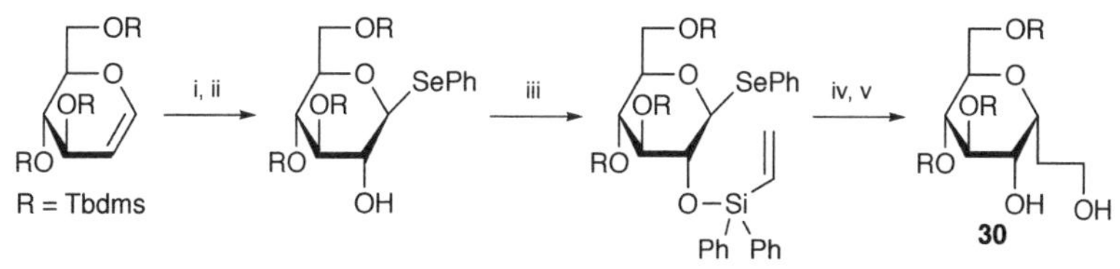

Reagents: i, dimethyl dioxirane; ii, PhSeH, Et$_3$N; iii, Si(CH=CH$_2$)Ph$_2$Cl; iv, Bu$_3$SnH, AIBN; v, H$_2$O$_2$, NaHCO$_3$, MeOH, H$_2$O

Scheme 6

affords the anhydro-octitol intermediate **30** bearing a stereoselectively (α) introduced C$_2$ unit.

1.3   Monomeric Cyclic Imino-alditols. – As sugar mimetics able to competitively and selectively inhibit glycosidases and glycosyltransferases, polyhydroxylated pyrrolidine and piperidine alkaloids and their synthetic analogues have again been the focus of considerable attention and a number of reviews of the literature in this area have appeared. The preparation of polyhydroxy pyrrolidines, piperidines and azepanes (imino-pentitols, -hexitols and -heptitols) from D-mannitol and their functionalization to yield non-peptide mimics of somatostatin and sandostatin has been covered.[30] A number of reports have been published which consider the biological activities of imino-alditols as anti-infectives,[31] glycosidase inhibitors[32] and as inhibitors of glycolipid biosynthesis.[33] The stereoselective synthesis of piperidines has been the subject of a recent review,[34] and a report on the use of anionic conjugate addition reactions for the preparation of substituted mono- and di-saccharides includes examples of aza-sugars such as 1-deoxynojirimycin.[35]

For convenience, the syntheses of cyclic imino-alditols are now categorized broadly according to whether they stem from carbohydrate or non-carbohydrate sources. Compounds derived from carbohydrate sources are further categorized according to their method of synthesis.

1.3.1   Cyclization by Displacement of Leaving Groups in Carbohydrate Substrates by Amines. The synthesis of 1,2,4-trideoxy-1,4-imino-D-*erythro*-pentitol (**31**) has been achieved in eight steps, in 42% overall yield, from 2,5-di-*O*-p-toluenesulfonyl-D-ribono-1,4-lactone according to Scheme 7. The correct stereochemistry in the product was achieved by epimerization at C-4 in a 5-*O*-tosylated 2-deoxy lactone intermediate; base action on the lactone produced a 4,5-epoxide of an open-chain carboxylate derivative which subsequently ring-closed by intramolecular attack of the carboxylate anion on the epoxide to produce the diastereomeric lactone.[36]

D-Arabinose has been used as starting material for an eight step preparation of 1,4-dideoxy-1,4-imino-L-xylitol. Ring closure was effected by a double displacement of the mesylate groups in the intermediate 2,3,5-tri-*O*-benzyl-1,4-di-*O*-methanesulfonyl-D-arabinitol with benzylamine.[37]

Galacto-isofagomine analogue **32**, a modest inhibitor of β-galactosidase, was also synthesized from D-arabinose in 12 steps *via* chain extension at C-2 and

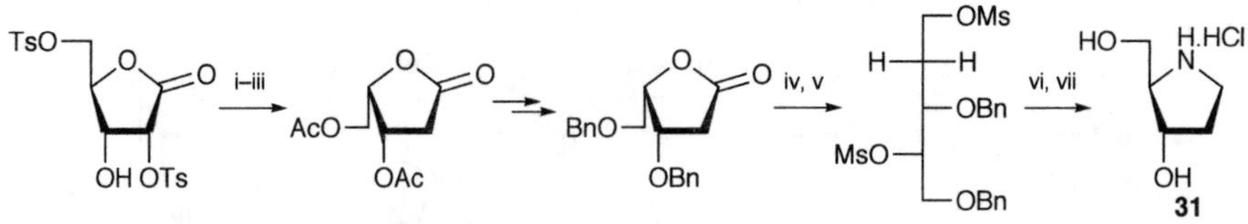

Reagents: i, NH$_2$NH$_2$, Br$_2$, dioxane; ii, aq. KOH then conc. HCl; iii, Ac$_2$O, cat. HClO$_4$; iv, LiAlH;
v, MsCl, py; vi, BnNH; vii, H$_2$, Pd/C

Scheme 7

reaction of the advanced intermediate dimesylate illustrated in Scheme 8 with hydrazine, followed by acid deprotection.[38]

In enantiospecific syntheses, (+)- (**34**) and (−)-1-azafagomine have been prepared from 2,3,5-tri-O-benzyl-D- and -L-xylose respectively (*e.g.* as depicted in Scheme 9 for the D-xylose). As a consequence of these syntheses, the glucosidase activity of (±)-1-azafagomine was found to be entirely due to the (−)-enantiomer. A more direct route to **34**, requiring a double displacement of the mesylate groups in **33** with hydrazine, was attempted. However, N-amino-1,4-dideoxy-1,4-imino-L-arabinitol **35** was the only product isolated.[39]

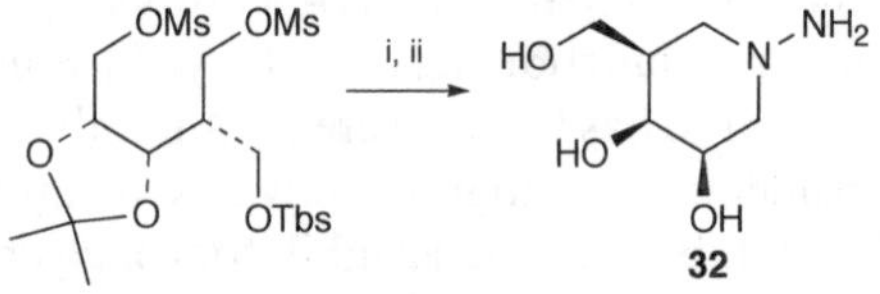

Reagents: i, NH$_2$NH$_2$.H$_2$O; ii, H$^+$

Scheme 8

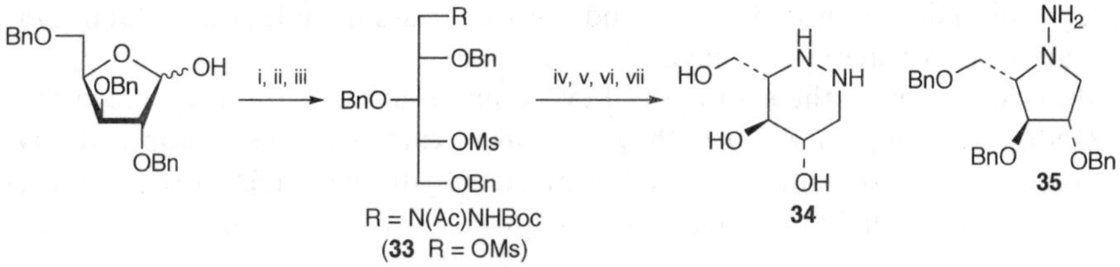

Reagents: i, NaBH$_3$CN, BocNHHN$_2$; ii, Ac$_2$O; iii, MsCl; iv, TFA; v, Diisopropylethylamine, MeNO$_2$;
vi, H$_2$, Pd/C, HCl, EtOH; vii, aq. HCl, 100 °C

Scheme 9

1,4-Dideoxy-1,4-imino-L-xylitol and 1,5-dideoxy-1,5-imino-L-iditol have been synthesized in roughly 50% yield from the common intermediate, 2-azido-2-deoxy-L-idonate derivative **36**, derived from L-gulonolactone. The key step in the transformation involves iodine-promoted one-pot cyclization (by displacement of a mesylate group) and is followed by selective removal of the 9-phenylfluoren-9-yl and isopropylidene groups (Scheme 10).[40]

A rapid route to N-substituted trihydroxypiperidines (1,5-dideoxy-1,5-imino-pentitols) by nucleophilic opening of *bis*-epoxides derived from D- and L-arabinitol, D-xylitol and related compounds has been reported. The 3-O-benzyl-*bis*-epoxides, generated by tosylation of the primary hydroxyl groups followed by treatment with sodium hydride and benzyl bromide, were treated with primary amines to yield piperidine/pyrrolidine mixtures, whose composition was found to be dependent on the nature of the alditol precursor, the amine and the reaction

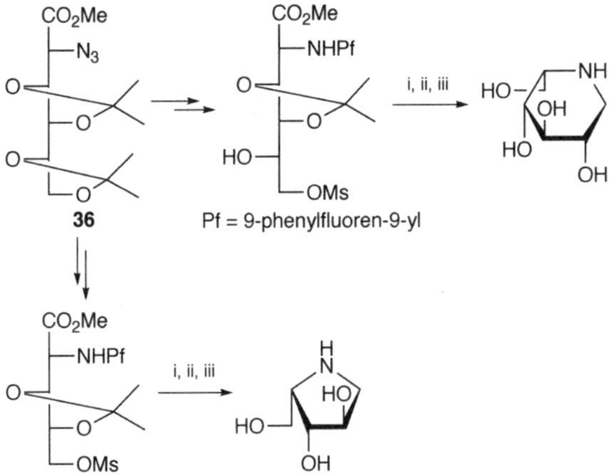

Reagents: i, LiAlH₄; ii, I₂, MeOH, heat; iii, Dowex-50W-X8, heat

Scheme 10

conditions used. For example, the *bis*-epoxide derived from D-arabinitol was treated with 12 equivalents of benzylamine in the presence of 6 equivalents of 60% perchloric acid to afford a 3:1 mixture of isomeric piperidine:pyrrolidine.[41] In a similar vein, the ring-opening of bis-cyclic sulfates by allylamine has resulted in the heterocyclization of various alditols. bis-Cyclic sulfates were prepared from erythritol and a range of pentitols and hexitols by treating the terminal vicinal hydroxyl pairs with thiocarbonyldiimidazole followed by sodium periodate and catalytic ruthenium(III) chloride. In this way *N*-allyl-imino-erythritol, -pentitols and -hexitols were produced in moderate yield.[42] Various isomers of 2,5-di(hydroxymethyl)-3,4-dihydroxypyrrolidine (DMDP or 2,5-dideoxy-2,5-imino-hexitols) have been analogously synthesized from different stereoisomers of bis-epoxides and cyclic sulfates derived from hexoses.[43]

The well known, D-mannitol-derived bis-aziridine **37** has been used to prepare 6-amino-6-deoxy-2,5-imino-D-glucitol and -1,5-imino-D-mannitol derivatives **38** and **39** as potential glycosidase inhibitors. Treatment of **37** with acetic acid, or alternatively allyl alcohol in the presence of an ytterbium catalyst, afforded mono-*O*-acetylated or -allylated glucitol/mannitol mixtures in 2:1 and 1:2 ratios respectively.[44]

A novel synthesis of chiral *N*-hydroxypyrrolidines, namely 1,4-dideoxy-1,4-imino-D-*galacto*-, -*gluco*- and -*talo*-hexitols, involving the reductive cyclization (with inversion of stereochemistry at C-4) of the oxime derivatives of protected D-gluco-, -galacto-, and -manno-pyranoses respectively, has been reported. *E/Z*-Benzyloximes were produced by treating the *O*-benzoyl protected hemiacetals with benzylhydroxylamine hydrochloride, and these were subsequently mesylated at O-5 and reductively cyclized by treatment with dimethylphenyl-silane in trifluoroacetic acid. A mechanistic cascade of neighbouring group participation was proposed (Scheme 11). Conversely, reductive cyclization of the oxime derived from the 6-deoxy L-rhamnose furnished a piperidine (1,5,6-trideoxy-1,5-imino-D-*gulo*-hexitol) as the only product.[45]

En route to a Sialyl Lewis X analogue containing 1,4,5,6-tetradeoxy-1,4-

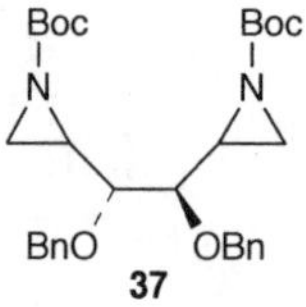

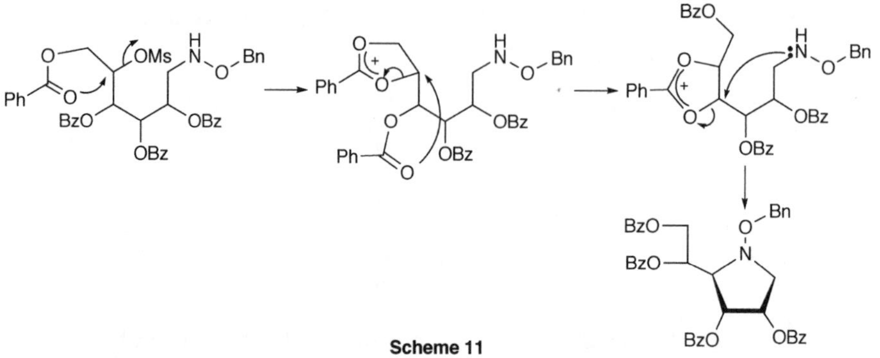

Scheme 11

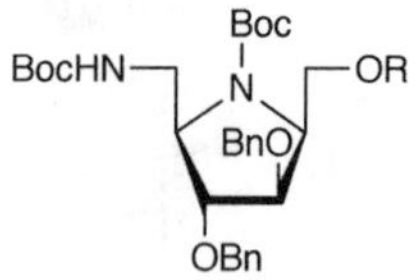

imino-D-*ido*-hexitol in place of an *N*-acetylgalactosamine residue (see also Section 1.5), the aforementioned imino-hexitol was synthesized in a protected form using the method of Bernotas (see Vol. 24, p. 198, ref. 36) which involves an intramolecular displacement by nitrogen under Mitsunobu conditions.[46]

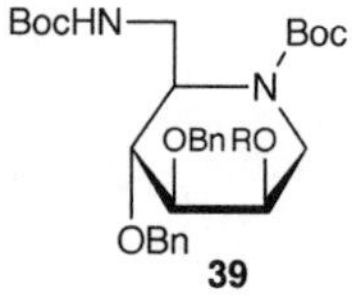

Full details of the synthesis of 6-azido-1,6-dideoxynojirimycin **41** (see Vol 28, p. 227, ref. 52) from 1-amino-1-deoxy-D-glucitol *via* the intermediate azido sugar **40** have been reported together with its conversion to various analogues **42** of DNJ, and the bicyclic iminosugars castanospermine and kifunensine.[47]

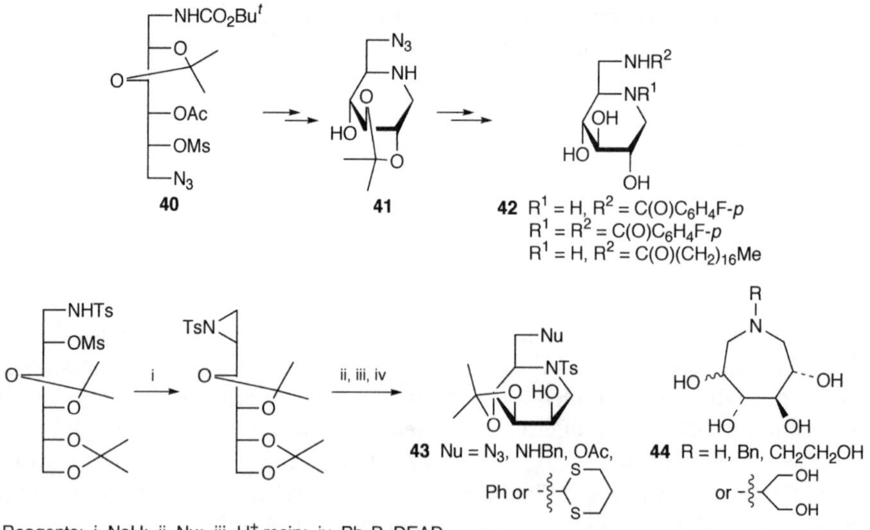

Reagents: i, NaH; ii, Nu; iii, H⁺ resin; iv, Ph₃P, DEAD

Scheme 12

1-Deoxymannojirimycin analogues **43** have been prepared from 1-amino-1-deoxy-D-glucitol according to Scheme 12. Various functionalities were incorporated at C-6 (in the product) by treating the aziridine precursor with a range of nucleophiles prior to ring-closure by intramolecular displacement of triphenylphospine oxide by nitrogen under Mitsunobu conditions.[48]

Bis-epoxides derived from D-glucitol and L-iditol have been used to create 1,6-dideoxy-1,6-imino-D-*gluco*- and -L-*ido*-hexitols (polyhydroxy azepanes) **44** by their reaction with various primary amines in dioxane at 100 °C followed by appropriate deprotection. These azepanes were found to be modest inhibitors of almond β-glucosidase and moderate inhibitors of various cancer cells.[49]

The synthesis of 2,5-dideoxy-2,5-imino-D-*altro*-hexitols **45** and 3,6-dideoxy-3,6-imino-D-*glycero*-L-*altro*-heptitols **46** from 2,3,5-tri-*O*-benzyl-D-lyxose has been reported. Chain extension was achieved using Wittig chemistry while nitrogen was introduced by azide displacement of a mesylate group. Reduction of the azide under Staudinger conditions proceeded with concomitant ring-closure by intramolecular ring-opening of an epoxide. The imino-hexitols **45** were obtained from 4,5,7-tri-*O*-benzylated **46** by lead tetraacetate-mediated cleavage of the α-diol followed by Pd/C catalysed hydrogenation.[50]

45 R^1 = H, Me

46 R = H, R^1 = H, Me

47 R = H, (CH$_2$)$_6$OH, (CH$_2$)$_3$NH$_2$
48 R = C(O)Me, C(O)(CH$_2$)$_5$OH, C(O)(CH$_2$)$_4$Me

49 R = Bn, R^1 = All, $\ce{->[\text{O}]}$
50 R = H, R^1 = Pr

51 $R^1 = R^3 = R^5 = R^6 = R^7$ = H, $R^2 = R^4 = R^8$ = OH
52 $R^1 = R^3 = R^5 = R^6 = R^8$ = H, $R^2 = R^4 = R^7$ = OH
53 $R^1 = R^4 = R^5 = R^6 = R^7$ = H, $R^2 = R^3 = R^8$ = OH
54 $R^2 = R^3 = R^5 = R^6 = R^8$ = H, $R^1 = R^4 = R^7$ = OH
55 $R^1 = R^3 = R^6 = R^8$ = H, $R^2 = R^4 = R^5 = R^7$ = OH
56 $R^1 = R^3 = R^6 = R^8$ = OH, $R^2 = R^4 = R^5 = R^7$ = H

1.3.2 Cyclization by Reductive Amination of Carbohydrate Substrates. Various 1-amino derivatives of 2,5-dideoxy-2,5-imino-D-mannitol (**47**) have been prepared in excellent yields in two steps from 5-azido-5-deoxy-D-glucofuranose. Reaction of this material with amines in the presence of acetic acid results in a 1-amino-fructopyranose intermediate *via* an Amadori rearrangement and catalytic hydrogenation of this ketose using Pearlman's catalyst yields the imino-mannitols by simultaneous reduction of the azido group and cyclization by reductive amination. 1-*N*-Acylated derivatives **48** proved to be very strong inhibitors of β-glucosidase from *Agrobacterium* sp.[51]

A new procedure for the preparation of deoxynojirimycin analogues bearing lipophilic substituents at C-1 has been reported starting from 1,2,3,4,6-penta-benzyl-α/β-D-glucosylamine. Stereoselective introduction of an allylic append-age at C-1 was achieved by reaction with allylmagnesium bromide. Subsequent pyridinium chlorochromate oxidation at C-5 and reductive amination using sodium triacetoxyborohydride in acetic acid afforded the α-allylic DNJ ana-logue (49 R^1 = All) with very high diastereoselectivity. Simple transformations of the allyl group provided access to propyl (50) and methyl ketone (49 R^1 = C(O)Me) derivatives.[52]

Trihydroxyazepanes 51–54 have been synthesized as potential glycosidase inhibitors from 6-azido-3,6-dideoxy-D-gluco-, -D-manno-, -D-galacto- and -L-galacto-pyranoses respectively by Pd/C-catalysed hydrogenation in methanol,[53] while tetrahydroxyazepanes 55 and 56 were generated from D- and L-*chiro*-inositol, respectively, utilizing a strategy involving isopropylidene protection of the *cis*-vicinal hydroxyls, periodate cleavage of the remaining *trans*-diol and double reductive amination of the product using aminodiphenylmethane and sodium cyanoborohydride.[54]

1.3.3 Cyclization by Addition of Amines to Multiple Bonds of Carbohydrate Substrates. A versatile unsaturated aldehydic intermediate derived from D-ribose has been used to synthesize a number of potential glycosidase inhibitors (Scheme 13). Addition of hydroxylamine and then bromine to the enal resulted in the formation of nitrone 57 with 66% diastereoselectivity, and this in turn was readily transformed into various imino-alditols (*e.g.* 58–60) by nucleophilic attack followed by reduction. A number of these compounds proved to be good inhibitors of α-L-fucosidases.[55]

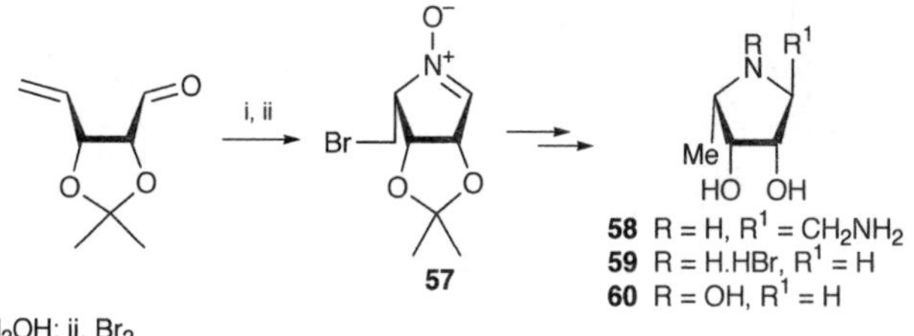

Reagents: i, NH$_2$OH; ii, Br$_2$

58 R = H, R^1 = CH$_2$NH$_2$
59 R = H.HBr, R^1 = H
60 R = OH, R^1 = H

Scheme 13

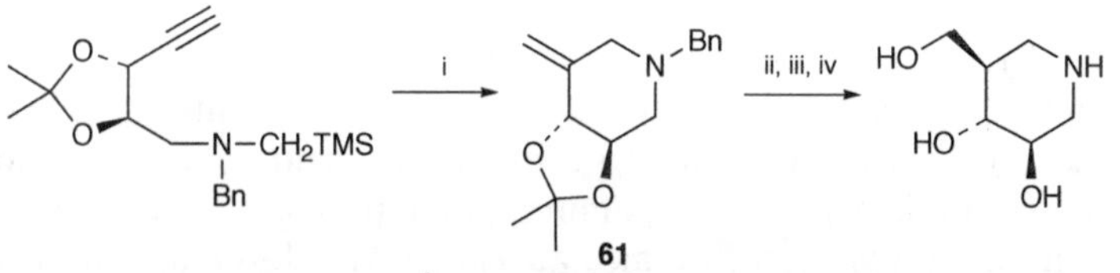

Reagents: i, *hv*, 1,4-dicyanonapthalene, Pr^{i}OH; ii, 9-BBN then NaOH, H$_2$O$_2$; iii, HCl, MeOH; iv, Pd(OH)$_2$/C, H$_2$

Scheme 14

A novel route to (+)-isofagomine from D-tartaric acid has been reported. In the presence of 1,4-dicyanonapthalene and light an α-trimethylsilylmethylamine radical cation is generated in an advanced alkynitol intermediate (Scheme 14).

This radical cation adds intramolecularly to the terminal triple bond to produce ring-closed **61** which is subsequently hydroborated and reduced. Enantiomeric (−)-isofagomine may be obtained if the alkynitol derived from L-tartrate is substituted.[56]

1-Deoxymannojirimycin has been synthesized with excellent diastereoselectivity in a process based upon a palladium(II)-catalysed cyclization of an allylic alcohol **62** derived from D-mannitol.[57] C-6 Homologues of 1-deoxynojirimycin and 1-deoxy-L-idonojirimycin (1,5,6-trideoxy-1,5-imino-D-*gluco*- and -L-*ido*-heptitols) have also been prepared using palladium chemistry. In these reports the protected aminohex-1-enitol **63**, derived from methyl α-D-glucopyranoside, was amino-carbonylated in the presence of palladium(II) chloride to afford a mixture of cyclized products whose composition was dependent upon the reaction conditions utilized. Two of these products, **64** and **65**, were reductively transformed into the DNJ and 1-deoxy-L-idonojirimycin homolgues respectively.[58,59]

A new, shorter synthesis of miglitol [*N*-(2-hydroxyethyl)-1-deoxynojirimycin] and its conversion to lipophilic prodrugs by coupling with cholesterol derivatives have been reported. Miglitol was synthesized from methyl 2,3,4-tri-*O*-benzyl-6-bromo-6-deoxy-α-D-glucopyranoside *via* mercuric trifluoroacetate/potassium bromide mediated intramolecular addition of a benzyloxyethylamino group to the terminal double bond in the advanced intermediate, 6-(benzyloxyethylamino)-3,4,5-tri-*O*-benzyl-1,2,6-trideoxy-L-*xylo*-hex-1-enitol. Chromatographic separation of the desired D-*gluco*- from the undesired L-*ido*-bromomercuric imino-hexitol followed by coupling with cholesterol hemisuccinate and deprotective hydrogenolysis afforded the lipophilic conjugate **66**.[60]

As general approach to dideoxy- and trideoxy-iminoalditols, the synthesis of 1,5-dideoxy- and 1,5,6-trideoxy-1,5-imino-D-glucitol and -galactitol from β-D-glycosides has been reported. Chromium trioxide oxidation of per-*O*-acetylated methyl β-D-gluco- and -galactosides to give 5-ulosonic acid esters, followed by amination using hydroxylamine yields the (*E*/*Z*)-oximes which in turn cyclize to form aldurono-lactams (**67**→**68**) under the conditions of catalytic hydrogenation. Borane reduction and final deprotection gives rise to the trideoxyiminoalditols. Interestingly, reductive cleavage of the acetoxy group at C-6 was observed during the hydrogenation step and afforded the 6-deoxy-lactam. If, however, the primary hydroxyl is unprotected and the ester group converted into an acyl hydrazide prior to hydrogenation by reaction with hydrazine, no such deoxygenation was observed.[61]

In a search for new glycosidase inhibitors an imino-*exo*-glycal, *N*-benzyloxycarbonyl-1,2,5-trideoxy-2,5-imino-D-*arabino*-hex-1-enitol (**69**) has been prepared from 2,3,5-tri-*O*-benzyl-D-arabinose. Wittig methylenation followed by a double Mitsunobu process, the second using phthalimide, generated (after deprotection) a 5-amino-hex-1-enitol with overall retention of configuration at C-5. This was protected as a carbamate and cyclized using *N*-iodosuccinimide followed by DBU.[62]

1.3.4 Modification of Azasugars. Boron trifluoride-promoted reaction of allyl trimethylsilane with 2,3-*O*-isopropylidene protected pyrrolidine sugar analogues has been shown to proceed with 2,3-*trans*-stereoselectivity to exclusively produce 2-allyl *C*-glycoside analogues in moderate to high yields (47–92%). As an example, 1,2-*O*-isopropylidene protected 4-amino-4-deoxy-L-arabinose derivative **70** was converted to the imino-arabinitol **71** in 60% yield and with 100% β-selectivity. The substitution was found to occur from the *exo*-face of the bicyclic aminal derivatives only, and the resulting 1,2-*trans*-stereoselectivity was independent of the configuration of the substituents in the ring.[63]

5-*O*-*tert*-Butyldimethylsilyl-1,4-dideoxy-1,4-imino-2,3-*O*-isopropylidene-D-ribitol has been used to prepare the 'Immucillins', aza-*C*-nucleoside analogues that are potent inhibitors of purine nucleoside phosphorylases. The inosine analogue **72** for example, bearing a 9-deazapurine, has a K_i of 0.07 nM against the human enzyme, and was synthesized by the addition of lithiated acetonitrile to C-1 of an imine intermediate derived from the ribitol precursor with subsequent step-wise construction of the deazapurine substituent using conventional chemistry. 5'-Deoxy-, 5'-deoxy-5'-fluoro- and 2'-deoxy deazainosine analogues of **72**, as well as the corresponding azasugar analogues of guanosine, were also prepared.[64]

1,5-Dideoxy-1,5-imino-D-*erythro*-pentulose hydrate **74** was produced as an Amadori product during the mild acid (pH > 5) hydrolysis of sulfur-linked pseudo-disaccharide **73** which has an iminosugar at the non-reducing end.[65] The synthesis of **73** is covered in Chapters 9 and 11.

O-Acetylated and non-acetylated *N*-(3'-iodobenzyl)- and *N*-(3'-iodo-2'-propenyl)-deoxynojirimycin derivatives, along with their corresponding ^{125}I-labelled analogues, were prepared for brain uptake studies in rats by simple treatment of 1-deoxynojirimycin with the appropriate alkylating agent.[66] Glycopeptidomimetic nojirimycinyl *C*-(L)-serine **76** was prepared with high diastereoselectivity from known α-*C*-allyl nojirimycin **75** by simple transformation of the allyl group. Ozonolysis followed by chain-extension using a Wittig–Horner reaction with a glycine-derived phosphonate and final hydrogenolysis/hydrogenation afforded the product as a diastereoisomeric mixture with an L/D ratio of 6:1.[67]

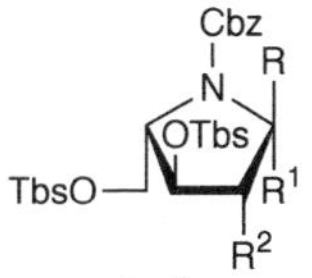

70 R = H, R^1,R^2 = O(CMe$_2$)O
71 R = Allyl, R^1 = H, R^2 = OH

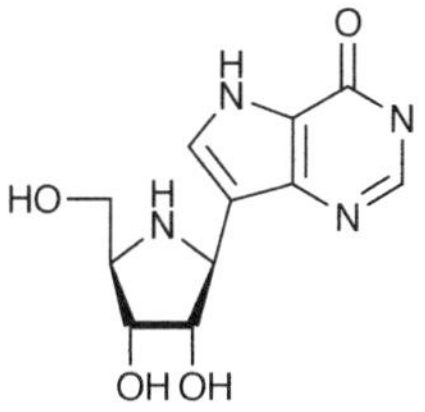

72

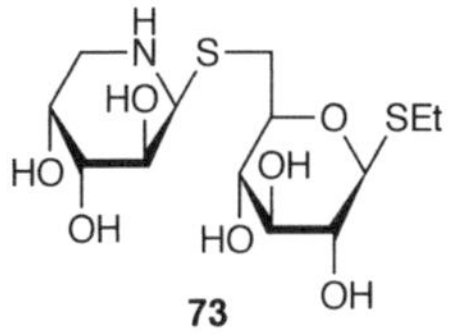

73

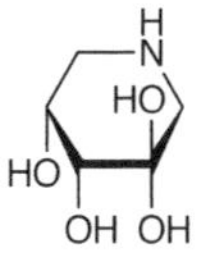

74

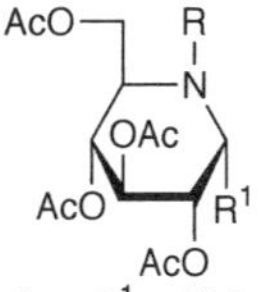

75 R = Boc, R^1 = Allyl
76 R = H, R^1 = $\sim$CO$_2$Me, NHBoc

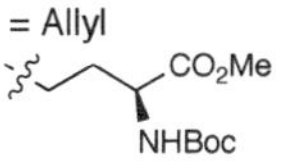

1.3.5 Other Carbohydrate-based Methods. *N*-Hydroxy-1,4-dideoxy-1,4-imino-arabinitol, **77**, has been prepared in nine steps from (±)-3-*O*-benzylglyceraldehyde. A transketolase mediated reaction was used to establish a pentulose (5-*O*-benzyl-D-xylulose) with correct absolute stereochemistry, and a 1,2-oxazine was the unexpected product of the acid-catalysed reaction of an aldehydic intermediate with triethylorthoformate (Scheme 15). Reduction of this oxazine with sodium cyanoborohydride in acetic acid, did not effect cleavage of the N-O bond, and yielded the *N*-hydroxypyrrolidine as a single diastereoisomer.[68]

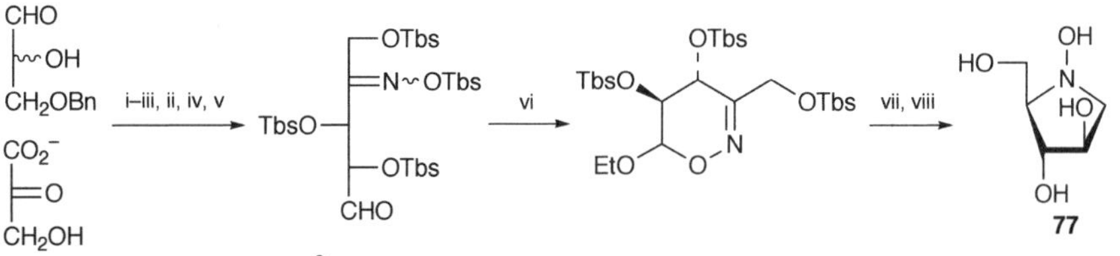

Reagents: i, transketolase, Mg^{2+}, pH 7; ii, TbsTf; iii, NH$_2$OH; iv, H$_2$, Raney Ni; v, Swern; vi, (EtO)$_3$CH, PTSA; vii, NaCNBH$_3$; viii, HF

Scheme 15

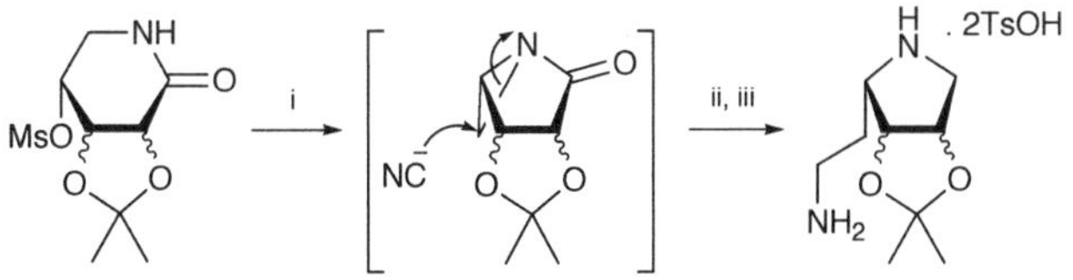

Reagents: i, Bu$_4$NCN, DMF, 100 °C; ii, LiAlH$_4$, THF; iii, TsOH, MeOH

Scheme 16

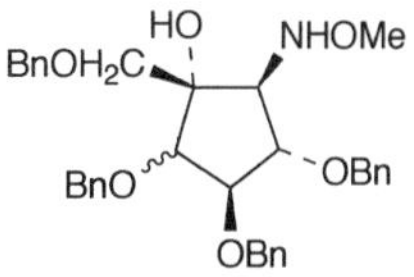

An unusual stereospecific ring contraction has been observed upon attempted substitution of 4-*O*-activated pentono-1,5-lactams with cyanide. The reaction of 4-*O*-methanesulfonyl-2,3-*O*-isopropylidene-D-ribo- or -D-lyxo-1,5-lactams with tetrabutylammonium cyanide gave 4-amino-5-*C*-cyano-4,5-dideoxy-2,3-*O*-iso-propylidene-L-lyxo- or -L-ribo-1,4-lactams respectively (Scheme 16), instead of the expected products of simple S$_N$ displacement of mesylate by cyanide. Reduction and deprotection afforded the corresponding 6-amino-1,4,5,6-tetradeoxy-1,4-imino-L-*lyxo*- and -L-*ribo*-hexitols, of which the latter was found to be a moderate inhibitor of α-L-fucosidase.[69]

1-Deoxynojirimycin (DNJ) and 1-deoxygalactostatin were reportedly prepared from *trans*-alkoxyaminocyclopentitols **78**, albeit in poor yield, by reductive ring expansion using lithium aluminium hydride in THF to afford the per-*O*-benzyl derivatives as minor (<30%) products, followed by hydrogenolytic debenzylation in ethanolic HCl.[70] The syntheses of **78** from D-glucose and -galactose are covered in Section 2.1.

As exemplifications of a new route to polyhydroxypiperidines, the 6-amino-1-deoxynojirimycin (DNJ) analogues **79** and **80** were synthesized as shown in Scheme 17 from a *xylo*-pentadialdose (obtained by the action of periodate on 1,2-*O*-isopropylidene-β-D-glucofuranose) by a novel condensative cyclization with (*R*)-(+)-phenylglycinol and potassium cyanide in the presence of zinc bromide to afford the thermodynamically more stable diastereoisomeric bicycles illustrated (45% and 5–15% respectively). Catalytic hydrogenation of these compounds resulted in the simultaneous reduction of the cyano group and hydrogenolysis of the chiral appendage.[71]

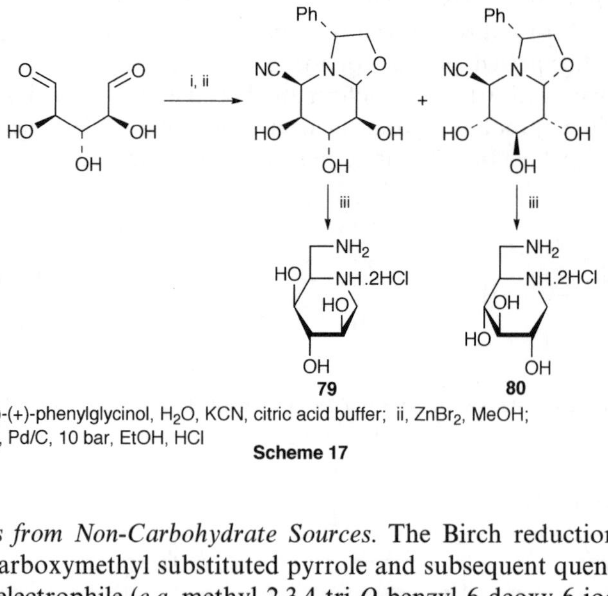

Reagents: i, (R)-(+)-phenylglycinol, H₂O, KCN, citric acid buffer; ii, ZnBr₂, MeOH;
iii, H₂, Pd/C, 10 bar, EtOH, HCl

Scheme 17

1.3.6 Syntheses from Non-Carbohydrate Sources. The Birch reduction of an *N*-protected 2-carboxymethyl substituted pyrrole and subsequent quench with an appropriate electrophile (*e.g.* methyl 2,3,4-tri-*O*-benzyl-6-deoxy-6-iodo-α-D-glucopyranoside) affords a simple route to *N*-protected 2,2-disubstituted dihydropyrroles **81**. These can be further elaborated into the corresponding 4-substituted imino-ribitols **82** in good overall yield by a simple strategy of reduction, acetylation, *syn*-dihydroxylation and deprotection.[72]

A [5-^{14}C]-labelled analogue of 1,4-dideoxy-1,4-imino-D-arabinitol, a potent α-glucosidase inhibitor, was conveniently prepared in a one-pot synthesis by ^{14}C-cyanosilylation of (3*R*,4*R*)-3,4-dibenzoyloxy-3,4-dihydro-2*H*-pyrrole **83** using an inexpensive mixture of potassium cyanide and chlorotrimethylsilane.

An epimeric mixture of 2-cyanopyrrolidines resulted and upon separation of the desired isomer, the arabinitol derivative was produced by acid hydrolysis and borane-dimethyl sulfide reduction of the so-formed carboxylic acid.[73]

Two total syntheses of pyrrolidine alkaloids have been reported. (−)-Codonopsinine (**85**), an imino-pentitol which exhibits antibiotic and hypotensive activity, has been synthesized in seven steps in 16% overall yield from dihydropyrrole **84**.[74] The enantioselective synthesis of the potent antifungal agent (+)-preussin (**87**) has also been achieved from L-phenylalanine derivative **86**. The pyrrolidine skeleton was established by hydrogenolysis of an intermediate oxazoline and subsequent diastereoselective reductive cyclization of the resultant aminoketone using Pearlman's catalyst.[75]

R = OAc, 6-deoxy-6-yl-α-D-Glc$_p$OMe

Two syntheses of iminoalditols from chiral 1,2-oxazines by catalytic reduction have appeared. A novel asymmetric hetero-Diels–Alder reaction of a 1-p-tolyl-sulfinyl-1,3-diene with benzyl nitrosoformate provides access to **89**, an optically pure 1,4-imino-L-ribitol derivative. The cycloaddition yields 1,2-oxazine **88** with complete regio- and diastero-selectivity, and in a very neat single synthetic step, a dihydroxylated derivative of this adduct is converted into an enatiomerically pure pyrrolidine by reduction under Pd/C. This 'reduction' encompasses hydrogenolysis of the benzyl group to yield a carbamic acid, decarboxylation, hydrogenolysis of the N–O bond, cleavage of the sulfonyl auxiliary with concomitant formation of an unstable aminoaldehyde, cyclization and final hydrogenation of the resulting imine.[76] Full details of the conversion of chiral 1,2-oxazines (*e.g.* **90**), derived from hetero-Diels–Alder cycloaddition of an α-chloronitroso-D-mannofuranosyl derivative with sorbaldehyde dimethyl acetal (see Vol. 31, p. 229, ref. 50) into 1,2,5-trideoxy-2,5-imino-D-alditols such as **91** have also been reported. Compound **91** is a potent α-D-galactosidase and α-L-fucosidase inhibitor with K_i values of 9 and 5 μM respectively.[77]

Several papers which detail the syntheses of isofagomine, its stereoisomers and various 3-susbstituted analogues by heterocycle modification have appeared. (±)-Isofagomine (**92**) and its (3,4-) stereoisomers (5-hydroxymethyl-3,4-dihydroxypiperidines) have been prepared in short syntheses from arecoline. Osmylation or epoxidation and hydrolytic ring-opening (mCPBA or methyltri-

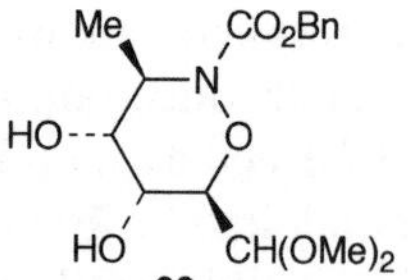

90 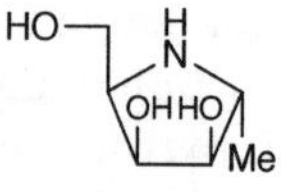   91

fluoromethyldioxirane/perchloric acid) carried out on the tetrahydropyridine intermediate **93** provided access to the four stereoisomers.[78] Piperidinone **94** has been used as a precursor to 3-substituted isofagomine analogues. A series of twelve 3,4,5-trisubstituted piperidines were synthesized *via* three key intermediates as depicted in Scheme 18. The free acids **95** were obtained by acid treatment, while the 3-deoxy-3-hydroxymethyl derivatives **96** were obtained by lithium borohydride reduction prior to acid treatment. Reaction of the free acids with diphenylphosphoryl azide produced highly reactive isocyanides, *via* Curtius rearrangement of the initially formed acyl azides, which underwent spontaneous intramolecular reaction to produce the cyclic carbamates **97**. Hydrolyses of these carbamates in refluxing HCl gave the corresponding 3-amino-isofagomine derivatives **98**.[79] In a succeeding report 3-deoxy- and 3-*C*-hydroxymethyl-isofagomines **99** and **100**, respectively, were also synthesized from **94**.[80] The cyclopentenoid building block **101**, accessed from the fragmentation of a norbornyl derivative, has been elaborated into isofagomine analogues **102** and **103** by

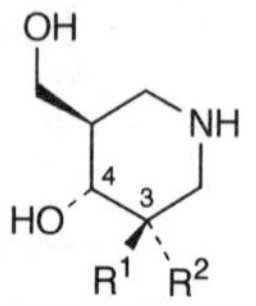

92 R^1 = OH, R^2 = H
99 R^1 = R^2 = H
100 R^1 = OH, R^2 = CH$_2$OH

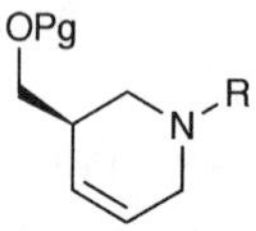

93

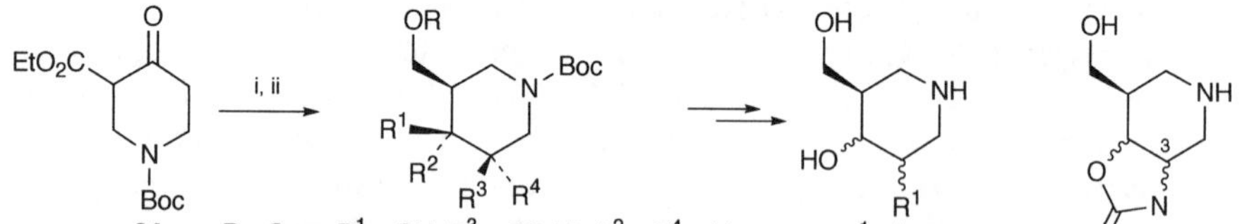

94
R = Sem, R^1 = OH, R^3 = CO$_2$Et, R^2 = R^4 = H
R = Sem, R^2 = OH, R^4 = CO$_2$Et, R^1 = R^3 = H
R = Sem, R^2 = OH, R^3 = CO$_2$Et, R^1 = R^4 = H

95 R^1 = CO$_2$H
96 R^1 = CH$_2$OH
96 R^1 = CH$_2$OH

97

Reagents: i, LDA, 2-(trimethylsilyl)ethoxymethyl chloride (SemCl); ii, H$_2$, Pd/C

Scheme 18

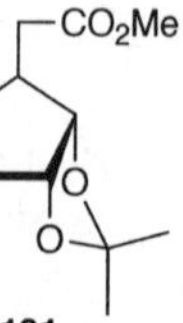

101

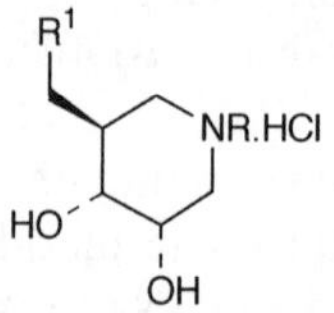

102 R = H, R^1 = OH
103 R = Bn or H, R^1 = H

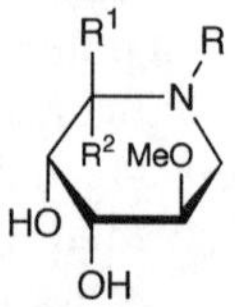

104 R = Bn or H, R^1 = CH$_2$OH, R^2 = H
105 R = Bn or H, R^1 = H, R^2 = CH$_2$OH

oxidative cleavage of the double bond followed by a strategy incorporating a double displacement of a dimesylate with *p*-toluenesulfonamide (→**102**) or a double reductive amination with benzylamine and sodium cyanoborohydride (→**103**).[81] The same building block **101** was also transformed into *altro-* and *galacto*-deoxynojirimycin analogues **104** and **105**, respectively, using similar methodology. The glycosidase inhibitory activity of the latter compounds was reported to be moderate-strong.[82]

1.3.7 Natural Products. Four new broussonetines, M, O, P and Q (**106–109**), pyrrolidine alkaloids with β-glycosidase inhibitory properties, have been isolated from branches of *Broussoneta kazinoki*.[83]

6-*C*-Butyl-DMDP (dihydroxymethyl dihydroxy pyrrolidine) derivative **110**, with an IC$_{50}$ of 68 μm against almond β-glucosidase, and α-1-*C*-ethyl-fagomine **111**, with an IC$_{50}$ of 29 μm against bovine liver β-galactosidase, were isolated from *Adenophora triphylla* sp., along with four known imino-sugars (DMDP, D-AB1, DNJ and DMJ).[84]

Five new homonojirimycin derivatives (**112–116**), including two glucosylated *pseudo*-disaccharides (**113** and **115**, see also Section 1.5), and deoxygalactonojirimycin derivative **117** have been extracted from the roots of *Adenophora* sp.[85]

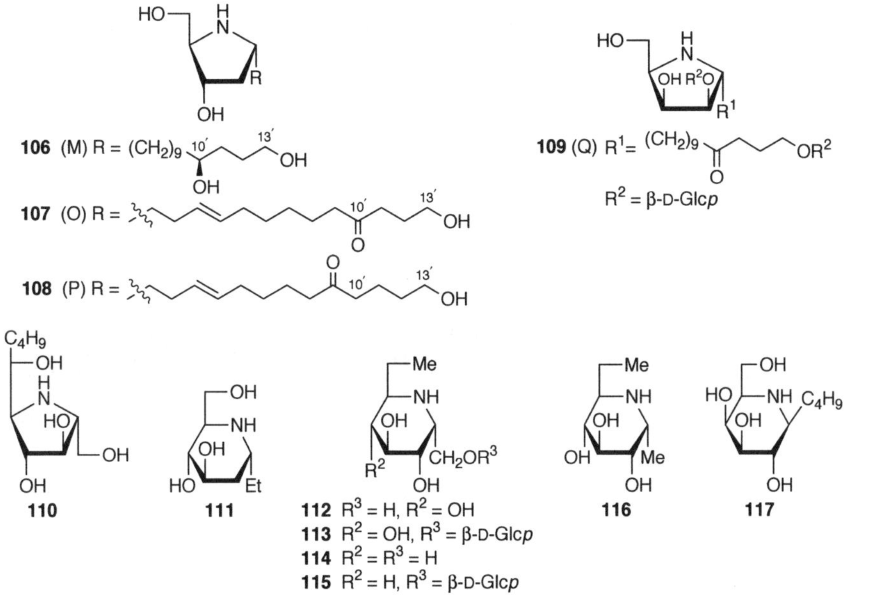

106 (M) R = (CH$_2$)$_9$ ...

107 (O) R = ...

108 (P) R = ...

109 (Q) R^1= (CH$_2$)$_9$ ...

R^2 = β-D-Glc*p*

110

111

112 R^3 = H, R^2 = OH
113 R^2 = OH, R^3 = β-D-Glc*p*
114 R^2 = R^3 = H
115 R^2 = H, R^3 = β-D-Glc*p*

116

117

1.4 Fused-ring and Bicyclic Azasugars. – Four new polyhydroxypyrrolizidines, the Hyacinthacines A$_1$, A$_2$, A$_3$ and B$_3$ (**118–121** respectively) were isolated from the bulbs of *Muscari armeniacum*, and their structures elucidated.[86]

Bis-aziridine **37** (see Section 1.3.1), derived from D-mannitol, was converted into oxazolidinone **122** by treatment with dilithium nickel tetrabromide followed by silver nitrate in DMF at elevated temperature.[44]

5-Isoxazolidinone **124** was synthesized from dehydroamino acid **123**, which has a pentose sugar moiety at its side-chain, by displacement of the terminal mesylate by hydroxylamine with spontaneous cyclization by intramolecular Michael-like addition. Interestingly, when a tetrose-derived dehydroamino acid was substituted, no cyclization was observed.[87]

Improved syntheses of (+)-lentiginosine (**125**) and (7R)-7-hydroxylentiginosine (**126**), inhibitors of amyloglucosidases, have been reported. In multi-step syntheses based upon a 1,3-dipolar cycloaddition of a pyrrolidine N-oxide derived from L-tartaric acid to 3-buten-1-ol, **125** and **126** were prepared in 25% and 37% overall yields from a common indolizidine intermediate. The enantiomers of these compounds could be similarly prepared using D-tartaric acid as a starting material.[88]

Castanospermine analogues **127** and **128** have been synthesized in many steps from a chirally unresolved pyrrolidinone.[89]

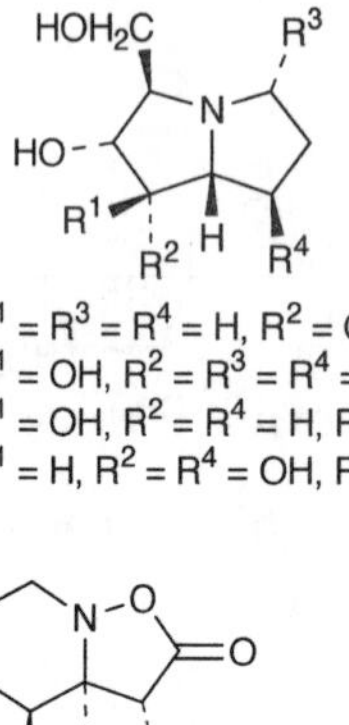

118 $R^1 = R^3 = R^4 = H$, $R^2 = OH$
119 $R^1 = OH$, $R^2 = R^3 = R^4 = H$
120 $R^1 = OH$, $R^2 = R^4 = H$, $R^3 = Me$
121 $R^1 = H$, $R^2 = R^4 = OH$, $R^3 = Me$

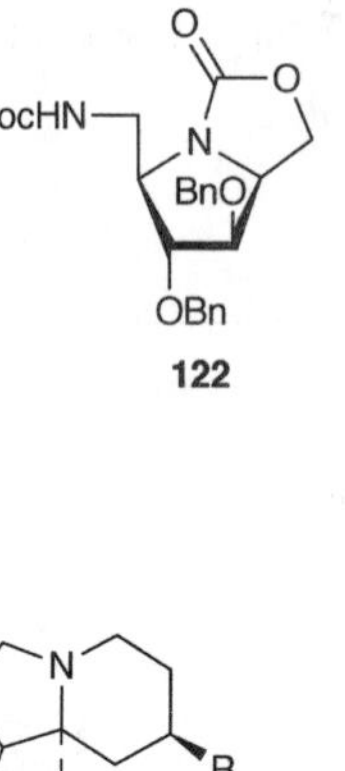

122

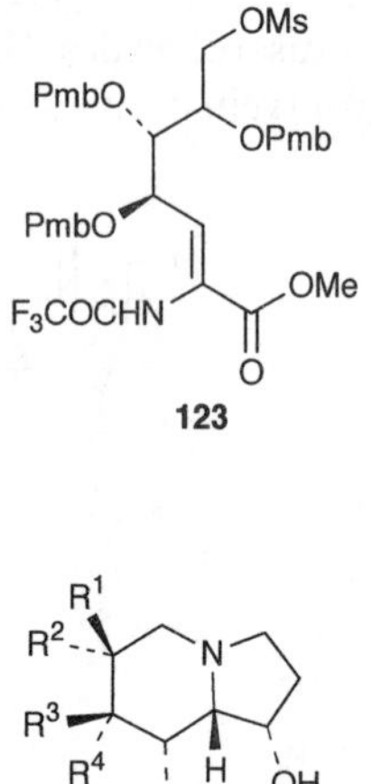

123

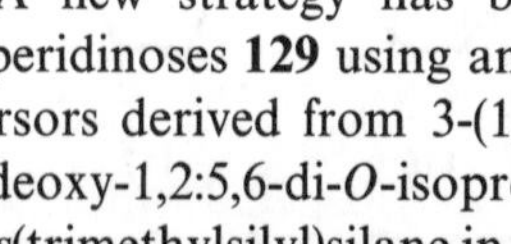

124

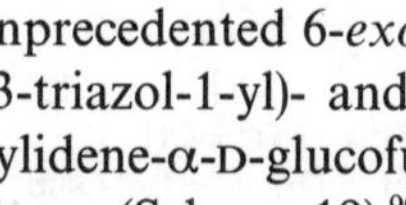

125 R = H
126 R = OH

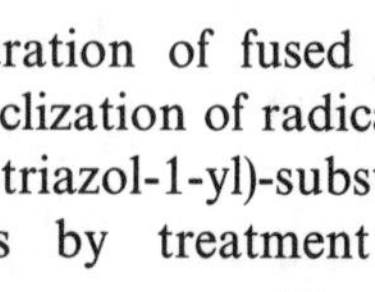

127 $R^1 = R^4 = OH$, $R^2 = R^3 = H$
128 $R^1 = R^4 = H$, $R^2 = R^3 = OH$

A new strategy has been reported for the preparation of fused azole-piperidinoses **129** using an unprecedented 6-*exo*-trig cyclization of radical precursors derived from 3-(1,2,3-triazol-1-yl)- and -(1,2,4-triazol-1-yl)-substituted 3-deoxy-1,2:5,6-di-O-isopropylidene-α-D-glucofuranoses by treatment with *tris*(trimethylsilyl)silane in toluene (Scheme 19).[90]

Bicyclic polyhydroxylated isoquinuclidines with conformations that mimic the boat-form of pyranosides have been prepared in multiple steps from an achiral pyridinone. These compounds proved to be strong and selective inhibitors of snail β-mannosidase (**130** R = H, K_i 20 μm; R = Bn, K_i 0.17 μm).[91]

1.5 Azasugar-containing Di- and Tri-saccharides. – A number of glycosylated iminoalditols have been isolated from plant extracts. Glycosylated homo-

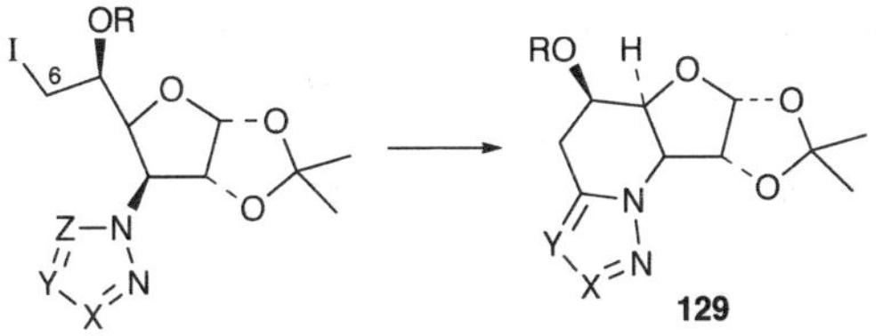

X = N, Y = Z = CHCO$_2$Et $\xrightarrow{\text{i}}$ X = N, Y = CHCO$_2$Et

X = CH, Y = N, Z = CH $\xrightarrow{\text{ii}}$ X = CH, Y = N

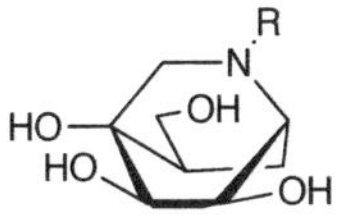

Reagents: i, [(CH$_3$)$_3$Si]$_3$SiH, Tol (0.02 M), 72% (R = H), 94% (R = Ac)
ii, [(CH$_3$)$_3$Si]$_3$SiH, Tol (0.02 M), 50% (R = H)

Scheme 19

nojirimycin **131** has been isolated from the plant *Lobelia sessilifolia*,[92] while **113** and **115** were found in the roots of *Adenophora* sp.[92] Broussonetine **109** was isolated from branches of *Broussonetia kazinoki*[93](see Section 1.3.3).

The *N*-linked pseudo-disaccharide **132** has been prepared from imino-D-*altro*-hexitol **45** (Section 1.3.1) and this compound was found to be a potent inhibitor of β-galactosidase (K_i 3.3 μm).[94]

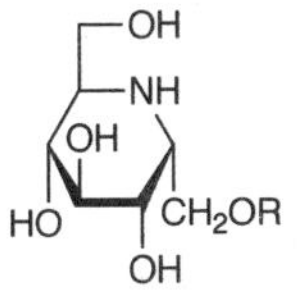

131 R = β-D-Glc*p* 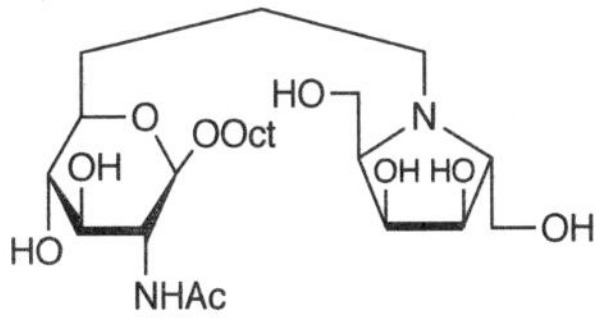
132 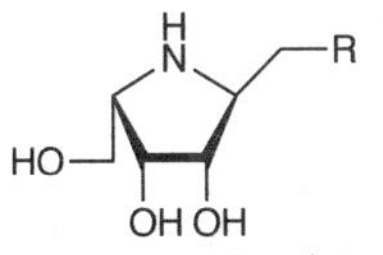
133 R = 6-yl-β-D-Gal*p*
134 R = 5-yl-β-D-Rib*f*

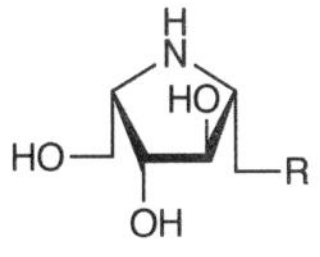

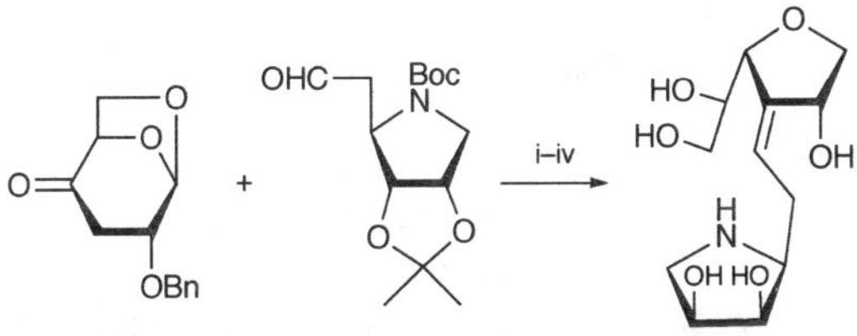

135

Reagents: i, BuLi, (Me$_3$Si)$_2$NH; ii, LiAlH$_4$; iii, Pd(OH)$_2$/C, H$_2$; iv, TFA

Scheme 20

Wittig condensation of *N*-benzyl-2,3,5-tri-*O*-benzyl-1,4-dideoxy-1-formyl-1,4-imino-L-*lyxo*-and -L-*xylo*-pentitols with protected D-galactopyranose-6- or D-ribofuranose-5-phosphoranes gave rise to the pyrrolidine-based (1→6)- and (1→5)-aza-*C*-disaccharides **133** (L-*lyxo*) and **134** (L-*xylo*) respectively after hydrogenation and deprotection.[95] The (1→3)-aza-*C*-disaccharide **135**, on the other hand, was synthesized as shown in Scheme 20 using a cross-aldol reaction between a 6-formyl-iminoribitol and an anhydro-keto sugar derived from isolevoglucosenone.[96] In a novel approach to disaccharide analogues of this type, stereoselective cycloaddition reactions between functionalized cyclic nitrones and sugar alkenes have also been successfully employed. Exemplified in Scheme 21, a nitrone derived in two steps from 2,3-*O*-isopropylidene-D-lyxose underwent a 1,3-dipolar cycloaddition with a protected α-D-*manno*- or -*galacto*-hept-6-

enopyranose to form an isoxazolidine cycloadduct which was reductively cleaved and deoxygenated to produce aza-*C*-disaccharides **136**.[97]

In the synthesis of a Sialyl Lewis X analogue, 1,4,5,6-tetradeoxy-1,4-imino-D-*ido*-hexitol (for synthesis see Section 1.3.1) was α-L-fucosylated at the less hindered (C-2) hydroxyl, then β-D-galactosylated at O-3 to produce a non-sulfated precursor to the *pseudo*-trisaccharide **137**.[46]

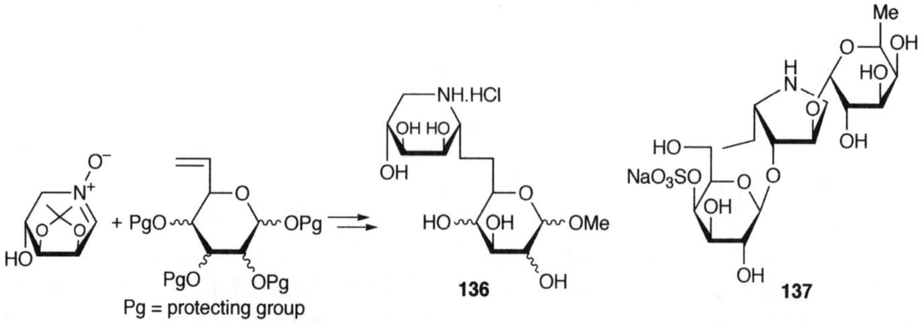

Pg = protecting group

Scheme 21

2   Cyclitols and Derivatives.

– A review on the application of the Pd(II)-catalysed Ferrier (II) carbocyclization in the synthesis of β-glucosidase inhibitors, cyclophellitol, all the diastereoisomers of inositol, and D-*myo*-inositol phosphates has been published.[98]

2.1   Cyclopentane Derivatives. – Carbapentafuranoses have been prepared in high yield from a variety of *O*-protected hexopyranosides, including α-D-gluco-, β-D-allo-, α-D-manno- and α-D-galacto-pyranosides, *via* cobalt-catalysed oxygenative radical cyclization of 6-iodo-hex-1-enitol derivatives. For example, **138**, derived from methyl 2,3,4-tri-*O*-benzyl-6-deoxy-6-iodo-α-D-glucopyranoside, was cyclized with very high diastereoselectivity to the cyclitol **139** using a catalytic Co(salen) complex in the presence of air (Scheme 22).[99] The same iodo-glucoside starting material has also been used to make the bicycle **140**. Zinc reduction followed by reaction with a hydrazine carboxylate is thought to afford the transient species **141**, the product of a prototropic shift, which cyclizes to the bicycle spontaneously under the reaction conditions.[55]

Ring-closing olefin metathesis has been used to effect cyclitol formation. The *carba*-β-D-arabinofuranose **143**[100] and *carba*-β- and -α-D-arabinofuranosides **144**[101] were synthesized by ring-closing metathesis of hept-1,6-dienitols derived from D-arabinose (**142**, P = Pmb) and D-mannose (**142**, P = Mom) respectively, by employing Schrock's catalyst followed by stereoselective hydrogenation of the resulting cyclopentene.

The syntheses of *carba*-β-D-xylofuranose and -xylofuranosylamine have been reported starting from 2,3-*O*-isopropylidene-D-glyceraldehyde and either 2-[(*tert*-butyldimethylsilyl)oxy]furan or *N*-(*tert*-butoxycarbonyl)-2-[(*tert*-butyldimethylsilyl)oxy]-1*H*-pyrrole respectively. The key reactions in the syntheses (depicted in Scheme 23) are the highly stereoselective boron trifluoride promoted nucleophilic addition of the furan or pyrrole to the aldehyde, and later, the

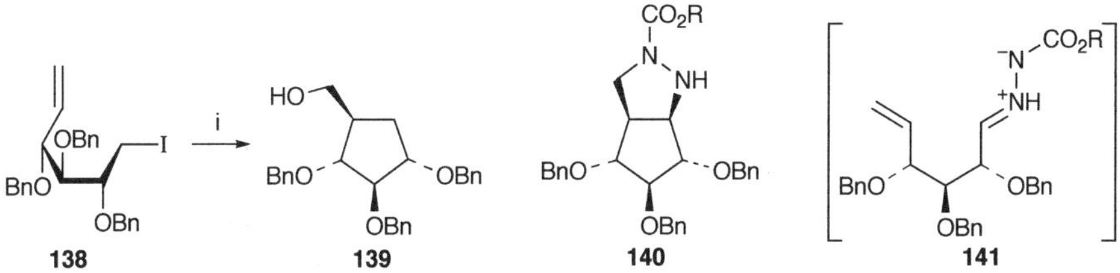

138   **139**   **140**   **141**

Reagents: i, cat. Co(salen) complex, air

Scheme 22

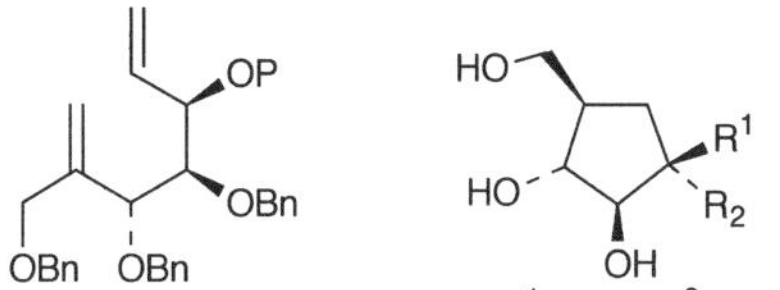

142

143 $R^1 = OH, R^2 = H$
144 $R^1 = OMe, R^2 = H$;
$\quad\quad R^1 = H, R^2 = OMe$

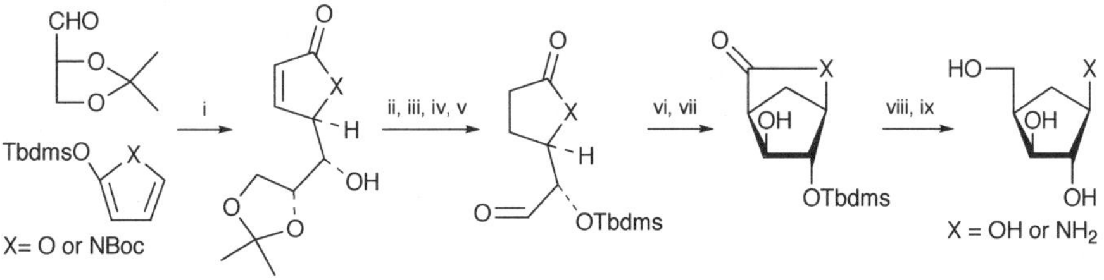

X= O or NBoc

X = OH or NH_2

Reagents: i, BF_3; ii, H_2, Pd/C; iii, TbdmsOTf; iv, H_3O^+; v, $NaIO_4$; vi, Swern; vii, LDA; viii, $LiAlH_4$; ix, H_3O^+

Scheme 23

diastereoselective cycloaldolization reaction that generates the cyclopentane ring in the form of a bicyclic lactone or lactam. This methodology was also used with slight modification to produce a *carba*-pyranose and -pyranosylamine (see Section 2.4).[102]

The synthesis and base-mediated intramolecular alkylation/cyclization of 6-deoxy-6-iodo-2,3:4,5-di-*O*-isopropylidene-D-glucose leading to the polyfunctionalised cyclopentane **145** has been reported. The reduction and acetylation of this compound to generate a 4a-*carba*-α-D-arabinofuranoside derivative was also described.[103]

The stereoselective construction of polyoxygenated carbocycles from carbohydrate templates using a samarium(II) iodide-promoted reductive coupling has been studied. The configuration of the hydroxy groups in the template (*e.g.* octa-6-enopyranuronate **146**) determines the stereo-outcome to a large extent (Scheme 24).[104] Carbocycles **147** (66% plus 21% epimer at*) and **148** (95% plus 5% di-epimer at*) were produced as the major diastereoisomeric components in one-step samarium(II) iodide-promoted and palladium(0)-catalysed ring contractions of methyl 2,3,4-tri-*O*-benzyl-6,7-dideoxy-α-D-*gluco*-hept-6-enopyranoside and methyl 2-*O*-benzyl-6,7-dideoxy-3,4-*O*-isopropylidene-β-D-*galacto*-hept-6-ynopyranoside respectively.[105]

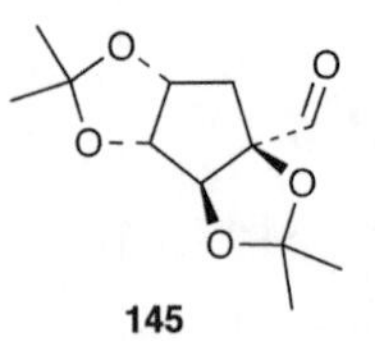

145 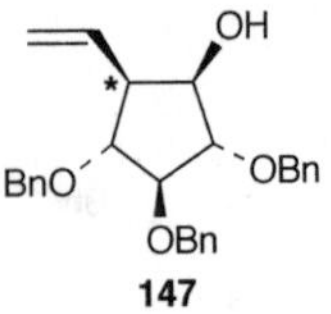   **147** 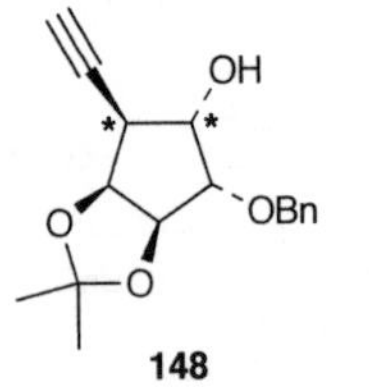   **148**

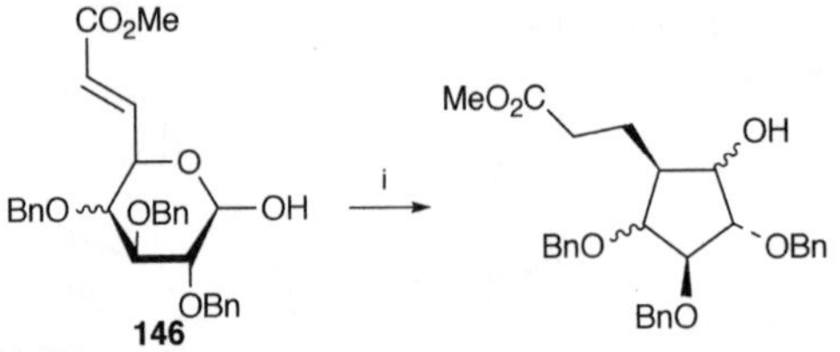

146

Reagents: i, SmI$_2$, THF/MeOH

Scheme 24

The 4-amino-5-(hydroxymethyl)cyclopentane-1,2,3-triol **149** was prepared in five steps from 2,3,4,6-tetra-*O*-benzyl-D-glucose for glycosidase inihibition studies. The results suggested that the hydrochloride salt of **149** acts in the enzyme active site as an analogue of a protonated (at the glycosidic oxygen) β-gluco-side.[106] Related aminocyclitols **150** have also been prepared, one of which (X = OMe) was found to be a potent α-mannosidase inhibitor.[107] The synthesis of alkoxyaminocyclitols **78**, **152** and **153** by stannyl radical addition of sugar (D-glucose or -galactose) derived oxime ethers **151** has been reported. The ratio of these diastereoisomeric compounds formed was found to vary with the substituent R. Both *gluco-* and *galacto-*derived cyclitols **78** were observed to undergo reductive ring-expansion to afford DNJ or 1-deoxygalactostatin respectively (see Section 1.3.5).[70]

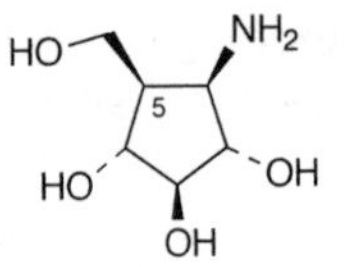

149 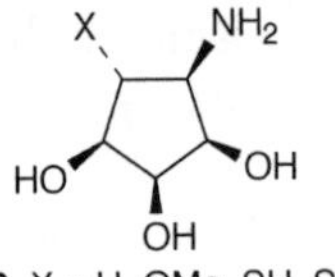   **150** X = H, OMe, SH, SEt 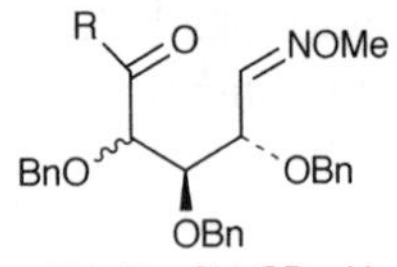   **151** R = CH$_2$OBn, H

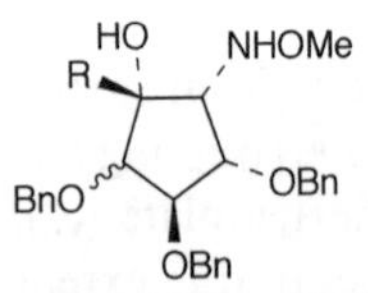

152   R = CH$_2$OBn, H 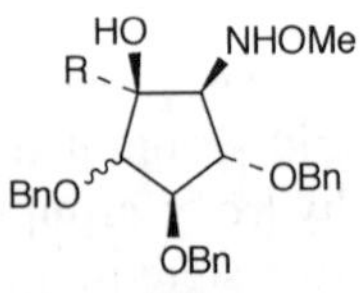   **153** 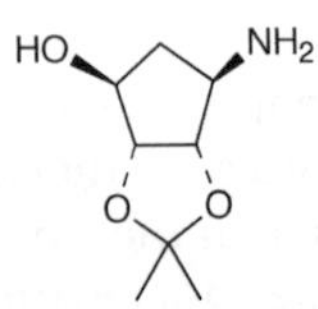   **154** 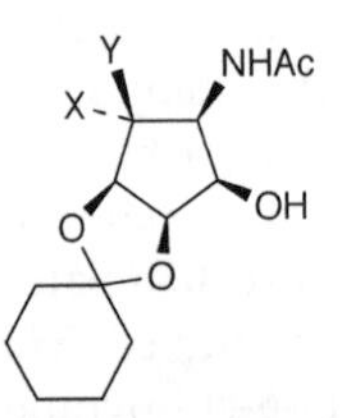   **155** X = H, Y = OH
156 X = SMe, Y = H

Carbocyclic nucleoside precursor **154** was synthesized in 36% overall yield from an *N*-hydroxy-D-alanine derivative by a reaction pathway involving oxidation, hetero-Diels–Alder cycloadditon of the nitroso intermediate to cyclopentadiene to give a chiral 1,2-oxazine, *syn*-dihydroxylation, and reductive cleavage.

The enantiomer of **154** can be made by using the corresponding L-alanine derivative as the chiral auxiliary. Compound **154** is also a precursor to the antiviral carbocyclic nucleoside, neplanocin A, and the antitumour agent, 5'-noraristeromycin.[108]

Three deoxy-analogues of mannostatin A have been synthesized conventionally from known precursors. 1-Deoxy- and 3-deoxymannostatin A were produced from precursor **155**, while a 2-deoxy-analogue was elaborated from **156**. These compounds displayed weak or no inhibition of α-mannosidase.[109]

The preparation of optically pure carbocycles from bicylic lactam templates has been described in a review looking at the use of chiral bicyclic lactams in auxiliary-based asymmetric reactions.[110]

2.2 Inositols and Related Compounds. – Labelled *myo*-[1-^{13}C]-inositol was synthesized from D-[6-^{13}C]-glucose for the purpose of investigating the biosynthetic pathway for *myo*-inositol in the protozoan parasite, *Leishmania donovani*.[111]

The formation of (±)-1,2-ethylidene-*myo*-inositol as a significant (33%) and previously unreported product of the orthoformylation of *myo*-inositol, using triethyl orthoformate and *p*-toluenesulfonic acid in DMSO, has been described and a mechanism proposed.[112] The migration of a benzoyl protecting group from the 4-position in 4-*O*-benzoyl-*myo*-inositol 1,3,5-orthoformate, the kinetic product of benzoylation using sodium hydride in DMF, to the thermodynamically favoured 2-position has also been reported. This novel 1(axial)→3(equatorial) intramolecular acyl migration occurs in the presence of excess sodium hydride.[113] The usefulness of the 1,3,5-orthoformate as a synthetic intermediate was demonstrated by the synthesis of *myo*-inositol 4,6-carbonate which was made effectively on a gram-scale in three steps by a sequence of regioselective *tert*-butyldimethylsilylation at O-2, reaction with carbonyldiimidazole to form the cyclic 4,6-carbonate and final deprotection by treatment with aqueous TFA. This compound readily forms complexes with cations.[114]

Three *myo*-inositol derivatives, (±)-3,6-di-*O*-valproyl-1,2:4,5-di-*O*-isopropylidene-*myo*-, (±)-3,6-di-*O*-valproyl-4,5-*O*-isopropylidene-*myo*- and (±)-3,6-di-*O*-valproyl-*myo*-inositol have been prepared as valproic acid (2-propylpentanoic acid) prodrugs for anticonvulsant studies. The last compound, resulting from acid hydrolysis of either of the first, was found to exhibit an anticonvulsant activity four times higher than that reported for valproic acid.[115]

An interesting and novel transformation was observed during the attempted fluorination of D-1,2-*O*-isopropylidene-3,4,5-tri-*O*-benzyl-*myo*-inositol using diethylaminosulfur trifluoride (DAST). The bicyclic anhydro-inositol **157** was isolated in 94% yield, presumably the result of fluoride-induced debenzylative intramolecular anhydride formation as depicted in Scheme 25.[116]

Ferulic acids isolated from rice bran have been coupled to partially protected *myo*-inositols by standard means to produce seven different feruloyl-*myo*-inositols, *e.g.* **158**. These feruloyl-inositols were subsequently tested for antioxidation and anticarcinogenesis activity with the result that **158** was the most active compound of the group.[117]

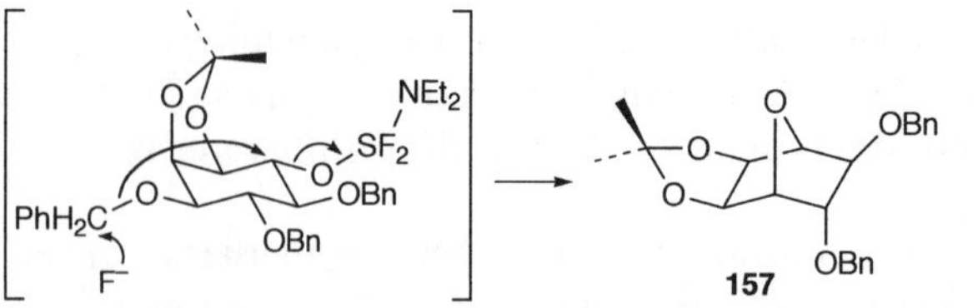

Scheme 25

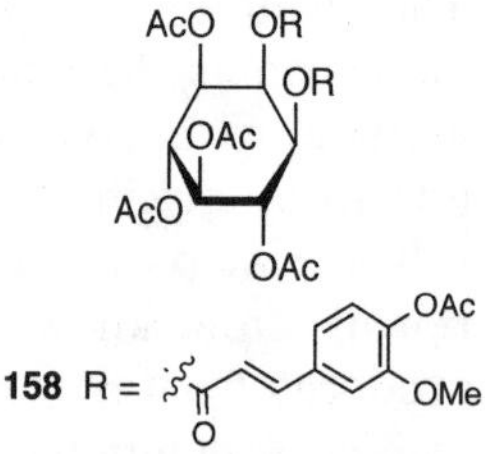

158 R = [structure shown]

As a development in the field of solution phase polymer-supported organic synthesis, 'xanthate transfer technology' has been used to prepare polymer-bound protected sugar derivatives which are soluble in organic solvents. For example, protected D-*chiro*-inositol **159** was elaborated into polystyrene bound **160** by radical polymerisation with 15 equivalents of styrene in the presence of 12 mol% lauroyl peroxide.[118]

Cyclization of known 2,6-di-O-benzyl-L-*arabino*-aldohexos-5-ulose with DBU afforded the inosose **161** in 60% yield by intramolecular aldol condensation. Hydrogenolytic removal of the O-benzyl groups followed by stereoselective reduction with sodium borohydride produced *epi*-inositol in high yield.[119]

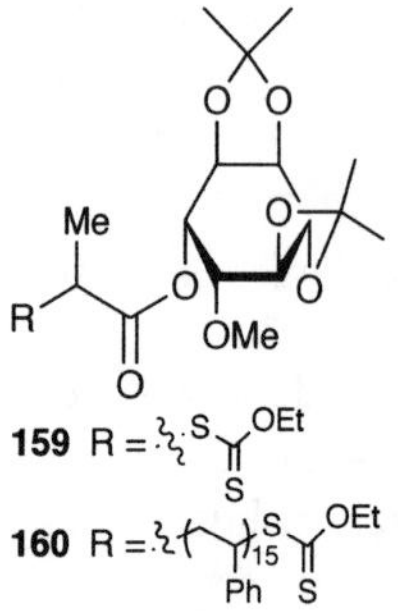

159 R = [structure]

160 R = [structure]

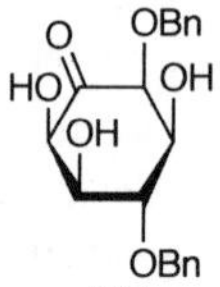

161 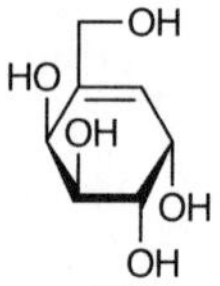   **162** 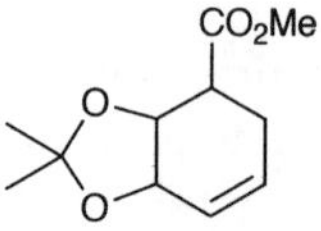   **163**

The syntheses of *neo*- and *allo*-inositol, conduritol E and the patented conduritol herbicide MK 7607 (**162**) have been reported from non-carbohydrate sources. *neo*-Inositol was efficiently synthesized from bromobenzene in a six step process that made use, *inter alia*, of an enzymatic *syn*-dihydroxylation with toluene dioxygenase.[120] On the other hand, *allo*-inositol and the conduritols were all made from **163** which was itself prepared by the fragmentation (C-1/C-7 bond scission) of a 7-norbornanone derivative with methoxide in methanol. Precursor **163** was also used to prepare a range of *carba*-sugars (see Section 2.4).[121]

The diene **164** was prepared from D-galactitol *via* Tebbe methylenation of the corresponding dialdehyde and subsequently subjected to ring-closing metathesis in the presence of Grubbs' (ruthenium) catalyst. Conduritol A derivative **165** was produced in 90% yield, and the methodology was equally successful in generating the isomeric conduritols E and F by reaction of the dienes derived from D-mannitol and -glucitol respectively.[122] (−)-Conduritols E (**166**) and F (**167**) have also been synthesized by Ramberg–Bäcklund reaction of the C-2-epimeric sulfones derived from 1,6-thio-D-mannitol and -glucitol (Scheme 26). 1,4-Di-O-methyl-(+)-conduritol B (**168**) was obtained analogously from the corresponding L-iditol derivative.[123] Conduritol F has been extracted as a natural product along with its 3-*O*- and 4-*O*-β-D-glucosides from the leaves of *Cynanchum liukiuense*. The regioisomeric 2-*O*-glucoside was found in a related species.[124]

Hitherto unreported (−)-conduramine E (**170**) was synthesized from **169** in *ca.* 50% overall yield as outlined in Scheme 27. Enzyme (esterase) induced monodeacetylation was followed by Mitsunobu-type inversion with an *N*-nucleophile in the presence of acetic acid. *endo*-Cyclization with concomitant epoxide-opening produced a urethane bearing the correct stereochemistry at the four chiral centres. Base treatment revealed the (−)-conduramine [for (+)-conduramine E see Vol. 29, p. 240].[125]

In the interest of devising simple routes to starting materials commonly used in synthesis from sustainable resources, catechol has been made chemoenzymatically in two steps from glucose in reasonable yield. 2-Deoxy-*scyllo*-inosose (**171**) was formed by incubation of D-glucose with recombinant 2-deoxy-*scyllo*-inosose synthase and a hexokinase, and this material was reductively dehydrated to catechol by treatment with HI in acetic acid.[126]

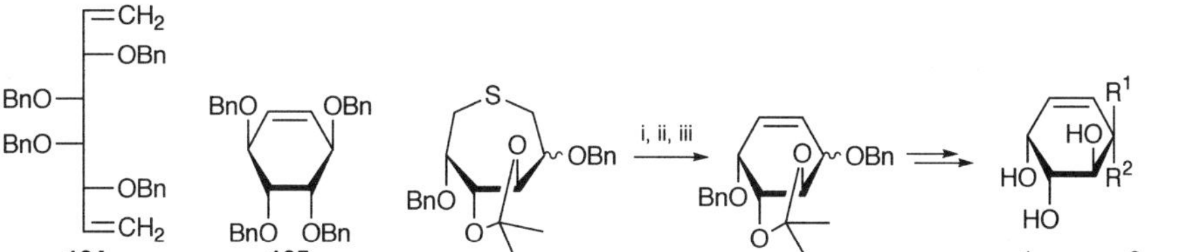

Reagents: i, mCPBA; ii, KOH, CCl₄, Bu^t^OH, H₂O; iii, SnCl₄, DCM

Scheme 26

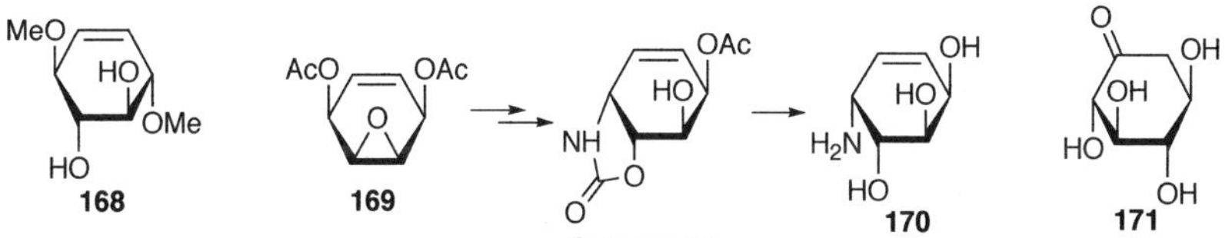

Scheme 27

2.3 Inositol Phosphates and Derivatives.

– A mixture of all possible *myo*-inositol monophosphates was obtained by lyophilizing a solution of *myo*-inositol and inorganic phosphate and heating the residue in a sealed tube at 66 °C for six days. The resulting mixture was separated by anion-exchange chromatography into four fractions containing 2-, 5-, racemic 1- plus 3-, and racemic 4- plus 6-phosphates in the ratio 3:1.5:1:3.[127] The first polyethylene glycol-linked dimers of *myo*-inositol 1,4,5-trisphosphate **172** have been prepared for use as multivalent ligands for inositol 1,4,5-trisphosphate receptors in the endoplasmic reticulum,[128] and the first total syntheses of *myo*-inositol 1,2,3,4- and 1,2,3,6-tetrakisphosphates from *p*-benzoquinone have been completed, taking advantage of the C₂-symmetry of conduritol B intermediates.[129] The dephosphorylation pathway of the hexakisphosphate of *myo*-inositol (phytic acid) by phytases from wheat bran of *Triticum aestivum* has also been examined.[130]

Following the trend of recent years attention has again been directed towards the synthesis of phosphatidylinositol and inositolphosphoglycan derivatives. A synthesis of 1-*O*-phosphatidyl-D-*myo*-inositols amenable to large scale production has been reported. It involves the treatment of a mixture of 2,3,4,5,6-penta-*O*-benzyl-D-*myo*-inositol and a diacylglycerophosphoric acid with 2,4,6-triisopropylbenzenesulfonyl chloride in pyridine followed by catalytic (Pd/C) hydrogenolysis.[131] The phosphatidyl-*myo*-inositols **173** were prepared and studied as substrates for phosphatidylinositol 3-kinase. Only those with short chain fatty acids, that are perhaps able to mimic the folded natural arachadonic acid group at this position, were phosphorylated at the 3-hydroxyl by ATP. Surprisingly, the enzyme could accept either the natural or unnatural *sn*-epimer.[132] The syntheses of 1-*O*-(1,2-di-*O*-pentadecanoyl-L-phosphatidyl)-D-*myo*-inositol 3- and 5-phosphates and 3,5-bisphosphate from (±)-2,4-di-*O*-benzyl-*myo*-inositol 1,3,5-orthoformate[133] and (±)-4,6-di-*O*-benzyl-3,5-di-*O*-(4-methoxybenzyl)-*myo*-inositol[134] by resolution with (1*R*)-(+)-camphor dimethylacetal have been described. In an investigation of signalling pathways by another group, the same 5-phosphate was made by the selective phosphorylation of 3,4-*O*-(tetraisopropyldisilox-1,3-diyl)-6-*O*-levulinoyl-D-*myo*-inositol 5-(dibenzyl phosphate) at O-1.[135] A resin-bound fluorescently labelled derivative **174** was also prepared in this study for use in affinity chromatography.

172

173 R = CO(CH$_2$)$_n$Me; n = 0, 2, 6, 10, 14, 16   **174** R = CO(CH$_2$)$_5$NHCO(CH$_2$)$_5$NH—[benzofurazan]—NO$_2$

Lengthy (>20 step) syntheses of 1-*O*-(1,2-di-*O*-pentadecanoylphosphatidyl)-D-*myo*-inositol 3,4- and 4,5-bisphosphates and 3,4,5-trisphosphate starting from L-(−)-quebrachitol have also been reported.[136]

The *myo*-inositol acceptor 1-*O*-allyl-2,3,4,5-tetra-*O*-benzyl-D-*myo*-inositol was synthesized from *meso*-2,5-di-*O*-benzoyl-*myo*-inositol for use in the total synthesis of a heptaose mimetic, a GPI anchor compound of *T. brucei* (see Vol. 32, p.71. ref. 156). The inositol was desymmetrized using the bis(spiroketal) protected intermediate **175** to enable the stepwise glycosylation at O-6 and attachment of a phosphoglyceride moiety at O-1. Cleavage of the 'dispoke' was effected by oxidizing the thiophenyl substituents with mCPBA and then treating the product disulfone with lithium hexamethyldisilazide in THF.[137] In a similar synthesis

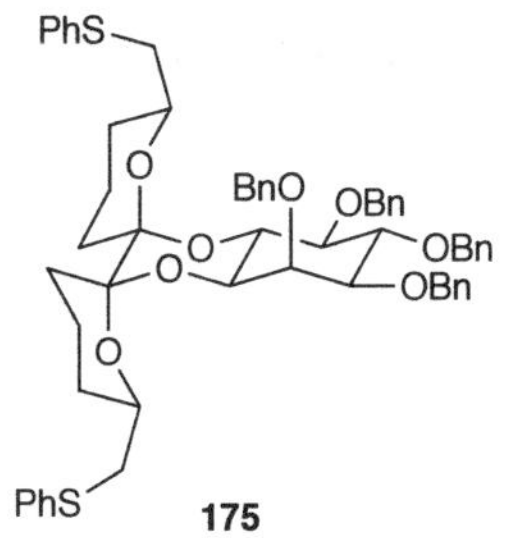

Reagents: i, Et$_3$B-BuCO$_2$BEt$_2$, hexane; ii, Bu$_2$Sn(acac)$_2$, toluene; iii, MntCOCl, NMI

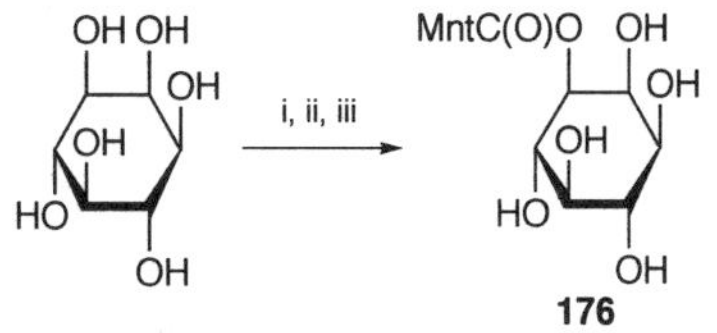

Scheme 28

of a pseudohexasaccharide IPG, a *myo*-inositol building block was again prepared with a view to regioselective glycosylation at O-6 and phosphorylation at O-1 (see also Chapter 4). Of interest was the use of a boron–tin exchange reaction to achieve a key selective acylation: the 1-*O*-(−)-menthoxycarbonyl derivative **176** was prepared from *myo*-inositol according to Scheme 28 to overcome the insolubility problem of *myo*-inositol in most organic solvents.[138]

A *myo*-inositol-containing IPG pseudopentasaccharide bearing the conserved linear structure of GPI anchors was synthesized for intracellular signalling studies. The structure of this IPG is detailed in Chapter 4; the key building blocks were a mannotriosyl donor and an inositol-containing O-6-glucosylated pseudodisaccharide acceptor with the C-1 and C-2 hydroxyls accessible for cyclo-phosphorylation.[139] Similarly the *myo*-inositol phosphoglycan fragments **177** found in *Leishmania* parasites have been synthesized by coupling an inositol-containing disaccharide with a mannobiosyl donor.[140]

The antigenic core of the glycoconjugate lipoarabinomannan (LAM) from *Mycobacterium tuberculosis*, 2,6-dimannosyl-*myo*-inositol 1-phosphate, was synthesized by stepwise regioselective mannosylation at the free C-2- and then at the C-6-hydroxyl in **178**. Removal of the 1-acetate provided for phosphorylation at the correct site.[141]

New preparative syntheses of 6-*O*-(2-amino-2-deoxy-α-D-glucopyranosyl)-D-*chiro*-inositol 1-phosphate and a 1,2-cyclic phosphate derivative have been reported. Differential protection was achieved by initially locking the four equatorial hydroxyls of D-*chiro*-inositol as cyclic Tipds ethers and then generating a *cis*-epoxide from the two remaining *trans*-diaxial hydroxyls under Mitsunobu conditions. Lewis acid catalysed *trans*-diaxial opening with allyl alcohol allowed for selective protection at O-1 and O-6.[142]

3-(Hydroxymethyl)-bearing phosphatidylinositol ether lipid analogues **179** (X = P) and carbonate surrogates (X = C) block the phosphatidylinositol 3-kinase (PI3-K) enzyme and inhibit cancer cell growth. Chain extension was effected by (Swern) oxidation at C-3, subsequent Wittig methylenation and hydroboration using 9-BBN to afford a separable C-3-epimeric mixture of (hydroxymethyl)inositols.[143] In another lengthy synthesis from L-(−)-quebrachitol (*ca.* 20 steps) the 3,4-dideoxy variant **180** of **179** (X = P) was made with the incorporation of two Barton deoxygenation reactions. This dideoxy analogue was found to be 18-fold more potent than its (mono) 3-deoxy counterpart in the inhibition of PI3-K.[144]

177 R = R^{1}O–P(=O)(OH)→6)-α-D-Man*p*-(1→3)-α-D-Man*p*-(1→4)-α-D-Glc*p*-NH$_2$

178

179 R = CH$_2$OH, R^1 = H, R^2 = OH, X = P(OH)
     R = H, R^1 = CH$_2$OH, R^2 = OH, X = P(OH)
     R = CH$_2$OH, R^1 = H, R^2 = OH, X = C
     R = H, R^1 = CH$_2$OH, R^2 = OH, X = C
180 R = CH$_2$OH, R^1 = R^2 = H, X = P(OH)

2.4   Carba-sugars. – Carba-furanoses and carba-furanosides are covered in Section 2.1. The first direct transformation of 1-*S*-, -*Se*- and -*C*-hex-5-enopyranosides into carbocycles with retention of the aglycon has been reported. Phenyl 2,3,4-tri-*O*-benzyl-6-deoxy-1-thio- and -1-seleno-β-D-*gluco*-hex-5-enopyranosides were converted into the corresponding cyclitols **181** in over 80% yield by treatment with five equivalents of triisobutylaluminium (TIBAL) in toluene at 50 °C. The 2,4,6-trimethoxyphenyl β-*C*-glycoside gave, on the other hand, a mixture of **182** and **183** in a 1.5:1 ratio under identical conditions.[145] Applying the same chemistry to an α-*C*-(2-furanyl)-glycoside afforded the cyclitol **184** and this was elaborated further into methyl 5a-*carba*-β-D-idopyranoside by methylation, ozonolytic cleavage of the furan moiety and subsequent reduction of the resulting ester group with concomitant O-debenzylation. The enantiomeric 5a-*carba*-β-L-idopyranoside was obtained *via* a different route from methyl 2,3,4-tri-*O*-benzyl-6-deoxy-β-D-*gluco*-hex-5-enopyranoside by reaction with isopropoxytitanium(IV) chloride and subsequent elaboration of the resulting carbocycle **185**. TIBAL-promoted rearrangement followed by oxidation of the newly formed secondary hydroxyl was a reaction motif used in the generation of the 5′a-*carba*-disaccharide **186**[146] as well as (1→4) ether-linked di- and tri-saccharide mimetics **188** from unsaturated thioglycoside di- and tri-saccharide precursors **187**.[147,148]

2,5-Di-*O*-benzyl-3,4-*O*-isopropylidene-D-mannitol was ring closed to afford a 9:1 *cis/trans* mixture of **189** in high yield by (Swern) oxidation and radical cyclization of the dialdehyde using samarium iodide in *tert*-butanol. The *cis*-diastereoisomer was selectively converted into a *cis*-cyclic sulfate which underwent cleavage upon treatment with potassium *tert*-butoxide to afford a cyclo-

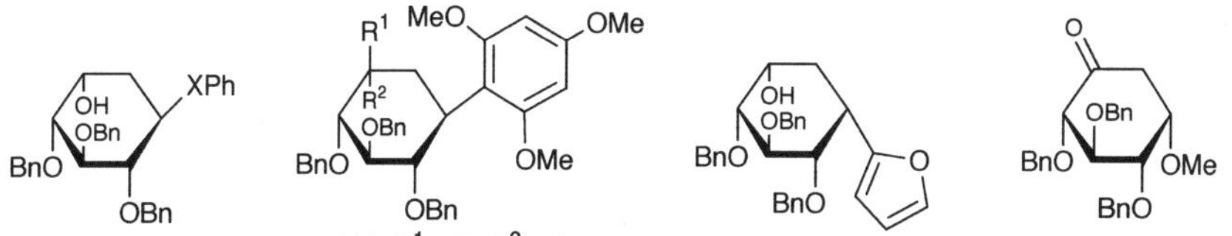

181 X = S, Se

182 R¹ = H, R² = OH
183 R¹ = OH, R² = H

184

185

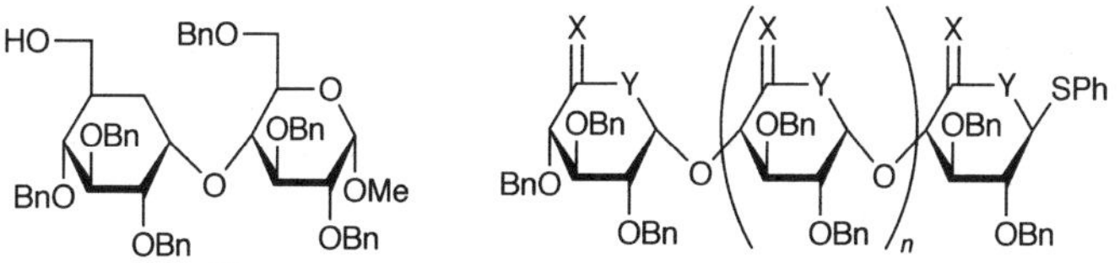

186

188 *n* = 0, 1; X = CH₂, Y = O
187 *n* = 0, 1; X = O, Y = CH₂

hexanone derivative. Further development of this intermediate, as shown in Scheme 29, allowed the production of either a 5a-*carba*-L-galactose (>9:1 L-/D-) or a 5a-*carba*-L-fucopyranose derivative.[149]

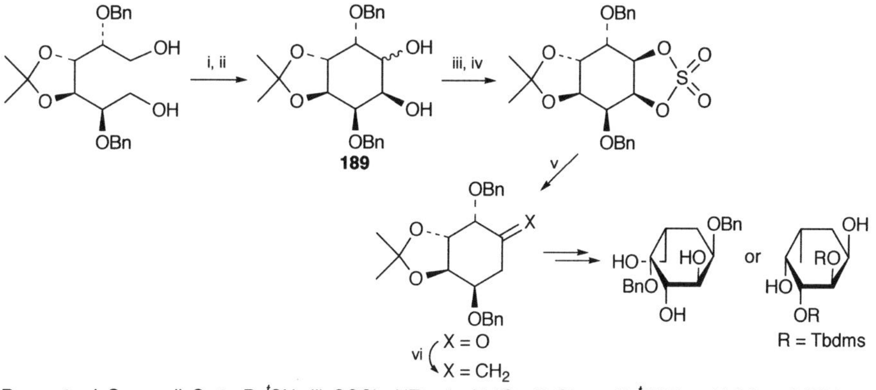

189

vi (X = O
   (X = CH₂

Reagents: i, Swern; ii, SmI₂, Bu^{t}OH; iii, SOCl₂, NEt₃; iv, NaIO₄, RuCl₃; v, Bu^{t}OK then H₂SO₄; vi, Wittig.

Scheme 29

The pseudo-sugars 6a-*carba*-α- (**191**) and -β-D-fructopyranose (**192**) have been prepared in six steps from an enzymatically resolved homochiral cyclohexane-triol building block *via* a common olefinic intermediate **190** by varying the dihydroxylation strategy (osmium tetroxide *vs. m*-CPBA respectively).[150] Another non-carbohydrate based route to 5a-*carba*-sugars has been reported which utilizes a 7-norbornanone-derived cyclohexenoid intermediate. The preparation of *carba*-α- and -β-galacto-, -α-talo- and -α-fucopyranose derivatives were described.[151]

D-Glyceraldehyde has been used as a starting material for the synthesis of pseudo β-D-gulopyranose and -gulopyranosylamine. As detailed in section 2.1 the method shown in Scheme 23 for the synthesis of *carba*-β-D-xylofuranose and -xylofuranosylamine was modified by the omission of a C-C cleavage to result in the formation of the above cyclohexane derivatives.[102]

The pseudo-sugars 5a-*carba*-α-DL-fucopyranosylamine and -galactopyranosylamine were made as racemates from known 2,3,4,6-tetra-*O*-acetyl-5a-*carba*-α-L-glucopyranosyl bromide by standard methods; the amino group was introduced *via* azide displacement of the bromide, inversion at C-4 was effected by displacement of a mesylate group and deoxygenation at C-6 (in the case of the former) by treatment of a 6-iodo intermediate with tributyltin hydride. The former *carba*-sugar was found to be a strong inhibitor of bovine kidney α-fucosidase.[152]

New syntheses of (+)-valienamine (**193**) and (+)-valiolamine (**194**), pseudo-aminosugars found in valiolamycins, acarbose and trestatines, have been described starting from a derivative of D-xylono-1,4-lactone in twelve or more steps,[153] and using much the same methodology the same group has reported the first total synthesis of pyralomycin 1c (**195**) from an L-arabinono-1,4-lactone derivative.[154] The synthesis of an α-hydroxymethyl-substituted α,β-unsaturated cyclohexenone intermediate common to all these compounds (with the exception of the stereochemistry of the hydroxyl groups) has been previously described in Vol. 32, p. 357. A preparation of *carba*-sugars from sugar lactones *via* spiro aldonic acid orthoesters has been reported. The cyclitol derivative **196**, a versatile synthon for the synthesis of valiolamine and related compounds, was synthesized in high overall yield (70% over 4 steps) from 2,3,4,6-tetra-*O*-benzyl-D-glucono-1,5-lactone according to Scheme 30. Enol ether formation from the sugar ortho ester with trimethylaluminium and the zinc chloride-promoted intramolecular aldol cyclization were the key steps in the synthesis.[155] Identical methodology used on the corresponding D-mannono-1,5-lactone gave rise to **197** as the major product (together with 9% of the C-5 epimer) and reduction of this cyclohexanone with sodium borohydride produced a roughly 1:1 mixture of the 'anomeric' 5-hydroxy-*carba*-sugars.[156]

A difluorinated *carba*-sugar **198** has been synthesized as a potential herbicide in 22% overall yield from D-ribose. The key step is an intramolecular 1,3-dipolar nitrone cycloaddition to generate an isoxazolidine as shown in Scheme 31.[157] Another cycloaddition strategy has been used to generate 1,4-glycosidically linked 'mono-*carba*-disaccharides'. A dieneophilic aglycon underwent stereospecific Diels–Alder cycloaddition to maleic anhydride and the resulting cycloadduct **199** was elaborated into a 5a-*carba*-α-L-idopyranose unit to form the pseudo-disaccharide **200**.[158]

In addition to three known gabosines, three new gabosines (L, N and O: **201–203** respectively) have been detected as secondary metabolites of *Streptomyces* strains.[159]

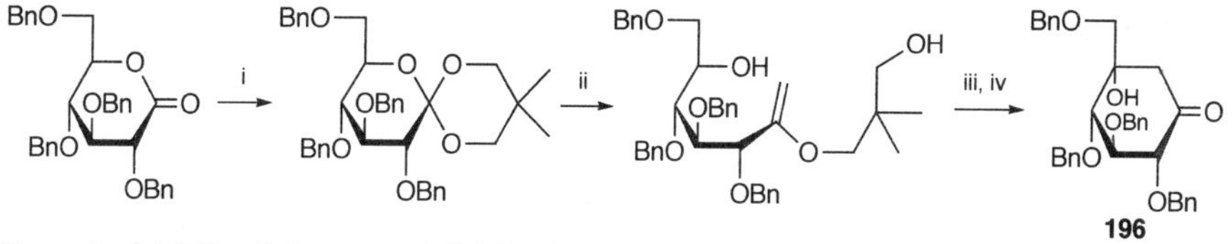

Reagents: i, 2,2-Dimethylpropane-1,3-diol, TmsOMe, TmsOTf, Toluene; ii, AlMe₃; iii, Ac₂O, DMSO; iv, ZnCl₂, THF/H₂O

Scheme 30

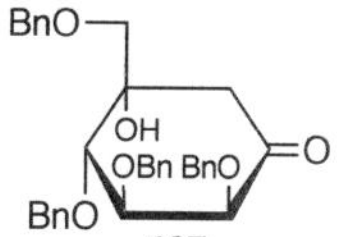

197

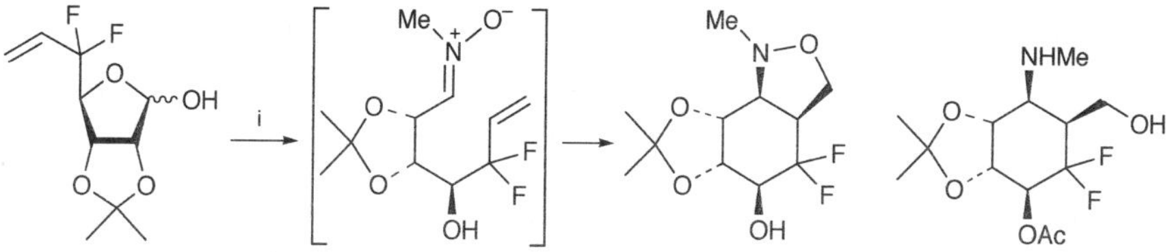

Reagents: i, MeNHOH.HCl, pyridine

Scheme 31

198

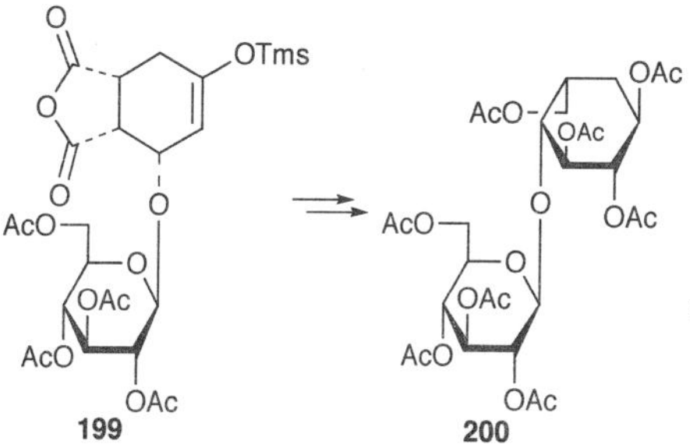

199   200 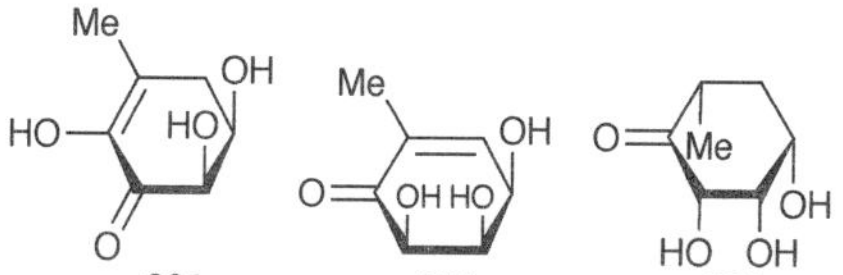   201   202   203

2.5 Aminoglycoside Antibiotics. – In a single study fourteen different amino-glycosides of the kanamycin/gentamycin, neomycin and astromycin classes were acetylated with aminoglycoside acetyltransferases from *Actinomycete* sp. in order to determine the degree of activity retained after acetylation.[160] The acetylation of istamycin and miconomycin antiobiotics at the 6′-methylamino group by a novel aminoglycoside 6′-acetyltransferase from *Actinomycete* sp. was also reported.[161] See also Chapter 20.

Neamine analogues have been prepared in several studies in order to explore their potential as small molecule antitumour and anti-HIV agents, as well as bacterial enzyme inhibitors. 5-*O*-Alkylated neamines with polyamine functionality in the side-chain have been synthesized from 5-*O*-allylated precursors and these were found to exhibit high binding affinity for oncogene fusion proteins.[162] The acylation of neamines at N-6′ with aromatic units was carried out to explore the effect of these substituents on the interaction of neamine aminoglycosides with HIV RNA. Pyrene substituents were found to impart the most effective level of binding (sub-micromolar) amongst those groups studied.[163] The tethering of a neamine aminoglycoside to an adenosine residue by the incorporation of a 6- or 7-carbon (methylene) spacer between O-3′ and O-5″(of the ribofuranose moiety) resulted in high binding affinities for bacterial 3′-phosphotransferases. These enzymes are known to inactivate aminoglycoside antibiotics, and their inhibition would serve to depress bacterial resistance to this class of compounds.[164]

Various analogues of neomycin B have also been synthesized in an effort to develop inhibitors of viral RNA. Compounds **204** and **205** were made from neamine, again *via* a 5-*O*-allylated precursor. Oxidative cleavage and reduction of the resulting aldehyde with sodium borohydride gave the acceptor **206** which was then glycosylated with the appropriately protected glycosyl donors (see also Chapter 3). Neither **204** nor **205** were found to be effective as anti-HIV agents.[165] Two analogues of neomycin B, each lacking one of the 2,6-diamino-sugars were prepared for binding studies by ribofuranosylation (at O-5) of a 4-*O*-amino-glycosyl inositol 'disaccharide' obtained from paromomycin, and by the coupling of a cyclitol derivative (at O-5) with an appropriate 3-*O*-diaminoglycosyl-ribofuranosyl acetate.[166] The neomycin-acridine conjugate **208** was found to be a potent inhibitor of Rev-Rev Response Element (RRE) binding. Synthesized from neomycin B in 5 steps according to Scheme 32, this compound has definite anti-HIV potential.[167]

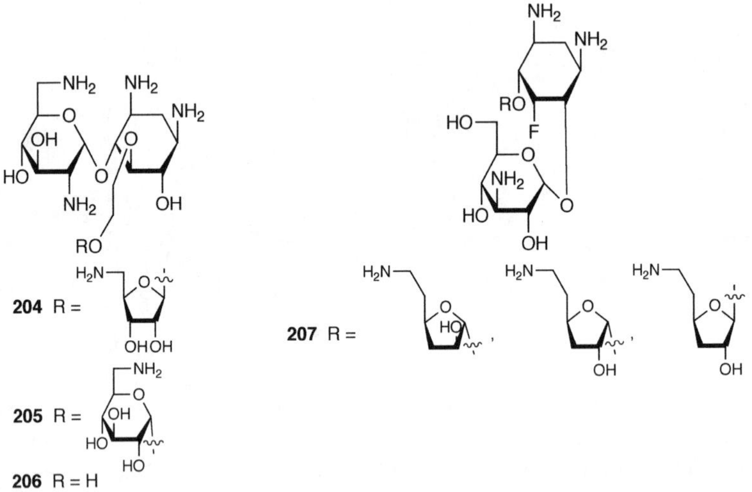

Kanamycin A has also received some attention. Three analogues **207** each containing a 6-amino-6-deoxyglycofuranose moiety have been prepared by coupling with appropriate glycosyl chlorides and fluorides. All three compounds, however, were inactive in antibacterial screens.[168]

A new guanidinylation reagent, *N*,*N*-di-Boc-*N*′-triflylguanidine, has been used to efficiently convert multiamine-containing aminoglycoside antibiotics into fully guanidinylated analogues in the presence of water under mild conditions. 'Guanidinylglycoside' analogues of kanamycin A and B, paromomycin, tobramycin and neomycin B were prepared in this way; the last two were found

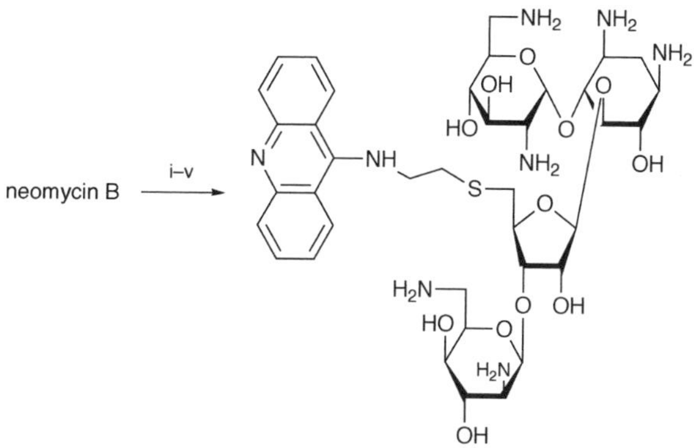

208

Reagents: i, Boc$_2$O, DMF, Et$_3$N, H$_2$O; ii, triisopropylbenzenesulfonyl chloride, pyr; iii, 2-aminoethanethiol, EtOH, EtONa; iv, 9-phenoxyacridine, phenol; v, 4M HCl, dioxane, 1,2-ethanedithiol

Scheme 32

to possess HIV inhibitory activities 100-fold greater than their parent amino-glycosides.[169] The chemistry of the guanidinylation reaction is covered in more detail in Chapters 9 and 10.

2.6 Quinic Acid Derivatives. – Two quinic acid derivatives have been isolated from natural sources. 1,3-Di-*O*-*trans*-feruoylquinic acid (**209**) was extracted from the roots of the grass *Brachiaria*[170] and 3,5-dicaffeoyl-*muco*-quinic acid (**210**) was isolated from aerial parts of the Korean culinary vegetable *Aster scaber*. The latter exhibited potent inhibition of HIV-1 integrase (IC$_{50}$ of 7 µg mL^{-1}).[171]

Quinic acid has been used as a starting material in the preparation of various enzyme inhibitors. Compounds **211** were made in multi-step syntheses as putative inositol monophosphatase inhibitors.[172] The phosphitamidites **212–214** were also made in lengthy reaction sequences from quinic acid and subsequently coupled to a 5′-*O*-unprotected cytidine derivative. Oxidation at phosphorus followed by deprotection gave *e.g.* phosphate diester **215** from **214**. These donor substrate analogues proved to be good inhibitors of α-(2→6)-sialyl transferases (K_i 10^{-4} – 10^{-5}).[173]

The vicinal diol **216**, prepared in 10 steps from quinic acid, was elaborated into *N*-alkylated 2-*epi*-valienamines **217** in variable (16–98%) yield by reaction with Viehe's salt followed by Pd(0)-catalysed coupling with a range of primary (*e.g.* R^1 = H; R^2 = Et, Bu, Hept, Oct, cyclohexyl, Bn) and secondary amines (*e.g.* R^1 = R^2 = Bu, cyclohexyl, *i*Pr).[174] Carbocyclic influenza neuraminidase inhibitors **219** (*n* = 3–8) with a cyclic amine side-chain have also been prepared *via* Pd(0)-catalysed coupling of acetate **218** with the corresponding cyclic amines.[175]

The use of pentafunctional quinic acid as a polyoxygenated scaffold for combinatorial synthesis has also been described.[176]

2.7 Other Cyclitol Derivatives. – The distribution of products in the gas-phase acid-induced ring-opening with methanol, as well as condensed phase methanolysis, of cyclohexene oxides (*e.g.* **220**) bearing remote O-functionality has been studied. Products and ratios varied widely depending on the reaction phase, the acid and in the case of gas-phase reaction, on the pressure used.[177]

Several polycyclic cyclitol derivatives have been reported. The decalinic compound **221** was synthesized from 3,4,6-tri-*O*-acetyl-D-glucal. As shown in Scheme 33, intramolecular Diels–Alder cycloaddition of the triene set up a tricyclic hex-5-enopyranoside derivative which underwent Ferrier rearrangent with mercuric sulfate and aqueous sulfuric acid to afford **221**.[178] Exhaustive osmylation of

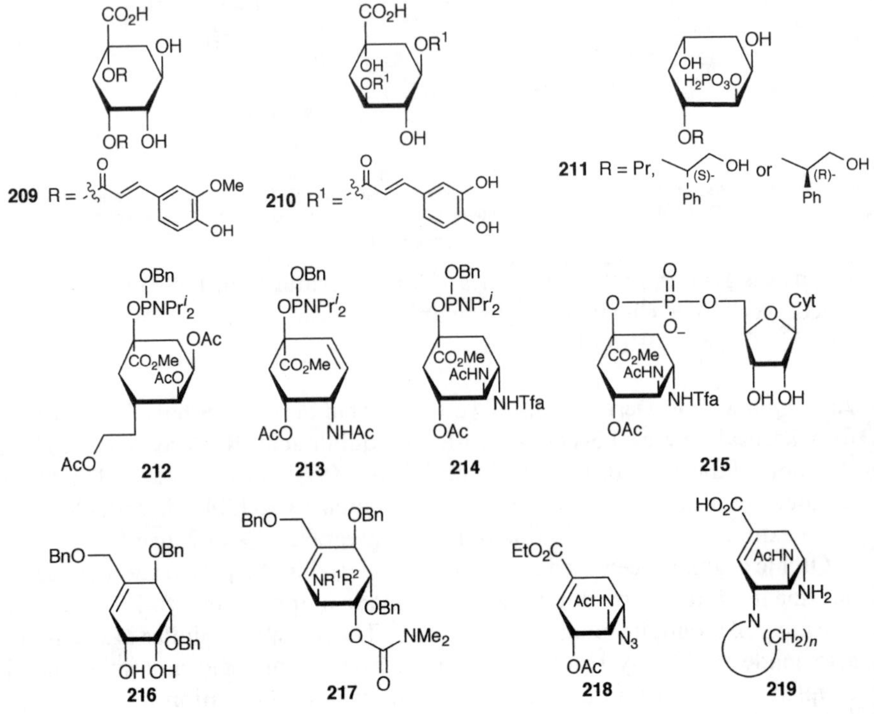

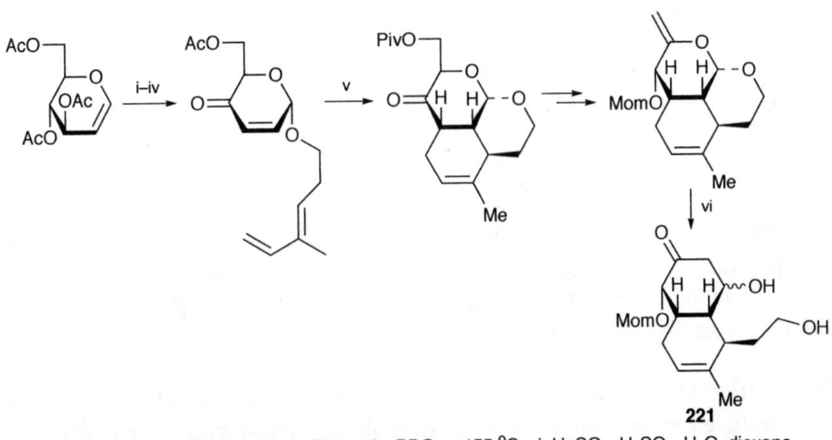

Reagents: i, BF₃.Et₂O; ii, NaOMe; iii, PivCl, pyr; iv, PDC; v, 155 °C; vi, HgSO₄, H₂SO₄, H₂O, dioxane

Scheme 33

tricyclic diene **222** was observed to give rise exclusively to the *exo*-hydroxylation product **223** in 66% yield. This compound was elaborated further to give, for example, the decahydronaphthalene-1,2,3,4,5,6,7,8-octaol **224** which proved to be a selective and potent α-glucosidase inhibitor.[179]

D-Glucose-derived radical precursors **225** and **226** have been subjected to carbocyclization reactions to yield cyclic polyol derivatives in enantiomerically pure form. They underwent 6-*exo-trig* and 7-*exo-dig* radical cyclizations upon treatment with tributyltin hydride and AIBN to give the cyclo-hexitol and -heptitols, **227** and **228** respectively, in moderate yields.[180]

Diastereoisomeric 'cyclooctanic *carba*-sugars' have been prepared in thirteen steps from D-glucose. The key step in the synthesis was a TIBAL-promoted Claisen rearrangement of the intermediate 2,6-anhydro-3,4,5-tri-*O*-benzyl-1,7,8-trideoxy-D-*gluco*-oct-1,7-dienitol (Scheme 34) which created the cyclooctene ring. Hydroboration of the double bond with borane-THF followed by chain extension by oxidation, methylenation and a second hydroboration, gave rise to a separable mixture of cyclooctano *carba*-sugars with D-*glycero*-D-*ido*- and L-*glycero*-D-*ido*-configurations.[181]

220 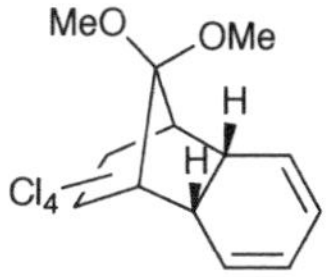   **222** 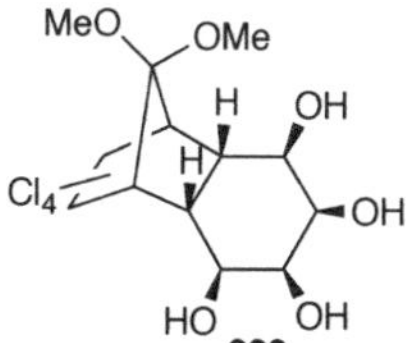   **223** 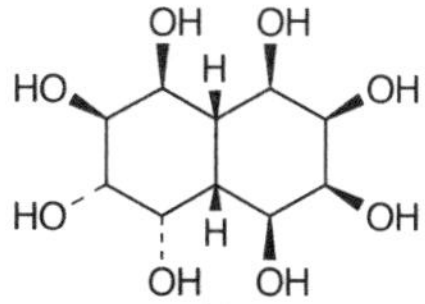   **224**

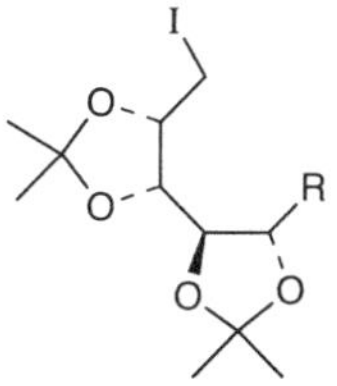

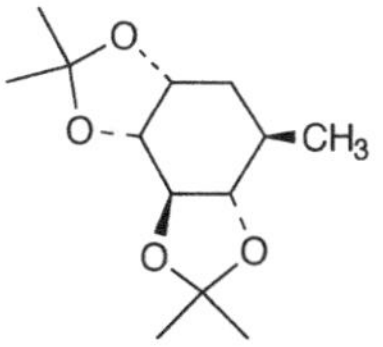

225 R = CH$_2$=CH$_2$

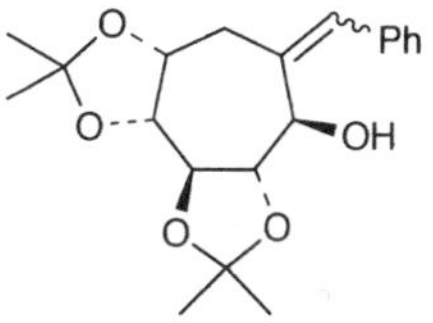

226 R = HO— —Ph

227   **228**

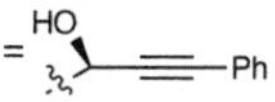

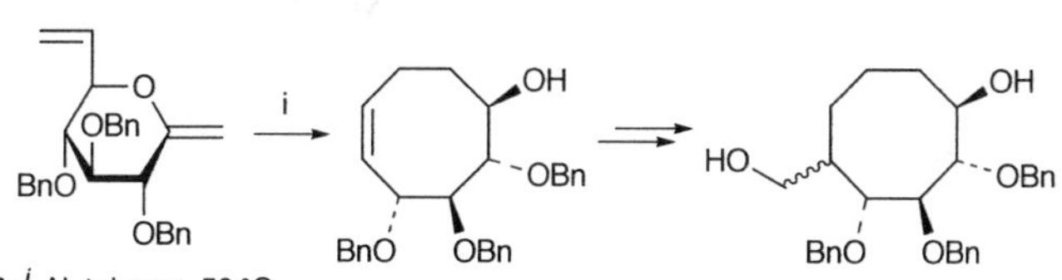

Reagents: i, Bu$^i_3$Al, toluene, 50 °C

Scheme 34

References

1. N. Maezahi, Y.-X. Li, K. Ohkubo, S. Goda, C. Iwata and T. Tanaka, *Tetrahedron*, 2001, **56**, 4405.
2. J. F. Hoeffler, C. Pale-Grosdemange and M. Rohmer, *Tetrahedron*, 2000, **56**, 1485.

3.   I. Sakamoto, K. Ichimura and H. Ohrui, *Biosci. Biotechnol. Biochem.*, 2000, **64**, 1915.
4.   T. Duvold and M. Rohmer, *Tetrahedron Lett.*, 2000, **41**, 3875.
5.   M. A. Lucas, O. T. K. Nguyen, C. H. Schiesser and S.-L. Zheng, *Tetrahedron*, 2000, **56**, 3995.
6.   V. Bouchez, I. Stasik and D. Beaupere, *Carbohydr. Res.*, 2000, **323**, 213.
7.   V. P. Roldan, V. A. Daier, B. Goodman, M. I. Santoro, J. C. Gonzalez, N. Calisto, S. R. Signorella and L. F. Sala, *Helv. Chim. Acta*, 2000, **83**, 3211.
8.   A. W. Heinen, J. A. Peters and H. van Bekkum, *Carbohydr. Res.*, 2000, **328**, 449.
9.   Y. Zhang and Z. Li, *Xiandai Huagong*, 1999, **19**, 26.
10.   J.-C. Thiery, C. Frechou and G. Demailly, *Tetrahedron Lett.*, 2000, **41**, 6337.
11.   M. J. Arevalo, M. Avalos, R. Babiano, P. Cintas, M. B. Hursthouse, J. L. Jimenez, M. E. Light, I. Lopez and J. C. Palacios, *Tetrahedron*, 2000, **56**, 1247.
12.   F. Leclercq, C. Dubertret, B. Pitard, D. Scherman and J. Herscovici, *Bioorg. Med. Chem. Lett.*, 2000, **10**, 1233.
13.   Y. Hirata and T. Nakagawa, *Bull. Chem. Soc. Jpn.*, 2000, **73**, 1905.
14.   A. Fragoso, M. L. Kahn, A. Castineiras, J.-P. Sutter, O. Kahn and R. Cao, *Chem. Commun.*, 2000, 1547.
15.   B. T. Cho, Y. S. Chun and W. K. Yang, *Tetrahedron: Asymm.*, 2000, **11**, 2149.
16.   Y. Le Merrer, L. Gauzy, C. Gravier-Pelletier and J.-C. Depezay, *Bioorg. Med. Chem.*, 2000, **8**, 307.
17.   M. Kurszewska, E. Skorupa, R. Kasprzykowska, P. Sowinski and A. Wisniewski, *Carbohydr. Res.*, 2000, **326**, 241.
18.   M. E. Maier and C. Hermann, *Tetrahedron*, 2000, **56**, 557.
19.   G. A. Freeman, D. W. Selleseth, J. L. Rideout and R. J. Harvey, *Nucleosides, Nucleotides, Nucleic Acids*, 2000, **19**, 155.
20.   C. M. El Hilali, T. A. Lamosse and G. R. Gray, *Carbohydr. Res.*, 2000, **329**, 189.
21.   S. M. Andersen, I. Lundt and J. Marcusson, *J. Carbohydr. Chem.*, 2000, **19**, 717.
22.   S. K. Readman, S. P. Marsden and A. Hodgson, *Synlett*, 2000, 1628.
23.   R. Persky and A. Albeck, *J. Org. Chem.*, 2000, **65**, 5632.
24.   T. K. Chakraborty, S. Ghosh, M. H. V. Ramana Rao, A. C. Kunwar, H. Cho and A. K. Ghosh, *Tetrahedron Lett.*, 2000, **41**, 10121.
25.   T. K. Chakraborty, *Tetrahedron Lett.*, 2001, **65**, 6441.
26.   C. Brown, R. W. Morston, P. F. Quigley and S. M. Roberts, *J. Chem. Soc. Perkin Trans. 1*, 2000, 1809.
27.   S. Chatti, M. Bortolussi and A. Loupy, *Tetrahedron Lett.*, 2000, **41**, 3367.
28.   D.-P. Pham-Huu, M. Petruskova, J. N. BeMiller and L. Petrus, *J. Carbohydr. Chem.*, 2000, **19**, 93.
29.   S. Shuto, Y. Yahiro, S. Ichikawa and A. Matsuda, *J. Org. Chem.*, 2000, **65**, 5547.
30.   Y. LeMerrer, L. Poitout and J.-C. Depezay, *Methods Mol. Med.*, 1999, **23**, 227 (*Chem. Abstr.*, 2000, **132**, 293 925).
31.   H. Hausler, R. P. Kawakami, E. Mlaker, W. B. Severn, T. M. Wrodnigg and A. E. Stutz, *J. Carbohydr. Chem.*, 2000, **19**, 435.
32.   N. Asano, R. J. Nash, R. J. Molyneux and G. W. J. Fleet, *Tetrahedron: Asymm.*, 2000, **11**, 1645.
33.   T. D. Butters, R. A. Dwek and F. M. Platt, *Chem. Rev.*, 2000, **100**, 4683.
34.   S. Laschat and T. Dickner, *Synthesis*, 2000, 1781.
35.   B. Becker, *J. Carbohydr. Chem.*, 2000, **19**, 253.
36.   B. M. Malle, I. Lundt and R. H. Furneaux, *J. Carbohydr. Chem.*, 2000, **19**, 573.
37.   Z. Chen, R. W. Wang, L. L. Zhu and X. T. Liang, *Chin. Chem. Lett.*, 1999, **10**, 651

(*Chem. Abstr.*, 2000, **132**, 208 058).

38. H. Søhoel, X. Liang and M. Bols, *Synlett*, 2000, 347.
39. B. V. Ernholt, I. B. Thomsen, A. Lohse, I. W. Plesner, K. B. Jensen, R. G. Hazell, X. Liang, A. Jakobsen and M. Bols, *Chem. Eur. J.*, 2000, **6**, 278.
40. B. W. Lee, I.-Y. Jeong, M. S. Yang, S. U. Choi and K. H. Park, *Synthesis*, 2000, 1305.
41. R. D. Smith and N. R. Thomas, *Synlett*, 2000, 193.
42. V. Glaçon, M. Benazza, D. Beaupère and G. Demailly, *Tetrahedron Lett.*, 2000, **41**, 505.
43. B. B. Lohray, V. Bhushan, M. Chatterjee, Y. Jayamma and G. Prasuna, *Res. Chem. Intermed.*, 1999, **25**, 887 (*Chem. Abstr.*, 2000, **132**, 152 063).
44. I. McCort, S. Fort, A. Duréault and J.-C. Depezay, *Bioorg. Med. Chem.*, 2000, **8**, 135.
45. C. W. Holzapfel, R. Crous, H. F. Greyling and G. H. Verdoorn, *Heterocycles*, 1999, **51**, 2801 (*Chem. Abstr.*, 2000, **132**, 180 786).
46. E. Dechaux, P. Savy, S. Bouyain, C. Monneret and J.-C. Florent, *J. Carbohydr. Chem.*, 2000, **19**, 485.
47. A. Kilonda, F. Compernolle, K. Peeters, G. J. Joly, S. Toppet and G. J. Hoornaert, *Tetrahedron*, 2000, **56**, 1005.
48. G. J. Joly, K. Peeters, H. Mao, T. Brossette, G. J. Hoornaert and F. Compernolle, *Tetrahedron Lett.*, 2000, **41**, 2223.
49. B. B. Lohray, V. Bhushan, G. Prasuna, Y. Jayamma, M. A. Raheem, P. Papireddy, B. Umadevi, M. Premkumar, N. S. Lakshmi and K. Narayanareddy, *Indian J. Chem.*, 1999, **38B**, 1311 (*Chem. Abstr.*, 2000, **132**, 308 572).
50. C. Saotome, Y. Kanie, O. Kanie and C.-H. Wong, *Bioorg. Med. Chem.*, 2000, **8**, 2249.
51. T. M. Wrodnigg, W. Gaderbauer, P. Greimel, H. Häusler, F. K. Sprenger, A. E. Stütz, C. Virgona and S. G. Withers, *J. Carbohydr. Chem.*, 2000, **19**, 975.
52. L. Cipolla, B. La Ferla, F. Peri and F. Nicotra, *Chem. Commun.*, 2000, 1289.
53. S. M. Andersen, C. Ekhart, I. Lundt and A. E. Stütz, *Carbohydr. Res.*, 2000, **326**, 22.
54. G. F. Painter and A. Falshaw, *J. Chem. Soc. Perkin Trans. 1*, 2000, 1157.
55. V. Jäger, L. Bierer, H.-Q. Dong, A. M. Palmer, D. Shaw and W. Frey, *J. Heterocycl. Chem.*, 2000, **37**, 455.
56. G. Pandey and M. Kapur, *Tetrahedron Lett.*, 2000, **41**, 8821.
57. H. Yokoyama, K. Otaya, H. Kobayashi, M. Miyazawa, S. Yamaguchi and Y. Hirai, *Org. Lett.*, 2000, **2**, 2427.
58. P. Szolcsányi, T. Gracza, M. Koman, N. Prónayová and T. Liptaj, *Chem. Commun.*, 2000, 471.
59. P. Szolcsányi, T. Gracza, M. Koman, N. Prónayová and T. Liptaj, *Tetrahedron: Asymm.*, 2000, **11**, 2579.
60. S. Fouace and M. Therisod, *Tetrahedron Lett.*, 2000, **41**, 7313.
61. G. Pistia and R. I. Hollingsworth, *Carbohydr. Res.*, 2000, **328**, 467.
62. A. Tatibouët, P. Rollin and O. R. Martin, *J. Carbohydr. Chem.*, 2000, **19**, 641.
63. P. de Armas, F. Garcia-Tellado and J. J. Marrero-Tellado, *Org. Lett.*, 2000, **2**, 3513.
64. G. B. Evans, R. H. Furneaux, G. J. Gainsford, V. L. Schramm and P. C. Tyler, *Tetrahedron*, 2000, **56**, 3053.
65. K. Suzuki and H. Hashimoto, *Carbohydr. Res.*, 2000, **323**, 14.
66. Y. Xu, S.-R. Choi, M.-P. Kung and H. F. Kung, *Nucl. Med. Biol.*, 1999, **26**, 833 (*Chem. Abstr.*, 2000, **132**, 180 782).

67.   T. Fuchss and R. R. Schmidt, *Synthesis*, 2000, 259.
68.   A. J. Humphrey, S. F. Parsons, M. E. B. Smith and N. J. Turner, *Tetrahedron Lett.*, 2000, **41**, 4481.
69.   M. Godskesen, I. Lundt and I. Sotofte, *Tetrahedron: Asymm.*, 2000, **11**, 567.
70.   T. Kiguchi, K. Tajiri, I. Ninomiya and T. Naito, *Tetrahedron*, 2000, **56**, 5819.
71.   E. Poupon, B.-X. Luong, A. Chiaroni, N. Kunesch and H.-P. Husson, *J. Org. Chem.*, 2000, **65**, 7208.
72.   F. Schieweck and H.-J. Altenbach, *J. Carbohydr. Chem.*, 2000, **19**, 647.
73.   B. M. Folkersen, I. Lundt, C. Foged and J. S. Valsborg, *J. Labelled Compd. Radiopharm.*, 1999, **42**, 1145 (*Chem. Abstr.*, 2000, **132**, 180 787).
74.   E. A. Severino and C. R. D. Correia, *Org. Lett.*, 2000, **2**, 3039.
75.   K.-Y. Lee, Y.-H. Kim, C.-Y. Oh and W.-H. Ham, *Org. Lett.*, 2000, **2**, 4041.
76.   C. Arribas, M. C. Carreno, J. L. Garcia-Ruano, J. F. Rodriguez, M. Santos and M. A. Sanz-Tejedor, *Org. Lett.*, 2000, **2**, 3165.
77.   T. Sifferlen, A. Defoin, J. Streith, D. Le Nouen, C. Tarnus, I. Dosbaa and M.-J. Foglietti, *Tetrahedron*, 2000, **56**, 971.
78.   S. U. Hansen and M. Bols, *J. Chem. Soc. Perkin Trans. 1*, 2000, 911.
79.   A. Lohse, H. H. Jensen, P. Bach and M. Bols, *J. Chem. Soc. Perkin Trans. 1*, 2000, 659.
80.   H. H. Jensen, A. Lohse, B. O. Petersen, J. O. Duus and M. Bols, *J. Chem. Soc. Perkin Trans. 1*, 2000, 667.
81.   G. Mehta and N. Mohal, *Tetrahedron Lett.*, 2000, **41**, 5747.
82.   G. Mehta and N. Mohal, *Tetrahedron Lett.*, 2000, **41**, 5741.
83.   M. Shibano, D. Tsukamoto, R. Fujimoto, Y. Masui, H. Sugimoto and G. Kusano, *Chem. Pharm. Bull.*, 2000, **48**, 1281.
84.   N. Asano, M. Nishida, M. Miyauchi, K. Ikeda, M. Yamamoto, H. Kizu, Y. Kameda, A. A. Watson, R. J. Nash and G. W. J. Fleet, *Phytochemistry*, 2000, **53**, 379.
85.   K. Ikeda, M. Takahashi, M. Nishida, M. Miyauchi, H. Kizu, Y. Kameda, M. Arisawa, A. A. Watson, R. J. Nash, G. W. J. Fleet and N. Asano, *Carbohydr. Res.*, 2000, **323**, 73.
86.   N. Asano, H. Kuroi, K. Ikeda, H. Kizu, Y. Kameda, A. Kato, I. Adachi, A. A. Watson, R. J. Nash and G. W. J. Fleet, *Tetrahedron: Asymm.*, 2000, **11**, 1.
87.   J. Choi, E. N. Jeong and K. H. Chun, *Bull. Korean Chem. Soc.*, 1999, **20**, 1123 (*Chem. Abstr.*, 2000, **132**, 194 586).
88.   F. Cardona, A. Goti, S. Picasso, P. Vogel and A. Brandi, *J. Carbohydr. Chem.*, 2000, **19**, 585.
89.   Y. St-Denis and T. H. Chan, *Can. J. Chem.*, 2000, **78**, 776.
90.   J. Marco-Contelles and C. Alhambra Jiménez, *Revista de la Sociedad de México*, 1999, **43**, 83 (*Chem. Abstr.*, 2000, **132**, 108 204).
91.   E. Lorthiois, M. Meyyappan and A. Vasella, *Chem. Commun.*, 2000, 1829.
92.   K. Ikeda, M. Takahashi, M. Nishida, M. Miyauchi, H. Kizu, Y. Kameda, M. Arisawa, A. A. Watson, R. J. Nash, G. W. J. Fleet and N. Asano, *Carbohydr. Res.*, 2000, **323**, 73.
93.   M. Shibano, D. Tsukamoto, R. Fujimoto, Y. Masui, H. Sugimoto and G. Kusano, *Chem. Pharm. Bull.*, 2000, **48**, 1281.
94.   C. Saotome, Y. Kanie, O. Kanie and C.-H. Wong, *Bioorg. Med. Chem.*, 2000, **8**, 2249.
95.   A. Dondoni, P. P. Giovannini and A. Marra, *Tetrahedron Lett.*, 2000, **41**, 6195.
96.   F. Cardona, I. Robia and P. Vogel, *J. Carbohydr. Chem.*, 2000, **19**, 555.

97. F. J. Duff, V. Vivien and R. H. Wightman, *Chem. Commun.*, 2000, 2127.
98. H. Takahashi, H. Kittaka and S. Ikegami, *Yuki Gosei Kagaku Kyokaishi*, 2000, **58**, 120 (*Chem. Abstr.*, 2000, **133**, 4 852s).
99. J. Desiré and J. Prandi, *Eur. J. Org. Chem.*, 2000, 3075.
100. M. Seepersaud and Y. Al-Abed, *Tetrahedron Lett.*, 2000, **41**, 7801.
101. C. S. Callam and T. L. Lowary, *Org. Lett.*, 2000, **2**, 167.
102. G. Rassu, L. Auzzas, L. Pinna, L. Battistini, F. Zanardi, L. Marzocchi, D. Acquotti and G. Casiraghi, *J. Org. Chem.*, 2000, **65**, 6307.
103. J. Marco-Contelles and E. de Opazo, *Tetrahedron Lett.*, 2000, **41**, 5507.
104. T. Kan, S. Nara, T. Ozawa, H. Shirahama and F. Matsuda, *Angew. Chem. Int. Ed. Engl.*, 2000, **39**, 355.
105. J. M. Aurrecoechea, B. López and M. Arrate, *J. Org. Chem.*, 2000, **65**, 6493.
106. O. Boss, E. Leroy, A. Blaser and J.-L. Reymond, *Org. Lett.*, 2000, **2**, 151.
107. S. Ogawa and T. Morikawa, *Bioorg. Med. Chem. Lett.*, 2000, **10**, 1047.
108. B. T. Shireman and M. J. Millar, *Tetrahedron Lett.*, 2000, **41**, 9537.
109. S. Ogawa and T. Morikawa, *Eur. J. Org. Chem.*, 2000, 1759.
110. M. D. Groaning and A. I. Meyers, *Tetrahedron*, 2000, **56**, 9843.
111. P. Sahai, M. Chawla and R. A. Vishwakarma, *J. Chem. Soc. Perkin Trans. 1*, 2000, 1283.
112. W. Zhu and Z. Li, *Synth. Commun.*, 2000, **30**, 3823.
113. K. M. Sureshan and M. S. Shashidhar, *Tetrahedron Lett.*, 2000, **41**, 4185.
114. S. J. Angyal, *Carbohydr. Res.*, 2000, **325**, 313.
115. N. Bodor, S. C. Moon and L. Bruno-Blanch, *Pharmazie*, 2000, **55**, 184 (*Chem. Abstr.*, 2000, **133**, 4 854u).
116. S.-K. Chung, Y.-U. Kwon, Y.-H. Ahn, T.-H. Jeong and Y.-T. Chang, *Bull. Korean Chem. Soc.*, 2000, **21**, 274 (*Chem. Abstr.*, 2000, **132**, 334 690).
117. A. Hosoda, E. Nomura, A. Murakami, K. Koshimizu, H. Ohigashi, K. Mizuno and H. Taniguchi, *Bioorg. Med. Chem. Lett.*, 2000, **10**, 1439.
118. B. Quiclet-Sire, A. Wilczewska and S. Z. Zard, *Tetrahedron Lett.*, 2000, **41**, 5673.
119. V. Pistarà, P. L. Barili, G. Catelani, A. Corsaro, F. D'Andrea and S. Fisichella, *Tetrahedron Lett.*, 2000, **41**, 3253.
120. T. Hudlicky, N. Restrepo-Sanchez, P. D. Kary and L. M. Jaramillo-Gomez, *Carbohydr. Res.*, 2000, **324**, 200.
121. G. Mehta and S. Lakshminath, *Tetrahedron Lett.*, 2000, **41**, 3509.
122. L. Ackermann, D. El Tom and A. Fürstner, *Tetrahedron*, 2000, **56**, 2195.
123. V. Cerè, G. Mantovani, F. Peri, S. Pollicino and A. Ricci, *Tetrahedron*, 2000, **56**, 1225.
124. F. Abe, T. Yamauchi, K. Honda and N. Hayashi, *Chem. Pharm. Bull.*, 2000, **48**, 1090.
125. D. Spielvogel, J. Kammerer, M. Keller and H. Prinzbach, *Tetrahedron Lett.*, 2000, **41**, 7863.
126. K. Kakinuma, E. Nango, F. Kudo, Y. Matsushima and T. Eguchi, *Tetrahedron Lett.*, 2000, **41**, 1935.
127. E. Tarelli, *J. Carbohydr. Chem.*, 2000, **19**, 233.
128. A. M. Riley and B. V. L. Potter, *Chem. Commun.*, 2000, 983.
129. O. Plettenburg, S. Adelt, G. Vogel and H.-J. Altenbach, *Tetrahedron: Asymm.*, 2000, **11**, 1057.
130. T. Nakano, T. Joh, K. Narita and T. Hayakawa, *Biosci. Biotechnol. Biochem.*, 2000, **64**, 995.
131. R. Aneja and S. G. Aneja, *Tetrahedron Lett.*, 2000, **41**, 847.

132. N. Morisaki, K. Morita, A. Nishikawa, N. Nakatsu, Y. Fukui, Y. Hashimoto and R. Shirai, *Tetrahedron*, 2000, **56**, 2603.

133. J. R. Falck, U. M. Krishna, K. R. Katipally, J. H. Capdevila and E. M. Ulug, *Tetrahedron Lett.*, 2000, **41**, 4271.

134. J. R. Falck, U. M. Krishna and J. H. Capdevila, *Bioorg. Med. Chem. Lett.*, 2000, **10**, 1711.

135. Y. Watanabe and H. Ishikawa, *Tetrahedron Lett.*, 2000, **41**, 8509.

136. L. Qiao, Y. Hu, F. Nan, G. Powis and A. P. Kozikowski, *Org. Lett.*, 2000, **2**, 115.

137. D. K. Baeschlin, A. R. Chaperon, L. G. Green, M. G. Hahn, S. J. Ince and S. V. Ley, *Chem. Eur. J.*, 2000, **6**, 172.

138. M. Martin-Lomas, M. Flores-Mosquera and J. L. Chiara, *Eur. J. Org. Chem.*, 2000, 1547.

139. M. Martín-Lomas, N. Khiar, S. García, J.-L. Koessler, P. M. Nieto and T. W. Rademacher, *Chem. Eur. J.*, 2000, **6**, 3608.

140. K. Ruda, J. Lindberg, P. J. Garegg, S. Oscarson and P. Konradsson, *Tetrahedron*, 2000, **56**, 3969.

141. G. Anilkumar, M. R. Gilbert and B. Fraser-Reid, *Tetrahedron*, 2000, **56**, 1993.

142. M. Martin-Lomas, M. Flores-Mosquera and N. Khiar, *Eur. J. Org. Chem.*, 2000, 1539.

143. Y. Hu, L. Qiao, S. Wang, S. Rong, E. J. Meuillet, M. M. Berggren, A. Gallegos, G. Powis and A. P. Kozikowski, *J. Med. Chem.*, 2000, **43**, 3045.

144. Y. Hu, E. J. Meuillet, L. Qiao, M. M. Berggren, G. Powis and A. P. Kozikowski, *Tetrahedron Lett.*, 2000, **41**, 7415.

145. M. Sollogoub, J.-M. Mallet and P. Sinaÿ, *Angew. Chem. Int. Ed. Engl.*, 2000, **39**, 362.

146. M. Sollogoub, A. J. Pearce, P. Sinaÿ and P. Sinaÿ, *Tetrahedron: Asymm.*, 2000, **11**, 283.

147. A. J. Pearce, J.-M. Mallet and P. Sinaÿ, *Heterocycles*, 2000, **52**, 819 (*Chem. Abstr.*, 2000, **132**, 279 409).

148. A. J. Pearce, R. Chevalier, J.-M. Mallet and P. Sinaÿ, *Eur. J. Org. Chem.*, 2000, 2203.

149. M. Carpintero, C. Jaramillo and A. Fernández-Mayoralas, *Eur. J. Org. Chem.*, 2000, 1285.

150. I. Hubrecht, E. van der Eycken and J. van der Eycken, *Synlett*, 2000, 971.

151. G. Mehta, N. Mohal and S. Lakshminath, *Tetrahedron Lett.*, 2000, **41**, 3505.

152. S. Ogawa, R. Sekura, A. Maruyama, H. Yuasa and H. Hashimoto, *Eur. J. Org. Chem.*, 2000, 2089.

153. K. Tatsuta, H. Mukai and M. Takahashi, *J. Antibiotics*, 2000, **53**, 430.

154. K. Tatsuta, M. Takahashi and N. Tanaka, *J. Antibiotics*, 2000, **53**, 88.

155. H. Ohtake and S. Ikegami, *Org. Lett.*, 2000, **2**, 457.

156. H. Ohtake, X.-L. Li, M. Shiro and S. Ikegami, *Tetrahedron*, 2000, **56**, 7109.

157. S. Jiang, G. Singh and A. S. Batsanor, *Tetrahedron: Asymm.*, 2000, **11**, 3873.

158. N. S. Trotter, D. S. Larsen, R. J. Stoddley and S. Brooker, *Tetrahedron Lett.*, 2000, **41**, 8957.

159. Y.-Q. Tang, C. Maul, R. Höfs, I. Sattler, S. Grabley, X.-Z. Feng and R. Thiericke, *Eur. J. Org. Chem.*, 2000, 149.

160. K. Hotta, A. Sunada, Y. Ikeda and S. Kondo, *J. Antibiotics*, 2000, **53**, 1168.

161. A. Sunada, Y. Ikeda, S. Kondo and K. Hotta, *J. Antibiotics*, 2000, **53**, 1416.

162. S. J. Sucheck, W. A. Greenberg, T. J. Tolbert and C.-H. Wong, *Angew. Chem. Int. Ed. Engl.*, 2000, **39**, 1080.

163. K. Hamasaki, M.-C. Woo and A. Veno, *Tetrahedron Lett.*, 2000, **41**, 8327.

164. M. Liu, J. Haddad, E. Azucena, L. P. Kotra, M. Kirzhner and S. Mobashery, *J. Org. Chem.*, 2000, **65**, 7422.
165. N. Nishizono and V. Nair, *Nucleosides, Nucleotides, Nucleic Acids*, 2000, **19**, 283.
166. Y. Ding, E. E. Swayse, S. A. Hofstadler and R. H. Griffey, *Tetrahedron Lett.*, 2000, **41**, 4049.
167. S. R. Kirk, N. W. Luedtke and Y. Tor, *J. Am. Chem. Soc.*, 2000, **122**, 980.
168. Y. Kobayashi, T. Ohgami, K. Ohtsuki and T. Tsuchiya, *Carbohydr. Res.*, 2000, **329**, 325.
169. T. J. Baker, N. W. Luedtke, Y. Tor and M. Goodman, *J. Org. Chem.*, 2000, **65**, 9054.
170. P. Wenzl, A. L. Chaves, J. E. Mayer, I. M. Rao and M. G. Nair, *Phytochemistry*, 2000, **55**, 389.
171. H. C. Kwon, C. M. Jung, C. G. Shin, J. K. Lee, S. U. Choi, S. Y. Kim and K. R. Lee, *Chem. Pharm. Bull.*, 2000, **48**, 1796.
172. J. Schulz, M. W. Beaton and D. Gani, *J. Chem. Soc. Perkin Trans. 1*, 2000, 943.
173. C. Schaub, B. Müller and R. R. Schmidt, *Eur. J. Org. Chem.*, 2000, 1745.
174. S. H. L. Kok and T. K. M. Shing, *Tetrahedron Lett.*, 2000, **41**, 6865.
175. W. Lew, H. Wu, X. Chen, B. J. Graves, P. A. Escarpe, H. L. MacArthur, D. B. Mendez and C. V. Kim, *Bioorg. Med. Chem. Lett.*, 2000, **10**, 1257.
176. C. W. Phoon and C. Abell, *J. Comb. Chem.*, 2000, **1**, 485 (*Chem. Abstr.*, 2000, **132**, 64 468).
177. F. Marianucci, G. Renzi, G. Amici and G. Roselli, *Tetrahedron*, 2000, **56**, 7513.
178. C. Taillefumier and Y. Chaplew, *Can. J. Chem.*, 2000, **78**, 708.
179. G. Mehta and S. S. Ramesh, *Chem. Commun.*, 2000, 2429.
180. J. Marco-Contelles and E. de Opazo, *Tetrahedron Lett.*, 2000, **41**, 5341.
181. W. Wang, Y. Zhang, M. Sollogoub and P. Sinaÿ, *Angew. Chem. Int. Ed. Engl.*, 2000, **39**, 2466.

19
Nucleosides

1 General

A review has discussed selective biocatalytic modifications of conventional nucleosides, carbocyclic nucleosides and C-nucleosides, including such topics as selective acylation of nucleosides, regioselective deacylation, and enzymic resolution of precursors of carbocyclic and C-nucleosides.[1] The enzymatic synthesis of antiviral nucleosides has also been reviewed, in Chinese.[2] A survey has been given of synthetic approaches to nucleosides in the unnatural L-series, and their potential use as antiviral and anticancer agents.[3]

A review on the synthesis of deuterionucleosides discusses the synthesis of deuteriated sugars, particularly ribose and 2-deoxyribose, prior to attachment of the base, and also deuteriation at the nucleoside level. Both single-site and multiple-site deuteriation is covered.[4]

An account from Wengel's laboratory covers the synthesis of bicyclic nucleosides and their use in conformational restriction of resultant oligonucleotides. Structural types such as 3'-C, 5'-C-linked bicyclic nucleosides, as in bicyclo-DNA, and 2'-O,4'-C-linked systems, as in 'Locked Nucleic Acid' (LNA) are reviewed to the end of 1999.[5]

De Clercq has reviewed the use of guanosine analogues as anti-herpes agents. Although mostly acyclonucleosides are covered, there are discussions of various types of carbocyclic nucleoside analogues, from cyclopropanes to cyclohexenyl compounds.[6] A review has also appeared covering the synthesis of coenzyme A analogues and derivatives, and their applications as mechanistic probes of enzymes utilizing CoA esters.[7]

2 Synthesis

The synthesis of nucleoside analogues with high β-selectivity from glycofuranosyl chlorides has been reported. Thus, for example, reaction of the stable crystalline 1 with the appropriate base and KOBu^t in DMSO gave the β-nucleoside 2 almost entirely. The use of DMSO seems to be essential, and it was speculated that an intermediate sulfonium species is involved. Although the process was not applied to β-D-ribofuranosyl nucleosides, several other cases were reported, such as the synthesis of **3** in 71% yield and with 32:1 β-selectivity.[8]

The D-erythrofuranosyl benzimidazole **4**, various related 2-substituted analogues and some compounds additionally halogenated at C-4 have been prepared by base–sugar coupling. The enantiomer of **4** was also made, since the analogous L-ribofuranosyl compound has anti-HCMV activity. Some of the novel compounds also proved effective against HCMV.[9]

The ribavirin analogue **5** has been prepared by conventional coupling using the silylated base, and was converted to its 5′-triphosphate for use in studies of an engineered protein kinase that does not use ATP as phosphate donor.[10] In connection with studies of compounds with Type 1 cytokine-inducing activity, the L-enantiomer **6** of ribivirin has been prepared from L-ribose; the 2′-deoxy-compound was made from **6** by deoxygenation, and the 5′-deoxy-L-*ribo*-compound, L-*xylo*-ribavirin and its 3′-deoxy-derivative were also all made by base–sugar condensation, with **6** displaying best potency.[11] The pyrimidinone **7** was prepared conventionally, for use in further chemistry (Section 13).[12] Standard coupling (silylated base, ribofuranosyl acetate, TmsOTf) was used to prepare **8**, convertible on ammonolysis to the 2-thiocytosine derivative **9**. Similar methods were used to make pyranosyl compounds such as **11** (Ar = Ph, *p*-tolyl, 2-naphthyl; D-*galacto*- and D-*xylo*-compounds also made),[13] but when condensations between salts of the bases and acetylated glycosyl bromides were used, the major products were bis-glycosylated compounds of type **10**, produced together with the normal nucleoside derivatives. However, ammonolysis of **10** produced products of type **11** in high yield.[13,14]

Standard coupling was used to prepare the 1-deazaadenosine derivative **12**, convertible to compounds of type **13**, used in studies of adenosine receptor binding.[15] Various pyrrolo[2,3-*d*]pyrimidinone nucleosides have been prepared and evaluated for their ability to enhance Type 2 and suppress Type 1 cytokines in human T cells.[16] The pyrido[2,3-*d*]pyrimidine nucleoside **14** has been prepared by base–sugar condensation; the 2′-deoxy-species was made by deoxygenation, the arabinofuranosyl analogue by a redox sequence on a derivative of **14**, and the xylofuranosyl compound by base–sugar condensation. The compounds were studied for anticancer activity, **14** being the best proliferation inhibitor, but the *xylo*-compound had selective activity against cancer cells.[17] 1-Deazaguanosine (**15**) has been prepared by modification of AICA riboside.[18]

The trisubstituted indole nucleosides **16** (X = Cl, Br) have been made as

4

5

6

7

8 **9**

10 **11**

12 **13**

14

15

analogues of TCRB, and similar 2′-deoxy- and 2′,3′-didehydro-2′,3′-dideoxy-compounds were also described, all having less activity against HCMV than does TCRB.[19] There has been a further report on 'stretched' tricyclic nucleosides related to TCRB, of type **17** (X = CH, Y = NHPr^i, NHcyclopropyl, SBn, SH, Cl) (see Vol. 32, p. 257), together with a range of similar analogues **17** (X = N) and the N^3-ribosylated species (see also Vol. 33, p. 276–277).[20]

16

17

18 **19**

20

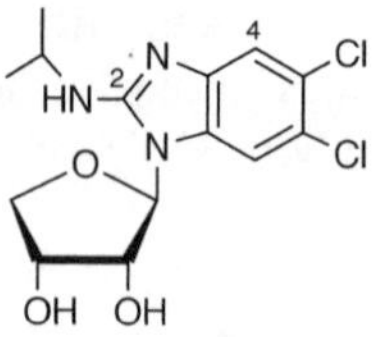

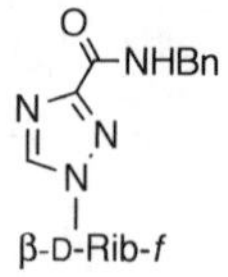

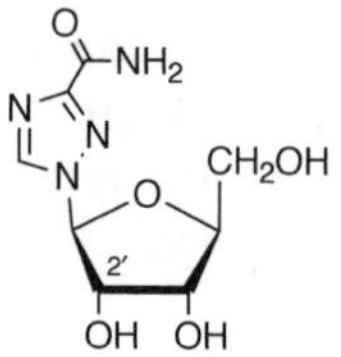

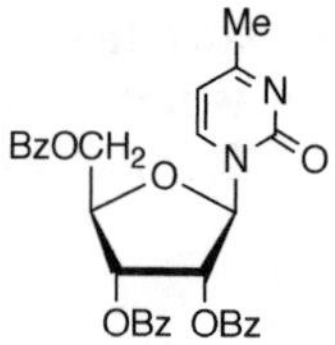

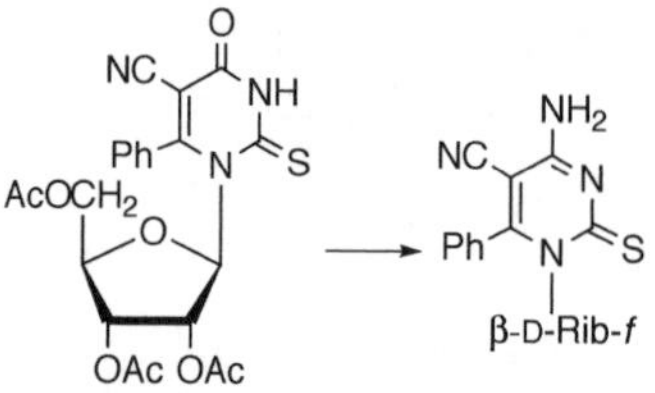

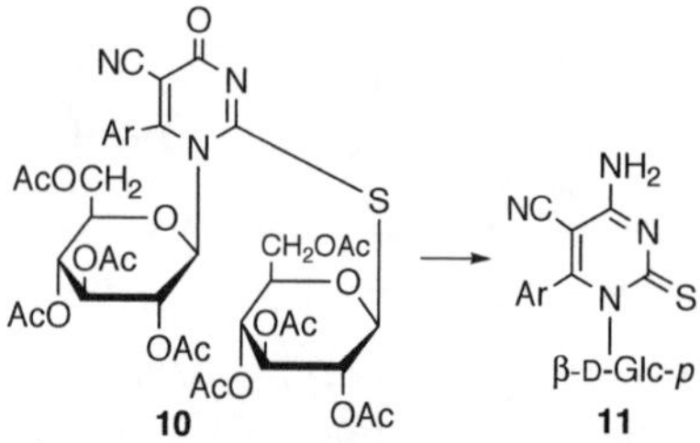

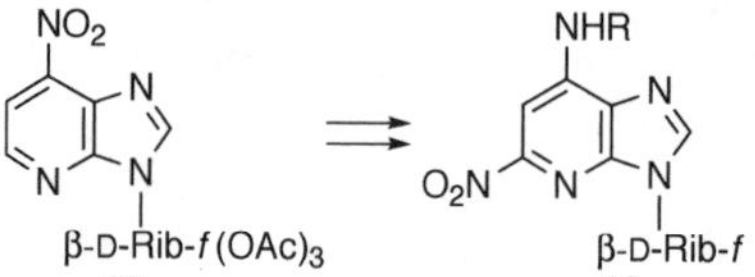

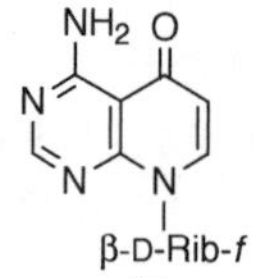

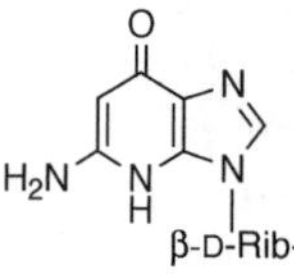

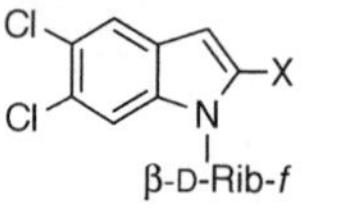

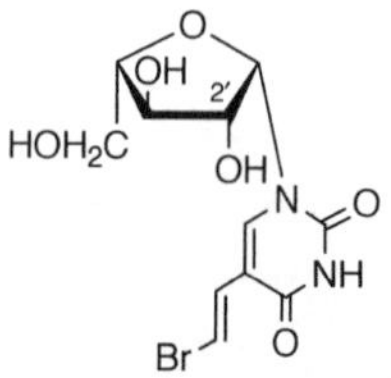

Reaction of 1-amino-1-deoxy-D-fructose or its *N*-substituted derivatives with peracylated glycosyl isothiocyanates has led to condensation products of type **18**. In the case of the ribofuranosyl compound subsequent treatment with acidic resin in ethanol gave the spironucleosides **19**.[21]

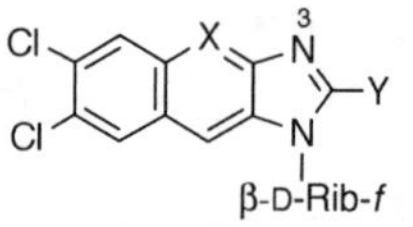

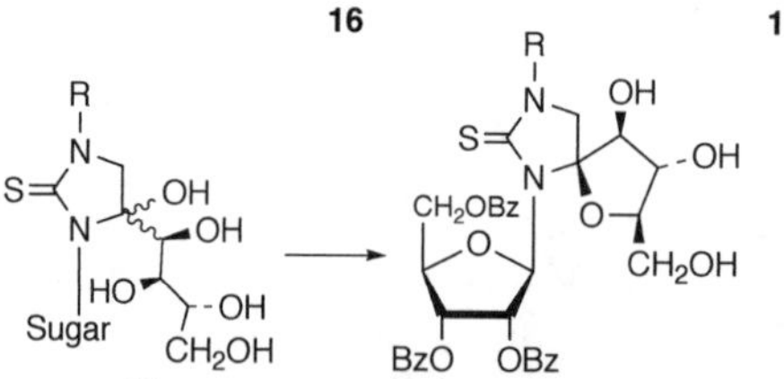

The L-arabinofuranosyl nucleoside **20** has been prepared from L-arabinose. Also reported were the 2′-deoxycompound (L-BVDU), and both 2′-deoxy-2′-fluoro-epimers.[22]

1,2-Epoxides were used as intermediates in the chemistry outlined in Scheme 1 for the formation of the differentially-protected β-D-xylofuranosylthymine **23** from the known tosylate **21** *via* the D-*lyxo*-isomer **22**.[23]

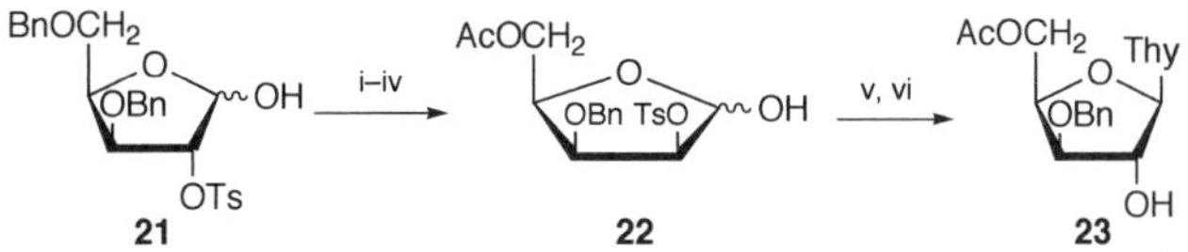

Reagents: i, NaOH, MeOH; ii, TsCl, py; iii, AcOH, Ac₂O, H₂SO₄; iv, NH₃, ether; v, KOBuᵗ; vi, (Tms)₂Thy, CH₂Cl₂

Scheme 1

Cyclization of a D-glucopyranosylhydrazine derivative was used to prepare the 1,2,4-triazole nucleoside **24**, and the analogous α-D-galactofuranosyl compound was synthesized similarly.[24] A range of 1-β-D-xylopyranosyl derivatives of 2-substituted-5-fluorouracils have been prepared by phase transfer catalysis or Koenigs–Knorr reactions.[25]

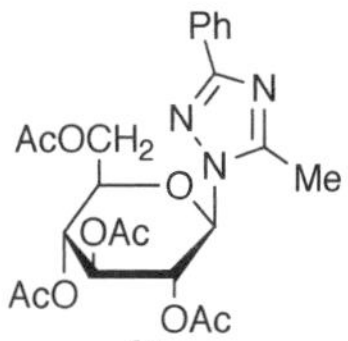

3   Anhydro- and Cyclo-nucleosides

Treatment of 9-(β-D-arabinofuranosyl)adenine (ara-A) with TPP and DEAD led in high yield to the D-*lyxo*-epoxide **25**, whilst a similar reaction on adenosine gave the 2′,3′-O-triphenylphosphorane.[26] The anhydronucleoside **26**, in which the furanose ring is restricted to a 2′-*endo*- (*S*-type) conformation, has been prepared by cyclization of a 1′-tosylate, and a 2,1′-anhydronucleoside, presumed to be an intermediate, could be isolated at shorter reaction times. The cytosine analogue of **26** was also made, and both products were devoid of anti-HIV activity in MT-4 cells.[27]

The 6,5′-anhydronucleoside **27** was obtained by reaction of 2,3-O-isopropylidene-D-ribofuranosylamine with 3-isothiocyanatopropanal, and on treatment with aqueous acetic acid, **27** was converted to the 6,2′-anhydro-system **28**. Crystal structures were reported for both **27** and **28**.[28]

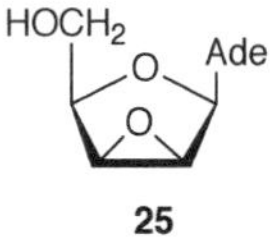

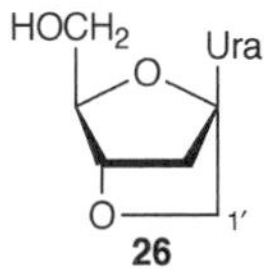

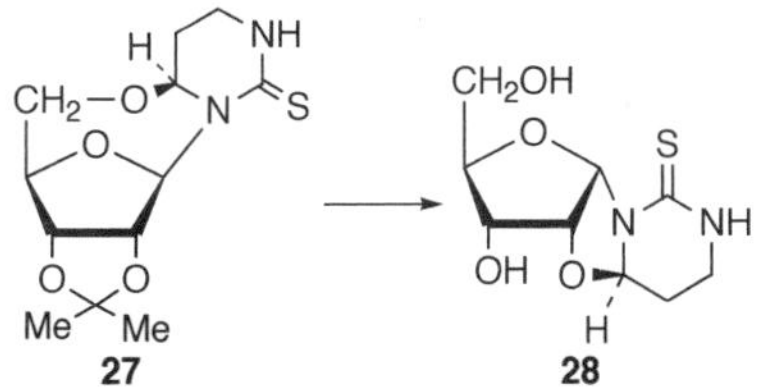

The use of 1,1′-sulfonyldiimidazole has been advocated for the synthesis of anhydronucleosides, as for example in the preparation of the 2,3′-anhydrocompound **29** from 2′-deoxy-5′-*O*-Tbdms-uridine in over 90% yield.[29] Compounds of type **30** (R = Pr^i, CH$_2$Pr^i, CH$_2$CH$_2$Pr^i) have been obtained by base-catalysed cyclization of 3′-mesylates.[30] When the 2,3′-anhydrothymidine derivative **31** was treated in DMF at 150 °C with either *O,O*-diethylphosphate or *O,O*-diethylphosphorothioate anions, the isomeric N^3-nucleoside **32** was obtained. It was speculated that the reaction proceeds by nucleophilic attack at C-1′ of **31**, followed by recyclization. The structures of both **31** and **32** were determined by X-ray methods.[31]

The key step in a synthesis of cyclo-5,6-dihydro-2′-deoxyuridine, a major product of gamma irradiation of deoxygenated aqueous solutions of deoxycytidine, is the cyclization of aldehyde **33** using Bu$_3$SnH and AIBN, to give **34** of stereochemistry as indicated. The method was previously used for the thymidine analogue (*J. Chem. Soc., Perkin Trans. 1*, 1999, 1257). Both the uridine and thymidine analogues were incorporated into oligodeoxynucleotides, where they acted as blocks for DNA polymerases.[32]

A number of examples of the use of anhydronucleosides as intermediates in syntheses of other types of analogue can be found elsewhere in this chapter.

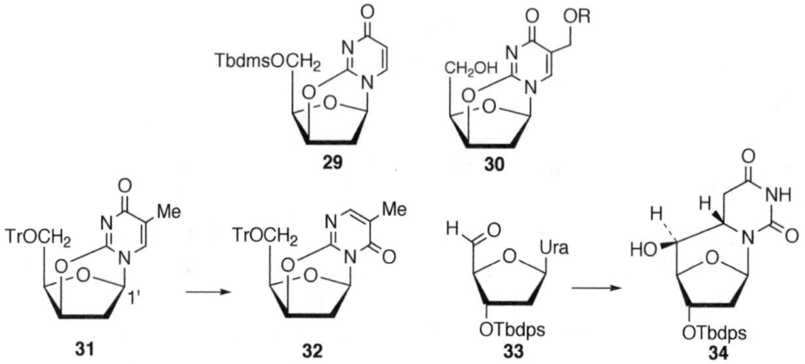

4   Deoxynucleosides

There has been a thorough account of work in Rizzo's laboratory concerning the synthesis of deoxynucleosides by deoxygenation using photoinduced electron transfer, with carbazole derivatives as photosensitizers. In addition to the photolysis of 2′-*m*-(trifluoromethyl)benzoates to produce β-2′-deoxynucleosides (see Vol. 30, p. 272–273), the method has been employed for the selective removal of benzoyloxy groups, as in the formation of the α-2′-deoxysystems **36** (B = Ura, Thy, *N*-AcGua) by photolysis of esters **35** in aqueous isopropanol in the presence of 3,6-dimethyl-9-ethylcarbazole and magnesium perchlorate, and for the synthesis of 3′-deoxy compounds **38** (B = Ura, Thy, Ade, *N*-AcGua) from precursors **37** under the same conditions. 2′,3′-Dideoxynucleosides **40** (B = Ura, Thy, Hx) are also accessible by photolysis of the triesters **39** [Ar = *m*-(trifluoromethyl)phenyl].[33]

An alternative to Barton–McCombie deoxygenation, restricted to cases with an electronegative group, particularly fluorine, β- to the radical centre, involves the thermolysis of xanthates in diglyme, as in the conversion of **41** to **42**. The reaction was thought to involve adventitious peroxides in the solvent, since their removal slowed the process considerably. Diglyme was found to be the hydrogen donor.[34] A synthesis of 2'-deoxy-2-fluoroadenosine, a potential prodrug for 2-fluoroadenine, involves deoxygenation by reduction of the 2'-thiocarbonylimidazolide with (Tms)₃SiH as hydrogen donor.[35]

A chemo-enzymatic approach has been used for the synthesis of thymidine specifically labelled with ^{13}C in the deoxyribose unit. Labels could be specifically placed at C-1' or C-2' starting from acetaldehyde labelled at C-1 or C-2 respectively, doubly-labelled acetate gave thymidine labelled at C-3' and C-4', whilst ^{13}C at C-5' had its origin in [^{13}C]diazomethane.[36] [*sugar*-^{13}C$_5$]Thymidine, additionally labelled stereoselectively with deuterium at C-2' and C-5', has been prepared from [^{13}C6]glucose, *via* a selectively deuteriated ribose derivative prepared by stereocontrolled reductions. The labelled sugar unit could then be transferred to other bases using purine and pyrimidine nucleoside phosphorylases.[37]

Standard base–sugar linking procedures have been used to make the benzimidazole nucleoside **43**, converted to the cross-linked bisnucleoside **44**,[38] the pyrroles **45** (R = H, CONH₂), which were made into triphosphates which exhibited a preference for incorporation into oligonucleotides with Klenow polymerase in place of either A or C,[39] and the pyrazoles **46** (X = OMe, NH₂) and their 2',3'-dideoxy-analogues.[40]

6-Methylpurine-2'-deoxyribonucleoside has been prepared by deoxygenation,[41] and, since the L-ribonucleoside has anti-HCMV activity, the L-2'-deoxycompound **47** and some related amines have been made, again using radical deoxygenation, and **47** was found to be active.[42] The enantiomer of **47**, and other related compounds such as the 2'-deoxyanalogue **48** of TCRB, have also been prepared, using either base–sugar coupling or exchange of 5,6-dichloroimidazole with the base of 2'-deoxyuridine, catalysed by *N*-deoxyribofuranosyl transferase.[43] The 2'-deoxyanalogue **49** of tricyribine, made by base–sugar coupling, has been reported, along with the 3'- and 5'-deoxycompounds, prepared from toyocamycin. The 2',3'-*ribo*-epoxide, the 2',3'-

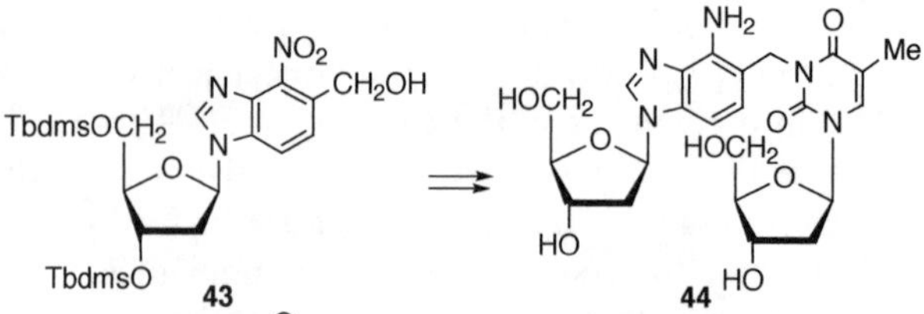

dideoxycompound and d4 systems were also made from triciribine itself, all these analogues being less effective as antivirals or antiproliferatives.[44]

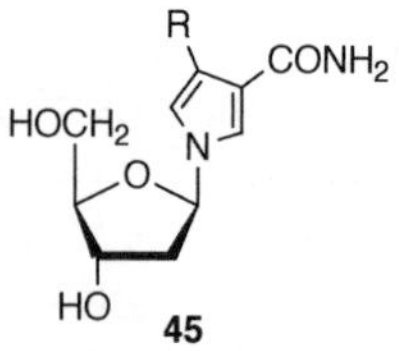

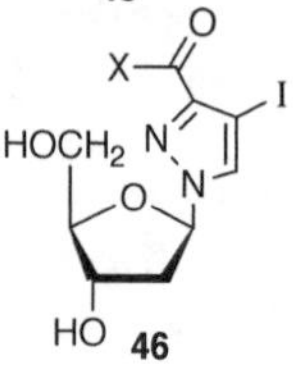

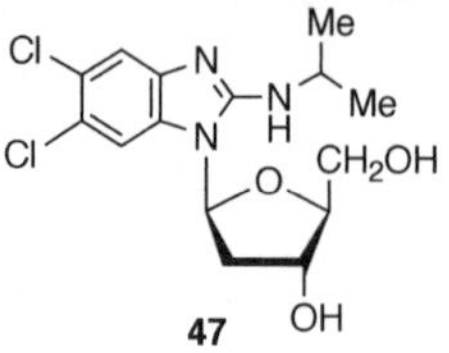

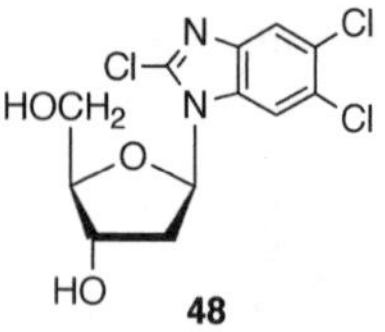

In connection with the synthesis of an RNA nonamer in both enantiomeric forms, in addition to the required L-ribonucleosides, L-2′-deoxythymidine was also prepared, to act as an anchoring derivative in the synthesis, from L-xylose.[45] The β-L-compound **50** (X = OH) has also been prepared from L-xylose, and was converted into the fluoro- and azido-analogues **50**, (X = F, N₃).[46]

5′-Deoxy-5-fluorocytidine has been prepared by base–sugar coupling, and used to make an orally-available prodrug of 5-fluorouracil.[47] The 5-iodotubercidin analogues **51** (R = Me, vinyl, Et), and some related species, have been synthesized as potential adenosine kinase inhibitors,[48] and the same group have also reported related 4-N-aryl-5-aryl-pyrrolo[2,3-d]pyrimidines.[49]

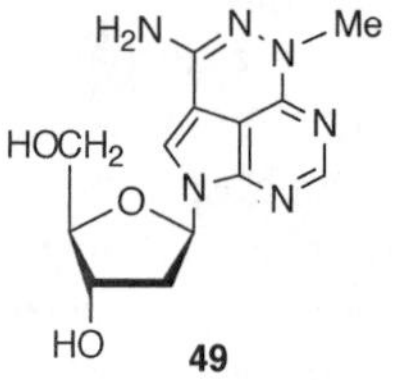

In the area of 2′,3′-didehydro-2′,3′-dideoxycompounds (d4 systems), a previous method for their synthesis from 5′-protected 2′,3′-di-O-mesyl-nucleosides by treatment with arylselenyl anions (Vol. 31, p. 272) has now been modified by the use of bis(4-perfluorohexylphenyl) diselenide and sodium borohydride, which permits the use of the diselenide in catalytic quantities, and also its ready recovery. The method was used for the synthesis of d4-uridine.[50] Analogues of d4T with potential linker arms at C-5, for attachment of either a fluorescent tag or a non-nucleoside reverse transcriptase inhibitor, have been prepared either from 5-(hydroxymethyl)uridine[51] or from the 2,2′-anhydronucleoside of 5-(methoxycarbonylmethyl)uridine (Vol. 28, p. 265–266).[52] Addition of iodine at C-2′ and C-3′ of protected pyrimidine nucleosides under Arbuzov reaction conditions has led to a new route to d4 systems,[53] and d4-uridine has been prepared from 2′-deoxyuridine using elimination from the 3′,5′-dimesylate.[54]

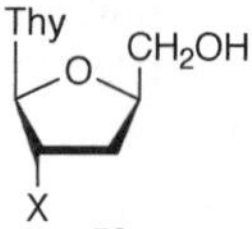

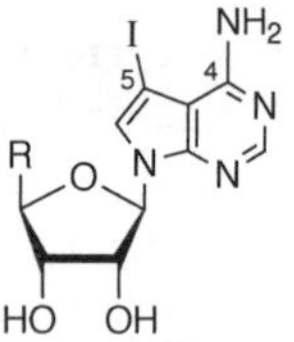

5 Halogenonucleosides

As part of an issue of *Carbohydrate Research* devoted to reviewing the whole area of fluorinated sugars, Pankiewicz has discussed nucleosides fluorinated in the sugar moiety, covering both synthesis and biological activity.[55] Another review has also covered recent strategies for the synthesis of fluoronucleosides, with some consideration of structure–activity relationships.[56]

Syntheses of the 2'-deoxy-2'-fluoronucleosides **52** have been carried out using the condensation of silylated 2,6-dichloropurine with 1,3,5-tri-*O*-benzoyl-2-deoxy-2-fluoro-α-D-arabinofuranose as a key step. Conformations of the nucleosides were analysed using the PSEUROT program, and they were incorporated into oligodeoxynucleotides which had practically the same affinity to both complementary DNA and RNA as did the parent unmodified systems.[57] The same sugar precursor was used in a synthesis of the fluorinated analogue **53** of TCRB, and the 3'-deoxy-3'-fluorocompound **54** was prepared from TCRB itself, through tritylation to give mostly the 2',5'-di-*O*-trityl compound, thus permitting introduction of fluorine at C-3' with inversion of configuration. The analogues **55** (X = Br, NHPr^i) were made by a sequence involving enzymic sugar exchange between 2'-deoxy-2'-fluorouridine and 5,6-dichlorobenzimidazole. All these compounds had less antiviral activity and more toxicity than TCRB.[58] A method for the synthesis of 2',3'-dideoxy-2'- or 3'-fluoronucleosides such as **42** was mentioned earlier,[34] and some 3'-amino-2',3'-dideoxy-2'-fluoronucleosides are referred to in the next section. 2',3'-Dideoxy- 3'-fluoro-L-ribonucleosides **56** (B = Cyt, Ura, Ade, Gua) have been prepared from D-glucitol, which was converted to methyl L-xylofuranoside, which was deoxygenated at C-2 prior to the introduction of fluorine at C-3 with inversion, and base–sugar coupling.[59] Other workers have also described the guanosine analogue (**56**, B = Gua) by a similar sequence, but with base–sugar coupling to an L-xylose derivative as an early stage.[60]

A procedure for the conversion of uridine to 5'-chloro-5'-deoxyuridine without the need for protection of O-2' or O-3' involves the treatment of the nucleoside with *N*-chlorodiisopropylamine and TPP, followed by hydrolysis of the resultant 5'-chlorinated 2',3'-*O*-triphenylphosphorane.[61]

6 Nucleosides with Nitrogen-substituted Sugars

Treatment of iodocompounds such as **57**, made from furanoid glycals (Vol. 31, p. 269–270), with Bu$_2$SnO gives rise to anhydronucleosides such as **58** (Scheme 2), treatment of which with azide ion gave the 2'-azido-2'-deoxynucleoside **59**. A better yield of **59** could be obtained by hydrolysis of **58** to the D-lyxonucleoside, followed by conversion to a triflate and azide displacement.[62]

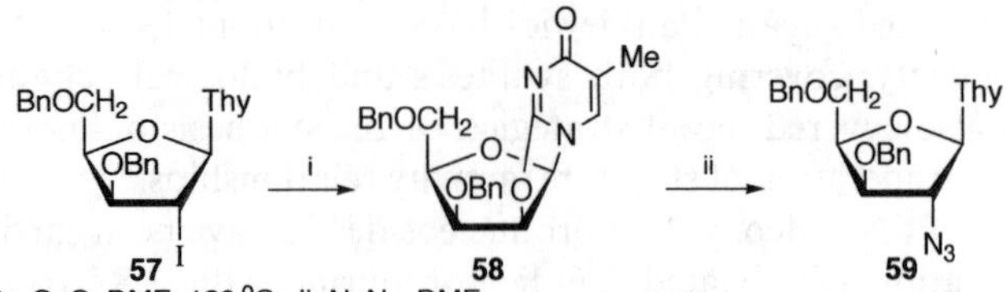

Reagents: i, Bu$_2$SnO, DMF, 120 °C; ii, NaN$_3$, DMF

Scheme 2

A similar displacement of a 2'-triflate was used in the synthesis of amides such as **60**, related to a known inhibitor of trypanosomal glyceraldehyde-3-phosphate dehydrogenase, from ara-A.[63] 2'-Amino-2'-deoxyuridine has been linked through an amide to a coumarin; the resultant derivative was incorporated into oligo-nucleotides to act as a fluorescence energy donor to DNA.[64]

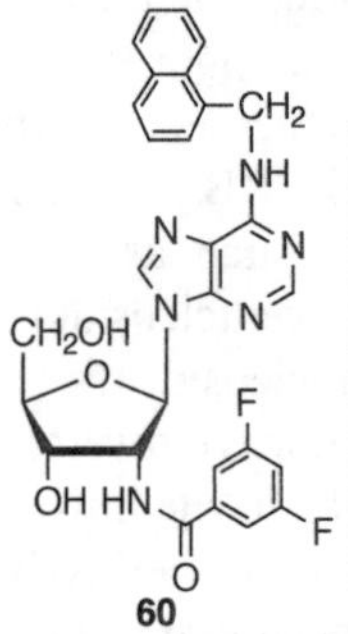

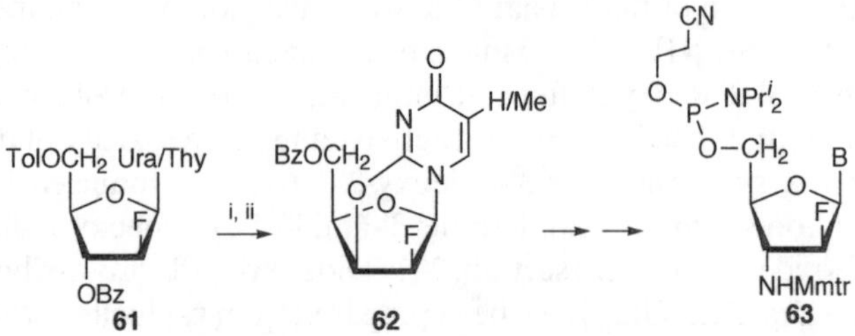

Reagents: i, NH$_3$, MeOH; ii, Ph$_3$P, DIAD, PhCO$_2$H

Scheme 3

The nucleosides **61** were formed with good β- stereoselectivity by condensation between the silylated bases and the α-glycosyl bromide. Conversion to cyclonucleosides **62** (Scheme 3) was followed by conversion to the *arabino*-configured 3'-amino-2'-fluoro-systems **63** (B = Ura, Thy, Cyt). These were incorporated into oligonucleotides linked by phosphoramidate bonds, and these were found to have high binding affinity to complementary nucleic acids (although not as high as for the corresponding 2'-*ribo*-fluoro-phosphoramidates; see *Nucleic Acids Res.*, 1996, **24**, 2966), as well as greater acid stability than the non-fluorinated phosphoramidate-linked oligomers.[65] The same group has also reported the synthesis of the building block **64**, made by base–sugar coupling, and a similar 2'-deoxycompound, made from AZT using chemical base exchange. These were incorporated into phosphoramidate-linked oligomers, which showed a considerable increase in stability for complexes with RNA or DNA, relative to adenosine-containing counterparts.[66]

AZT was used as a precursor to prepare a compound in which fluorescein was linked to the amino group of 3'-amino-3'-deoxythymidine through a thioamide, a link resistant to enzymatic degradation, and a spacer. The product was converted to its 5'-triphosphate for use as a chain terminator for DNA dye-terminator sequencing.[67] 3'-Amino-3'-deoxythymidine has been joined in hybrid structures, in which the two components are linked by urea units, to erythromycin (*via* its 9-amino-derivative) and azithromycin.[68] AZT has also been converted to the aminodiacid **65**, and a similar structure was prepared from 5′-amino-5′-deoxythymidine. These were incorporated into oligonucleotides at terminal positions, in an effort to preorganize high-affinity metal-binding sites.[69] Conjugates of nucleoside and non-nucleoside reverse transcriptase inhibitors have been prepared. These include species in which AZT is linked through spacers to the NNRTI *via* N^3, or *via* a side-chain replacing the methyl group, and compounds involving ddC were also made. Some of the conjugates displayed anti-HIV activity, but no synergistic effects were found.[70]

The L-enantiomer of 3′-azido-2′,3′-dideoxyguanosine has been prepared from L-xylose, but no significant antiviral activity was observed.[60]

Uridine has been converted in seven steps into 5′-amino-3′,5′-dideoxyuridine (**66**), related to the 4′,5′-unsaturated structure found in the mureidomycin antibiotics.[71] Some 5′-Amino-5′-deoxy-5-iodo-pyrrolopyrimidine nucleosides such as **67** (X = Cl, NH₂) have been prepared by sodium salt glycosylation procedures, and shown to be powerful adenosine kinase inhibitors.[48]

In connection with the synthesis of glucosamine-based oligonucleotide analogues (see Vol. 31, p. 293–295), a study was made of the regioselectivity of glycosylation of guanine derivatives with the acylated glucosamine **68**. Use of 2-*N*-acetyl-6-*O*-diphenylcarbamoylguanine, which has been used with high N-9 selectivity in other cases, gave a mixture of regioisomers, and indeed only the N-7 isomer under some conditions. Use of 2-*N*-acetyl-6-*O*-benzylguanine, however, gave only the N-9 substituted product **69** in moderate yield.[72] Addition reactions of *N*-heterocycles to the nitrogalactal **70** proceeded stereoselectively to give, after further manipulation, β-D-galactosaminyl nucleosides such as **71**.[73] Reaction of silylated thymine with di-*O*-acetyl-L-rhamnal gave a 2′-enopyranosyl nucleoside which could be converted to the AZT analogue **72**.[74]

7 Thio- and Seleno-nucleosides

The 1′-phenylselenyl derivative **73** was obtained, but with only low diastereoselectivity, by reaction of the enolate with PhSeCl. Treatment of **73** with NaBH₄ in the presence of CeCl₃ gave stereoselective reduction to alcohol **74**.[75]

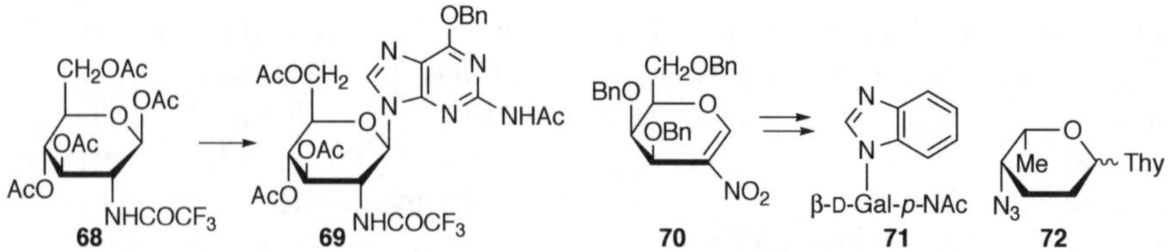

Opening of 2,2'-anhydrouridine with 2-(Tms)ethanethiol gave the thioether **75**, which could be converted to the methyldisulfide **76** by reaction with dimethyl(methylthio)sulfonium tetrafluoroborate.[76] *S*-Hexyl-2'-thiouridine has been synthesized, again by opening of a 2,2'-anhydro-system, and was incorporated into oligodeoxyribonucleotides, which had decreased duplex stability with both complementary RNA and DNA.[77] In an extension of earlier work (Vol. 33, p. 283), the β-D-*ribo*-thioglycoside **77** was converted under Mitsunobu conditions in the presence of 3-benzoylthymine into the α-D-arabinofuranosyl nucleoside **78** with good stereocontrol, whilst the α-D-*arabino*-compound **79** gave the β-D-ribonucleoside **80**, again in good yield. Phenylselenyl glycosides behaved similarly.[78]

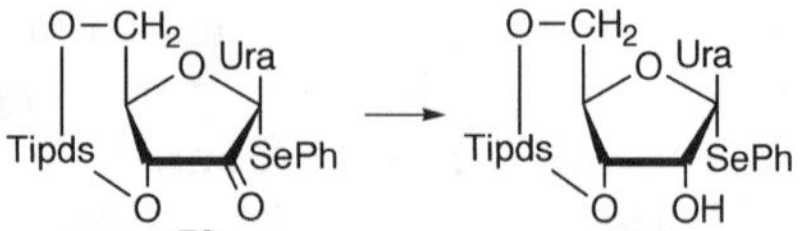

Reaction of a thiol with 5'-*O*-Dmtr-2,3'-anhydrothymidine was used to prepare *S*-(2-aminoethyl)-3'-thiothymidine, which was converted to its 5'-triphosphate and coupled to an oxazine dye (see also ref. 67 above).[79]

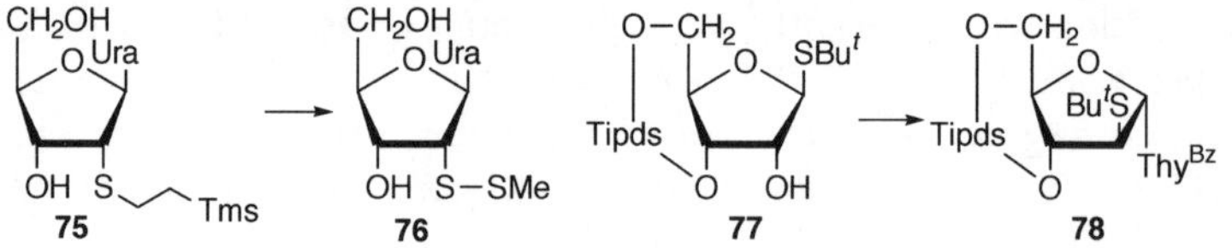

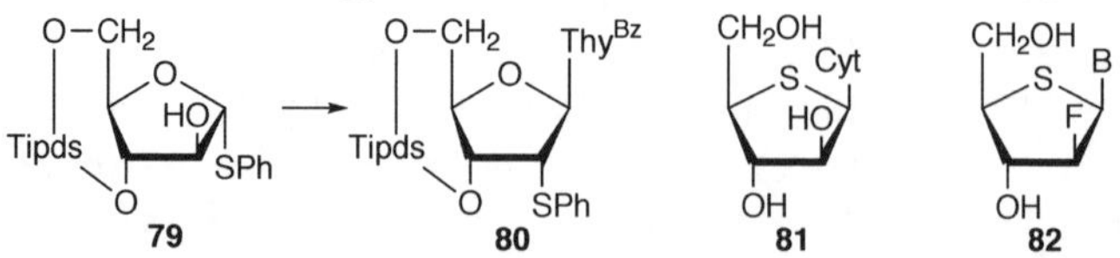

A review has been given of the synthesis and biological activity of 4'-thio-nucleosides, which also embraces oxathiolane analogues.[80] The 4'-thio-analogue **81** of ara-C has been prepared by base–sugar condensation, using a 4'-thio-arabinofuranose building block previously used to make related purines (Vol. 32, p. 270), and was evaluated against human tumours, with ara-C proving generally more cytotoxic.[81] A range of 2'-deoxy-2'-fluorocompounds of type **82**, with both purine and pyrimidine bases, have been prepared by base–sugar condensations in which significant amounts of the α-anomers were also formed. In the pyrimidine series, a number of the 4'-thionucleosides showed potent anti-HSV activity, and the cytosine and 5-fluorocytosine derivatives had potent anti-tumour activity.[82] The (4-thio-L-arabinofuranosyl)-5-halopyrimidines **83** (X = F, Cl, Br) have been prepared from a sugar unit accessible from D-xylose.[83]

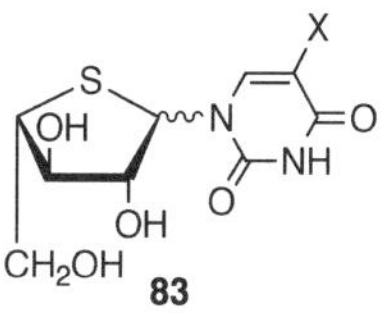

Reagents: i, ICl, 2,6-di-Me-4-Bu^t–pyridine; ii, ICl, (Tms)$_2$-5-Et-Ura;
iii, Bu$_3$SnH, AIBN; iv, BBr$_3$, DCM, –78 °C

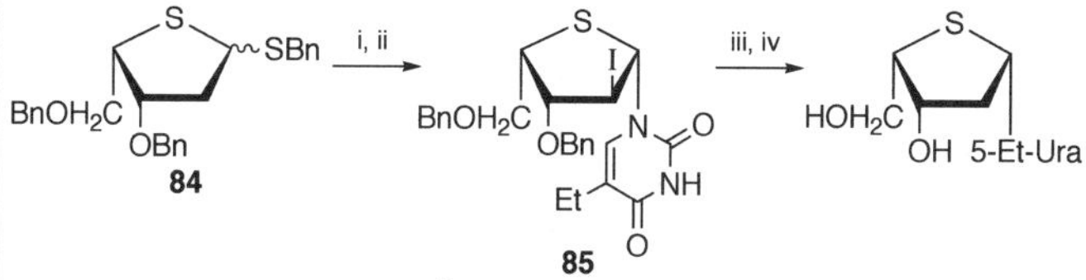

Scheme 4

The method of Scheme 4 has been developed for the synthesis of 2′-deoxy-4′-thionucleosides. In the conversion of thioglycoside **84** to **85**, a thiofuranoid 1,2-glycal was formed as an intermediate, and could be isolated, although the one-pot procedure, which gives good stereocontrol, was superior.[84] The same procedure could be applied to the formation of **86** (R = H, Et), although stereocontrol was poor, and a comparable amount of the alternative 1,2-*trans*-isomer was obtained.[84,85] The products **87** (X = Cl, N$_3$) could be obtained from **86**, with 2,2′-anhydronucleosides as intermediates.[85] An alternative route to 2′-deoxy-4′-thionucleosides developed in the same laboratory involves the synthesis of thioglycoside **88** from L-ascorbic acid (ascorbate carbons numbered) in a procedure (see Chapter 11) which compares favourably with an alternative approach to a similar compound (see Vol. 25, p. 138), and then linkage with the base with little stereocontrol. The β-anomer of **89** has good antiherpes activity.[86] Some 4′-thio-*C*-nucleosides are mentioned in Section 10, and a 4′-thionucleoside antibiotic in Section 12.

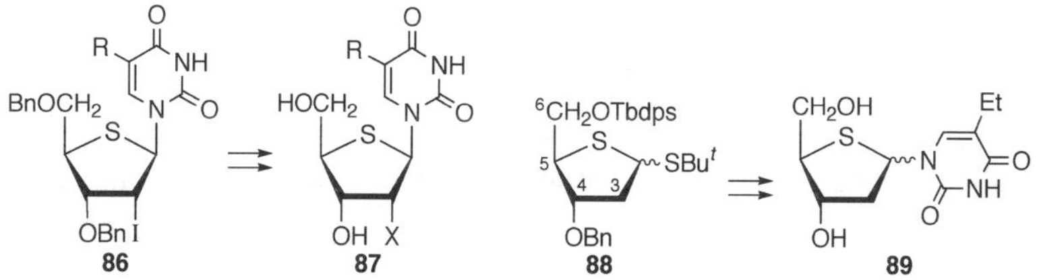

5′-Acetylthio-2′,3′-*O*-isopropylideneadenosine has been made from iso-propylideneadenosine by a Mitsunobu reaction, giving a route to 5′-thioadenosine, and various *S*-alkylated derivatives such as the *S*-adenosyl-methionine analogues **90** (R = NH$_2$, CO$_2$H).[87]

The nucleosides **91** have been prepared by coupling of the silylated bases with the methyl glycoside of the selenoanhydrofuranoside (Chapter 11), in the presence of TmsOTf.[88]

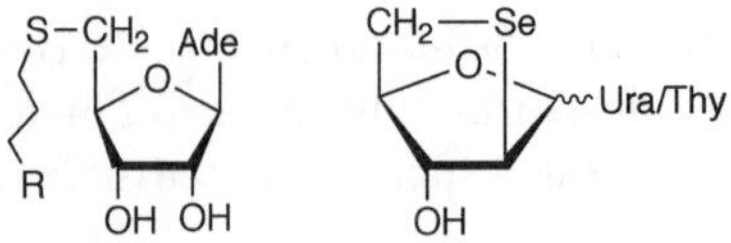

90   **91**

8   Nucleosides with Branched-chain Sugars

The 2'-deoxy-2'-methylene nucleoside **92** has been prepared using a Wittig reaction, and a number of other 4-amino-5-oxo-pyrido[2,3-*d*]pyrimidine ribonucleosides with chain branches at C-2', C-3' and C-4' were also reported.[17]

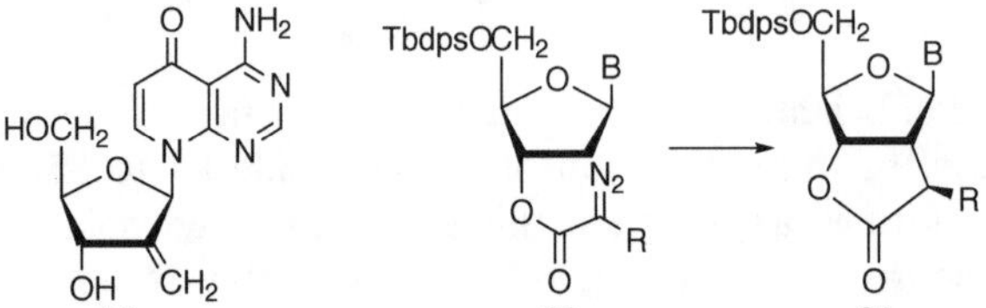

92   **93**   **94**

Rhodium-catalysed carbenoid insertion has been used to make fused γ-lactones, as in the conversion of diazoesters of type **93** into products **94** (R = H, COMe, CO_2Me), in which the indicated *exo*-isomer strongly predominated.[89] When the readily-accessible uridine derivative **95** is treated as indicated in Scheme 5, the lactone **96** is obtained, and this can be used to make nucleosides **97** containing other bases, either purines or pyrimidines.[90]

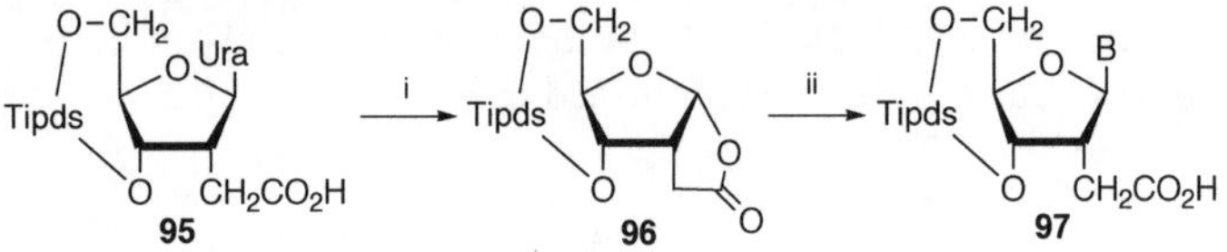

95   **96**   **97**

Reagents: i, (Tms)$_2$NH, (NH$_4$)$_2$SO$_4$; ii, base, HMDS, TmsCl, TmsOTf, MeCN

Scheme 5

The antitumour agent 2'-deoxy-2'-methylenecytidine has been conjugated (through the 4-amino-group) with folic acid, since the cell-surface folate receptor is overexpressed in some tumours and is therefore a target for drug delivery.[91] Pyrimidine 2'-deoxynucleosides chain-branched at C-2' have been prepared as outlined in Scheme 6. Slow addition of Bu$_3$SnH and AIBN to **98** led to the formation of the product with a 2-hydroxyethyl group at C-2' by rearrangement of the initially-formed radical prior to H-abstraction, but the yield was only moderate.[92] A paper discussing 2'-deoxy-2'-substituted-1',2'-unsaturated uridines is mentioned in the next section.

A route to 2',3'-dideoxy-2'-trifluoromethylpyrimidine nucleosides involves the synthesis of the lactone **99**, and its separable *cis*-isomer, from isopropylidene-D-glyceraldehyde, which provides C-3 to C-5. Reduction of **99** and glycosylation gave, with moderate stereoselectivity, the nucleosides **100** (B = Ura, Thy, Cyt), and the *cis*-isomer of **99** could also be converted to 1',2'-*trans*-nucleosides.[93] An alternative route to **100** (B = Ura) proceeds through the difluoromethylene

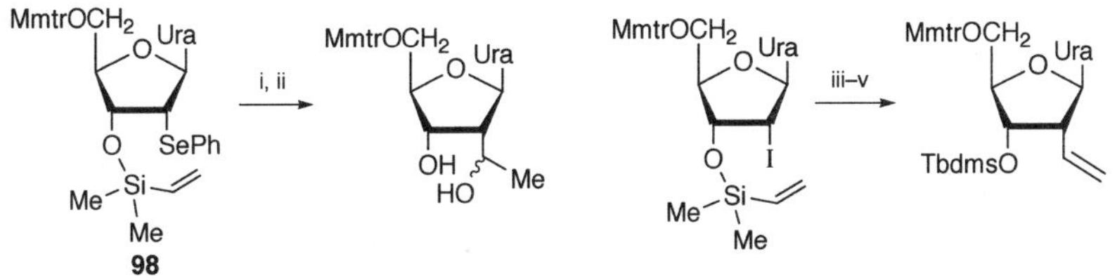

Reagents: i, Bu$_3$SnH, AIBN; ii, H$_2$O$_2$, KF, KHCO$_3$; iii, (Me$_3$Sn)$_2$, AIBN; iv, TBAF; v, TbdmsCl

Scheme 6

compound **101**, prepared using a Wittig-type reaction. Treatment of **101** with TBAF leads to the trifluoromethyl alkene **102**, and a mechanism was proposed in which loss of ethylene from the Sem protecting group was followed by intramolecular delivery of fluoride to the difluoromethylene unit. Reduction and deprotection of **102** gave **100** (B = Ura). Similar chemistry starting from a 3′-keto-nucleoside gave the isomeric nucleoside **103** (B = Ura).[94] An alternative route to 3′-deoxy-3′-trifluoromethylpyrimidine nucleosides has also been described, with the sugar unit **104**, made from D-xylose (Chapter 14) as an intermediate. This was converted to nucleosides **105**, and conventional radical deoxygenation was used to make the 2′-deoxycompounds **103** (B = Thy, Cyt). Elimination applied to a 2′-O-mesyl derivative was also used to give access to the alkene **106**.[95]

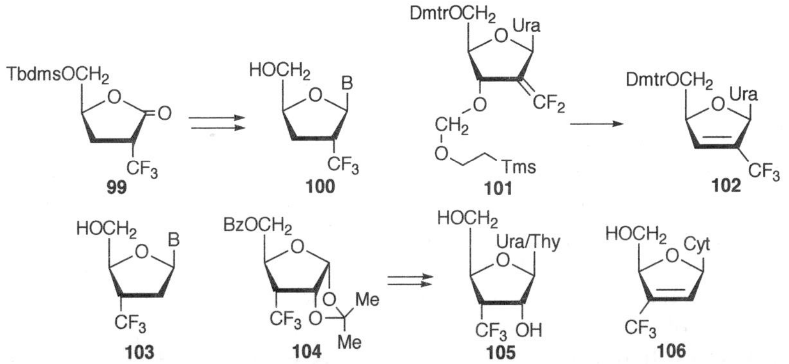

2′,3′-*endo*-Methylene nucleosides **107**, involving all the main nucleobases, have been prepared in a stereoselective manner from isopropylidene-D-glyceraldehyde. The bases were introduced with stereocontrol using an α-glycosyl chloride. The enantiomers **108** were also made using similar chemistry, starting from L-gulono-γ-lactone.[96]

A review on HIV-1 specific reverse transcriptase inhibitors has appeared, with special emphasis on TSAO-T (**109**) and related compounds.[97] Some new analogues of TSAO-T and its N^3-methyl derivative have been reported, with ethers and esters replacing the Tbdms group at O-5′, and also with aminogroups at C-5′, and some of the ethers and amines had moderate activity.[98]

A route to the 3′-carboxymethyl-3′-deoxyribonucleosides **110** (X = N$_3$, ODmtr) has been reported, in which the sugar unit is prepared from diisopropylidene-α-D-glucofuranose, and the base is attached at a late stage.[99] A synthesis has been described in detail of the amide **111**, together with its X-ray structure and that of the corresponding 5'-O-Tbdps carboxylic acid.[100] Epoxy-compounds of type **112** (X = NH$_2$, NHOH, NHNH$_2$) have been prepared through Darzens reactions on 3'-ketonucleosides. Ring opening of one such compound gave the chlorohydrin **113**, and a similar reaction occurred with azide ion. Treatment of **112** (X = NHNH$_2$, Ura series) with DBU in methanol gave the reduction product **114**, a mechanism for the reaction being proposed.[101] The triphosphate of 3'-(2-aminoethyl)-3'-deoxythymidine and a similar compound with a longer spacer arm have been prepared, and linked through their amino-groups to an oxazine derivative, but these compounds failed as terminators in DNA sequencing.[102] 3'-Deoxy-3'-hydroxymethyl-L-ribonucleosides of the main nucleobases have been prepared from L-xylose, but they did not show significant antiviral activity.[103]

The epoxide **115**, made from (S)-glycidol, gave, on treatment with lithium hexamethyldisilazide, the furanoid glycal **116**. This could be converted stereoselectively into nucleoside analogues of type **117**, involving all the main nucleobases, by a sequence involving reductive removal of the sulfone, formation of an α-epoxide, and coupling of this with the appropriate base. In the py-rimidine series, related compounds in which the secondary alcohol was replaced with retention of configuration by fluoride and azide, or inverted, were made *via* anhydronucleoside intermediates, and the 2'-deoxycompound (B = Ura) was made by convertional radical deoxygenation.[104] Both anomers of **118** were made, as racemates, by base–'sugar' linkage.[105] The C-vinyl compound **119** was con-verted, by hydroxylation, base-catalysed cyclization, coupling to silylated thymine, and debenzylation, into the tricyclic nucleoside **120**.[106] A range of nucleosides **121** have been prepared by base–sugar coupling, with the quaternary carbon being established by chirality transfer using a Claisen rearrangement of an intermediate derived from isopropylidene-D-glyceraldehyde.[107] Radical cycl-ization of **122** gave the 3'-methylene nucleoside **123**, as well as the product **124** of 1,6-H transfer derived from the other diastereoisomer of **122**.[108] Some references mentioning 3'-branched nucleosides are discussed in Section 14 below.

Some new 3'- and 4'-C-branched compounds have been made and incorpor-ated into oligonucleotides **125** (X = H, OMe) and **126** (X = F, OMe). The presence of units **126** increased affinity towards complementary RNA and DNA, whilst the systems **125** led to duplexes of unchanged or lower stability.[109] A report from Matsuda's laboratory has described the preparation of 4′α-*C*-(2-aminoethyl)thymidine and its incorporation into oligodeoxynucleotides which were significantly resistant to snake venom phosphodiesterase and endonuclease.[110] Compounds in which the aminoethyl group is linked to a lipophilic group such as a fatty acid have also been made and incorporated into oligodeoxynucleotides, with a view to increasing membrane permeability.[111]

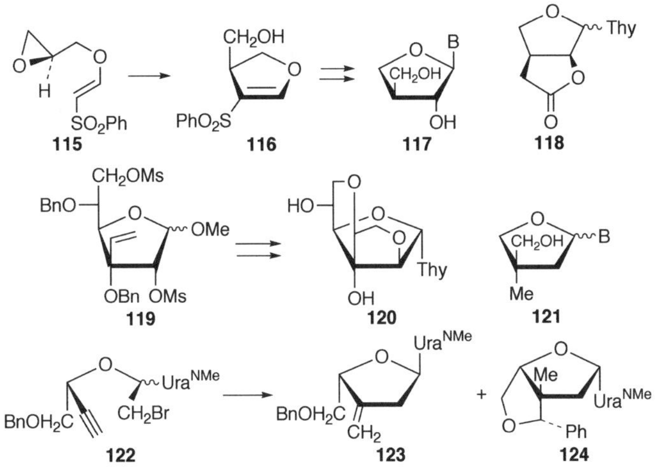

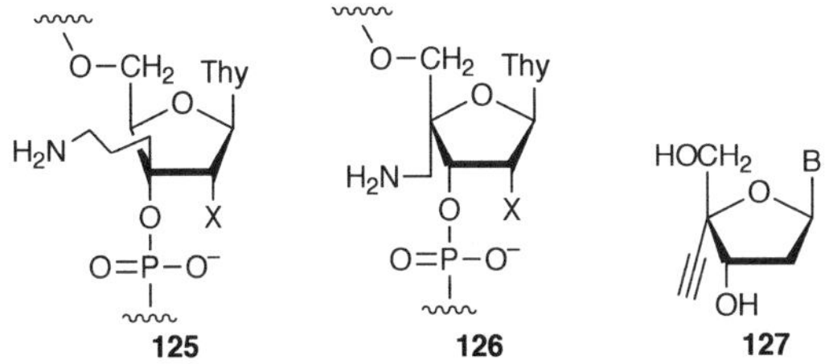

There has been a further report (see Vol. 33, p. 292) on the preparation of 4′-*C*-alkynyl-2′-deoxynucleosides **127**, including cases with both purine and pyrimidine bases, and, in the pyrimidine series, arabinofuranosyl nucleosides were also prepared through anhydronucleoside intermediates. Some of these compounds had good anti-HIV activity.[112]

Interest continues in conformationally-locked bicyclic nucleosides and their oligomers. A report from Wengel's laboratory describes details of the synthesis of the α-L-LNA (locked nucleic acid) nucleoside **129**, in which a key step is the treatment of ditosylate **128** with NaOH in aqueous ethanol to establish the bicycle, a reaction thought to involve a 2,2′-anhydronucleoside as an intermediate. Also reported is the similar conversion of **130** into the α-L-*xylo*-LNA nucleoside **131**.[113] The effect on RNA binding of the incorporation of these two compounds, and the previously-prepared β-D-*xylo*-isomer and LNA nucleoside itself, into oligonucleotides has been studied; the behaviour of the other four stereoisomers of LNA, enantiomeric with those synthesized, was also studied indirectly by using *ent*-RNA, derived from L-ribonucleosides. Increased binding efficiencies towards RNA, as compared to the DNA reference, was found for six of the eight stereoisomeric LNAs, and particularly strong binding was found for LNA itself and α-L-LNA, the oligomer of 129.[114] Full details have been given of the synthesis of the abasic LNA monomer 132 (Vol. 33, p. 292), and the synthesis of the ring-cleaved analogues 133 (X = OH, H) were made by adaption of earlier chemistry. On incorporation into oligomers, 132 had similar effects to a normal abasic nucleoside, whereas the monocyclic species 133 caused destabilization of duplexes.[115] The structures of LNA:RNA duplexes have been studied by NMR; it was found that the introduction of three modified units induces major conformational changes in the remaining unmodified units in the DNA strand, changing all except the terminal ones to *N*-type conformations, and leading to speculation as to how to tune an LNA:RNA duplex so that it would become a substrate for RNAse H.[116]

Imanishi and co-workers have shown that one LNA modification in a pyrimidine deoxyoligonucleotide could promote triplex formation with a DNA duplex in a highly sequence-selective manner.[117] A further synthesis of the 2'-*O*-,4'-*C*-methylene analogue of AZT has been reported (see Vol. 33, p. 292–293), and the α-L-*xylo*-compound 134 was also prepared, using a trimesylate analogous to 130 as precursor. Neither compound displayed anti-HIV activity.[118] Some related *C*-nucleosides are mentioned in Section 10.

9   Nucleosides of Unsaturated Sugars and Dialdoses

As in earlier volumes, 2',3'-didehydro-2',3'-dideoxynucleosides (d4 compounds) are discussed with their saturated counterparts in Section 4.

There has been a full account of the synthesis of the 2'-stannylated alkene 135 (X = SnBu₃) by base-induced stannyl migration from C-6 (see Vol. 32, p. 275), and the application of this compound to the preparation of the alkenyl halides 135 (X = Cl, Br, I), and products with carbon substituents at C-2' through Stille couplings.[119] Reaction of di-*O*-acetyl-L-rhamnal with silylated thymine gave the 2'-enopyranosyl nucleoside by allylic rearrangement, as a mixture of anomers.[74] A paper discussing a glycal substituted at C-3 with a nucleobase is mentioned in Chapter 10, and a 3'-ene derived from thymidine is mentioned in Section 17.

The bromoalkene 136 is produced stereoselectively when the corresponding

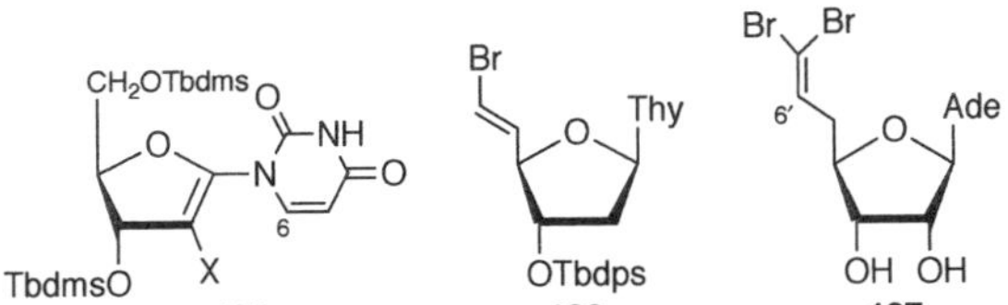

135   **136**   **137**

gem-dibromide is reduced with dimethyl phosphite and triethylamine.[120] The doubly-homologated dibromoalkene **137** has been prepared by coupling adenine to a sugar unit prepared (Chapter 2) from di-isopropylideneglucose. Treatment of **137** with BuLi gave the corresponding 6'-alkyne, and the *gem*-bromofluoride was also reported. The effect of these compounds on *S*-adenosylhomocysteine hydrolase was studied, with time- and concentration-dependent inactivation being observed, as well as partial reduction of the enzyme-bound NAD⁺.[121] Some related alkynyl nucleosides are mentioned in Section 17.

In further investigations of transglycosidic tethers for conformational restriction of pyrimidine nucleosides, it has been found that 6-formyluridine-5'-carboxaldehyde exists in aqueous solution as the bis-hydrate **138**, whereas the isopropylidene derivative of 6-formyluridine-5'-carboxaldehyde formed the bridged spiro-fused system **139** (X = OH), the structure of which has been previously confirmed by X-ray crystallography (Vol. 29, p. 335). The isopropylidene derivative of 6-(hydroxymethyl)uridine-5'-carboxaldehyde also cyclized to give **139** (X = H), whilst once again the compound without the isopropylidene group showed no tendency to form a transannular link.[122]

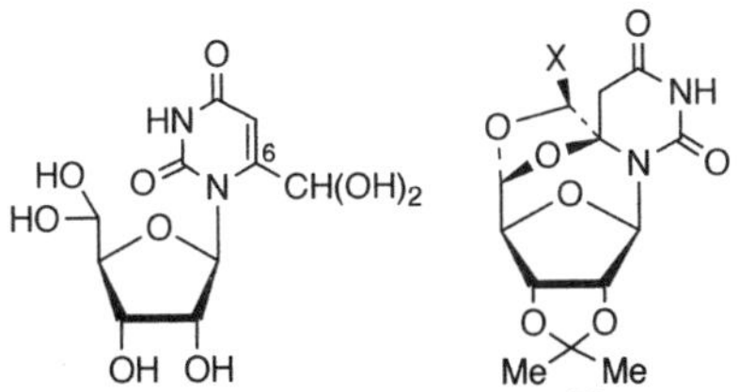

138   **139**

10   *C*-Nucleosides

The oxadiazole **140**, related to ribavirin, has been prepared by cyclization of a known sugar-protected amidoxime.[123] The 4'-thio-analogues **141** (X = O, S) of furanfurin and thiophenfurin (Vol. 29, p. 283), themselves analogues of ribavirin, have been prepared by condensation between the heterocycles, as their ethyl esters, with *O*-protected derivatives of 4'-thioribofuranosyl acetate. The thiophene compound ('thiophenthiofurin') had cytotoxicity towards cancer cell lines, but less so than thiophenfurin.[124] A previously-known 2,5-anhydroglucose derivative **142** has been converted to the pyrazole **143**, and some related compounds.[125] Some pyrazole iso-*C*-nucleosides are mentioned in Chapter 2. Various 3-cyano-2-(β-D-ribofuranosyl)-1,5-benzodiazepines have been prepared by a novel ring transformation of 5-(tri-*O*-benzoyl-β-D-ribofuranosyl)isoxazole-4-carbaldehyde.[126]

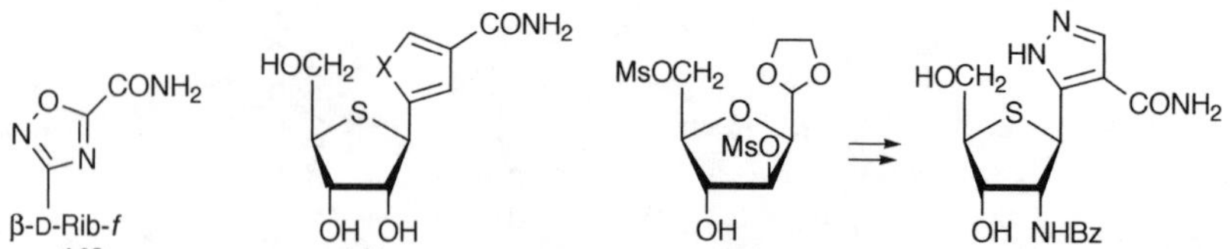

A range of 2'-deoxypyrimidinyl-*C*-nucleosides **144** have been prepared by parallel synthesis from acetylenic ketones and amidines; the anomers could be obtained as separated isomers, although anomerization had occurred during the pyrimidine-forming step.[127] The pyrimidinyl *C*-nucleoside **145** has been prepared using Pd-catalysed coupling of 2-amino-5-iodopyrimidine to a furanoid glycal as a key step. It was incorporated into oligonucleotides, as was a related pyridine *C*-nucleoside, for use in studies of base-triplet formation involving C–G and G–C base pairs.[128]

The 1,2,3-triazole **146**, and the α-anomer, have been prepared by cyclization of acylic precursors.[129]

Conformationally-locked *C*-nucleosides such as **147** have been reported from Imanishi's laboratory. These were prepared by formation of the C-1'–O bond in Mitsunobu reactions, the necessary diols being formed by stereoselective addition of Grignard derivatives of the heterocycles to an aldehyde. Use of lithiated heterocycles gave substantially more of the other epimers of the diols, thus permitting access to the α-anomers after Mitsunobu reaction.[130] The oxazole **147** and the compound without the phenyl group were incorporated into oligonucleotides, and the triplex-forming ability of the these towards a purine sequence of duplex DNA was studied.[131]

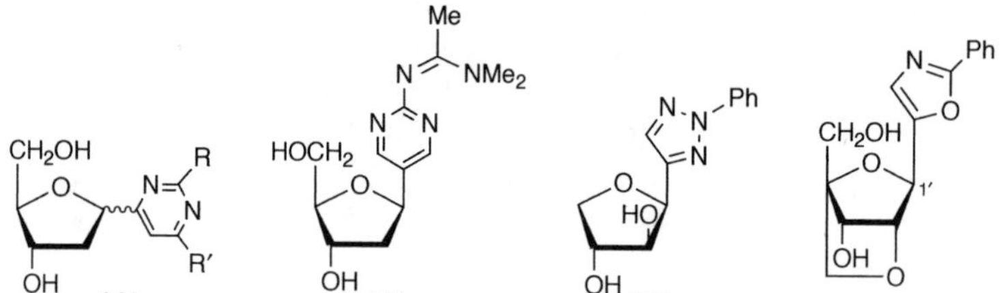

The aza-*C*-nucleosides **149** (X = H, NH₂) are potent inhibitors of purine nucleoside phosphorylase. These compounds, termed 'immucillins', have been prepared from the acetonitrile derivative **148**, which could itself be made by addition of lithiated acetonitrile to the cyclic imine. Various 5'-deoxy-, 5'-deoxy-5'-fluoro-, and 2'-deoxy-analogues were also prepared.[132]

11   Carbocyclic Nucleosides

A review has appeared covering the chemistry of carbocyclic nucleosides published over the period 1994–1998, discussing compounds with ring sizes from three to six, and also bicyclic systems.[133]

148 **149**

A report has discussed at length work carried out in Trost's laboratory on the synthesis of carbocyclic nucleosides such as carbovir (**150**) and abacavir (**151**) from *cis*-3,5-dibenzoyloxycyclopent-2-ene, using Pd(0)-catalysed introduction of the nucleobase and a precursor of the hydroxymethyl group, and introducing asymmetry by the use of chiral ligands on palladium (see Vol. 26, p. 247). The work was extended to the synthesis of 2',3'-dihydroxylated compounds, including (−)-aristeromycin and (−)-neplanocin A (see Vol. 31, p. 261).[134] In an alternative approach to the same compounds, the *syn*-aldol product **152** was converted using Grubbs' catalyst into the cyclopentene **153**, and Pd(0) chemistry was again used to attach the base units in syntheses of carbovir (**150**) and abacavir (**151**).[135] A similar approach using olefin metathesis was also used in a synthesis of carbocyclic ribavirin (**154**) but, due to a low yield in the step in which the base (as its ethyl ester) was linked to a 1',2'-epoxide, a more linear sequence in which the heterocycle was assembled stepwise, and with enantiopure 2-aza-bicyclo[2.2.1]hept-5-en-3-one as starting material, proved more satisfactory. Linkage of the intact base unit to a different carbocyclic synthon gave a route to the N²-linked analogue of **154**.[136]

150 **151** **152** **153** **154**

150, 151

Enantiopure 2-azabicyclo[2.2.1]hept-5-en-3-one has also been used as starting material in an efficient synthesis of abacavir (**151**), amenable to large-scale production.[137] Carbocyclic 5'-norcytidine (**156**), which showed activity against the Epstein–Barr virus, has been prepared, with the alkene **155** as the immediate precursor. The enantiomers of both of these compounds were also described.[138] The uridine analogues of **155** and **156**, and their enantiomers, have also been prepared, using Pd(0)-catalysed coupling of the sodium salt of uracil with enantiopure allylic acetates.[139] A paper describing a route to either enantiomer of a hydroxylated cyclopentylamine useful for making carbocyclic nucleosides is mentioned in Chapter 18. Racemic homocarbovir (**157**) has been prepared from norbornadiene,[140,141] and the analogue **158** of carbovir, and the related abacavir derivative, have also been synthesized as racemates, starting from cyclopenta-diene.[142]

155   156   157   158

(−)-Carvone was used as a precursor for the synthesis of the cyclohexene **159**, and this was used to make the cyclohexenyl nucleoside analogue **160**. The same chiral precursor was also manipulated to give the enantiomers of **159** and **160**. Both enantiomers of **160** showed potent antiherpetic activity. Molecular modelling of the binding of both compounds to the active site of HSV-1 thymidine kinase was carried out, and a model for the binding of both enantiomers was proposed.[143] An analogue of carbocyclic 2′-deoxyuridine, conformationally-restricted due to a 6,6′-oxido-link (Vol. 32, p. 278), has now been reported in optically-active form, in the L-series.[144]

An intermediate prepared from isopropylidene-D-glyceraldehyde *via* a Claisen rearrangement and previously used for the synthesis of some branched-chain fluorinated nucleosides (Vol. 32, p. 272–273), has now been used to make a fluorinated cyclopentane unit, to which was linked 6-chloropurine, thus leading to carbocyclic 2′,3′-dideoxy-4′-fluoro-L-adenosine (**161**).[145] The same synthetic sequence could be modified to generate, *via* a metathesis step, an enantiomeric cyclopentane unit, and hence the enantiomer of **161**, and equivalent structures with the other nucleobases.[146]

There has been a further report on the synthesis of racemic 2′,3′-dideoxy-2′,3′-methanoadenosine (see Vol. 33, p. 297).[147] 4′-Hydroxymethyl-carbocyclic nucleosides **162** (B = Ade, Thy) have been made as racemates, along with their 3′-epimers,[148] and the compounds **163** (B = Ade, Hx) have been described.[149] References to the carbocyclic analogues of oxanosine and oxetanocin are mentioned in the next section.

159   160   161   162   163

12   Nucleoside Antibiotics

Oxanosine (**164**, X = O) has been converted into its 2′-deoxy-, 2′,3′-dideoxy- and 2′,3′-didehydro-2′,3′-dideoxy-analogues, and into its 5′-monophosphate, by conventional manipulations. The carbocyclic analogue **164** (X = CH_2) of oxanosine was also prepared from (−)-2-azabicyclo[2.2.1]hept-5-en-3-one, and, by variations on this route, the 2′,3′-dideoxy- and 2′,3′-didehydro-2′,3′-dideoxy-compounds were synthesized.[150] Derivatives of bredinin (**165**) have been prepared, in which the antibiotic is deoxygenated, phosphorylated or carbamoylated at O-5′. To make these it was necessary to carry out a photochemical ring opening of the isopropylidene derivative of bredinin, giving **166**, on which the modifications could be performed, followed by reformation of the imidazole ring.[151]

A synthesis has been reported of phosmidosine A (**167**), the *N*-acylphosphoramidate being assembled by reaction of a 5′-phosphoramidite of 8-oxoadenosine with prolinamide, protected at the ring nitrogen with the base-labile 4,4′,4″-tris(benzoyloxy)trityl group, followed by oxidation at phosphorus and deprotection under non-acidic conditions.[152]

The glycosylated uridine **168** has been prepared, using Crich's procedure for making β-mannosides (Vol. 30, p. 18). This compound is an analogue of tunicamycin, in which the disaccharide mimics the undecose tunicamine in the antibiotic, and which is capable of orthogonal derivatization on the L-mannose unit at positions corresponding to further units of the tunicamycin structure, with a view to producing analogues with more selective bioactivity. The analogue **169**, incorporating a fatty acid chain similar to that in tunicamycin, was made from **168**.[153]

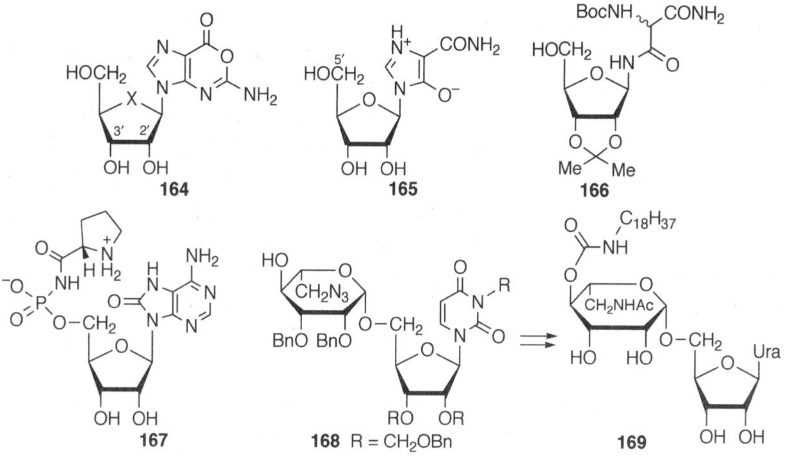

Ribosylation of isopropylideneuridine and subsequent manipulations led to the synthesis of **170** (R = H), which constitutes a part-structure of the liposidomycin class of antibiotics. The two isomers of **170** (R = CH₂OH) were also prepared in synthetic sequences that involved ribosylation of D-allofuranose and L-talofuranose derivatives at O-5, with introduction of uracil at a late stage. Molecular modelling was carried out of both liposidomycins and tunicamycin with the UDP-*N*-acetylmuramic acid-pentapeptide that is the substrate for the enzyme (translocase) in bacterial cell wall biosynthesis that the antibiotics inhibit, and, in accordance with the predictions, only the *S*-isomer of **170** (R = CH₂OH) was a good inhibitor.[154]

A number of nikkomycin analogues of type **171** have been prepared by acylation of uracil polyoxin C with aminoacid units, using *N*-hydroxysuccinimidyl esters as intermediates. The best inhibitory activity against fungal chitin synthase was found for cases in which the aminoacid was an *S*-arylmethyl derivative of L-cysteine or penicillamine.[155]

The final stages in the biosynthesis of the peptidyl-nucleoside antifungal agent blasticidin S, produced by *Streptomyces griseochromogenes*, have been elucidated. The organism exports blasticidin S in the form of an *N*-leucyl derivative as a self-protective mechanism.[156]

A new synthesis of the inositol 1,4,5-trisphosphate receptor agonist adenophostin A (**172**) has been reported, in which a suitably-protected adenosine is linked to a glucosyl 1-phosphite in an α-selective coupling, and where the manipulation between the coupled product and the final target **172** was minimized.[157,158] The method was also applied to the synthesis of analogues in which the glucose bisphosphate unit is replaced by either a D-mannose bisphosphate or a D-xylose bisphosphate.[158] A synthetic approach in which attachment of the base is a late step was used to make the adenophostin analogues with the adenine unit replaced by either an imidazole ring or purine itself (*i.e.* lacking the 6-aminogroup). This latter analogue had similar potency to adenophostin A, whilst the imidazole derivative was approximately equipotent with inositol 1,4,5-trisphosphate.[159] A similar synthetic approach has been used to prepare the 5'-amino-5'-deoxy-analogue of adenophostin A, which was then linked as an amide to a benzophenone derivative to give an adenophostin derivative with an attached photoaffinity label. Another photoaffinity derivative was also prepared, in which the benzophenone was linked *via* a tether as an aglycone replacing the adenine unit.[160]

Staurosporine (**173**) has been converted to the 4',5'-alkene *via* amine oxide pyrolysis; the alkene was hydroxylated, and also subjected to hydroboration–oxidation to give regioselectively the 5'-α-alcohol. Various other subsequent manipulations at the 4'- and 5'-positions were also reported.[161] 4'-*N*-Methyl-5'-hydroxystaurosporine and 5'-hydroxystaurosporine are new indolocarbazoles that have been isolated from a marine *Micromonospora* strain.[162]

In the area of carbocyclic nucleoside antibiotics, a full account of syntheses of (−)-aristeromycin and (−)-neplanocin A (see Vol. 31, p. 261) was mentioned above.[134] The known intermediate **174**, prepared from D-ribose, has been used for the first synthesis of neplanocin C (**175**), a minor component of the neplanocin family. The diastereomeric epoxide was also obtained.[163] In efforts to prepare prodrugs, the oxetanocin analogue **176**, an antiviral agent (lobucavir), has been selectively aminoacylated with L-valine on either of the hydroxymethyl groups using enzymic methods.[164] A synthesis of the cyclohexenyl nucleoside antibiotic pyralomycin 1c is mentioned in Chapter 18.

An inhibitor of seryl t-RNA synthetase, SB-217452, isolated from a *Streptomyces* species, has been identified as the 4'-thionucleoside **177**. The nucleoside moiety seems to be identical to one obtained as an enzymic cleavage product of albomycin δ2 (Vol. 18, p. 182).[165]

A paper on the total synthesis of spicamycin aminonucleoside is mentioned in Chapter 10.

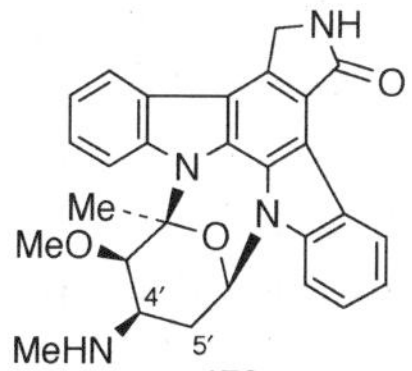

13   Nucleoside Phosphates and Phosphonates

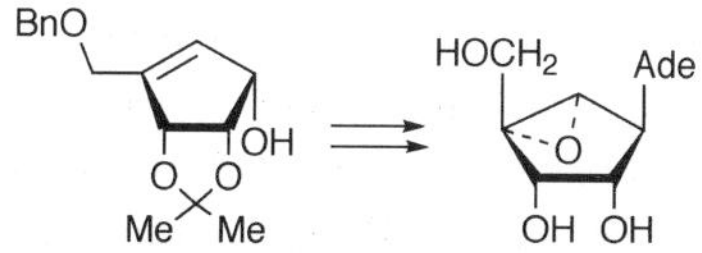

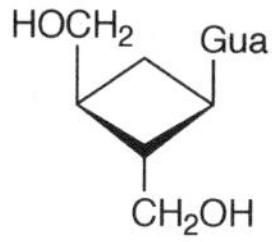

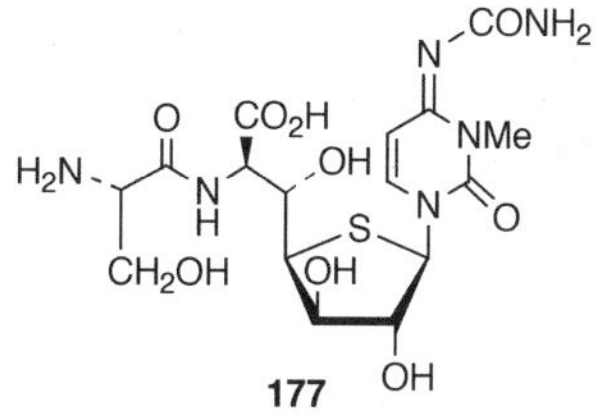

13.1 Nucleoside Mono- and Di-phosphates, Related Phosphonates and Other Analogues. – The reagent **178** has been developed for the phosphorylation of the 3′- or 5′-hydroxy-groups of otherwise-protected deoxynucleosides. The reagent incorporates a Dmtr group to help monitor the progress of phosphorylation, and the nucleoside phosphate is produced by treatment of the initial phosphodiester with NaOH in pyridine–ethanol, which can be done during work-up of the reaction.[166]

A route has been developed for the synthesis of ribonucleoside 2′-phosphates **179**, which could be incorporated into oligonucleotides. The *t*-butyl esters could be removed under mild acidic conditions without cleavage of internucleotidic links.[167]

The mesylate **180**, the synthesis of which from the tertiary alcohol required the preparation of the sulfinate followed by oxidation, gave **181** on treatment with TBAF, by an unusual phosphonate-to-phosphate rearrangement of unknown mechanism.[168]

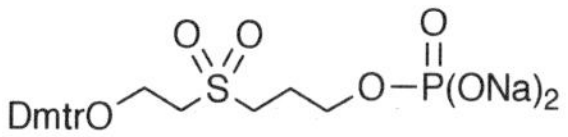

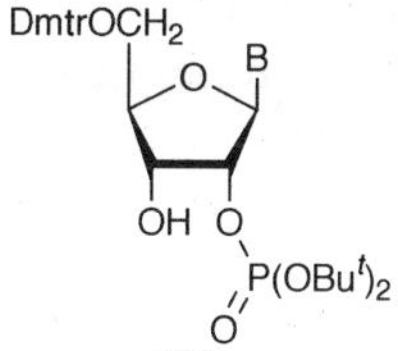

Triesters of type **182** ($n = 2–5$) have been prepared from the nucleoside 3′-H-phosphonate, *via* the cyclic phosphites. Methods were also developed for making H-phosphonates and phosphodiesters involving just one of the hydroxy-groups of the diols.[169]

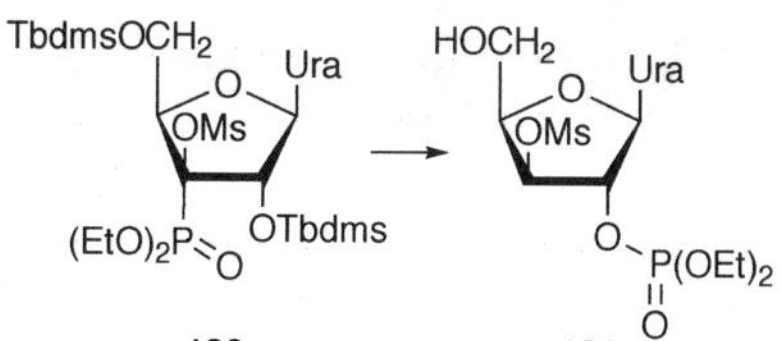

A report from Stec's laboratory describes the preparation and separation of the diastereoisomers of triethylammonium 5′-*O*-Tbdms-thymidine 3′-*O*-methanephosphonothioate, and their reaction with *p*-nitrophenylsulfenyl chloride to give the disulfide **183**. Treatment of **183** with TPP gave the pyrophosphonate derivative **184**, whereas the use of TPP in the presence of excess *p*-nitrophenyl-sulfenyl chloride gave the methanephosphonodithioate **185** in a stereospecific manner.[170] *S*-Phenyl methanephosphonothioates **186**, the diastereomers of which can be separated, have been prepared by reaction of the protected nucleosides with methanephosphonyl dichloride, followed by thiophenol. They could be used for the preparation of methanephosphonyl dinucleotides, but the procedure did not prove to be optimal.[171] Procedures have been developed for the conversion of nucleoside 3′-H-phosphonate monoesters into H-phosphonothioates **187** (X = O) or H-phosphonodithioates **187** (X = S) by reaction of aryl diesters of bis(aryl) phosphites respectively with 1,1,1,3,3,3,-hexamethyl-disilathiane.[172]

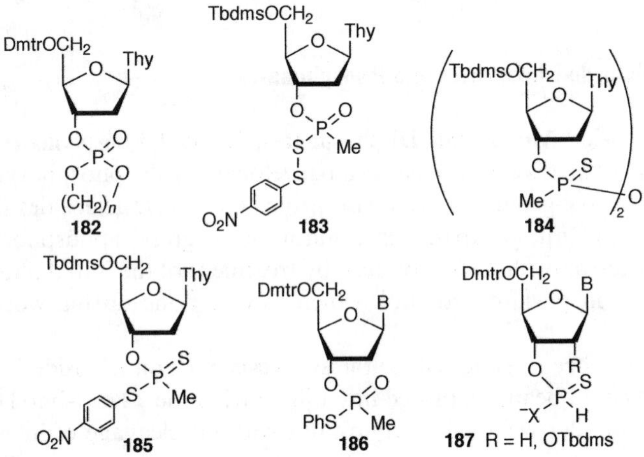

2′-Deoxynucleoside 3′-*C*-phosphonates **188** have been prepared by reaction of 3′-ketonucleosides with tris(Tms) phosphite. Some of the chemistry of these geminal hydroxyphosphonates was also explored.[173] 3′-Methylene-H-phosphonates **189** [X = H, OMe, F, O(CH$_2$)$_2$OMe], and also 5′-*O*-Dmtr species of use for oligonucleotide synthesis by the H-phosphonate method, have been prepared by reaction of iodomethylene compounds with bis(trimethylsilyoxy)phosphine.[174]

GMP has been prepared from guanosine by an enzymic method,[175] and an improved synthesis of guanosine 5′-monothiophosphate from 2′,3′-*O*-iso-propylideneguanosine, which avoids the need for chromatography, has been described.[176] The pyrimidinone **7** has been converted into the 5′-phosphate **190**, for use in studies of covalently-linked base pairs.[12] Nucleotide libraries of type **191** (X = O, S) have been assembled using solid-phase techniques.[177] The uridine-derived 5′-oxyphosphorane **191a** has been prepared from the 5′-phosphite and *o*-chloranil.[178]

A range of deoxyadenosine bisphosphates with modified ribose units have been prepared and evaluated as ligands at the P2Y$_1$ receptor. The modications included deoxy-compounds, carbocyclic analogues, anhydrohexitol nucleosides, and morpholino-compounds.[179] Improved routes have been reported to the

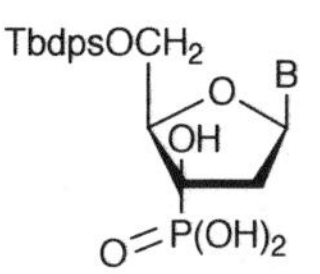

188 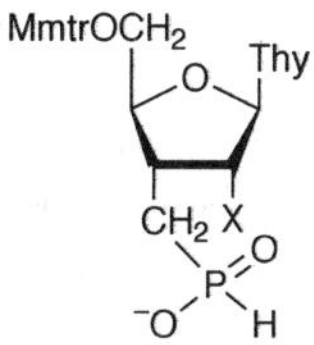 **189** 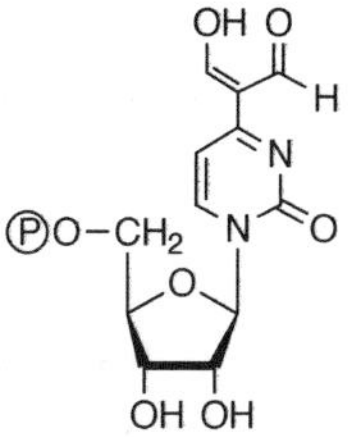 **190**

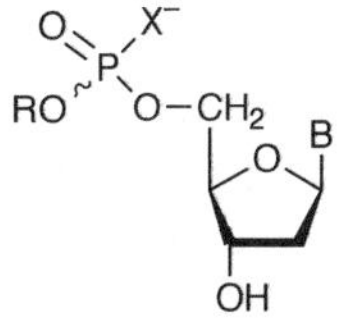

191 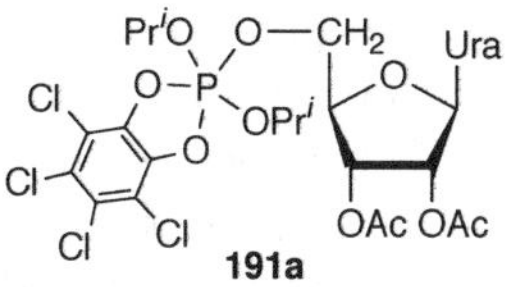 **191a**

3′,5′-bisphosphates of 4-thiouridine, 6-thioinosine and 6-thioguanosine, for use as donor molecules in RNA ligation.[180]

An extensive paper from Sekine's laboratory has discussed the synthesis of aminoacylamido-derivatives of AMP (**192**), analogues of aminoacyl adenylates. The syntheses, which were successfully completed using a number of α-amino acids, used as a key step the reaction of 5′-*O*-phosphoramidite derivatives of adenosine with the amides of the suitably-protected amino acid.[181] The application of similar chemistry in the synthesis of phosmidosine was mentioned above.[152]

2′-*O*-(5-*O*-Phosphoryl-β-D-ribofuranosyl)adenosine has been prepared from a previously-described ribosylated adenosine (Vol. 31, p. 296).[182]

There have been further reports on potential prodrugs of AZT and its monophosphate. Phosphoramidates **193** (X = Y = O) have been prepared using H-phosphonate chemistry,[183] and both phosphoramidothioates **193** (X = S, Y = O) and phosphoramidodithioates **193** (X = Y = S), with phenylalanine and tryptophan as amino acids, have been synthesized using 1,3,2-oxathia- or -dithia-phospholane chemistry.[184] *S*-Acyl-2-thioethyl aryl phosphotriester derivatives of AZT have been prepared as mononucleotide prodrugs. They acted as such in thymidine kinase-deficient cells, and it was proposed that activation involves successive esterase and phosphodiesterase hydrolysis.[185] H-Phosphonate diesters **194** have been prepared; esters of primary or secondary alcohols were degraded in serum or phosphate buffer to give AZT, whereas esters of tertiary alcohols gave AZT-5′-hydrogen phosphonate. Similar findings were reported for ddA and d4T.[186]

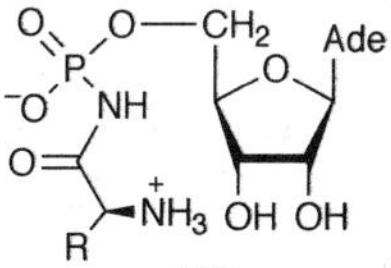

192 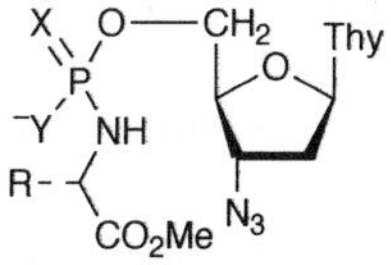 **193** 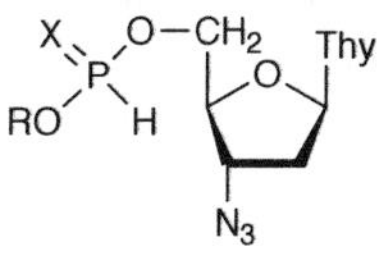 **194**

The novel 5-fluoro-2′-deoxyuridine phosphoramidate prodrug **195** has been prepared, along with the compound with only one bromoethyl group. The mechanism of conversion of **195** to FdUMP was thought to be *via* aziridinium species, after initial hydrolysis of the *N*-hydroxybenzotriazolyl ester.[187] The mechanism of hydrolysis of some related thymidine-derived phosphoramidates was investigated, with aziridinium intermediates being involved in the case of *N*-(2-bromoethyl) compounds, whilst a piperidino-group on phosphorus was hydrolysed by an endogenous phosphoramidase.[188] Some 5'-phosphodiesters of 1-(β-D-arabinofuranosyl)-2-thiocytosine and a long-chain alkanol were synthesized and evaluated *in vivo* as antitumour agents, with some compounds showing promising properties.[189]

5'-Vinylphosphonates of cytosine, uracil and ara-C have been prepared by Wittig reactions, and intermediates in these syntheses were hydroxylated using AD-mix-α to give the product **196** and the related derivatives of uridine and cytosine.[190] Geminal hydroxyphosphonates **197** (R = Me or H, and with all four nucleobases) have been prepared by addition of dimethyl phosphite or tris-Tms phosphite to 5'-aldehydes. Similar products were obtained from 2',3'-*O*-isopropylidene ribonucleosides, except in the case of adenosine. NMR studies indicated that the major diastereoisomer formed in each case had 5'-*R*-stereochemistry, and in some cases this was produced in up to 90% d.e.[191]

The 5'-(P^α-boranodiphosphate) analogues **198** of ADP and GDP have been prepared, by reaction of 2',3'-di-*O*-acetyl-5'-phosphoramidites with the borane complex of Hünig's base, followed by deacetylation and coupling with phosphate. The diastereoisomers were separable by RP-HPLC. The products, more lipophilic than the diphosphates and resistant to nucleases, could have application in boron neutron-capture therapy.[192]

13.2 Cyclic Monophosphates and Their Analogues. – A new preparative route to cyclic phosphorothioates of type **199** involves the reaction of 5'-*O*-Dmtr-nucleosides with diphenyl H-phosphonate in the presence of pyridine, followed by treatment of the resultant cyclic H-phosphonates with sulfur and final deprotection with acetic acid. The method was used with all four ribonucleobases, and gave the products as 1:1 mixtures of diastereoisomers.[193]

The phosphinic acid analogue **200** of cyclic AMP has been prepared by linking adenine to a cyclic methyl phosphinate derived from di-*O*-isopropylidene-glucose (Chapter 17).[194] Thymidine cyclic 3',5'-phosphorofluoridate and its sulfur analogue (**201**, X = O or S), as 1:1 mixtures of diastereomers, have been prepared by oxidation or sulfurization of the cyclic phosphorofluoridite.[195] 7-Deoxypaclitaxel has been linked to cyclic AMP as a phosphotriester; the product showed enhanced cytotoxicity against human cancer cells.[196]

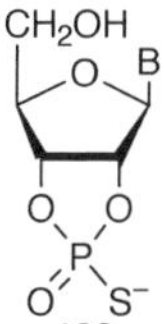

199 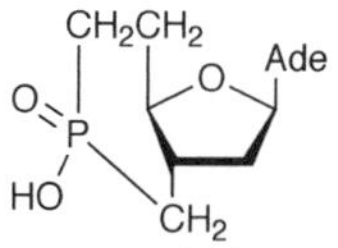 **200** 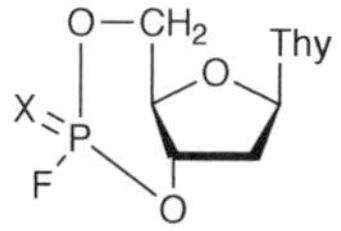 **201**

13.3 Nucleoside Triphosphates and Their Analogues. – A review has been given of the most useful methods for the synthesis of nucleoside triphosphates.[197] 5′-Triphosphates of 8-(alkylthio)adenosines have been prepared as inhibitors of nucleoside triphosphate diphosphohydrolase,[198] and the triphosphates **202** and that derived from 2,2′-anhydrouridine have been made as agonists for P2X$_2$-purinoceptors, but they showed lesser potencies than the parent nucleosides.[199] Derivatives of ATP, UTP and CTP have been prepared in which methyl ketone groups are attached *via* spacers to the base units, in order to permit interaction with fluorescent probes after enzymic incorporation into oligonucleotides. The triphosphate was assembed by Eckstein's procedure, in which a 2′,3′-*O*-iso-propylidene nucleoside is treated sequentially with salicyl phosphorochloridite, pyrophosphate and an oxidant.[200]

Reduction of ATP to the 2′-deoxy-compound has been carried out on a preparative scale using recombinant *Lactobacillus leichmannii* ribonucleotide triphosphate reductase.[201]

The 3′-*O*-phosphonomethyl compound **203** has been prepared by alkylation of O-3′ of thymidine, and was converted into the triphosphate analogue **204** by reaction with tributylammonium pyrophosphate, previously activated by carbonyldiimidazole. Initial reaction of **203** with carbonyldiimidazole led to the formation of a 'dimeric' pyrophosphonate of **203**. Similar chemistry was also carried out on α-thymidine, and the substrate properties of the products **203** and its α-anomer towards DNA polymerases were studied.[202]

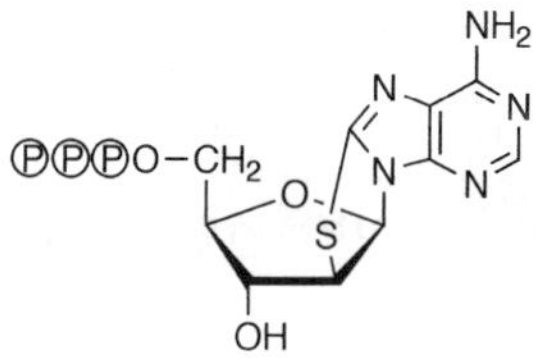

202 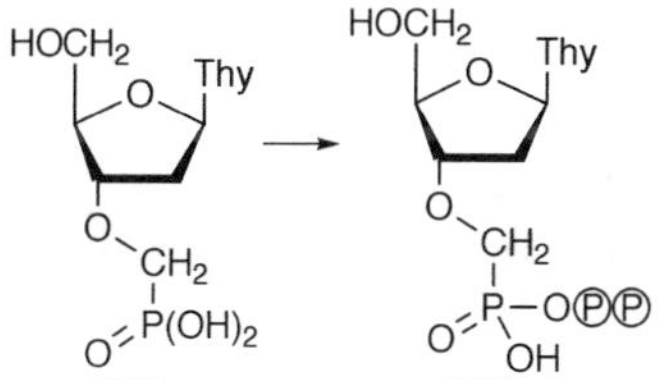 **203** **204**

13.4 Nucleoside Mono- and Di-phosphosugars and Their Analogues. – A synthesis of UDP-GlcNAc from UMP and GlcNAc has been developed which uses an engineered mutant.[203] The analogue **205** of UDP-Gal has been prepared as a potential transferase inhibitor; the synthesis involved the formation of a *C*-hydroxymethyl compound *via* the reaction of tetra-*O*-benzyl methyl galactoside with propargyl trimethylsilane and BF$_3$, followed by ozonolyis and reduction. Similar analogues of UDP-GlcNAc and UDP-GalNAc were also described.[204]

A range of analogues of CMP-NeuNAc have been reported from Schmidt's laboratory. These include compounds in which the cytosine unit is replaced by a methoxy group or a *C*-linked resorcinol unit, and cases where the sialic acid moiety is modified in the *N*-acyl group, is replaced by a KDN unit, or modified at C-8 and C-9. All were made by reaction of sialyl phosphites with ribosyl phosphates, using phosphate–phosphate exchange. The compounds with the cytosine unit replaced were not substrates for α(2-6)-sialyltransferase, whereas the modifications at the 5-,8- and 9-position of the neuraminic acid residue were tolerated.[205] The *S*-linked analogue of CMP-NeuNAc has been prepared; in the key step, a cytidine-5′-phosphoamidite reacted with a sialyl thiol. The product was a substrate for α-(2-3) sialyltransferase.[206]

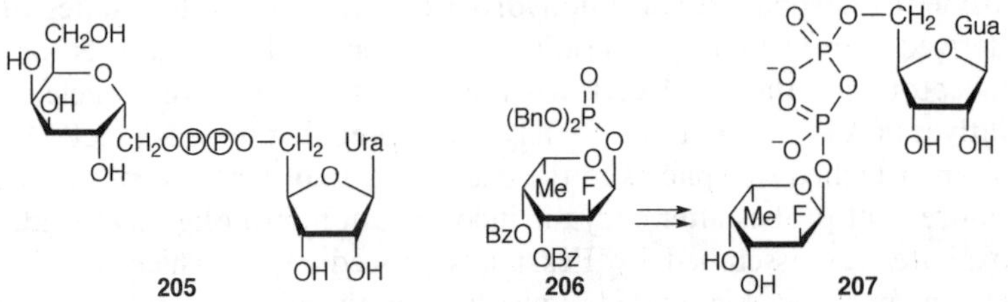

Three fluorinated sugar nucleotide analogues have been reported. L-Galactose was used to prepare GDP-6-deoxy-6-fluoro-β-L-fucose ,and the glycosyl phosphate **206**, prepared from the glycal using Selectfluor and dibenzyl phosphate, was used to make the α-anomer **207** of GDP-2-deoxy-2-fluoro-L-fucose, the pyrophosphate link being constructed chemically. A similar procedure was used to synthesize UDP-2-deoxy-2-fluoro-α-D-galactose, the fluorinated glycosyl phosphate being coupled enzymically to the UMP unit. The effects of the compounds on fucosyltransferases and sialyltransferases were studied, with which in general they acted as competitive inhibitors and formed tight complexes with the enzymes.[207]

In a route to biosynthetically-significant nucleotides of ketosugars, oxidation of partially-protected glycosyl H-phosphonates such as **208** (Scheme 7) with RuO$_4$ gives ulosyl phosphates in good yield, and these can then be converted conventionally to nucleoside diphosphosugars such as **209**.[208] TDP-3,6-Dideoxy-β-L-*arabino*-hexopyranose (TDP-β-L-ascarylose) has been produced from TDP-3-deoxy-α-D-glucose by the rhamnose-synthesizing enzyme system isolated from *Salmonella entericum* LT2.[209]

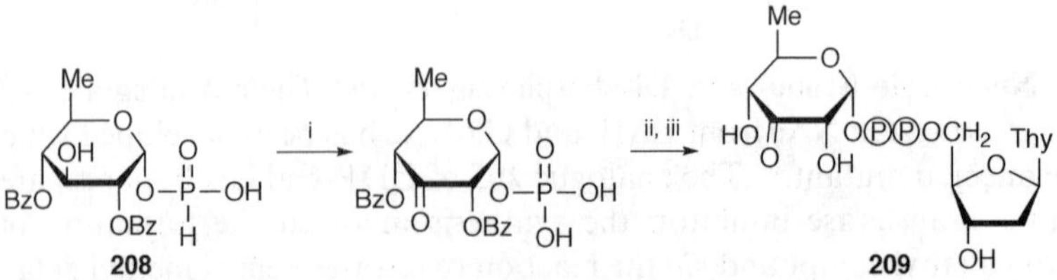

Reagents: i, RuO$_2$, NaIO$_4$, CH$_2$Cl$_2$/H$_2$O; ii, dTMP-morpholidate, MeCN; iii, LiOH aq.

Scheme 7

The stereochemistry of the ADP-heptose which acts as the glycosyl donor for the heptose sugar units in bacterial lipopolysaccharides has previously been undefined, with regard to both anomeric configuration and side-chain stereochemistry. The synthesis of both anomers of ADP-L-*glycero*-D-*manno*-heptose, and of the D-*glycero*-isomer, has now been accomplished. Evaluation in an *in vitro* system in which two Kdo residues and one heptose were transferred sequentially to a synthetic lipid A analogue showed that ADP-L-*glycero*-β-D-*manno*-heptose (**210**) is the natural sugar donor for the heptosyltransferases from *E. coli*, although the D-*glycero*-β-D-*manno*- isomer was also accepted at a considerably slower rate.[210]

The CDP-derivative **211** of 2-*C*-methyl-D-erythrito, an intermediate in the non-mevalonate pathway to isoprenoids, has been synthesized from 2-*C*-methyl-D-erythritol-4-phosphate and CTP, using the transferase which produces **211** naturally.[211] Enzymatic routes have also been used to produce **211** with multiple specific ¹³C-labels, and also 2-¹⁴C-labelled material.[212]

The boranodiphosphate analogue **212** of ADP-glucose has been synthesized, in a procedure in which a 5′-phosphoramidite of adenosine was treated with a borane–amine complex, and the product then deprotected and coupled with α-D-glucopyranose-1-phosphate.[213]

210 **211** **212**

13.5 Small Oligonucleotides and Their Analogues. – *13.5.1 3′→5′-Linked systems; methodology and modified internucleotidic links.* New protecting groups have been developed for the internucleosidic phosphate links in oligonucleotide synthesis. The phosphoramidite **213**, containing the 2-[(1-naphthyl)carbamoyloxy] ethyl (NCE) group has been used in conventional solid-phase synthesis, and the NCE group could be removed from the assembled phosphate or phosphorothioate oligonucleotide using aqueous ammonia.[214] Other workers have studied a number of β-(heteroaryl)ethyl groups, such as in **214**, where a broad range of sensitivity towards base could be achieved depending on the nature of the heteroaromatic unit.[215] Phosphoramidites containing the 2-cyano-1-*t*-butylethyl protecting group have also been used for oligonucleotide synthesis.[216] Phosphoramidites of 6-methyluridine and 5,6-dimethyluridine have been made and used to incorporate the bases site-selectively into oligonucleotides.[217]

Ring-closing metathesis was applied to the phosphotriester **215** to give the conformationally restricted dinucleotide **216**, as a mixture of all four diastereomers.[218] There has been a fuller report on the synthesis and properties of a uridylyl-(3′→5′)-thymidine dinucleotide with a link between O-2′ of the uridine and the methyl group of the thymine base, designed to mimic the hydrogen-bonded situation in certain tRNAs (Vol. 32, p. 286–287). The unit was incorporated into oligonucleotides, where a bending of the structure was demon-

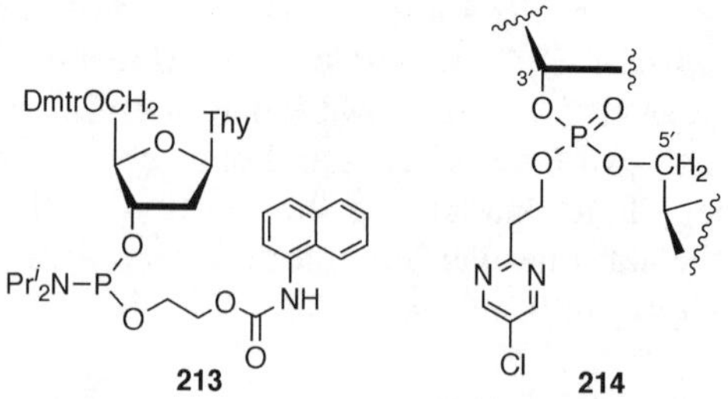

213   **214**

strated by physical measurements.[219] The same group has also prepared the cyclic phosphate **217**, in which the ribose ring has a 3′-*endo*-geometry, and incorporated it into into oligonucleotides at the 5′-terminal site.[220] This work was extended to the synthesis of the dinucleotide **218**, the isomers of which, separated by chromatography, were assigned absolute configurations by CD and computational methods. Detailed conformational studies were carried out, the 'top' ribose unit adopting a 2′-*endo*-geometry and the 'lower' one a 3′-*endo*-conformation. The analogous P=S species was also studied.[221]

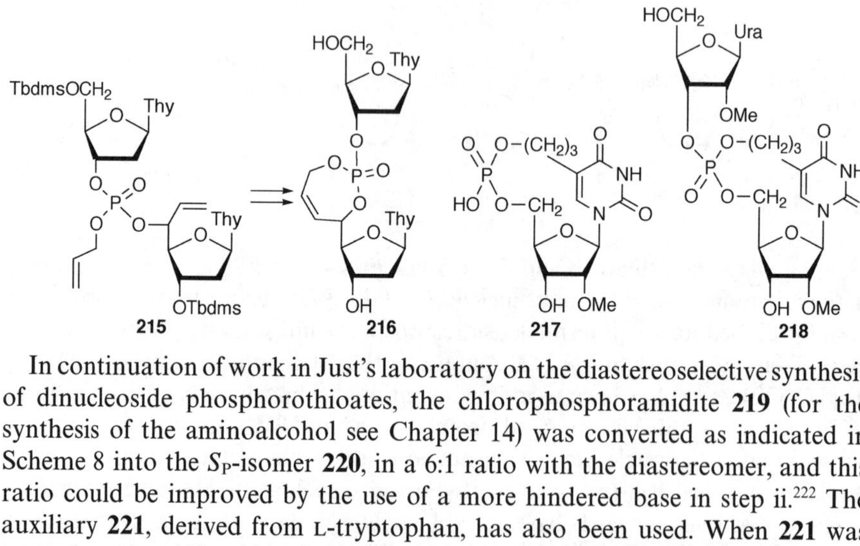

215   **216**   **217**   **218**

In continuation of work in Just's laboratory on the diastereoselective synthesis of dinucleoside phosphorothioates, the chlorophosphoramidite **219** (for the synthesis of the aminoalcohol see Chapter 14) was converted as indicated in Scheme 8 into the S_P-isomer **220**, in a 6:1 ratio with the diastereomer, and this ratio could be improved by the use of a more hindered base in step ii.[222] The auxiliary **221**, derived from L-tryptophan, has also been used. When **221** was treated sequentially with 5′-*O*-Tbdms-thymidine, 3′-*O*-Tbdms-thymidine and Beaucage's reagent, a phosphorothioate triester was produced. The auxiliary could be removed as the aminomethyl compound by treatment with ammonia, to give the R_P-isomer of **220**, in 40:1 excess.[223]

The influence of diastereomeric ratios of deoxyribonucleoside phosphoramidites on the synthesis of phosphorothioate oligonucleotides has been investigated. Almost diastereomerically-pure phosphoramidites were found to give an almost equal mixture of diastereomers of a monophosphorothioate decamer.[224] A facile synthetic route to dimeric phosphorothioate building blocks **222** has been developed. Dinucleoside phosphite triesters were obtained in a one-pot procedure by sequential coupling to PCl₃ of protected nucleosides with

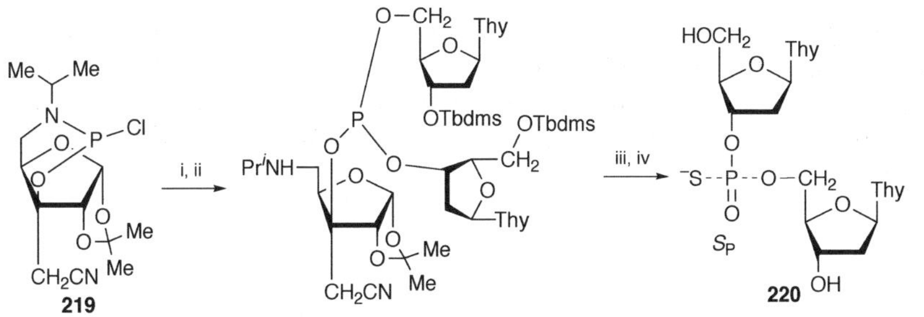

Reagents: i, 5'-O-Tbdms-thymidine, Et₃N; ii, 3'-O-Tbdms-thymidine, 2-bromo-4,5-dicyanoimidazole; iii, Beaucage's reagent; iv, NH₃ aq., then TBAF

Scheme 8

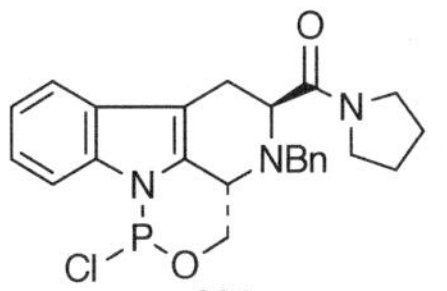

free 5'- and 3'-hydroxy groups, followed by cyanoethanol. Subsequent sulfuriz-ation gave the products **222**.[225] Diastereomeric S-methyl compounds **223** have been made by reaction of the dinucleoside phosphorothioates with MeI and pyridine.[226]

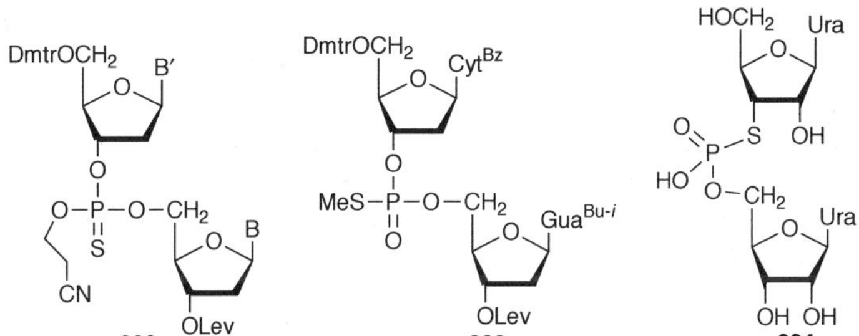

There has been a fuller account of the preparation of 3'-thiouridine dinucleo-tide **224** (Vol. 30, p. 280–281), and a detailed study was made of its stability in acid and base. It was less stable in acid than the parent dinucleotide, and the initial product of treatment either with acid or at pH 10 was 3'-thiouridine 2',3'-cyclic phosphorothioate.[227]

A previously-reported bicyclic 3'-aminonucleoside has now been converted to the dinucleotide building block **225**, which caused a reduction in T_m values when incorporated into oligonucleotides.[228] There has been an extended and expanded account of the preparation of dideoxynucleoside boranophosphates through the intermediacy of H-phosphonate diesters. The conversion of these to the boranophosphates proceeds with retention of configuration.[229]

Workers at Isis Pharmaceuticals have developed methods for the synthesis of chimeric oligonucleotides containing blocks of phosphodiester, phos-phorothioate and phosphoramidate internucleosidic links. H-Phosphonate methodology was used, and methods were developed for conversion of a block of H-phosphonate diester links into the required final internucleosidic links without affecting those of other types already in place. The installation of PO groups in the presence of PS and PN units required oxidation with trithylamine in CCl_4/pyridine/water.[230]

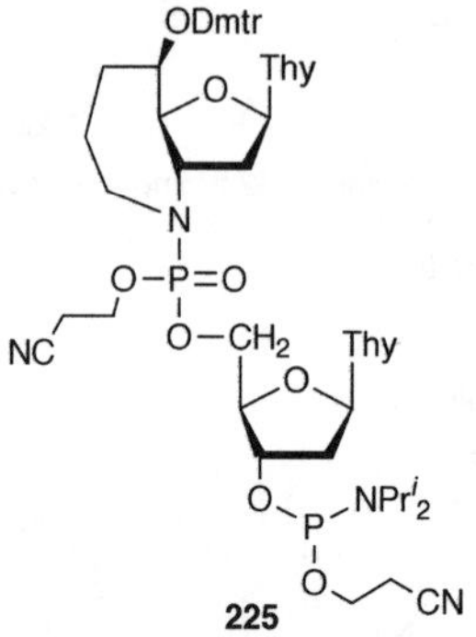

225 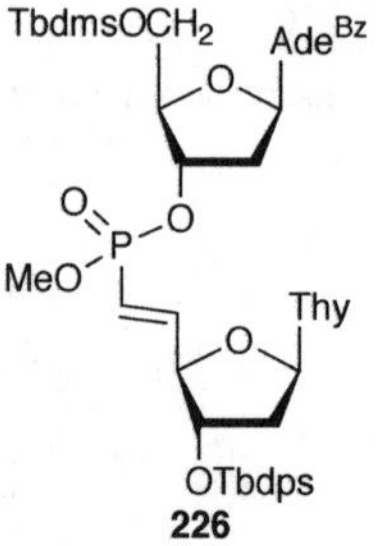   **226**

An improved procedure has been developed for the preparation of vinylphosphonate-linked dinucleotide analogues such as **226**. The method involves the coupling of an H-phosphonate diester at O-3′ of the 'top' nucleoside with a β-bromovinyl derivative of thymidine, using palladium catalysis. Optimization studies were carried out, and under the best conditions, yields were high.[231]

The L-enantiomers of deoxycytidine and deoxyguanosine have been incorporated into the middle of a decadeoxynucleotide, and the effects on the conformation were studied.[232]

A number of references to the incorporation of modified nucleosides into oligonucleotides are given elsewhere in this chapter, and some relevant chemistry of protecting groups in discussed in Section 15. A reference to altritol nucleic acids is mentioned in Section 16.

13.5.2 5′→5′-Linked systems. Treatment of inosine 5′-methylenephosphonate with Im_3PO gave the [bis(inosine-5′)]-tetraphosphate analogue **227**, with two methylene replacements, together with the [bis(inosine-5′)]-pentaphosphate analogue.[233]

An improved route has been developed for the synthesis of the carbocyclic analogue **228** of cyclic IDP-ribose (see Vol. 32, p. 289). In the new approach, the intramolecular formation of the pyrophosphate link was carried out efficiently using an S-phenyl phosphorothiate at O-5′ of the ribose unit, activated by iodine, and there was no need to install a bromine atom at C-8 of the hypoxanthine in order to bias the conformation in favour of cyclization.[234] This strategy of conformational restriction was used in chemistry *en route* to cyclic ADP-carbocyclic ribose, in which the pyrophosphoryl macrocycle was formed, but final deprotection was not reported.[235] Some chemistry directed towards the synthesis of cyclic IDP-glucose has been described, but the pyrophosphate link was not established.[236]

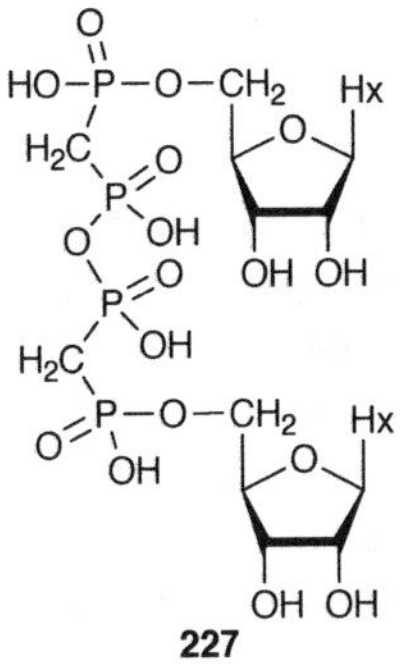

227 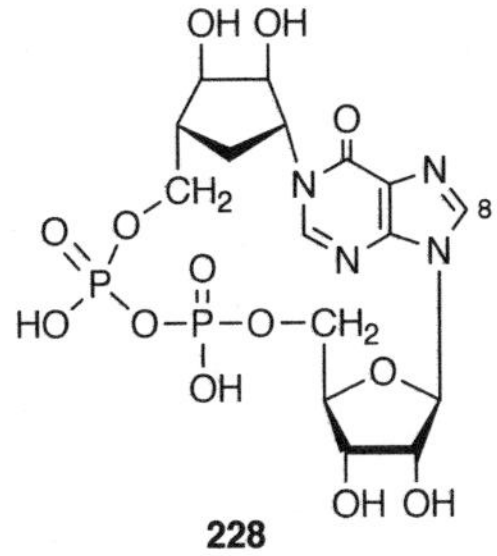   228

14 Oligonucleotide Analogues with Phosphorus-free Linkages

In the area of amide replacements, Robins's group have reported the synthesis of the active ester **229** and its reaction with the amine **230** (Scheme 9) to give the trinucleotide analogue **231**, capable of further extension using **229** after reduction of the azide to an amine. This iterative procedure was used to produce a pentanucleotide analogue, and alternatively coupling of two dimeric units was used to make a tetramer with a 3'-terminal hydroxy group.[237] Other workers have reported the synthesis of C-T dinucleotide analogue with the same amide replacement and its incorporation at the cleavage position of a hammerhead ribozyme substrate. The analogue had high affinity for the ribozyme but did not undergo hydrolysis.[238] A similar U-G dinucleotide analogue has also been made, the amide bond being introduced by reaction of a 5'-amino-5'-deoxyguanosine with a 2',3'-butyrolactone on the uridine unit. This unit was incorporated into a hammerhead ribozyme to give a catalytically active species which was stable to RNase A degradation.[239]

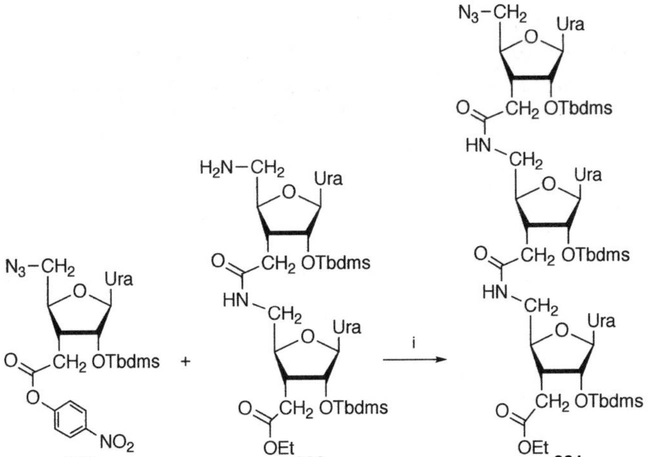

229   +   230   i →   231

Reagents: i, diglyme, 65 °C   **Scheme 9**

The thymidine dimers **232** (R = Me, Bn), with *O*-alkylhydroxamate links, have been prepared.[240] When the *O*-methyl compound was incorporated into oligonucleotides, only minor changes in T_m values were observed for duplexes with either complementary RNA or DNA, but significant decreases were found using the *O*-benzyl species. The very similar *N*-benzyloxycarbamate **233** has also been synthesized, and up to three replacement linkages were incorporated into a thymidine 16-mer, from which the benzyl groups could be removed by hydrogenolysis. The oligomers with *N*-benzyloxy groups showed significant decreases in T_m values for duplexes with poly-dA, but the decrease in stability was less for the *N*-hydroxy-species. The presence of the *N*-hydroxycarbamate links increased stability of the oligonucleotides towards nuclease S1, and the oligomers formed Fe(III) complexes.[241] The bicyclic aminonucleoside which was used in the formation of **225** has also been combined with a thymidine unit to give a carbamate-linked dinucleotide analogue, incorporation of which into oligonucleotides caused significant decreases in duplex stability.[228]

Workers at Novartis have prepared structures **234** (B = Thy, Cyt, Ade) from D-hydroxyproline, and have assembled them into amide-linked oligonucleotide analogues by conventional solid-state peptide synthesis. A homopyrimidine oligomer bound to complementary RNA with significant affinity, whilst no binding was observed with complementary DNA.[242]

A new approach to 3'-deoxy-3'-*C*-formylribonucleosides involves (Scheme 10) the synthesis of **235** from D-xylose; stereoselective reduction with LiAlH$_4$ to give the saturated dithiane was followed by attachment of the base and liberation of the aldehyde **236**. This was converted as indicated into the methylene(methylimino)-linked dimer **237**, with 2'-hydroxy and methoxy groups. This was used to make oligonucleotides with alternating phosphodiester and methylene(methylimino) links, which showed good increases in T_m against RNA complements.[243]

The formacetal-linked dimer **238** was prepared, using the 3'-methylthiomethyl ether as an intermediate, and activating it for coupling using NBS. Photolysis of **238** gave the analogue **239** of a DNA photodecomposition product. This analogue was incorporated into oligonucleotides for comparison with the phosphate-linked material.[244]

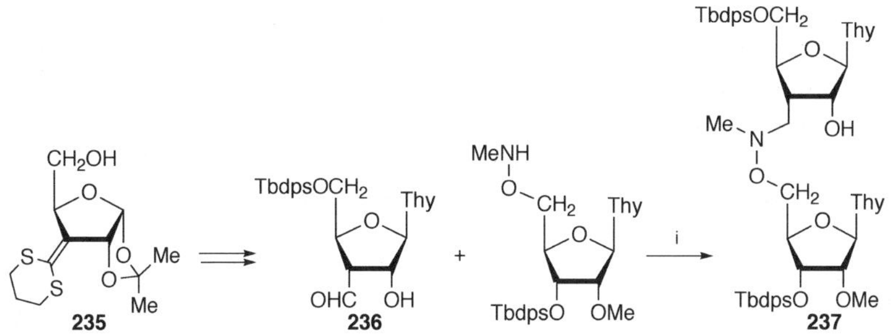

235

Reagents: i, PPTS, then BH$_3$.py

Scheme 10

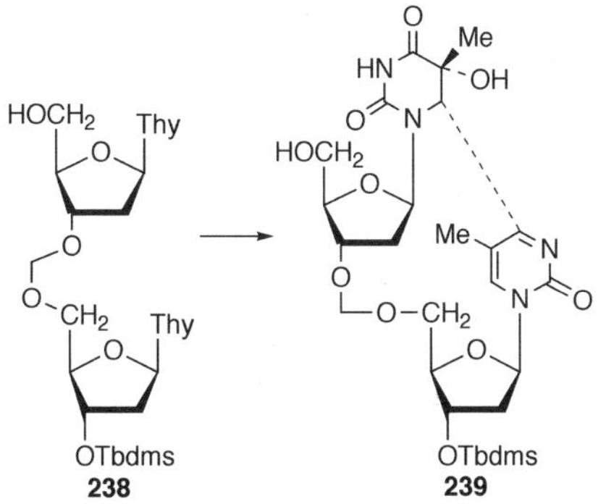

238 **239**

15 Ethers, Esters and Acetals of Nucleosides

15.1 Ethers. – A large-scale synthesis of 2'-*O*-methyluridine has been reported, involving the formation of the 2,2'-anhydronucleoside and its subsequent ring-opening with magnesium methoxide. The method could also be applied to the synthesis of other 2'-*O*-alkyluridines.[245] Several improved procedures have been described for making 2'-*O*-methyl-adenosine and -guanosine, and some *N*-acyl derivatives. These include the preparation of 2'-*O*-methyladenosine by trans-glycosylation of a 2'-*O*-methylcytidine derivative, itself accessible *via* an anhydro-nucleoside, and the selective 2'-*O*-methylation of 2-amino-6-chloropurine ribo-side followed by base modification. Also, 3',5'-Tipds-2,6-diaminopurine riboside can be methylated at O-2', followed by modification of the base, and selective 2'-*O*-methylation of 1-benzylinosine could be accomplished using trimethylsul-fonium hydroxide.[246]

Reaction of *N*-benzyloxymethyl-5-methyluridine with cyclic sulfates gave the 2'-*O*-alkylated compounds **240** (*n* = 2, 3), in a *ca.* 3:1 ratio with the 3'-regioisomer. The sulfate could then be displaced with dimethylamine to give after further manipulation the intermediates **241** (*n* = 2, 3), and similar (methylthio)alkyl ethers could be made using methanethiol.[247] The 3-(dimethylamino)propyl com-pound **241** (*n* = 3) has been incorporated into oligonucleotides, which showed very high nuclease resistance, as did the previously prepared 3-aminopropyl systems, and maintained high binding affinity to target RNA when a few of the modifications were dispersed throughout the oligonucleotide.[248] The same team, from Isis Pharmaceuticals, have also made the related species **242** (R = Me, Et) *via* the intermediacy of a 2'-*O*-(2-hydroxyethyl)-nucleoside and the *N*-alkoxyphthalimide derived from it by Mitsunobu reaction. Again, these modifications, when introduced into oligonucleotides, gave high binding affinity to complementary RNA (but not DNA) and good nuclease stability.[249] Other workers have prepared 2'-*O*-methoxycarbonylmethyl- and 2'-*O*-(2,3-dibenzoyloxypropyl)-derivatives of nucleosides by alkylation of 3',5'-Tipds derivatives, and these were used to make oligonucleotides with carboxymethyl, 2,3-dihydroxypropyl and 2-oxoethyl ethers at *O*-2'.[250]

240   241   242   243

The bisubstrate inhibitors **243** of catechol-*O*-methyltransferase have been prepared by base–sugar coupling, after attachment of the aminoethyl spacer to a ribose derivative. The adenosine analogue was ten times more effective than the compound in which the nucleoside unit was replaced by a methyl group.[251] In somewhat similar fashion, O-3' of the aminoglycoside antibiotic neamine has been linked to O-5' of adenosine by chains of between five and eight methylene units, in order to make bisubstrate inhibitors for aminoglycoside 3'-phosphotransferase, an enzyme which phosphorylates the antibiotic using ATP and thus conveys bacterial resistance. The six- and seven-carbon tethers proved most effective.[252]

The 3-bromopropyl-substituted nucleoside phosphoramidite **244** has been prepared, with a view to its incorporation into oligonucleotides which would permit post-synthetic functionalization of the sugar moiety on the solid support by reaction with appropriate nucleophiles.[253]

Ceric ammonium nitrate on silica gel has been used for the rapid cleavage of trityl, Mmtr and Dmtr ethers from primary alcohol groups of nucleoside and nucleotide derivatives. Primary Tbdms and Tips ethers were also cleaved, the silica gel adding considerably to the catalytic activity.[254] The selective hydrolysis of a Tbdms protecting group at O-5' of multisilylated nucleosides has been carried out in high yields using a mixture of TFA, water and THF in the proportions 1:1:4, at 0 °C.[255] The same method also liberates the 5'-hydroxyl group selectively when it is applied to 3',5'-Tipds-protected nucleosides, with **245** being produced in 95% yield.[256]

15.2 Esters. – A *Pseudomonas* lipase deposited on ceramic particles or on diatomite can acetylate 2'-deoxynucleosides regioselectively at O-3' by acetyl

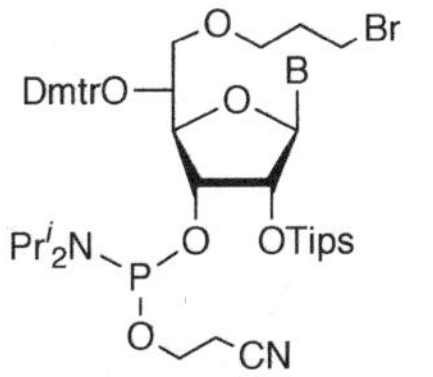

244 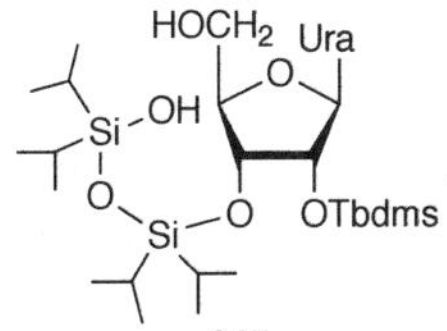   **245**

transfer from various *O*-acetyl-aldoximes or -ketoximes.[257] Amano A lipase will selectively remove the three *O*-acetyl groups from 4-*N*-acetyl-2′,3′,5′-tri-*O*-acetylcytidine, whereas *Burkholderia cepacia* esterase will remove selectively the 5′-*O*-acetyl group from the same substrate.[258]

The ester **246** has been prepared as a potential bioreductively-activated prodrug of 5-fluoro-2′-deoxyuridine (FUDR). It was resistant to human serum esterases due to the steric hindrance of the α-methyl groups, but chemical reduction led to rapid release of FUDR by intramolecular aminolysis.[259]

When the ester **247** was treated with lauroyl peroxide in the presence of 15 equivalents of styrene, the polymer **248** was obtained. This was soluble in many common organic solvents but which could be precipitated with methanol, making the technique potentially useful for parallel synthesis.[260] 5′-*O*-Methacryloyl-uridine and -adenosine have been polymerized onto a silica support, using Cu(I) mediated radical polymerization, under conditions favouring products of narrow polydispersity.[261]

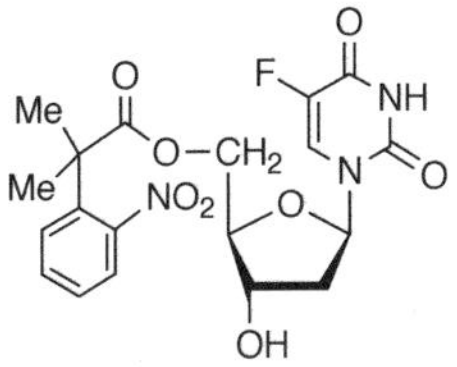

246 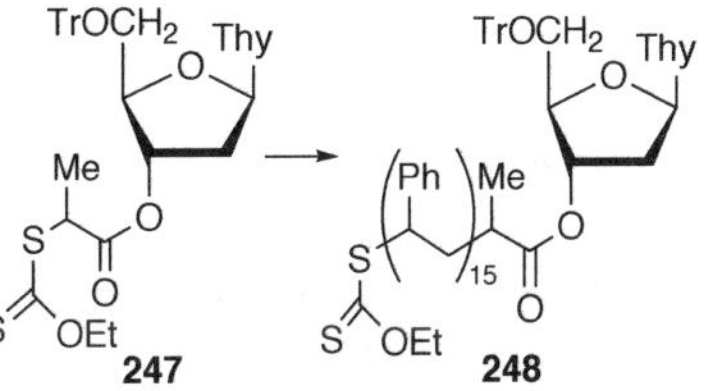   **247**   **248** 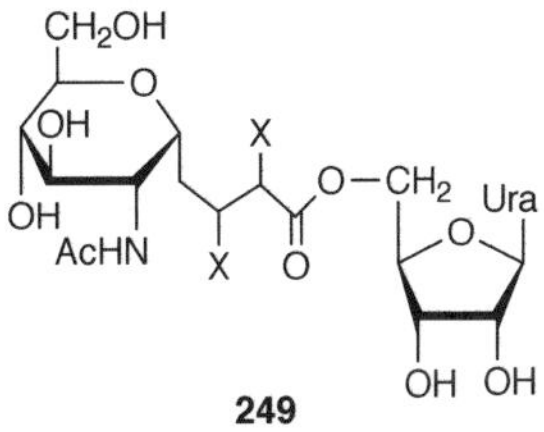   **249**

Analogues **249** (X = H, OH) of UDP-GlcNAc have been prepared as potential inhibitors of chitin synthetases. The synthetic route involved a *C*-allyl derivative of GlcNAc, which was elaborated by ozonolysis, Wittig reaction and coupling with the nucleoside, followed by hydrogenation or hydroxylation as appropriate. The corresponding amides were also prepared from 5′-amino-5′-deoxyuridine.[262]

Lipophilic amino acid methyl esters and methylamides have been coupled to O-5′ of AZT by carbamate links. The products showed anti-HIV activity, but this was not due to carbamate hydrolysis or to direct inhibition of reverse transcriptase, and the mechanism of action may be one not previously observed with nucleoside antivirals.[263]

Adenosine has been converted regioselectively to its 2′-tosylate in 90% yield by treatment with TsCl and Et$_3$N in acetonitrile, in the presence of sub-stoichiometric amounts of organotin reagents such as Bu$_2$SnCl$_2$ or Bu$_2$SnO.[264]

15.3 Acetals. – In continuation of work on protecting groups suitable for use at O-2' during oligoribonucleotide synthesis, Reese and co-workers have studied the properties of a range of 1-aryl-4-alkoxypiperidin-4-yl groups, leading to the development of the 1-(4-chlorophenyl)-4-ethoxypiperidin-4-yl (Cpep) group (as in **250**). This group has greater stability at pH 0.5 and greater lability at pH 3.75 than a number of alternatives such as the previously-used Ctmp and Fpmp groups.[265]

The 2'-*O*-(3-bromopropoxy)methyl group has been introduced at O-2' of ribonucleosides, permitting post-synthetic functionalization of the sugar units of oligonucleotides on the solid support by nucleophilic substitution.[253]

Ribonucleoside 2'- and 4'-*O*-methylthiomethyl derivatives have been synthesized directly from selectively protected nucleosides by reaction with DMSO–Ac$_2$O–AcOH mixtures.[266]

To study the structural requirements of the binding of aminoglycosides to nucleic acids, the conjugate **251**, related to the naturally-occurring nucleoside J, has been prepared and incorporated into oligonucleotides. This led to stabilization of duplexes, which was most marked with complementary RNA.[267]

250   251

16   Other Types of Nucleoside Analogue

A review has been given of the synthesis and biological activity of isonucleosides, analogues in which the heterocyclic base occupies a non-anomeric position.[268] There have been reports on the synthesis of isonucleosides of type **252**, with R = H[269,270] or CH$_2$OH.[270] Isonucleosides related to bioactive benzimidazole nucleosides have been prepared, including **253**, made by displacement of a tosylate with 2-bromo-5,6-dichlorobenzimidazole under phase-transfer conditions, followed by reaction with isopropylamine. The 3'- and 5'-deoxy-analogues of **253** were also reported.[271] Chain-extended compounds **254** have been synthesized from D-glucose, the base (Ade, Ura, 5-fluoro-Ura, Thy) being introduced by reaction with an epoxide at a late stage.[272] The species **255** (B = Ade, Ura, 5-fluoro-Ura, Thy) have also been prepared; again the base was introduced by reaction with an epoxide, itself prepared from D-glucosamine through the intermediacy of 2,5-anhydro-D-mannitol.[273] Compounds enantiomeric with **255** have been reported previously from the same laboratory (Vol. 31, p. 128). Isonucleosides **256** (B = *e.g.* Ade, 8-aza-Ade, Cyt) with an exocyclic methylene group have been prepared, using a keto-sugar derived from D-xylose as an intermediate, and introducing the nucleobases using Mitsunobu reactions. The compounds did not have significant anti-HIV activity.[274] The iso-*C*-nucleosides **257** (R = Ph, Me) have been

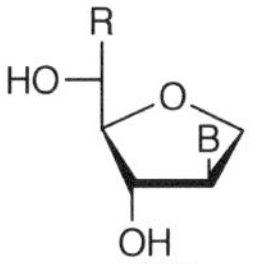

252 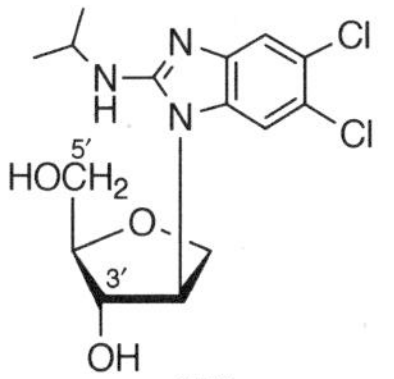   253 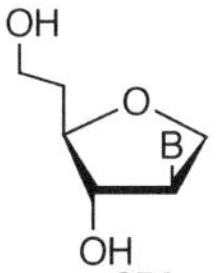   254 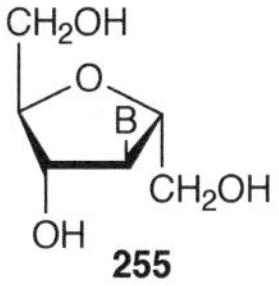   255 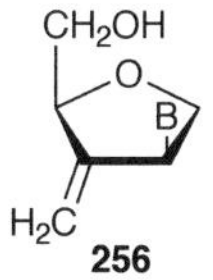   256 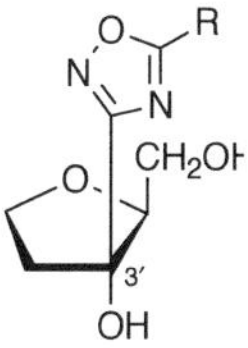   257

synthesized, again using D-xylose as precursor, and the 3'-epimers were also produced.[275]

There has been a further report on altritol nucleic acids (ANA, 258) (see Vol. 33, p. 316), which notes that they are superior to the corresponding DNA, RNA or hexitol nucleic acids (as 258, but lacking the hydroxy groups) in supporting efficient non-enzymatic template-directed synthesis of complementary RNAs from nucleoside 5'-phosphoro-2-methylimidazolides.[276]

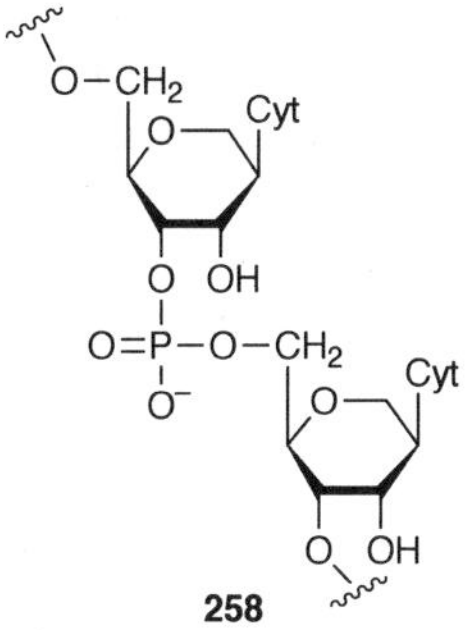

258

Further reports have appeared concerning dioxolane and oxathiolane nucleoside analogues. Compounds of these types, in both enantiomeric series, have been reported with 5-(2-halovinyl)uracils as bases, and the dioxolanes in the L-series showed potent anti-VZV activity.[22] Dioxolane analogues with 5-azacytosine and 6-azathymine as base have been prepared in both enantiomeric series, and as α- and β-anomers. The compound 259 of the L-series showed significant activity against HBV, whilst the enantiomer of 259 has potent anti-HIV activity.[277] Similar oxathiolanes in the L-series, with 6-azacytosine, 5-azacytosine, and various 3-deazapyrimidines as base, have been prepared by chemical transglycosylation of 3-TC.[278] Racemic oxaselenolanes 260 have been made, along with the α-anomers. The enantiomers of 260 were separated by chiral HPLC. Anti-HIV and anti-HBV activity was found for the cytidine and 5-fluorocytidine analogues, and this was greatest for the (−)-isomers, as in the case for 3-TC, but activities were significantly less than for 3-TC.[279] New prodrugs for 3-TC have been reported, in which polyaminated sidearms are attached at either O-5' or N-4, or both positions, with a view to using the polyamine transport system for incorporation of the drug. Some compounds reported had good antiviral activity.[280] The dioxolane and oxathiolane derivatives 261 (X = O, S; R = H, Me) have been prepared, along with the trans-isomers, but did not show significant antiviral activity.[281]

The isoxazolidine **262** was the major isomer formed by cycloaddition of vinyl acetate with a nitrone derived from isopropylidene-D-glyceraldehyde. Further chemistry then led to the L-nucleoside analogue **263**. Reaction of the same nitrone with 1-vinylthymine also gave diastereoselectively a *cis*-disubstituted isoxazolidine which could be converted to the same enantiomer **263**.[282] An alternative route to **263** and its uracil analogue involves the BF$_3$-catalysed reaction of the nitrone with the Tbdms ketene acetal of methyl acetate, which proceeds with the same facial selectivity as the cycloadditions.[283] In a comprehensive paper, the use of serine-derived starting materials has led to the synthesis of isoxazolidinyl analogues of the polyoxins. Thus, for example, the isoxazolidinone **264** was prepared stereoselectively through Michael addition of *N*-benzylhydroxylamine to an enoate derived from D-serine, and could be converted to the nucleoside analogues **265**. Additions of silyl ketene acetals to nitrones, and cycloadditions of nitrones with vinyl acetate, were also used to make oxazolidines, and appropriate choice of protecting groups and chirality of serine permitted access to all possible stereochemistries.[284]

A series of papers from Vasella's laboratory has discussed the synthesis of oligonucleotide analogues that possess a 'nucleobase-including' backbone, with the 3′,5′-phosphodiester link replaced by an acetylenic link between C-5′ of the ribose and C-6 of uridine or C-8 of adenosine. The bis-uridine **266** was made using Sonogashira coupling of a C-5′-alkyne with a 6-iodouridine.[285] The 8-iodoadenosine derivative **267** was prepared by addition of Tms-acetylide to a C-5′ aldehyde.[286] Other workers have reported a stereocontrolled route to a similar non-iodinated adenosine derivative through diastereoselective reduction of a ribofuranosyl acetylenic ketone, followed by introduction of the adenine unit.[287] A bis-adenine unit was made by Sonogashira coupling,[288] and an adenosine tetramer **268** was obtained by coupling an iodinated dimer with an ethynylated dimer.[289]

3′-*O*-Thiocarbamoylthymidine has been converted into various species with heterocycles joined to thymidine through an ether link.[290] Water-soluble C-linked nucleoside-porphyrins have been made by reaction of 2′,3′-*O*-isopropylideneuridine-5′-carboxaldehyde with pyrrole to give a dipyrromethane, which was treated with *p*-fluorobenzaldehyde to give after oxidation a porphyrin with uridine units at opposite meso-positions, into which Pd(II) could be inserted. The Pd(II) metalloporphyrin was an efficient reagent for selective cleavage of double-stranded DNA in visible light.[291]

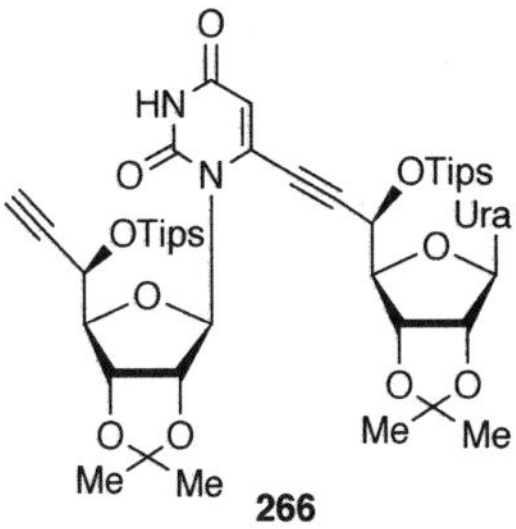

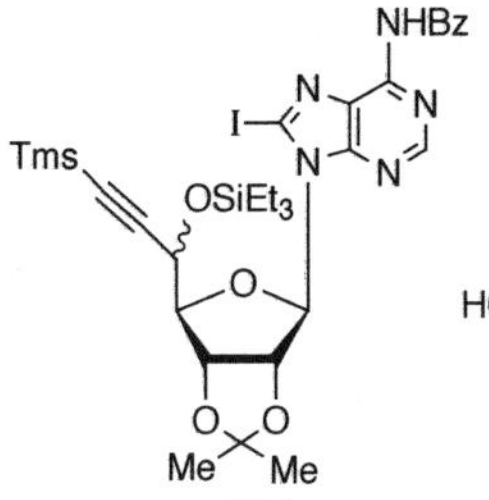

Nucleoside analogues **269** (B = Ade, Gua, Thy) have been made as racemates.[292] The antiviral activity of synadenol has led to the synthesis of the related cyclopropanes **270** and its *E*-isomer, which did not have useful antiviral activity.[293] The cyclobutane **271**, and its *E*-isomer, and the cyclohexenes **272** and **273** have also been described.[294] Various pyrimidine nucleosides of type **274** have been made (for related purines, see Vol. 32, p. 297), and the BVDU analogue shown had the best anti-VZV activity.[295] The cyclopropanes **275**, involving all five main nucleobases, have been made in chiral form, the Ade and Gua compounds having moderate anti-HIV activity.[296] The bicyclo[2.1.1]hexane nucleoside analogues **276** have been reported,[297] and cyclopentane nucleoside analogues such as **277** have been made from the iridoids geniposide and aucubin.[298]

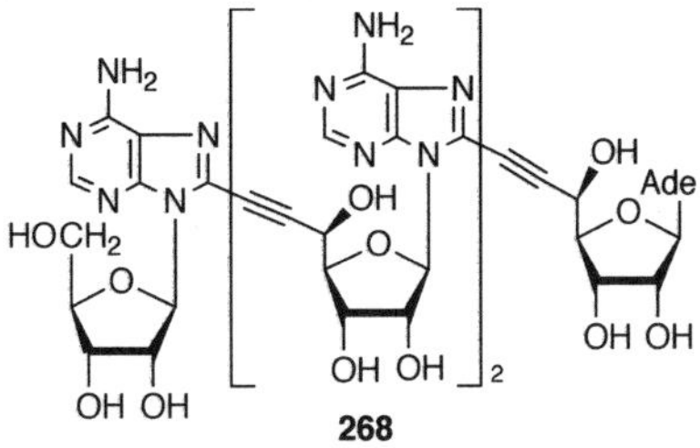

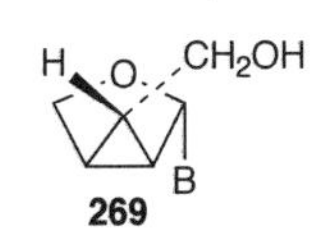

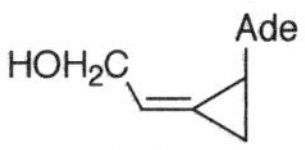

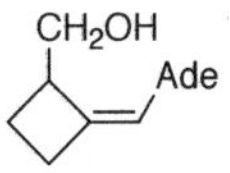

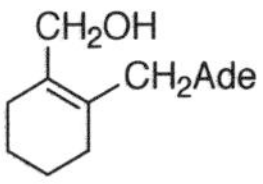

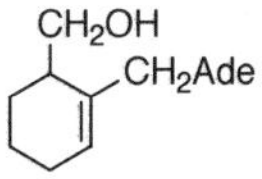

17 Reactions

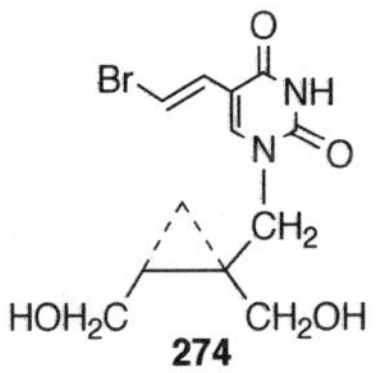

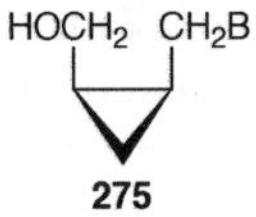

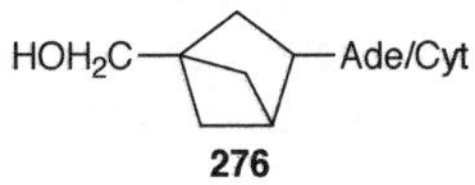

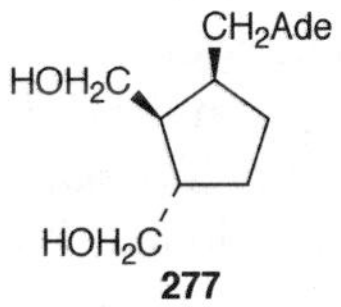

The reactions of some 5'-protected pyrimidine ribonucleosides with phosphoryl chloride and pyridine in dichloromethane, followed by reaction with excess of an alcohol, have been studied. It was claimed that the first step of this procedure led to a mixture of the 2'- and 3'-phosphorodichloridates (*e.g.* **278** and its regioisomer). Since the rate of the subsequent alcoholysis to give a mixture of the 2'- and 3'-phosphotriesters was considerably faster that the corresponding cases using 2'-deoxynucleosides, it was thought that the adjacent *cis* vicinal hydroxyl group was involved in hydrogen bonding to the P=O unit, thereby accelerating the displacements of chloride by electrophilic catalysis.[299] Other workers, however, have reinterpreted the NMR data on which this theory was based, and have concluded that the intermediates in this reaction sequence are the two diastereomers of the cyclic phosphorochloridates such as **279**.[300]

The hydrolysis of the the 3'-thioinosine derivative **280** has been studied over a

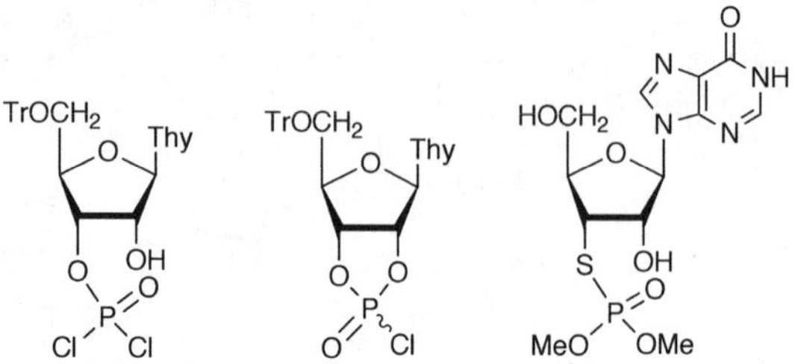

wide acidity range. At pH >3, only hydroxide ion catalysed isomerization to the 2-phosphotriester is observed, whereas under more acidic conditions hydrolysis to the 2'-monomethylphosphate and 3'-S-monomethylphosphorothiolate competed. The latter was the only product to accumulate in very acidic solutions, and mechanisms were discussed.[301] Uridylyl-(3',5')-8-carbomethoxyaminoadenosine has been synthesized, and its transesterification to uridine 2',3'-cyclic phosphate has been studied in the presence and absence of Zn^{2+} ions. The carboxylate in the vicinity of the phosphodiester bond accelerated the cleavage in the presence of Zn(II) ions, but not when the metal was not present.[302]

In connection with mechanisms for the alkaline hydrolysis of DNA, the kinetics of hydrolysis of thymidine 3'-phosphodiesters with alcohols such as 2,2,2-trichloroethanol, and also of thymidinyl-(3'-5')-thymidine, have been studied. With the trifluoroethyl phosphate, the 3',5'-cyclic phosphate is an intermediate in alkaline hydrolysis, but this was not the case with the dinucleotide. Studies using density functional theory calculations were also reported.[303]

The hydrolysis of the cyclic phosphorothiolate **281** has been studied between pH −2 and pH 7.6. Below pH 2, the products **282** and **283** were obtained with **283** predominating, whereas between pH 2 and pH 5, the cyclic phosphorothiolate **282** was the main product. Above pH 5, **283** and **284** were obtained, with the latter predominating by 2:1. Mechanisms were discussed, with the uncatalysed reaction to give **282** occurring through attack of water on the methyl group, with C–O bond cleavage.[304]

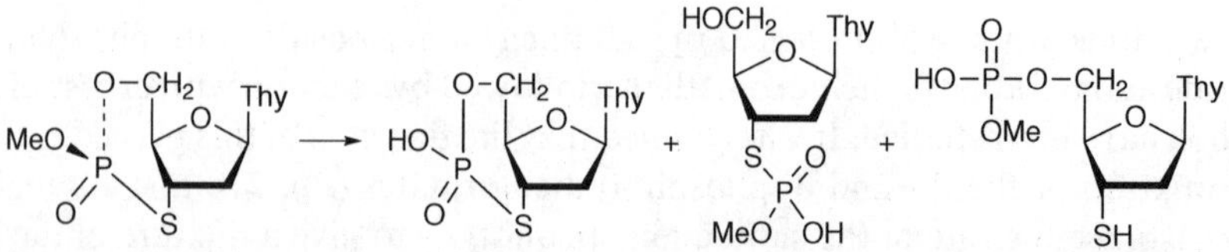

Radicals generated by reaction of cytidine with the sulfate radical-anion in aqueous solution have been characterized by EPR spectroscopy. Radicals were generated either *in situ* in a continuous-flow system, or the sulfate anion radical was generated in the presence of 2-methyl-2-nitrosopropane. In the continuous-flow system, the sugar radical **285** was observed, but the intensity of this decreased in the presence of phosphate dianion, with a base radical of indeterminate structure being formed. In the spin-trapping experiments the persistent radical **286** was detected in the absence of phosphate, but with phosphate dianions present, spin-trapped radicals on the base were observed. The results were interpreted in terms of competition between reaction of the intermediate base radical-cation with phosphate dianion and radical transfer to the sugar.[305] The radical **287** was made by photolysis of a benzylcarbonyl precursor. Its radical recombination and hydrogen abstracting characteristics suggested that it could abstract a hydrogen atom from deoxyribose and thus transfer radical character to the sugar of a neighbouring nucleotide in DNA.[306] A model compound has been used to study the effects of thiiyl radicals on nucleosides, of relevance to radical damage to DNA. The work supported the proposal that thiiyl radicals can add at C-6 to pyrimidine nucleosides.[307] An investigation has been made of the mechanism of direct strand cleavage induced by anaerobic irradiation of DNA containing 5-bromouracil. An oligonucleotide fragment containing 2-(adenin-9-yl)furan at the 3'-terminus was found as a major photoproduct.[308]

285   286   287   288

Photolysis of 1-(2'-deoxyribosyl)-7-nitroindole was presumed to give the diradical **288**, and hence, by radical recombination and fragmentation, the observed photoproducts 2-deoxyribonolactone and 7-nitrosoindole. When the 7-nitroindole nucleoside was incorporated into oligodeoxynucleotides, protolysis led to the formation of abasic positions.[309]

When oligonucleotide chains incorporating the 4'-C-pivaloylthymidine unit **289** are irradiated in the absence of radical traps, the cation-radical **290** (Scheme 11) is produced, which is capable of undergoing reduction by electron transfer from a nearby guanosine unit to give **291**, which on enzymic digestion gives the 3'-ene **292**. Variation of the nucleotide sequence demonstrated a strong distance dependence of the electron transfer rate, which was more efficient with 8-oxoguanosine as electron donor. The enol ether **292** was independently synthesized using reductive elimination from a 4'-phenylselenyl thymidine derivative to install the double bond.[310]

The degradation of hexopyranosyl cytosine nucleosides in buffers of acid, neutral and alkaline pH has been followed by HPLC. The compounds were found to degrade by hydrolysis to cytosine and/or deamination to the corresponding uracil nucleosides.[311]

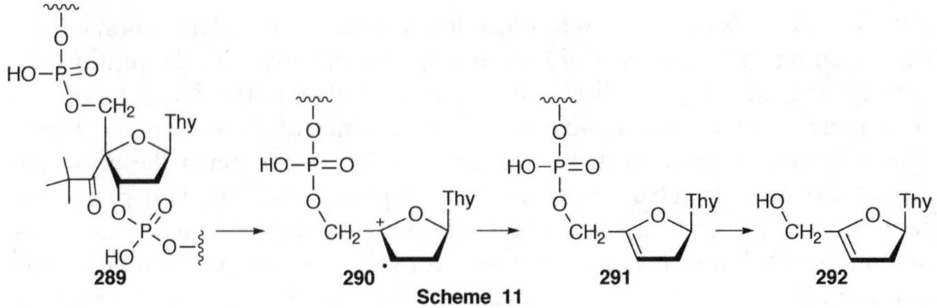

Scheme 11

References

1. M. Ferrero and V. Gotor, *Chem. Rev.*, 2000, **100**, 4319.
2. W.-R. Qiu and Q.-B. Ding, *Zhonguo Yiyao Gongye Zazhi*, 1999, **30**, 474 (*Chem. Abstr.*, 2000, **132**, 137 616).
3. J.-C.G. Graciet and R.F. Schinazi, *Adv. Antiviral Drug Des.*, 1999, **3**, 1 (*Chem. Abstr.*, 2000, **132**, 108 161).
4. A. Földesi, A. Trifonova, M.K. Kundu and J. Chattopadhyaya, *Nucleosides, Nucleotides, Nucleic Acids*, 2000, **19**, 1615.
5. M. Meldgaard and J. Wengel, *J. Chem. Soc., Perkin Trans. 1*, 2000, 3539.
6. E. De Clercq, *Nucleosides, Nucleotides, Nucleic Acids*, 2000, **19**, 1531.
7. P.K. Mishra and D.G. Drueckhammer, *Chem. Rev.*, 2000, **100**, 3283.
8. R.W. Scott, D.E. Fox and D.K. Williams, *Tetrahedron Lett.*, 2000, **41**, 8207.
9. K.S. Gudmundsson, J. Tidwell, N. Lippa, G.W. Koszalka, N. van Draanen, R.G. Ptak, J.C. Drach and L.B. Townsend, *J. Med. Chem.*, 2000, **43**, 2464.
10. S.M. Ulrich, O. Buzko, K. Shah and K.M. Shokat, *Tetrahedron*, 2000, **56**, 9495.
11. K.S. Ramasamy, R.C. Tam, J. Bard and D.R. Averett, *J. Med. Chem.*, 2000, **43**, 1019.
12. K. Gao and L.E. Orgel, *Nucleosides, Nucleotides, Nucleic Acids*, 2000, **19**, 935.
13. I.M. Abdou and L. Strekowski, *Tetrahedron*, 2000, **56**, 8631.
14. L. Strekowski, I.M. Abdou, A.M.E. Attia and S.E. Patterson, *Tetrahedron Lett.*, 2000, **41**, 4757.
15. M.J. Wanner, J.K. von Frijtag Drabbe Künzel, A.P. IJzerman and G.-J. Koomen, *Bioorg. Med. Chem. Lett.*, 2000, **10**, 2141.
16. G.Y. Wang, R.C. Tam, E. Gunic, J.F. Du, J. Bard and B. Pai, *J. Med. Chem.*, 2000, **43**, 2566.
17. J.-L. Girardet, E. Gunic, C. Esler, D. Cieslak, Z. Pietrzkowski and G. Wang, *J. Med. Chem.*, 2000, **43**, 3704.
18. N. Kojima, N. Minakawa and A. Matsuda, *Tetrahedron*, 2000, **56**, 7909.
19. J.J. Chen, Y. Wei, J.C. Drach and L.B. Townsend, *J. Med. Chem.*, 2000, **43**, 2449.
20. Z. Zhu, B. Lippa, J.C. Drach and L.B. Townsend, *J. Med. Chem.*, 2000, **43**, 2430.
21. C. Gasch, M.A. Pradero, B.A.B. Salameh, J.L. Molina and J. Fuentes, *Tetrahedron: Asymmetry*, 2000, **11**, 435.
22. Y. Choi, L. Li, S. Grill, E. Gullen, C.S. Lee, G. Gumina, E. Tsujii, Y.C. Cheng and C.K. Chu, *J. Med. Chem.*, 2000, **43**, 2538.
23. J. Ning and F. Kong, *Carbohydr. Res.*, 2000, **326**, 235.
24. E.S.H. El Ashry, L.F. Awad and M. Winckler, *J. Chem. Soc., Perkin Trans. 1*, 2000, 829.
25. J. Zhang, D. Wang, G. Zhang, C. Sun and Y. Qi, *Quingdao Haiyang Daxue Xuebao*,

1999, **29**, 229 (*Chem. Abstr.*, 2000, **132**, 208 075).

26. A.F. Barrero, E.J. Alvarez-Manzaneda and R. Chahboun, *Tetrahedron Lett.*, 2000, **41**, 1959.

27. L. Kværnø, C. Nielsen and R.H. Wightman, *J. Chem. Soc., Perkin Trans. 1*, 2000, 2903.

28. A.D. Shutalev, V.E. Zavodnik and G.V. Gurskaya, *Nucleosides, Nucleotides, Nucleic Acids*, 2000, **19**, 1831.

29. Z. No, D.S. Shin, B.J. Song, M. Ahn and D.-C. Ha, *Synth. Commun.*, 2000, **30**, 3873.

30. A. E.-S. Abdel-Megied, *J. Chem. Res.*, 2000, (*S*) 152, (*M*) 0501.

31. X.-B. Yang, K. Misiura, W.J. Stec, M.J. Potrzebowski, S. Kazmierski, M. Wieczorek, W.R. Majzner and G.D. Bujacz, *Nucleosides, Nucleotides, Nucleic Acids*, 2000, **19**, 1657.

32. E. Muller, D. Gasparutto, M. Jaquinod, A. Romieu and J. Cadet, *Tetrahedron*, 2000, **56**, 8689.

33. Z. Wang, D.R. Prudhomme, J.R. Buck, M. Park and C.J. Rizzo, *J. Org. Chem.*, 2000, **65**, 5969.

34. M.A. Siddiqui, J.S. Driscoll, E. Abushanab, J.A. Kelley, J.J. Barchi, Jr. and V.E. Marquez, *Nucleosides, Nucleotides, Nucleic Acids*, 2000, **19**, 1.

35. A.E.A. Hassan, A.T. Shortnacy-Fowler, J.A. Montgomery and J.A. Secrist III, *Nucleosides, Nucleotides, Nucleic Acids*, 2000, **19**, 559.

36. N. Ouwerkerk, J.H. van Boom, J. Lugtenburg and J. Raap, *Eur. J. Org. Chem.*, 2000, 861.

37. Y. Oogo, K. Nonaka, A. Ono, A. Ono and M. Kainosho, *Nucleic Acids Symp. Ser.*, 1999, **42**, 123 (*Chem. Abstr.*, 2000, **132**, 279 443).

38. Y.-L. Qiu, H.-Y. Li, G. Topalov and Y. Kishi, *Tetrahedron Lett.*, 2000, **41**, 9425.

39. D. Loakes, M.J. Guo, D.M. Brown, S.A. Salisbury, C.L. Smith, I. Felix, S. Kumar and S. Nampalli, *Nucleosides, Nucleotides, Nucleic Acids*, 2000, **19**, 1599.

40. S. Manfredini, P.G. Baraldi, R. Bazzanini, E. Darini, S. Vertuani, A. Pani, T. Marceddu, F. Demontis, L. Vargiu and P. La Colla, *Nucleosides, Nucleotides, Nucleic Acids*, 2000, **19**, 705.

41. A.E.A. Hassan, R.A.I. Abou-Elkair, J.A. Montgomery and J.A. Secrist III, *Nucleosides, Nucleotides, Nucleic Acids*, 2000, **19**, 1123.

42. J.H. Chen, S.D. Chamberlain, K.K. Biron, M.G. Davis, R.J. Harvey, D.W. Sellerseth, R.E. Dornsife, E.H. Park, L.W. Frick, L.B. Townsend, J.C. Drach and G.W. Koszalka, *Nucleosides, Nucleotides, Nucleic Acids*, 2000, **19**, 101.

43. R. Zou, E. Kawashima, G.A. Freeman, G.W. Koszalka, J.C. Drach and L.B. Townsend, *Nucleosides, Nucleotides, Nucleic Acids*, 2000, **19**, 125.

44. A.R. Porcari, R.G. Ptak, K.Z. Borysko, J.M. Breitenbach, S. Vittori, L.L. Wotring, J.C. Drach and L.B. Townsend, *J. Med. Chem.*, 2000, **43**, 2438.

45. E. Moyroud, E. Biala and P. Strazewski, *Tetrahedron*, 2000, **56**, 1475.

46. C. Mathé and G. Gosselin, *Nucleosides, Nucleotides, Nucleic Acids*, 2000, **19**, 1517.

47. N. Shimma, I. Umeda, M. Arasaki, C. Murasaki, K. Masubuchi, Y. Kohchi, M. Miwa, M. Ura, N. Sawada, H. Tarahara, I. Kuruma, I. Horii and H. Ishitsuka, *Bioorg. Med. Chem.*, 200, **8**, 1697.

48. B.G. Ugarkar, J.M. DaRe, J.J. Kopcho, C.E. Browne III, J.M. Schanzer, J.B. Wiesner and M.D. Erion, *J. Med. Chem.*, 2000, **43**, 2883.

49. B.G. Ugarkar, A.J. Castellino, J.M. DaRe, J.J. Kopcho, J.B. Wiesner, J.M. Schanzer and M.D. Erion, *J. Med. Chem.*, 2000, **43**, 2894.

50. D. Crich, S. Neelamkavil and F. Sartilo-Piscil, *Org. Lett.*, 2000, **2**, 4029.

51. D. Gavriliu, C. Fossey, G. Fontaine, S. Benzaria, A. Ciurea, Z. Delbederi, B. Lelong,

D. Ladurée, A.M. Aubertin and A. Kirn, *Nucleosides, Nucleotides, Nucleic Acids*, 2000, **19**, 1017.

52.   Z. Delbederi, C. Fossey, G. Fontaine, S. Benzaria, D. Gavriliu, A. Ciurea, B. Lelong, D. Ladurée, A.M. Aubertin and A. Kirn, *Nucleosides, Nucleotides, Nucleic Acids*, 2000, **19**, 1441.

53.   V.G. Kostina, A.S. Shalamay, L.S. Usenko and V.A. Gladkaya, *Biopolim. Kletka*, 1998, **14**, 246 (*Chem. Abstr.*, 2000, **132**, 122 843).

54.   I. Liu, Y. Chen and X. Wen, *Zhongguo Yaowu Huaxue Zashi*, 2000, **10**, 62 (*Chem. Abstr.*, 2000, **133**, 105 240).

55.   K.W. Pankiewicz, *Carbohydr. Res.*, 2000, **327**, 87.

56.   F. Viani, *Enantiocontrolled Synth. Fluoro-Org. Compd.*, 1999, 419 (*Chem. Abstr.*, 2000, **132**, 166 395).

57.   T Tennilä, E. Azhayeva, J. Vepsäläinen, R. Laatikainen, A. Azhayev and I.G. Mikhailopulo, *Nucleosides, Nucleotides, Nucleic Acids*, 2000, **19**, 1861.

58.   K.S. Gudmundsson, G.A. Freeman, J.C. Drach and L.B. Townsend, *J. Med. Chem.*, 2000, **43**, 2473.

59.   B.K. Chun, R.F. Schinazi, Y.-C. Cheng and C.K. Chu, *Carbohydr. Res.*, 2000, **328**, 49.

60.   A. Marchand, C. Mathé, J.-L. Imbach and G. Gosselin, *Nucleosides, Nucleotides, Nucleic Acids*, 2000, **19**, 205.

61.   Y.-S. Zhou, Z.-W. Miao and Y.-F. Zhao, *Synlett*, 2000, 671.

62.   R. Robles, C. Rodríguez, I. Izquierdo, M.T. Plaza, A. Mota and L.A. de Cienfuegas, *Tetrahedron: Asymmetry*, 2000, **11**, 3069.

63.   A. Golisade, S. Van Calenbergh and A. Link, *Tetrahedron*, 2000, **56**, 3167.

64.   T. Mitsui, H. Nakano and K. Yamana, *Tetrahedron Lett.*, 2000, **41**, 2605.

65.   R.G. Schultz and S.M. Gryaznov, *Tetrahedron Lett.*, 2000, **41**, 1895.

66.   T. Matray, S. Gamsey, K. Pongracz and S. Gryaznov, *Nucleosides, Nucleotides, Nucleic Acids*, 2000, **19**, 1553.

67.   C. Wojczewski, K. Schwarzer and J.W. Engels, *Helv. Chim. Acta*, 2000, **83**, 1268.

68.   A.M. Costa and J. Vilarrasa, *Tetrahedron Lett.*, 2000, **41**, 3371.

69.   M.D. Jonklass and R.R. Kane, *Tetrahedron Lett.*, 2000, **41**, 4035.

70.   R. Pontikis, V. Dollé, J. Guillaumel, E. Dechaux, R. Note, C.H. Nguyen, M. Legraverend, E. Bisagni, A.-M. Aubertin, D.S. Grierson and C. Monneret, *J. Med. Chem.*, 2000, **43**, 1927.

71.   D.M. Bender, D.D. Hennings and R.M. Williams, *Synthesis*, 2000, 399.

72.   A.W.-H. Cheung, A. Sidduri, L.M. Garofalo and R.A. Goodnow, *Tetrahedron Lett.*, 2000, **41**, 3303.

73.   G.A. Winterfeld, J. Das and R.R. Schmidt, *Eur. J. Org. Chem.*, 2000, 3047.

74.   F. Sztaricskai, A. Csorvasi, A. Horvath, G. Batta and Z. Dinya, *J. Carbohydr. Chem.*, 2000, **19**, 1223.

75.   T. Kodama, S. Shuto, M. Nomura and A. Matsuda, *Tetrahedron Lett.*, 2000, **41**, 3643.

76.   S. Chambert, I. Gautier-Luneau, M. Fontecave and J.-L. Décout, *J. Org. Chem.*, 2000, **65**, 249.

77.   H. Ozaki, Y. Sato, S. Azuma and H. Sawai, *Nucleosides, Nucleotides, Nucleic Acids*, 2000, **19**, 593.

78.   A. Viso, N. Poopeiko and S. Castillón, *Tetrahedron Lett.*, 2000, **41**, 407.

79.   C. Wojczewski and J.W. Engels, *Synthesis*, 2000, 149.

80.   M. Yokoyama, *Synthesis*, 2000, 1637.

81.   K.N. Tiwari, A.T. Shortnacy-Fowler, L. Cappellacci, W.B. Parker, W.R. Waud, J.A.

Montgomery and J.A. Secrist III, *Nucleosides, Nucleotides, Nucleic Acids*, 2000, **19**, 329.

82. Y. Yoshimura, K. Kitano, K. Yamada, S. Sakata, S. Miura, N. Ashida and H. Machida, *Bioorg. Med. Chem.*, 2000, **8**, 1545.
83. J. Wirsching, J. Voss, J. Balzarini and E. De Clercq, *Bioorg. Med. Chem. Lett.*, 2000, **10**, 1339.
84. J.A. Miller, A.W. Pugh and G.M. Ullah, *Tetrahedron Lett.*, 2000, **41**, 3265.
85. J.A. Miller, A.W. Pugh and G.M. Ullah, *Nucleosides, Nucleotides, Nucleic Acids*, 2000, **19**, 1475.
86. J.A. Miller, A.W. Pugh, G.M. Ullah and C. Gutteridge, *Tetrahedron Lett.*, 2000, **41**, 10099.
87. M. Pignot, G. Pljevaljcic and E. Weinhold, *Eur. J. Org. Chem.*, 2000, 549.
88. G. Adiwidjaja, O. Schulze, J. Voss and J. Wirsching, *Carbohydr. Res.*, 2000, **325**, 107.
89. J. Lim, D. Choo and Y.H. Kim, *Chem. Commun.*, 2000, 553.
90. V. Fehring, M.Y. Chan, I.A. O'Neil and R. Cosstick, *J. Chem. Soc., Perkin Trans. 1*, 2000, 3185.
91. M. Nomura, S. Shuto and A. Matsuda, *J. Org. Chem.*, 2000, **65**, 5016.
92. M. Sukeda, S. Shuto, I. Sugimoto, S. Ishikawa and A. Matsuda, *J. Org. Chem.*, 2000, **65**, 8988.
93. X. Zhang, F.-L. Qing and Y.Yu, *J. Org. Chem.*, 2000, **65**, 7075.
94. P.J. Serafinowski and C.A. Brown, *Tetrahedron*, 2000, **56**, 333.
95. P.K. Sharma and V. Nair, *Nucleosides, Nucleotides, Nucleic Acids*, 2000, **19**, 757.
96. B.K. Chun, S. Olgen, H. Hong, M.G. Newton and C.K. Chu, *J. Org. Chem.*, 2000, **65**, 685.
97. M.J. Camarasa, A. San-Félix, M.J. Pérez-Pérez, S. Velázquez, R. Alvarez, C. Chamorro, M.L. Jimeno, C. Pérez, F. Gago, E. De Clercq and J. Balzarini, *J. Carbohydr. Chem.*, 2000, **19**, 451.
98. A. San-Félix, C. Chamorro, M.J. Pérez-Pérez, S. Velázquez, E. De Clercq, J. Balzarini and M.J. Camarasa, *J. Carbohydr. Chem.*, 2000, **19**, 635.
99. M.J. Robins, B. Doboszewski, V.A. Timoshchuk and M.A. Peterson, *J. Org. Chem.*, 2000, **65**, 2939.
100. J.Y. Kim and B.H. Kim, *Nucleosides, Nucleotides, Nucleic Acids*, 2000, **19**, 637.
101. J.M.J. Tronchet, I. Kovacs, M. Seman, P. Dilda, E. De Clercq and J. Balzarini, *Nucleosides, Nucleotides, Nucleic Acids*, 2000, **19**, 775.
102. T. Schoetzau, U. Koert and J.W. Engels, *Synthesis*, 2000, 707.
103. J.S. Cooperwood, V. Boyd, G. Gumina and C.K. Chu, *Nucleosides, Nucleotides, Nucleic Acids*, 2000, **19**, 219.
104. M.E. Jung and A. Toyota, *Tetrahedron Lett.*, 2000, **41**, 3577.
105. C. Lescop, P.-P. Nguyen-Kim and F. Huet, *Tetrahedron Lett.*, 2000, **41**, 3057.
106. P. Nielsen, M. Petersen and J.P. Jacobsen, *J. Chem. Soc., Perkin Trans. 1*, 2000, 3706.
107. J.H. Hong, M.-Y. Gao, Y. Choi, Y.-C. Cheng, R.F. Schinazi and C.K. Chu, *Carbohydr. Res.*, 2000, **328**, 37.
108. N. Baret, J.-P. Dulcere, J. Rodriguez and J.-M. Pons, *Eur. J. Org. Chem.*, 2000, 1507.
109. H.M. Pfundheller, T. Bryld, C.E. Olsen and J. Wengel, *Helv. Chim. Acta*, 2000, **83**, 128.
110. M. Kanazaki, Y. Ueno, S. Shuto and A. Matsuda, *J. Am. Chem. Soc.*, 2000, **122**, 2422.
111. Y. Ueno, K. Tomino, I. Sugimoto and A. Matsuda, *Tetrahedron*, 2000, **56**, 7903.

112. H. Ohrui, S. Kohgo, K. Kitano, S. Sakata, E. Kodama, K. Yoshimura, M. Matsuoka, S. Shigeta and H. Mitsuya, *J. Med. Chem.*, 2000, **43**, 4516.
113. A.E. Håkansson, A.A. Koshkin, M.D. Sørensen and J. Wengel, *J. Org. Chem.*, 2000, **65**, 5161.
114. V.K. Rajwanshi, A.E. Håkansson, H.D. Sørensen, S. Pitsch, S.K. Singh, R. Kumar, P. Nielsen and J. Wengel, *Angew. Chem. Int. Ed. Engl.*, 2000, **39**, 1656.
115. L. Kværnø, R. Kumar, B.M. Dahl, C.E. Olsen and J. Wengel, *J. Org. Chem.*, 2000, **65**, 5167.
116. K. Bondensgaard, M. Petersen, S.K. Singh, V.K. Rajwanshi, R. Kumar, J. Wengel and J.P. Jacobsen, *Chem. Eur. J.*, 2000, **6**, 2687.
117. S. Obika, Y. Hari, T. Sugimoto, M. Sekiguchi and T. Imanishi, *Tetrahedron Lett.*, 2000, **41**, 8923.
118. A.G. Olsen, V.K. Rajwanshi, C. Nielsen and J. Wengel, *J. Chem. Soc., Perkin Trans. 1*, 2000, 3610.
119. H. Kumamoto, S. Shindoh, H. Tanaka, Y. Itoh, K. Haraguchi, E. Gen, A. Kittaka, T. Miyasaka, M. Kondo and K.T. Nakamura, *Tetrahedron*, 2000, **56**, 5363.
120. S. Abbas, C.J. Hayes and S. Worden, *Tetrahedron Lett.*, 2000, **41**, 3215.
121. S.F. Wnuk, C.A. Valdez, J. Khan, P. Moutinho, M.J. Robins, X. Yang, R.T. Borchardt, J. Balzarini and E. De Clercq, *J. Med. Chem.*, 2000, **43**, 1180.
122. M.P. Groziak and R. Lin, *Tetrahedron*, 2000, **56**, 9885.
123. R. Pratap and V.N. Yarovenko, *Nucleosides, Nucleotides, Nucleic Acids*, 2000, **19**, 845.
124. P. Franchetti, S. Marchetti, L. Cappellacci, H.N. Jayaram, J.A. Yalowitz, B.M. Goldstein, J.-L. Barascut, D. Dukhan, J.-L. Imbach and M. Grifantini, *J. Med. Chem.*, 2000, **43**, 1264.
125. M. Popsavin, L. Torovic, S. Spaic, S. Stankov and V. Popsavin, *Tetrahedron Lett.*, 2000, **41**, 5737.
126. N. Nishimura, H. Hisamitsu, M. Sugiura and I. Maeba, *Carbohydr. Res.*, 2000, **329**, 681.
127. R.M. Adlington, J.E. Baldwin, G.J. Pritchard and K.C. Spencer, *Tetrahedron Lett.*, 2000, **41**, 575.
128. D.L. Chen and L.W. McLaughlin, *J. Org. Chem.*, 2000, **65**, 7468.
129. M.A.E. Sallam, F.F. Louis and J.M. Cassady, *Nucleosides, Nucleotides, Nucleic Acids*, 2000, **19**, 941.
130. S. Obika, Y. Hari, K.-i. Morio and T. Imanishi, *Tetrahedron Lett.*, 2000, **41**, 215.
131. S. Obika, Y. Hari, K.-i. Morio and T. Imanishi, *Tetrahedron Lett.*, 2000, **41**, 221.
132. G.B. Evans, R.H. Furneaux, G.J. Gainsford, V.L. Schramm and P.C. Tyler, *Tetrahedron*, 2000, **56**, 3053.
133. X.F. Zhu, *Nucleosides, Nucleotides, Nucleic Acids*, 2000, **19**, 651.
134. B.M. Trost, R. Madsen, S.D. Guile and B. Brown, *J. Am. Chem. Soc.*, 2000, **122**, 5947.
135. M.T. Crimmins, B.W. King, W.J. Zuercher and A.L. Choy, *J. Org. Chem.*, 2000, **65**, 8499.
136. R. Kuang, A.K. Ganguly, T.-M. Chan, B.N. Pramanik, D.J. Blythin, A.T. McPhail and A.K. Saksena, *Tetrahedron Lett.*, 2000, **41**, 9575.
137. S.M. Daluge, M.T. Martin, B.R. Sickles and D.A. Livingston, *Nucleosides, Nucleotides, Nucleic Acids*, 2000, **19**, 297.
138. V.R. Hegde, K.L. Seley and S.W. Schneller, *J. Heterocycl. Chem.*, 2000, **37**, 1361.
139. V.R. Hegde, K.L. Seley and S.W. Schneller, *Nucleosides, Nucleotides, Nucleic Acids*, 2000, **19**, 269.

140. H. Rhee, D. Yoon and M.E. Jung, *Nucleosides, Nucleotides, Nucleic Acids*, 2000, **19**, 619.
141. W. Kim, H. Kim and H. Rhee, *Heterocycles*, 2000, **53**, 219.
142. G. An and H. Rhee, *Nucleosides, Nucleotides, Nucleic Acids*, 2000, **19**, 1111.
143. J. Wang, M. Froeyen, C. Hendrix, G. Andrei, R. Snoeck, E. De Clercq and P. Herdewijn, *J. Med. Chem.*, 2000, **43**, 736.
144. H. Urata, H. Miyagoshi, T. Yumoto, K. Mori, R. Teramichi and M. Agaki, *Nucleic Acids Symp. Ser.*, 1999, **42**, 45 (*Chem. Abstr.*, 2000, **132**, 279 442).
145. G. Gumina, Y. Chong, Y. Choi and C.K. Chu, *Org. Lett.*, 2000, **2**, 1229.
146. Y. Chong, G. Gumina and C.K. Chu, *Tetrahedron: Asymmetry*, 2000, **11**, 4853.
147. Y. Yamatoya, M. Ishikura and N. Katagiri, *Nucleic Acids Symp. Ser.*, 1999, **42**, 23 (*Chem. Abstr.*, 2000, **132**, 279 439).
148. H. Hrebabecky and A. Holy, *Collect. Czech. Chem. Commun.*, 1999, **64**, 1485 (*Chem. Abstr.*, 2000, **132**, 23 163).
149. C. Teran, L. Santana, M. Teijeira, E. Uriarte and E. De Clercq, *Chem. Pharm. Bull.*, 2000, **48**, 293.
150. Y. Saito, M. Nakamura, T. Ohno, C. Chaicharoenpong, E. Ichikawa, S. Yamamura, K. Kato and K. Umezawa, *J. Antibiotics*, 2000, **53**, 309.
151. S. Shuto, K. Haramuishi, M. Fukuoka and A. Matsuda, *J. Chem. Soc., Perkin Trans. 1*, 2000, 3603.
152. T. Moriguchi, N. Asai, T. Wada, K. Seio, T. Sasaki and M. Sekine, *Tetrahedron Lett.*, 2000, **41**, 5881.
153. D.J. Silva and M.J. Sofia, *Tetrahedron Lett.*, 2000, **41**, 855.
154. C. Dini, P. Collette, N. Drochon, J.C. Guillot, G. Lemoine, P. Mauvais and J. Aszodi, *Bioorg. Med. Chem. Lett.*, 2000, **10**, 1839.
155. K. Obi, J. Uda, K. Iwase, O. Sugimoto, H. Ebisu and A. Matsuda, *Bioorg. Med. Chem. Lett.*, 2000, **10**, 1451.
156. Q. Zhang, M.C. Cone, S.J. Gould and T.M. Zabriskie, *Tetrahedron*, 2000, **56**, 693.
157. R.D. Marwood, V. Correa, C.W. Taylor and B.V.L. Potter, *Tetrahedron: Asymmetry*, 2000, **11**, 397.
158. R.D. Marwood, A.M. Riley, D.J. Jenkins and B.V.L. Potter, *J. Chem. Soc., Perkin Trans. 1*, 2000, 1935.
159. R.D. Marwood, D.J. Jenkins, V. Correa, C.W. Taylor and B.V.L. Potter, *J. Med. Chem.*, 2000, **43**, 4278.
160. M. de Kort, J. Luijendijk, G.A. van der Marel and J.H. van Boom, *Eur. J. Org. Chem.*, 2000, 3085.
161. Z. Li, T. Sunazuka, R. Yamada, Y. Kato, A. Enomoto, M. Hayashi, Y. Harigaya and S. Omura, *J. Antibiotics*, 2000, **53**, 426.
162. L.M. Cañedo Hernandez, J.A. de la Fuente Blanco, J. Pérez Baz, J.L. Fernández Puentes, F. Romero Millán, F. Espliego Vázquez, R.I. Fernández-Chimeno and D. García Grávalos, *J. Antibiotics*, 2000, **53**, 895.
163. M.J. Comin and J.B. Rodriguez, *Tetrahedron*, 2000, **56**, 4639.
164. R.L. Hanson, Z. Shi, D.B. Brzozowski, A. Banerjee, T.P. Kissick, J. Singh, A.J. Pullockaran, J.T. North, J. Fan, J. Howell, S.C. Durand, M.A. Montana, D.R. Kronenthal, R.H. Mueller and R.N. Patel, *Bioorg. Med. Chem.*, 2000, **8**, 2681.
165. A.L. Stefanska, M. Fulston, C.S.V. Houge-Frydrych, J.L. Jones and S.R. Warr, *J. Antibiotics*, 2000, **53**, 1346.
166. M. Taktakishvili and V. Nair, *Tetrahedron Lett.*, 2000, **41**, 7173.
167. R. Kierzek, M.A. Steiger, S.L. Spinelli, D.H. Turner and E.M. Phizicky, *Nucleosides, Nucleotides, Nucleic Acids*, 2000, **19**, 917.

168. T.J. Baker and D.F. Wiemer, *Tetrahedron*, 2000, **56**, 3127.
169. M. Sobkowski, M. Wenska, A. Kreszewski and J. Stawinski, *Nucleosides, Nucleotides, Nucleic Acids*, 2000, **19**, 1487.
170. A. Chworos, L.A. Wozniak and W.J. Stec, *Tetrahedron Lett.*, 2000, **41**, 1219.
171. J. Pysowski, A. Chworos, L.A. Wozniak and W.J. Stec, *Tetrahedron Lett.*, 2000, **41**, 1223.
172. J. Cieslak, J. Jankowska, J. Stawinski and A. Kraszewski, *J. Org. Chem.*, 2000, **65**, 7049.
173. S. Králíková, M. Budesínsky, M. Masojídková and I. Rosenberg, *Nucleosides, Nucleotides, Nucleic Acids*, 2000, **19**, 1159.
174. H. An, T. Wang and P.D. Cook, *Tetrahedron Lett.*, 2000, **41**, 7813.
175. H. Kawasaki, Y. Usuda, M. Shimaoka and T. Utagawa, *Biosci. Biotechnol. Biochem.*, 2000, **64**, 2259.
176. E.J. Behrman, *J. Chem. Res.*, 2000, (S) 446.
177. W. Zhou, A. Roland, Y. Jin and R.P. Iyer, *Tetrahedron Lett.*, 2000, **41**, 441.
178. Y.-F. Zhao and Y.-S. Zhou, *Synth. Commun.*, 2000, **30**, 2769.
179. E. Nandanan, S.-Y. Jang, S. Moro, H.O. Kim, M.A. Siddiqui, P. Russ, V.E. Marquez, R. Busson, P. Herdewijn, T.K. Harden, J.L. Boyer and K.A. Jacobsen, *J. Med. Chem.*, 2000, **43**, 829.
180. M. Kadokura, T. Wada, K. Seio and M. Sekine, *J. Org. Chem.*, 2000, **65**, 5104.
181. T. Moriguchi, T. Yanagi, M. Kunimori, T. Wada and M. Sekine, *J. Org. Chem.*, 2000, **65**, 8229.
182. A.A. Rodionov, E.V. Efimtseva, S.N. Mikhailov, J. Rosenski, I. Luyten and P. Herdewijn, *Nucleosides, Nucleotides, Nucleic Acids*, 2000, **19**, 1847.
183. V.V. Iyer, G.W. Griesgraber, M.R. Radmer, E.J. McIntee and C.R. Wagner, *J. Med. Chem.*, 2000, **43**, 2266.
184. J. Baraniak, R. Kaczmarek and W.J. Stec, *Tetrahedron Lett.*, 2000, **41**, 9139.
185. N. Schlienger, S. Peyrottes, T. Kassem, J.-L. Imbach, G. Gosselin, A.-M. Aubertin and C. Périgaud, *J. Med. Chem.*, 2000, **43**, 4570.
186. A.L. Khandazhinskaya, E.A. Shirokova, I.L. Karpenko, N.F. Zakirova, N.B. Tarussova and A.A. Krayevsky, *Nucleosides, Nucleotides, Nucleic Acids*, 2000, **19**, 1795.
187. C.L. Freel Meyers, L. Hong, C. Joswig and R.F. Borch, *J. Med. Chem.*, 2000, **43**, 4313.
188. C.L. Freel Meyers and R.F. Borch, *J. Med. Chem.*, 2000, **43**, 4319.
189. T. Kawaguchi, T. Ichikawa, T. Hasegawa, M. Saneyoshi, T. Wakayama, H. Kato, A. Yukita and T. Nagata, *Chem. Pharm. Bull.*, 2000, **48**, 454.
190. K.-Y. Jung, R.J. Hohl, A.J. Wiemer and D.F. Wiemer, *Bioorg. Med. Chem.*, 2000, **8**, 2501.
191. S. Králíková, M. Budesínsky, M. Masojídková and I. Rosenberg, *Tetrahedron Lett.*, 2000, **41**, 955.
192. J. Lin, K. He and B. Ramsay Shaw, *Helv. Chim. Acta*, 2000, **83**, 1392.
193. J. Jankowska, M. Wenska, M. Popenda, J. Stawinski and A. Kraszewski, *Tetrahedron Lett.*, 2000, **41**, 2227.
194. A.C. Regan, N. Sciammetta and P.I. Tattersall, *Tetrahedron Lett.*, 2000, **41**, 8211.
195. W. Dabkowski, I. Tworowska, J. Michalski and F. Cramer, *Nucleosides, Nucleotides, Nucleic Acids*, 2000, **19**, 1779.
196. Q. Cheng, T. Oritani and T. Horiguchi, *Tetrahedron*, 2000, **56**, 1667.
197. K. Burgess and P.D. Cook, *Chem. Rev.*, 2000, **100**, 2047.
198. F.-P. Gendron, E. Halbfinger, B. Fischer, M. Duval, P. D'Orléans-Juste and A.R.

Beaudion, *J. Med. Chem.*, 2000, **43**, 2239.

199. G. Tusa and J.K. Reed, *Nucleosides, Nucleotides, Nucleic Acids*, 2000, **19**, 805.
200. E. Trevisol. E. Defrancq, L. Lhomme, A. Laayoun and P. Cros, *Tetrahedron*, 2000, **56**, 6501.
201. A. Brunella, M. Abrantes and O. Ghisalba, *Biosci. Biotechnol. Biochem.*, 2000, **64**, 1836.
202. I.A. Mikhailopulo, T.I. Kulak, O.V. Tkachenko, S.L. Sentyureva, L.S. Victorova, H. Rosemeyer and F. Seela, *Nucleosides, Nucleotides, Nucleic Acids*, 2000, **19**, 1885.
203. K. Okuyama, T. Hamamoto, K. Ishiga, K. Takenouchi and T. Noguchi, *Biosci. Biotechnol. Biochem.*, 2000, **64**, 386.
204. A. Schäfer and J. Thiem, *J. Org. Chem.*, 2000, **65**, 24.
205. G. Dufner, R. Schwörer, B. Müller and R.R. Schmidt, *Eur. J. Org. Chem.*, 2000, 1467.
206. S.B. Cohen and R.L. Halcomb, *J. Org. Chem.*, 2000, **65**, 6145.
207. M.D. Burkhart, S.P. Vincent, A. Düffels, B.W. Murray, S.V. Ley and C.-H. Wong, *Bioorg. Med. Chem.*, 2000, **8**, 1937.
208. A. Naundorf, S. Natsch and W. Klaffke, *Tetrahedron Lett.*, 2000, **41**, 189.
209. W. Klaffke and S. Chambron, *Tetrahedron: Asymmetry*, 2000, **11**, 639.
210. A. Zamyatina, S. Gronow, C. Oertelt, M. Puchberger, H. Brade and P. Kosma, *Angew. Chem. Int. Ed. Engl.*, 2000, **39**, 4150.
211. T. Kuzuyama, M.Takagi, K. Kaneda, T. Dairi and H. Seto, *Tetrahedron Lett.*, 2000, **41**, 703.
212. F. Rohdich, C.A. Schuhr, S. Hecht, S. Herz, J. Wungsintaweekul, W. Eisenreich, M.H. Zenk and A. Bacher, *J. Am. Chem. Soc.*, 2000, **122**, 9571.
213. J. Lin and B. Ramsay Shaw, *Tetrahedron Lett.*, 2000, **41**, 6701.
214. A.P. Guzaev and M. Manohavan, *Tetrahedron Lett.*, 2000, **41**, 5623.
215. T. Reiner, E. Kvasyuk and W. Pfleiderer, *Helv. Chim. Acta*, 2000, **83**, 3053.
216. A. Kitamura, Y. Horie and T. Yoshida, *Chem. Lett.*, 2000, 1134.
217. E. Sochacka, G. Czerwinska, R. Guenther, R. Cain, P.F. Agris and A. Malkiewicz, *Nucleosides, Nucleotides, Nucleic Acids*, 2000, **19**, 515.
218. A.M. Sørensen and P. Nielsen, *Org. Lett.*, 2000, **2**, 4217.
219. K. Seio, T. Wada and M. Sekine, *Helv. Chim. Acta*, 2000, **83**, 162.
220. M. Sekine, O. Kurasawa, K. Shohda, K. Seio and T. Wada, *J. Org. Chem.*, 2000, **65**, 3571.
221. M. Sekine, O. Kurasawa, K. Shohda, K. Seio and T. Wada, *J. Org. Chem.*, 2000, **65**, 6515.
222. Y. Lu and G. Just, *Tetrahedron Lett.*, 2000, **41**, 9223.
223. Y. Lu and G. Just, *Angew. Chem. Int. Ed. Engl.*, 2000, **39**, 4521.
224. Z.S. Cheruvallath, H. Sasmor, D.L. Cole and V.T. Ravikumar, *Nucleosides, Nucleotides, Nucleic Acids*, 2000, **19**, 533.
225. T. Miyashita, K. Yamada, K. Kondo, K. Mori and K. Shinozuka, *Nucleosides, Nucleotides, Nucleic Acids*, 2000, **19**, 955.
226. D. Machytka, G. Sági, M. Kajtar-Peredy and E. Gács-Baitz, *Nucleosides, Nucleotides, Nucleic Acids*, 2000, **19**, 903.
227. X. Liu and C.B. Reese, *J. Chem. Soc., Perkin Trans. 1*, 2000, 2227.
228. G. Wang and V. Stoisavljevic, *Nucleosides, Nucleotides, Nucleic Acids*, 2000, **19**, 1413.
229. Z.A. Sergueeva, D.S. Sergueev, A.A. Ribeiro, J.S. Summers and B. Ramsay Shaw, *Helv. Chim. Acta*, 2000, **83**, 1377.
230. M.A. Maier, A.P. Guzaev and M. Manoharan, *Org. Lett.*, 2000, **2**, 1819.

231. S. Abbas and C.J. Hayes, *Tetrahedron Lett.*, 2000, **41**, 4513.
232. S. Vichier-Guerre, F. Santamaria and B. Rayner, *Tetrahedron Lett.*, 2000, **41**, 2101.
233. A.V. Shipitsyn, N.B. Tarussova, E.A. Shirokova and A.A. Krayevsky, *Nucleosides, Nucleotides, Nucleic Acids*, 2000, **19**, 881.
234. M. Fukuoka, S. Shuto, N. Minakawa, Y. Ueno and A. Matsuda, *J. Org. Chem.*, 2000, **65**, 5238.
235. Y. Sumita, M. Shirato, Y. Ueno, A. Matsuda and S. Shuto, *Nucleosides, Nucleotides, Nucleic Acids*, 2000, **19**, 175.
236. A. De Capua, L. De Napoli, G. Di Fabio, A. Messere, D. Montesarchio and G. Piccialli, *Nucleosides, Nucleotides, Nucleic Acids*, 2000, **19**, 1289.
237. M.J. Robins, B. Doboszewski, B.L. Nilsson and M.A. Peterson, *Nucleosides, Nucleotides, Nucleic Acids*, 2000, **19**, 69.
238. F. Burlina, A. Faure, J.-L. Fourrey and M. Thomas, *Eur. J. Org. Chem.*, 2000, 633.
239. M. Dunkel and V. Reither, *Nucleosides, Nucleotides, Nucleic Acids*, 2000, **19**, 749.
240. K.S. Ramasamy, L. He, V. Stoisavljevic, B. Harpham and W. Seifert, *Tetrahedron Lett.*, 2000, **41**, 4317.
241. H. Li and M.J. Miller, *Tetrahedron Lett.*, 2000, **41**, 4323.
242. K.-H. Altmann, D. Hüsken, B. Cuenoud and C. García-Echeverría, *Bioorg. Med. Chem. Lett.*, 2000, **10**, 929.
243. M. Prhavc, G. Just, B. Bhat, P.D. Cook and M. Manoharan, *Tetrahedron Lett.*, 2000, 41, 9967.
244. K. Satou, Y. Komatsu, T. Torizawa, K. Kato, I. Shimada, O. Nikaido and E. Ohtsuka, *Tetrahedron Lett.*, 2000, **41**, 2175.
245. S.K. Roy and J. Tang, *Org. Process Res. Dev.*, 2000, **4**, 170.
246. L. Beigelman, P. Haeberli, D. Sweedler and A. Karpeisky, *Tetrahedron*, 2000, **56**, 1047.
247. A.S. Fraser, A.M. Kawasaki, M.E. Jung and M. Manoharan, *Tetrahedron Lett.*, 2000, **41**, 1523.
248. T.P. Prakash, M. Manoharan, A.S. Fraser, A.M. Kawasaki, E.A. Lesnik and S.R. Owens, *Tetrahedron Lett.*, 2000, **41**, 4855.
249. T.P. Prakash, M. Manoharan, A.N. Kawasaki, E.A. Lesnik, S.R. Owens and G. Vasquez, *Org. Lett.*, 2000, **2**, 3995.
250. A.V. Kachalova, T.S. Zatsepin, E.A. Romanova, D.A. Stetsenko, M.J. Gait and T.S. Oretskaya, *Nucleosides, Nucleotides, Nucleic Acids*, 2000, **19**, 1693.
251. B. Masjost, P. Ballmer, E. Borroni, G. Zürcher, F.K. Winkler, R. Jacob-Roetne and F. Diederich, *Chem. Eur. J.*, 2000, **6**, 971.
252. M. Liu, J. Haddad, E. Azucena, L.P. Kotra, M. Kirzhner and S. Mobashery, *J. Org. Chem.*, 2000, **65**, 7422.
253. X. Wu and S. Pitsch, *Helv. Chim. Acta*, 2000, **83**, 1127.
254. J.R. Hwu, M.L. Jain, F.-Y. Tsai, S.-C. Tsay, A. Balakumar and G.H. Hakimelahi, *J. Org. Chem.*, 2000, **65**, 5077.
255. X.-F. Zhu, H.J. Williams and A.I. Scott, *J. Chem. Soc., Perkin Trans. 1*, 2000, 2305.
256. X.-F. Zhu, H.J. Williams and A.I. Scott, *Tetrahedron Lett.*, 2000, **41**, 9541.
257. R.V. Nair and M.M. Salunkhe, *Synth. Commun.*, 2000, **30**, 3115.
258. A. Kuboki, T. Ishihara, E. Kobayashi, H. Ohta, T. Ishii, A. Inoue, S. Mitsuda, T. Miyazaki, Y. Kajihara and T. Sugai, *Biosci. Biotechnol. Biochem.*, 2000, **64**, 363.
259. L. Hu, B. Lui and D.R. Hacking, *Bioorg. Med. Chem. Lett.*, 2000, **10**, 797.
260. B. Quiclet-Sire, A. Wilczewska and S.Z. Zard, *Tetrahedron Lett.*, 2000, **41**, 5673.
261. A. Marsh, A. Kahn, M. Garcia and D.M. Haddleton, *Chem. Commun.*, 2000, 2083.
262. J. Grugier, J. Xie, I. Duarte and J.-M. Valéry, *J. Org. Chem.*, 2000, **65**, 979.

263. S. Chang, G. Griesgraber, T.W. Abraham, T. Garg, H. Song, C.L. Zimmerman and C.R. Wagner, *Nucleosides, Nucleotides, Nucleic Acids*, 2000, **19**, 87.
264. M. Kawana, M. Tsujimoto and S. Takahashi, *J. Carbohydr. Chem.*, 2000, **19**, 67.
265. W. Lloyd, C.B. Reese, Q. Song, A.M. Vandersteen, C. Viscontin and P.-Z. Zhang, *J. Chem. Soc., Perkin Trans. 1*, 2000, 165.
266. A.E. Pechenov, S.G. Zavgorodny, V.I. Shvets and A.I. Miroshnikov, *Russ. J. Bioorg. Chem.*, 2000, **26**, 327 (*Chem. Abstr.*, 2000, **133**, 177 398).
267. R. Toña, R. Bertolini and J. Hunziker, *Org. Lett.*, 2000, **2**, 1693.
268. K. Walczak, *Pol. J. Chem.*, 1999, **73**, 1613 (*Chem. Abstr.*, 2000, **132**, 12 448).
269. H. Zhang, M. Zhang, Z. Piao, L. Ma and L. Zhang, *Yaoxue Xuebao*, 1999, **34**, 363 (*Chem. Abstr.*, 2000, **132**, 64 474).
270. B.-H. Yang and X.-J. Wu, *Chin. J. Chem.*, 2000, **18**, 118 (*Chem. Abstr.*, 2000, **132**, 222 789).
271. G.A. Freeman, D.W. Selleseth, J.L. Rideout and R.J. Harvey, *Nucleosides, Nucleotides, Nucleic Acids*, 2000, **19**, 155.
272. X.B. Tian, J.M. Min and L.H. Zhang, *Tetrahedron: Asymmetry*, 2000, **11**, 1877.
273. Z. Lei, J.M. Min and L.H. Zhang, *Tetrahedron: Asymmetry*, 2000, **11**, 2899.
274. S. Bera and V. Nair, *Helv. Chim. Acta*, 2000, **83**, 1398.
275. W.D. Wu, L.T. Ma, L.H. Zhang, Y. Lu, F. Guo and Q.T. Zheng, *Tetrahedron: Asymmetry*, 2000, **11**, 1527.
276. I.A. Koslov, M. Zielinski, B. Allart, L. Kerremans, A. Van Aerschot, B. Busson, P. Herdewijn and L.E. Orgel, *Chem. Eur. J.*, 2000, **6**, 151.
277. M.-Z. Luo, M.-C. Liu, D.E. Mozdziesz, T.-S. Lin, G.E. Dutschmann, E.A. Gullen, Y.-C. Cheng and A.C. Sartorelli, *Bioorg. Med. Chem. Lett.*, 2000, **10**, 2145.
278. M.-C. Liu, M.-Z. Luo, D.E. Mozdziesz, T.-S. Lin, G.E. Dutschman, E.A. Gullen, Y.-C. Cheng and A.C. Sartorelli, *Nucleosides, Nucleotides, Nucleic Acids*, 2000, **19**, 603.
279. C.K. Chu, L. Ma, S. Olgen, C. Poerra, J. Du, G. Gumina, E. Gullen, Y.-C. Cheng and R.F. Schinazi, *J. Med. Chem.*, 2000, **43**, 3906.
280. N. Mourier, M. Camplo, G. S. Della Bruna, F. Pellacini, D. Ungheri, J.-C. Chermann and J.-L. Kraus, *Nucleosides, Nucleotides, Nucleic Acids*, 2000, **19**, 1057.
281. N. Nguyen-Ba, N. Lee, L. Chan and B. Zacharie, *Chem. Commun.*, 2000, 2311.
282. P. Merino, E.M. del Alamo, S. Franco, F.L. Merchan, A. Simon and T. Tejero, *Tetrahedron: Asymmetry*, 2000, **56**, 2995.
283. P. Merino, E.M. del Alamo, M. Bona, S. Franco, F.L. Merchan, T. Tejero and O. Vieceli, *Tetrahedron Lett.*, 2000, **41**, 9239.
284. P. Merino, S. Franco, F.L. Merchan and T. Tejero, *J. Org. Chem.*, 2000, **65**, 5575.
285. S. Eppacher, N. Solladié, B. Bernet and A. Vasella, *Helv. Chim. Acta*, 2000, **83**, 1311.
286. H. Gunji and A. Vasella, *Helv. Chim. Acta*, 2000, **83**, 1311.
287. T. Umino, N. Minakawa and A. Matsuda, *Tetrahedron Lett.*, 2000, **41**, 6419.
288. H. Gunji and A. Vasella, *Helv. Chim. Acta*, 2000, **83**, 2975.
289. H. Gunji and A. Vasella, *Helv. Chim. Acta*, 2000, **83**, 3229.
290. G. Prestat, D. Dubreuil, A. Adjou, J.P. Pradère, J. Lebreton, M. Evers and Y. Henin, *Nucleosides, Nucleotides, Nucleic Acids*, 2000, **19**, 735.
291. M. Cornia, M. Menozzi, E. Ragg, S. Mazzini, A. Scarafoni, F. Zanardi and G. Casiraghi, *Tetrahedron*, 2000, **56**, 3977.
292. C. Lescop and F. Huet, *Tetrahedron*, 2000, **56**, 2995.
293. Y.-L. Qiu, M.B. Ksebati and J. Zemlicka, *Nucleosides, Nucleotides, Nucleic Acids*, 2000, **19**, 31.

294. H.-P. Guan, M.B. Ksebati, E.R. Kern and J. Zemlicka, *J. Org. Chem.*, 2000, **65**, 5177.

295. T. Onishi, C. Mukai, R. Nakagawa, T. Sekiyama, M. Aoki, K. Suzuki, H. Nakazawa, N. Ono, Y. Ohmura, S. Iwayama, M. Okunishi and T. Tjuji, *J. Med. Chem.*, 2000, **43**, 278.

296. C. Pierra, S. Olgen, S.C.H. Cavalcanti, Y.-C. Cheng, R.F. Schinazi and C.K. Chu, *Nucleosides, Nucleotides, Nucleic Acids*, 2000, **19**, 253.

297. G. Wang, *Tetrahedron Lett.*, 2000, **41**, 7139.

298. H. Franzyk and F.R. Stermitz, *J. Nat. Prod.*, 1999, **62**, 1646 (*Chem. Abstr.*, 2000, **132**, 122 846).

299. C.D. Roussev, G.D. Ivanova, E.K. Bratovanova and D.D. Petrov, *Angew. Chem. Int. Ed. Engl.*, 2000, **39**, 779.

300. P.M. Cullis, M.J.P. Harger and M. Lee, *Angew. Chem. Int. Ed. Engl.*, 2000, **39**, 4245.

301. M.I. Elzagheid, E. Mäki, U. Kaukinen, M. Oivanen and H. Lönnberg, *Nucleosides, Nucleotides, Nucleic Acids*, 2000, **19**, 827.

302. S. Mikkola, M. Oivanen, K. Neuvonen, S. Piitari, K. Kekomäki and H. Lönnberg, *Nucleosides, Nucleotides, Nucleic Acids*, 2000, **19**, 1675.

303. N. Takeda, M. Shibata, N. Tajima, K. Hirao and M. Komiyama, *J. Org. Chem.*, 2000, **65**, 4391.

304. M.I. Elzaghheid, K. Mattila, M. Oivanen, B.C.N.M. Jones, R. Cosstick and H. Lönnberg, *Eur. J. Org. Chem.*, 2000, 1987.

305. H. Niehaus and K. Hildenbrand, *J. Chem. Soc., Perkin Trans. 2*, 2000, 947.

306. A.A. Anderson, J.-T. Hwang and M.M. Greenberg, *J. Org. Chem.*, 2000, **65**, 4648.

307. K.N. Carter, T. Taverner, C.H. Schiesser and M.M. Greenberg, *J. Org. Chem.*, 2000, **65**, 8375.

308. K. Fujimoto, Y. Ikeda and I. Saito, *Tetrahedron Lett.*, 2000, **41**, 6455.

309. M. Kotera, Y. Roupioz, E. Defrancq, A.-G. Bourdat, J. Garcia, C. Coulombeau and J. Lhomme, *Chem. Eur. J.*, 2000, **6**, 4163.

310. E. Meggers, A. Dussy, T. Schäfer and B. Giese, *Chem. Eur. J.*, 2000, **6**, 485.

311. G.N. Thoithi, A. Van Schepdael, R. Busson, G. Janssen, A. Van Aerschot, P. Herdewijn, E. Roets and J. Hoogmartens, *Nucleosides, Nucleotides, Nucleic Acids*, 2000, **19**, 189.

20
Enzymes in Mono- and Oligo-saccharide Chemistry

1 General

Chapters 3 and 4 contain many references to the use of enzymes in the synthesis of glycosides, and di- and higher saccharides.

Specific reviews have described both the traditional method of enzymatic glycosylation and some novel approaches which were designed to improve the efficiency of reactions involving poorly water soluble alcohols,[1] and the enzymatic synthesis of oligosaccharides on a preparative scale (greater than milligram quantities).[2] Several general reviews on recent developments in enzyme-catalysed syntheses of carbohydrates have appeared.[3–5] Those that directly relate to specific subjects are detailed in the appropriate following sections.

2 Enzymes in Synthesis

2.1 Aldolases and Ketolases. – Kajimoto has reviewed aldolase-catalysed synthesis of various azasugars (45 refs.) using fructose-1,6-diphosphate-aldolase and other dihydroxyacetone phosphate (DHAP)-dependent aldolases in the key step.[6] A recent review of enzymatic preparations of monosacccharides, oligosaccharides and related compounds based on aldolases has been published.[7] Protecting group-free chemoenzymatic syntheses of the tetrahydroxylated pyrrolizidines australine (**1**) and 7-epialexine (**2**) have been achieved by fructose 1,6-diphosphate aldolase-catalysed reaction of dihydroxyacetone phosphate with appropriate aldehydes.[8] Similar procedures involving coupling of DHAP with ω-phosphonylated α-hydroxyaldehydes have been used to prepare a set of unnatural ω-phosphonic sugars **3** (Scheme 1).[9]

Short syntheses of L-fucose analogues have been achieved (3 enzymatic steps) by aldol condensation of DHAP with various α-hydroxyaldehyde derivatives (using L-fuculose phosphate aldolase), followed by phosphate hydrolysis (alkaline phosphatase) and subsequent ketol isomerization (L-fucose isomerase).[10] Selective ^{13}C-labelling of thymidine in all positions of the 2′-deoxyribose ring has been achieved using an aldolase-catalysed reaction with appropriately ^{13}C-labelled DHAP/acetaldehyde substrates. The labelled sugar was then converted

Carbohydrate Chemistry, Volume 34
© The Royal Society of Chemistry, 2003

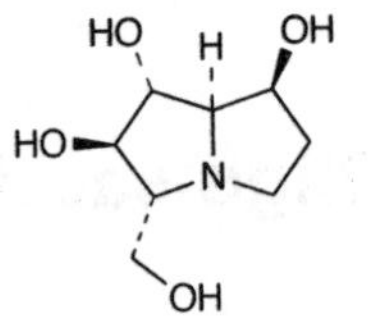

Reagents: i, Fructose 1,6-diphosphate aldolase; ii, Phosphatase

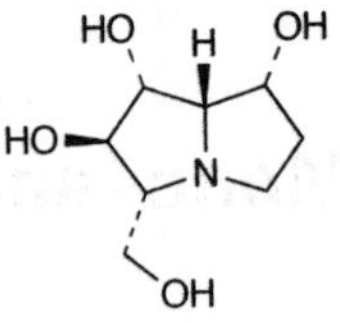

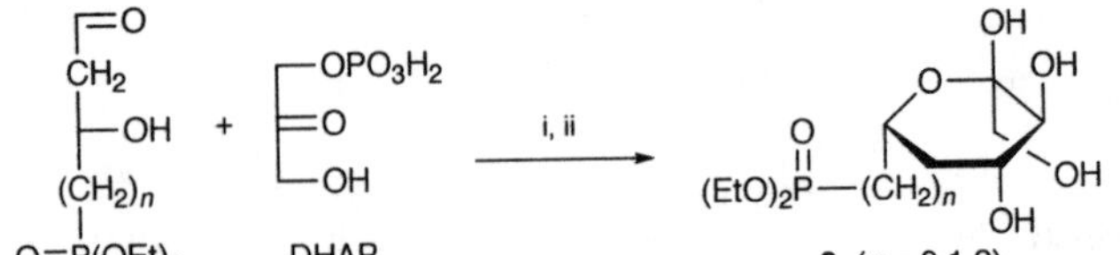

Scheme 1

to the O-protected ribosyl chloride prior to addition of the thymidine nucleobase.[11] N-Hydroxypyrrolidine derivatives **6** has also been prepared (on a 2–3 g scale) by an *E. coli.* transketolase-mediated reaction of ($\pm$)-3-O-benzyl glyceraldehydes **4** with hydroxypyruvate to set up the required stereochemistry on the triol building block **5** (Scheme 2).[12]

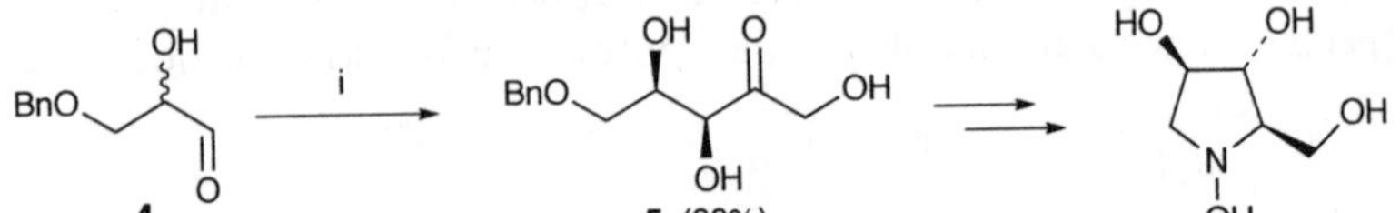

Reagents: i, Transketolase, TPP, hydroxypyruvate, pH 7.0

Scheme 2

2.2 Glycosidases. – Several reviews on the enzymatic preparation of glycosides and oligosaccharides using glycosidase-catalysed *trans*-glycosidation reactions have appeared.[13-17] Another review discusses the inulase-catalysed mass production of difructose anhydrides **7** and **8**.[18] The α-glucosidase-catalysed reaction of maltose and glycerol has been shown to give a mixture of the three glyceryl α-D-glucoside regio-isomers.[19] A hydrolytically inactive E358A mutant of an *Agrobacterium sp.* β-glucan hydrolase has been used to construct β-1,4-linked di-, tri- and tetrasaccharides **9** with a C-glycoside at the 'reducing' terminal end using α-D-glucopyranosyl fluoride donor and various β-C-glucosides as the 'reducing' end acceptor molecules.[20] A new β-glucosidase from Thai rosewood has been shown to catalyse the synthesis of di- and tri-saccharides from glucose. The major di-saccharide product was gentiobiose, whilst starting from a mixture of D-fucose and D-glucose the main product obtained was β-D-Fuc-(1→6)-D-Glc.[21] A partially purified β-glucosidase from *Trichoderma viride* cellulase catalyses 1,6-transglycosylation reactions with cellobiose, laminaribiose and gentiobiose, but with poor yields.[22] Glycoside synthesis from aqueous starch solution and hydroxylbenzyl alcohols (*o, m* and *p*), using amyloglucosidase from *Rhizopus sp.*, has been shown to yield hydroxybenzyl α-glucosides.[23] There have been two reports on the synthesis of novel hetero-branched β-cyclodextrins by the reverse action of *Aerobacter aerogenes* pullulanase on β-cyclodextrin and various tri-/tetra-saccharides.[24,25]

A comparative study of transglycosylation reactions using various newly cloned α-galactosidases from *B. stearothermophilus*, *Therus. brockianus*, *Streptococcus mutans* and *E. coli*. has been carried out and all displayed the capacity for regioselective synthesis of α-(1→6)-disaccharides using a *p*-nitrophenyl β-D-galactopyranoside as donor.[26] The β-gluco-/galacto-sides of *cis*- and *trans*-2-*p*-methoxybenzyl cyclohexanol have been made using almond β-glucosidase and β-galactosidase from *E. coli*, respectively.[27] Transglycosylation of lactose using *Streptococcus thermophilus* led to 3'-β-D-galactopyranosyl-lactose as the major product (16%), which was isolated on a 1 g scale.[28] The transglycosylating properties of a β-galactosidase from *Aspergillus aculeatus* have been characterized using lactose as a substrate, and its activities towards exopolysaccharides from *Lactococcus lactis* have been studied.[29]

Cycloinulo-oligosaccharide fructanotransferase (CFTase)-catalysed decomposition of inulin in the presence of D-mannose and D-sorbose acceptors has been used to obtain inulotriosyl derivatives **10** and **11** on a preparative scale.[30] Transfructosylation of 2-mercaptoethanol using sucrose and β-fructosidases gives the *S*-glycoside product after long reaction times (20 h), whereas the *O*-glycoside accumulates during shorter reaction times and then decays suggesting that the thioglycoside is relatively resistant to the hydrolytic action of the fructofuranosides.[31] The formation and retention of cyclomaltodextrins inside starch granules has been achieved by reaction of cyclomaltodextrin glucanosyltransferase with solid granules.[32] There are also two reports on the use of cyclomaltodextrin-glucanosyltransferase in the preparation of pyridoxine β-glycosides, **12** and **13**, from pyridoxin (vitamin B6) and maltodextrin.[33,34]

A bovine testes β-galactosidase has shown a surprising lack of regiospecificity in β-galactosylation of chito-oligosaccharides giving an equimolar ratio of terminal β-(1→4) and β-(1→3) galactosylated products.[35] The enzymatic synthesis of *N*-acetyl-lactosamine has been investigated with a variety of galactose donors, cosolvents and β-galactosidases.[36] Pig liver β-galactosidase has been used in

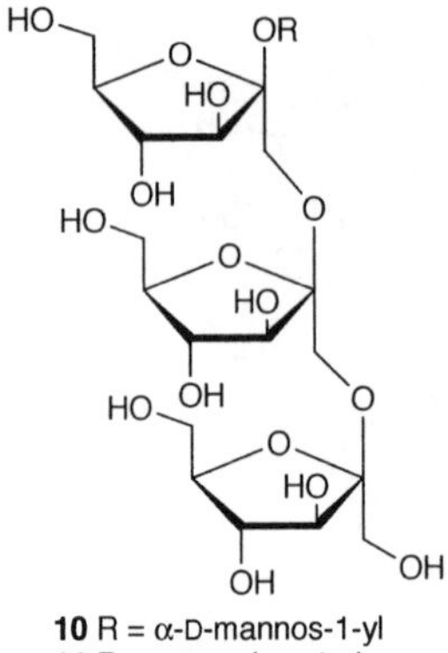

10 R = α-D-mannos-1-yl
11 R = α-L-sorbos-1-yl   **12** 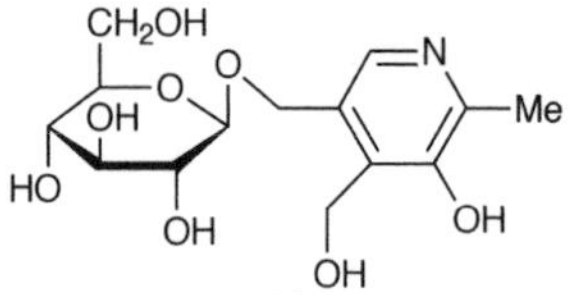   **13**

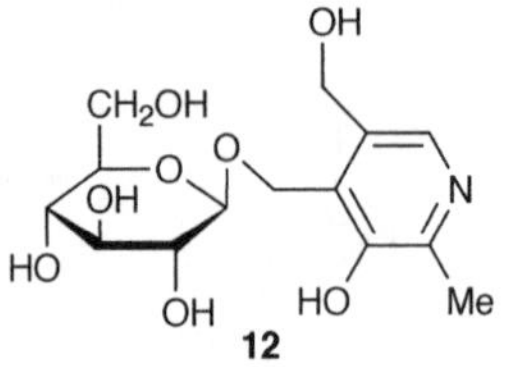

β-(1→6)-disaccharide syntheses with *p*-nitrophenyl galactoside derivatives as
donors and *p*-nitrophenyl glycosides of Gal, GalNAc and Glc as acceptors to
obtain, for example, β-Gal-(1→6)-α-Gal-OPNP and β-Gal-(1→6)-β-GlcNAc-
OPNP. With β-D-galactosidase from *Bacillus circulans* the respective (1→3)-
linked disaccharides were obtained.[37] Both enzymes displayed a higher activity
for α-glycoside acceptors. Regiospecific transglycosidic syntheses of β-Gal-
(1→4)-GlcNAc and α-NeuAc-(2→3)-β-Gal-(1→4)-GlcNAc have been reported.[38] Galactosylation of 2-mercaptoethanol using β-galactosides from *Penicillium multicolour* or *Aspergillus lorysae* has been shown to give thioglycoside **14**
as the major product (Scheme 3).[39]

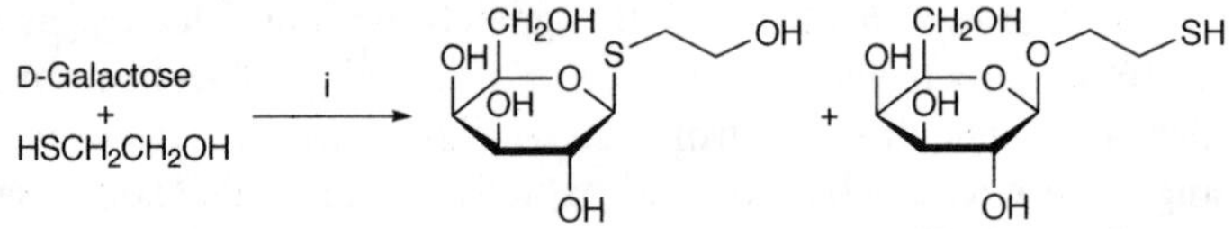

Reagents: i, β-galactosidases from various sources

Scheme 3

β-D-Glc, α- and β-D-Man and β-Lac have been transferred from various
donors to methanol and various glycosides and glycosyl fluorides by use of a
crude filtrate from *Thermoascus aurantiacus*.[40] Synthesis of α-(1→2) and α-(1→3)
linked disaccharides of mannofuranose using the reverse hydrolysis activity of
partially purified α-mannosidases from almond meal (*Prunus amygdalus*) and
limpets (*Patella rulgatal*) has been reported.[41] *Penicillium multicolour* α-L-
fucosidase has been used to prepare a variety of 1,3-linked fucose-containing
disaccharides.[42] Fucosylated epitopes such as Le[X] and Lac(NAc)₂ have been
prepared enzymatically.[43] Sugar oxazoline donors have been used in the
chitinase-catalysed synthesis of GlcNAc-containing oligosaccharides.[44] Incubation of *Trichoderma reesei* cellulase with xyloglycan or cellulose donors in an
aqueous solution of various acceptor alcohols gave high yields of alkyl β-
glycosides.[45]

Concentrated plasticised glass mixtures of acceptors and donors have been
shown to enable the glycosidase-catalysed syntheses of a functionally diverse

range of *O*-glycosides such as **16** and **17** at rates up to an order of magnitude greater than for dilute systems.[46] There have been several reports on the sialidase-catalysed synthesis of α-sialylated oligosaccharides using α-NeuNAcOPNP as the donor.[47,48] Notably, various terminally sialylated di- and trisaccharides have been prepared by combination of sialidase-catalysed transglycosylation with polyethylene glycol ω-monomethyl ether-based glycoside acceptors.[49]

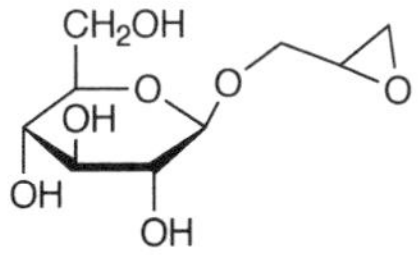

16 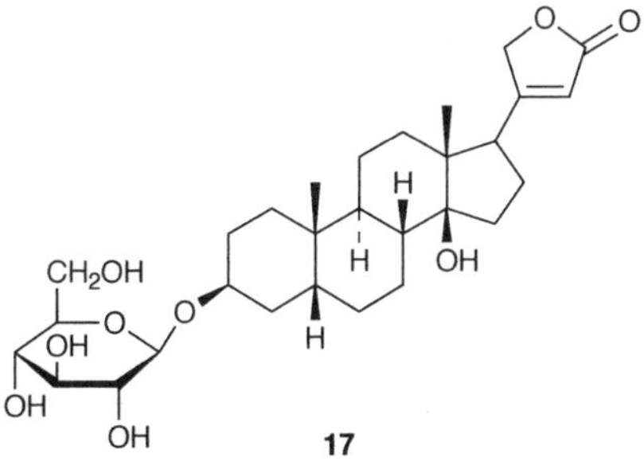 **17**

A novel thermophilic glycosynthase which effects branching glycosylation has been obtained by mutagenesis (D387G) of the active site nucleophile of a glycosidase from *Sulfolobus solfataricus,* and by use of α-hexosyl fluoride donors and *o*- or *p*-nitrophenyl hexoside acceptors various di-, tri- and tetra-saccharides have been prepared.[50] Another glycosynthase has been derived from *Agrobacterium sp.* β-glucosidase in a similar manner.[51] The synthesis of β-(1→4)-oligosaccharides using a glycosynthase derived from *Humicola insolens,* α-lactosyl fluoride and a variety of acceptors has been studied.[52]

2.3 Glycosyltransferases. – Three different transferases have been used in the chemoenzymatic synthesis of 'bisecting-type' deca- and dodeca-oligosaccharides of the bisecting-*N*-glycans.[53,54] Dendritic polyethylene glycols have been used as soluble supports for acceptor substrates. For reactions which fail to reach completion the partially substituted support is recycled after simple dialysis, and treated with fresh enzymes and donors to improve the yields.[55] A range of 5a-carba-sugars, including both anomers of 5a-carba-D-GlcNAc **18** (Scheme 4), have been tested as acceptors for β-(1→4)-galactosyl transferase-catalysed reactions with UDP-Gal as donor.[56] Isomers **18**, which give **19**, are the best acceptors of the compounds studied.

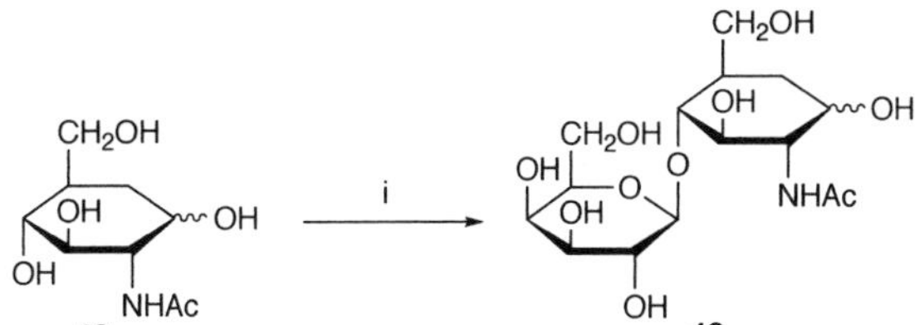

18 **19**

Reagents: i, UDP-Gal, galactosyl β-(1,4)-transferase

Scheme 4

L-Glycosylamine derivatives **20** and **21** act as acceptor substrates for bovine β-(1→4)-galactosyltransferase which, surprisingly, catalyses transfer to the 3-

hydroxyl groups.[57] The two human blood group A and B glycosyltransferases utilize UDP-GalNAc and UDP-GlcNAc donors, respectively. A hybrid enzyme has been constructed that can utilize both these donors.[58] A synthetic tri-lactosamine has been further elaborated to sialyl-trimeric-Le^X by enzymatic addition of the sialyl and fucose units.[59] Fucosyltransferase V only added two fucose units to the sialyl-tri-lactosamine precursor, whereas fucosyltransferase VI either completed the fucosylation or was used alone to put on all three fucose units. α-(1→3)-Fucosyltransferase has also been used to transfer L-galactose moieties in the synthesis of a dimeric sialyl Le^X analogue.[60]

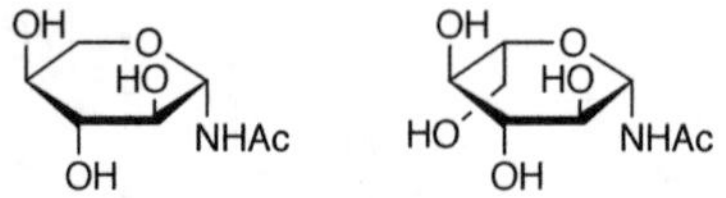

Recombinant hexa-histidine-tagged human α-(1→3/4) fucosyltransferase III immobilized on nickel-agarose, exhibits marked stability which has been exploited in the synthesis of Le^A and Le^X trisaccharides with the enzyme still retaining 50% activity after a 10 day incubation at 37 °C.[61] Fucosyltransferase III has also been used to prepare glycal derivatives of SiaLe^X.[62] Studies of a 2,3-sialyltransferase (v-ST3Gal I), obtained from myxoma virus infected RK13 cells, have revealed its unique substrate specificity amongst previously characterized sialyltransferases.[63] In addition to catalysing sialyl transfer from CMP-NeuAc to type I, II and III acceptors, this viral enzyme also transfers sialic acid to the fucosylated acceptors Le^X and Le^A. UDP-Glucuronyltransferases from ovine[64] and human liver microsomes[65] have been used to prepare lysergol β-D-glucuron-ide and 3-O-benzylmorphine-6-glucuronide, respectively. An intein-mediated protein ligation strategy has been used to incorporate glycopeptides **22** at the C-terminus of proteins, and the GlcNAc unit of the ligation product **23** served as a substrate for subsequent β(1→4)-galactosyltransferase- catalysed modification (Scheme 5).[66]

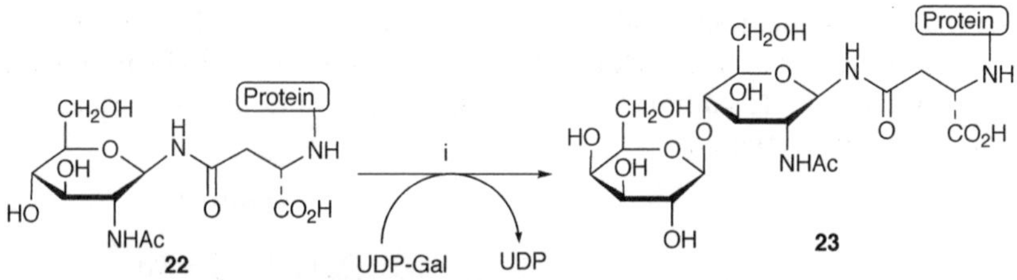

Reagents: i, Galactosyl β-(1,4)-transferase

Scheme 5

2.4 Lipases and Acyl Transferases. – A recent review of synthetic applications of enzymes in organic solvents details examples of lipases and proteases exploited in the chemoenzymatic O-substitution of sugar derivatives.[67] Lipase from *Humicola lanuginosa*, together with various vinyl esters, has been used for specific and high yielding (70–90%) 6-O-acylation of the non-reducing-terminal glucose moieties of maltose, maltotriose and dodecyl α,β-maltosides.[68] A mixture of

1,3,4-trideoxy-5,6:7,8-di-O-isopropylidene-β-D-*manno*- and *gluco*-non-5-ulo-5,9-pyranose [(R,S)-**24**] have been separated by stereoselective acylation using vinyl acetate and Chirazyme® L-2 to give the *manno*-acetate **25** (Scheme 6).[69]

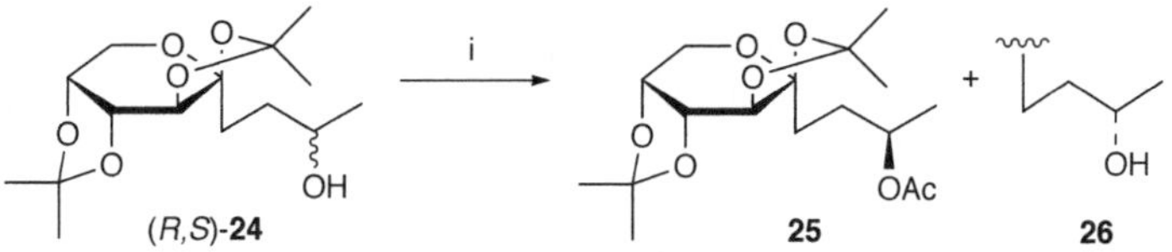

Reagents: i, Vinyl acetate, Chirazyme® L-2

Scheme 6

Lipase from *Candida antartactica* has been used in the preparation of various acylated maltooligosaccharides,[70] ascorbic acids,[71] disaccharides[72] and human milk trisaccharides[73] and also for the regioselective 1,6-co-polymerization of D-glucitol and di-vinyl sebacate.[74] A novel aminoglycoside 6'-acetyltransferase from *Actinomycete sp.* has been used for the acetylation of a large selection of aminoglycoside antibiotics.[75,76] A *Pseudomonas* lipase on ceramic particles and on diatomite has been used to regioselectively acylate nucleosides and 2'-deoxynucleosides at O-3' by transfer from various O-acetyl aldehyde or ketone oximes in 70–90% yields.[77]

Chitin deacetylase has been used in the synthesis of partially and fully N-deacetylated 4-methyl umberlliferyl chitobiose derivatives as fluorogenic substrates for chitinase.[78] Conversely, the reverse acylating action of chitin deacetylase has been used to prepare a partially N-acetylated chitosamine tetramer.[79] The preparation of a 4-O-acetyl sialic acid derivative by a lipase OF-catalysed deacetylation of the peracetylated precursor has been reported.[80]

Regioselectively protected cytidine derivatives have been prepared from the peracetylated nucleoside using lipase and esterase reactions.[81] There have been reports on the use of bacterial proteases for the selective acylation of sucrose to produce a variety of different acyl[82] and acrylate esters.[83] 8-Aminooctyl 5-S-coniferyl-5-thio-α-L-arabinofuranoside (**27**) attached to sepharose proved to be a selective affinity ligand for feruloyl esterase *A. niger*.[84]

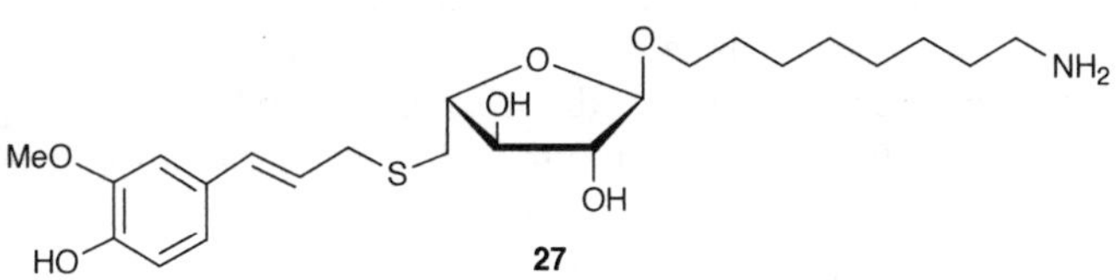

27

2.5 Sulfotransferases. – A recent study of keratan sulfate Gal-6-sulfotransferase (KSGal6ST) has demonstrated its ability to sulfate O-6 of the Gal residue of NeuAc-(2→3)-β-Gal-(1→4)-GlcNAc (3'SLN), fetuin oligosaccharides and their desialylated derivatives.[85] The relative sulfation rate of some of the compounds was much higher than for keratan sulfate.

2.6 Coupled, Multi-Enzymatic and Whole Cell-based Syntheses. – A recent review discusses preparations of antiviral nucleosides by the sequential use of nucleoside phosphorylase and *N*-deoxyribosyltransferase enzymes.[86] A detailed review of the integration between the enzymatic epimerization of GlcNAc to nNAc and the Neu5Ac aldolase-catalysed bio-transformation to Neu5Ac has been reported.[87] Wong has reviewed complex carbohydrate synthesis tools that are accessible to glycobiologists, with particular focus on enzymatic and computer based one-pot approaches for the preparation of complex carbohydrates/glycoconjugates.[88]

The tetra-saccharide β-D-Gal-(1→4)-β-D-GlcNAc-(1→3)-β-D-Gal-(1→4)-Glc has been made by three methods of increasing efficiency: (i) conventional organic synthesis, using a lactose acceptor (30%); (ii) similarly with an MPEG-supported lactose acceptor (53%); (iii) enzymatically (96%).[89] A sequential one-pot synthesis of UDP-2-deoxy-2-fluoro-D-galactose from D-galactal using selectfluoride, galactokinase and galactose-1-phosphate uridyltransferase has been reported.[90] The five per-*O*-acetylated disaccharide glycones of **28–32**, structurally related to the glycans of vertebrate mucins, have been evaluated for their ability to prime oligosaccharide synthesis, inhibit glycoprotein synthesis and alter adhesion to E-selectin expressed in endothelial cells. All these compounds served as substrates for the enzymic synthesis of oligosaccharides.[91]

28 Gal-β-(1-4)-GlcNAc-β-OR
29 GlcNAc-β-(1-3)-Gal-β-OR
30 Gal-β-(1-3)-GalNAc-α-OR   R = [naphthalenylmethyl structure]
31 GlcNAc-β-(1-3)-GalNAc-β-OR
32 GlcNAc-β-(1-6)-GalNAc-β-OR

A preparative synthesis of GDP-β-L-fucose from GDP-D-mannose using recombinant enzymes from enterobacterial sources has been achieved.[92] The enzymatic conversion of D-glucose to nucleotides by the stepwise addition of sequential enzymes in the glycolytic and pentose phosphate pathway has been used for the preparation of ^{2}H- and ^{13}C-labelled RNA.[93] TDP-3,6-dideoxy-β-L-*arabino*-hexopyranose has been produced on a 200 mg scale from TDP-3-deoxy-α-D-glucose by the L-rhamnose synthesizing system isolated from *Salmonella entericum LT*2.[94] Various ^{13}C-labelled 4-*O*-diphosphocytidyl-2-*C*-methyl-D-erythritols have been prepared from sodium pyruvate and labelled D-glucose in a one-pot procedure by use of a mixture of five recombinant enzymes from the nonmevalonate isoprenoid pathway.[95] The uptake of a 12-azidododecyl β-lactoside primer for sialylation by B16 melanoma cells resulted in the 3′-glycosylation product which is a glycoside of ganglioside GM3.[96]

A one-pot, two-step enzymatic ^{13}C-labelling method for the preparation of sialic acid analogues **35** has been developed (Scheme 7).[97] Multi-enzyme approaches have been used to prepare bisecting-type oligosaccharides from bisecting-*N*-glycan saccharides.[53] The lipophilic benzyloxycarbonylaminoaniline group has been used to derivatize reducing oligosaccharides, to give products such as **36** following *N*-acetylation, which can be further elaborated by multistep enzymatic glycosylation. The lipophilic moiety enables facile purification by C18 reverse phase chromatography after each reaction.[98] The UDP-2- and

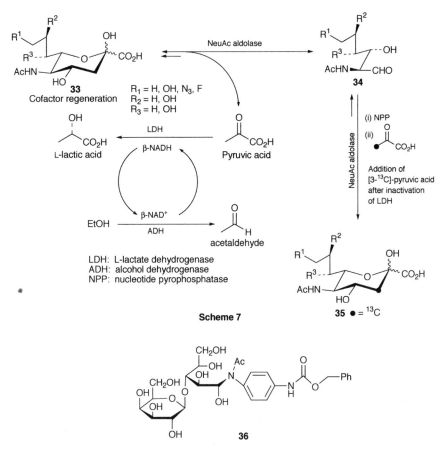

Scheme 7

3-fluorogalactopyranose sugar nucleotides have been prepared from 2- and 3-deoxy-2- and 3-fluoro-D-galactose using a one-pot cocktail of nine commercial enzymes.[99] Both these compounds were used as mechanistic probes for UDP-Gal$p$-mutase.

Starting from UDP-Glc, a variety of α-D-Gal-(1→4)-β-D-Gal derivatives have been prepared using a UDP-Glc-4-epimerase-α-1,3-Gal-transferase fusion enzyme. The fusion protein enhanced reaction rates by 300% compared to the same reactions carried out with equivalent amounts of the two separate enzymes.[100] A one-pot route to ketoses, such as **38**, from glycerol and pyrophosphate, *via* their phosphorylated precursors **37** has been developed using phytase (Scheme 8).[101]

Glycopeptides derived from P-selectin glycoprotein PSGL-1 have been prepared from peptidic disaccharide precursors using the appropriate glycosyl transferases.[102] Manipulation of the biosynthetic machinery in *Streptomyces venezuelae*, which normally produces macrolides containing D-desosamine, has resulted in L-rhamnose being appended instead.[103] A GlcNAc kinase gene (yqgR) from *Bacillus subtilis* has been added to a yeast-based reaction system to improve the whole cell synthesis of UDP-GlcNAc.[104]

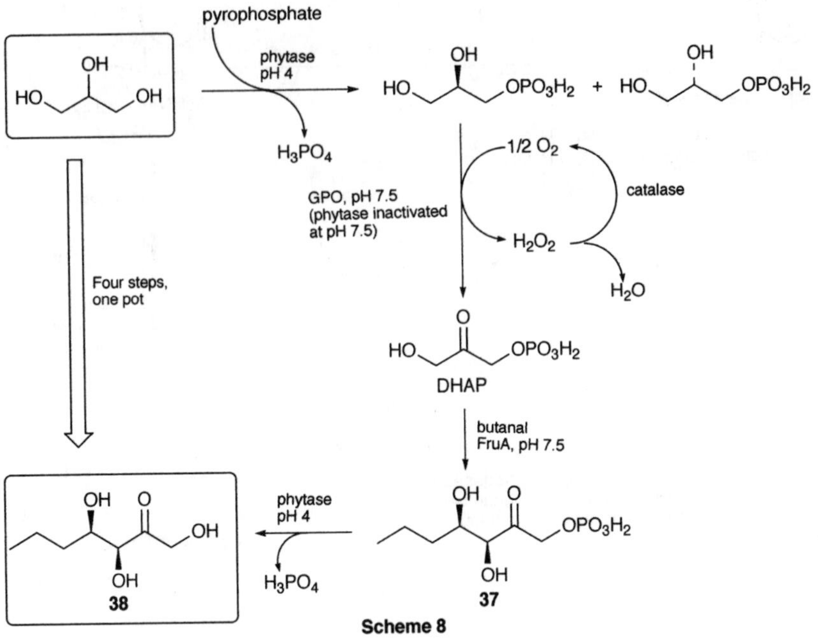

Scheme 8

3 Enzyme Mechanisms

A review of glycosidase mechanisms has recently appeared, which primarily deals with the hydrolytic mechanisms of β-glucosidases.[105] Two other review papers cover a broad range of novel enzymatic mechanisms in carbohydrate metabolism such as epimerase reactions, 1,2-carbonyl rearrangements (*e.g.* with D-xylose isomerase), C–O bond cleavage reactions (excluding glycosidases), C–C bond formation in branched chain sugars and reactions involved in the formation/rearrangement of hexose skeletons (*e.g.* DHQ synthase).[106,107] A review of mechanism-based inhibitors of Kdo8P synthase has also appeared (see Chapter 16).[108]

   Photoreactive aryl glycosides bearing an azido, a diazonium salt or a diazirine substituted aryl group as the photoprobe have been developed to study the sugar binding site of melbiose permease.[109] All possible monomethyl ethers of *p*-nitrophenyl α-D-gluco-, α-D-galacto and α-D-mannopyranosides have been made and used to evaluate the hydrolytic activities of various glycosidases. *Mortierella vinacea* α-galactosidase and almond and jack bean α-mannosidases were found to hydrolyse the glycosidic bonds of the respective 6-*O*-methyl ether derivatives whereas the other monomethyl ethers were unreactive.[110] The substrate properties of the 2-, 3-, 4- and 6-deoxy-α-D-galactopyranosides of *p*-nitrophenol have been studied with α-galactosidases from several sources. The 3-, 4- and 6-hydroxyl groups were essential for some of the enzymes, and the 2-deoxyglycoside was a substrate for all the enzymes studied.[111] 5-Deoxy-5-fluoro-α-D-galactosyl fluoride has been prepared and used to trap an intermediate of green coffee bean α-galactosidase reaction in order to identify the catalytic aspartate nucleophile.[112] The glycosidase-catalysed hydrolysis of isoquinolinium salts **39** and **40** have been studied.[113] Both are enzyme substrates, but **39** binds more tightly to the enzyme than does the 2-deoxy compound **40**, and the enzyme-catalysed hydrolysis rate of the former is 10^4 times higher. The authors concluded that the interaction of the 2-OH group with the α-glucosidase generates a relative transition state stabilization of about 23.5 kJ mol^{-1}.

39 R = OH
40 R = H   **41** n = 0-2

1-*O*-(*p*-Hydroxybenzoyl) β-galactose was synthesized to examine the hydrolytic activity of the β-galactosyl ester linkage by β-galactosidases. Of particular interest, the enzyme from *Penicillium multicolor* hydrolysed this substrate in the presence of ^{18}O-labelled water to liberate galactose containing ^{18}O suggesting that C-O bond cleavage occurs at the anomeric carbon rather than the ester carbonyl carbon centre.[114] A computational method has been presented for predicting the alcohols which can be glucosylated with β-glucosidases.[115] Experiments with *p*-nitrophenyl β-D-fucofuranoside have shown it not to be a substrate for the exo-β-D-galactofuranosidase from *P. fellutanum* which indicates that the C-6 hydroxyl group is essential for activity.[116] ^1H NMR spectroscopy has been used to investigate the transfer of sialic acid from donor to acceptor molecules using *trans*-sialidase from *T. cruzi*.[117] A separate study of this enzyme by ^1H NMR has shown that its stereoselectivity is identical to that of bacterial, viral and mammalian sialidases, suggesting a similar active-site architechture.[118] The action pattern of human salivary amylase (HSA) has been examined by HPLC analysis of the product pattern and cleavage frequency of model maltooligosaccharides of dp 4–8.[119] Results suggest the binding region in HSA is longer than the five subsites usually considered in the literature, and imply differences between the 3D structures of HSA and porcine pancreatic α-amylase. The 2', 3' and 4'-modified deoxy and *O*-methylfucopyranose derivatives of α-L-Fuc*p*-(1→2)-β-D-Gal*p*-1-*O*-octyl have been made as acceptor analogues for the human blood group A and B Gal- and GalNAc-transferases. Experiments with all these compounds have shown that the 2'-OH is an essential feature for recognition by these enzymes.[120] Both enantiomers of 1,2-anhydro-*myo*-inositol have been prepared and assessed as glycosidase inhibitors. The 1-D-isomer was shown to be active against β-glucosidase but neither isomer was active against α-glucosidase or β-galactosidase.[121] A series of C-4' branched deoxynucleotides **41** have been prepared and incorporated into oligodeoxynucletides the resistance of which to snake venom phosphodiesterase and DNAase is notably increased by the modifications.[122]

NMR studies have shown that the conformation of D-*gluco*-dihydroacarbose bound in the active site of glucoamylase resembles that found in the crystal structure rather than the unbound solution conformation.[123] Both α-fluorophosphonate analogues of glucose 6-phosphate have been made and studied as substrate mimics for glucose 6-phosphate dehydrogenase. Surprisingly, the k_{cat}/K_m value of the 7(S)-diastereomer is an order of magnitude greater than that of the 7(R)-diastereomer, which is explained in the context of the known phosphate binding pocket of this enzyme.[124]

4   Other Enzymatic Modifications

An extensive review on the chemoenzymatic radiolabelling of gangliosides has been reported.[125] Tritium labelling has been achieved by C-6 oxidation of the galactose moieties using galactose oxidase [EC 1.1.3.9] followed by chemical reduction of the aldehyde products with tritium-labelled sodium borohydride. The selective biocatalytic modifications of conventional, carbocyclic and *C*-nucleosides have also been reviewed.[126] L-Rhamnose isomerase of *Pseudomonas sp.* LL172 immobilized on Chitopearl beads has been used to produce L-talose and D-gulose from L-tagatose and D-sorbose, respectively.[127] Eleven synthetic D-glucuronans have been epimerized to varying extents with *Azobacter vinelandii* poly-β-D-mannuronic acid C-5-epimerase to give copolymers of D-glucuronic acid and L-iduronic acid with different viscocities, acid susceptibilities and capacity to form calcium-gels which were compared with those of the unepimerized D-glucuronans.[128] An enzyme involved in the biosynthesis of the 4-epivancosamine substituents of a vancomycin group antibiotic has been expressed and its role as a TDP-4-keto-6-deoxyglucose-3,5-epimerase has been demonstrated.[129] The stereo-/regio-specific syntheses of Gal-β-(1→3)-GlcNAc derivatives using the β-1,3-galactosyl-*N*-acetylhexosamine phosphorylase enzyme from *Bifidobacterium bifidum* DSM 20083, which catalyses the reversible phosphorolytic cleavage of β-(1→3)-galactooligosaccharides, have been reported.[130] The preparation of 2'-deoxyribonucleosides with identically ^{2}H/^{13}C-labelled sugar residues has been achieved by the enzymatic transglycosylation reactions of purine/pyrimidine nucleoside phosphorylases with isotopically labelled thymine (the preparation of which is described in Chapter 19).[131]

Sucrose phosphorylase from *Leuconstroc mesentroides* has been used for the enzymatic transglucosylation from sucrose to 4-hydroxy-3(2H)-furanone derivatives in the preparation of potential antioxidant compounds **42** and **43**.[132] A

42 R^1 = Me, R^2 = Et
43 R^1 = Et,   R^2 = Me

procedure for the preparation of GMP using guanosine-inosine kinase from *Exiguobacterium acetylicum* coupled with ATP regeneration has been reported.[133] A range of fluoro- and/or azido-derivatives of D-fructose have been prepared by the polyol dehydrogenase catalysed isomerization of the corresponding D-glucose substrates.[134]

An efficient double oxidation of D-xylose to D-*glycero*-aldopentos-2,3-diulose (2,3-diketo-D-xylose, **45**) has been achieved (80% yield) using pyranose dehydrogenase from the mushroom *Agaricus bisporus* (Scheme 9).[135] The gram scale 2′-deoxygenation of ATP has been achieved using a recombinant ribonucleoside triphosphate reductase from *Lactobacillus leichmannii*.[136]

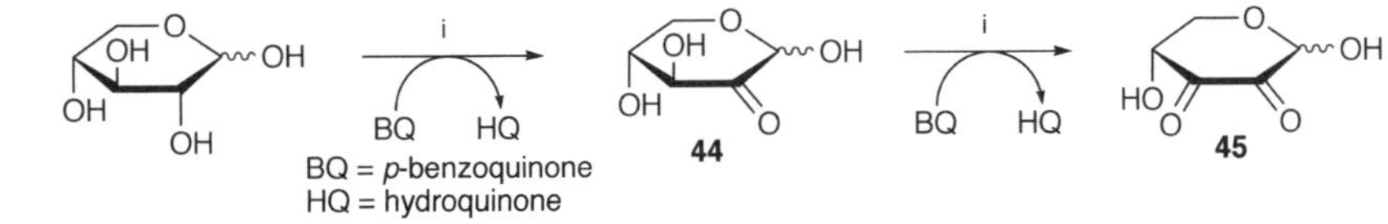

BQ = *p*-benzoquinone
HQ = hydroquinone

Reagents: i, Pyranose dehydrogenase

Scheme 9

5 Miscellaneous

A potential two-step process to convert D-glucose to catechol (**48**) *via* D-glucose-6-phosphate (**46**) has been developed. This comprises a one-pot incubation of the sugar with hexokinase and recombinant 2-deoxy-*scyllo*-inosose synthase, followed by chemical reductive dehydration of the resulting 2-deoxy-*scyllo*-inosose **47** with hydrogen iodide (Scheme 10).[137]

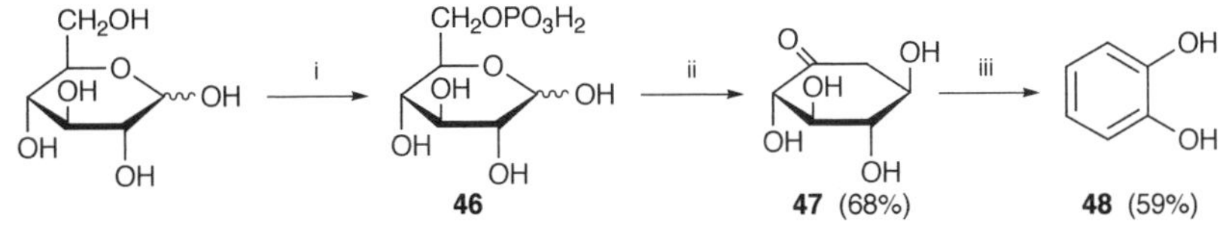

Reagents: i, Hexokinase; ii, 2-deoxy-scyllo-inosose synthase; iii, HI, HOAc

Scheme 10

Various phosphorylated pentose derivatives (*e.g.* D-arabinose 5-phosphate, **49**) have been condensed with mono-deuterated (Z) and (E)-phosphoenolpyruvates in the presence of *E. coli* 3-deoxy-D-*arabino*-heptulosonate 7-phosphate (DAH 7-P) synthase. The products were in complete agreement with the observed facial selectivity of DAH 7-P synthase with its native substrates (Scheme 11).[138]

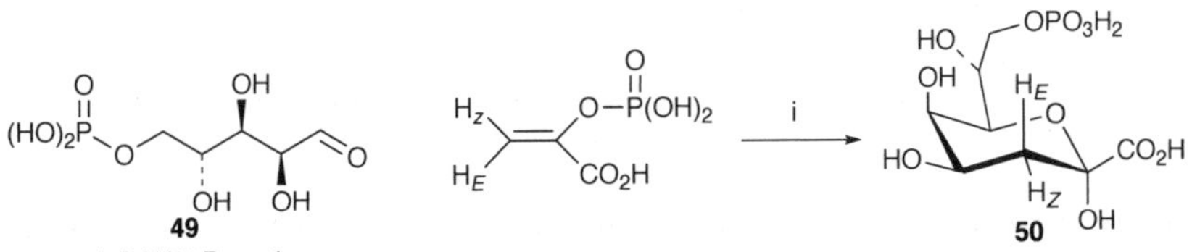

Reagents: i, DAH 7-P synthase

Scheme 11

A nonmevalonate pathway intermediate, 4-(cytidine 5'-diphospho)-2-*C*-methyl-D-erythritol is transformed to its 2-phospho-derivative in the presence of ATP by a novel *E. coli* enzyme, 4-(cytidine 5'-diphospho)-2-*C*-methyl-D-erythritol kinase.[139] The same authors report the enzymatic conversion of this product to 2-*C*-methyl-D-erythritol 2,4-cyclodiphosphate using a cyclodiphosphate synthase isolated from the same source.[140] Waldmann *et al.* have reported the use of the tetra-*O*-acetyl gluco-/galactopyranosyl oxycarbonyl unit as an *N*-terminal peptide protecting group. α/β Mixtures of the resulting carbohydrate-derived urethanes, *e.g.* **51**, are readily cleaved in a one-pot procedure involving enzymatic deacetylation followed by glycosidase-catalysed glycosidic bond fragmentation (Scheme 12).[141]

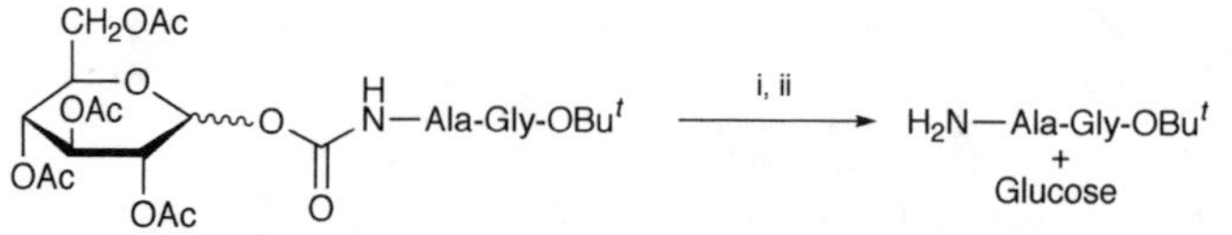

Reagents: i, Lipase WG, 5% MeOH, pH 6.0; ii, α/β glucosidase

Scheme 12

Various dermatan sulfate derived di-, tetra-, hexa-, octa-, deca- and dodeca-saccharide mixtures have been prepared and purified (on a semi-preparative scale) by controlled depolymerization of dermatan sulfate using chondroitin ABC lyase.[142] The pathways of InsP$_6$ hydrolysis by phytase from wheat bran of *Triticum aestivum* have recently been established.[143]

An improved ($\times$ 10) method for assaying α-glucosidases has been developed using aryl β-maltosides as substrates with β-glucosidase as an auxiliary enzyme.[144]

References

1. J.A. Khan, A. Millqvist-Fureby and E.N. Vulfson, *Methods Biotechnol.*, 1999, **10**, 313 (*Chem. Abstr.*, 2000, **132**, 108 149).
2. O. Hindsgaul, *Essent. Glycobiol.*, 1999, 509, Ed. A. Varki (*Chem. Abstr.*, 2000, **132**, 108 159).
3. K.M. Koeller and C.-H. Wong, *Chem. Rev.*, 2000, **100**, 4465.
4. M.M. Palcic, *Curr. Opin. Biotechnol.*, 1999, **10**, 616.
5. N. Wymer and E.J. Toone, *Curr. Opin. Chem. Biol.*, 2000, **4**, 110.
6. T. Kajimoto, *Yakugaku Zasshi*, 2000, **120**, 42 (*Chem. Abstr.*, 2000, **132**, 194 551).
7. M. Hendrix and C.-H. Wong, *Bioorg. Chem.: Carbohydr.*, 1999, **198**, 569 (*Chem. Abstr.*, 2000, **132**, 293 929).
8. A. Romero and C.-H. Wong, *J. Org. Chem.*, 2000, **65**, 8264.
9. G. Guanti, L. Banfi and M.T. Zannetti, *Tetrahedron Lett.*, 2000, **41**, 3181.
10. W.-D. Fessner, C. Grosse, G. Jaeschke and O. Eyrisch, *Eur. J. Org. Chem.*, 2000, 125.
11. N. Ouwerkerk, J.H. van Boom, J. Lugtenburg and J. Raap, *Eur. J. Org. Chem.*, 2000, 861.

12. A.J. Humphrey, S.F. Parsons, M.E.B. Smith and N.J. Turner, *Tetrahedron Lett.*, 2000, **41**, 4481.
13. S.J. Williams and S.G. Withers, *Carbohydr. Res.*, 2000, **237**, 27.
14. S. Shoda and M. Fujita, *Kagaku to Seibutsu*, 2000, **38**, 309 (*Chem. Abstr.*, 2000, **132**, 347 799).
15. S.G. Withers, *Spec. Publ. – R. Soc. Chem.*, 1999, **246**, 62 (*Chem. Abstr.*, 2000, **132**, 222 714).
16. C. Bucke, J. Packwood, S. Suwasono, R.A. Rastall, *Methods Biotechnol.*, 1999, **10**, 213 (*Chem. Abstr.*, 2000, **132**, 108 148).
17. D.H.G. Crout, P. Critchley, D. Muller, M. Scigelova, S. Singh and G. Vic, *Spec. Publ. – R. Soc. Chem.*, 1999, **246**, 15 (*Chem. Abstr.*, 2000, **132**, 222 713).
18. K. Saito and F. Tomita, *Biosci. Biotechnol. Biochem.*, 2000, **64**, 1321.
19. F. Takenaka and H. Uchiyama, *Biosci. Biotechnol. Biochem.*, 2000, **64**, 1821.
20. J.K. Fairweather, R.V. Stick and S.G. Withers, *Aust. J. Chem.*, 2000, **53**, 913.
21. C. Srisomsap, P. Subhasitanont, S. Techasakul, R. Surarit and J. Svasti, *Biotechnol. Lett.*, 1999, **21**, 947.
22. H. Kono, S. Kawano, T. Erata and M. Takai, *J. Carbohydr. Chem.*, 2000, **19**, 127.
23. H.K. Shin, J.Y. Kong, I.D. Lee and T.H. Lee, *Biotechnol. Lett.*, 2000, **22**, 321.
24. S. Kitahata, T. Tanimoto, Y. Okada, K. Tanska, H. Murakami, H. Nakano and K. Koizami, *Biosci. Biotechnol. Biochem.*, 2000, **64**, 2406.
25. S. Kitahata, T. Tanimoto, A. Ikuta, K. Tanaka, K. Fujita, H. Hashimoto, H. Murakani, H. Nakano and K. Koizumi, *Biosci. Biotechnol. Biochem.*, 2000, **64**, 1223.
26. P. Spangenberg, C. André, M. Dion, C. Rabiller and R. Mattes, *Carbohydr. Res.*, 2000, **329**, 65.
27. M. Zarevucka, M. Vacek, Z. Wimmer, C. Brunet and M.-D. Legoy, *Biotechnol. Lett.*, 1999, **21**, 785.
28. V. Perrin, B. Fenet, J.-P. Praly, F. Lecroix and C.D. Ta, *Carbohydr. Res.*, 2000, **325**, 202.
29. W.H.M. van Casteren, M. Eimermann, L.A.M. van den Brock, J.-P. Vincken, H. A. Schols and A.G.J. Voragen, *Carbohydr. Res.*, 2000, **329**, 75.
30. M. Kawamura, H. Nakai and T. Uchiyama, *Carbohydr. Res.*, 2000, **323**, 49.
31. H. Nakano, H. Murakami, M. Shizuma, T. Kiso, T.L. de Aragio and S. Kitahata, *Biosci. Biotechnol. Biochem.* 2000, **64**, 1472.
32. Y.-K. Kim and J.F. Robyt, *Carbohydr. Res.*, 2000, **328**, 509.
33. T. Yamamoto, T. Hosokawa, S. Kishihara, H. Tawaki, N. Wakiuchi and Y. Suzuki, *Seito Gijutsu Kenkyu Kaishi*, 1998, **46**, 49 (*Chem. Abstr.*, 2000, **132**, 108 169).
34. T. Yamamoto, T. Hosokawa, S. Kishihara, S. Fujii, H. Tawaki, N. Wakiuchi and Y. Suzuki, *Seito Gijutsu Kenkyu Kaishi*, 1998, **46**, 57 (*Chem. Abstr.*, 2000, **132**, 108 170).
35. U. Gambert, H.S. Conradt, M. Nimtz and J. Thiem, *J. Carbohydr. Chem.*, 2000, **19**, 621.
36. J.H. Yoon and J.S. Rhee, *Carbohydr. Res.*, 2000, **327**, 377.
37. X. Zeng, R. Yoshino, T. Murata, K. Ajisaka and K. Usui, *Carbohydr. Res.*, 2000, **325**, 120.
38. A. Vetere, M. Miletich, M. Bosco and S. Paoletti, *Eur. J. Biochem.*, 2000, **267**, 942.
39. H. Nakano, M. Shizuma, T. Kiso and S. Kitahata, *Biosci. Biotechnol. Biochem.*, 2000, **64**, 735.
40. J. Ortner, M. Albert, K. Terler, W. Steiner and K. Dax, *Carbohydr. Res.*, 2000, **327**, 483.
41. S. Singh, M. Scigelora and D.H.G. Crout, *Tetrahedron Asymm.*, 2000, **11**, 223.

42. E. Farkas, J. Thiem and K. Ajisaka, *Carbohydr. Res.*, 2000, **328**, 293.
43. A. van Remoortere, C.H. Hokke, G.J. van Dam, I. Van Die, A.M. Deelder and D. van den Eijinden, *Glycobiology*, 2000, **10**, 601.
44. S.-I. Shoda, T. Kiyosada, H. Mori and S. Kobayashi, *Heterocycles*, 2000, **52**, 599.
45. W.S. Yor and R. Hawkins, *Glycobiology*, 2000, **10**, 193.
46. I. Gill and R. Valivety, *Angew. Chem. Int. Ed. Engl.*, 2000, **39**, 3804.
47. S. Singh, M. Scigelova, M.L. Hallberg, O.W. Howarth, S. Schenkman and D.H.G. Crout, *Chem. Commun.*, 2000, 1013.
48. D. Schmidt, B. Sauerbrei and J. Thiem, *J. Org. Chem.*, 2000, **65**, 8518.
49. D. Schmidt and J. Thiem, *Chem. Commun*, 2000, 1919.
50. A. Trincone, G. Perugino, M. Rossi and M. Moracci, *Bioorg. Med. Chem. Lett.*, 2000, **10**, 365.
51. C. Mayer, D.L. Zechel, S.P. Reid, R.A.J. Warren and S.G. Withers, *FEBS Lett.*, 2000, **466**, 40.
52. S. Fort, V. Boyer, L. Greffe, G.J. Davies, O. Moroz, L. Christianse, M. Schiilein, S. Cotaz and H. Driguez, *J. Am. Chem. Soc*, 2000, **122**, 5429.
53. C. Unverzagt and J. Seifert, *Tetrahedron Lett.*, 2000, **41**, 4549.
54. V. Wittmann, A.K. Datta, K.M. Koeller and C.-H. Wong, *Chem. Eur. J.*, 2000, **6**, 162.
55. A. Lubinean, A. Malleron and C. Le Narvor, *Tetrahedron Lett.*, 2000, **41**, 8887.
56. Y. Kajihara, H. Hashimoto and S. Ogawa, *Carbohydr. Res.*, 2000, **323**, 44.
57. Y. Nishida, H. Tamakoshi, Y. Kitagawa, K. Kobayashi and J. Thiem, *Angew. Chem. Int. Ed. Engl.*, 2000, **39**, 2000.
58. M.O.L. Seto, C.A. Comston, A. Szpacenko and M.M. Palcic, *Carbohydr. Res.*, 2000, **324**, 161.
59. K.M. Koeller and C.-H. Wong, *Chem. Eur. J.*, 2000, **6**, 1243.
60. A. Duffels, L.G. Green, R. Lenz, S.V. Ley, S.P. Vincent and C.-H. Wong, *Bioorg. Med. Chem.*, 2000, **8**, 2519.
61. C. Augé, A. Malleron, H. Tahrat, A. Marc, J.-L. Goergen, M. Cerlutti, W.F.A. Steelant, P. Delannoy and A. Lublineau, *Chem. Commun.*, 2000, 2017.
62. B. Ernst, B. Wagner, G. Baisch, A. Katopadis, T. Wrinkler and R. Ohrlein, *Can. J. Chem.*, 2000, **78**, 892.
63. K. Sujino, R.J. Jackson, N.W.C. Chan, S. Tsuji and M.M. Palcic, *Glycobiology*, 2000, **10**, 313.
64. D.E. Stephenson, V. Kren, P. Halada and P. Sedmera, *Collect. Czech. Chem. Commun.*, 2000, **65**, 117.
65. G.N. Jenkins, A.V. Stachulski, F. Scheinmann and N.J. Turner, *Tetrahedron Asymm.*, 2000, **11**, 413.
66. T.J. Tolbert and C.-H. Wong, *J. Am. Chem. Soc.*, 2000, **122**, 5421.
67. G. Carrea and S. Riva, *Angew. Chem. Int. Ed. Engl.*, 2000, **39**, 2227.
68. M. Ferrer, M.A. Cruces, F.J. Plou, M. Bernabé and A. Ballesteros, *Tetrahedron*, 2000, **56**, 4043.
69. I. Izquierdo, M.T. Plaza, M. Rodriguez and J. Tamayo, *Tetrahedron Asymm.*, 2000, **11**, 1749.
70. P. Degn, K.L. Larsen, J.Ø. Duus, B.O. Petersen and W. Zimmermann, *Carbohydr. Res.*, 2000, **329**, 57.
71. Y. Yan, U.T. Bornsheuer and R.D. Schmid, *Biotechnol. Lett.*, 1999, **21**, 1051 (*Chem. Abstr.*, 2000, **132**, 251 346).
72. B. La Ferla, L. Lay, G. Russo and L. Panza, *Tetrahedron Asymm.*, 2000, **11**, 3647.

73. B. La Ferla, L. Lay, L. Poletti, G. Russo and L. Panza, *J. Carbohydr. Chem.*, 2000, **19**, 331.
74. H. Ugama, E. Klegraf, S. Wada and S. Kobayashi, *Chem. Lett.*, 2000, 800.
75. K. Hotta, A. Sunada, Y. Ikeda and S. Kondo, *J. Antibiotics*, 2000, **53**, 1168.
76. A. Sunada, Y. Ikeda, S. Kondo and K. Hotta, *J. Antibiotics*, 2000, **53**, 1416.
77. R.V. Nair and M.M. Salunkhe, *Synth. Commun.*, 2000, **30**, 3115.
78. Y. Honda, S. Tanimori, M. Kirihata, S. Kaneko, K. Tokuyasu, M. Hashimoto and T. Watanabe, *Bioorg. Med. Chem. Lett.*, 2000, **10**, 827.
79. K. Tekuyasu, H. Ono, M. Mitsutomi, K. Hayashi and Y. Mori, *Carbohydr. Res.*, 2000, **325**, 211.
80. T.-L. Shih, M.-C. Cheng and S.-H. Wu, *Tetrahedron Lett.*, 2000, **41**, 7921.
81. A. Kuboki, T. Ishihara, F. kabayashi, H. Ohta, T. Ishii, A. Inoue, S. Mitsuda, T. Miyazaki, Y. Kajihara and T. Sugai, *Biosci. Biotechnol. Biochem.*, 2000, **64**, 363.
82. P. Potier, A. Bouchu, G. Descotes and Y. Queneau, *Tetrahedron Lett.*, 2000, **41**, 3597.
83. H.G. Park and H.N. Chang, *Biotechnol. Lett.*, 2000, **22**, 39 (*Chem. Abstr.*, 2000, **132**, 265 380).
84. Y. Zhao and M.L. Sinnott, *Bioorg. Med. Chem.*, 2000, **8**, 917.
85. T. Torii, M. Fukuta and O. Habuchi, *Glycobiology*, 2000, **10**, 203.
86. W.-R. Qui and Q.-B. Ding, *Zhonguo Yiyao Gongye Zazhi*, 1999, **30**, 474 (*Chem. Abstr.*, 2000, **132**, 137 616).
87. S. Blayer, J.M. Woodley, M.J. Dawson and M.D. Lilly, *Biotechnol. Bioeng.*, 1999, **66**, 131.
88. K.M. Koeller and C.-H. Wong, *Glycobiology*, 2000, **10**, 1157.
89. F. Yan, W.W. Wakarchuk, M. Gilbert, J.C. Richards and D.M. Whitfield, *Carbohydr. Res.*, 2000, **328**, 3.
90. M.D. Burkart, S.P. Vincent, A.D. Duffels, B.W. Murray, S.V. Ley and C-H. Wong, *Bioorg. Med. Chem.*, 2000, **8**, 1937.
91. A.K. Sarkar, J.R. Brown and J.D. Esko, *Carbohydr. Res.*, 2000, **329**, 287.
92. C. Albermann, J. Distler and W. Piepersberg, *Glycobiology*, 2000, **10**, 875.
93. L.G. Scott and T.J. Tolbert, *Methods Enzymol.*, 2000, **317**, 18.
94. W. Kaffke and S. Chambon, *Tetrahedron Asymm.*, 2000, **11**, 639.
95. F. Rohdich, C.A. Schuhr, S. Hacht, S. Herz, J. Wungsintaweekul, W. Eisenreich, M.H. Zenk and A. Bacher, *J. Am. Chem. Soc.*, 2000, **122**, 9571.
96. M. Carmelita, Z. Kasuya, L.X. Wang, Y.C. Lee, M. Mitsuki, H. Nakajima, Y. Miura, T. Sato, K. Hatakana, S. Yamagata and T. Yamagata, *Carbohydr. Res.*, 2000, **329**, 755.
97. T. Mikazaki, H. Sato, T. Sakakibara and Y. Kajihara, *J. Am. Chem. Soc.*, 2000, **122**, 5678.
98. M. Båström, M. Bengsston, O. Blixt and T. Norberg, *Carbohydr. Res.*, 2000, **328**, 525.
99. J.N. Barlow and J.S. Blanchard, *Carbohydr. Res.*, 2000, **328**, 473.
100. J. Fang, X. Chen, W. Zhang, A. Janczuk and P.G. Wang, *Carbohydr. Res.*, 2000, **329**, 873.
101. R. Schoevaart, F. van Rantwijk and R.A. Sheldon, *J. Org. Chem.*, 2000, **65**, 6940.
102. K.M. Koeller, M.E.B. Smith and C.-H. Wong, *Bioorg. Med. Chem.*, 2000, **8**, 1017.
103. H. Yamase, L. Zhao and H. Lui, *J. Am. Chem. Soc.*, 2000, **122**, 12397.
104. Okuyama, T. Hamamoto, F. Ishige, K. Takenouchi and T. Noguchi, *Biosci. Biotechnol. Biochem.*, 2000, **64**, 386.
105. D.L. Zechel and S.G. Withers, *Acc. Chem. Res.*, 2000, **33**, 11.

106. X. He, G. Agrihotri and D.-W. Lui, *Chem. Rev.*, 2000, **100**, 4615.
107. D.A. Johnson and H.-W. Liu, *Biol. Chem. Interface*, 1999, 351 (*Chem. Abstr.*, 2000, **132**, 108 160).
108. T. Baasov and V. Belakhov, *Recent Res. Devel, Organic Chem.*, 1999, **3**, 195 (*Chem. Abstr.*, 2000, **132**, 249).
109. Y. Ambroise, C. Mioskowski, G. Leblanc and B. Rosseau, *Bioorg. Med. Chem. Lett.*, 2000, **10**, 1125.
110. W. Hakamata, T. Nishio, R. Sato, T. Mochizuki, K. Tsuchiya, M. Yasuda and T. Oku, *J. Carbohydr. Chem.*, 2000, **19**, 359.
111. W. Hakamaya, T. Nishio and T. Oku, *Carbohydr. Res.*, 2000, **324**, 107.
112. H.D. Ly, S. Howard, K. Shum, S. He, A. Zhu and S.G. Withers, *Carbohydr. Res.*, 2000, **329**, 529.
113. K.S.E. Tamaka, J. Huang, F. Lipari and A.J. Bennet, *Can. J. Chem.*, 2000, **78**, 577.
114. T. Kiso, H. Nakano, H. Nakajina, T. Terai, K. Okamoto and S. Kitahata, *Biosci. Biotechnol. Biochem.*, 2000, **4**, 1702.
115. B.M. de Roode, H. Zuilhof, M.C.R. Frannsen, A. Nan du Padt and A. de Groot, *J. Chem. Soc. Perkin Trans. 2*, 2000, 2217.
116. A. Chiocconi, C. Marino and R.M. de Lederkremer, *Carbohydr. Res.*, 2000, **323**, 7.
117. J.C. Wilson, M.J. Kiefel, S. Albouz-Abo and M. von Itzstein, *Bioorg. Med. Chem. Lett.*, 2000, **10**, 2791.
118. A.R. Todeschini, L. Mendorca-Previatol, J.O. Previato, A. Varki and H. van Halbeek, *Glycobiology*, 2000, **10**, 213.
119. L. Kandra and G. Gyemant, *Carbohydr. Res.*, 2000, **329**, 579.
120. A. Mukherjee, M.M. Palcic and Q. Hindsgaul, *Carbohydr. Res.*, 2000, **326**, 1.
121. A. Falshaw, J.B. Hart and P.C. Tyler, *Carbohydr. Res.*, 2000, **329**, 301.
122. M. Kanazaki, Y. Ueno, S. Shuto and A. Matasuda, *J. Am. Chem. Soc.*, 2000, **122**, 2422.
123. T. Weimar, B.O. Petersen, B. Svensson and B.M. Pinto, *Carbohydr. Res.*, 2000, **326**, 50.
124. D.B. Berkowitz, M. Bose, T.J. Pfannenstiel and T. Doukov, *J. Org. Chem.*, 2000, **65**, 4498.
125. S. Sonino, V. Chigorno and G. Tettamanti, *Methods Enzymol.*, 2000, **311**, 639.
126. M. Ferreo and V. Gotor, *Chem. Rev.*, 2000, **100**, 4319.
127. S.H. Bhuiyan, Y. Itami, G. Takada and K. Izumori, *J. Biosci. Bioeng.*, 1999, **88**, 567 (*Chem. Abstr.*, 2000, **132**, 222 721).
128. P.S. Chang, R. Mukerjea, D.D. Fulton and J.F. Robyt, *Carbohydr. Res.*, 2000, **329**, 913.
129. P.N. Kirkpatrick, W. Scaife, T.M. Hallis, H. Lui, J.B. Spencer and D.H. Williams, *Chem. Commun.*, 2000, 1565.
130. E. Farkas, J. Thiem, F. Krzewinski and S. Baiquelet, *Synlett*, 2000, 728.
131. Y. Oogo, K. Nonaka, A. Ono, A. Ono and M. Kainosho, *Nucleic Acids Symp. Ser.*, 1999, **42**, 123 (*Chem. Abstr.*, 2000, **132**, 279 443).
132. S. Kitao, T. Matsudo, T. Sasaki, T. Koga and M. Kawamura, *Biosci. Biotechnol. Biochem.*, 2000, **64**, 134.
133. H. Kawasaki, Y. Usuda, M. Shimoaka and T. Utagawa, *Biosci. Biotechnol. Biochem.*, 2000, **64**, 2259.
134. P. Hadwiga, P. Mayr, B. Nidetzky, A.E. Stütz and A. Tauss, *Tetrahedron Asymm.*, 2000, **11**, 607.
135. J. Volc, P. Sedmera, P. Halada, V. Prikrylova and D. Haltrich, *Carbohydr. Res.*, 2000, **329**, 219.

136. A. Brunella, M. Abantes and O. Ghisalbe, *Biosci. Biotechnol. Biochem.*, 2000, **64**, 1836.
137. K. Kakinuma, E. Nango, F. Kudo, Y. Matsushima and T. Eguchi, *Tetrahedron Lett.*, 2000, **41**, 1935.
138. A.K. Sundaram and R.W. Woodard, *J. Org. Chem.*, 2000, **65**, 5891.
139. T. Kuzuyama, M. Takagi, K. Kaneda, H. Watanabe, T. Dairi and H. Seto, *Tetrahedron Lett.*, 2000, **41**, 2925.
140. M. Takagi, T. Kuzuyama, M. Takagi, K. Kaneda, H. Watanabe, T. Dairi and H. Seto, *Tetrahedron Lett.*, 2000, **41**, 3395.
141. A.G. Gum, T. Kappes-Roth and H. Waldmann, *Chem. Eur. J.*, 2000, **6**, 3714.
142. H.O. Yang, N.S. Gunay, T. Toida, B. Kuberan, G. Yu, Y.S. Kim and R.J. Linhardt, *Glycobiology*, 2000, **10**, 1033.
143. T. Nakano, T. Joh, K. Narita and T. Hayakawa, *Biosci. Biotechnol. Biochem.*, 2000, **64**, 995.
144. Y.V. Voznyi, I.S. Lukomskaya, I.M. Lanskaya and E.I. Podkidysheva, *Vopr. Med. Khim.*, 1996, **42**, 348 (*Chem. Abstr.*, 2000, **132**, 237 265).

21
Structural and Quantitative Analytical and Separatory Methods

1   Computational Methods

An extensive review on structures, configurations and dynamics of bioactive oligosaccharides included a section on computational studies.[1]

Improvements in the quantum mechanical calculations of the potential energy surfaces of D-aldohexoses and D-ketohexoses have been reported.[2] The relative reactivities of the free hydroxyl groups in the partially protected methyl 4,6-O-benzylidene-α- and β-hexopyranosides **1** (D-*allo*-, D-*galacto*-, D-*gluco*- and D-*manno*-configuration) have been explored by conducting experiments with the molecular modelling program STR3DI.EXE and its molecular mechanics module QVBMM.[3]

Conformational energy maps for the furanosyl and pyranosyl rings of ribose and 2-deoxyribose in solution, generated with the MM3 force field, indicated the presence of several tautomers in multiple conformations for both compounds.[4] The preferred conformations and energy pseudorotational barriers of 2-deoxy-β-D-ribofuranosylamine in protonated and unprotonated form have been established by use of *ab initio* molecular orbital and density functional theory calculations.[5] Three-, two- and one-bond ^{1}H-^{13}C spin coupling constants in anhydrodeoxythymidines have been determined both by *ab initio* calculations and by NMR experiments.[6] Modelling experiments have been undertaken to compare the conformations of the seven-membered ring aza-sugars **2** (see Chapter 18 for synthesis), with those of DNJ and isofagamine.[7]

Further monosaccharide pyranosyl compounds to have been investigated by use of computational techniques were the anticancer agent etoposide in CD_3OD, dry $CDCl_3$ and wet $CDCl_3$,[8] and a series of ellagitannin models with axial chirality in the hexahydroxydiphenoyl moieties, such as compound **3**.[9]

Modifications have been made to the AMBER force field to improve the correlations between calculated and observed molecular properties of α-linked saccharides;[10] these led to refinements in solvation studies on maltose, α-, β- and γ-cyclodextrin and two larger cyclodextrins (DP 10 and 21).[11] A molecular dynamics simulation investigation of the solvation patterns of the model disaccharide **4** in aqueous DMSO defined regions in which competition exists between the solvents, whereas other parts of the molecule were preferentially solvated by one solvent.[12]

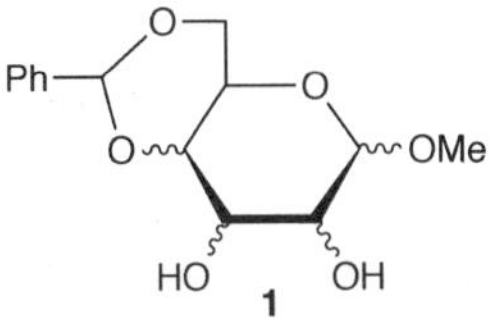

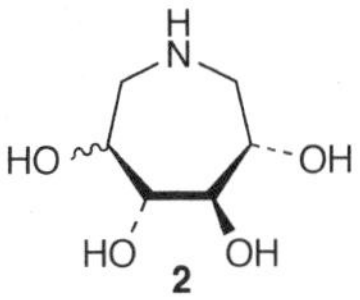

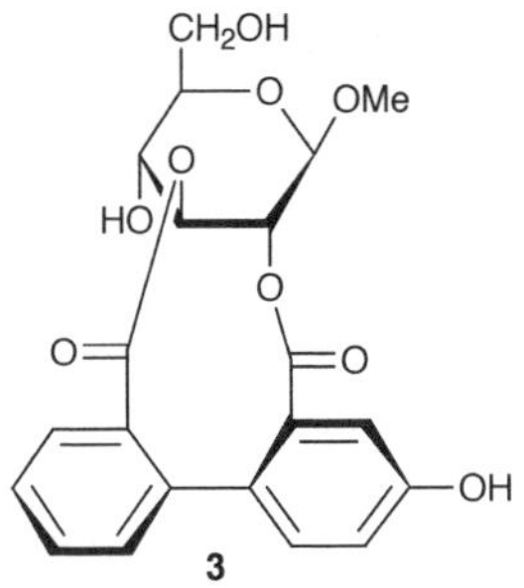

$$\alpha\text{-D-Man}p\text{-}(1\rightarrow3)\text{-}\beta\text{-D-Glc}p\text{OMe}$$
4

The decisive role of the *exo*-anomeric effect in the conformational behaviour of interglycosidic linkages has been demonstrated by molecular dynamics and molecular mechanics studies on the preferred solution conformations of eight *C*-linked D-glucosyl disaccharides,[13] of the *C*-linked disaccharide **5** together with its natural *O*-linked equivalent **6**,[14] and of the two isomeric hydroxymethylene-linked analogues of sialyl-α-(2→3)-D-galactose.[15]

$$\alpha\text{-X-D-Man}p\text{-}(1\leftrightarrow1)\text{-}\beta\text{-D-Gal}p$$
5 X = CH$_2$
6 X = O

The preferred solution conformations of a number of glycosyl *myo*- and *chiro*-inositols, *e.g.* compounds **7** and **8**, that are potential mimics of the putative inositolphosphoglycan mediators (IPGs), have been examined by computational methods as well as NMR spectroscopy.[16]

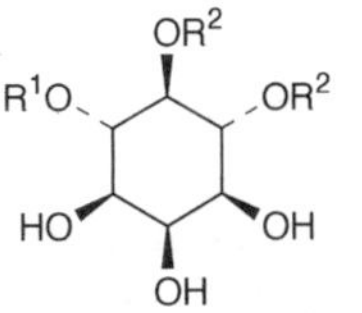

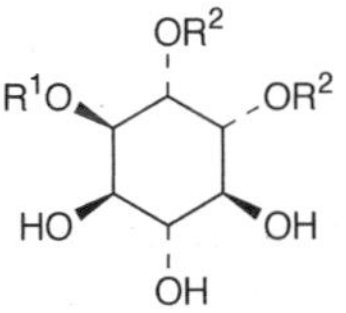

R^1 = β-D-Glc$p$NH$_2$, R^2 = H or
R^1 = β-D-Glc$p$NH$_3^+$, R^2,R^2 =

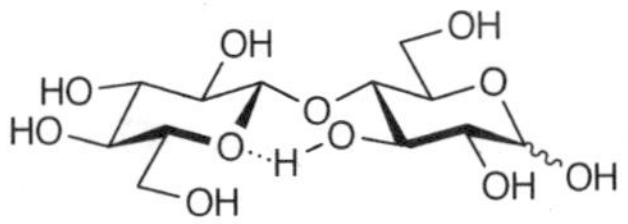

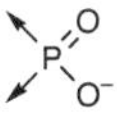

2   Spectroscopy

The Pr(II) and Nd(III) complexes of three pentoses, three hexoses and two disaccharides were characterized by various spectral and analytical techniques including FT IR, ^{13}C NMR, solution absorption and solid-state diffused reflectance spectroscopy, magnetic susceptibility and CD measurements, as well as cyclic voltammetry and thermal analysis.[17]

2.1   NMR Spectroscopy. – Several extensive reviews on the use of ^{1}H and ^{13}C NMR spectroscopy in the carbohydrate field have been published. One address-ing the general problem of structural assignments of complex carbohydrates covered traditional as well as new NMR techniques including computer-assisted evaluation of spectra,[18] another one focussed on blood-group oligosaccharides, Lewis X, sialyl Lewis X, fragments of glycosaminoglycans and structural oligosaccharide mimetics.[1] A third one dealt with the use of stable isotopes (^{13}C, ^{2}H) in conjunction with NMR spectroscopy in the evaluation of carbohydrate structure, conformation and reactivity.[19] An essay entitled 'NMR of Carbohy-drates, Lipids and Membranes' highlights the application of contemporary NMR techniques to solving microbiological, industrio-pharmaceutical and bio-medical problems.[20] In a review on recent advances in the *O*-sialylation of saccharides the NMR criteria for the anomeric assignment of sialosides are summarized.[21]

2.1.1   Technique. Mixtures of sugars with the same molecular weight (*e.g.* D-arabinose and D-ribose), as well as tautomeric forms of a single sugar (*e.g.* α-D-ribofuranose, β-D-ribofuranose, α-D-ribopyranose and β-D-ribopyranose) have been distinguished by diffusion-ordered NMR spectroscopy (DOSY) in the presence of lanthanide cations. This significant development is based on the different diffusion rates of complexed and non-complexed species.[22] A new, one-dimensional ^{13}C NMR approach for determining the configuration of *O*-methyl substituents in glycosides by use of non-refocussed INEPT experiments associated with numerical methods has been reported.[23]

   New procedures for suppressing the interfering methylene signals of protect-ing groups, such as those of benzyl ethers and cyclohexylidene acetals, in 2D NMR spectra of oligosaccharides,[24] and for eliminating spin diffusion effects from NOE measurements[25] have been established. A new strategy to overcome resonance overlap involving 3D-TOCSY and transfer-NOE experiments has been employed for the unambiguous identification of bioactive ligands in prod-uct mixtures; as an example, a library of fifteen mono-, di-, and tri-saccharides was successfully screened for compounds that bind specifically to *Aleuria auran-tia* agglutinin.[26]

   The active site of KDO 8-phosphate synthase complexed to its natural sub-strate and to an inhibitor has been identified by use of rotational-echo double-resonance (REDOR) solid state NMR spectroscopy.[27]

2.1.2   Conformational and Structural Analysis. Detailed ^{1}H NMR spectroscopic studies of hydrogen bonding in selected monosaccharides and inositols,[28] as well as in selected di- and oligo-saccharides (cellobiose, lactose, *N,N′*-diacetyl-chitobiose, agarose, *etc.*),[29] in CDCl$_3$ and DMSO-d$_6$ revealed that discrimina-tions can be made between intra- and inter-molecular hydrogen bonds, between weak and strong intramolecular hydrogen bonds and between hydroxyl groups acting as donors and acceptors. Simple formulae or rules for approximating δ(OH) of fully solvated secondary hydroxyl groups of monosaccharides in DMSO have been developed. Restricted conformational freedom associated with persistent C(3)-OH ···· O-C(5′) bonds was detected in a number of disaccharides, for example in cellobiose (**9**), lactose and *N,N′*-diacetylchitobiose.

By use of similar techniques, inter-residue hydrogen bonding has also been observed in disaccharides **10** and **11**,[30] but in the case of the Lewis b tetrasaccharide derivative **12** no evidence was found for persistent hydrogen bonds participating in the structural stabilization.[31]

β-D-Glc*p*NAc-(1→4)-β-D-Glc*p*NAc   β-D-Gal*p*-(1→3)-α-D-Gal*p*NAc-OMe
10   **11**

α-L-Fuc*p*
1
↓
6
α-L-Fuc*p*-(1→2)-β-D-Gal*p*-(1→3)-β-D-Glc*p*NAc-O(CH₂)₂NHCOCHCH₂
12

A study on intramolecular hydrogen bonding in *N*-(2-amino-2-deoxy-β-D-glucopyranosyl)-*N′*-carbamoyl-L-dipeptidylester focussed mainly on the conformations of the peptide moieties.[32]

2.1.2.1 Monosaccharides. *Erythro*-Hexo-2,5-diulose has been shown by ¹H NMR spectroscopy to exist as a mixture of the acyclic and the four pyranose forms in DMSO. In water, three furanose forms and four hydrated pyranose forms were also present.[33]

A variety of NMR spectroscopic methods have been applied to structural and conformational analyses of the following furanose systems: sugar *gem*-dimethyl substituted alkenyl ethers, *e.g.* α-D-xylofuranose derivatives **13** (see Chapter 5 for synthesis),[34] the diastereomeric 6-chloro-6-deoxy-1,2-*O*-isopropylidene-α-D-gulofuranose cyclic 3,5-piperididomonophosphates **14**,[35] several 4′-thionucleosides,[36] 6-formyl- and 6-(hydroxymethyl)-uridine 5′-carboxaldehydes **15** and **16**, respectively, and their 2′,3′-*O*-isopropylidene derivatives,[37] two ¹³C-²H double-labelled 2′-deoxynucleosides complexed to deoxycytidine kinase,[38] and various nucleoside 5′-triphosphates (AZTTP, dTTP and dATP) bound to HIV-RT in the ground state, *i.e.*, prior to the conformational changes to form catalytic complexes.[39] Three-, two- and one-bond ¹H-¹³C spin coupling constants in anhydrodeoxythymidines (2,3′-anhydro- and 2,5′-anhydro-1-(2′-deoxy-β-D-furanosyl)thymine) have been determined both by NMR experiments and by *ab initio* calculations.[6]

The H-C(1)-C(2)-H torsion angles of α- and β-D-glucopyranose and of methyl α-D-glucopyranoside in the crystal as well as in solution have been calculated on the basis of double-quantum heteronuclear local field NMR experiments using [1,2-¹³C₂]-D-glucopyranose and methyl [1,3-¹³C₂]-α-D-glucopyranoside.[40] Mono- and di-fatty acid esters of α-D-glucopyranose have been characterized on the basis of the substituent-induced chemical shift effects on the carbonyl carbon atoms.[41]

The chiralities at sulfur in epimeric sulfenyl glycosides, obtained by *m*CPBA oxidation of the corresponding ethyl 1-thio-β-glycopyranosides, were readily determined from NMR spectral data, since under the influence of the *exo*-anomeric effect the minor R_s-isomers, such as **17**, assume the preferred conformation shown, giving rise to very large (≥ 170 Hz) differences in the chemical shifts of H_{pro-S} and H_{pro-R}.[42]

Further monosaccharide pyranosyl compounds to have been investigated by NMR spectroscopic methods included α-and β-D-idopyranose,[43] methyl 2,6:3,4-dianhydro-α-D-altropyranoside (**18**),[44] methyl D-glucopyranuronate derivatives **19**,[45] ellagic acid derivatives of D-xylose and D-glucose,[46] the anticancer agent etoposide in CD$_3$OD, dry CDCl$_3$ and wet CDCl$_3$,[8] the major and minor conformational isomers (owing to restricted rotation about the amide bond) of the rare alkaloid casimiroedine (**20**) and its peracetate,[47] and the diacylmannosylerythritols **21**.[48]

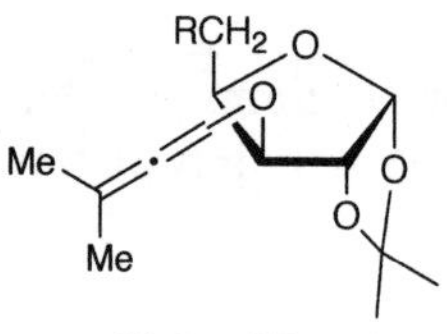

13 R = OTr or H 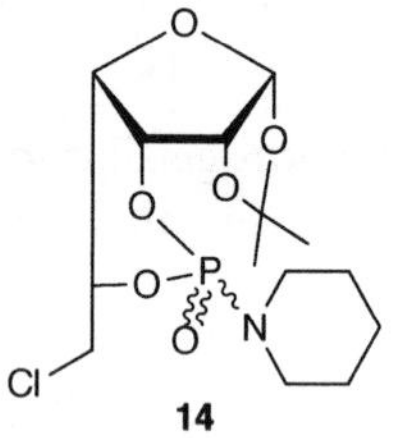   14 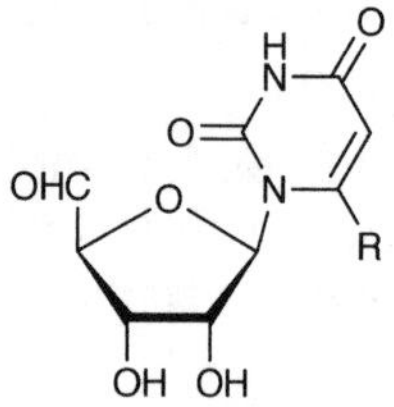   15 R = CHO
                                     16 R = CH$_2$OH

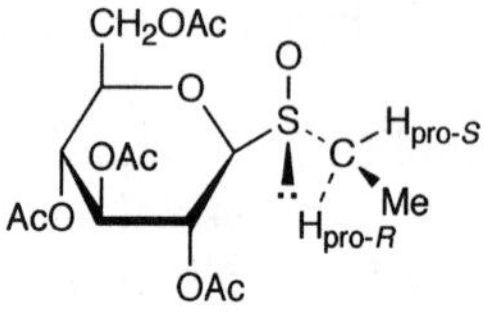

17 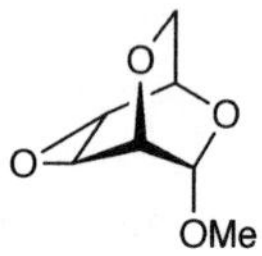   18 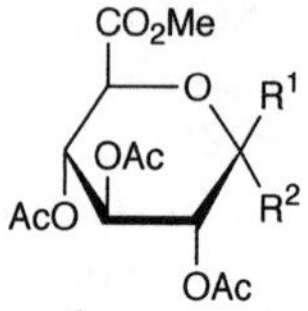   19 R^1 = H, R^2 = OH or R^1 = OAll, R^2 = H

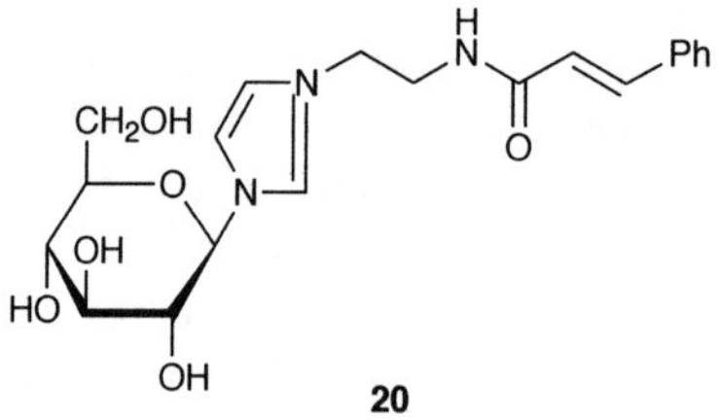

20 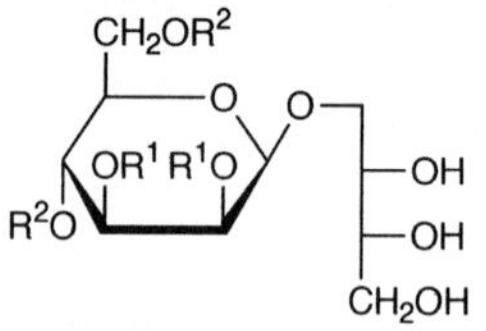   21 R^1 = C(O)(CH$_2$)$_n$Me, n = 6–10; R^2 = H or Ac

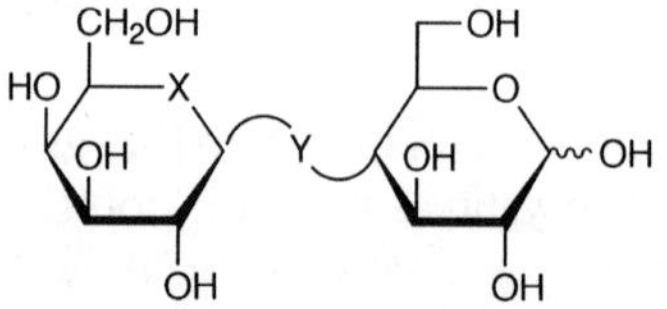

22 X = O, Y = S or X = CH$_2$, Y = O or X = CH$_2$, Y = NH

2.1.2.2 Disaccharides. Full ^{1}H and ^{13}C NMR assignments for thirteen glucosylated or galactosylated *N*-acetylhexosaminitols have been recorded to aid the structural analysis of oligosaccharide alditols related to *O*-linked protozoan glycans.[49] Four 2,3,4-trideuterio-α-L-fucose-containing disaccharides have been synthesized to minimize overlap of ¹H NMR signals and facilitate conformational analysis in saturation transfer difference- (STD-) as well as bioaffinity-NMR studies.[50]

NMR spectroscopic investigations have also been undertaken with the following disaccharides: crystalline sucrose moieties,[51] neocarrabiose and nine of its sulfated and/or pyruvylated derivatives,[52] a series of non-ionizable lactose analogues **22**,[53] cellobiose and five disaccharides containing β-D-glucopyranoside linked to a 6-deoxy- or 2,6-dideoxy-sugar, with the aim of rationalizing the substrate specificities of cardenolide glucohydrolases,[54] various uridine diphosphoglucose salts,[55] and UDPG.[56]

Some of the above studies included molecular mechanics and/or molecular dynamics calculations, and the computational studies on the conformational behaviour of *C*-linked disaccharides and of a number of glycosyl *myo*- and *chiro*-inositols referred to above (Refs. 14–16) were supplemented by NMR experiments.

2.1.2.3 Oligosaccharides and Complex Carbohydrates. The lactonization of α-*N*-acetylneuraminyl-(2→3)-lactose has been studied by NMR and FAB mass spectroscopy.[57]

A series of per-*O*-acetylated cellooligomers have been examined by CP/MAS ¹³C NMR spectroscopy to obtain a structural model for cellulose triacetate in the solid state.[58] Molecular modelling experiments with trisaccharides **23** and **24** indicated that the (1→6)-α-linked side chains did not interfere significantly with the stereochemistry of the α-(1→3) backbone of (1→3)-α-D-glucans.[59]

$$\alpha\text{-D-Glc}p$$
$$1$$
$$\downarrow$$
$$3$$

α-D-Glc*p*-(1→6)-α-D-Glc*p*   α-D-Glc*p*-(1→6)-α-D-Glc*p*-(1→3)-α-D-Glc*p*
23   **24**

It has been suggested on the basis of NMR NOE experiments that the conformation of D-*gluco*-dihydroacarbose bound in the active site of glucoamylase resembles that of the unbound inhibitor in the crystalline state rather than that in solution.[60] Conformational effects observed in a NMR ROESY/molecular dynamics simulation study on Man₅GlcNAc₂ and its conjugate with a pentapeptide indicated that the oligosaccharide stabilized the peptide in solution and the peptide influenced the oligosaccharide conformation.[61]

In addition, the following oligosaccharides and complex carbohydrates have been investigated: the tetrasaccharide **25** and its close analogues **26**,[62] a synthetic octasaccharide fragment of the O-specific polysaccharide of *Shigella dysenteriae* type 1,[63] the ganglioside headgroups GM1, GM2, 6′-GM2 and GM4,[64] and several analogues of the nucleoside antibiotic adenophostin A.[65]

Computational solvation studies of cyclodextrins are referred to in Part 1.1.1 above (Ref. 9).

2.1.3 NMR of Nuclei Other Than ¹H and ¹³C. A review article containing ¹⁹F α-NeuAc

2

↓

3

Hex-(1→4)-β-D-Gal*p*-(1→4)-β-D-Glc*p*NAc-O(CH₂)₅NH₂

25 Hex = β-D-Gal*p*NAc   **26** Hex = β-D-Glc*p*NAc or β-D-Gal*p*

NMR spectral data of fluorinated carbohydrates (δ- and *J*-data on *ca.* 130 compounds and including $J_{F,F}$ values of many more) has been published.[66] A new method for the simultaneous detection of different glycosidase activities in crude culture filtrates involved addition of a mixture of glycosyl fluorides (α-D-Man*p*F, β-D-Man*p*F, α-D-Glc*p*F, β-D-Glc*p*F, β-D-Gal*p*F, β-Cell F, β-D-LacF) to the crude culture filtrates and observation of the changes in the concentrations of individual fluorides by ¹⁹F NMR spectroscopy.[67]

²H NMR spectroscopy, in addition to ¹H and ¹³C methods, has been used in an analysis of octa-*O*-decanoylthio-β,β-trehalose in the solid state,[68] and in the elucidation of the three-dimensional structures of various oligosaccharides containing 2-deoxysugar residues.[69]

Borate esters of furanoid *cis*-1,2-diols were shown by combined ¹¹B NMR, ¹³C NMR and X-ray studies to form bis(diolato)borate anions, such as **27**.[70] ¹¹B NMR experiments have also been employed to investigate complexes between sugar-based bolaamphiphiles, such as the glucuronamide derivative **28**, and the aromatic boronate **29**.[71]

Glycosylmanganese pentacarbonyl complexes have been characterized on the basis of ¹H-, ¹³C- and ⁵⁵Mn-NMR spectral data,[72] and ten tungstate and seven molybdate complexes of D-gulonic acid in aqueous solutions have been identified by multinuclear NMR studies (¹H, ¹³C, ¹⁷O, ⁹⁵Mo, ¹⁵³W).[73]

2.2   IR Spectroscopy. – The FT IR spectra of galactaric acid and its K⁺, NH₄⁺, Ca²⁺, Ba²⁺ and La³⁺ salts have been recorded and interpreted.[74] The mid-IR spectra of some trivalent lanthanide complexes of inositols, *e.g.* PrCl₃·*myo*-inositol·9H₂O, were consistent with their crystal structures.[75] The FT IR spectra for the 2:1 1-*O*-α-D-glucopyranosyl-D-mannitol/ethanol adduct supported the X-ray crystallographic results referred to in Part 3.1 below.[76]

Hydrogen bonding in polycrystalline ribitol, xylitol, D-arabinitol, methyl α-D-manno-, methyl α-D-gluco- and methyl β-D-galacto-pyranoside has been studied by IR spectroscopy at 20–300 °K.[77]

The Raman spectra in the region 20–4000 cm⁻¹ of D-glucose and D-fructose in the crystalline state and in the amorphous state with varying water contents, as well as in solution with various sugar concentrations have been recorded. The Raman spectra of the amorphous sugars were found to be very similar to those of their aqueous solutions.[78]

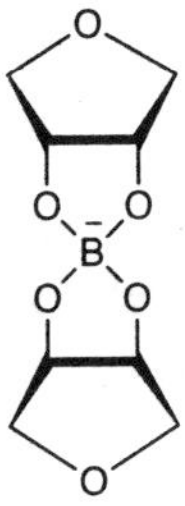

2.3 Mass Spectrometry. – A review on the use of mass spectrometry for ident-ifying flavonoid glycosides has been published.[79]

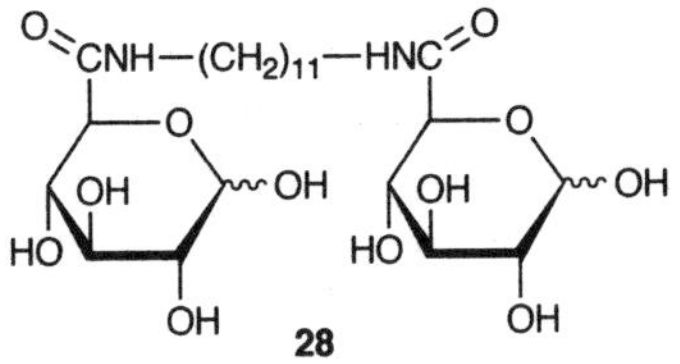

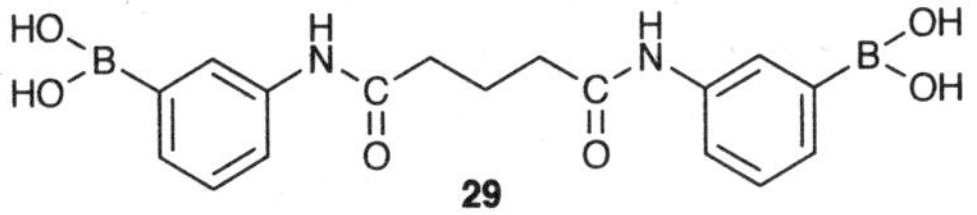

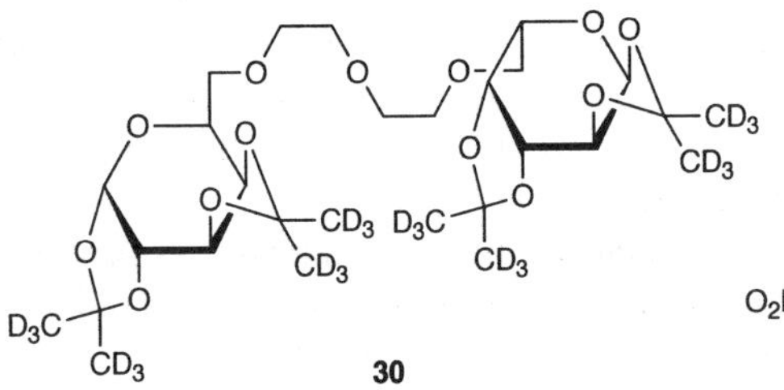

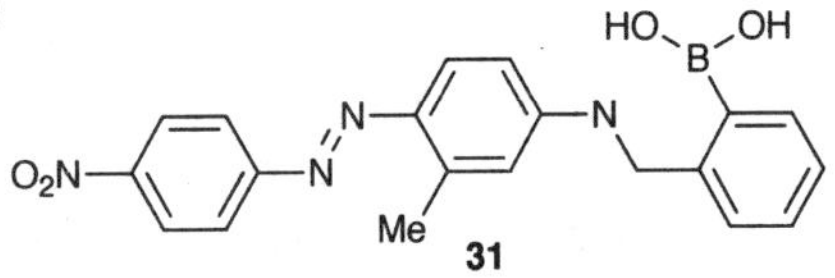

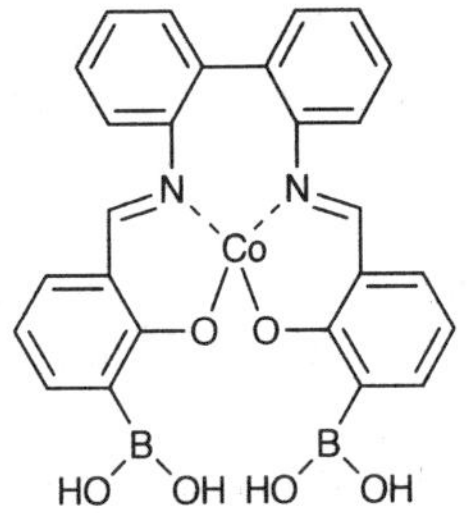

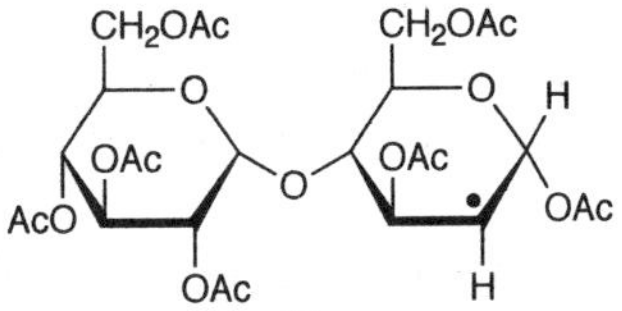

The malonitrile derivatives of oligosaccharides, which were developed for the separation and detection of multi-component oligosaccharide mixtures by nega-tive ion electrospray MS, have now been detected by positive MALDI at considerably lower concentrations, even without prior purification.[80] The lacton-ization of α-N-acetylneuraminyl-(2→3)-lactose has been studied by FAB MS and NMR specroscopy.[57] The O-isopropylidene and O-benzylidene acetals of several 2,6-anhydro-1-deoxy-1-nitroalditols (β-D-glycopyranosylnitromethanes) were studied by EI MS. Distinct differences in the fragmentation patterns of stereoisomers were observed.[81] Negative ion FAB MS gave a simple, abundant and informative fragmentation pattern for ziracin, one of the everninomycin class of oligosaccharide antibiotics.[82]

A method for determining the ee of organic amine salts, such as 1-naph-thylethylamine hydrochloride, by FAB MS relied on the differential complexation of the *R*- and *S*-enantiomers with a deuterium-labelled/unlabelled podand pair comprising the labelled LL-compound 'Gal2deg-d$_{24}$' (**30**) and its unlabelled DD-counterpart.[83] For investigating the site-specific glycosylation in glycoproteins, a strategy based on peptide mass fingerprinting using MALDI-TOF MS, following sequential digestion by a protease and a glycan-specific endoglycosidase, has been developed.[84]

2.4 Other Spectroscopic Methods. – A review (62 refs.) on the elucidation of structures of both natural and unnatural oligomers and polymers, and in particular *N*-acetylneuraminic acid-containing oligo- and poly-saccharides focussed on the use of circular dichroism.[85] Improvements in the calculational version of the MOLROT model for describing the optical rotation of simple saccharides have been illustrated by the excellent agreement between observed and calculated values. For example, 309 and 307 deg cm^{-1} dmol^{-1}, respectively, were the figures for the optical rotation of methyl α-D-glucopyranoside.[86] The influence of alkali chlorides and related chlorides (LiCl, NaCl, KCl, CsCl, NH$_4$Cl and Me$_4$NCl) on the rate of the benzoic acid-catalysed mutarotation of *N*-(*p*-chlorophenyl)-β-D-glucopyranosylamine in methanol has been investigated.[87]

The formation of cyclic esters of boronic acid derivative **31** with monosaccharide diols in aqueous solution causes a large visible colour change from purple to red suitable for use in a diagnostic test paper for glucose in body fluids.[88] The binding events of (salen)-Co(II) complexes with monosaccharides can be monitored by CD-spectral changes (see *Tetrahedron*, 1999, **55**, 9455), and the prochiral (salen)-Co(II) complex **32**, bearing two boronic acid groups, has now been proposed for use in sugar 'chirality' sensing at visible wavelengths.[89] The use of sugar borate complexes for UV-detection in HPLC is referred to in Part 3.1.2 below.

An ESR-spectrometric study showed that the anomeric radical formed on treatment of acetobromomaltose with Bu$_3$SnH at ambient temperature undergoes (2→1)-acetyl migration at elevated temperatures, giving rise to a second radical species **33**.[90]

3 Other Analytical Methods

3.1 X-Ray Crystallography. – No listing of published crystallographic analyses has been prepared this year. Readers are referred to electronic searching methods.

The crystal structures of four mannose-derivatives which might exist as either open-chain Schiffs bases or as glycosylamines showed that the hydroxylamine derivative is the oxime **34**, whereas the semicarbazide, aniline- and *p*-chloroaniline-derivatives are the β-glycosylamines **35**.[91]

X-Ray diffraction has been used to determine the molecular and crystal structure of 1-*O*-(α-D-galactopyranosyl)-*myo*-inositol dihydrate,[92] and to con-

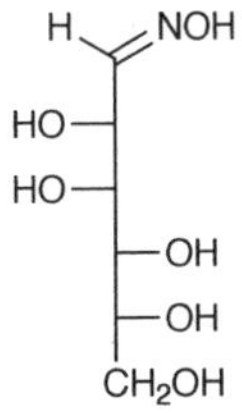

34

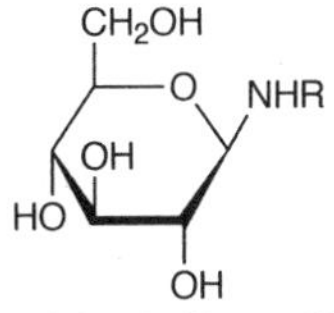

35 R = NHCONH₂, Ph or *p*Cl-C₆H₄

firm that 1-*O*-α-D-glucopyranosyl-D-mannitol crystallizes from ethanol as a 2:1 substrate solvent adduct, the structure being held together by a complex system of hydrogen bonds.[76] X-Ray analysis, in addition to DSC and thermogravimetric measurements, has been employed in a study of the thermotropic properties of crystalline long-chain-alkyl α-D-glucopyranosides and their hydrates.[93]

3.2 Physical Measurements. – The use of aryl boronic acid derivatives as sugar sensors is referred to in Part 2.4 above.

Partial solubility parameters for mannitol, sucrose and lactose in a wide variety of solvents have been extracted from solubility measurements by use of a modified, extended Hansen method. The solubilities were determined by use of evaporative light scattering detection HPLC.[94] The viscosities of glucose, glucitol and maltitol have been measured over eleven orders of magnitude by using capillary tube-, falling spheres-, as well as penetro-viscometers.[95] The adiabatic compressibilities, apparent molar compressibilities and solvation numbers of fructose and maltose have been calculated from measurements of their ultrasonic velocities in water and in 0.5 M NH₄Cl solution at various temperatures.[96] In addition to X-ray analysis, DSC and thermogravimetric measurements have been used in a study of the thermotropic properties of crystalline long-chain-alkyl α-D-glucopyranosides and their hydrates.[93]

The gelation abilities of four configurational diol sugar isomers, namely methyl 4,6-*O*-benzylidene-α-D-gluco-, allo- altro- and ido-pyranoside, in several organic solvents have been tested in an attempt to find rules for predicting the gelation abilities of hydrogen bond-based gelators.[97]

4   Separatory Methods

A review article on dehydroascorbic acid (72 refs.) covered, in addition to its chemistry, many methods for its separation and analysis.[98]

4.1   Chromatography. – Reviews of chromatographic methods (HPLC, GLC, capillary electrophoresis) for the quantitative determination of aminoglycoside antibiotics[99] and of sugars, amino acids and carboxylic acids in foods (175 refs.)[100] have been published.

4.1.1   Gas–Liquid Chromatography. Selectively substituted γ-cyclodextrins, *e.g.*

the octakis-(2-O-methyl-6-O-Tbdms)-derivative, have been introduced as stationary phases in capillary GC.[101]

GLC determination of reducing end sugar residues in oligo- and poly-saccharides following reduction, hydrolysis and acetylation has been recommended as an alternative to colorimetric and ^{1}H NMR methods.[102] The iridoid glycosides catalpol and aucubin have been quantified by use of a fast micellar electrokinetic capillary chromatographic method.[103]

4.1.2 High-pressure Liquid Chromatography. A simple and highly sensitive ion-exchange HPLC method with UV-detection for measuring small quantities (<100 pmol) of monosaccharides (Ara, Xyl, Glc, Gal, Man, Fuc, Rha) from pectic polysaccharides has been reported. The sugars were converted to their borate complexes and visualized by addition of ethanolamine to the mobile phase which gave intense absorption at 310 nm.[104] Reverse-phase HPLC analysis of the enantiomers of the β-adrenergic blocker atendol has been achieved after coupling *via* the hydroxyl group of the chiral centre with tetra-O-acetyl-β-D-glucopyranosyl isothiocyanate.[105]

HPLC quantification of 2′-deoxyuridine in deproteinated human plasma has been used to follow the effect of administering drugs which inhibit thymidylate synthase.[106] HPLC has also been applied to the quantitative determination of ascorbic acid in food, using monosodium L-glutamate[107] or disodium guanosine-5′-monophosphate[108] in the mobile phase, of sugar alcohols (xylitol, D-glucitol, D-mannitol, maltitol) in confectioneries, after *p*-nitrobenzylation,[109] and of carbohydrates in drinks, using a modified amino column and evaporative light scattering detection.[110] An HPLC method for determining the partial solubility parameters for mannitol, sucrose and lactose in a variety of solvents is referred to above (Ref. 94).

4.1.3 Column Chromatography. Alditols and free sugars have been determined by high pH anion-exchange chromatography with pulsed amperometric detection.[111] Analysis of lectin-oligosaccharide interactions by frontal affinity chromatography has been improved by incorporating contemporary LC techniques.[112] Brain neuro-chemicals, including ascorbic acid, have been monitored continuously *in vivo* by microdialysis coupled with LC.[113] LC has also been used to determine ascorbic acid and dehydroascorbic acid in fruit and vegetables [114] and inositol in infant formula and clinical products.[115]

4.2 Electrophoresis. – A review on the use of chiral glycosidic surfactants for the separation of enantiomers by capillary electrophoresis has been published.[116]

A considerable increase in the reliability of identifying compounds by capillary electrophoresis has been achieved by a reduction in the standard deviation of migration times of various compounds, including tetramethylrhodamine-labelled oligosaccharides, from >3% to <0.04%.[117]

Free sugars have been determined without derivatization by high performance capillary electrophoresis with refractometric detection.[118] Complex carbohydrates, such as O-glycosylated amino acids and glycopeptides, in urine have been investigated by off-line capillary electrophoresis/electrospray ionization tandem mass spectroscopy.[119]

Capillary electrophoresis has been applied to the analysis of enzymically glucosylated flavonoids[120] and of monosaccharides from glycosaminoglycans,[121] using UV detection and indirect laser-induced fluorescence detection, respectively, and the separation of the cyanogenic glycosides amygdalin and prunasin from their isomers has been achieved by a micellar capillary electrophoresis method.[122]

Analysis of carbohydrates in delignified wood and pulp involved enzymic hydrolysis (cellulase, hemicellulase), derivatization of the free sugars with ethyl *p*-aminobenzoate and capillary zone electrophoresis. All reducing sugars, *i.e.* neutral monosaccharides as well as uronic acids, were quantified in a single analytical run, and 4-deoxy-β-L-*threo*-hex-4-enopyranosyluronic acid present in the pulps was also detected.[123]

References

1. A. Imberty and S. Perez, *Chem. Rev.*, 2000, **100**, 4567.
2. J.-H. Lii, B. Ma and N.L. Allinger, *J. Comput. Chem.*, 1999, **20**, 1593 (*Chem. Abstr.*, 2000, **132**, 78 751).
3. V.G.S. Box and T. Evans-Lora, *J. Mol. Struct.*, 2000, **516**, 203 (*Chem. Abstr.*, 2000, **132**, 237 258).
4. M.K. Dowd, A.D. French and P.J. Reilly, *J. Carbohydr. Chem.*, 2000, **19**, 1091.
5. F. Cloran, Y. Zhu, J. Osborn, I. Carmichael and A.S. Seranni, *J. Am. Chem. Soc.*, 2000, **122**, 6435.
6. J. Czernek, J. Lang and V. Sklenar, *J. Phys. Chem. A*, 2000, **104**, 2788 (*Chem. Abstr.*, 2000, **132**, 322 061).
7. B.B. Lohray, V. Bhushan, G. Prasuna, Y. Jayamma, M.A. Raheem, P. Papireddy, B. Umadevi, M. Premkumar, N.S. Lakshmi and K. Narayanareddy, *Indian J. Chem.*, 1999, **38B**, 1311 (*Chem. Abstr.*, 2000, **132**, 308 572).
8. K. Zhang, A. Franz, G.C. Hill, C.M. McCallum and M.J. Minch, *Magn. Res. Chem.*, 1999, **37**, 788.
9. S. Immel and K. Khanbabaee, *Tetrahedron: Asymm.*, 2000, **11**, 2495.
10. F.A. Momany and J.L. Willett, *Carbohydr. Res.*, 2000, **326**, 194.
11. F.A. Momany and J.L. Willett, *Carbohydr. Res.*, 2000, **326**, 210.
12. A. Vishnyakov, G. Widmalm and A. Laaksonen, *Angew. Chem., Int. Ed. Engl.*, 2000, **39**, 140.
13. E. Mikros, G. Labrinidis and S. Pérez, *J. Carbohydr. Chem.*, 2000, **19**, 1319.
14. J.L. Asensio, F.J. Cañada, X. Cheng, N. Khan, D.R. Mootoo and J. Jiménez-Barbero, *Chem. Eur. J.*, 2000, **6**, 1035.
15. A. Poveda, J.L. Asensio, T. Polat, H. Bazin, R.J. Linhardt and J. Jiménez-Barbero, *Eur. J. Org. Chem.*, 2000, 1805.
16. M. Martin-Lomas, P.M. Nieto, N. Khiar, S. García, M. Flores-Mosquera, E. Poirot, J. Angulo and J.L. Muñoz, *Tetrahedron: Asymm.*, 2000, **11**, 37.
17. A. Mukhopadhyay, E. Kolehmainen and C.P. Rao, *Carbohydr. Res.*, 2000, **328**, 103.
18. J.O. Duus, C.H. Godfredsen and K. Bock, *Chem. Rev.*, 2000, **100**, 4589.
19. A.S. Serianni, *Bioorg. Chem.: Carbohydr.*, 1999, **244–312**, 575 (*Chem. Abstr.*, 2000, **132**, 293 930).

20. E.F. Hounsell, *Nucl. Magn. Reson.*, 2000, **29**, 406 (*Chem. Abstr.*, 2000, **133**, 208 040).
21. G.-J. Boons and A.V. Demchenko, *Chem. Rev.*, 2000, **100**, 4539.
22. M.D. Díaz and S. Berger, *Carbohydr. Res.*, 2000, **329**, 1.
23. L. Pouysegu, P. Norbert, D. Deffieux, B. De Jeso, J.-C. Lartigue, M. Petraud and M. Ratier, *Spectrochim. Acta, Part A*, 1999, **55A**, 2383 (*Chem. Abstr.*, 2000, **132**, 3504).
24. K.E. Kövér, K. Fehér, L. Szilágyi, A. Borbás, P. Herczegh and A. Lipták, *Tetrahedron Lett.*, 2000, **41**, 393.
25. G.J. Harris, N. Patel, B.J. Rawlings and T.J. Norwood, *J. Magn. Res.*, 1999, **140**, 504 (*Chem. Abstr.*, 2000, **132**, 12 469).
26. L. Herfurth, T. Weimar and T. Peters, *Angew. Chem., Int. Ed. Engl.*, 2000, **39**, 2097.
27. L. Kaustov, S. Kababya,. S. Du, T. Baasov, S. Gropper, Y. Shoham and A. Schmidt, *J. Am. Chem. Soc.*, 2000, **122**, 2649.
28. B. Bernet and A. Vasella, *Helv. Chim. Acta*, 2000, **83**, 995.
29. B. Bernet and A. Vasella, *Helv. Chim. Acta*, 2000, **83**, 2055.
30. L. Kindahl, C. Sandström, T. Norberg and L. Kenne, *J. Carbohydr. Chem.*, 2000, **19**, 1291.
31. S. Bekiroglu, C. Sandström, T. Norberg and L. Kenne, *Carbohydr. Res.*, 2000, **328**, 409.
32. M. Plass, M. Weychert, I. Wawer, B. Piekarska-Bartoszewicz and A. Temeriusz, *J. Carbohydr. Chem.*, 2000, **19**, 1059.
33. S. Freimund and A. Huwig, *Carbohydr. Lett.*, 2000, **3**, 419 (*Chem. Abstr.*, 2000, **133**, 120 535).
34. R. Lysek, P. Krajewski, Z. Urbanczyk-Lipkowska, B. Furman, Z. Kaluza, L. Kozerski and M. Chmielewski, *J. Chem. Soc., Perkin Trans. 2*, 2000, 61.
35. M.P. Koroteev, S.B. Khrebtova, E.S. Perfil'Eva, A.S. Shashkov, A.V. Ignatenko, K.A. Lyssenko and E.E. Nifantiev, *Carbohydr. Lett.*, 2000, **3**, 437 (*Chem. Abstr.*, 2000, **133**, 135 492).
36. M. Crnugelj, D. Dukhan, J.-L. Barascut, J.-L. Imbach and J. Plavec, *J. Chem. Soc., Perkin Trans. 2*, 2000, 255.
37. M.P. Groziak and R. Lin, *Tetrahedron*, 2000, **56**, 9885.
38. T. Maltseva, E. Usova, S. Eriksson, J. Milecki, A. Foldesi and J. Chattopadhayaya, *J. Chem. Soc., Perkin Trans. 2*, 2000, 2199.
39. G.R. Painter, C.W. Andrews and P.A. Furman, *Nucleosides, Nucleotides, Nucleic Acids*, 2000, **19**, 13.
40. S. Ravindranathan, X. Feng, T. Karlsson, G. Widmalm and M.H. Levitt, *J. Am. Chem. Soc.*, 2000, **122**, 1102.
41. D. Kuang, O.J. Obaje, A. Kassim, G.C.L. Ee and H. Suhaimi, *J. Am. Oil Chem. Soc.*, 2000, **77**, 43 (*Chem. Abstr.*, 2000, **132**, 265 359).
42. N. Khiar, *Tetrahedron Lett.*, 2000, **41**, 9059.
43. F.L. Tobiason, D.D. Swank, G. Vergoten and P. Legrand, *J. Carbohydr. Chem.*, 2000, **19**, 959.
44. B. Coxon, *Carbohydr. Res.*, 2000, **329**, 131.
45. A. Trynda, J. Madaj, A. Konitz and A. Wisniewski, *Carbohydr. Res.*, 2000, **329**, 249.
46. X.-C. Li, H.N. El Sohly, C.D. Hufford and A. Clark, *Magn. Res. Chem.*, 1999, **37**, 856.
47. W.F. Reynolds, R.G. Enriquez, G.A. Magos and D. Gnecco, *Magn. Res. Chem.*, 2000, **38**, 366.
48. D. Kitamoto, S. Ghosh, G. Ourisson and Y. Nakatani, *Chem. Commun.*, 2000, 861.
49. C. Jones, J.O. Previato and L. Mendonca-Previato, *Carbohydr. Res.*, 2000, **328**, 321.

50. W. Hellebrandt, T. Haselhorst, T. Köhli, E. Bäuml and T. Peters, *J. Carbohydr. Chem.*, 2000, **19**, 769.

51. A.D. French, A.-M. Kelterer, C.J. Cramer, G.P. Johnson and M.K. Dowd, *Carbohydr. Res.*, 2000, **326**, 305.

52. C.A. Stortz and A.S. Cerezo, *J. Carbohydr. Chem.*, 2000, **19**, 1115.

53. E. Montero, A García-Herrera, J.L. Asensio, K. Hirai, S. Ogawa, F. Santoyo-Gonzáles, F.J. Cañada and J. Jiménez-Barbero, *Eur. J. Org. Chem.*, 2000, 1945.

54. Y. Terada, R. Misoi, N. Watanabe, M. Hornberger and W. Kreis, *Chem. Pharm. Bull.*, 2000, **48**, 349.

55. C. Monteiro, S. Neyret, J. Leforestier and C. Hervé du Penhoat, *Carbohydr. Res.*, 2000, **329**, 141.

56. K.B. Kim, E.C. Behrman, C.E. Cottrell and E.J. Behrman, *J. Chem. Soc., Perkin Trans. 2*, 2000, 677.

57. T. Nakamura, W.A. Bubb, T. Saito, I. Arai and T. Urashima, *Carbohydr. Res.*, 2000, **329**, 471.

58. H. Kono, Y. Numata, N. Nagai, T. Erata and M. Takai, *J. Polym. Sci., Part A: Polym. Chem.*, 1999, **37**, 4100 (*Chem. Abstr.*, 2000, **132**, 35 969).

59. T. Yui, K. Goto, Y. Kawano and K. Ogawa, *Biosci. Biotechnol. Biochem.*, 2000, **64**, 52.

60. T. Weimar, B.O. Petersen, B. Svensson and B.M. Pinto, *Carbohydr. Res.*, 2000, **326**, 50.

61. D. Bailey, D.V. Renouf, D.G. Large, C.D. Warren and E.F. Hounsell, *Carbohydr. Res.*, 2000, **324**, 242.

62. J.L.J. Blanco, J.J.M. Van Rooijen, P.J.A. Erbel, B.R. Leeflang, J.P. Kamerling and J.F.G. Vliegenthart, *J. Biomol. NMR*, 2000, **16**, 59 (*Chem. Abstr.*, 2000, **132**, 35 969).

63. B. Coxon, N. Sari, G. Batta and V. Poszgay, *Carbohydr. Res.*, 2000, **324**, 53.

64. P. Brocca, A. Bernardi, L. Raimondi and S. Sonnino, *Glycoconjugate J.*, 2000, **17**, 283.

65. M. de Kort, A.D. Regenbogen, H.S. Overkleeft, R.A.J. Challiss, I. Iwata, S. Miyamoto, G.A. van der Marel and J.H. van Boom, *Tetrahedron*, 2000, **56**, 5915.

66. M. Michalik, M. Hein and M. Frank, *Carbohydr. Res.*, 2000, **327**, 185.

67. M. Albert, W. Repetschnigg, J. Ortner, J. Gomes, B.J. Paul, C. Illaszewicz, H. Weber, W. Steiner and K. Dax, *Carbohydr. Res.*, 2000, **327**, 395.

68. B. Stevensson, C. Hoeoeg, K. Ulfstedt-Jaekel, Z. Huang, G. Widmalm and A. Maliniak, *J. Phys. Chem. B*, 2000, **104**, 6065 (*Chem. Abstr.*, 2000, **133**, 177 372).

69. D. Deepak, A. Sethi, S.S. Srivastav, A. Khare and S. Srivastav, *Trends Carbohydr. Chem.*, 1999, **4**, 25 (*Chem. Abstr.*, 2000, **132**, 35 965).

70. K. Benner and P. Klüfers, *Carbohydr. Res.*, 2000, **327**, 287.

71. I. Nakazawa, S. Suda, M. Masuda, M. Asai and T. Shimizu, *Chem. Commun.*, 2000, 881.

72. P. DeShong, E.D. Soli, G.A. Slough, D.R. Sidler, V. Elango, P.J. Rybcynski, L.J.S. Vosejpka, T.A. Lessen, T.X. Le, G.B. Andersen, W. Von Philipsborn, M. Vohler, D. Rentsch and O. Zerbe, *J. Organomet. Chem.*, 2000, **593-4**, 49 (*Chem. Abstr.*, 2000, **132**, 251 310).

73. M.L. Ramos, M.M. Caldeira and V.M.S. Gil, *Carbohydr. Res.*, 2000, **329**, 387.

74. W. Tian, L.-M. Yang, Y.-Z. Xu, S.-F. Weng and J.-G. Wu, *Carbohydr. Res.*, 2000, **324**, 45.

75. L. Yang, Z. Wang, Y. Zhao, W. Tian, Y. Xu, S. Weng and J. Wu, *Carbohydr. Res.*, 2000, **329**, 847.

76. P. Perkkalainen, I. Pitkanen and J. Huuskonen, *J. Mol. Struct.*, 1999, **510**, 179 (*Chem. Abstr.*, 2000, **132**, 108 185).

77. M. Rozenberg, A. Loewenschuss and Y. Marcus, *Carbohydr. Res.*, 2000, **328**, 307.

78. S. Soderholm, Y.H. Roos, N. Meinander and M. Hotokka, *J. Raman Spectrosc.*, 1999, **30**, 1009 (*Chem. Abstr.*, 2000, **132**, 122 806).

79. M. Stobiecki, *Phytochemistry*, 2000, **54**, 237.

80. Y.H. Ahn and J.S. Yoo, *Rapid Commun. Mass Spectrom.*, 1999, **13**, 1985 (*Chem. Abstr.*, 2000, **132**, 50 178).

81. V. Kovacik, V. Patoprsty, E. Lattova, L. Petrus, V. Ovcharenko and K. Pihlaja, *J. Mass Spectrom.*, 1999, **34**, 1322 (*Chem. Abstr.*, 2000, **132**, 152 058).

82. B.N. Pramanik, P.A. Shipkova, P.L. Bartner, L. Heimark, Y.-H. Liu, P.R. Das, A.K. Saksena, O.Z. Sarre, V.M. Girijavallabhan and A.K. Ganguly, *J. Antibiotics*, 2000, **53**, 640.

83. M. Shizuma, H. Imamura, Y. Takai, H. Yamada, T. Takeda, S. Takahashi and M. Sawada, *Chem. Lett.*, 2000, 1292.

84. K. Mells, A.W. Johnson, O. Dietrich, P.T. Clayton and B.G. Winchester, *Tetrahedron: Asymm.*, 2000, **11**, 75.

85. K.D. McReynolds and J. Gervay-Hague, *Tetrahedron: Asymm.*, 2000, **11**, 337.

86. E.P. Stroyan and E.S. Stevens, *Carbohydr. Res.*, 2000, **327**, 447.

87. K. Smiataczowa and K. Maj, *Pol. J. Chem.* 1999, **73**, 429 (*Chem. Abstr.*, 2000, **132**, 322 052).

88. C.J. Ward, P. Patel, P.R. Ashton and T.D. James, *Chem. Commun.*, 2000, 229.

89. T. Mizuno, M. Yamamoto, M. Takeuchi and S. Shinkai, *Tetrahedron*, 2000, **56**, 6193.

90. A. Alberti, S. Bertini, M. Comoli, M. Guerrini, A. Mele and E. Vismara, *Tetrahedron*, 2000, **56**, 6291.

91. W.H. Ojala, J.M. Ostman and C.R. Ojala, *Carbohydr. Res.*, 2000, **326**, 104.

92. K. Noguchi, K. Okuyama, S. Ohno, T. Hidano, N. Wakiuchi, T. Tarui, H. Tamaki, S. Kishihara and S. Fujii, *Carbohydr. Res.*, 2000, **328**, 241.

93. B. Hoffmann, W. Milius, G. Voss, M. Wunschel, S. van Smaalen, S. Diele and G. Platz, *Carbohydr. Res.*, 2000, **323**, 192.

94. M.A. Peña, Y. Daali, J. Barra and P. Bustamante, *Chem. Pharm. Bull.*, 2000, **48**, 179.

95. A. Nakheli, A. Eljiazouli, M. Elmorabit, E. Ballouki, J. Fornazero and J. Huck, *J. Phys.: Condens. Matter*, 1999, **11**, 7977 (*Chem. Abstr.*, 2000, **132**, 78 778).

96. P.S. Nikam, H.R. Ansari and M. Hasan, *J. Mol. Liq.*, 2000, **84**, 169 (*Chem. Abstr.*, 2000, **132**, 251 302).

97. R. Luboradzki, O. Gronwald, M. Ikeda, S. Shinkai and D.N. Reinhoudt, *Tetrahedron*, 2000, **56**, 9595.

98. J.C. Deutsch, *J. Chromatogr., A*, 2000, **881**, 299.

99. D.A. Stead, *J. Chromatogr., B*, 2000, **747**, 69.

100. I. Molnár-Perl, *J. Chromatogr., A*, 2000, **891**, 1.

101. G. Cravotto, G. Palmisano, L. Panza and S. Tagliapietro, *J. Carbohydr. Chem.*, 2000, **19**, 1235.

102. C.M. Courtin, H. van den Broeck and J.A. Delcour, *J. Chromatogr., A*, 2000, **866**, 97.

103. J. Suomi, H. Sirén, K. Hartonen and M.-L. Riekkola, *J. Chromatogr., A*, 2000, **868**, 73.

104. A. Nakamura, C. Hatanaka and Y. Nagamatsu, *Biosci. Biotechnol. Biochem.*, 2000, **64**, 178.

105. X. Li, T.-W. Yao and S. Zeng, *J. Chromatogr., B*, 2000, **742**, 433.

106. F. Mitchell, S. Lynn and A.L. Jackman, *J. Chromatogr., B*, 2000, **744**, 351.
107. H. Iwase, *J. Chromatogr., A*, 2000, **881**, 317.
108. H. Iwase, *J. Chromatogr., A*, 2000, **881**, 327.
109. S. Nojiri, N. Taguchi, M. Oishi and S. Sukuki, *J. Chromatogr., A*, 2000, **893**, 195.
110. Y. Wei and M.-Y. Ding, *J. Chromatogr., A*, 2000, **904**, 113.
111. R. Andersen and A. Sorensen, *J. Chromatogr., A*, 2000, **897**, 195.
112. J. Hirabayashi, Y. Arata and K.-i. Kasai, *J. Chromatogr., A*, 2000, **890**, 261.
113. F.-C. Cheng, L.-L. Yang, D.-Y. Yang, T.-H. Tsai, C.-W. Lee and S.-H. Chen, *J. Chromatogr., A*, 2000, **870**, 389.
114. V. Gökmen, N. Kahraman, N. Demir and J. Acar, *J. Chromatogr., A*, 2000, **881**, 309.
115. E.G. Tagliaferri, G. Bonetti and C.J. Blake, *J. Chromatogr., A*, 2000, **879**, 129.
116. Z. El Rassi, *J. Chromatogr., A*, 2000, **875**, 207.
117. X.-F. Li, H. Ren, X. Le, M. Qi, I.D. Ireland and N.J. Dovichi, *J. Chromatogr., A*, 2000, **869**, 375.
118. A.R. Ivanov, I.V. Nazimov, A.P. Lobazov and G.B. Popkovich, *J. Chromatogr., A*, 2000, **894**, 253.
119. A. Zamfir, S. König, J. Althoff and J. Peter-Katalinc, *J. Chromatogr., A*, 2000, **895**, 291.
120. T. Akiyama, T. Yamada and T. Maitani, *J. Chromatogr., A*, 2000, **895**, 279.
121. V. Ruiz-Calero, L. Puignou and M.T. Galceran, *J. Chromatogr., A*, 2000, **873**, 269.
122. C. Campa, P. Schmitt-Kopplin, T.R.I. Cataldi, S.A. Bufo, D. Freitag and A. Kettrup, *J. Chromatogr., B*, 2000, **739**, 95.
123. O. Dahlman, A. Jacobs, A. Liljenberg and A.I. Olsson, *J. Chromatogr., A*, 2000, **891**, 157.

22
Carbohydrates in Chiral Organic Synthesis

1   Carbocyclic Compounds

The iodides **2** (R = Bz, Tbdps, R′ = H) can be prepared from di-*O*-iso-propylidene-α-D-glucofuranose (sugar carbons numbered) by a sequence involving deoxygenation at C-3 and introduction of iodine with inversion of configuration. Reductive cyclization (Scheme 1) of the diols **2** (R′ = H) gave predominantly the lactones **1** (R = Bz, Tbdps), whereas similar reactions on the bis(triethylsilyl) ethers gave mostly products **3** of alternative stereochemistry.[1] The product **3** (R = Tbdps), which was produced with slightly better stereoselectivity (75% of the isomer indicated) than was the corresponding benzoate, was subsequently used to prepare 4-(*RS*)-F$_{4t}$-isoprostane methyl ester (**4**).[2] There has been a fuller account of an earlier report (Vol. 32, p. 353) on related cyclizations of ω-iodo-enoates such as **5**, derived from 2-deoxy-D-ribose; use of SmI$_2$ to induce reaction gave a high yield of the all-*cis-meso*-compound **6**, whereas reactions using Bu$_3$SnH were much less diastereoselective.[3]

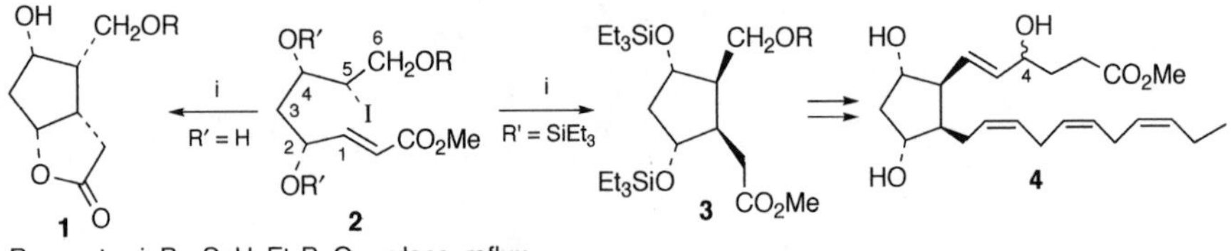

Reagents: i, Bu$_3$SnH, Et$_3$B, O$_2$, xylene, reflux

Scheme 1

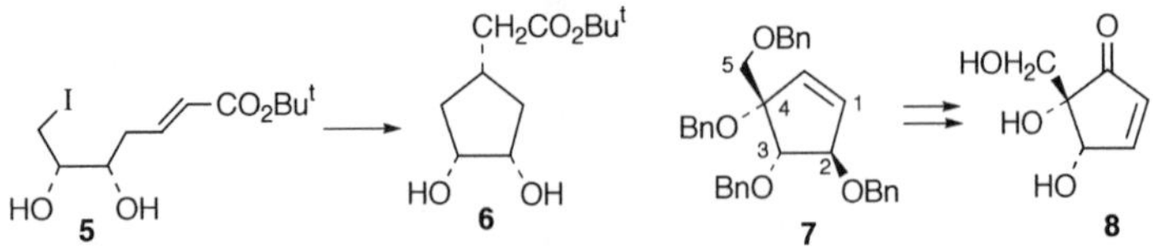

The cyclopentene **7** can be prepared from 2,3,5-tri-*O*-benzyl-D-arabinofuranose (sugar carbons numbered) in five steps and high overall yield (Vol. 33, p. 241; see also *Org. Lett.*, 1999, **1**, 1463). This intermediate has been used to prepare (−)-pentenomycin (**8**) in a further five steps, the first of which was a hydroboration–oxidation which occurred with 2:1 regioselectivity in the desired sense.[4]

A Claisen rearrangement was a key step in the preparation of **9** from di-*O*-

isopropylidene-D-glucose (Vol. 21, p. 142). This compound was used (Scheme 2) to prepare the methyl uronate 10, with introduction of the methoxy group at C-5 by displacement of a mesylate, and hence the cyclopentene 11. This was elaborated in further chemistry to 12, which represents the cyclopentene carboxylic acid present as a substructure of the macrolide-macrolactam viridenomycin.[5]

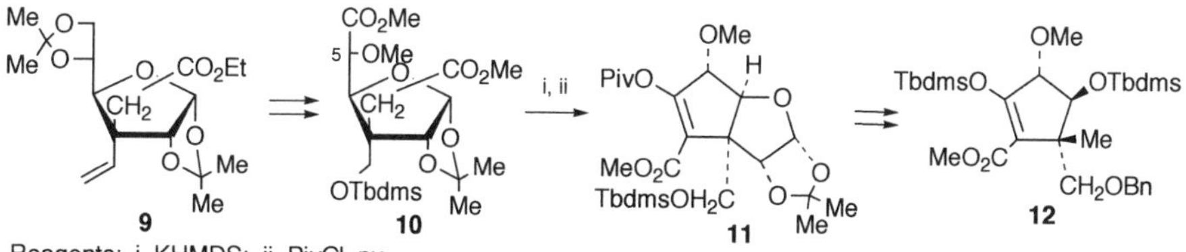

Reagents: i, KHMDS; ii, PivCl, py

Scheme 2

Some time ago, the conversion of 9 into the spirocyclic cyclopentenone 13 was reported (Vol. 22, p. 150). An investigation has now been made into the conjugate additions of organocopper reagents to 13, which give predominantly the epimers 14 (R = Me, *n*-Bu, *i*-Pr, vinyl), although in some cases this selectivity could be reversed under particular conditions. Compound 9 can also be converted into the allylic alcohol 15 (*J. Carbohydr. Chem.*, 1993, 12, 1187), and orthoester Claisen rearrangement of this gives the isomer 16 with high diastereoselectivity.[6]

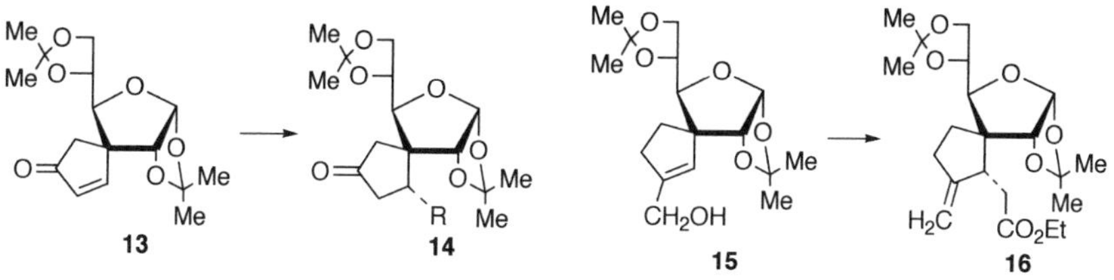

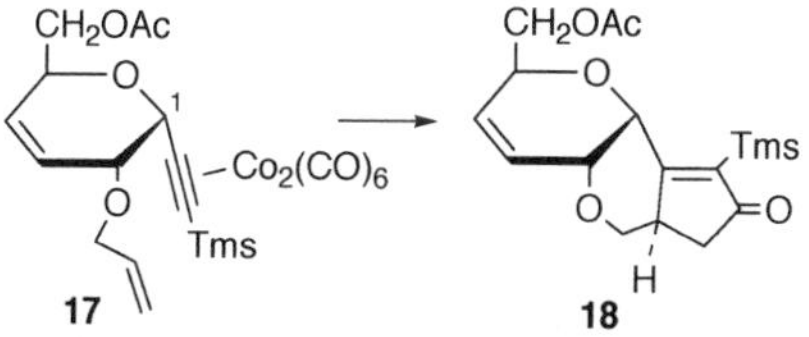

When the biscobalthexacarbonyl complex 17, prepared from 2-acetoxy-3,4,6-tri-*O*-acetyl-D-glucal, was treated with NMNO, a Pauson–Khand reaction occurred to give the fused cyclopentenone 18 in virtually quantitative yield. Treatment of 17 with triflic acid caused epimerization at C-1, and the stereoisomer also underwent Pauson–Khand cyclization efficiently.[7]

The D-*manno*-epoxide 19 has been converted (Scheme 3) into the enynes 20 (*n* = 0, 1, 2, sugar carbons numbered), and these were used to prepare the vitamin D$_3$ analogues 21, of which the two shortest-chain compounds had high affinity for the vitamin D receptor.[8]

In a synthesis of the CD ring unit of paclitaxel, the cyclohexenone 22 was prepared from triacetyl-D-glucal by a four-step sequence involving Ferrier carbocyclization (sugar carbons indicated), and was then converted as outlined in Scheme 3 into the CD unit 23.[9]

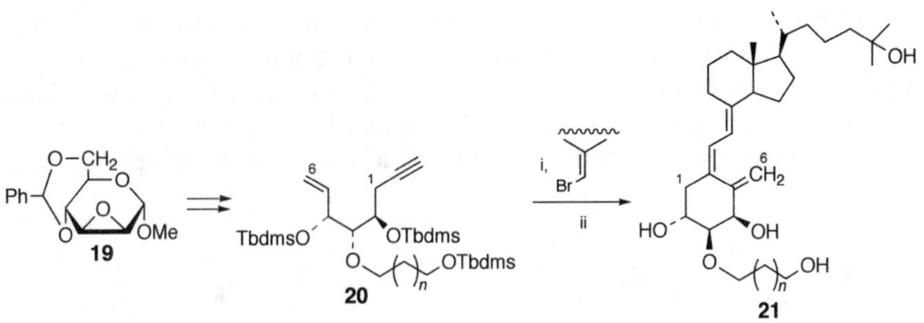

Reagents: i, (Ph₃P)₄Pd (cat.). Et₃N, toluene, reflux; ii, TBAF

Scheme 3

In an approach to polycyclic compounds by tin-mediated radical cyclizations, the dienyne **24** was treated (Scheme 5) with Bu₃SnH to give a mixture of the fused cyclohexene **25** and the cyclopentane **26**. A number of other similar sequences were investigated at different oxidation levels.[10]

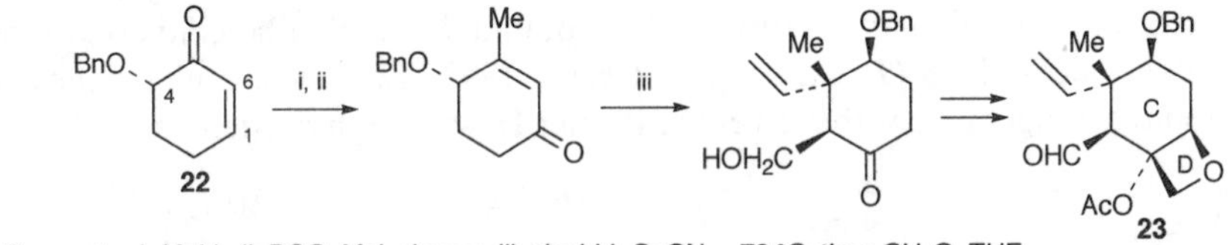

Reagents: i, MeLi; ii, PCC, Mol. sieves; iii, vinyl-Li, CuCN, −78 °C, then CH₂O, THF

Scheme 4

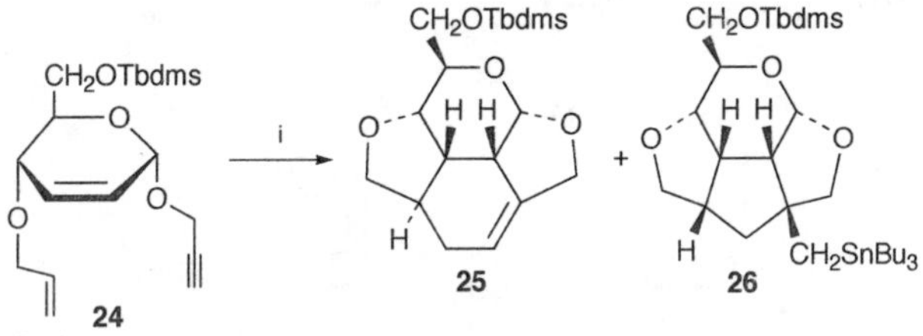

Reagents: i, Bu₃SnH

Scheme 5

In an interesting approach to bicyclo[4.3.0] systems, Wadsworth–Emmons reaction between the diene-aldehyde **27** (Vol. 28, p. 8; see also Section 6) and phosphonate **28** was followed by cycloaddition at room temperature to give **29** (Scheme 6) in high yield as a single stereoisomer. Similar products, with the same stereochemistry at the newly-created chiral centres, were obtained from diene-aldehydes of D-*lyxo*- and L-*arabino*- configurations, implying that stereocontrol was derived from the chirality of **28**. In support of this, a product **30** of opposite stereochemistry at the ring junction was obtained from **27** and a phosphonate, derived from isopropylidene-D-glyceraldehyde, of opposite chirality α-to the carbonyl group.[11] In an accompanying study, reactions of diene-phosphonates with sugar aldehydes were used to give *cis*-decalins, as in the example in Scheme 7, and again different stereochemistries in the diene component were investigated. Most of these reactions occurred with good stereoselectivity, which for diene-phosphonates of D-*lyxo*- and L-*arabino*-configuration was determined primarily by the configuration of the diene-phosphonate, whereas for the D-*xylo*-

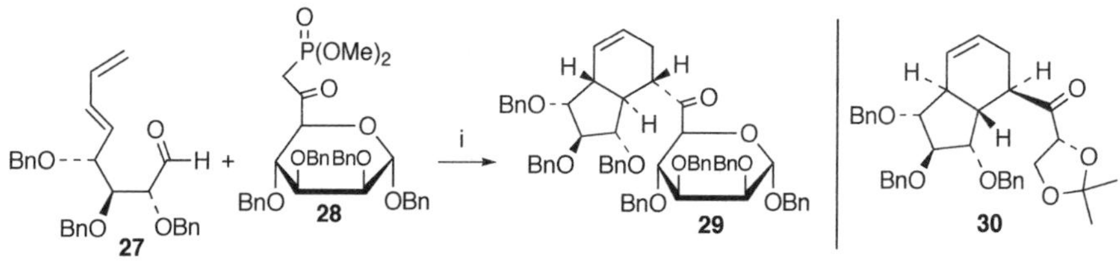

Reagents: i, K_2CO_3, 18-C-6, toluene, r.t.

Scheme 6

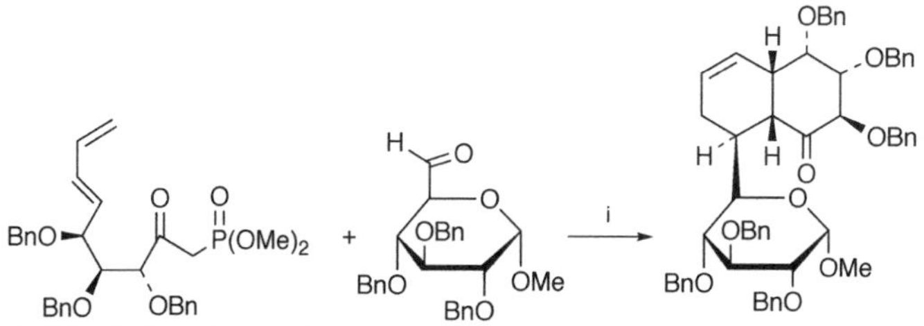

Reagents: i, K_2CO_3, 18-C-6, toluene, r.t.

Scheme 7

diene, the chirality of the aldehyde was dominant.[12] An alternative route to a *cis*-decalin from a sugar (Scheme 8) involves the synthesis of **31** by intramolecular Diels–Alder reaction, the required precursor being made from tri-*O*-acetyl-D-glucal and the diene-alcohol. Subsequent formation of a 5,6-ene **32** and Ferrier cyclization gave the *cis*-decalin **33**, an advanced intermediate for the synthesis of the hexahydronaphthalene portion of the mevinic acids.[13]

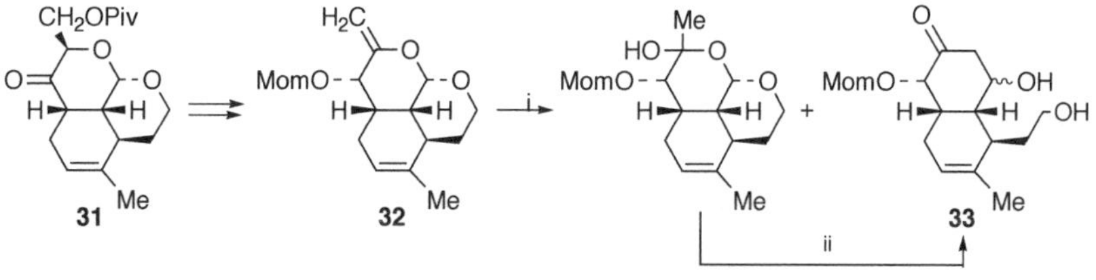

Reagents: i, HgSO$_4$, H$_2$SO$_4$, dioxane-H$_2$O, ii, Na$_2$CO$_3$, DBU, THF-MeOH

Scheme 8

The first enantioselective total synthesis of (–)-tetracycline (**41**) has been reported. The glucosamine derivative **34** was converted into the branched alkene **35** in a sequence in which the benzyloxymethyl unit was introduced using a Wittig-hydroboration sequence, and the 5-alkene by selenoxide elimination. Since the enone produced from **36** by elimination did not undergo the desired cycloaddition, an epimerization step was employed prior to elimination to give **37**. Diels–Alder reaction and subsequent oxidation then gave the decalin derivative **38**. Annelation to tetracycle **40** was followed by a multistep sequence to give (–)-tetracycline (**41**).[14]

There have been several reports of the use of ring-closing metathesis applied to sugar-derived compounds for the preparation of cycloheptene and cyclooctene derivatives. The mannose-derived dienes **42** (R = H, Ac) were converted in high yields to the carbocycles **43** using Grubbs' ruthenium-based catalyst, and the

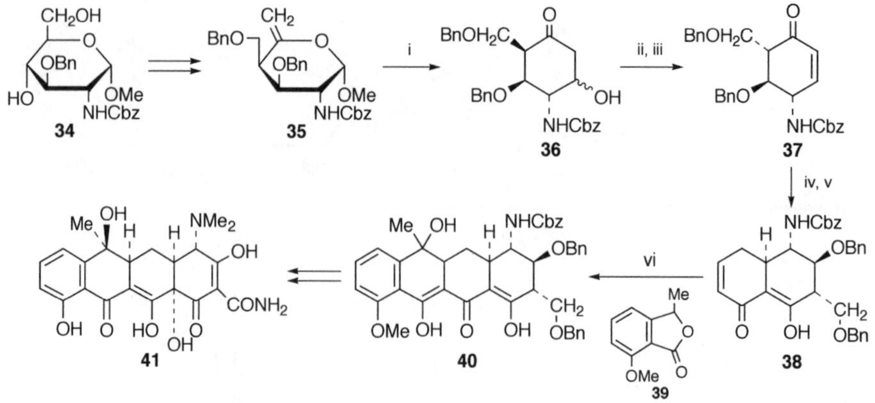

Reagents: i, HgCl₂; ii, DBU, toluene; iii, MsCl; iv, 1-TmsO-butadiene; v, Jones' reagent; vi, **39**, LDA
Scheme 9

C-1 epimers (sugar numbering, R = Ac, Bn) were similarly cyclized.[15,16] The presence of the isopropylidene ring is not essential, as evidenced, for example, by the conversion of **44** to **45** in 90% yield.[16] Others have reported similar results with dienes derived from D-galactose, leading to cycloheptenes and also to cyclooctenes, as in the case of the conversion of **46** (sugar carbons numbered) into **47** in 93% yield.[17] The fused cyclopentane **48** was one of a number of compounds made (with an alkene originally present between the two asterisked positions) by olefin metathesis carried out on sugars derivatives doubly-branched at C-2 and C-3 (see Chapter 14). On treatment with zinc **48** gave rise to the cyclohexanone **49** by fragmentation followed by a pinacol-type rearrangement.[18]

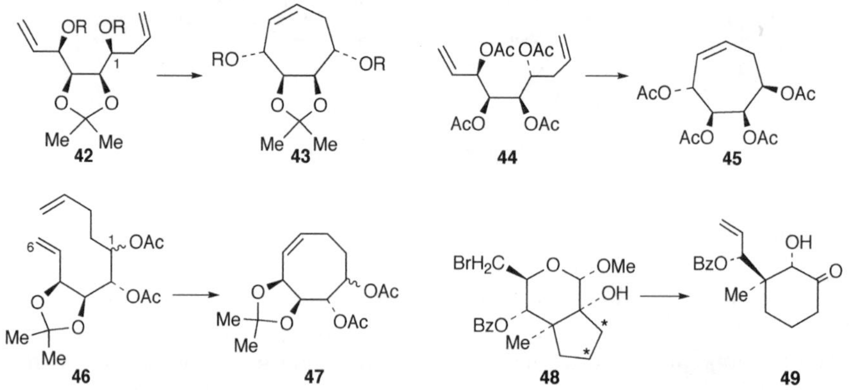

2   γ- and δ-Lactones

A review article has described routes to small, general-purpose chiral synthons from carbohydrates, including the preparation of D-erythronolactone and L-threonolactone from D-isoascorbic and L-ascorbic acids respectively, and the synthesis of lactone **50** and its enantiomer from disaccharides and L-arabinose respectively. The application of these units in synthesis was also discussed.[19] 3,5-Di-*O*-benzyl-D-xylose was used to make the lactone **51**, an immediate precursor of (+)-muricatacin.[20]

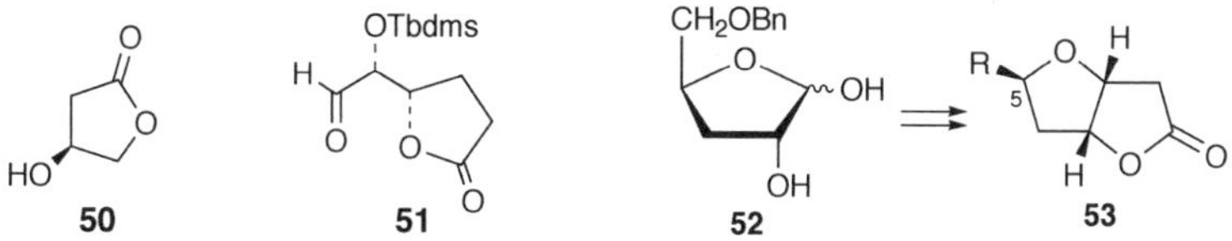

Diacetone-D-glucose can be converted by conventional manipulations into **52**, and this was used to synthesize the (5*R*)-Hagen's gland lactones **53** (R = *n*-C₄H₉, *n*-C₆H₁₃), the bicyclic system being established in one step by reaction of **52** with ethoxycarbonylmethylene triphenylphosphorane.[21] The same group has also described an alternative route (Scheme 10), in which the alkenyl diol **54**, prepared from D-mannose (sugar carbons numbered), was converted by oxidative cyclization to the bicyclic hemiacetal **55**, which could be further manipulated to give the lactones **53** and the (5*S*)-epimers **56** (R = *n*-C₄H₉, *n*-C₆H₁₃).[22]

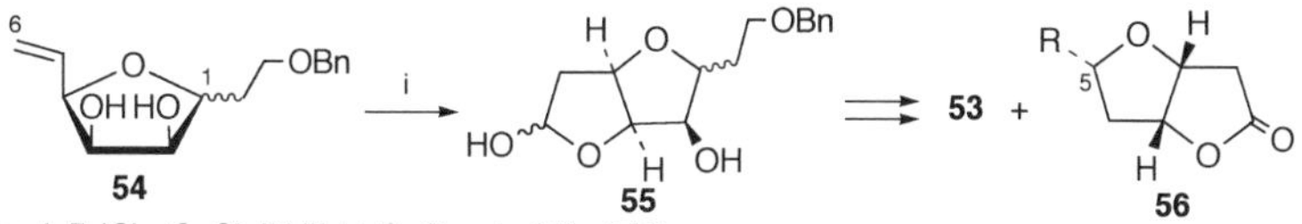

Reagents: i, PdCl₂, CuCl, DMF-H₂O, O₂, r.t., 24h, 84%

Scheme 10

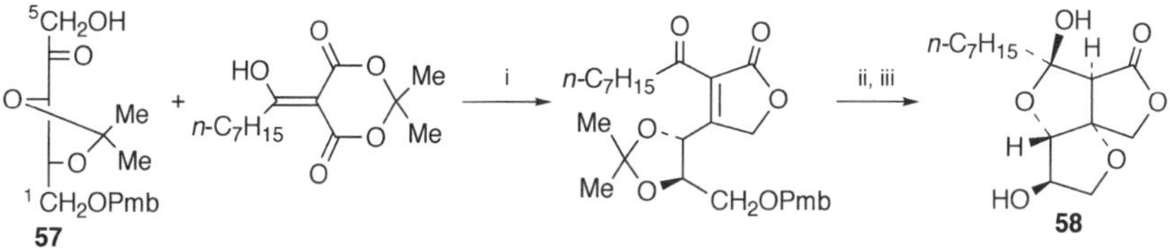

Reagents: i, THF, reflux, then SiO₂; ii, DDQ; iii, TsOH, H₂O-acetone

Scheme 11

(−)-Syringolide 2 (**58**), a bacterial signal molecule that elicits active defence in resistant plants, has been made (Scheme 11) from the protected ketose **57**, itself prepared from D-arabinose (sugar carbons numbered) in nine steps.[23] This sequence can be considerably shortened by use of a chemo-enzymatic approach, in which **57** was made in three steps by condensation of dihydroxyacetone phosphate with *O*-Pmb-glycolaldehyde, catalysed by fructose 1,6-diphosphate aldolase, followed by action of phosphatase and formation of the isopropylidene derivative.[24]

New formal enantioselective syntheses of (+)-avenaciolide and (+)-isoavenaciolide (the enantiomers of the natural fungal metabolites) have been reported. Ring contraction of the triflate **59** (see Vol. 29, p. 199) gave the hemiacetal **60** (Scheme 12). Reaction of this with *n*-octylmagnesium bromide gave a mixture of two separable diols in 3:1 ratio, and oxidation of the major one under Jefford's conditions gave the lactone **61** in good yield. Further oxidation using Grieco's method then gave the bis-lactone **62**, which is convertible by established methods to (+)-avenaciolide (the α-methylene-lactone). The minor diol could be converted similarly to (+)-isoavenaciolide, although in this case Grieco oxidation of the C-6 epimer of **61** gave a different product, explained on conformational grounds, and a two-step hydrolysis-PCC sequence was therefore used.[25]

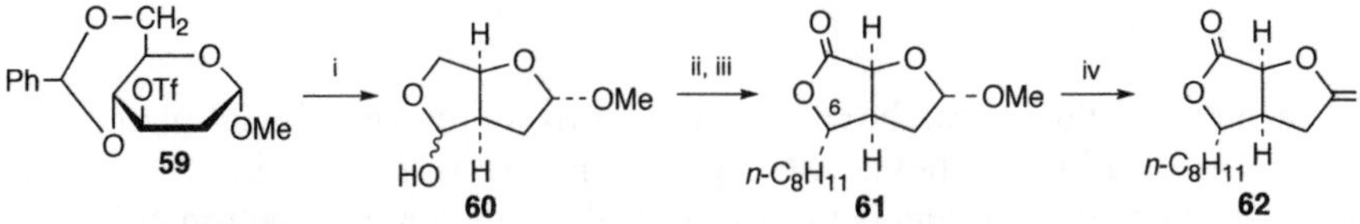

Reagents: i, (CF₃)₂CHOH, py, H₂O, reflux, 92%; ii, *n*-C₈H₁₇MgBr; iii, KMnO₄, CuSO₄.5H₂O, CH₂Cl₂; iv, *m*CPBA, BF₃.Et₂O, CH₂Cl₂

Scheme 12

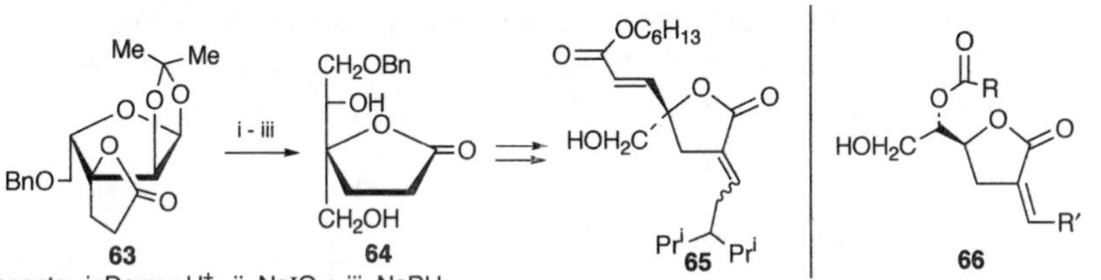

Reagents: i, Dowex H⁺; ii, NaIO₄; iii, NaBH₄

Scheme 13

There have been further reports from Marquez' laboratory on the synthesis of conformationally-restricted analogues of diacylglycerol. The spirolactone **63** (Scheme 13) was made by addition of an organometallic to a 3-ketone derived from L-xylose, and was converted *via* the lactone **64** into analogues such as **65**, some of these protein kinase C ligands having potent antitumour activity.[26] A previously-reported intermediate (Vol. 30, p. 358–9) has been converted into analogues of type **66**, in which the lactone carbonyl corresponds to the *sn*-1 carbonyl of diacylglycerol, in contrast to earlier analogues in which the lactone mimicked the *sn*-2 position. However, these new lactones proved less effective than the earlier analogues as protein kinase C ligands.[27]

Treatment of the 1,6-anhydromannose derivative **67** as indicated in Scheme 14 gave the product **68** of radical addition to the oxime ether, and **68** was then convertible to the α-methylene-γ-lactone **69**, which can be considered as a potential intermediate in a synthesis of tetrodotoxin. Interestingly, the *O*-silylated form of **67** did not undergo cyclization satisfactorily. A number of related radical annulations were also reported (see Section 4 for one case).[28]

The epoxide **70**, derived from D-mannitol, has been converted into the δ-lactone **71** *via* the key intermediates shown in Scheme 15. This compound is the dihydro-derivative of (+)-boranolide, the corresponding α,β-unsaturated lactone, and this work constitutes a formal synthesis of the natural product.[29] The enal **72**, derived from tri-*O*-acetyl-D-glucal, has been converted in a multistep process into the unsaturated lactone **73**, an analogue of the α,β-unsaturated lactone portion of a number of natural products.[30]

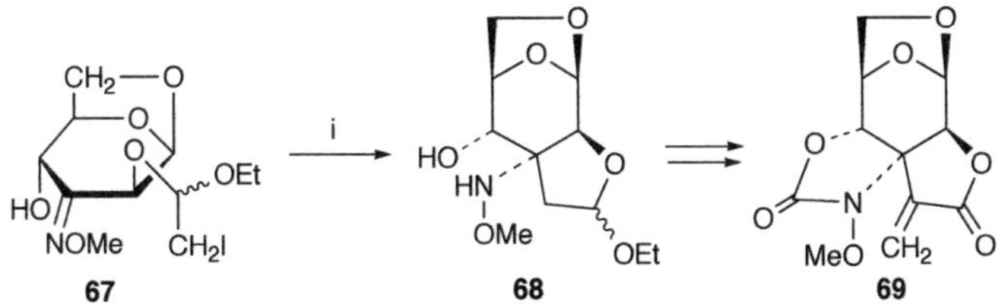

67 → **68** ⇒ **69**

Reagents: i, Ph₃SnH, AIBN, slow addition, toluene, 100 °C, 58%

Scheme 14

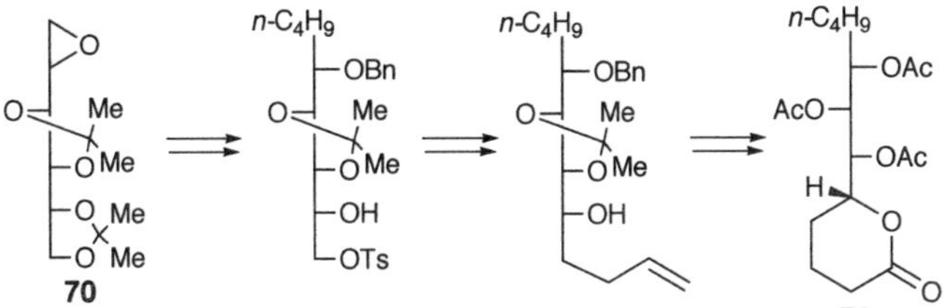

Scheme 15

70 ⇒ **71**

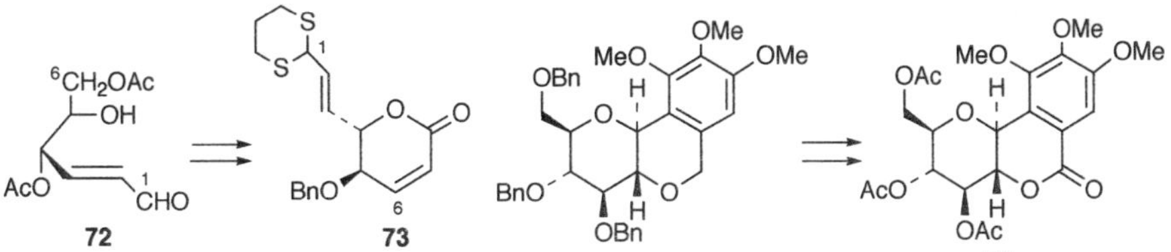

72 ⇒ **73** | **74** ⇒ **75**

In a route to bergenin-related natural products, the *trans*-fused tricyclic structure **74** was prepared by intramolecular *C*-glycosylation (Chapter 3), followed by isomerization of the initially-formed *cis*-fused system by BF₃.Et₂O. Exchange of the benzyl ethers for *O*-acetyl groups and oxidation with catalytic RuO₄ then gave the triacetate **75** of tri-*O*-methylbergenin.[31]

3   Macrolides and Their Constituent Segments

The stereostructure of mueggelone, an inhibitor of fish development, has been determined by synthesis. The building block **76** (Scheme 16), made from D-arabinose by known methods, was converted into the allylic alcohol **77** (sugar carbons numbered) and its epimer at the secondary alcohol centre, the appropriate chirality at that position being introduced by CBS reduction. Since it was known that mueggelone was a *trans*-epoxide, **77** and its epimer were converted into all four possible diastereomers by appropriate deprotection and epoxide formations **78**. It was shown by NMR and optical rotation measurements that mueggelone was the isomer of stereochemistry **78**.[32]

The diazoketone **79** was made from the furanuronic acid, and on Wolff rearrangement in the presence of methanol it gave the ester **80**, which represents the C-1 to C-6 fragment of carbonolide B.[33]

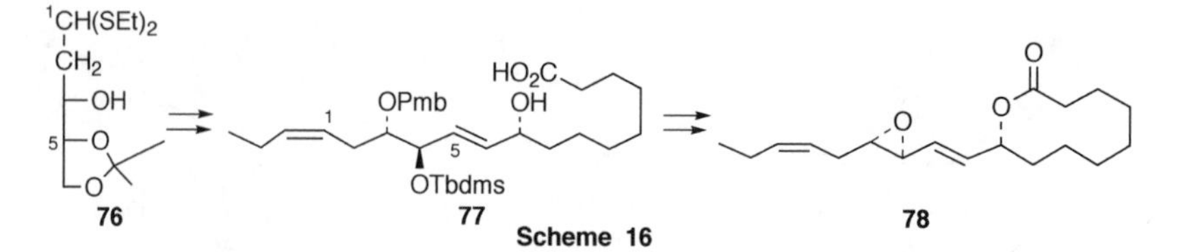

Scheme 16

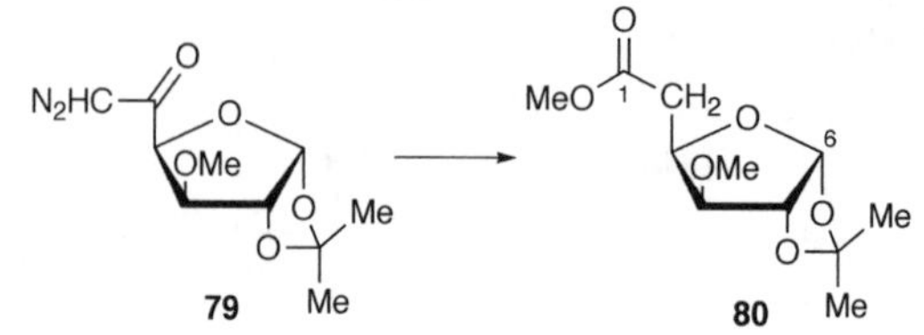

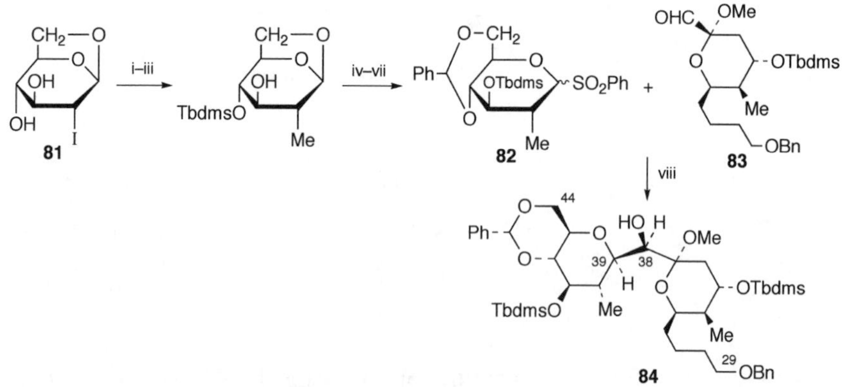

Reagents: i, TbdmsCl; ii, NaH, DMF; iii, Me₂CuLi; iv, PhSH, TsOH; v, PhCH(OMe)₂, mol. sieves; vi, TbdmsOTf, 2,6-lutidine; vii, *m*CPBA, CH₂Cl₂; viii, **82**, LDA, then **83**, then Li naphthalenide, then MeOH

Scheme 17

A sugar-based synthesis of the C29-C44 fragment of the spongipyran macrolides has been reported. The iodide **81**, derived from D-glucal (see Vol. 26, p. 127 and 150), was converted into the sulfone **82** as outlined in Scheme 17. This was then coupled *via* its anion to aldehyde **83**, prepared from simple aliphatic precursors by asymmetric aldol condensations, followed by reductive desulfonylation. The major product **84** had the indicated stereochemistry at C-38 and C-39 (macrolide numbering), and a minor product (4:1 ratio) was the epimer at C-38. The natural products have the configuration of this minor product, but the stereochemistry at C-38 of **84** could be inverted cleanly by an oxidation-reduction sequence.[34]

4   Other Oxygen Heterocycles

Di-*O*-isopropylidene-D-glucose can be converted into **85** by known procedures, and this has been used to prepare the lactone **86**, thus providing a new formal synthesis of (−)-*trans*-kumausyne (**87**).[21] In a synthetic approach to the laurenynes, secondary metabolites of red algae, **88**, derived from α-D-glucono-6,3-lactone, was converted by [2,3]-Wittig rearrangement into the alkenol **89**, produced exclusively as the Z-isomer. Further elaboration led to the synthesis of dienyne **90**.[35] The 5-lipoxygenase inhibitor CMI-977 (**91**) has been prepared from D-mannitol in 13 steps.[36]

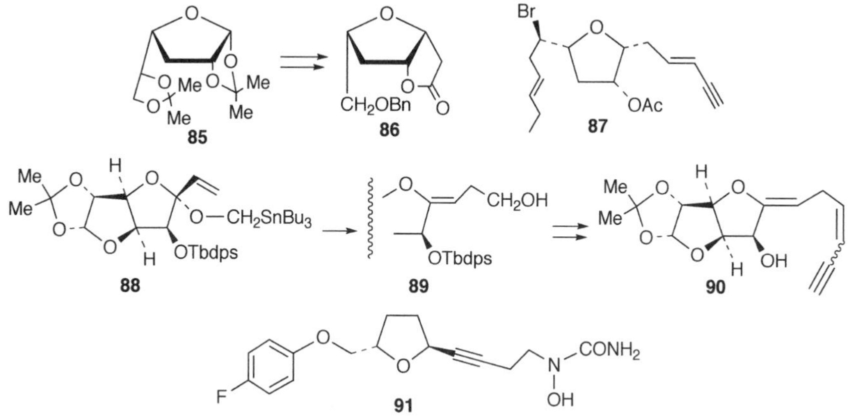

In a route to partially hydroxylated 2,5-disubstituted bis-tetrahydrofurans, related to annonaceous acetogenins, the glucose-derived lactone **92** (Scheme 18) was converted by routine operations into aldehyde **93**. On reaction with 3-butenylmagnesium bromide the major product formed (3:1 ratio) was the epimer **94**, which was cyclised to give **95** as a mixture of epimers at the new centre of chirality. Similar chemistry was also carried out on the epimer of **94**.[37]

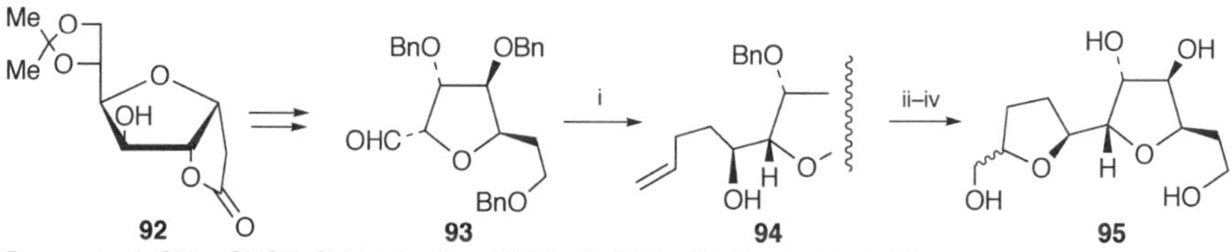

Reagents: i, CH$_2$=CHCH$_2$CH$_2$MgBr; ii, mCPBA; iii, CSA, CH$_2$Cl$_2$; iv, H$_2$, Pd/C

Scheme 18

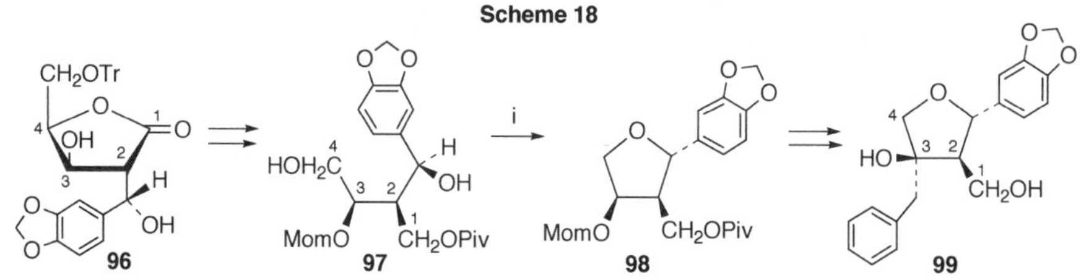

Reagent: i, TsOH

Scheme 19

In an approach to the olivil type of lignan (Scheme 19), the *erythro*-aldol product **96** was produced as a 1:1 mixture with the *threo*-isomer by reaction between piperonal and a lactone derived from D-xylose. Manipulation of **96** gave the diol **97** (sugar carbons numbered), which underwent acid-catalysed cyclization to give the tetrahydrofuran **98**. Interestingly, similar processing of the *threo*- aldol adduct also gave the same product **98**, which was converted to the lignan **99**.[38] The same workers have also reported the stereoselective formation of the *threo*-product **100** from piperonal and a lactone made from L-arabinose, and this was converted to the lignan isomer **101**, where again the sugar carbons are indicated.[39]

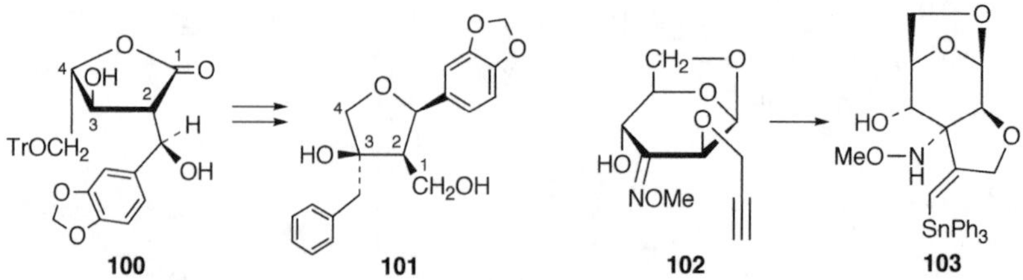

100   **101**   **102**   **103**

A tetrahydrofuran ring has been annelated onto a sugar by means of the radical cyclization of **102** which gave **103** on treatment with triphenylstannane and triethylborane.[28]

Two sugar-based syntheses of the novel neuroexcitotoxic amino acid (−)-dysiherbaine have been reported. In the first of these (Scheme 20), the epoxyalcohol **104** was prepared from di-*O*-acetyl-L-xylal by reaction with allyltrimethylsilane and $TiCl_4$ (Vol. 23, p. 40), followed by hydrolysis and hydroxyl-directed epoxidation. Further transformations *via* the key intermediates shown led to (−)-dysiherbaine (**105**), although the exocyclic chiral centre was not introduced stereoselectively.[40] In another approach (Scheme 21), the dianhydrohexose **106** was converted to **107**, the initial benzylation proceeding with epoxide migration. Further chemistry led to alkenyl iodide **108**. This was subjected to palladium-catalysed coupling with a unit derived from L-alanine, but the further steps that led to (−)-dysiherbaine (**105**) did not permit stereoselective establishment of the chirality at C-4. However the C-4 epimer which thus became accessible also possessed bioactivity, although at a lower level.[41]

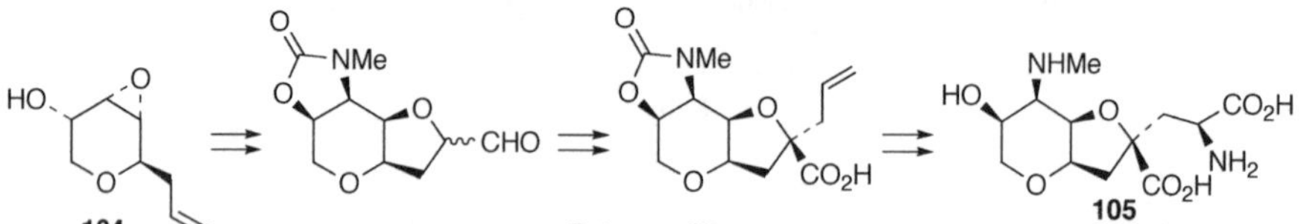

104   Scheme 20   **105**

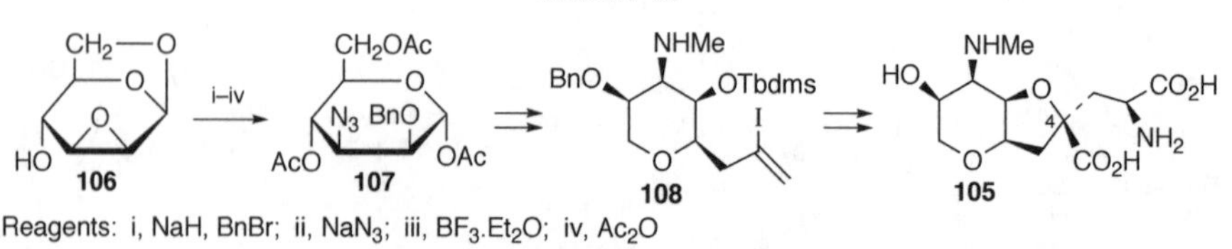

106   **107**   **108**   **105**

Reagents: i, NaH, BnBr; ii, NaN₃; iii, BF₃.Et₂O; iv, Ac₂O
Scheme 21

When the diazospecies **109**, prepared by diazo transfer fron mesyl azide, was treated with $Rh_2(OAc)_4$ in benzene under reflux, the two products **110** and **111** were formed in a ratio of 38:62. Use of the corresponding Pmb ether gave only the product of type **111**, whereas the 3-*O*-allyl ether gave just the analogue of **110**. The epimeric benzyl ether **112** gave, under the same conditions, the product **113** of formal carbenoid C-H insertion. Possible mechanisms and rationalizations of these findings were discussed.[42]

An interesting rearrangement to form a oxabicyclo[3.3.1]-system has been observed (Scheme 22). When the 5-hexityl-4-nitrocyclohexene **114** was treated with NaOMe, the deacetylated nitronate salt was formed, and this on reacetylation gave the cyclic nitronate **115**. A similar observation was made for the D-*manno*-analogue of **114**. When **115** was treated with catalytic NaOMe in methanol, the product was found to be the bicyclic oxime **116**, whereas the D-*arabino*-analogue of **115** (from the D-*manno*-analogue of **114**) did not rearrange. A mechanistic rationale was proposed.[43]

Reaction of the alkenyl carbene complex **117**, derived from tetra-*O*-benzyl-*aldehydo*-D-ribose, with cyclohexene oxide in the presence of [Cp₂TiCl]₂ gave the oxa-*trans*-decalin complex **118**, together with a smaller proportion of the epimer at the asterisked carbon. This work was extended to the use of sugar epoxide **119** and the carbene complex **120** to give the dioxabicyclo-system **121**.[44]

This last product is structurally similar to the polyether antibiotics, and a review has appeared, arranged chronologically, covering strategy and development of methodology for the total synthesis of polyether ionophore antibiotics. This includes carbohydrate-based syntheses as well as other approaches.[45] An interesting route to fused tetrahydropyran systems, indicated in Scheme 23, uses Hg(II)-catalysed transetherification-cyclization to establish the *trans*-fused ring system of **122**, but the same method could also be used to make the *cis*-fused analogue of **122**.[46]

In a route to the ABC-ring system of brevetoxin B, the known intermediate **123**, derived from 2-deoxy-D-ribose (sugar carbons indicated), was converted (Scheme 24) into **124**. This was elaborated into **125**, the A-ring being made by alkene metathesis, and the C-ring was closed by reductive cyclization to give **126**.[47] In an approach to the IJK-ring portion of the same target, intermediate **124** could be converted into **127** (Scheme 25: deoxyribose numbers shown), which was manipulated into **128**, the I ring being established using a 6-*endo*-

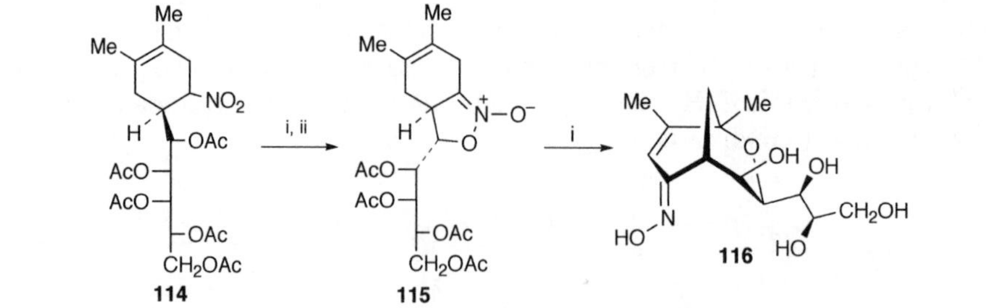

Reagents: i, NaOMe; ii, Ac₂O, py

Scheme 22

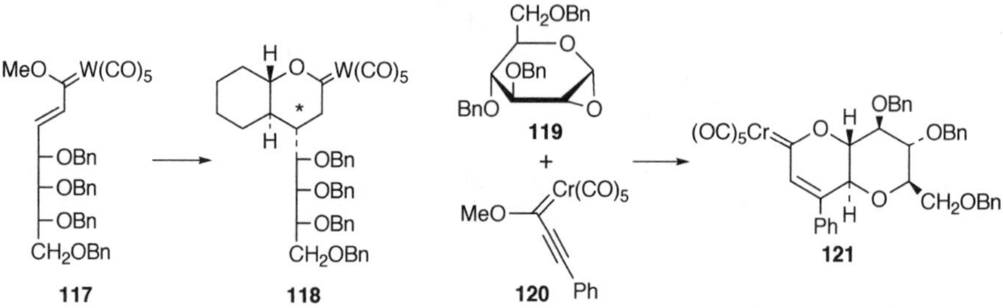

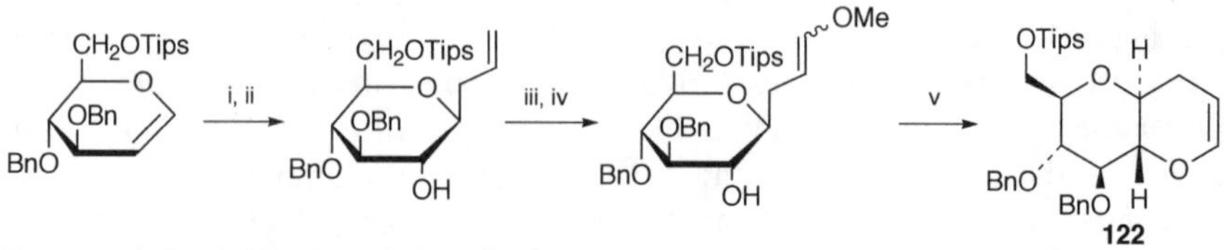

Scheme 23

Reagents: i, dimethyldioxirane; ii, CH₂=CH-CH₂MgCl; iii, O₃, then Ph₃P; iv, MeOCH=PPh₃;
v, Hg(OCOCF₃)₂ (0.05 eq.), THF, reflux, 24 h, 71%

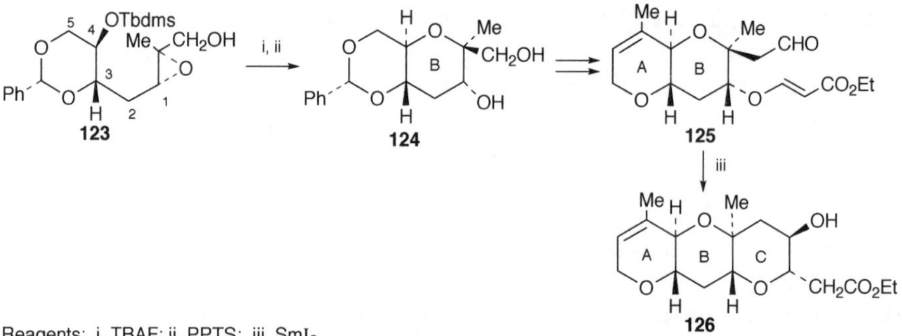

Reagents: i, TBAF; ii, PPTS; iii, SmI₂

Scheme 24

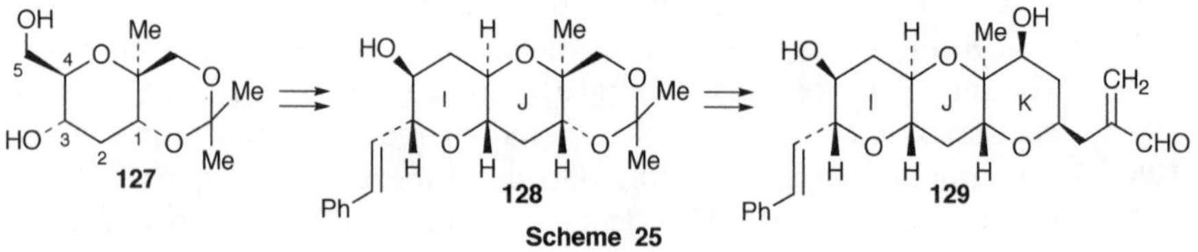

Scheme 25

epoxide cyclization. Further chemistry led to **129**, in which the side chain on the K-ring was introduced by reaction of an allyl silane with an α-acetoxy-compound.[48]

The FGH-ring system of the polyether toxin gambierol has been made as outlined in Scheme 26. Intermediate **130** was prepared from 2-deoxy-D-ribose in nine steps (sugar carbons numbered). Exocyclic alkene **131** was subjected to stereoselective hydroboration followed by a *B*-alkyl Suzuki coupling with enol phosphate **132**, also made from 2-deoxy-D-ribose, to give **133**, which was further transformed to the FGH tricycle **134**.[49]

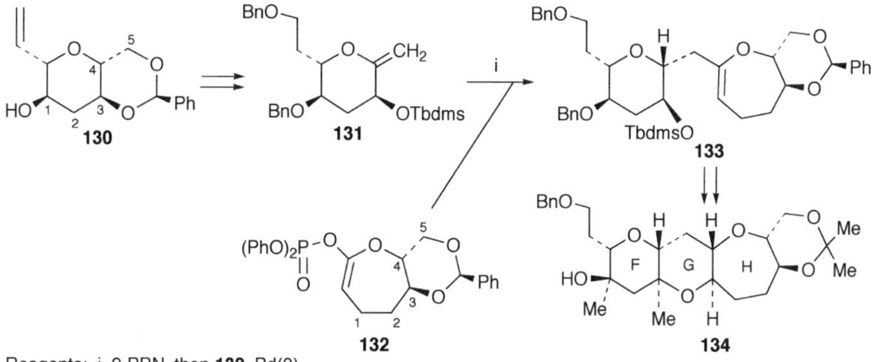

Reagents: i, 9-BBN, then **132**, Pd(0)

Scheme 26

In a new stereoselective total synthesis of zaragozic acid A, a key step was the [1,2]-Wittig rearrangement of the acetal **135** (Scheme 27), derived from L-arabinonolactone, to give the product **136**, having two contiguous quaternary chiral centres, with high diastereoselectivity. This was then transformed, using further ingenious chemistry, into the lactone **137** and hence the late intermediate **138**.[50]

The tricycle **139** (Scheme 28) has been previously prepared by intramolecular cyclization of a nitrile oxide with an allyl ether, and this compound has now been converted to **140**, and hence, by intramolecular nitrone cyclization, into the tetracycle **141**, containing an oxepinopyran ring system. A related cyclization of a nitrile oxide derived from **140** was also described, as were similar transformations from previously-reported cycloadducts with seven-membered rings (Vol. 28, p. 363), to give oxepinooxepanes.[51]

5   Nitrogen and Sulfur Heterocycles

In a new formal synthesis of (+)-thienamycin (Scheme 29), the 6-deoxy-glucosamine derivative **142** underwent base-catalysed rearrangement (see Vol. 16, p. 149), followed by oxidation, to give the separable isomers of **143**. Both of these could be transformed, by sequences involving base-catalysed epimerization, into the same all-*trans*-lactone **144**. Formation of a β-lactam and then oxidation by lead tetraacetate gave **145** (original sugar carbons indicated), which is an intermediate in a previous route to thienamycin.[52]

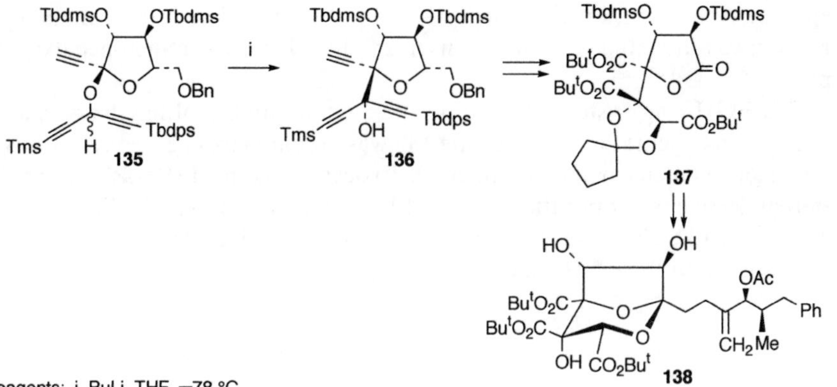

Reagents: i, BuLi, THF, −78 °C

Scheme 27

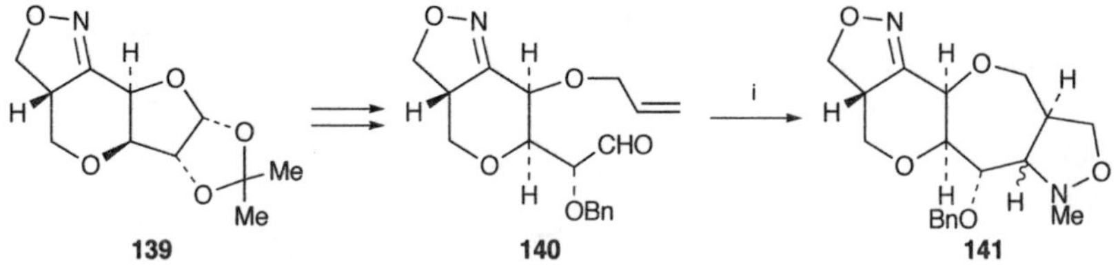

Reagent: i, MeNHOH

Scheme 28

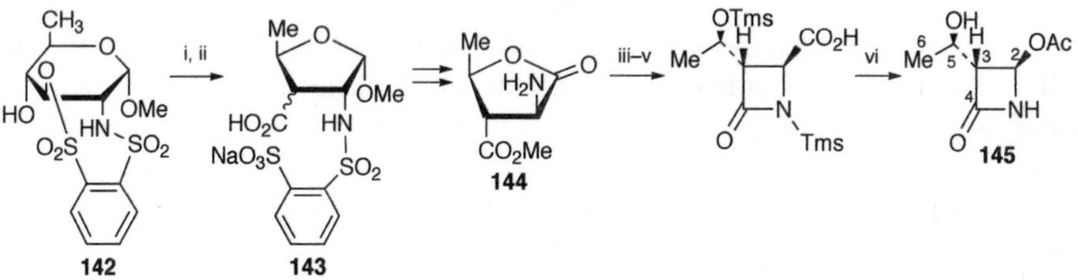

Reagents: i, KOBu^t, Bu^{t}OH; ii, NaClO$_2$; iii, NaOH, MeOH; iv, TmsCl, HMDS; v, Bu^{t}MgCl; vi, Pb(OAc)$_4$

Scheme 29

Also in the β-lactam area, [2+2]-cycloaddition of chlorosulfonyl isocyanate with a xylose-derived allenyl ether was used to make the adduct **146** with reasonable stereocontrol. Acidic detritylation, followed by sulfonylation and base-induced cyclization then gave the fused β-lactam **147**. The presence of the *gem*-dimethyl group was necessary to give good yields in the initial cycloaddition.[53] Others have used ketene-imine cycloadditions to give β-lactams such as **148**, which was converted to the simple α-alkoxy-β-lactam **149** by oxidative degradation of the sugar. This paper also describes the use of β-lactams produced in this way to make polyhydroxy amino acids, including more details of a synthesis of (−)-polyoxamic acid (see Vol. 27, p. 114), and the conversion of the adduct **150**, derived from isopropylidene-D-glyceraldehyde, into the lactone **151** on acid treatment.[54] A paper describing a β-lactam derived from a 3-amino-3-deoxyaldonic acid, and its subsequent polymerization, is mentioned in Chapter 16.

In an interesting approach to enantiopure hydroxylated piperidines (Scheme 30), the *meso*-compound **152**, made from 1,2-*O*-isopropylidene-α-D-glucopyranose by periodate cleavage followed by acid hydrolysis, was converted as indicated into the two bicycles **153** and **154**, with the former predominating (*ca.* 4:1). In both of these products, the cyano-group occupies an axial position, *trans*- to the nitrogen lone pair. These were converted into the deoxynojirimycin analogues **155** and **156** by reduction-hydrogenolysis.[55] When 5-bromo-5-deoxy-D-xylose (**157**) was treated with 3-amino-1-propanol, the bicyclic compounds **158** and **159** were obtained. The isomer **158** was produced about twenty times faster, but **159** was the more stable isomer. It was determined that *N*-alkylations of diastereomeric tetrahydro-1,3-oxazines were the rate-determining steps, and the selective formation of **158** was ascribed to the operation of an anomeric effect that favours the transition state for axial alkylation.[56]

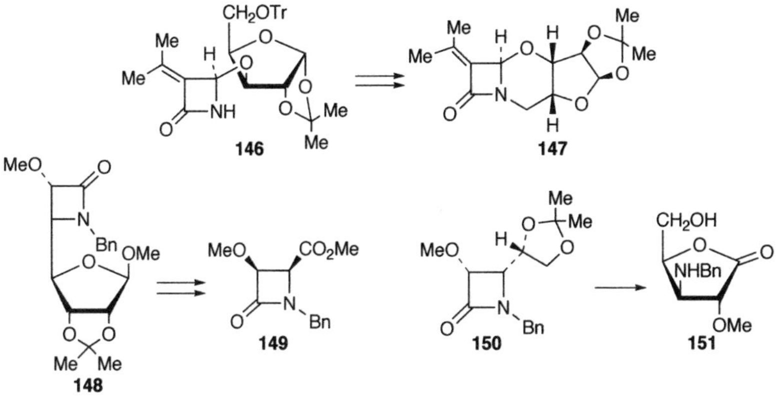

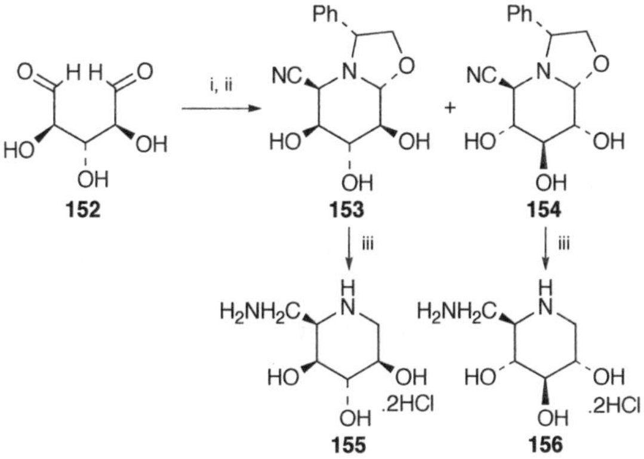

Reagents: i, (*R*)-PhCH(NH$_2$)CH$_2$OH, KCN; ii, ZnBr$_2$; iii, H$_2$, Pd/C, then EtOH, HCl

Scheme 30

Some simpler analogues of the bacterial tyrosyl tRNA synthetase inhibitor SB-219383 have been prepared. Reaction of the nitrone **160**, prepared from L-arabinose by intramolecular alkylation of an oxime, with the anion of Ph$_2$C=NCH$_2$CO$_2$Et gave the separable isomers of **161**. These were converted to **162** and its diastereoisomer, and compounds with opposite chirality in the piperidine ring were made from D-arabinose. Compound **162** had much greater bioactivity than the other three isomers, also displaying selectivity for the bacterial enzyme over the mammalian one, and it thus seems likely that SB-219383 has the stereostructure shown in **163**.[57]

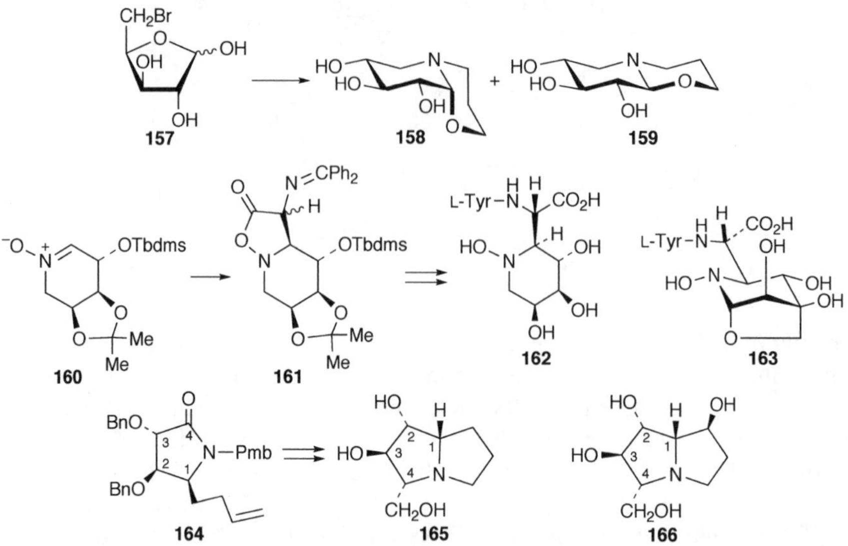

A number of papers have dealt with sugar-based syntheses of hydroxylated pyrrolizidine alkaloids. The glycosylamine formed from tri-*O*-benzyl-D-arabinofuranose and *p*-methoxy-benzylamine was treated with but-3-enylmagnesium bromide, followed by oxidative degradation using PCC to give the lactam **164**. This was convertible in a further nine steps to 7-deoxyalexine (**165**, sugar carbons numbered).[58] The same starting material was subsequently used by the same group in a somewhat similar appoach to (+)-alexine (**166**) itself.[59] Tri-*O*-benzyl-L-xylofuranose was used to make the azido-alkene **167** (Scheme 31), which was then converted (ozonolysis, Wittig reaction and oxidation with *m*CPBA) into the azido-epoxide **168** as a 2:1 mixture of two isomers. On ring closure and deprotection as indicated, both (+)-australine (**169**) and (–)-7-epi-alexine (**170**) were obtained.[60] A further synthesis of (+)-australine has also been reported, proceeding from isopropylidene-D-glyceraldehyde *via* anhydropentitol intermediates, and using olefin metathesis to establish an eight-membered nitrogen-containing ring, followed by transannular ring closure to the pyrrolizidine.[61]

A chemoenzymatic route to 7-epi-alexine (**170**) (Scheme 32) involves the formation of aldehyde **171** from non-sugar precursors, and its coupling with dihydroxyacetone phosphate using fructose 1,6-diphosphate aldolase to give, after acid phosphatase treatment, the ketose **172**. This was converted to 7-epi-alexine (**170**) as indicated, and related methods were also used to prepare (+)-australine (**169**) and 3-epi-australine (**173**).[62]

Reaction of ethyl 2-nitrosoacrylate, produced *in situ*, with alkene **174** gave the cycloadduct **175** (Scheme 33). Reduction of this with cyanoborohydride, followed by epimerization to the more favoured isomer **176**, was followed by Raney nickel reduction of the N-O bond and subsequent condensations and reductions to give the pyrrolizidine **177**.[63]

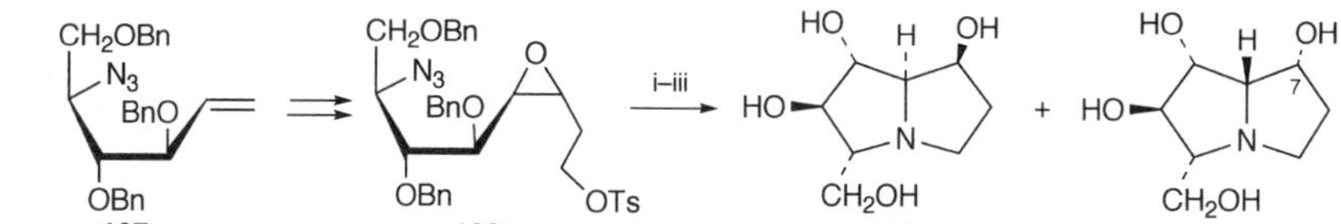

167 168 169 170

Reagents: i, H_2, Pd/C (5 wt.%), EtOH; ii, K_2CO_3; iii, H_2, Pd/C (300 wt.%), EtOH

Scheme 31

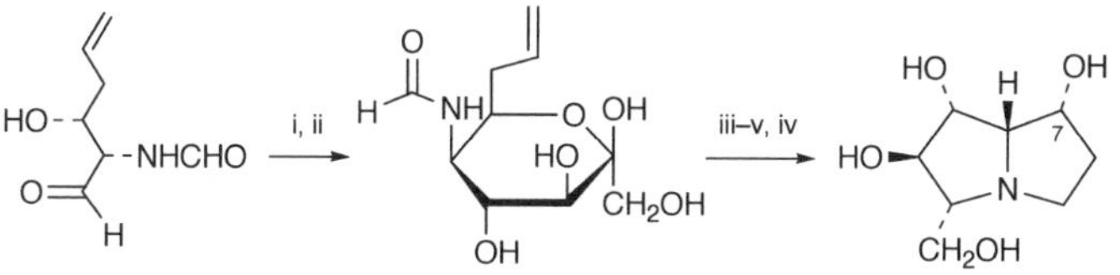

171 172 170

Reagents: i, Dihydroxyacetone phosphate, fructose 1,6-diphosphate aldolase; ii, acid phosphatase; iii, O_3; iv, H_2, Pd/C; v, HCl

Scheme 32

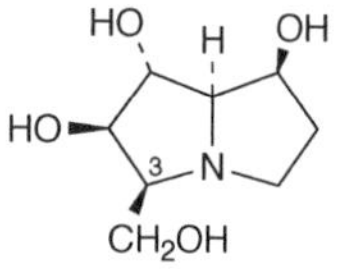

173

An account of a lecture by Jäger incorporates some new work in this area, including the synthesis of the cycloadduct **178** by cycloaddition of a glucose-derived nitrone with ethyl acrylate (sugar numbers indicated), and its subsequent conversion to the pyrrolizidine **179**.[64]

A review has discussed synthetic methods for the synthesis of the indolizidine swainsonine and its analogues, covering sugar-based methods as well as other approaches.[65] When the azido-nitrile **180**, prepared from L-sorbose with an inversion of configuration at C-4 (Vol. 33, p. 396), was treated firstly with aqueous TFA and then with Pd-C in methanol, the indolizidinone **181** was obtained.[66]

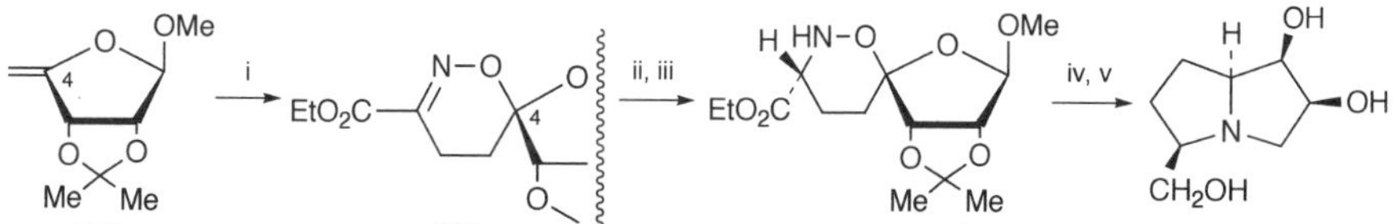

174 175 176 177

Reagents: i, $BrCH_2C(NOH)CO_2Et$, Na_2CO_3; ii, $NaBH_3CN$, AcOH; iii, Et_3N, $CHCl_3$; iv, Ra Ni, H_2, H_3BO_3, MeOH; v, $LiBH_4$

Scheme 33

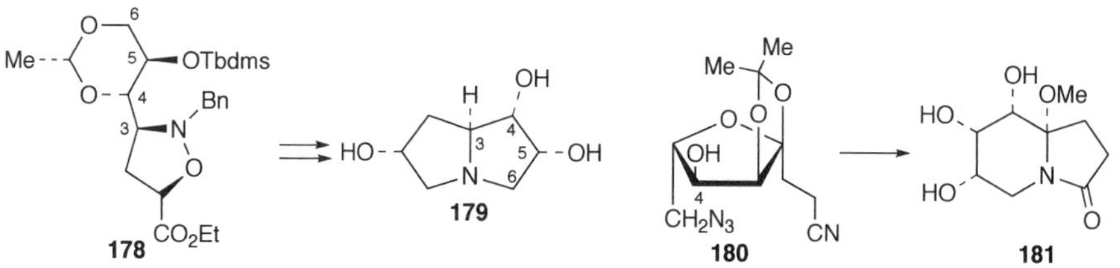

178 179 180 181

In a route to polyhydroxylated quinolizidines, condensation between the silyl enol ether **182** and aldehyde **183** (Scheme 34) gave the aldol product **184** with high diatereoselectivity. Conditions were also found to obtain diastereoselective condensations between the enantiomer of **183** and either **182** or its precursor ketone to give stereoisomers of **184**. Both **182** and **183** were obtained from furan.

Subsequent manipulation (including reduction of the ketone, selenoxide elimin-ation, hydroxylation, Baeyer–Villiger oxidation, and a ring contraction of the product) gave **185**, which gave the quinolizidine **186** on reduction and hydrolysis. Two stereoisomers of **186** were made similarly from stereoisomers of **184**.[67]

Four novel isomeric cyclic thiosulfinates, the zeylanoxides, have been isolated from the tropical weed *Sphenoclea zeylanica*. The structures and absolute stereochemistry of these compounds, which are plant growth inhibitors, have been determined by synthesis from the enantiomers of glucose. Use of L-glucose as outlined in Scheme 35 gave a mixture, separable by HPLC, of zeylanoxide A (**187**, α-sulfoxide), *epi*-zeylanoxide A (**187**, β-sulfoxide), zeylanoxide B (**188**, α-sulfoxide), and *epi*-zeylanoxide B (**188**, β-sulfoxide), with optical rotations in good agreement with those of the natural products.[68]

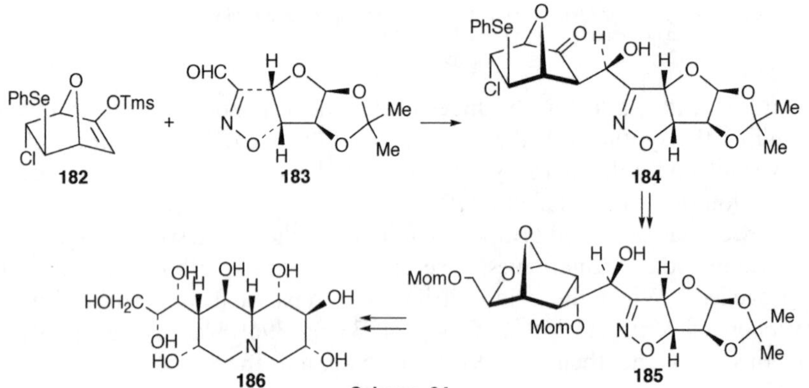

Scheme 34

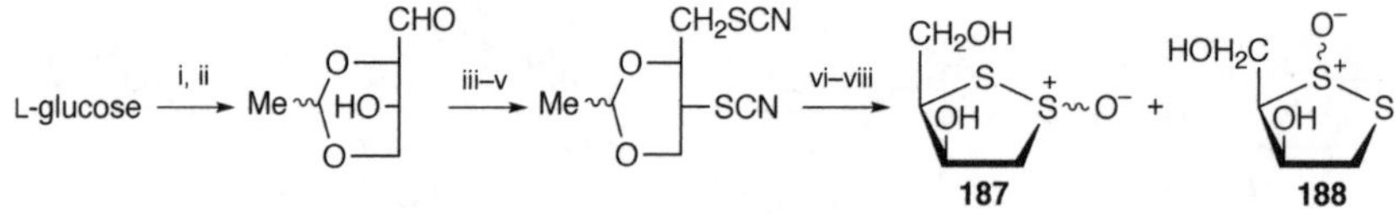

Reagents: i, paraldehyde; ii, NaIO$_4$; iii, LiBH$_4$; iv, TsCl, py; v, KSCN; vi, KOH; vii, HCl, H$_2$O–MeOH; viii, H$_2$O$_2$

Scheme 35

6   Acyclic Compounds

An improved method has been described for the conversion of D-arabinose into the aldehyde **189** (sugar carbons indicated), which was then used in a synthesis of a biotinylated 5-HETE for use in affinity chromatography.[69]

A full account has been given of the conversion of L-erythulose derivatives **190** into their di-cyclohexylboron enolates and reaction of these with aldehydes to give products of type **191**, of 2,4-*syn*-, 4,5-*syn*-stereochemistry (Vol. 33, p. 10). These compounds can be used, by treatment with periodic acid followed by diazomethane, to make mono-protected dihydroxyesters **192**.[70]

The vinyl epoxide **193** can be made from *N*-benzoyl-D-glucosamine in six steps and 38% overall yield (*Synth. Commun.*, 1997, **27**, 4255), and this has now been used for the synthesis of the sphingadienine-type glucocerebrosides **194**.

Methods were developed to make the 8,9-alkene stereoselectively with either *cis*- or *trans*-geometry, prior to glucosylation using tetra-*O*-benzoyl-α-D-glucopyranosyl bromide and AgOTf.[71]

There has been a further report (see Vol. 28, p. 8 for earlier) on the preparation of conjugated dienes of *trans*-configuration by treatment of sugar-derived allyl stannanes with ZnCl₂. Thus, for example, stannane **195** gave diene-aldehyde **196**, and **197** gave **198**.[72]

The C-5 to C-13 fragment **199** of the cytotoxic marine natural product myriaporone 4 has been synthesized from D-glucose, with C-5 to C-8 of myriaporone 4, and the one-carbon branch at C-8, being derived from C-6 to C-2 of the sugar, and a C-9 aldehyde being coupled with an iodoalkene corresponding to C-10 to C-13.[73]

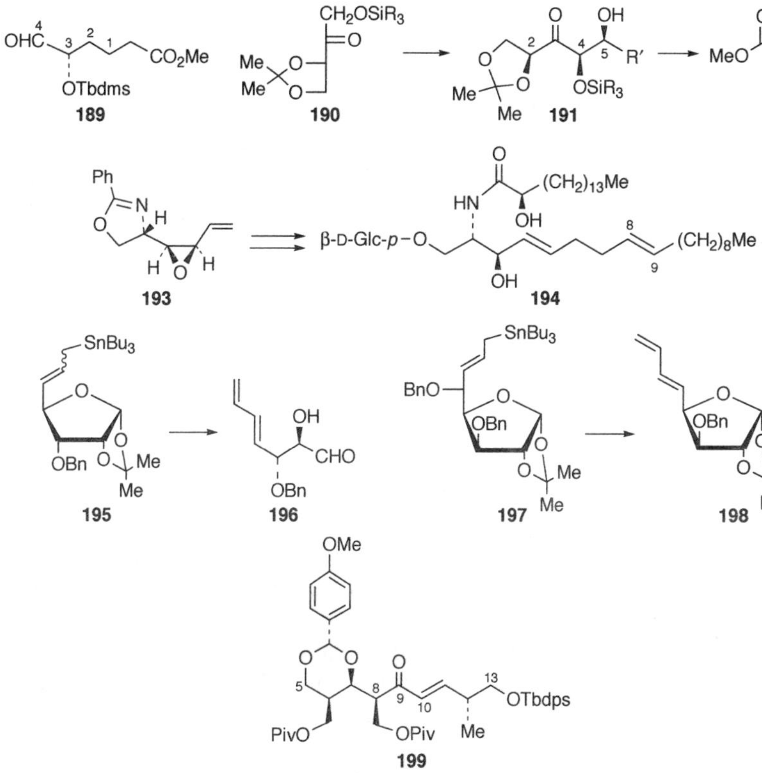

7   Carbohydrates as Chiral Auxiliaries and Catalysts

7.1   Carbohydrate-derived Auxiliaries. – Reaction of the fructose-derived acrylate **200** with cyclopentadiene in the presence of Et₂AlCl, followed by disconnection of the auxiliary by reduction, gave the cycloadduct **201** in high ee. Use of the same auxiliary in an intramolecular Diels–Alder reaction catalysed by Et₂AlCl gave, after reductive disconnection with LiAlH₄, the bicycle **202**.[74] It has previously been shown that **203** can be produced with high *endo*-selectivity and 90% de by reaction of cyclopentadiene and the isomannide-derived acrylate, catalysed by EtAlCl₂ (Vol. 30, p. 390–1). It has now been found that higher diastereoselectivities can be obtained if the benzyl group is converted into an η^6 arene chromium carbonyl complex prior to cycloaddition, and a de of 99% was obtained using the Cr(CO)₂[P(OEt)₂] complex, again with EtAlCl₂ as Lewis acid catalyst.[75]

High levels of diastereoselectivity can be achieved in conjugate radical additions of the type shown in Scheme 36, particularly when the group R′ is bulky, and where the protecting group R is an acyloxy or bulky silyl group. In the case where R = Piv and R′ = Pr^i, the product was obtained in 84% yield, and with a de of 86%.[76]

Two new glucose-derived oxazolidinones have been prepared, and converted to *N*-acyl derivatives of type **204** (R = Me or Piv). The dialkylboron enolates derived from **204** underwent aldol reactions to give *syn*-products **205**, with diastereomeric ratios between 8:1 and 16:1.[77] The same group has also made the oxazolidinone **206** from D-xylose. When this was treated with Mukaiyama's reagent (2-chloro-1-methylpyridinium iodide) in the presence of an imine, a Staudinger ketene-imine cyclization occurred to give a β-lactam such as **207**, the structure of which was confirmed by X-ray crystallography, in >98% de.[78]

In a synthesis of a potent selective inhibitor of Factor Xa, modified Ugi four-component condensation between tetra-*O*-pivaloyl-β-D-glucopyranosylamine, pyridine-4-carboxaldehyde, formic acid, and ethyl isocyanoacetate gave the product **208** in high yield and with 81% de. This was subsequently converted to the desired peptidomimetic **209**.[79]

A neat approach (Scheme 37) to the spiroketal **213**, a component of the pheromone of the olive fruit fly, involves a base-catalysed oxy-Michael addition between ketosugar **210** and **211** to give stereoselectively the adduct **212**, convertible as indicated to the target.[80]

Further examples of hetero-Diels–Alder reactions of carbohydrate-derived chloronitroso compounds have been reported. When the chloronitroso compound **214**, made from D-xylose, reacted with cycloheptadiene in the presence of some water, the cycloadduct **215**, which could be used in a synthesis of (–)-physoperuvine, was obtained in ≥96% ee, together with the sugar ketone which can be recycled to **214**. The pseudoenantiomeric species **216**, from L-sorbose, gave the dihydrooxazine **217** on reaction with cyclohexadiene, again in high e.e.[81] The chloronitroso compound **218**, easily accessible from D-ribose, underwent

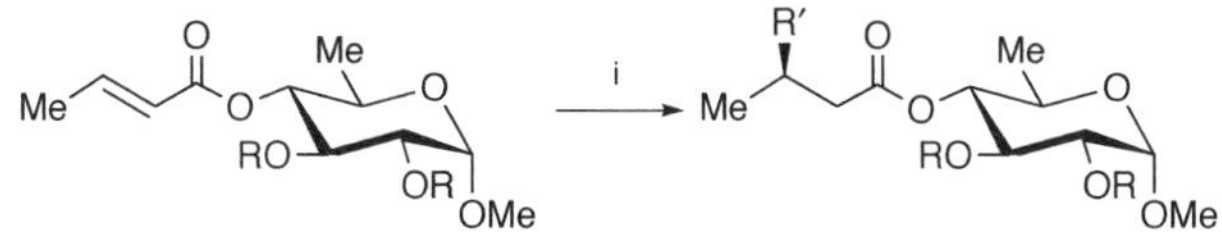

Reagents: i, R′I, Bu_3SnH, Et_3B/O_2, Et_2AlCl, CH_2Cl_2

Scheme 36

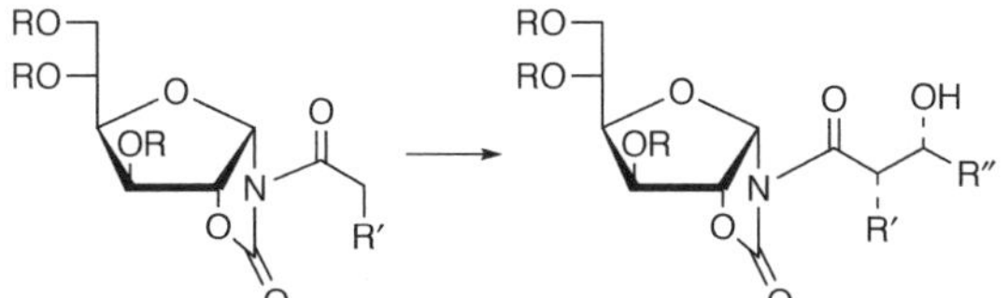

204 **205**

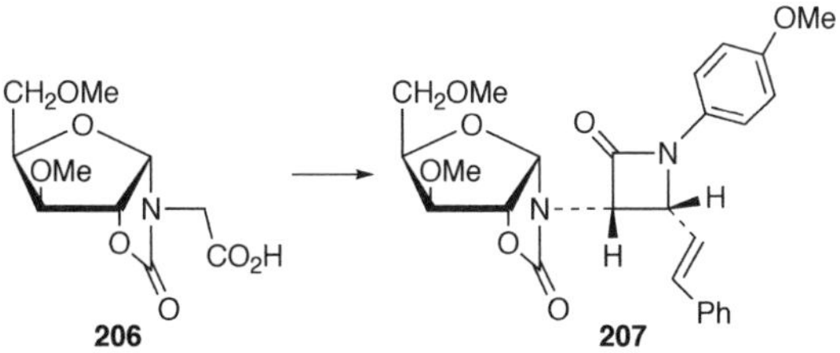

206 **207**

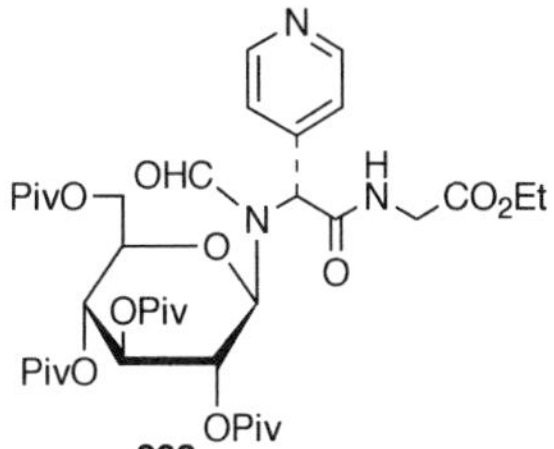

208

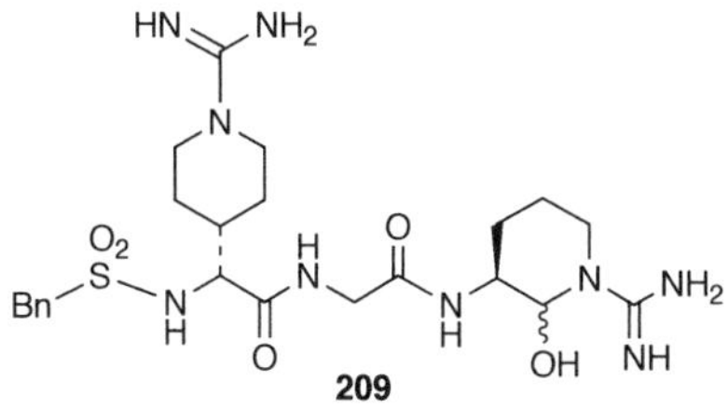

209

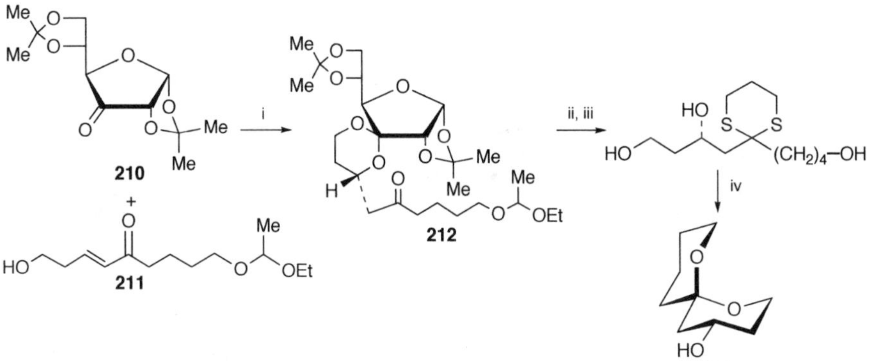

210 **211** **212** **213**

Reagents: i, $C_{16}H_{13}NMe_3OH$; ii, $HS(CH_2)_3SH$, $BF_3.Et_2O$; iii, HCl, THF iv, $Hg(ClO_4)_2$

Scheme 37

rapid cycloaddition with cyclohexadiene to give the enantiomer of **217** with 96% ee, and reacted more slowly with acyclic dienes to give, for example from sorbic acid, the adduct **219** with >99% ee, after derivatization.[82]

When the sulfinates **220** (R = Me, Et) were treated with the anions of racemic

3-methylisoxazolines, the products **221** were obtained as a $\sim$1:1 mixture of separable diastereoisomers. The enantiomers of **221** could be obtained by the use of the isomers of **220** epimeric at sulfur.[83]

In continuation of work in Just's laboratory on the diastereoselective synthesis of dinucleoside phosphorothioates (see Vol. 30, p. 290–1, Vol. 32, p. 287), the chlorophosphoramidite **222** (for the synthesis of the aminoalcohol from D-xylose see Chapter 14) was converted as indicated in Scheme 38 into the S_P-isomer **223**, in a 6:1 ratio with the diastereomer, and this ratio could be improved by the use of a more hindered base in step ii.[84]

A study has been reported on the photochemistry of tropolone alkyl ethers within α-, β-, and γ-cyclodextrins. Moderate asymmetric induction in the bicyclic products **224** were observed in some cases.[85]

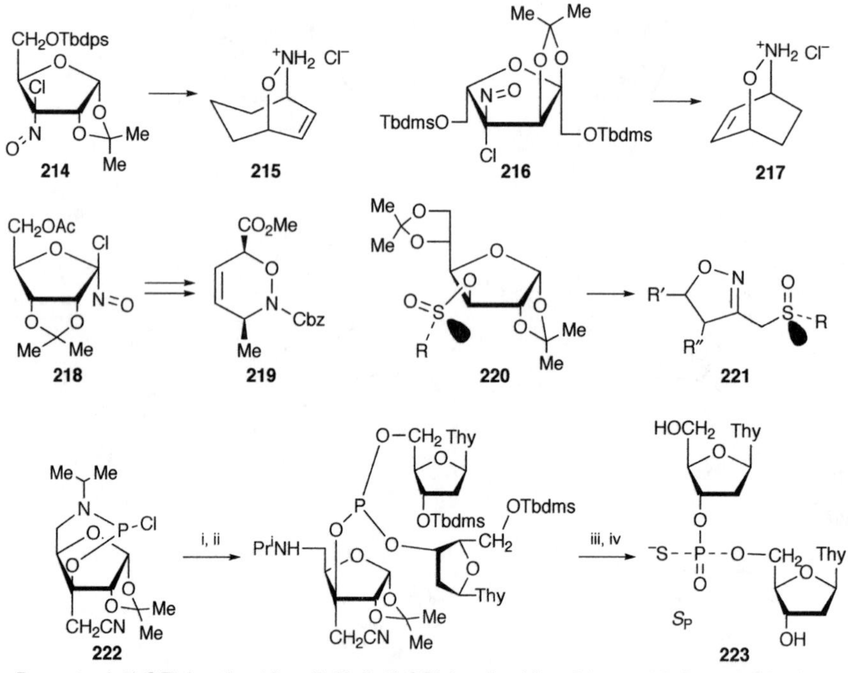

Reagents: i, 5'-*O*-Tbdms-thymidine, Et$_3$N; ii, 3'-*O*-Tbdms-thymidine, 2-bromo-4,5-dicyanoimidazole; iii, Beaucage's reagent; iv, NH$_3$ aq., then TBAF

Scheme 38

Some uses of sugars in novel resolutions of enantiomers have been reported. The ferrocene derivative **225** could be isolated from reaction of a mixture of *d,l*- and *meso*-isomers of the corresponding ferrocene dicarbonyl chloride with methyl 4,6-*O*-benzylidene-α-D-glucopyranoside. This could then be converted by methanolysis into the enantiomerically-pure (*R,R*)-ferrocene derivative **226**.[86] When the chiral pyridinium salt **227** was photolysed, equimolar amounts of the diastereomers **228** and **229** were obtained. Although no chirality transfer occurred, the two isomers could be separated after acetylation, and some further chemistry was performed on the fused aziridine ring of one of the diastereoisomers.[87] The acetal **230** was formed by reaction of phthaldehyde with monoisopropylidene-α-D-xylofuranose. On creating a new chiral centre in the form of **231**, the two isomers could be separated, and these isomers on methanolysis gave the methyl acetals **232** and the isomers at the asterisked carbon.[88]

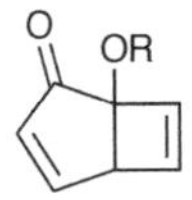

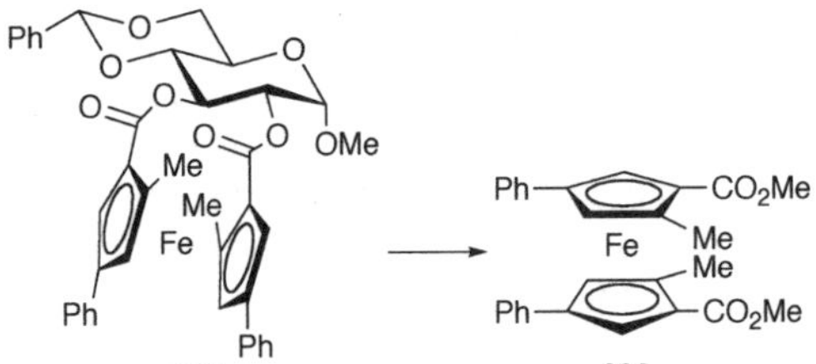

7.2 Carbohydrates as Chiral Catalysts. – A review on asymmetric epoxidation using chiral ketones as catalysts includes a section discussing the work of, in particular, Shi on the generation *in situ* of dioxiranes from sugar ketones and oxone, and their use in the asymmetric epoxidation of *trans*-alkenes.[89]

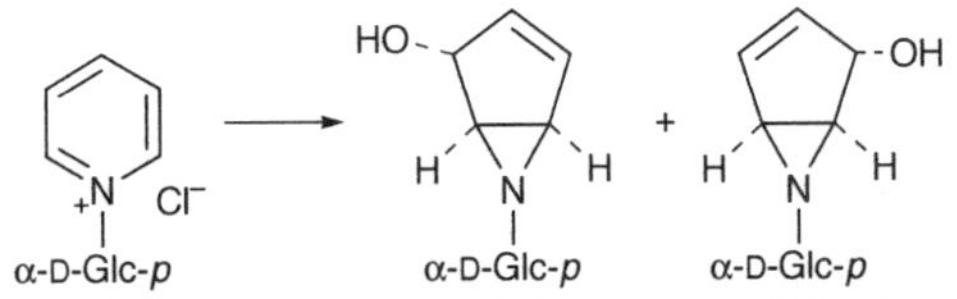

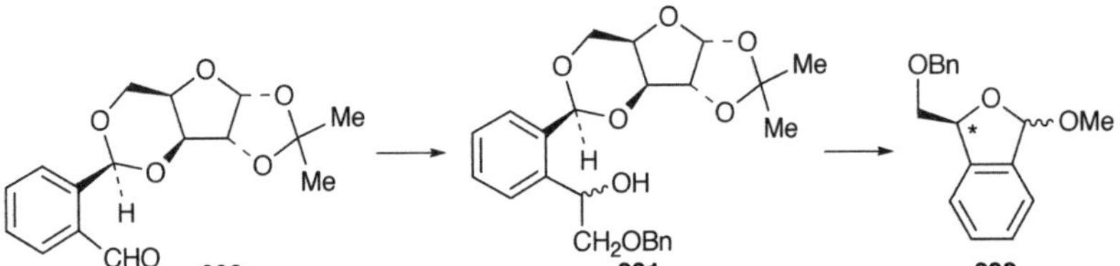

The new mannitol-based diphosphinite ligands **233** (R = Ph, cyclopentyl, cyclohexyl) have been prepared, and used in rhodium complexes for the asymmetric hydrogenation of prochiral ketones. The cases where R = cyclohexyl gave highest enantioselectivity, up to 86% for the reduction of methyl pyruvate to give methyl *R*-lactate.[90]

A review on carbohydrate complexes of the platinum-group metals includes discussion of the use of such complexes in asymmetric homogeneous hydrogenation of alkenes.[91] In this area, the new bisphosphine **234** ('xylophos') has been made from D-xylose, and converted to the complex $[Rh(cod)(xylophos)]^+BF_4^-$, which was used for homogeneous hydrogenation of acrylates. For the case of methyl 2-acetamidoacrylate and α-acetamidocinnamic acid, the *N*-acetyl-aminoacids were produced in high yields with ~90% ee in favour of the *S*-enantiomers.[92] The bidentate ligands **235** and **236** have also been made, from D-glucose, and used similarly. The reduction of methyl 2-acetamidoacrylate using the Rh(I) catalyst containing D-*gluco*-ligand **236** occurred with 100%

conversion to give the *S*-product with an improved 98% ee.[93] The bis-phosphite **237**, and two related species with similar diphenols, have also been used, as Rh(I) and Ir(I) cationic complexes, in asymmetric hydrogenation, but enantioselectivity in the reduction of acrylates was only moderate. Better enantioselectivity was observed when these complexes were used for the hydroformylation of styrene, with the Rh(I) complex containing **237** giving **238** of 55% ee, with the *R*-isomer predominant. Use of the D-*xylo*-analogue of **237** as ligand led to selective formation of the *S*-enantiomer of **238** (51% ee).[94] Synthesis of the D-*gluco*-bis-phosphite **239** led to much improved enantioselectivity in hydroformylation [using Rh(acac)(CO)₂ and **239**] of styrene and *p*-substituted styrenes, with styrene itself giving (*S*)-**238** with 90% ee.[95]

Ligands of type **240** (X = O, S) have been prepared from D-mannitol and used to catalyse the addition of dialkylzinc reagents to benzaldehyde and heptanal.

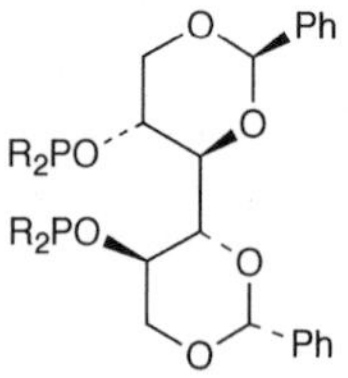

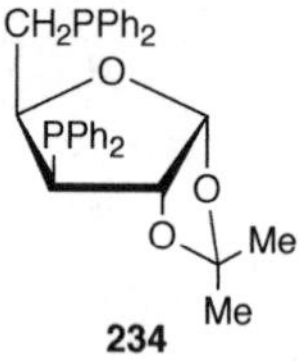

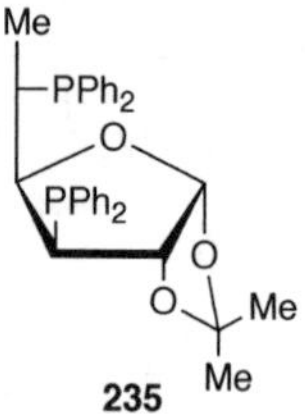

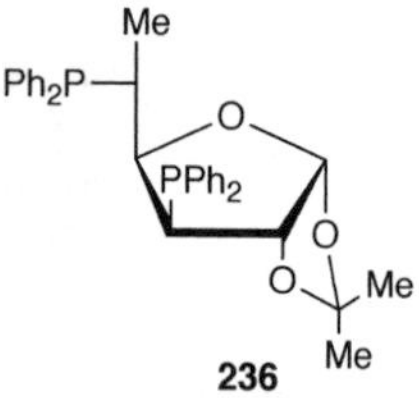

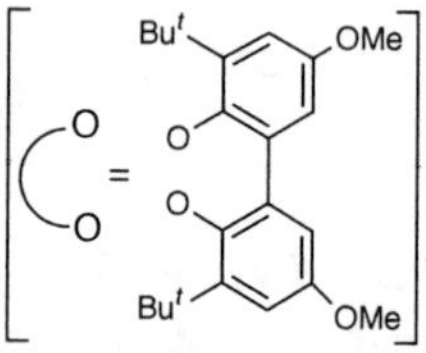

The ligand **240** (X = O, R = *n*-octyl) gave the best enantioselectivity (~90%) for the addition of diethylzinc to both aldehydes, with the (*R*)-enantiomer predominant.[96] The aminoalcohol **241**, previously used to catalyse the addition of diethylzinc to aldehydes (Vol. 30, p. 125) has now been used in the addition of di-isopropylzinc to a range of aldehydes, giving in each case predominantly the products **242** of addition to the *re* face of the aldehyde, in all except one case investigated with >90% ee.[97] Copper-catalysed conjugate addition of diethylzinc to cyclohexenone has been investigated in the presence of the ligand **243**, to give under the best conditions found 3-ethylcyclohexanone with ee 62%, favouring the (*S*)-enantiomer.[98] The bis-phosphite **244** was investigated more successfully for the same transformation, giving 3-ethylcyclohexanone of (*R*)-chirality in up to 81% ee.[99]

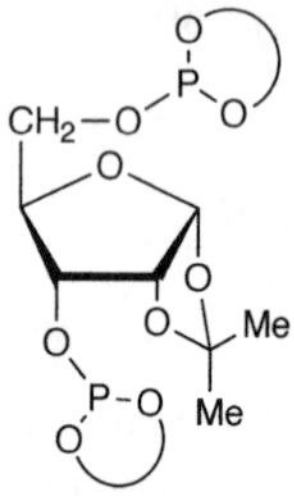

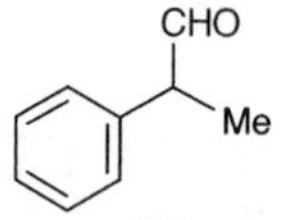

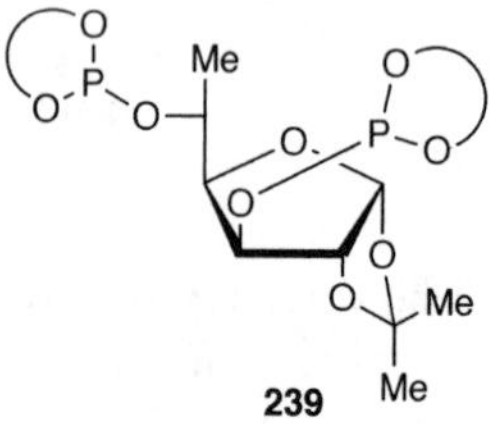

D-Glucosamine has been converted into η³-allyl species **245**, containing an amphiphilic chiral ligand, which permitted allylic substitutions to occur enantioselectively in aqueous media, or in aqueous-organic biphasic systems. Thus, for example, the racemic allylic acetate **246** could be converted into **247** in 95% yield and with 92% ee.[100]

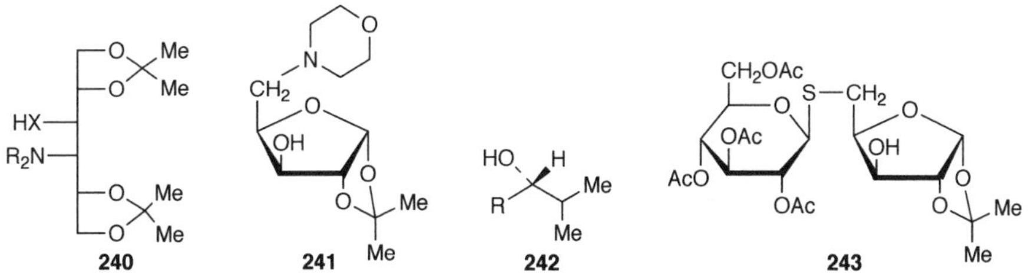

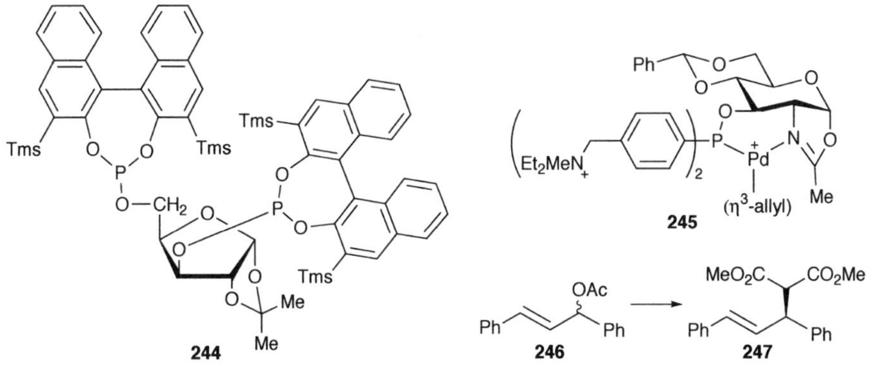

A sugar-based catalyst for catalytic asymmetric cyanosilylation has been developed. This catalyst, **248**, is derived from tri-*O*-acetyl-D-glucal *via* the intermediates shown in Scheme 39. It incorporates a Lewis acidic and a Lewis basic site within the molecule, and it was found that the conformational constraint induced by the phenyl group was necessary for good enantioselectivity. Treatment of benzaldehyde and TmsCN with catalytic quantities of **248** gave after acid hydrolysis the cyanohydrin **249** in 80% ee, and several other aldehydes behaved similarly.[101]

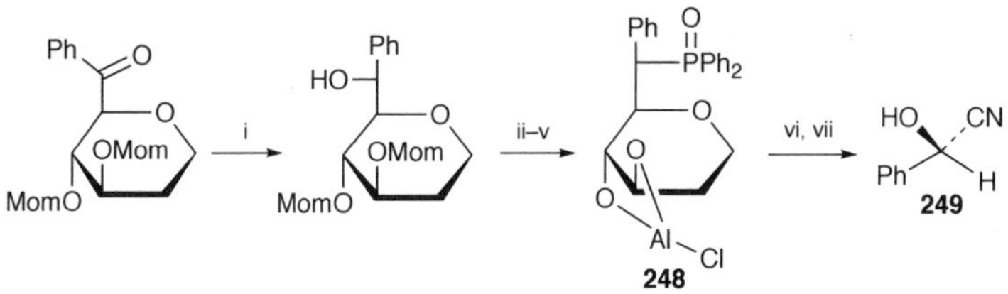

Reagents: i, NaBH₄, MeOH; ii, MsCl, py; iii, Ph₂PK, THF; iv, H₂O₂; v, Me₂AlCl; vi, TmsCN, PhCHO; vii, H⁺

Scheme 39

References

1. A. Roland, T. Durand, D. Egron, J.-P. Vidal and J.-C. Rossi, *J. Chem. Soc., Perkin Trans. 1*, 2000, 245.
2. T. Durand, A. Guy, J.-P. Vidal, J. Viala and J.-C. Rossi, *Tetrahedron Lett.*, 2000, **41**, 3859.
3. B.S.F. Salari, R.K. Biboutou and S.M. Bennett, *Tetrahedron*, 2000, **56**, 6385.
4. M. Seepersaud and Y. Al-Abed, *Tetrahedron Lett.*, 2000, **41**, 4291.
5. J. Ishihara, K. Hagihara, H. Chiba, K. Ito, Y. Yanagisawa, K. Totani and K. Tadano, *Tetrahedron Lett.*, 2000, **41**, 1771.
6. K. Takao, H. Saegusa, G. Watanabe and K. Tadano, *Tetrahedron: Asymm.*, 2000, **11**, 453.
7. M. Isobe and S. Takai, *J. Organomet. Chem.*, 1999, **589**, 122 (*Chem. Abstr.*, 2000, **132**, 93 554).
8. A. Kittaka, Y. Suhara, H. Takayanagi, T. Fujishima, M. Kurihara and H. Takayama, *Org. Lett.*, 2000, **2**, 2619.
9. T. Momose, M. Setoguchi, T. Fujita, H. Tamura and N. Chida, *Chem. Commun.*, 2000, 2237.
10. D.R. Kelly and M.R. Picton, *J. Chem. Soc., Perkin Trans. 1*, 2000, 1571.
11. S. Jarosz and S. Skóra, *Tetrahedron: Asymm.*, 2000, **11**, 1425.
12. S. Jarosz and S. Skóra, *Tetrahedron: Asymm.*, 2000, **11**, 1433.
13. C. Taillefumier and Y. Chapleur, *Can. J. Chem.*, 2000, **78**, 708.
14. K. Tatsuta, T. Yoshimoto, H. Gunji, Y. Okado and M. Takahashi, *Chem. Lett.*, 2000, 646.
15. J. Marco-Contelles and E. de Opazo, *Tetrahedron Lett.*, 2000, **41**, 2439.
16. J. Marco-Contelles and E. de Opazo, *J. Org. Chem.*, 2000, **65**, 5416.
17. I. Hanna and L. Ricard, *Org. Lett.*, 2000, **2**, 2651.
18. D.J. Holt, W.D. Barker, P.R. Jenkins, J. Panda and S. Ghosh, *J. Org. Chem.*, 2000, **65**, 482.
19. R.J. Hollingsworth and G. Wang, *Chem. Rev.*, 2000, **100**, 4267.
20. V. Popsavin, S. Grabez, M. Popsavin and J. Petrovic, *Carbohydr. Lett.*, 2000, **3**, 411.
21. H.B. Mereyala and R.R. Gadikota, *Tetrahedron: Asymm.*, 2000, **11**, 743.
22. H.B. Mereyala, R.R. Gadikota, K.S. Sunder and S. Shailaja, *Tetrahedron*, 2000, **56**, 3021.
23. R. Chenevert and M. Dasser, *Can. J. Chem.*, 2000, **78**, 275.
24. R. Chenevert and M. Dasser, *J. Org. Chem.*, 2000, **65**, 4529.
25. E. Alcázar, M. Kassou, I. Matheu and S. Castillón, *Eur. J. Org. Chem.*, 2000, 2285.
26. K. Nacro, B. Bienfait, J. Lee, K.-C. Han, J.-H. Kang, S. Benzaria, N.E. Lewin, D.K. Bhattacharyya, P.M. Blumberg and V.E. Marquez, *J. Med. Chem.*, 2000, **43**, 921.
27. H. Tamamura, B. Bienfait, K. Nacro, N.E. Lewin, P.M. Blumberg and V.E. Marquez, *J. Med. Chem.*, 2000, **43**, 3209.
28. B. Noya, M.D. Paredes, L. Ozores and R. Alonso, *J. Org. Chem.*, 2000, **65**, 5960.
29. M. Chandrasekhar, S. Raina and V.K. Singh, *Tetrahedron Lett.*, 2000, **41**, 4969.
30. H.H.A.M. Hassan and M.M.A. Rahman, *Synth. Commun.*, 2000, **30**, 201.
31. C. Rousseau and O.R. Martin, *Tetrahedron: Asymm.*, 2000, **11**, 409.
32. K. Ishigami, H. Motoyoshi and T. Kitahara, *Tetrahedron Lett.*, 2000, **41**, 8897.
33. J.N. Tilekar, N.T. Patel and D.D. Dhavale, *Synthesis*, 2000, 395.
34. M. Samadi, C. Munoz-Letelier, S. Poigny and M. Guyot, *Tetrahedron Lett*, 2000, **41**, 3349.
35. H. Sugimura, Y. Hasegawa amd K. Osumi, *Heterocycles*, 2000 **52**, 99.

36. M.S. Chorghade, K. Sadalapure, S. Adhikari, S.V.S. Lalitha, A.M.S. Murugaiah, P.R. Krishna, B.S. Reddy and M.K. Gurjar, *Carbohydr. Lett.*, 2000, **3**, 405.
37. R. Bruns, J. Köpf and P. Köll, *Chem. Eur. J.*, 2000, **6**, 1337.
38. S. Yamauchi and Y. Kinoshita, *Biosci. Biotechnol. Biochem.*, 2000, **64**, 1563.
39. S. Yamauchi and Y. Kinoshita, *Biosci. Biotechnol. Biochem.*, 2000, **64**, 2320.
40. B.B. Snyder and N.A. Hawryluk, *Org. Lett.*, 2000, **2**, 635.
41. M. Sasaki, T. Koike, R. Sakai and K. Tachibana, *Tetrahedron Lett.*, 2000, **41**, 3923.
42. V.N. Desai, N.N. Saha and D.D. Dhavale, *J. Chem. Soc., Perkin Trans. 1*, 2000, 147.
43. M.V. Gil, E. Román and J.A. Serrano, *Tetrahedron Lett.*, 2000, **41**, 3221.
44. K.H. Dötz and E. Gomes da Silva, *Tetrahedron*, 2000, **56**, 8291.
45. M.M. Faul and B.E. Huff, *Chem. Rev.*, 2000, **100**, 2407.
46. D.S. Tan and S.L. Schreiber, *Tetrahedron Lett.*, 2000, **41**, 9509.
47. G. Matsuo, H. Matsukura, N. Hori and T. Nakata, *Tetrahedron Lett.*, 2000, **41**, 7673.
48. H. Matsukura, N. Hori, G. Matsuo and T. Nakata, *Tetrahedron Lett.*, 2000, **41**, 7681.
49. H. Fuwa, M. Sasaki and K. Tachibana, *Tetrahedron Lett.*, 2000, **41**, 8371.
50. K. Tomooka, M. Kikuchi, K. Igawa, M. Suzuki, P.-H. Keong and T. Nakai, *Angew. Chem. Int. Ed. Engl.*, 2000, **39**, 4502.
51. A. Pal, A. Bhattacharjya and R. Mukhopadhyay, *Tetrahedron Lett.*, 2000, **41**, 10135.
52. K. Tatsuta, M. Takahashi, N. Tanaka and K. Chikauchi, *J. Antibiotics*, 2000, **53**, 1231.
53. R. Lysek, B. Furman, Z. Kaluza, J. Frelek, K. Suwinska, Z. Urbanczyk-Lipkowska and M. Chmielowski, *Tetrahedron: Asymm.*, 2000, **11**, 3131.
54. A.K. Bose, B.K. Banik, C. Mathur, D.R. Wagle and M.S. Manhas, *Tetrahedron*, 2000, **56**, 5603.
55. E. Poupon, B.-X. Luong, A. Chiaroni, N. Kunesch and H.-P. Husson, *J. Org. Chem.*, 2000, **65**, 7208.
56. D.A. Berges, J. Fan, S. Devinck and K. Mower, *J. Org. Chem.*, 2000, **65**, 889.
57. J.M. Berge, R.C.B. Copley, D.S. Eggleston, D.W. Hamprecht, R.L. Jarvest, L.M. Mensah, P.J. O'Hanlon and A.J. Pope, *Bioorg. Med. Chem. Lett.*, 2000, **10**, 1811.
58. H. Yoda, F. Asai and K. Takabe, *Synlett*, 2000, 1001.
59. H. Yoda, H. Katoh and K. Tanabe, *Tetrahedron Lett.*, 2000, **41**, 7661.
60. W.H. Pearson and J.V. Hines, *J. Org. Chem.*, 2000, **65**, 5785.
61. J.D. White and P. Hrnciar, *J. Org. Chem.*, 2000, **65**, 9129.
62. A. Romero and C.-H. Wong, *J. Org. Chem.*, 2000, **65**, 8264.
63. J.K. Gallos, V.C. Sarli, T.V. Koftis and E. Coutouli-Argyropoulou, *Tetrahedron Lett.*, 2000, **41**, 4819.
64. L. Fisera, V. Ondrus, J. Kubán, P. Micúch, I. Blanáriková and V. Jäger, *J. Heterocycl. Chem.*, 2000, **37**, 551.
65. A.E. Nemr, *Tetrahedron*, 2000, **56**, 8579.
66. I. Izquierdo, M.T. Plaza, R. Robles and A.J. Mota, *Eur. J. Org. Chem.*, 2000, 2071.
67. C. Schaller and P. Vogel, *Helv. Chim. Acta*, 2000, **83**, 193.
68. N. Hirai, S. Sakashita, T. Sano, T. Inoue, H. Ohigashi, C. Premasthira, Y. Asakawa, J. Harada and Y. Fujii, *Phytochemistry*, 2000, **55**, 131.
69. S.P. Knanapure, G. Saha, W.S. Powell and J. Rokach, *Tetrahedron Lett.*, 2000, **41**, 5807.
70. M. Carda, J. Murga, E. Falomir, F. González and J.A. Marco, *Tetrahedron*, 2000, **56**, 677.

71. T. Murakami, T. Shimizu and K. Taguchi, *Tetrahedron*, 2000, **56**, 533.
72. S. Jarosz, S. Skóra and K. Szewczyk, *Tetrahedron: Asymm.*, 2000, **11**, 1997.
73. B.-Z. Zheng, M. Yamauchi, H. Dei and O. Yonemitsu, *Chem. Pharm. Bull.*, 2000, **48**, 1761.
74. E.J. Enholm and S. Jiang, *J. Org. Chem.*, 2000, **65**, 4756.
75. G.B. Jones and M. Guzel, *Tetrahedron Lett.*, 2000, **41**, 4695.
76. R. Munakata, K. Totani, K. Takao and K. Tadano, *Synlett*, 2000, 979.
77. M. Stöver, A. Lützen and P. Köll, *Tetrahedron: Asymm.*, 2000, **11**, 371.
78. R. Saul, J. Kopf and P. Köll, *Tetrahedron: Asymm.*, 2000, **11**, 423.
79. S.Y. Tamura, O.E. Levy, T.H. Uong, J.E. Reiner, E.A. Goldman, J.Z. Ho, C.R. Cohen, P.W. Bergum, R.F. Nutt, T.K. Brunck and J.E. Semple, *Bioorg. Med. Chem. Lett.*, 2000, **10**, 745.
80. H. Watanabe, D. Itoh and T. Kitahara, *Synthesis*, 2000, 1925.
81. A. Hall, P.D. Bailey, D.C. Rees, G.M. Rosair and R.H. Wightman, *J. Chem. Soc., Perkin Trans. 1*, 2000, 329.
82. A. Defoin, M. Joubert, J.-M. Heuchel, C. Strehler and J. Streith, *Synthesis*, 2000, 1719.
83. J.A. López-Sastre, J.D. Martín-Ramos, J.F. Rodríguez-Amo, M. Santos-García and M.A. Sanz-Tejedor, *Tetrahedron: Asymm.*, 2000, **11**, 4791.
84. Y. Lu and G. Just, *Tetrahedron Lett.*, 2000, **41**, 9223.
85. S. Koodanjeri, A. Joy and V. Ramamurthy, *Tetrahedron*, 2000, **56**, 7003.
86. S. Shirakami and T. Itoh, *Tetrahedron: Asymm.*, 2000, **11**, 2823.
87. F. Glarner, B. Acar, I. Etter, T. Damiano, E.A. Acar, G. Bernardinelli and U. Burger, *Tetrahedron*, 2000, **56**, 4311.
88. D.F. Ewing, C. Len, G. Mackenzie, G. Ronco and P. Villa, *Tetrahedron: Asymm.*, 2000, **11**, 4995.
89. M. Frohn and Y. Shi, *Synthesis*, 2000, 1979.
90. S. Naili, I. Suisse, A. Mortreux, F. Agbossou, M.A. Ali and A. Karim, *Tetrahedron Lett.*, 2000, **41**, 2867.
91. D. Steinborn and H. Junicke, *Chem. Rev.*, 2000, **100**, 4283.
92. O. Pàmies, G. Net, A. Ruiz and C. Claver, *Eur. J. Inorg. Chem.*, 2000, 2011.
93. M. Diéguez, O. Pàmies, A. Ruiz, S. Castillón and C. Claver, *Tetrahedron: Asymm.*, 2000, **11**, 4701.
94. O. Pàmies, G. Net, A. Ruiz and C. Claver, *Tetrahedron: Asymm.*, 2000, **11**, 1097.
95. M. Diéguez, O. Pàmies, A. Ruiz, S. Castillón and C. Claver, *Chem. Commun.*, 2000, 1607.
96. B.T. Cho, Y.S. Chun and W.K. Yang, *Tetrahedron: Asymm.*, 2000, **11**, 2149.
97. W.K. Yang and B.T. Cho, *Tetrahedron: Asymm.*, 2000, **11**, 2947.
98. O. Pàmies, G. Net, A. Ruiz, C. Claver and S. Woodward, *Tetrahedron: Asymm.*, 2000, **11**, 871.
99. O. Pàmies, M. Diéguez, G. Net, A. Ruiz and C. Claver, *Tetrahedron: Asymm.*, 2000, **11**,4377.
100. T. Hashizume, K. Yonehara, K. Ohe and S. Uemura, *J. Org. Chem.*, 2000, **65**, 5197.
101. M. Kanai, Y. Hamashima and M. Shibasaki, *Tetrahedron Lett.*, 2000, **41**, 2405.

Author Index

In this index the number in parenthesis is the chapter number of the citation and this is followed by the reference number of the relevant citations within that Chapter.